W0037762

OXYGEN TRANSPORT TO TISSUE XXIV

ADVANCES IN EXPERIMENTAL MEDICINE AND BIOLOGY

Recent Volumes in this Series

Volume 523
ADVANCES IN MODELLING AND CLINICAL APPLICATION OF INTRAVENOUS ANAESTHESIA
Edited by Jaap Vuyk and Stefan Schraag

Volume 524
DIPEPTIDYL AMINOPEPTIDASES IN HEALTH AND DISEASE
Edited by Martin Hildebrandt, Burghard F. Klapp, Torsten Hoffmann, and Hans-Ulrich Demuth

Volume 525
ADVANCES IN PROSTAGLANDIN, LEUKOTRIENE, AND OTHER BIOACTIVE LIPID RESEARCH: Basic Science and Clinical Applications
Edited by Zeliha Yazıcı, Giancarlo Folco, Jeffrey M. Drazen, Santosh Nigam, and Takao Shimizu

Volume 526
TAURINE 5: Beginning the 21st Century
Edited by John B. Lombardini, Stephen W. Schaffer, and Junichi Azuma

Volume 527
DEVELOPMENTS IN TRYPTOPHAN AND SEROTONIN METABOLISM
Edited by Graziella Allegri, Carlo V. L. Costa, Eugenio Ragazzi, Hans Steinhart, and Luigi Varesio

Volume 528
ADAMANTIADES-BEHÇET'S DISEASE
Edited by Christos C. Zouboulis

Volume 529
THE GENUS *YERSINIA*: Entering the Functional Genomic Era
Edited by Mikael Skurnik, José Antonio Bengoechea, and Kaisa Granfors

Volume 530
OXYGEN TRANSPORT TO TISSUE XXIV
Edited by Jeffrey F. Dunn and Harold M. Swartz

Volume 531
TROPICAL DISEASES: From Molecule to Bedside
Edited by Sangkot Marzuki, Jan Verhoef, and Harm Snippe

Volume 532
NEW TRENDS IN CANCER FOR THE 21ST CENTURY
Edited by Antonio Llombart-Bosch and Vicente Felipo

A Continuation Order Plan is available for this series. A continuation order will bring delivery of each new volume immediately upon publication. Volumes are billed only upon actual shipment. For further information please contact the publisher.

OXYGEN TRANSPORT TO TISSUE XXIV

Edited by

Jeffrey F. Dunn
Harold M. Swartz
Dartmouth Medical School
Hanover, New Hampshire

Springer Science+Business Media, LLC

Library of Congress Cataloging-in-Publication Data

International Society on Oxygen Transport to Tissue. Meeting (27th: 1999: Hanover, N.H.)
Oxygen transport to tissue XXIV/edited by Jeffrey F. Dunn and Harold M. Swartz.
p. ; cm. — (Advances in experimental medicine and biology; v. 530)
"Presentations made at the 27th Annual Meeting of the International Society on Oxygen Transport to Tissue (ISOTT) .... held in Hanover, NH, USA, at Dartmouth Medical School"—Pref.
Includes bibliographical references and index.

ISBN 978-1-4613-4912-9 ISBN 978-1-4615-0075-9 (eBook)
DOI 10.1007/978-1-4615-0075-9

1. Oxygen—Physiological transport—Congresses. 2. Tissue respiration—Congresses. I. Title: Oxygen transport to tissue 24. II. Title: Oxygen transport to tissue twenty-four. III. Dunn, Jeffrey F. (Jeffrey Frank). 1956- IV. Swartz, Harold M. V. Title. VI. Series.
[DNLM: 1. Oxygen Consumption—physiology—Congresses. 2. Biological Transport—physiology—Congresses. 3. Oxygen—metabolism—Congresses. WF 110 161o 2004]
QP99.3.O915 1999
572'.47—dc21

2003050632

This is a proceedings of the 27th annual meeting of the International Society on Oxygen Transport to Tissue (ISOTT), held at Dartmouth Medical School, Hanover, New Hampshire, from August 28–September 2, 1999.

ISBN 0-306-47774-2

Originally published by Kluwer Academic / Plenum Publishers, New York in 2003

http://www.wkap.nl/

10 9 8 7 6 5 4 3 2 1

A C.I.P. record for this book is available from the Library of Congress

INTERNATIONAL SOCIETY ON OXYGEN TRANSPORT TO TISSUE

1999

Officers

President (1999)	Harold M. Swartz, USA
Past President (1998)	Andras Eke, Hungary
President-Elect (2000)	Berend Oeseburg, The Netherlands
Secretary	David F. Wilson, USA
Treasurer	Peter E. Kiepert, USA
Chairman of the Knisely Award Committee	Duane F. Bruley, USA

Executive Committee

David Benaron, USA
Duane Bruley, USA
Andras Eke, Hungary
Simon Faithfull, USA
David Harrison, UK
Louis Hoofd, The Netherlands
Peter Kiepert, USA
Josef Moravec, France
Paul Okunieff, USA
Berend Oeseburg, The Netherlands
Oliver Thews, UK
David Wilson, USA

Scientific Program Committee

International

Britton Chance, USA
David Delpy, UK
David Harrison, UK
P.W. Hochachka, Canada
Antal Hudetz, USA
Paul Okunieff, USA
Peter Vaupel, Germany

Local

Ted Abraham
Donald Bartlett
Jeff F. Dunn
Richard W. Dow
Michael Gazzaniga
David Glass
Marsh Tenney
Bernard Trumpower
Keith Paulsen
Peter Spiegel
Harold M. Swartz

TECHNICAL EDITOR
Laraine Visser-Isles, Erasmus University Rotterdam, Holland

VOLUME ADMINISTRATORS
Elena Jmourova, Dartmouth College
Virginia S. ("Dinny") Carreiro, Dartmouth College

SPONSORS
The organizers of ISOTT99 are particularly grateful for the sponsorship of the following organizations:

Allos Therapeutics
Alliance Pharmaceutical Corporation
Baxter Healthcare Corporation
Biopure Corporation
Bruker Instruments, Inc.
Johnson & Johnson
Oxford Optronix
Reming Bioinstruments Co.
Department of Neurology, Dartmouth Medical School
Department of Radiology, Dartmouth Medical School

PREFACE

This volume contains refereed manuscripts prepared from presentations made at the 27th annual meeting of the International Society on Oxygen Transport to Tissue (ISOTT). The meeting was held in Hanover, NH, USA, at Dartmouth Medical School, the 3rd oldest medical school in the USA. ISOTT attempts to produce high quality publications on cutting edge topics relating to oxygen in living systems. The goal is to allow contributors to contribute original data, as with a main-stream journal article, but also to voice individual opinions and ideas in a more relaxed scientific forum.

The meeting brought together an international group of scientists who share a common interest in the measurement and role of oxygen in living systems. The organizers of ISOTT99 made a special effort to bring together people from industry, medicine, and basic sciences in order to improve the links in the chain of discovery through to application. As a result, this volume contains publications on a range of subjects. There are contributions from companies on modifiers of oxygen carrying capacity (allosteric modifiers of hemoglobin and infusible oxygen carriers or blood substitutes); technical reports on oxygen measurement devices including advances in near-infrared spectroscopy and imaging, oxygen electrodes, magnetic resonance spectroscopy and imaging, and fluorescence based measurements. There are medically related sections on modifying and measuring tumor oxygenation in order to improve therapy, assessment and interpretation of oxygenation in the central nervous system, and general issues relating oxygen to pathological conditions. There are special sections dealing with the issue of critical pO_2 in tissues, oxygen related gene activation, and protein C engineering.

We trust that you will find the volume as exciting and interesting as we have done in the preparation of this volume.

Jeffrey F. Dunn
Harold M. Swartz

CONTENTS

METHODS FOR ASSESSING OXYGEN IN TISSUES

OXYGEN IN TUMORS

CNS OXYGENATION

MUSCLE OXYGENATION AND METABOLISM

CRITICAL pO_2

OXYGEN RELATED GENE ACTIVATION

Chapter 1

MEASUREMENTS OF OXYGEN IN TISSUES: OVERVIEW AND PERSPECTIVES ON METHODS

Harold M. Swartz & Jeff F. Dunn

EPR Center for Study of Viable Systems, and Biomedical NMR Laboratory, Department of Diagnostic Radiology, Dartmouth Medical School, 7785 Vail Building, Hanover, NH 03755 USA

Abstract: The goal of this manuscript is to provide a summary of the major techniques that currently are being applied to measure oxygenation of tissues *in vivo.* Oxygen is one of the key components of metabolism. Oxygen is also a major variable in many diseases, both with respect to the pathophysiological processes and influencing the efficacy of treatment. Unfortunately, however, the measurement of tissue oxygenation is non-trivial. Consequently many different methods have been developed to try to make this measurement. This paper presents a summary, largely in tabular form, of most of the current methods for assessing tissue oxygenation. The table attempts to cover the most pertinent aspects of the techniques and their applications, including their potential niche, limitations, and advantages. Citations are given for each method to point the reader in the direction of relevant literature.

Key words: oxygen, methods, pO_2, *in vivo*

1. INTRODUCTION

While it is widely recognized that the amount of oxygen in tissues is a critically important factor in both physiology and pathophysiology, it is only recently that it has been possible to measure it directly *in vivo*. This new capability has led to important new knowledge. It also has made less direct measurements more interpretable and more meaningful. The goal of this manuscript is to provide a summary of the major techniques that currently are being applied to measure oxygenation of tissues *in vivo*. The summary is presented in the form of a table that provides short descriptions of key aspects of the various methods to make such measurements.

Any such summary inevitably involves decisions that are likely to be idiosyncratic and potentially arbitrary. In an attempt to minimize this, input has been solicited from a wide range of experts in the various techniques and we are very grateful for the valuable advice that we have received. Undoubtedly, in spite of this input, many short-comings remain in this summary. The reader is requested to provide additional input to the authors so that future versions of such a summary may be more complete and more accurate. There have been several previous overviews of the various techniques, and these remain very valuable summaries of the potential advantages and limitations of many of the techniques that are considered here (22,36,40).

2. DEFINITIONS OF COLUMNS IN THE TABLE

2.1 Method

The choices here were made on a practical basis, in some cases keeping together under a single heading two or more approaches that could have been listed separately, with the rationale that most of the features of the grouped methods had similar principles, etc. There is a special emphasis on methods based on magnetic resonance because those are the ones with which we are most familiar and therefore most likely to view as being distinct approaches. The table attempts to list all of the major methods that provide *in vivo* data either intracellularly, interstitially, or in the microvasculature. It does not consider some less direct methods such as measurement of blood gases.

2.2 Parameter measured

This considers the type of parameter that is measured by each technique. There are essentially three inter-related types of measurements of oxygen in tissues:

1. direct measurement of the partial pressure of oxygen (pO_2),
2. direct measurement of the concentration of oxygen ($[O_2]$),
3. measurements that indirectly provide data on one or both of these parameters.

In order to reach valid conclusions it may be important to recognize which of these is being measured by each technique. It could be argued that the pO_2 relates to diffusion gradients and the $[O_2]$ to enzyme and reaction kinetics. Due to our limited knowledge about the variations in solubility across the cell and around enzymes, however, the two measurements are often interconverted using a single assumption about solubility. The precise solubility in a given tissue is dependent on factors such as salinity, temperature and lipid content. This makes it difficult to come up with an accurate value. For instance, at 37°C, solubility in a tumor model was reported to be 0.93μM/mmHg (14), while in serum a value of 1.52μM/mm Hg has been reported (13). A common conversion for tissue solubility is 1.35μM/mm Hg (17), which is the solubility in pure water at 37°C.

It is especially important to understand the basis of the more indirect measurements, in order to know when the extrapolation to the more fundamental values is valid. But as noted below, even methods that are considered to "directly" measure pO_2 or $[O_2]$ often have intrinsic assumptions, which could influence the validity of the measurement. For example, polarography records a current, which is not pO_2, but rather a parameter that is expected to directly reflect this variable and therefore any factor that affects the current would affect the validity of the data.

2.3 Mechanism for measurement

This describes the principle by which the parameters are measured. It should be noted that in virtually all cases the parameter really is not directly measured. Usually a physical property is measured (e.g. the amount of current generated at the electrode) that under appropriate conditions is a good measure of the parameter. When the information is crucial, the validity of the assumptions should be supported by additional studies.

2.4 Method for making the measurement

These descriptions are intended to provide an indication as to how the actual measurements are made. Such brief descriptions are necessarily incomplete.

2.5 Time required

Measurement times usually vary depending on the precision that is being sought. Also, as techniques continue to develop, the time required for measurement with a particular precision is likely to decrease. Assuming that technology can improve indefinitely, with the exception of polarography and substances that localize in hypoxic regions, the intrinsic limit for time resolution for all of the techniques is milliseconds or less.

2.6 Site of measurement

The site of measurement is important to recognize because this can have a direct bearing on the interpretation of the measurement. Several of the methods involve an assumption as to the localization of the substances that are being probed. An important related consideration is the variant of the technique that is used for the actual measurement, because this may affect the region that is sampled. For example, quite different regions may be sampled depending on whether non-invasive or invasive variants of the same technique are used. In many of the methods, especially those using magnetic resonance, the detection system will have a limited volume that it probes; the sensitivity of detection within that volume may vary, which will affect the actually site that is measured.

2.7 Minimum volume directly sampled

The minimum volume that is sampled may be of more theoretical than practical interest. It is quite complex and uncertain to try to go from this fairly well defined aspect to a conclusion on the volume of tissue for which this measurement provides information. The latter depends greatly on the homogeneity of the pO_2 in the region (42).

2.8 Approximate resolution (given as μM $[O_2]$)

The $[O_2]$ that can be resolved by the method is of great interest. Unfortunately, however, it depends greatly on a number of factors, and therefore a particular value cannot be assigned without defining the conditions of the measurement, including the time that is allocated for making the measurement. Therefore, the values that are given are approximate and represent what is usually achievable by current techniques. As noted above, different methods convert to $[O_2]$ using limited knowledge of solubility.

2.9 Perturbation caused from measurement

In principle, the perturbations from the measurements can be divided into three broad categories:

1. none - those techniques that use an intrinsic probe (e.g. deoxyhemoglobin) and a non-perturbing measurement technique (e.g. NIRS at levels that do not cause local heating);
2. minimal - techniques that use non-perturbing querying techniques but require the administration of a substance;

3. invasive - many of the techniques require some degree of invasiveness, usually in the form of the acute insertion of the measuring device.

Invasiveness per se is not necessarily a significant drawback for routine clinical use, especially if the information that is obtained provides critically needed information. Perhaps the more important question is how the invasiveness affects the parameter that is being measured. This is a very important and incompletely resolved question that now should be able to be resolved by the combined use of several methods under appropriate conditions. It also should be noted that many of the potentially minimal or non-invasive techniques have invasive variants that have been developed to enhance sensitivity and provide better localization of the measurement.

2.10 Limitations

It is very unlikely that any technique will be free of limitations. Therefore, an understanding of the limitations of the techniques is essential for the proper selection of approaches and interpretation of the results that are obtained. The requirement for the administration of a sensing material is unlikely to be a significant limitation if there is not significant toxicity from it. The degree of invasiveness that is required may be potentially limiting in some circumstances, but again this is not likely to be a major drawback. The most important limitations are related to the region that can be measured and potential perturbations in pO_2 caused by the method. For some applications, poor sensitivity may also be a strong limitation. Problems with interpreting the measurements in terms of the information that is sought also can be a significant limitation for the use of some techniques. This section is probably the most incomplete due to the complexities of measurement conditions and the various requirements for different applications.

2.11 Potential niche & advantages

At this time tissue pO_2 is not measured routinely in the clinical setting, although it is used clinically in a limited number of sites. The value of making such measurements has been demonstrated, especially in Oncology (18). As the value and feasibility of making measurements of oxygen in tissues becomes more appreciated, it is likely that the development of clinically applicable methodology will accelerate rapidly. It is likely that the developments will proceed in two different but complementary directions. One direction is to improve the feasibility and reliability of making the measurements. This often will require increasingly complex and perhaps costly technology. The other direction is to use methods that can be widely applied, even if these methods do not provide the high quality information that the first direction will allow. The relative weighting of these two factors will vary with the clinical application. For example, precise and repeated measurements of oxygen in tumors may have important clinical implications and therefore for this use more sophisticated methodology may be especially appropriate. Similar considerations may apply for the intraoperative measurement of oxygen in tissues. On the other hand, for clinical conditions that affect large numbers of patients whose clinical needs for such measurements is likely to persist for many years for each patient (e.g. peripheral vascular disease), simpler and less precise methodology may be more appropriate.

2.12 Citations

These have been chosen to provide a start for accessing the pertinent literature and in themselves do not necessarily provide a sufficient background for understanding a particular technique. In some cases, the field is moving very fast and so a selection of recent citations was warranted. This certainly fits the field of hypoxia markers, which are injected and localized to hypoxic areas. In other more established fields, such as that of electrodes, some historical references have been included, specifically that of the Clark and Whalen electrodes. The Clark electrode included a cellophane cover (6), which minimized artifacts caused by biological materials including red cells. The Whalen electrode is a good example of a needle electrode (50). The field of BOLD (blood oxygen level dependent) MR imaging (34) is also moving fast, fueled by the potential to map brain function with this non-invasive method. This is a very complex field. Some of the assumptions are outlined in (10, 48) and include variation in sensitivity associated with variations with field strength (12) and vessel diameter (20) and orientation (34).

3. CONCLUSIONS

The appropriateness and value of a particular type of measurement of oxygen in tissues *in vivo* depends on the experimental and clinical needs for the measurement and the feasibility of applying various approaches. While it is quite likely that a particular type of measurement will be the method of choice under a particular set of circumstances, it is unlikely that any method will be intrinsically superior for most uses. The most useful data are likely to be obtained by judicious selection of at least one and preferably more techniques that provide the type of data needed for a particular use, combined with careful interpretation of the results in terms of the type of information that is obtained.

Table 1. Characteristics of Methods to Measure Oxygen in Tissues

	METHOD	PARAMETER MEASURED	MECHANISM FOR MEASUREMENT	METHOD FOR MAKING THE MEASUREMENT
A	MICRO ELECTRODE	pO_2	Current generated by the electrolytic decomposition of dioxygen	Insertion of polarographic needle directly into tissue, with periodic measurements at different depths.
B	NEAR INFRARED MONITORING OF HEMOGLOBIN, MYOGLOBIN	Physiological parameter relative or absolute changes in saturation	Amount or fraction of hemoglobin [or myoglobin] and relative saturation	Usually by optodes on the surface; invasive variant using catheter.
C	NEAR INFRARED MONITORING OF CYTOCHROMES	Physiological parameter relative changes in cytochrome oxidation	Redox state of cytochromes	Usually by optodes on the surface; invasive variant using catheter.
D	PHOSPHORESCENT AND FLUORESCENT METHODS BASED ON REDOX STATES OF INTERMEDIATES	Physiological parameter based on redox potential	Ratio of reduced and oxidized states of redox couples	Usually by surface probes; invasive variants with optical; detector on fiberoptic catheter.
E	PHOSPHORESCENT AND FLUORESCENT METHODS BASED ON QUENCHING BY OXYGEN	$[O_2]$	Change in lifetimes of the excited state	Surface probes with injected fluorescent material or; invasive variants with material on fiberoptic.
F	NMR PERFLUORO-CARBON RELAXATION	$[O_2]$	Effect on relaxation rates of fluorine nuclei	Injection of emulsion with specific fluorine containing agent, directly into site of interest.
G	SUBSTANCES THAT LOCALIZE IN HYPOXIC AREAS (e.g., 2-nitroimidazoles)	Physiological parameter	Amount of material that localizes in the tissue; related to per-fusion and $[O_2]$ at time of administration	Administration of tagged material and subsequent biopsy or with gamma emitting radioactive tag, assay by nuclear medicine techniques.
H	EPR OXIMETRY BASED ON SOLID PARTICLES	pO_2	Effect on linewidth of EPR spectrum	Insertion of material into sites of interest with subsequent moni-toring from surface; invasive variant with insertion of catheter directly into site of interest.
I	EPR OXIMETRY BASED ON SOLUBLE MATERIALS	$[O_2]$	Effect on linewidth of EPR spectrum or relaxation rates	Administration of soluble paramagnetic material with subsequent monitoring from surface.
J	NMR SPECTROSCOPY	Physiological parameter metabolic correlates with oxygen	Concentrations of metabolites which change with oxidative status of cells	Localized NMR spectroscopy from surface using naturally occurring intermediates.
K	PROTON NMR SPECTRA OF MYOGLOBIN	Physiological parameter relative or absolute change in oxymyoglobin	Relative concentrations of deoxy and/or oxymyoglobin	Localized NMR spectroscopy from surface over myoglobin-rich regions.
L	NMR OVERHAUSER EFFECT	$[O_2]$	Relaxation rates of protons that couple to free radicals	Administration of free radicals with subsequent measurements using combined EPR and NMR frequencies; may field cycle magnet.
M	NMR "BOLD" EFFECT	Physiological parameter	Amount of deoxyhemoglobin in the voxels	MRI with pulse sequences that are sensitive to the paramagnetism associated with the deoxy state of hemoglobin.

(Table 1 cont'd)

	TIME REQUIRED	SITE OF MEASUREMENT	MINIMUM VOLUME DIRECTLY SAMPLED	APPROX. RESOLUTION (μM [O_2])	PERTURBATION CAUSED FROM MEASUREMENT
A	0.1 sec. – 5 min.	Interstitial volume in contact with the tip.	μl	0.01-1	Acute tissue damage; electrode consumes O_2.
B	0.1 sec. – 10 sec.	Location of the proteins. In the vascular system with non-linear weighting to vessel diameter for Hb methods. In myocytes for myoglobin.	ml's	2μM, but only over the sensitive range of Hb or Mb	None if optode is superficial; acute damage if optode is implanted.
C	Seconds to minutes	Intracellular cytochromes.	5 ml's	0.1-1	As above.
D	0.1 to 60 sec.	Sites of the redox intermediates (usually intracellular).	μl's	0.1-10	As above.
E	0.1 sec. – 60 sec.	Sites of the introduced probe molecules, intravascular or at a catheter tip.	μl's	0.1-10	As above; may require injection of material.
F	Minutes	Sites of the introduced emulsion.	μl-ml's	2-10	Requires injection of emulsion.
G	Minutes	Tissues where substances localize.	< 10μ in biopsy	Not applicable in biopsy	Requires biopsy.
H	Seconds to minutes	Sites of the particles (usually interstitial).	100 μl	1-5	Minimal with chronic implants; invasive with implanted resonator.
I	Seconds to minutes	Sites of the soluble molecules (usually throughout the tissues).	~1 ml	10 - 20	Minimal heating of tissue.
J	4 sec. – 30 min.	Sites of metabolites.	~1 ml 25μl-ml's	Relates to critical pO_2 for particular metabolic events	None.
K	13-30 minutes	Muscle (myoglobin).	~1 ml μl-ml's	1-10	None.
L	5-10 minutes	Sites of the soluble free radicals (usually throughout the tissues).	Potential for resolution of MRI	uncertain	Minimal heating of tissue.
M	0.35-10- minutes	Vascular system with a non-uniform weighting to vascular diameters.	<0.1 ml μl-ml's	<5	None.

(Table 1 cont'd)

	LIMITATIONS	POTENTIAL NICHE & ADVANTAGES	CLINICAL STATUS	CITATIONS
A	Invasive, may perturb system. Difficult to make repeated measurements.	Reference for other techniques. Provides distribution of pO_2 and unambiguous measurement of pO_2 under most conditions.	The "gold" standard. Has been used mostly in brain and tumors.	6,9,21,23,29, 50
B	For Hb, measurement is in blood. For Mb, it is in muscle cells. Relation between saturation and pO_2 can be complex. Frequency resolved methods are more quantitative but instrumentally demanding.	Based on intrinsic probe. Instrumentation could be widely available, very simple and wearable.	Routine monitoring of hemoglobin saturation. Many clinical trials in progress.	4,8,24,30
C	Very difficult to resolve from much larger absorbencies from hemoglobin or myoglobin.	Based on intrinsic probe. Measures key physiological parameters. Reflects intracellular conditions.	Could be developed for intraoperative and repeated studies.	35,44
D	Sensitivity *in vivo* may be limited. Interpretation of data as pO_2 may be complex.	Based on intrinsic probe. Measures key physiological parameters. Reflects intracellular conditions.	Same as above.	32,37
E	Requires administration of material, either through injection or implantation of fiber optode with material bonded on the tip.	Can make accurate measurements to very low pO_2. Technology could be fast.	Commercial fiberoptic probes in use.	3,7,28,49
F	Requires perfluorocarbon injection and knowledge of localization site. Sensitivity may be limiting.	Uses widely available instrument.	Potential for early use if sensitivity is demonstrated.	31,39
G	Requires injection of drug. May be difficult to quantify.	Can be implemented for use in tumors in near future. Relatively specific for hypoxia.	Clinical studies in progress for isotopic labeled compounds and for biopsied samples.	1,19,45,47
H	Requires administration of probe. Non-invasive variant has limited depth for sensitive measurements.	Useful for repeated measurement and for low pO_2 from same site and time. One probe (India ink) can now be used clinically. Instrumentation is fairly simple.	Early clinical studies of complete system in process.	41,42,43
I	Requires administration of a probe. Resolution of low pO_2 may be difficult.	Potential for dynamic imaging of pO_2 distribution. Repeated studies potentially feasible.	Early clinical studies of oxygen-sensitive materials in process.	15,16,26
J	Information is only indirectly linked to pO_2. Sensitivity for pO_2 may be low.	Widespread availability of instruments in clinic. Based on intrinsic probes. May provide multiple parameters.	Potentially implementable on machines already used widely for patients.	33,38
K	Sensitivity may be low. Only used on tissues where myoglobin is present. May be difficult to quantify.	Provides evidence for radiologically significant hypoxia. Based on intrinsic probe. Intracellular data.	Potentially implementable on machines already used widely for patients.	5,25,30
L	May not be sufficiently sensitive. Requires administration of probe. Indirect measure of pO_2. Requires special complex instrumentation.	Potentially could provide map of pO_2 in tissue, using the capabilities of MRI.	Requires development of new instrumentation.	2,11,27
M	Not directly related to pO_2. May be difficult to quantify. Sensitivity may be low. Measures only deoxy Hb.	Can be measured with existing clinical MRI instruments. Based on intrinsic probe.	Instrumentation widely available.	10,34,46

ACKNOWLEDGEMENTS

This study was supported by NIH Grants PO1 GM51630 and R01 CA67431, and used the facilities of the EPR Center for Viable Systems, supported by NIH Grant P41 RR11602.

REFERENCES.

1. Aboagye EO, Maxwell RJ, Horsman MR, Lewis AD, Workman P, Tracy M, et al. The relationship between tumour oxygenation determined by oxygen electrode measurements and magnetic resonance spectroscopy of the fluorinated 2-nitroimidazole SR-4554. Brit J Cancer 1998;77:65-70.
2. Ardenkjaer-Larsen JH, Laursen I, Leunbach I, Ehnholm G, Wistrand LG, Petersson JS, et al. EPR and DNP properties of certain novel single electron contrast agents intended for oximetric imaging. J Magn Reson 1998;133:1-12.
3. Buerk DG, Tsai AG, Intaglietta M, Johnson PC. *In vivo* tissue pO_2 measurements in hamster skinfold by recessed pO_2 microelectrodes and phosphorescence quenching are in agreement. Microcirc 1998;5:219-225.
4. Chance B, Luo Q, Nioka S, Alsop DC, Detre JA. Optical investigations of physiology: a study of intrinsic and extrinsic biomedical contrast. Phil Trans R Soc Lond B 1997;352:707-716.
5. Chen W, Cho Y, Merkle H, Ye Y, Zhang Y, Gong G, et al. In vitro and *in vivo* studies of 1H NMR visibility to detect deoxyhemoglobin and deoxymyoglobin signals in myocardium. Magn Reson Med 1999;42:1-5.
6. Clark LC, Jr., Wold D, Granger D, Taylor Z. Continous recording of blood oxygen tensions by polargraphy. J Appl Physol 1953;6:189-193.
7. Collingridge DR, Young WK, Vojnovic B, Wardman P, Lynch EM, Hill SA, et al. Measurement of tumor oxygenation: a comparison between polarographic needle electrodes and a time-resolved luminescence based optical sensor. Rad Res 1997;147:329-334.
8. Delpy DT, Cope M. Quanitification in tissue near-infrared spectroscopy. Phil Trans R Soc Lond B 1997;352:649-639.
9. Dings J, Meixensberger J, Jager A, Roosen K. Clinical experience with 118 brain tissue oxygen partial pressure catheter probes. Neurosurg 1998;43:1082-95.
10. Dunn JF, Zaim Wadghiri Y, Pogue BW, Kida I. BOLD MRI vs. NIR spectrophotometry: will the best technique come forward? Adv Exp Med Biol 1999;454:103-114.
11. Foster MA, Seimenis I, Lurie DJ. The application of PEDRI to the study of free radicals *in vivo*. Physics Med Biol 1998;43:1893-1897.
12. Gati JS, Menon RS, Ugurbil K, Rutt BK. Experimental determination of the BOLD field strength dependence in vessels and tissue. Magn Reson Med 1997;38:296-302.
13. Groebe K, Vaupel P. Evaluation of oxygen diffusion distances in human breast cancer xenografts using tumor-specific *in vivo* data: role of various mechanisms in the development of tumor hypoxia. Int J Radiat Biol Phys 1988;15:691-697.
14. Grote J, Susskind R, Vaupel P. Oxygen diffusivity in tumor tissue (DS-carcinosarcoma) under temperature conditions within the range of 20-40° C. Pflugers Arch 1977;372:37-42.

15. Halpern HJ, Yu C, Peric M, Barth E, Grdina DJ, Teicher BA. Oxymetry deep in tissues with low-frequency electron paramagnetic resonance. Proc Nat Acad of Sci USA 1994;91:13047-13051.
16. He G, Shankar RA, Chzhan M, Samouilov A, Kuppusamy P, Zweier JL. Noninvasive measurement of anatomic structure and intraluminal oxygenation in the gastrointestinal tract of living mice with spatial and spectral EPR imaging. Proc Nat Acad of Sci USA 1999;96:4586-4591.
17. Hitchman ML. Measurement of dissolved oxygen. New York: John Wiley and Sons; 1978.
18. Hockel M, Knoop C, Schlenger K, Vorndran B, Baussmann E, Mitze M, et al. Intratumoral pO_2 predicts survival in advanced cancer of the uterine cervix. Radiother Oncol 1993;26:45-50.
19. Hodgkiss RJ. Use of 2-nitroimidazoles as bioreductive markers for tumour hypoxia. Anti-Cancer Drug Design 1998;13:687-702.
20. Hoppel BE, Weisskoff RM, Thulborn KR, Moore JB, Kwong KK, Rosen BR. Measurement of regional blood oxygenation and cerebral hemodynamics. Magn Reson Med 1993;30:715-723.
21. Kallinowski F, Zander R, Hoeckel M, Vaupel P. Tumor tissue oxygenation as evaluated by computerized-pO_2-histography. Int J Rad Oncol Biol Phys 1990;19:953-961.
22. Kavanagh MC, Tsang V, Chow S, Koch C, Hedley D, Minkin S, et al. A comparison in individual murine tumors of techniques for measuring oxygen levels. Int J Rad Oncol Biol Phys 1999;44:1137-1146.
23. Kiening KL, Unterberg AW, Bardt TF, Schneider GH, Lanksch WR. Monitoring of cerebral oxygenation in patients with severe head injuries: brain tissue pO_2 versus jugular vein oxygen saturation. J Neurosurg 1996;85:751-757.
24. Kleinschmidt A, Obrig H, Requardt M, Merboldt KD, Dirnagl U, Villringer A, et al. Simultaneous recording of cerebral blood oxygenation changes during human brain activation by magnetic resonance imaging and near-infrared spectroscopy. J Cereb Blood Flow Metab 1996;16:817-826.
25. Kreutzer U, Jue T. Critical intracellular O_2 in myocardium as determined by ^1H nuclear magnetic resonance signal of myoglobin. Am J Physiol 1995;268(Heart Circ Physiol 37):H1675-H1681.
26. Kuppusamy P, Shankar RA, Zweier JL. *In vivo* measurement of arterial and venous oxygenation in the rat using 3D spectral-spatial electron paramagnetic resonance imaging. Phys Med Biol 1998;43:1837-1844.
27. Leunbach I. On a novel MRI technique (OMRI) for the determination of tissue parameters. Acta Anaesthes Scand Supp 1997;110:121-122.
28. Lo LW, Vinogradov SA, Koch CJ, Wilson DF. A new, water soluble, phosphor for oxygen measurements *in vivo*. Adv Exp Med Biol 1997;428:651-656.
29. Lubbers DW, Baumgartl H, Zimelka W. Heterogeneity and stability of local pO_2 distribution within the brain tissue. Adv Exp Med Biol 1994:567-574.
30. Mancini DM, Wilson JR, Bolinger L, Li H, Kendrick K, Chance B, et al. *In vivo* magnetic resonance spectroscopy measurement of deoxymyoglobin during exercise in patients with heart failure. Demonstration of abnormal muscle metabolism despite adequate oxygenation. Circulation 1994;90:500-508.
31. Mason RP, Rodbumrung W, Antich PP. Hexafluorobenzene: a sensitive 19F NMR indicator of tumor oxygenation. NMR Biomed 1996;9:125-134.
32. Mayevsky A, Manor T, Meilin S, Doron A, Ouaknine GE. Real-time multiparametric monitoring of the injured human cerebral cortex--a new approach. Acta Neurochir Suppl 1998;71:78-81.

33. Nioka S, Chance B, Smith DS, Mayevsky A, Reilly MP, Alter C, et al. Cerebral energy metabolism and oxygen state during hypoxia in neonate and adult dogs. Pediatric Res 1990;28:54-62.
34. Ogawa S, Lee TM, Kay AR, Tank DW. Brain magnetic resonance imaging with contrast dependent on blood oxygenation. Proc Natl Acad Sci U S A 1990;87:9868-9872.
35. Quaresima V, Springett R, Cope M, Wyatt JT, Delpy DT, Ferrari M, et al. Oxidation and reduction of cytochrome oxidase in the neonatal brain observed by *in vivo* near-infrared spectroscopy. Bioenergetics 1998;1366:291-300.
36. Raleigh, J A, Dewhirst M W and Thrall D E Measuring tumor hypoxia. In Hypoxia and Its Clinical Significance: Raleigh JA, editor1996; 37-45
37. Rampil IJ, Litt L, Mayevsky A. Correlated, simultaneous, multiple-wavelength optical monitoring *in vivo* of localized cerebrocortical NADH and brain microvessel hemoglobin oxygen saturation. J Clin Monitor 1992;8:216-225.
38. Rofstad EK, DeMuth P, Fenton BM, Ceckler TL, Sutherland RM. ^{31}P NMR spectroscopy and HbO_2 cryospectrophotometry in prediction of tumor radioresistaance caused by hypoxia. Int J Rad Oncol Biol Phys 1989;16:9119-923.
39. Sotak CH, Hees PS, Huang HN, Hung MH, Krespan CG, Raynolds S. A new perfluorocarbon for use in fluorine-19 magnetic resonance imaging and spectroscopy. Magn Reson Med 1993;29:188-195.
40. Stone HB, Brown J M, Phillips T L and Sutherland R M Summary of workshop on measuring tumor hypoxia, November 19-20, 1992, National Cancer Institute, Bethesda, MD, Radiat Res 1993;422-434
41. Swartz HM, Clarkson RB. The measurement of oxygen *in vivo* using EPR techniques. Phys Med Biol 1998;43:1957-1975.
42. Swartz HM, Dunn JF, Grinberg OY, O'Hara JA, Walczak T. What does EPR oximetry with solid particles measure--and how does this relate to other measures of pO_2? Adv Exp Med Biol 1996;428:663-670.
43. Swartz HM, Walczak T. Developing *in vivo* EPR oximetry for clinical use. Adv Exp Med Biol 1998;454:243-252.
44. Sylvia AL, Piantadosi CA, Jobsis-VanderVliet FF. Energy metabolism and *in vivo* cytochrome c oxidase redox relationships in hypoxic rat brain. Neurol Res 1985;7:81-88.
45. Thrall DE, Rosner GL, Azuma C, McEntee MC, Raleigh JA. Hypoxia marker labeling in tumor biopsies: quantification of labeling variation and criteria for biopsy sectioning. Radioth Oncol 1997;44:171-176.
46. Thulborn K, Waterton J, Mathews P, Radda G. Oxygenation dependence of the transverse relaxation time of water protons in whole blood at high field. Biochim Biophys Acta 1982;714:265-270.
47. Urtasun RC, Parliament MB, McEwan AJ, Mercer JR, Mannan RH, Wiebe LI, et al. Measurement of hypoxia in human tumours by non-invasive spect imaging of iodoazomycin arabinoside. Brit J Cancer - Suppl 1996;27:S209-S212.
48. van Zijl PC, Eleff SM, Ulatowski JA, Oja JM, Ulug AM, Traystman RJ, et al. Quantitative assessment of blood flow, blood volume and blood oxygenation effects in functional magnetic resonance imaging. Nature Med 1998;4:159-167.
49. Vinogradov SA, Wilson DF. Phosphorescence lifetime analysis with a quadratic programming algorithm for determining quencher distributions in heterogeneous systems. Biophys J 1994;67:2048-2059.
50. Whalen WJ, Spande JI. A hypodermic needle pO_2 electrode. J Appl Phys Resp Environ Exercise Physiol 1980;48:186-187.

Chapter 2

QUANTITATIVE BRAIN TISSUE OXIMETRY, PHASE SPECTROSCOPY AND IMAGING THE RANGE OF HOMEOSTASIS IN PIGLET BRAIN

Britton Chance, Hong Yan Ma & Shoko Nioka

Department of Biochemistry and Biophysics, Johnson Research Foundation, University of Pennsylvania, and Center for Biomedical Optics, University City Science Center, Philadelphia, PA USA

Abstract: The quantification of tissue oxygen by frequency or time domain methods has been discussed in a number of prior publications where the meaning of the tissue hemoglobin oxygen saturation was unclear and where the CW instruments were unsuitable for proper quantitative measurements [1, 2]. The development of the IQ Phase Meter has greatly simplified and made reliable the difficult determination of precise phase and amplitude signals from brain. This contribution reports on the calibration of the instrument in model systems and the use of the instrument to measure tissue saturation (S_tO_2) in a small animal model. In addition, a global interpretation of the meaning of tissue oxygen has been formulated based on the idea that autoregulation will maintain tissue oxygen at a fixed value over a range of arterial and venous oxygen values over the range of autoregulation. Beyond that range, the tissue oxygen is still correctly measured but, as expected, approaches the arterial saturation at low metabolic rates and the venous saturation at high metabolic rates of mitochondria.

Key words: Oximetry , spectroscopy, imaging, brain, homeostasis

Abbreviations: S_tO_2 Tissue oxygen saturation; S_aO_2 Arterial oxygen saturation; S_sO_2 Saggital sinus oxygen saturation; S_vO_2 Venous oxygen saturation; IQ in-phase and quadrature detection

1. INTRODUCTION

1.1 O_2 gradients

We are studying the tissue oxygen levels by Near Infrared (NIR) optical measurement of the capillaries, arterioles, and venules that deliver oxygen to the cortical neurons, to tumors or muscles [1,2]. A steep O_2 gradient from arteriolar to venolar (10 to 15 torr) facilitates oxygen diffusion to the tissue (Figure 1). The O_2 gradient has been simulated by a number of workers in previous ISOTT volumes [3]. The O_2 sink is cytochrome oxidase and the O_2 source is oxyhemoglobin.

The spectrum of the two forms of hemoglobin in the NIR region is well known but the baseline for this measurement is very important and other absorbers are present, water, lipid, etc. A perfusion of the dog brain [3,4] with fluorocarbon removes the blood, yet there remains a very significant baseline (S. Nioka, pers. Commun.), part of which is due to water, part lipid, part unknown. Thus, five wavelengths are needed for brain O_2 saturation studies, as used here [5]. We chose the immature, poorly differentiated piglet brain to be nearly homogeneous, recognizing the layered structure of the human brain would require data at various source-detector separations as well.

1.2 Site of measurement

NIRS does not receive signals from the larger blood vessels as does NMR to which the vessels are transparent. Only the small vessels, for example, and the pial mater of the cortex are seen by NIR, particularly vessels < 50 μ in diameter. NIR affords an unique measurement of oxygen concentrations at the site where extraction of O_2 from blood to tissue occurs which we term (S_tO_2).

The instrument that we use is contained in a very simple little circuit box [3], the contents of which can be placed on a PC card with surface mount technology. In order to maximize simplicity, we do not use a heterodyne system, but a homodyne system which detects at the same rf frequency, the real and imaginary parts of the phase angle φ and the amplitude (A) from which hemoglobin saturation can be displayed on a PC. Up to 5 wavelengths are time-shared (690, 720, 760, 805, 850 nm). The absorption (μ_a) and scattering (μ_s') are calculated from φ, and A data by the usual equations; saturation values are calculated based upon standard values of extinction coefficients for Hb, HbO_2, H_2O and lipid, and other background features [5-8].

The system is calibrated by a model containing a scatterer, blood and yeast. Oxygenation is varied from 100% to 10% saturation. Based upon the calculation of absorption and scattering from equations based upon the diffusion theory, we calculate absolute values rather than relative quantities as do continuous light oximeters and the average saturation value found for piglet brain is from 20-75% in the normoxia range (F_iO_2 = 21% -12%) and the average deviation is a few % (5).

2. METHODS AND RESULTS

The anesthetized piglet is the most convenient for *in vivo* calibration of brain oximetry because it accepts O_2 saturation perturbations from near zero in hypoxia to near 100% in hyperemia. In a typical study (Figure 1) the dots are the saturation of the arterial blood (S_aO_2), the solid line is the saturation of tissue as determined by IQ NIR (S_tO_2), and the triangles are the saturation of venous blood (S_vO_2). It is clear that arterial (S_aO_2) and venous (S_vO_2) bloods do not give the same values as the tissue (StO2) values. So tissue hemoglobin saturation (S_tO_2) is a novel and important measure. When tissue metabolism is active, tissue saturation (S_tO_2) will lie below arterial and above as in Figure 2. These studies have been repeated on 16 piglets (or 200 measures).

3. DISCUSSION

In this study, the FiO2 was cycled between air to 20% to 5%, then to 100% and then to 20%, etc. The tissue value varies from below arterial in normoxia and may reach the venous or saggital sinus value in hypoxia. In hyperemia it approaches the arterial value particularly in recovery from a hypoxic stress. Thus, the consequences of imbalances and balances of oxygen delivery and oxygen uptake are measured as S_tO_2 in the capillary bed, the site of extraction of O_2 from hemoglobin. The classical measures of arterial and venous bloods cannot directly tell what occurs in the tissue and what the saturation of hemoglobin may be at the anaerobic threshold unless we find some fixed relationship between these values:

1) One extreme is hypothermia and minimal O_2 uptake: S_tO_2 will approach the arterial level and minimal O_2 gradients will be present.

2) The other extreme is maximal uptake such as uncoupling or maximal function; S_tO_2 will reach the venous or saggital sinus level (Figure 1).

3) In-between the extremes, presumably, there is a homeostatic balance between delivery and utilization of O_2 under microvascular control. This is

identified by the nearly horizontal line of proportional changes at S_tO_2, S_aO_2 and S_tO_2 (S_vO_2), from 20 to 80% S_tO_2 (Figure 2). In order that hyperoxia be on the left and hypoxia be on the right, we plot I-S_tO_2 vs the fraction of arterial signal/arterial + venous signals over the available range for 16 piglets. Critical hypoxia can now be defined by the failure of homeostasis as the tissue value falls toward the venous or the saggital sinus blood values.

This could be a signal for aggressive therapeutic measures in neonate or adult human brain. While this relationship has therapeutic values, it can also be used to evaluate oximeter function in the 20 to 80% range.

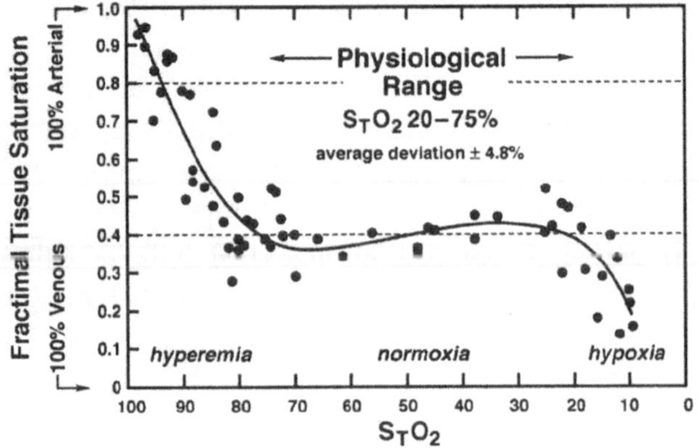

Experimental relationship between tissue saturation and arterial/venous bloods (HYM 32)

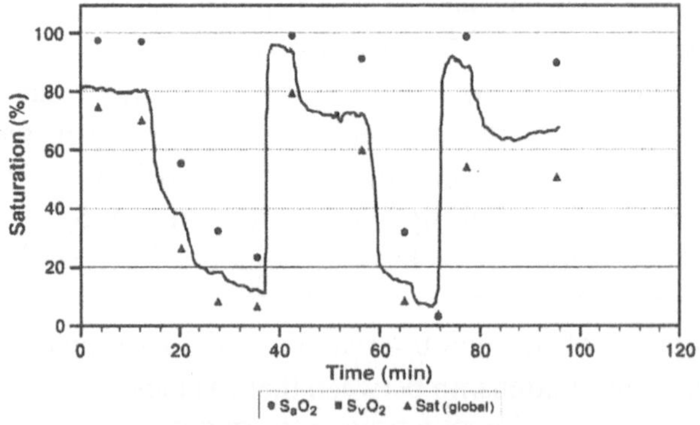

Piglet hypoxia test, IQ oximetry 12/17/98 (HYM 34)

Figure 1. A typical experiment of two variations in the piglet brain with respect to arterial, venous and saggital sinus bloods. The data are calculated from the μ_a, μ_s' seen to clearly to lie between the arterial and venous bloods, approaching arterial in reactive hyperemia and approaching saggital sinus venous blood in severe hypoxia

3.1 Imaging

There are some cases where single site oximetry may not be adequate and imaging of hypoxia is necessary to determine the "worst case" localization of hypoxic stress. Also, failure of functional activation may also provide evidence of critical hypoxia. Such responses may be very important in stroke because the absence of a functional response may indicate a critical low S_tO_2 Alternatively, pre-frontal activation of blood volume changes in response to standard tests can be used to ensure adequate O_2 delivery (N-back, anagram, etc.).

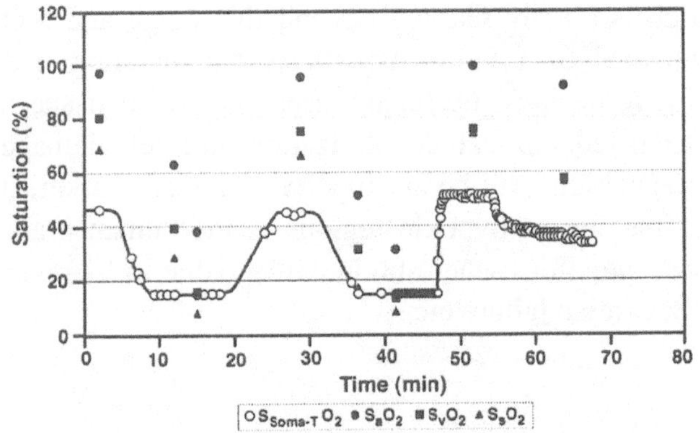

Piglet hypoxia test (Somanetics 5/12/99) (HYM 35)

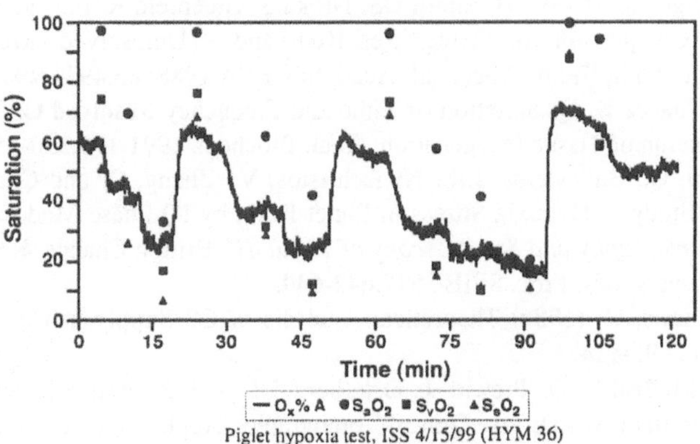

Piglet hypoxia test, ISS 4/15/99 (HYM 36)

Figure 2. An analysis of the meaning of S_tO_2 by comparing the S_tO_2 values to the arterial and venous values illustrating that the tissue value is rigorously maintained at about 40% of the difference between arterial and venous signals in the range from 20 to 80% S_tO_2 where our explanation suggests that autoregulation is maintaining this relationship while beyond this range autoregulation fails and hypometabolism reaches the arterial value or it hypermetabolism can reach the saggital sinus or venous value.

4. SUMMARY

4.1.1 Quantitative measurements of tissue oxygen saturation

The S_tO_2 now reveal hitherto predictable but unverified relationships between tissue, arterial and venous/saggital sinus oxygen saturations. The concept that autoregulation will establish an average value for S_tO_2 between arterial and venous and that the capillary system lies between the two values is strongly suggested by these data.

This relationship appears to be maintained by autoregulation over a wide range of values, i.e. from 20 to 80%. Thus, autoregulation may sense the capillary bed value rather than the arterial/venous values. Furthermore, loss of autoregulation is clearly marked beyond this range and S_tO_2 can reach arterial or venous values. It is our hypothesis that signaling of the failure of autoregulation may be most useful in cases of cortical tissue hypoxia that may lead to localized cortical depolarization and cell damage. Since the failure of autoregulation may be an important clinical marker, it is strongly suggested that imaging of hypoxic regions in the human brain be used to monitor all those situations where brain damage due to hypoxia may occur, i.e. stroke, CPB, cardiac failure etc.

REFERENCES

1. Chance B, Leigh *JS,* Miyake H, Smith DS, Nioka S, Greenfeld R, Finander M, Kaufmann K, et. al. Comparison of Time Resolved and Unresolved Measurements of Deoxyhemoglobin in Brain. Proc. Natl. Acad. Sci. USA 1988; 85:4971-4975.
2. Sevick EM, Chance B. Quantitation of Time-and Frequency-Resolved Optical Spectra for the Determination of Tissue Oxygenation. Anal. Biochem. 1991;195:330-351
3. Ma, H.Y., Xu, Q., Ballesteros, J.R., Ntziachristos, V., Zhang, Q. and Chance, B. (1999) Quantitative Study of Hypoxia Sttress in Piglet Brain by IQ Phase Modulation Oximetry. In *Optical Tomography and Spectroscopy of Tissue III,* Britton Chance, Robert R. Alfano, Bruce J. Tromberg, eds. Proc. SPIE 3597:642-649.
4. Groebe K, Thews, G. (1986) Theoretical Analysis of O2 Supply in Contracted Muscle. ISOTT , 1986; 495-514.
5. Pogue BW, McBride TO, Prewitt J, Osterberg UL and Paulsen KD Spatially Variant Regularization Improves Diffuse Optical Tomography. Appl. Optics 1999; 38:2950-2961.
6. Cope M. The development of a Near-Infrared System (dissertation) University College London, 1991
7. Nioka S, Smith DS, Chance, B. Control of Respiration in Uncoupling Mitochondria in *In Vivo* Brain. This volume.
8. Chance B, Anday E, Nioka S, Zhou S, Long H, Worden K, Li C, Turray T, Ovetsky Y, Pidikiti D, Thomas R. A Novel Method for Fast Imaging of Brain Function, Non-Invasively, with Light. Optics Express 1998;2:411-423.

Chapter 3

TUMOR OXIMETRY: COMPARISON OF ^{19}F MR EPI AND ELECTRODES

Ralph P. Mason, Sandeep Hunjan, Anca Constantinescu, Yulin Song, Dawen Zhao, Eric W. Hahn, Peter P. Antich, and Peter Peschke[+].
U.T. Southwestern Medical Center, Dallas, TX 75390, USA and [+]DKFZ, Heidelberg, Germany

Abstract: We recently described a novel approach to measuring regional tumor oxygen tension. This approach is based on ^{19}F pulse burst saturation recovery NMR echo planar imaging relaxometry of hexafluorobenzene or "FREDOM" (Fluorocarbon Relaxometry using Echo planar imaging for Dynamic Oxygen Mapping). We have now compared oxygen tension measurements using FREDOM with a traditional polarographic method (the Eppendorf Histograph) in a group of size matched Dunning prostate rat tumors R3327-AT1. We also compare MR and electrode approaches to monitoring dynamic changes with respect to interventions and demonstrate extension of the MR technique to rat breast tumors.

Key words: echo planar imaging, electrode, MRI, oxygen, prostate, tumor

Abbreviations: ARDVARC (Alternated Relaxation Delays with Variable Acquisitions to Reduce Clearance effects); EPI (echo planar imaging); FREDOM (Fluorocarbon Relaxometry using Echo planar imaging for Dynamic Oxygen Mapping); HFB (hexafluorobenzene); i.t (intra tumoral)

1. INTRODUCTION

It is widely appreciated that tumor oxygenation may significantly influence therapeutic success. In particular, the efficacy of radiotherapy [1], photodynamic therapy [2] and hypoxia selective chemotherapeutic agents [3] depends on pO_2. It had been suggested that the ability to measure tumor oxygenation in patients could allow therapy to be individualized and

Oxygen Transport to Tissue XXIV, edited by
Dunn and Swartz, Kluwer Academic/Plenum Publishers, 2003

optimized [4], and indeed, several recent studies have found significant prognostic value based on the Eppendorf Histograph in assessing clinical tumors [5-7]. While electrodes may be considered a "gold standard", they have certain shortcomings, such as invasiveness and limited sampling, and there is clearly a need for alternative methods [8]. We have been developing a new approach based on ^{19}F NMR of perfluorocarbons [9-12] and believe the method can now provide useful measurements of tumor oxygen dynamics *in vivo*.

The FREDOM approach exploits the exceptional response of the ^{19}F NMR spin lattice relaxation rate, R1, of fluorocarbons to changes in oxygen tension. Since fluorocarbons act as ideal liquids the solvation of gases is directly proportional to the partial pressure of the gas (Henry' law). Oxygen (O_2) is paramagnetic and induces relaxation in solution directly proportional to its concentration, and hence, pO_2 [13]. The highly hydrophobic nature of fluorocarbons ensures both high solubility of gases, providing molecular amplification, and minimal solvation of other materials (*e.g.*, metal ions) minimizing interference from other environmental factors. We, and others, have explored the use of numerous PFC reporter molecules and various routes of administration [14]. We believe that direct intra tumoral (i.t.) injection of HFB provides an optimal approach to tumor oximetry, and should provide measurements comparable to those obtained using electrodes. In addition, this minimally invasive approach facilitates mapping of dynamic changes in pO_2 with respect to interventions.

2. METHODS

Dunning prostate adenocarcinomas (R3327-AT1) were implanted in male Copenhagen rats (~250 g), as described in detail previously [15]. Tumors were divided into two groups and allowed to grow to either ~2 cm^3 or > 3.5 cm^3 volume. For MR investigations the rats were placed under general gaseous anesthesia with 33% inhaled O_2 (0.3 dm^3/min O_2, 0.6 dm^3/min N_2O, and 0.5% methoxyflurane. Hexafluorobenzene (25 - 40 μl) was injected directly into the tumors in both central and peripheral regions using a Hamilton syringe with a custom made fine sharp needle (32 G). A fiber optic probe was placed rectally to monitor core temperature. NMR experiments were performed using an Omega CSI 4.7 Tesla horizontal bore magnet system with actively shielded gradients with a tunable ($^1H/^{19}F$) single turn size-matched solenoid coil placed around the tumor. Following traditional imaging to establish the distribution of HFB, tumor oxygenation was estimated on the basis of ^{19}F PBSR EPI relaxometry of the HFB [10] with a typical 1.25 mm in plane resolution. For initial work three consecutive R1 measurements were made over a period of 1 hour to investigate reproducibility, and stability of the system when the rats breathed 33% O_2 (baseline). Since R1 is a linear function of pO_2 at constant temperature, pO_2 was estimated on a voxel by voxel basis using the relationship pO_2 (torr) = (R1 - 0.074)/0.0016 [10]. The inhaled gas was then altered to 100% oxygen, and relaxation measurements (three) were immediately repeated over a period of 1 hour. Finally, the gas was switched back to the baseline state and three further pO_2 determinations were immediately performed over 1 hour. Our initial studies required 20 mins to produce a pO_2 map, but more recent introduction of the ARDVARC acquisition protocol [12] provides enhanced maps in 8 mins. Breast 13762 NF adenocarcinomas were examined similarly.

Histography was applied to groups of size matched tumors, which did not receive HFB. Anesthesia was induced with halothane. Using the Eppendorf Histograph 100 to 200 individual pO_2 determinations were made in each tumor, as recommended by the manufacturer. For dynamic measurements a Diamond General micro-electrode (700 μm) was inserted to a specific location. Baseline pO_2 was measured and the inhaled gas altered to 100% O_2 or carbogen (95%O_2/5%CO_2) for 30 mins. At this stage pO_2 was again measured. Following a series of measurements with different gases at one location, the needle was moved and the gases cycled again.

Statistical significance of changes in oxygenation was assessed using analysis of variance (ANOVA) on the basis of Fisher PLSD. Experiments were approved by the Institutional Animal Care and Advisory Committee conducted in accordance with National Laws.

3. RESULTS

Both FREDOM and electrode methods indicated similar oxygen tension distributions for the AT1 tumors (Fig. 1). Moreover, both techniques showed that tumors with volume > 3.5 cm^3 were significantly ($p < 0.0001$) less well oxygenated than smaller tumors (volume < 2 cm^3). For the large tumors FREDOM indicated median pO_2 = 2 torr and fraction < 10 torr (HF_{10}) = 82 %, while the Eppendorf electrode indicated median pO_2 = 3 torr and HF_{10} = 84%. For the small tumors the match was less good with median = 15 v 8 torr and HF_{10} = 44 versus 66% for FREDOM and electrode, respectively. Examination of the MR images showed that for 1 small tumor most of the HFB resided very close to the tumor edge and may have biased the apparent pO_2. Indeed, if this tumor was excluded there was no significant difference between the respective pO_2 distributions.

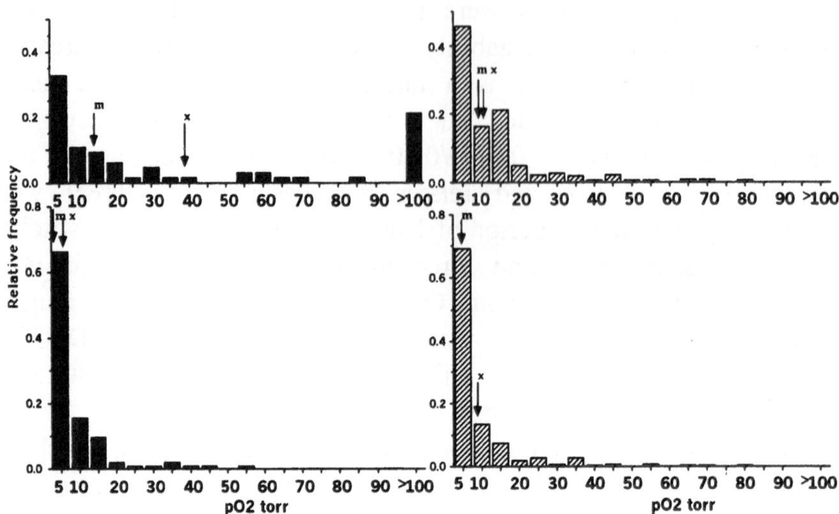

Figure 1. Comparison of oxygenation in size-matched groups of AT1 tumors based on ^{19}F MR EPI relaxometry (left) and electrode polarography (right), when rats inhaled 33% O_2. Small tumors are shown at top (volume < 2 cm^3) and large tumors below (volume > 3.5 cm^3). Each method shows a significant difference in tumor oxygenation for small versus large tumors ($p < 0.0001$).

Using the FREDOM approach we also examined response to respiratory challenge. Increasing the concentration of inspired oxygen from 33% to 100% O_2 produced a significant increase ($p < 0.0001$) in tumor oxygenation for the group of small tumors. In contrast no change was observed in the mean pO_2 for the group of large tumors. A strength of the FREDOM approach is the ability to follow individual tumor regions, with respect to intervention, in this case respiratory challenge. Six representative regions were selected from a single tumor (Fig. 2a). Three regions, which were initially well oxygenated (pO_2 > 10 torr) showed rapid and significant increases within 8 minutes of switching from 33% O_2 to 100% O_2. Changes in relatively poorly oxygenated regions were much slower, although 2 of 3 regions did show a significant change in pO_2 after 24 mins.

Electrode investigation of dynamic changes in pO_2 also showed 3 of six regions with significant changes in switching from 33% O_2 to 100% O_2, but only 1 region was also significantly different with carbogen (Fig. 2b).

In a representative large breast tumor (~ 4 cm^3) we found significant changes in pO_2 ($p < 0.0001$) with respect to respiratory challenge with baseline mean $pO_2 = 40 \pm 3$ rising to mean $pO_2 = 99 \pm 4$, when rat inhaled carbogen and mean $pO_2 = 145 \pm 4$ for oxygen inhalation.

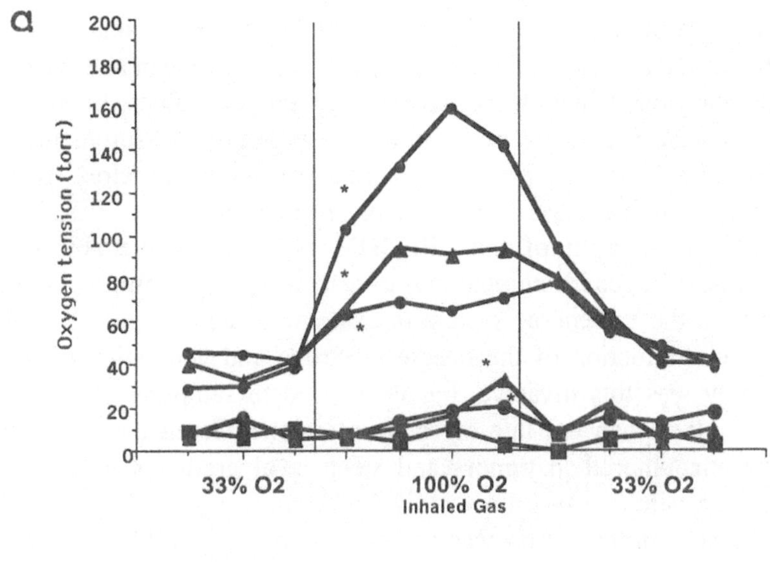

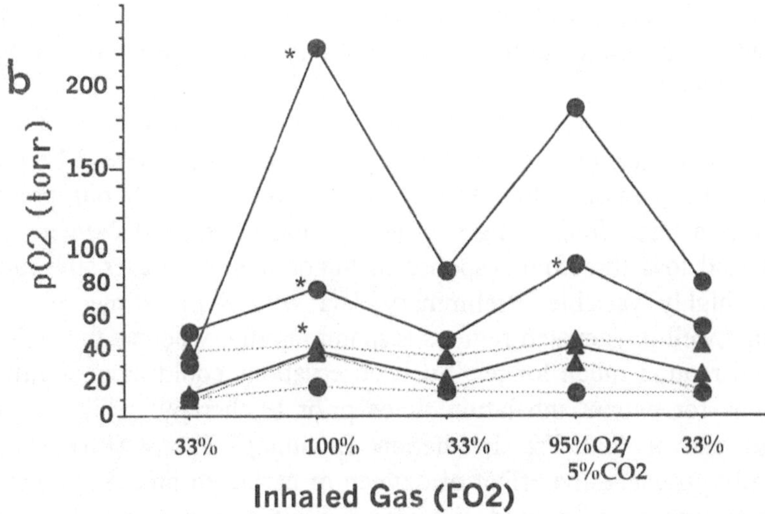

Figure 2. a) Dynamic changes in pO_2 of six specific regions of an AT1 tumor. The three high pO_2 regions had significantly different pO_2 (* $p< 0.05$)) from those with low pO_2 at each time point. Within 8 mins of elevating inspired O_2 the three high pO_2 voxels had significantly increased pO_2 ($p< 0.05$) while the low pO_2 voxels required > 24 mins to show significant changes. All six regions were observed simultaneously using the FREDOM approach.
b) Dynamic changes in pO_2 of six specific regions of an AT1 tumor. The electrode was placed in one location at a time and inhaled gases cycled for subsequent locations.

4. DISCUSSION

These results demonstrate the similarity of measurements obtained using traditional electrodes or the new FREDOM approach to tumor oximetry. In each case there was a significant difference in pO_2 observed in small versus large AT1 tumors. For larger tumors the hypoxic fraction, mean and median were very similar, together with the range of typical pO_2 values. In smaller tumors MR suggested a larger range with a number of measurements in excess of 100 torr. This may have arisen from measurements close to the tumor periphery, which are less common using electrodes.

A significant strength of the FREDOM approach is the ability to monitor dynamic changes in regional pO_2 in response to acute interventions. Others have used the Eppendorf system to examine acute changes [16], but this required reintroduction of the needle electrode and generation of new tracks. Not only was this invasive, but it also led to sampling of parallel tissue regions rather than the fate of specific regions. Given the extensive heterogeneity encountered in tumors and steep local gradients in pO_2 we believe it will be valuable to follow individual tumor regions. Historically, regional response to intervention was assessed by placing an electrode at a specific location and monitoring changes in pO_2 [17]. We have now performed such experiments with a micro electrode and found a range of baseline pO_2 values and response to respiratory challenge similar to those seen using MRI.

We have now shown both that there is distinct intra tumoral heterogeneity in baseline oxygenation in the Dunning prostate AT1 tumor and also in the response to intervention. In common with our previous observations a three fold change in the fraction of inhaled oxygen (FO_2) seems to lead to a threefold response in tumor pO_2. However, the rate of change is highly variable. Preliminary data with 8 min time resolution suggest that well oxygenated regions respond rapidly, whereas those poorly oxygenated require much longer. Such observations could have significant implications for patient inhalation times prior to therapy: while previous work had shown that Pre Irradiation Breathing Times (PIBT) could substantially influence the effect of oxygen or carbogen breathing [18], the differential response of individual tumor regions may not have been fully appreciated.

In developing a new technique it is important to demonstrate its reliability, robustness and general application. We and several other groups have now applied the FREDOM approach to tumor oximetry. Initially investigators favored intra venous or intra peritoneal administration of emulsions of fluorocarbons. While material became trapped in tumors and could be used to report pO_2 [19-22], it became increasingly apparent that material delivered via the vasculature tended to bias measurements towards well perfused tumor regions [22]. Indeed, recent measurements by Griffiths *et al.* have confirmed such a bias [23]. Furthermore, the use of emulsions to carry the PFCs tends to lead to extensive uptake by the reticuloendothelial system with hepatomegaly. Intra tumoral administration is minimally invasive provided that a fine sharp needle is applied, as we have used here. We have now extended our work from the Dunning prostate R3327-AT1 tumor, which is poorly differentiated, has only microscopic necrotic foci and is firm, to the 13762 breast tumor, which has less structure and considerable cystic fluid. Here, we have simply reported the ability to measure dynamic changes in the breast tumor oxygenation, but in the accompanying work (Song *et al.*, this volume), we show more extensive results.

Since the MR and electrode approaches appear to give similar results one may debate their relative merits. Clearly, MR is very expensive, with a typical imaging system costing upwards of $1 M, compared with $60 000 for the Eppendorf system and < $5 000 for a laboratory micro electrode system. However, MR facilitates the simultaneous measurement of dynamic changes over extensive tumor regions in response to intervention, which is particularly valuable for tumors exhibiting a high degree of heterogeneity. While we were able to follow changes in pO_2 at specific regions using a needle electrode with placement at sequential locations accompanied by cycling of the intervention, such an approach would be less satisfactory for other interventions, and even here, may have led to some conditioning or hysteresis. The FREDOM approach may be readily combined with other measurements such as blood flow/perfusion [24], pH [25], or metal ions by infusion of appropriate reporter molecules [26].

As a reporter molecule HFB has many advantageous properties. It is cheap, readily available, and exhibits minimal acute toxicity (LD_{50} > 25 g/kg) [27]. No signs of renal or hepatic toxicity have been found [2] and others have tested doses as high as 50 g/kg (twice weekly) orally in rats over 35 weeks [29]. We use ~ 250 mg/kg and typically find substantial clearance from tumors within 24 h, though this does limit our measurements to acute response to interventions [12]. High symmetry within the molecule leads to a single ^{19}F MR resonance providing optimal SNR. The R1 (=1/T1) is highly sensitive to pO_2 while showing little response to temperature [9]. Long T1s up to 14 s appear to make HFB less efficient for spin lattice relaxometry, but use of the pulse burst saturation recovery approach minimizes the length of the experiment [10] and a large range of T1 values is a requisite for sensitivity to changes in pO_2. The long transverse relaxation time (T2) is ideally suited to echo planar imaging.

The ultimate value of a novel technique will depend on its adoption by multiple laboratories, and the significance of the results that can be generated. We believe that the FREDOM approach is versatile, as demonstrated by increasing applications, and thus, we foresee expanded future application of the technique.

ACKNOWLEDGMENTS

This work was supported in part by The American Cancer Society (RPG-97-116-010CCE; RPM), NCI RO1 79515 (RPM), DOD Breast Cancer Initiative (YS), Verein zur Forderung der Krebserkennung and Krebshandlung e.V. Heidelberg (PP) and the NIH BRTP Facility #5-P41-RR02584.

REFERENCES

1. Hall EJ. The oxygen effect and reoxygenation. In: Hall EJ, ed. Radiobiology for the Radiologist. 3 ed. Philadelphia: Lippincott, J. B., 1994;133-152.
2. Chapman JD, Stobbe CC, Arnfield MR, Santus R, Lee J, McPhee MS. Oxygen Dependency of Tumor Cell Killing *In Vitro* by Light Activated Photofrin II. Radiat Res 1991;126:73-79.
3. Brown JM, Giaccia AJ. Tumor hypoxia: the picture has changed in the 1990s. Int J Radiat Biol 1994;65:95-102.
4. Vaupel PW, Höckel M. Oxygenation status of human tumors: a reappraisal using computerized pO_2 histography. In: Vaupel PW, Kelleher DK, Günderoth M, eds. Tumor Oxygenation. Stuttgart: Gustav Fischer, 1995;219-232. (Thews G, ed. Funktionsanalyse biologischer Systeme; vol 24).
5. Fyles AW, Milosevic M, Wong R, et al. Oxygenation predicts radiation response and survival in patients with cervix cancer. Radiother Oncol 1998;48:149-156.
6. Höckel M, Schlenger K, Aral B, Mitze M, Schäffer U, Vaupel P. Association between tumor hypoxia and malignant progression in advanced cancer of the uterine cervix. Cancer Res 1996;56:4509-4515.
7. Nordsmark M, Overgaard M, Overgaard J. Pretreatment oxygenation predicts radiation response in advanced squamous cell carcinoma of the head and neck. Radiother Oncol 1996;41:31-40.
8. Stone HB, Brown JM, Phillips T, Sutherland RM. Oxygen in human tumors: correlations between methods of measurement and response to therapy. Radiat Res 1993;136:422-434.
9. Mason RP, Rodbumrung W, Antich PP. Hexafluorobenzene: a sensitive ^{19}F NMR indicator of tumor oxygenation. NMR in Biomed 1996;9:125-134.
10. Le D, Mason RP, Hunjan S, Constantinescu A, Barker BR, Antich PP. Regional tumor oxygen dynamics: ^{19}F PBSR EPI of hexafluorobenzene. Magn Reson Imaging 1997;15:971-981.
11. Mason RP, Constantinescu A, Hunjan S, et al. Regional tumor oxygenation and measurement of dynamic changes. Radiat Res 1999;152:239.
12. Hunjan S, Mason RP, Constantinescu A, Peschke P, Hahn EW, Antich PP. Regional tumor oximetry: ^{19}F NMR spectroscopy of hexafluorobenzene. Int J Radiat Oncol Biol Phys 1998;40:161-171.

13. Delpuech J-J, Hamza MA, Serratice G, Stébé M-J. Fluorocarbons as oxygen carriers. I. An NMR study of oxygen solutions in hexafluorobenzene. J Chem Phys 1979;13:399.
14. Mason RP. Non-invasive physiology: ^{19}F NMR of perfluorocarbon. Art. Cells, Blood Sub & Immob Biotech 1994;22:1141-1153.
15. Hahn EW, Peschke P, Mason RP, Babcock EE, Antich PP. Isolated tumor growth in a surgically formed skin pedicle in the rat: a new tumor model for NMR studies. Magn Reson Imaging 1993;11:1007-1017.
16. Laurence V, Ward R, Bleehen N. Tumor pO_2 distribution in patients treated with the combination of nicotinamide and carbogen breathing. In: P. W. Vaupel, D. K. Kelleher, M. Günderoth, eds. Tumor Oxygenation. Stuttgart: Gustav Fischer, 1995;185-193.
17. Cater D, Silver I. Quantitative measurements of oxygen tension in normal tissues and in the tumors of patients before and after radiotherapy. Acta Radiol 1960;53:233-256.
18. Chaplin DJ, Horsman MR, Siemann DW. Further evaluation of nicotinamide and carbogen as a strategy to reoxgenate hypoxic cells *in vivo*: importance of nicotinamide dose and pre-irradiation breathing time. Br J Cancer 1993;68:269-273.
19. Dardzinski BJ, Sotak CH. Rapid tissue oxygen tension mapping using ^{19}F Inversion-recovery Echo-planar imaging of Perfluoro-15-crown-5-ether. Magn Reson Med 1994;32:88-97.
20. Baldwin NJ, Ng TC. Oxygenation and metabolic status of KHT tumors as measured simultaneously by ^{19}F magnetic resonance imaging and ^{31}P magnetic resonance spectroscopy. Magn Reson Imaging 1996;14:514-551.
21. Fishman JE, Joseph PM, Carvlin MJ, Saadi-Elmandjra M, Mukherji B, Sloviter HS. *In vivo* measurements of vascular oxygen tension in tumors using MRI of a fluorinated blood substitute. Invest Radiol 1989;24:65-71.
22. Mason RP, Antich PP, Babcock EE, Constantinescu A, Peschke P, Hahn EW. Non-invasive determination of tumor oxygen tension and local variation with growth. Int J Radiat Oncol Biol Phys 1994;29: 95-103.
23. McIntyre DJO, McCoy CL, Griffiths JR. Tumour oxygenation measurements by ^{19}F MRI of perfluorocarbons. Curr Sci 1999;76:753-762.
24. Brown SL, Ewing JR, Kolozsvary A, Butt S, Cao Y, Kim JH. Magnetic Resonance Imaging of perfusion in rat cerebral 9L tumor after nicotinamide administration. Int J Radiat Oncol Biol Phys 1999;43:627-633.
25. Mason R, Hunjan S, He S, et al. Tumor trans membrane pH gradient and regional oxygen tension measured by fluorine magnetic resonance. In: Moraes M, Brentani R, Bevilacqua R, eds. 17th International Cancer Congress. Rio de Janeiro: Monduzzi, 1998;1627-1631. vol 2).
26. Mason RP. Transmembrane pH gradients *in vivo*: measurements using fluorinated vitamin B6 derivatives. Curr Med Chem 1999;6:533-551.
27. Lancaster. Material Safety Data Sheet. In: Lancaster Synthesis Inc., 1998.
28. Hall LW, Jackson SRK, Massey GM. Hexafluorobenzene in veterinary anaesthesia. In: Arias A, Llaurado R, Nalda MA, Lunn JN, eds. Recent Progress in Anaesthesiology and Resuscitation. Oxford: Excerpta Medica, 1975;201-204.
29. Rietjens IMCM, Steensma A, den Besten C, et al. Comparative biotransformation of hexachlorobenzene and hexafluorobenzene in relation to the induction of porphyria. Eur J Pharmacol 1995;293:292-299.

Chapter 4

MAPPING CEREBRAL GLUTAMATE ^{13}C TURNOVER AND OXYGEN CONSUMPTION BY *IN VIVO* NMR

Fahmeed Hyder [#,%], Peter Brown, Terennce W. Nixon, [*]Kevin L. Behar
Magnetic Resonance Research Center, Departments of [#]Diagnostic Radiology, [%]Biomedical Engineering, and [*]Psychiatry, Yale University School of Medicine, New Haven, CT, USA

Abstract: Regional rates of ^{13}C incorporation from glucose to glutamate were detected in anesthetized rat brain *in vivo* at 7T with high temporal and spatial resolution using NMR method ICED PEPSI (*i*n vivo *c*arbon *e*dited *d*etection with *p*roton *e*cho *p*lanar *s*pectroscopic *i*maging). Time courses of regional glutamate ^{13}C turnover were fitted by a metabolic model to obtain regional tri-carboxylic acid (TCA) cycle flux and cerebral metabolic rate of oxygen consumption (CMR_{O2}) in each voxel (8 μL) of rat cortex. CMR_{O2} maps obtained for rats under either α-chloralose or morphine anesthesia revealed average cortical values of 1.5±0.2 (n = 3) and 3.2±0.3 (n = 4) μmol/g/min, respectively. These values of CMR_{O2} are in good agreement with previous cortical measurements with coarser spatial resolution. The heterogeneity within each map, which depicted predominantly gray and white matter differences, was significantly greater under morphine (higher cortical activity) than under α-chloralose (lower cortical activity) anesthesia. The regional variations in the basal awake state, which are expected to be even greater, should be considered to avoid partial-volume artifacts in functional activation studies of awake subjects.

Key words: glucose - lactate - glutamine - Kreb's cycle - metabolism - echo planar imaging

1. INTRODUCTION

In vivo ^{13}C NMR spectroscopic detection in conjunction with infused ^{13}C-labeled stable isotopes of glucose in the blood stream can provide important information on brain energy metabolism [1]. A variety of techniques are available for this purpose, e.g., direct ^{13}C NMR detection [2] or indirect ^{13}C NMR detection using single quantum (^{1}H observed ^{13}C editing (POCE); [3]) or multiple quantum techniques [4], each with specific advantages and disadvantages in terms of sensitivity and resolution. The flow of ^{13}C label from glucose to glutamate, glutamine, and γ-amino butyric acid (GABA) can be used to calculate several metabolic fluxes [5] including the tri-carboxylic acid (TCA) cycle flux (V_{TCA}), cerebral metabolic rates of glucose (CMR_{glc}) and oxygen (CMR_{O2}) consumption, and rate of neurotransmitter cycling (V_{cyc}).

The *in vivo* ^{13}C data is richer in biochemical information in comparison to data obtained by other methods, e.g., ^{14}C-2-deoxyglucose autoradiography [6], because of the ability to detect turnover of the ^{13}C label from the intrinsic substrate to one (or more) metabolite(s) at different carbon positions within the molecule(s) in real time. In the autoradiography method a ^{14}C-labeled analogue of glucose, 2-deoxyglucose, is infused into the blood stream in trace amounts. The radioactive analog crosses the blood-brain-barrier and is phosphorylated much like glucose. Since the ^{14}C-labeled phosphate cannot be metabolized further, the amount of the trapped radioactive derivative is then used to interpret CMR_{glc}. While ^{13}C methods provide superb metabolic information, the ^{14}C autoradiographic method provides exceptional spatial information of CMR_{glc}. Improved spatial resolution in ^{13}C detection would be of significant value for quantitative neurobiological research.

Because of the greater sensitivity advantage of ^{1}H NMR, regional metabolic ^{13}C turnover is generally obtained by indirect ^{13}C detection in conjunction with chemical shift imaging (CSI) requiring long acquisition times (e.g., ref. [4]). A viable method for improving the temporal resolution of indirect ^{13}C CSI is *p*hase *e*ncoded echo *p*lanar *s*pectroscopic *i*maging or PEPSI [7]. Recently, we described our preliminary results in the development of *i*n vivo *c*arbon *e*dited *d*etection with *p*roton *e*cho *p*lanar *s*pectroscopic *i*maging or ICED PEPSI, which combines the PEPSI sequence with ^{13}C-^{1}H *J*-editing and semi-selective water suppression of POCE to permit imaging of ^{13}C-labeled metabolite turnover [8]. Here we apply the ICED PEPSI technique to measure regional turnover of C4-glutamate from [1,6-^{13}C]glucose in 8 μL voxels of the rat cortex under two different anesthetics.

2. METHODS

2.1 Animals and materials

Adult, male, Sprague-Dawley rats (*n* = 5; 105-165 g; fasted >16 hours) were tracheotomized (0.7-1.2% halothane anesthesia) and mechanically ventilated (70%/30% ratio of N_2O/O_2). A femoral artery was cannulated for continuous blood pressure monitoring and periodic blood sampling of pCO_2, pO_2, pH, and glucose. A femoral vein was cannulated for the intravenous infusion of *D*-[1,6-^{13}C]glucose (99 atom %; Cambridge Isotopes, Andover, MA). An intra peritoneal (i.p.) catheter was placed for delivery of either α-chloralose or morphine and *D*-tubocurarine chloride (initial 0.5 mg/kg; supplemental 0.25 mg/kg/30 min). The scalp above the bregma was clipped, the dura was cleared along with all fatty tissues on the skull, and the exposed wound was cleaned and secured with a layer of Saran Wrap. The rat's torso was covered with a warm water pad to maintain body temperature at ~37 °C and then placed prone in a cradle with a bite-bar to secure the head. Foam cushions were tightly fixed on either side of the head to prevent head movement. The radio-frequency (RF) surface-coil receiver was centered above the bregma and the rat's head was positioned at the magnet isocenter. Halothane anesthesia was then discontinued and thereafter, anesthesia was maintained with either morphine sulfate (initial 50 mg/kg; supplemental 30 mg/kg/30min; i.p.) or α-chloralose (initial 80 mg/kg; supplemental 20 mg/kg/½ hour; i.p.).

2.2 *In vivo* NMR measurements

NMR data were acquired on an extensively modified 7T Bruker Biospec I horizontal-bore spectrometer (Bruker, Billerica, MA), operating at 300.4 and 74.6 MHz for ^{1}H and ^{13}C, respectively. The RF probe consisted of one ^{13}C and two ^{1}H coils. The two ^{1}H RF coils, a transmit resonator (8 cm, diameter) and a receive surface-coil (10 mm, diameter), were oriented orthogonally such that the homogenous region of the two RF coils were coincident and the surface-coil receiver experienced a negligible loss in signal-to-noise ratio. A concentric ^{13}C RF surface-coil (20 mm, diameter) was used for transmission and decoupling. Localized static magnetic field homogeneity was optimized prior to data acquisition. Gradient echo coronal images were acquired with an image matrix of 128×128 pixels, field of view of 20 mm, repetition time (TR) of 250 ms and echo time (TE) of 16 ms.

The ICED PEPSI pulse sequence has been described [8]. The sequence was separated into four segments: (i) water suppression with chemical shift selective excitation of water followed by crusher gradients, (ii) selection of a 2D rectangular box in the *x*, *z* dimension (14.0×7.5 mm2) using 2D *i*mage-*s*elected *i*n vivo *s*pectroscopy (ISIS; [9]), (iii) 1D outer volume suppression by *y* slice excitation followed by crushing of signals from the surface lipid and deeper structures [10] to allow selection of a 3D rectangular box in the *x*, *y*, and *z* dimensions (14.0×7.5×5.0 mm^3), and (iv) heteronuclear editing with echo planar acquisition. A sinc ^1H pulse was used in segments (i) and (iii) and an adiabatic fast passage hyperbolic secant ^1H pulse was used in segment (ii). An incremented *y* gradient (G_y) was used in segment (iv) for spatial phase-encoding and periodic switching of the *x* gradient (G_x) throughout the acquisition period was used for spatial read-out convolved with the free-induction decay (FID). Segment (iv) consisted of a spin-echo with TE of 16 ms and TR of 2 s. A ^1H-90° hard pulse (α_{+x}) followed by a ^1H-180° hard 2-τ-2 semi-selective pulse (β_{+x} β_{-x}; τ = 676 μs) was used for spin-echo excitation and additional water suppression. Two balanced crusher gradients were used in each half of the spin-echo sequence to eliminate non-refocused magnetization. A ^{13}C-180° phase-cycled hard pulse (γ_{+x} $\gamma_{\pm x}$) was centered at 1/2*J* (= 3.9 ms; *J* 125 Hz) from the ^1H-90° pulse to inhibit the ^1H-^{13}C *J* modulation. The ^{13}C inversion pulse resulted in 95% inversion with a bandwidth of ±2.5 ppm (in ^1H chemical-shift). Each FID was acquired in the presence of a broadband (±10 ppm in ^{13}C chemical-shift) ^{13}C decoupling [3] where the receiver phase was alternated (i.e., δ_{+x} δ_{+x} δ_{-x} δ_{-x}). The phase-cycling scheme, shown for 4 consecutive scans, resulted in an edited POCE spectrum (without any subtraction) after summing of 8 consecutive scans ($[\alpha_{+x}]_1$, $[\alpha_{+x}]_2$, $[\alpha_{+x}]_3$, $[\alpha_{+x}]_4$; $[\beta_{+x}\ \beta_{-x}]_1$, $[\beta_{+x}\ \beta_{-x}]_2$, $[\beta_{+x}\ \beta_{-x}]_3$, $[\beta_{+x}\ \beta_{-x}]_4$; $[\gamma_{+x}\ \gamma_{+x}]_1$, $[\gamma_{+x}\ \gamma_{-x}]_2$, $[\gamma_{+x}\ \gamma_{+x}]_3$, $[\gamma_{+x}\ \gamma_{-x}]_4$; $[\delta_{+x}]_1$, $[\delta_{+x}]_2$, $[\delta_{-x}]_3$, $[\delta_{-x}]_4$).

The first echo in the train was carefully adjusted for the spin-echo. In each echo train, 128 consecutive echoes were acquired with switching G_x with a period of 640 μs, which allowed for a spectral bandwidth of 5.20 ppm. Each echo, which provided spatial information in the *x* dimension, contained 32 complex points with a dwell time of 10 μs. The total acquisition time for all echoes was 40.96 ms. Phase encoding was implemented in the *y* direction. The sequence was run with 16 phase-encode steps, guided by the coordinates of the anatomical image, to provide a 16×16 imaging grid with spatial resolution of 1250×1250×5000 μm^3 per voxel in *x*, *y*, and *z* dimensions. A minimum of 16 averages per phase-encode were used with 4 dummy scans per phase-encode step. All data were processed off-line on an Indy Silicon Graphics (Mountain View, CA) workstation with home-written algorithms in MATLAB (Natick, MA). Each data set of 64 echos, was separated into odd and even echoes and the even echoes were reversed in the time domain. The time-forward odd and time-reversed even echo trains were then corrected for sequential sampling [11]. The 16×16×64 data matrix per echo train was zero-filled to a 16×16×1024 points. No window function or filter was applied to spatial dimensions prior to Fourier transformation. An additional 5-10 Hz exponential line broadening was applied in the chemical shift dimension. A small zero order phase correction was applied and no first order phase correction nor baseline correction were necessary. The spectral data from odd and even echoes were added to improve spectral sensitivity. Limited integral bandwidth (0.3 ppm) was applied to create 2D CSI maps either from the time-forward odd or time-reversed even echo trains. Small zero- and first-order baseline corrections were necessary to match the baselines of the two spectra before they were added. Prior to the start of [1,6-^{13}C]glucose infusion, one set of PEPSI data were collected and processed in the same manner as above. Integrated peaks of C4-glutamate from 2D CSI maps of ICED PEPSI and PEPSI data were compared to obtain a ^{13}C fractional enrichment (FE) map of C4-glutamate *in vivo*. ^{13}C FE of C1,6-glucose at 15 minute intervals were obtained from acid extracts of frozen plasma samples [2].

2.3 The metabolic model

The metabolic model has been described [12-15]. Plasma and brain glucose exchange via Michaelis-Menten kinetic parameters K_m (13.9 mM) and V_{max} (1.16 μmol/g/min) which corresponds to V_{max}/CMR_{glc} = 5.8 for resting α-chloralose anesthetized rats. The brain C1,6-glucose concentration contributes to a time lag before the ^{13}C label reaches the C4-glutamate pool via the TCA cycle. The fits performed with V_{max}/CMR_{glc} ranging from 2.5 to 25, which represent either slow or rapid glucose transport, resulted in V_{TCA} values which were within 15% of values obtained with V_{max}/CMR_{glc} of 5.8. This indicates that glucose transport is not the rate limiting factor in glucose metabolism *in vivo* [14]. The ^{13}C label flows at a rate $2CMR_{glc}$ through the glycolytic intermediates (negligible concentrations) and arrives at C3-pyruvate and C3-lactate, which are in rapid exchange, represented by a single pool L (1.5 μmol/g). Two sources of unlabeled carbon into the TCA cycle are the exchange of blood-brain pyruvate and lactate, V_{ex}, and the ketone body flux, V_{ket}. Both of these sources of dilution for the acetyl CoA pool, represented as V_{dil}, could be represented either with total dilution source from ketone bodies (i.e., $V_{ket}=V_{dil}$) or total dilution source due to pyruvate (and lactate) blood-brain exchange (i.e., $V_{ex}=V_{dil}$). The dilution source may have large contributions from ketone bodies since the rats in this study and in our previous studies [8,14-16] were fasted for >16 hours. Similarly, the pyruvate (and lactate) blood-brain exchange may have made significant contributions of unlabeled carbons to the TCA cycle via the acetyl CoA pool. From our previous studies [14] it was determined that the major source of dilution to the acetyl CoA pool is the pyruvate (and lactate) blood-brain exchange (i.e., $V_{ex}=V_{dil}$ and $V_{ket}=0$) which is based on the finding that the ^{13}C FE of C4-glutamate is slightly less than that of C1,6-glucose. The ^{13}C label enters the TCA cycle and labels C4-α-ketoglutarate and C4-glutamate (12.0 μmol/g). These two pools are in very rapid isotopic exchange, V_x, where $V_x/V_{TCA} >> 1$.

One turn of the TCA cycle, V_{TCA}, is approximately coupled with CMR_{O2}. In the first pass of the TCA cycle the ^{13}C label arrives at C4-glutamate and in the second and subsequent passes of the cycle the ^{13}C label arrives at C2- and C3-glutamate. There is an exchange between C4-glutamate and C4-glutamine (6.2 μmol/g) at a rate of V_{gln} (0.08 μmol/g/min). The value of V_{gln} contributes to a time lag between the turnover data in C4- and C3-glutamate ^{13}C labeling. The fits performed with V_{gln} ranging from 0.01 to 0.44 μmol/g/min, which represent either slow or rapid glutamate-glutamine exchange, resulted in V_{TCA} values which were within 20% of values obtained with V_{gln} of 0.08 μmol/g/min [14]. This indicates that glutamate-glutamine exchange is not the rate limiting factor in the ^{13}C turnover data of C4-glutamate. The efflux at L, V_{out}, causes some ^{13}C label to be lost to the blood. This source of ^{13}C label loss does not contribute to dilution of the acetyl CoA pool. A maximum value of V_{out} was estimated to be in the range of V_{TCA} [15].

The model has inputs consisting of the time courses of measured plasma glucose concentrations and ^{13}C FEs of plasma C1,6-glucose and brain C4-glutamate. A set of coupled differential equations are used to describe the model and an iterative method is used to fit the model to the C4-glutamate data. Data from ICED PEPSI experiments were fitted to obtain best fitted V_{TCA} values. CMR_{O2} was calculated as $3V_{TCA} - 0.75V_{ket}$ (where V_{ket} was found to be negligible), and the value of total dilution of the acetyl CoA pool, V_{dil}, was determined as V_{TCA}[1 – (C4-glutamate ^{13}C FE/½C1,6-glucose ^{13}C FE)]. All data are presented as mean ± standard deviation.

3. RESULTS

3.1 CMR_{O2} maps

Typical CMR_{O2} maps of the sensorimotor cortex of rats under α-chloralose (A) and morphine (B) anesthesia, along with a representative anatomical image (C) are show in Fig. 1. The nominal spatial resolution of these images is 8 μL per voxel. Metabolic heterogeneity was detected in deep and central structures in the CMR_{O2} map, which is consistent with the differences in metabolic rates between gray and white matter reported for these regions using ^{14}C-2-deoxyglucose autoradiography [6]. Poorer RF sensitivity and shimming in the lower portions of the brain and along the midline compared to the dorsal surface may also contribute to some of the heterogeneity detected in each map [8]. The heads of the rats in (A) and (B) were oriented such that the brain dipped lower in front and higher in rear along the slice dimension. Since the slice thickness in the metabolic data was approximately four times the in-plane resolution, most of the partial volume in the data came from the slice dimension. Specific pixels in (A) and (B) have been identified as gray and white matter where ever possible based on the multiple anatomical images that were obtained for each rat. CMR_{O2} averaged from voxels in the cortex for data in (A) and (B) were 1.5±0.2 and 3.1±0.5 μmol/g/min, respectively. Averaged CMR_{O2} values from a group of rats for α-chloralose and morphine anesthesia were 1.5±0.2 (n = 3) and 3.2±0.3 (n = 4) μmol/g/min, respectively. These cortically localized values of CMR_{O2} are in excellent agreement with localized POCE measurements under identical conditions [14-16].

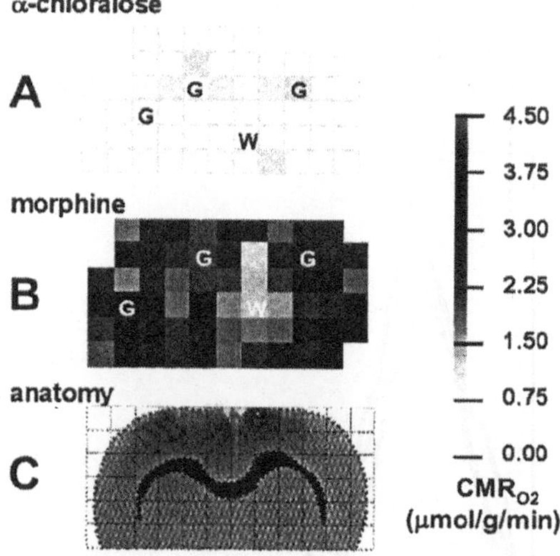

Figure 1. CMR_{O2} maps of two separate rats under α-chloralose (A) and morphine (B) anesthesia with nominal spatial resolution of 8 μL per voxel. In each case, the map is oriented coronally with respect to anatomy where the cortical surface is located in the top of each coronal map and the midline is located approximately in the middle of each map. In (C) a high resolution anatomical image reveals the approximate location of the 16×16 grid for each map. CMR_{O2} averaged from voxels in the cortex in (A) and (B) were 1.5±0.2 and 3.1±0.5 μmol/g/min, respectively. The slice thickness in the metabolic data was approximately four times the in-plane resolution. Some pixels in (A) and (B) have been identified as gray (G) and white (W) matter based on multiple anatomical images. Averaged CMR_{O2} values from a group of rats for -chloralose and morphine anesthesia were 1.5±0.2 (n = 3) and 3.2±0.3 (n = 4) μmol/g/min, respectively, which are in excellent agreement with localized POCE measurements under identical conditions [14-16].

3.2 Distributions of CMR_{O2} with anesthesia

Fig. 2 shows the distribution of CMR_{O2} values in α-chloralose and morphine anesthetized rats represented as histograms. In morphine anesthetized rats CMR_{O2} values span a significantly wider metabolic range (i.e., 0.4 to 4.0 μmol/g/min vs. 0.2 to 2.2 μmol/g/min) than in α-chloralose anesthetized rats. However, the dynamic range (i.e., ratio of maximum-to-minimum) of CMR_{O2} values for both α-chloralose and morphine anesthesia is approximately the same, 10 versus 8 times, respectively. This indicates that while the absolute metabolic rates may vary with type and/or depth of anesthesia, the same dynamic range is approximately maintained which is a strong gauge for maintenance of autoregulation in these two anesthetized conditions [14-16]. This comparison reveals that at higher levels of activity, metabolic heterogeneity is greater, thus rendering lower spatial resolution methods of CMR_{O2} measurement more susceptible to partial-volume errors. Since CMR_{O2} in the basal awake state is higher than the anesthetized conditions investigated in this study [17], the regional variation in CMR_{O2} is expected to be greater. Therefore, regional variations in metabolism in the basal awake state should be considered in order to avoid partial-volume artifacts in functional activation studies of awake subjects (e.g., humans).

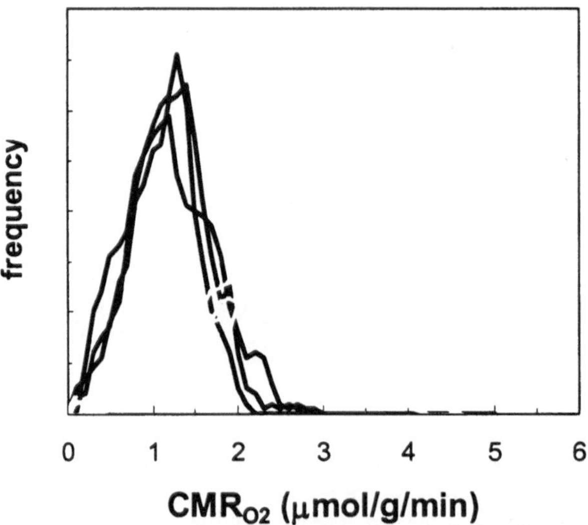

Figure 2. Distribution of CMR_{O2} values in α-chloralose (solid line) and morphine (dotted line) anesthetized rats. In morphine anesthetized rats CMR_{O2} values span a significantly wider metabolic range (i.e., 0.4 to 4.0 μmol/g/min; data from two other rats not shown) than in α-chloralose anesthetized rats (0.2 to 2.2 μmol/g/min; data from one other rat not shown). In these studies, peak values of CMR_{O2} for α-chloralose and morphine anesthetized rats were 1.2±0.3 and 2.7±0.3 μmol/g/min, respectively, which are significantly lower than the mean cortical values of 1.5±0.2 and 3.2±0.3 μmol/g/min (see Fig. 1).

4. DISCUSSION

In this study, the spectroscopic data did not suffer from chemical shift artifacts [18] because slice excitation pulses were not used for heteronuclear editing and outer volume signals were suppressed by combination of 2D ISIS and 1D OVS. The advantage of ICED PEPSI over conventional CSI is the lesser time required for data acquisition. In the current modified editing and localization scheme, only 8 averages were required to produce an edited spectrum per phase-encode step; along with 8 repetitions to generate a 16×16 grid, the ICED PEPSI sequence required ~11 minutes. In contrast to our approach, generation of a 16×16 grid using conventional CSI would require ~68 minutes. Reduced *k* space acquisition methods [19] in conjunction with single shot editing [20] and localization [21], and spatial filtering [22] could lead to isotropic 2 μL voxels in 5 minutes with ICED PEPSI for a 16×16 grid. Thus, we anticipate that the ICED PEPSI method will be useful for imaging turnover of ^{13}C-labeled metabolites (e.g., glutamate, glutamine, and GABA) at high temporal and spatial resolution.

Irrespective of the time resolution of each spectroscopic image used to depict the regional ^{13}C turnovers, the metabolic maps obtained by this approach represents metabolism at steady-state. A metabolic map obtained by the NMR approach is similar to a map obtained by autoradiography because both methods require long times for the tracer to be turned over or trapped. Therefore, metabolic maps obtained by ICED PEPSI cannot be used to detect acute changes in metabolism. However, these spatially resolved metabolic maps would be ideal for autoregulation and functional activation studies where changes in steady-state metabolism are measured (see ref. [6] for autoradiography and refs. [14-16] for POCE). The ICED PEPSI method will be useful for imaging ^{13}C turnover at high temporal and spatial resolution.

5. CONCLUSION

A method for imaging of ^{13}C turnover in brain with ICED PEPSI has been described. CMR_{O2} maps obtained for rats under either α-chloralose or morphine anesthesia revealed significantly different averaged cortical values. Since the heterogeneity in CMR_{O2} maps was greater at higher metabolic activity levels (i.e., morphine anesthesia), regional variations in the non-anesthetized state are expected to be even greater. As presently demonstrated, the ICED PEPSI method provides sufficient sensitivity to image ^{13}C turnover at spatial resolution relevant for studying activation in the rat cortex [23,24].

ACKNOWLEDGEMENTS

The authors thank Drs. Graeme F. Mason and Douglas L. Rothman for helpful discussions, and S. McIntyre and B. Wang for technical support. Supported by National Institutes of Health: NS-034813 (KLB), HD-032573 (KLB), NS-037203 (FH), DC-003710 (FH), MH-067528 (FH); National Science Foundation: DBI-9730892 (FH) and DBI-0095173 (FH).

REFERENCES

1. Shulman RG, Rothman DL, Hyder F. Stimulated changes in localized cerebral energy consumption under anesthesia. Proc Natl Acad Sci USA 1999;96:3245-325.
2. Behar KL, Petroff OAC, Prichard JW, Alger JR, Shulman RG. Detection of metabolites in rabbit brain by ^{13}C NMR spectroscopy following administration of [1-^{13}C]glucose. Mag Reson Med 1986;3:911-920.
3. Fitzpatrick SM, Hetherington HP, Behar KL, Shulman RG. The flux from glucose to glutamine in the rat brain *in vivo* as determined by ^{1}H-observed, ^{13}C-edited NMR spectroscopy. J Cereb Blood Flow Metab 1990;10:170-179.
4. Inubushi T, Morikawa S, Kito K, Arai T. ^{1}H-detected *in vivo* ^{13}C NMR spectroscopy and imaging at 2T magnetic field: efficient monitoring of ^{13}C-labeled metabolites in the rat brain derived from [1-^{13}C]glucose. Biochem Biophys Res Commun 1993;191:866-872.
5. Rothman DL, Sibson NR, Hyder F, Shen J, Behar KL, Shulman RG. *In vivo* nuclear magnetic resonance spectroscopy studies of the relationship between the glutamate-glutamine neurotransmitter cycle and functional neuroenergetics. Phil Trans R Soc Lond B 1999;354:1165-1177.
6. Sokoloff L, Reivich M, Kennedy C, des Rosiers MH, Patlak CS, Pettigrew KD, Sakurada O, Shinohara M. The [^{14}C]deoxyglucose method for the measurement of local cerebral glucose utilization: theory, procedure, and normal values in the conscious and anesthetized albino rat. J Neurochem 1977;28:897-916.
7. Matsui S, Sekihara K, Kohno H. High-speed spatially resolved high-resolution NMR spectroscopy. J Am Chem Soc 1985;107:2817-2818.
8. Hyder F, Renken R, Rothman DL. *I*n vivo *c*arbon *e*dited *d*etection with *p*roton *e*cho *p*lanar *s*pectroscopic *i*maging (ICED PEPSI): [3,4-$^{13}CH_2$]glutamate/glutamine tomography in rat brain. Magn Reson Med 1999;42:997-1003.
9. Ordidge RJ, Connelly A, Lohman AB. Image-selected *in vivo* spectroscopy (ISIS). A new technique for spatially selective NMR spectroscopy. J Magn Reson 1986;66:283-294.
10. Hyder F, Petroff OA, Mattson RH, Rothman DL. Localized ^{1}H NMR measurements of 2-pyrrolidinone in human brain *in vivo*. Magn Reson Med 1999;41:889-896.
11. Hyder F, Rothman DL, Blamire AM. Image reconstruction of sequentially sampled echo-planar data. Magn Reson Imaging 1995;13:97-103.
12. Mason GF, Rothman DL, Behar KL, Shulman RG (1992) NMR determination of the TCA cycle rate and -ketoglutarate/glutamate exchange rate in rat brain *J Cereb Blood Flow Metab* 12:434-447
13. Mason GF, Behar KL, Rothman DL, Shulman RG. NMR determination of glucose concentration and transport kinetics in rat brain. J Cereb Blood Flow Metab 1992;12:448-455.
14. Hyder F, Chase JR, Behar KL, Mason GF, Siddeek M, Rothman DL, Shulman RG. Increase tricarboxylic acid cycle flux in rat brain during forepaw stimulation detected with ^{1}H [^{13}C] NMR. *Proc Natl Acad Sci USA* 1996;93:7612-7617.

15. Hyder F, Rothman DL, Mason GF, Rangarajan A, Behar KL, Shulman RG. Oxidative glucose metabolism in rat brain during single forepaw stimulation: a spatially localized ^{1}H [^{13}C] nuclear magnetic resonance study. J Cereb Blood Flow Metab 1997;17:1040-1047.
16. Hyder F, Kennan RP, Kida I, Mason GF, Behar KL, Rothman DL. Dependence of oxygen delivery on blood flow in rat brain: A 7 Tesla NMR study. J Cereb Blood Flow Metab 2002;20:485-498.
17. Hyder F, Shulman RG, Rothman DL. A model for the regulation of cerebral oxygen delivery. J Appl Physiol 1998;85:554-564.
18. Brown TR, Kincaid BM, Ugurbil K. NMR chemical shift imaging in three dimensions. Proc Natl Acad Sci USA 1982;79:3523-3526.
19. Maudsley AA, Matson GB, Hugg JW, Weiner WM. Reduced phase encoding in spectroscopic imaging. Magn Reson Med 1994;31:645-651
20. Garwood M, Merkle H. Heteronuclear spectral editing with adiabatic pulses. J Magn Reson 1991;94:180-185
21. de Graaf RA, Nicolay K, Garwood M. Single-shot, B1-insensitive slice selection with a gradient-modulated adiabatic pulse. BISS-8 *Mag Reson Med* 1996;35:652-657
22. Hetherington HP, Pan JW, Mason GF, Adams D, Vaughn MJ, Tweig DB, Pohost GM. Quantitative ^{1}H spectroscopic imaging of human brain at 4.1T using image segmentation. Magn Reson Med 1996;36:21-29
23. Hyder F, Rothman DL, Shulman RG. Total neuroenergetics support localized brain activity: Implications for the interpretation of fMRI. Proc Natl Acad Sci USA 2002;99:10771-10776.
24. Smith AJ, Blumenfeld H, Behar KL, Rothman DL, Shulman RG, Hyder F. Cerebral energetics and spiking frequency: The neurophysiological basis of fMRI. Proc Natl Acad Sci USA 2002;99:10765-10770.

Chapter 5

MONTE-CARLO SIMULATION OF LIGHT TRANSPORT FOR NIRS MEASUREMENTS IN TUMORS OF ELLIPTIC GEOMETRY

Mojca Pavlin*, Tomaž Jarm & Damijan Miklavčič
Univ.of Ljubljana, Faculty of Electrical Eng., Trzaska 25, SI-1000 Ljubljana, SLOVENIA
** Tel. +386 61 1768 264, Fax +386 61 1264 658, E-mail: mojca@svarun.fe.uni-lj.si*

Abstract: Propagation of light in a highly scattering medium such as biological tissue is difficult to study. For complex geometry and multilayer structures computer simulation has to be used for light transport analysis. A Monte Carlo model of light propagation in tissue has been applied for the purpose of better understanding of the results of near-infrared spectroscopy (NIRS) measurements in experimental tumors. The major objective was to determine the percentage and location of the illuminated area in tumor and to estimate fraction of NIRS signal originating from the underlying tissues. Values of optical parameters used in the model were taken from literature. Tumor shape was approximated with a rotational ellipsoid. Computer simulations were made for two positions of optodes: reflectance and transmittance mode. Results of simulations indicate that in both configurations the majority of signal originates from tumor and not from surrounding tissue. In reflectance mode collected light comes from limited area near the optode whereas in transmittance mode the collected light illuminate almost whole tumor. This difference between the two modes is valid for all tissue parameters.

Key words: Near-infrared spectroscopy, Monte-Carlo model, light transport, tumor

1. INTRODUCTION

Propagation of light in tissue is a complex phenomenon due to strong scattering, insufficiently known optical parameters and heterogeneous structure of tissue. Even in simple geometry it is hard to solve this problem theoretically and impossible for more complex geometry. For this reason different numerical approaches are used, where one of most commonly used is Monte-Carlo method: Understanding and quantitative analysis of light transport in tissue is important for medical applications of different optical methods. One of the recently developed optical methods is near--infrared spectroscopy (NIRS), which enables noninvasive measurement of oxygenation and perfusion in soft tissue *in vivo*. NIRS method uses a difference in the light absorption spectra of oxyhemoglobin and deoxihemoglobin in near-infrared spectrum to measure concentrations of oxy- and deoxyhemoglobin. From relative change of light attenuation concentration changes of the absorbers can be determined.

The major problem related to quantification of NIRS data arises from the fact that light attenuation in tissue originates not just from absorption but also from light scattering. Light loss due to scattering depends on tissue parameters and measurement geometry and results in a non-linear relationship between absorption changes and attenuation changes. One of the problems of experiments with NIRS method is also nonhomogeneous structure of tissue. As a result of tissue inhomogenity signals may actually be obtained from different tissue types, in different physiological states [1]. Another complication follows from the fact that due to strong scattering of light in tissue the pathlength is greater than the distance d(cm) between emitting and receiving optode. The traveled average distance d_{opt}(cm) - optical path is larger by a differential pathlength factor B, defined as: $B=d_{opt}/d$. It was proven for the attenuation of light of certain wavelength in tissue that it agrees with modified Beer Lamber's law [1]:

$$A=\log(J/J_0)=\Sigma_i=\alpha_i c_i B\, d + G \tag{1}$$

where J_0 (W/m^2), and J (W/m^2) represent intensity of inward and outward light flux, A (OD) is attenuation, B is differential pathlength factor, α (μM^{-1}cm^{-1}) specific extinction coefficient, c (μM^{-1}) is concentration of the absorber and index i stands for different absorbers. Parameter G represents unknown light losses due to scattering, undefined absorption and other undefined losses.

Our work was related to measurements of oxygenation with NIRS method on small subcutaneous tumors [2], experimental configuration of these measurements is presented in figure 1a. The main problem in this kind of measurements is unknown fraction of light originating from tumor itself due to its small size and position of the optodes. Therefore our main objective was to estimate how much signal emerges from underlying tissues and to determine the illuminated area of tumor tissue for reflectance and transmittance modes.

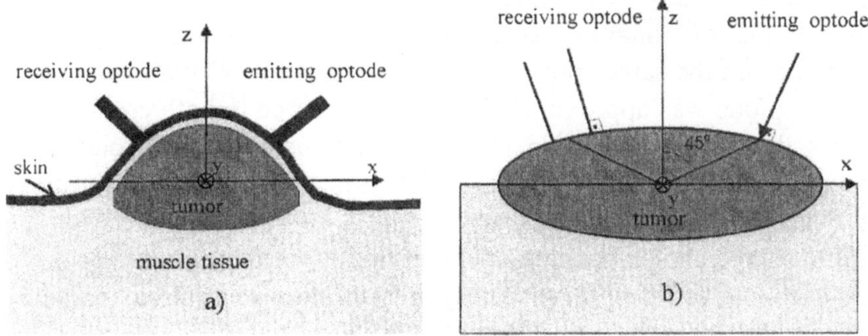

Figure 1. Position of optical fibers-optodes in transmittion mode a) for NIRS measurements in tumor b) in simulations.

A Monte-Carlo model of light propagation in tissue has been applied. Values of optical parameters for near-infrared wavelengths have been selected in the range of values reported in literature. We have also studied how changing the values of parameters affects attenuation and differential pathlength factor of photons.

2. METHODS

Monte-Carlo method has been widely applied in radiation transport studies. It is a stochastic method that simulates photon propagation as a random walk from one interaction to another. Transport of light in the tissue can be described by computer simulation of appropriately weighted random absorption and scattering interactions [3,4,5]. A three-dimensional Monte-Carlo model based on transport equation [5] was developed in our study. Parameters for the model are the following;

- Parameter *g* - anisotropy factor which is defined as the average scattering angle $g=\langle cos\theta \rangle$, where larger g means more scattering in the forward direction. For scattering in tissue it was shown that it is mainly forward scattering and can be approximated with Henyey-Greenstein function.
- Total attenuation coefficient μ_t $(\mu_t = \mu_a + \mu_s)$; the sum of scattering coefficient μ_s and absorption coefficient μ_a .
- Albedo, defined as: $w = \mu_s/(\mu_a + \mu_s)$.
- Values of optical parameters used in the model were taken from literature [6] and were for albedo (*w*) in range from 0.994 to 0.998, for scattering coefficient (μ_s) from 160 to 410 cm^{-1} and for mean cosine of the scattering angle (*g*) from 0.93 to 0.97.

Tumor shape was approximated with rotational ellipsoid with orthogonal diameters being a=8mm, b=8mm and c=3 mm long (volume = 96 mm^3). In transmittance mode two optodes were placed symmetrically at angle 45 degrees to the rotational axis (Figure 1b). In reflectance mode both optodes were placed on the same spot at angle 45 degrees to the rotational axis. The emitting optode was approximated by a point source in both configurations, whereas the receiving optode had finite dimensions of the one used in measurements. Fiber diameter was 1586 μm and numerical aperture was 0.55. The basic simulation steps were:

1. Point source photon generation of a normally incident beam.
2. At each interaction point scattering or absorption was chosen randomly, proportionally to the probability of each.
3. Pathway generation. The distance between two successive interactions was calculated as: $l=1/\mu_t(-\ln RN)$, where RN is a random number from a sequence of independent uniformly distributed random numbers. With this formula we generated an exponential distribution of interaction path lengths with mean free path $s=1/\mu_t$.
4. New direction was randomly selected from appropriate distribution function. The azimuth angle Φ and the cosine of the scattering angle θ with respect to the previous direction were generated by:

$$\Phi = 2\pi RN , \tag{3}$$

$$\cos\theta=[(1+g^2) - (1- g^2)^2 - (1- g + 2g\, RN)^{-2}][2g]^{-1} , \tag{4}$$

   where Eq. (4) is derived from the Henyey-Greenstein phase function.
5. At each point we checked whether the photon crossed the fiber tip at an angle smaller than numerical aperture. If so, all scattering points for this photon were stored. Typically 100,000 photons were generated to detect 200 photons in transmittance mode and 1000 photons in reflectance mode.

We used Fresnel's laws to calculate the possibility of reflection and refraction for each collision of a photon with an interface between air and tissue [7]. Reflections at tissue-fiber surfaces were ignored due to small difference in refractive index. The tissue volume was defined with the sum of ellipsoid volume and semi-infinite space defined by $z<0$.

3. RESULTS

All scattering points of detected photons were stored. To present light distribution graphically we took two projections of all scattering points in *xz* and *yz* plane. We separated the planes into array of fields 55μm×55μm large and counted the number of appropriately weighted interactions in each field.

Figures 2 and 3 represent light distribution drawn in planes *xz* and *yz* of simulation with parameters being *w*=0.998, *g*=0.96 and *s*=25μm.

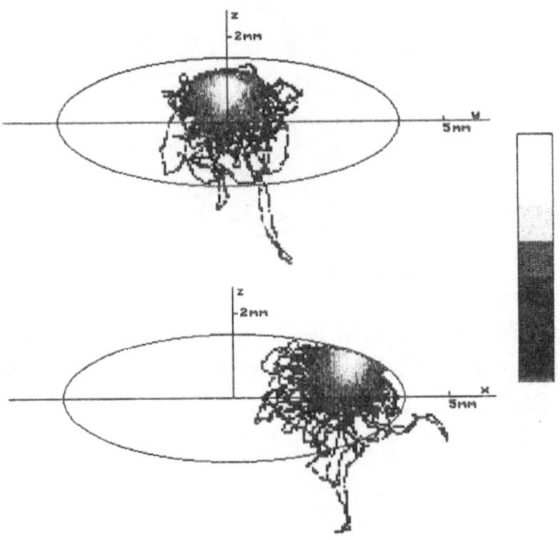

Figure 2. Light intensity distribution for reflectance mode in a) *xz* plane and b) *yz* plane. Gray scale is logarithmic and relative to maximum of intensity, brighter shading represents higher intensity. Parameters of simulation are *g*=0.96, *w*=0.998 and average path μ_s^{-1}= *s*=25μm.

Density of interactions represents measure of light intensity at specific place and is represented in a gray scale. The scale is logarithmic and relative to the maximum of intensity, where brighter shading represent higher intensity.

We can see the difference in illuminated volume between the two modes. In the reflectance mode the collected light originates from a limited area (approx. 5 mm^3) near the optode. In contrast in the transmittance mode the collected light illuminates approximately 65 mm^3 of ellipsoid. It can also be seen from figure 3 that in the transmittance mode a fraction of photons travels outside ellipsoid.

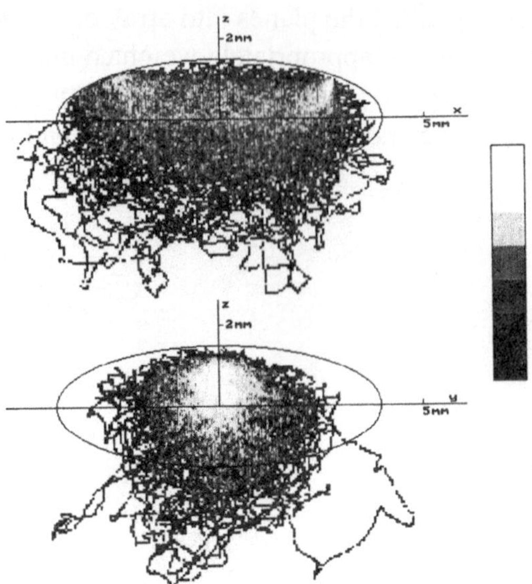

Figure 3. Light intensity distribution for trasmittance mode in a) *xz* plane and b) *yz* plane. Gray scale is logarithmic and relative to maximum of intensity, brighter shading represent higher intensity. Parameters of simulations are *g*=0.96, *w*=0.998 and average path μ_t^{-1}=*s*=25µm.

Table 1. Percentage of the signal originating outside the ellipsoid (1) and fraction of illuminated volume (2) for reflectance and transmittance mode.

%		g=0.93	g=0.94	g=0.95	g=0.96	g=0.97	g=0.97	
		s=60µ m					s=25µ m	s=35µ m
		w=0.994					w=0.998	
1	ref.	1.5	1.5	3.3	5.5	5.4	3.0	5.8
	tran	12.5	12.5	15.5	15.6	18.7	15.8	26.3
2	refl.	5.0	5.0	5.0	4.9	4.1	5.1	4.9
	tran	60	77	72	57	58	56	60
%		g=0.97		g=0.95				
		s=45µ m	s=60µ m	s=60 µ m				
		w=0.998		w=0.995	w=0.996	w=0.997	w=0.998	
1	ref	9.2	12.8	4.2	5.4	5.4	6.1	
	tran	22.8	36.1	20.5	21.9	23.8	26.7	
2	ref	4.3	3.6	4.8	4.9	4.5	3.7	
	tran	62	60	68	69	72	67	

In table 1 percentages of signal (photon interactions) originating outside the ellipsoid (1) and fraction of illuminated volume (2) for reflectance and transmittance mode for 13 different sets of parameters are given. The results demonstrate that 2-9% and 15-37% of detected signal for the reflectance mode and the transmittance mode respectively originates from the tissue outside the tumor.

To estimate the illuminated volume we counted number of fields with at least one interaction in both planes. With this we calculated that in reflectance mode only 4-5% of ellipsoid volume was illuminated in contrast to transmittance mode where 56-72% of volume was illuminated.

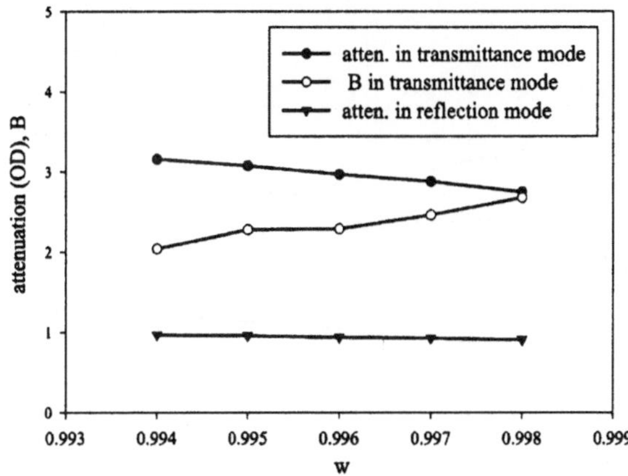

Figure 4. Attenuation (OD) in transmittance and reflectance mode and differential pathlength factor- B in transmittance mode at different values of albedo w. Parameters of simulations are g=0.95 and s=60μm.

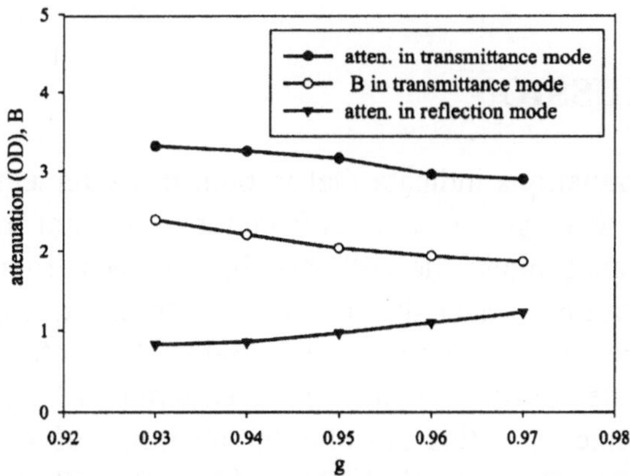

Figure 5. Attenuation (OD) in transmittance and reflectance mode and differential pathlength factor - B in transmittance mode at different values of parameter g. Parameters of simulations are w=0.994 and s=60μm.

In figure 5 dependency on anisotropy parameter g is shown. Larger parameter g means less scattering and more signal in forward and less in backward direction. This results in smaller attenuation and smaller B for transmittance mode and larger attenuation in reflectance mode. In figure 6 dependency of attenuation and parameter B on average free path is shown. If average free path is increased there is less collisions and less light loss, which results in smaller attenuation in transmission mode. Average free path has no significant impact on differential pathlength factor. For all simulations counting errors are 0.07 OD in the transmittance mode and 0.03 OD in the reflectance mode.

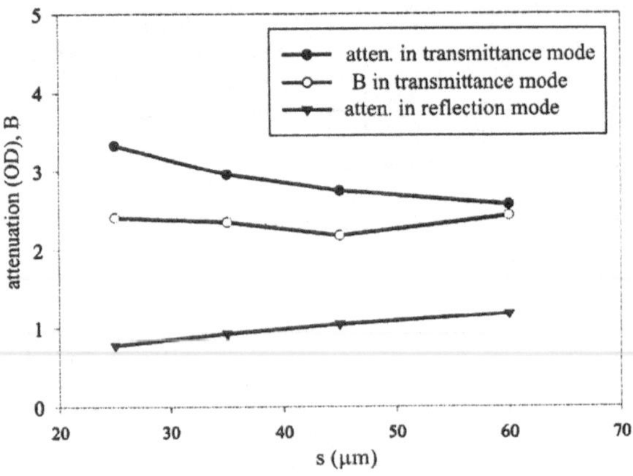

Figure 6. Attenuation (OD) in transmittance and reflectance mode and differential pathlength factor- B in transmittance mode at different values of average path μ_t^{-1}=s(μm). Parameters of simulations are w=0.998 and g=0.97.

4. DISCUSSION

Results of simulations indicate that in both measurement configurations the majority of NIRS signal originates from tumor and not from surrounding tissue. In reflectance mode the collected light comes from a limited area (approx. 5% of volume) near the optode; therefore, we should measure at more different positions to obtain valid results. In contrast to this in the transmittance mode collected light illuminated almost whole tumor (approx. 65% of the volume) and this is valid for all range of tissue parameters. Although the transmittance mode signal is few orders of magnitude smaller than the reflectance mode signal, the first mode gives more relevant information of overall tumor oxygenation and perfusion. Therefore, for all but smallest tumors the transmittance mode seems to be more appropriate.

Attenuation and average path of photons do not depend strongly on the values of optical parameters, but much more so on the position of the optodes. The results support our assumptions that most of NIRS signal recorded from small subcutaneous tumors originate from the tumor itself and are not significantly contaminated by the underlying tissues. Since values of parameters vary for different tissue types, we should consider a multilayer structure as a more realistic model.

REFERENCES

1. Delpy DT, Cope M. Quantification in tissue near-infrared spectroscopy. Phil Trans R Soc Lond B 1997; 649-657.
2. Jarm T, Wickramasinghe YABD, Deakin M, Cemazar M, Miklavcic D, Sersa G, Elder J, Rolfe P, Vodovnik L: Blood Perfusion and Oxygenation of Tumours Following Electrotherapy - A Study by Means of Near Infrared Spectroscopy and Patent Blue Staining. Proceedings of the 9th International Conference on Mechanics in Medicine and Biology, Ljubljana, Slovenia, 1996;339-342
3. Wilson BC, Adam G. A Monte-Carlo model for the absorption and flux distribution of light in tissue. Med Phys 1983;10:824-830.
4. Groenhuis RAJ, Ferwenda HA, Ten Bosch JJ. Scattering and absorption of turbid materials determined from reflection measurments. 1: Theory. Appl Opt 1983;22(16):2456-2462.
5. Ishimaru A. Diffusion of light in turbid media. Appl Opt 1989;28:2210-2215.
6. Beek JF, Blokland P, Posthumus P, Aalders M, Pickering JW, Sterenborg HJCM, van Gemert MJC. In vitro double-integrating-sphere optical properties of tissues between 630 and 1064 nm. Phys Med Biol 1997;42:2255-2261.
7. Bolin FP et al. Refractive index of some mammalian tissues using a fiber optic cladding method. Appl Opt 1989;28(12):24297-2303.

Chapter 6

ARTERIAL PULSATIONS ARE PRESENT IN ONE THIRD OF THE HUMAN CRANIAL VASCULAR VOLUME PENETRATED BY NEAR INFRARED LIGHT

Robert Stingele[1], Emanuela Keller[2], Benedicht P. Wagner[3], Thorsten Steiner[1], Karoline Stingele[1], Werner Hacke[1]

[1]Department of Neurology, University of Heidelberg, Germany, [2]Department of Neurosurgery, University of Zürich, Switzerland, [3]Department of Pediatrics, University of Berne, Switzerland

Abstract: The percentage of cranial vascular volume that undergoes arterial pulsations is estimated by 18 paired measurements in nine healthy volunteers using near infrared spectroscopy (NIRS, 10 Hz sampling frequency, 769 nm). The NIR absorption is decomposed in pulsatile and non-pulsatile components by digital filtering. The time course of absorption changes of these components after sudden intravenous injection of indocyanine-green (ICG) is used to estimate the arterial fraction of vascular volume (f_{art}). Approximately 28% of the vascular volume within the optical field of the NIR spectroscope was arterial. The range of values was 6 - 43%. Variance of f_{art} was 4.8 times higher between subjects than within subjects, indicating that the variability observed is not due to imprecise estimation but to optode position and subject.

Key words: Blood volume; humans; indocyanine-green; NIRS

ABBREVIATIONS

$A_{tot}, A_{art}, A_{ven}, A_p$: NIR absorptions originating in volumes with corresponding indices

$\overline{A}_p$: instantaneous power of pulsatile absorption

A_{np}: non-pulsatile absorption originating in V_{ven} and $\overline{V}_{art}$

f_{art}: arterial fraction of arterial vascular volume

ICG: indocyanine-green

RMS: ROOT MEAN SQUARE VALUE

V_{tot}, V_{art}, V_{ven}, V_p: Total, arterial, venous, and pulsatile vascular volume

$\overline{V}_{art}$: sum of the cylindrical core volumes of pulsatile vessels

$\overline{V}_p$: instantaneous amplitude of pulsatile volume (= instantaneous root mean square value)

1. INTRODUCTION

The arterial fraction of cranial vascular volume is of interest because oxy- and deoxy-hemoglobin, the principal chromophores investigated by NIRS, are present in different concentrations in arterial and venous vessels due to oxygen extraction along the vessel. Therefore, NIRS signals are affected not only by the amount of hemoglobin (Hb) and the overall oxygenation of Hb, but also by f_{art} in the optical field at a specific optode position. We present a method allowing non-invasive estimation of f_{art} by fast-scanning NIRS and intravenous injection of ICG.

2. METHODS

The study was approved by our local ethical committee. Nine healthy volunteers were investigated using NIRS. In each subject, two paired measurements with unchanged optode positions approximately 30 minutes apart were performed. Subjects were breathing spontaneously. Optodes were placed over the lateral aspect of the frontal lobe on either side at a distance of 45 or 55 mm. NIR absorption was recorded at 10 Hz and 769 nm using an Oxymon system (University of Nijmegen, Netherlands). After a stable baseline reading was reached, ICG was injected into the cubital vein (ICG Pulsion, total dose 10-30 mg).

Both ICG and Hb are NIR-chromophores with a strict intravascular volume of distribution. NIR absorption due to these chromophores is proportional to their respective intravascular concentrations and the vascular volumes inside the optical field containing the chromophores. The vascular volume in the optical field is subject to periodic changes due to changing transmural blood pressure and therefore NIR absorption is pulsatile. In the present method, pulsatility of blood volume at heart frequency is used as a tag for arterial volume portions. The sampling frequency of 10 Hz allows to follow pulsatile absorption changes at heart frequency with a large margin of oversampling. To analyze the recorded absorption data, the following model is used. The total volume of blood vessels inside the optical field (V_{tot}) is composed of vessels that pulsate at heart frequency (V_{art}) and vessels that do not pulsate in that frequency band (V_{ven}). The arterial volume V_{art} is decomposed in two parts. Part one is the sum of the cylindrical core volumes of the arterial vessels ($\overline{V}_{art}$). This part is constant and it can be pictured as the volume that the arterial vessels would occupy, if there would be no changes in blood pressure. The second part (V_p) is the difference between $\overline{V}_{art}$ and V_{art} that is due to changes in transmural blood pressure with the heart cycle. V_p is positive in systole, negative in diastole and zero on average. V_p can be pictured as a cylindrical layer added to the pulsating part of the vessels the thickness of which is changing with blood pressure. In the model used here, the NIR-absorptions A_{tot}, A_{ven}, A_{art}, and A_p are attributed as originating from the presence of chromophore in the volumes V_{tot}, V_{ven}, V_{art}, and V_p. The pulsatile absorption A_p is positive in systole, negative in diastole and zero on average, just as the corresponding volume V_p. Before injection of ICG, A_p is due to presence of Hb in the arterial part of the vessel and the amplitude of A_p is constant because the concentration of Hb is constant. The instantaneous power of the A_p signal ($\overline{A}_p$) is constant before ICG appears. When ICG arrives in the optical field for the first time after injection, proportionally filling the two arterial volumes (V_p and $\overline{V}_{art}$), the amplitude of A_p and thus $\overline{A}_p$ increase proportional to the increase of the ICG concentration. For convenience non-pulsatile absorption (A_{np}) is defined as the lumped absorption originating in non-pulsatile volumes $\overline{V}_{art}$ and V_{ven}.

If the concentration of a chromophore is equal in any two compartments in the optical field, then the ratio of the absorptions due to presence of the chromophore in these compartments is equal to the ratio of their volumes. This observation follows from Beer's law because all other factors that determine absorption, such as optical path length, differential path length factor and absorption coefficient simplify when the ratio of absorptions is calculated. Equal concentrations of ICG for two vascular compartments can be assumed in two conditions. Condition 1: In a short period when ICG arrives in the optical field for the first time, the concentration of ICG can assumed to be equal in V_p and $\overline{V}_{art}$, the two volumes that make up the arterial vessel. During this period of time, the ICG concentration in V_{ven} is zero because ICG particles have not yet arrived in the venous volume. The ratio $\overline{A}_p / A_{np} = \overline{A}_p / \left(\overline{A}_{art} + A_{ven}\right) = \overline{A}_p / \overline{A}_{art}$ is equal to the volume ratio $r_1 = \overline{V}_p / \overline{V}_{art}$. This ratio indicates the average fraction of the arterial volume that the arterial vessels are inflated and deflated by during the heart cycle.

Condition 2: Many circulation times after the injection of ICG thorough mixing over the entire vascular volume can be assumed and the concentration of ICG is the same in all the volumes mentioned. Calculation of the ratio $\overline{A}_p / A_{np} = \overline{A}_p / (\overline{A}_{art} + A_{ven}) = \overline{A}_p / \overline{A}_{tot}$ for this condition yields $r_2 = \overline{V}_p / V_{tot}$, the ratio indicating the average percentage of volume the arterial vessels are inflated and deflated during the heart cycle in relation to

$$\frac{r_2}{r_1} = \frac{\overline{V}_p / V_{tot}}{\overline{V}_p / \overline{V}_{art}} = \frac{\overline{V}_{art}}{V_{tot}} = f_{ar}$$

the total vascular volume. Finally, by calculating the quotient one obtains the desired arterial volume fraction.

The steps of data analysis were as follows:

Digital filters were applied to the raw absorption time series at 769 nm to obtain A_p and A_{np} (A_{np}: lowpass filter design with edge frequency below heart rate; A_p: bandpass filter design with edge frequencies around the heart rate; filters were of Chebyshev type II; the minimum filter order leading to at least 40 dB of stopband attenuation and less than 0.3 dB of passband ripple were used).

To obtain an estimate for the time course of the instantaneous power of the A_p signal, the signal envelope was calculated (the signal envelope is

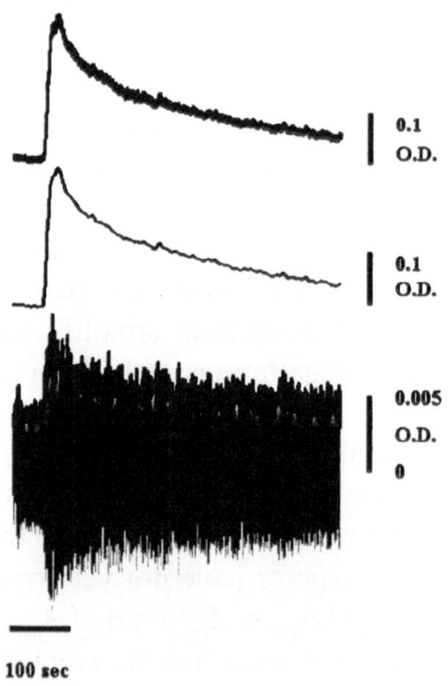

Figure 1. Raw NIR absorption data at 769 nm sampled at 10 Hz (top). Lowpass filtered data (middle), and bandpass filtered data (bottom, in gray). The black line overlaid to the bottom curve is the envelope of the bandpass filtered data. Note the different scale of the ordinate in the bottom panel.

obtained by taking the absolute value of the analytical signal of A_p, which in turn is calculated from A_p by a Hilbert transform, for details see [1]) (Figure 1).

To remove the contribution of Hb to Anp and the envelope of A_p , the average values of Anp and of the envelope of A_p before injection of ICG were subtracted from the respective time series.

Figure 2 is a parametric plot of the changes in absorption due to ICG only. The abscissa plots non-pulsatile absorption (A_{np}^{ICG}), the ordinate plots the instantaneous RMS amplitude of A_p ($\tilde{A}_p^{ICG}$), both in optical densities (O.D.). Before injection of ICG, both absorptions are zero. As ICG appears at the arterial extreme of the vessels there is a steep increase of both pulsatile and non-pulsatile absorption. As the first particles of ICG spill over into the non-pulsatile section of the vessels, the non-pulsatile absorption increases relatively more than does the pulsatile absorption. This results in a downward concavity of the initial part of the parametric plot. The initial slope of that section of the parametric plot corresponds to the ratio $r_1 = \overline{V}_p / \overline{V}_{art}$. After many circulation times, the concentration of ICG will be equal in the entire vascular volume and thus the amount of ICG entering and leaving the vessels in the optical field during any instant of time will be equal. As ICG is eliminated from circulation the concentration decreases towards zero by the same rate in both pulsatile and non-pulsatile part of the vessels. This results in a linear slope of the parametric plot after about 2 minutes of time have elapsed. The slope of this line corresponds to the ratio $r_2 = \overline{V}_p / V_{tot}$. As stated above, the quotient of the two slopes gives the desired arterial volume fraction f_{art}.

3. RESULTS

All values reported are means ± SD. Figure 1 (top) shows the raw absorption data of a representative measurement. The amplitude of pulsatile absorption changes is higher after ICG has been injected (see Figure 1, bottom). The results of filtering the data are shown in Figure 1 (middle and bottom). The envelope of A_p , shown in black, linearly reflects the concentration of ICG in the pulsatile section of the vessels.

The maximal change from baseline of the non-pulsatile absorption A_{np} was 0.25 ± 0.08 O.D., the maximal increase of the RMS-amplitude of the pulsatile absorption A_p was 0.0032 ±0.001 O.D. The parametric plot of RMS-pulsatile versus non-pulsatile absorption changes due to ICG is shown in Figure 2.

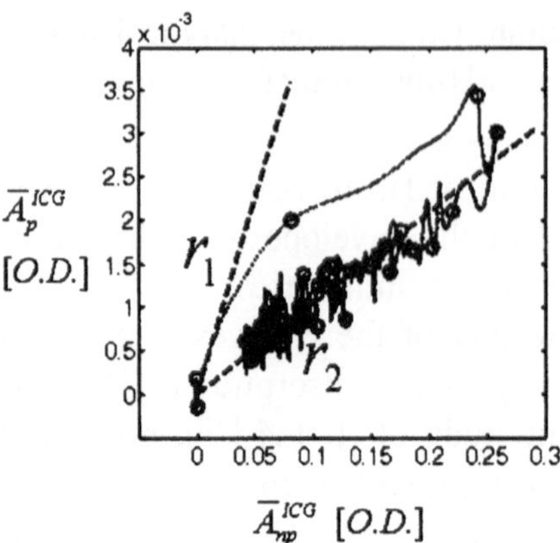

Figure 2. Parametric plot of pulsatile (ordinate) versus non-pulsatile (abscissa) absorption after injection of ICG. Small data points are 0.1 sec apart. The large circles are 10 sec apart. For explanation of the slopes r_1 and r_2 see text.

Both absorptions are zero before ICG is injected. As ICG appears, the parametric plot shows a large increase of both absorptions with initial slope r_1. The values of r_1 were 0.022 ± 0.012, indicating that the RMS-pulsatile volume change due to blood pressure changes is 2.2% of the arterial volume $\overline{V}_{art}$ on average. As ICG starts to fill non-pulsatile vessels, the parametric plot follows a downward concavity. This can be explained because indicator particles leaving the pulsatile compartment by entering the non-pulsatile compartment will continue to contribute to non-pulsatile absorption but will be lost to detection in the pulsatile compartment. On the other hand, during any instant of time both compartments will be filled by the same number of newly arriving particles. The result is that, in this phase of the parametric plot, the increase of absorption is lower for Ap than for A_{np}. After approximately two minutes, both absorptions start to decrease towards zero at the same rate, resulting in a linear slope r_2 of the parametric plot. The value of r_2 was 0.0056 ± 0.0036 O.D., indicating that the pulsatile RMS-amplitude change was 0.56% of the total vascular volume. The quotient $r_2/r_1 = f_{art}$ was 0.28 ± 0.11 for the 18 measurements. Therefore, the fraction of arterial volume was about 28% of the total vascular volume in the optical field.

Figure 3 shows the 18 measurements in 9 subjects with paired measurements connected by a line. The range of observed values of f_{art} was large (6 – 43%). However, the variance within single subjects is smaller than the variance between subjects by a factor of 4.8 (One-way ANOVA for repeated measurements).

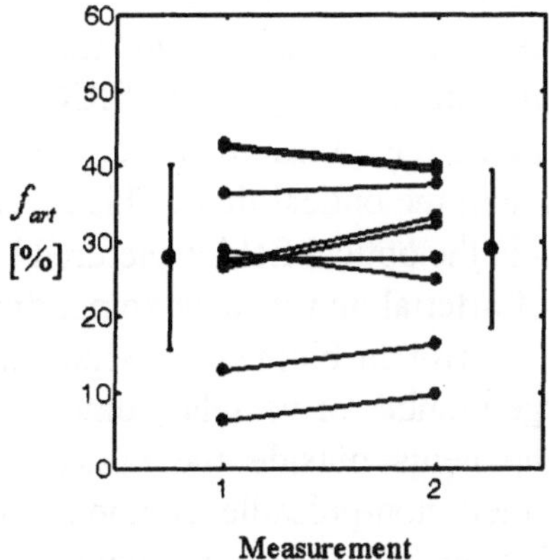

Figure 3. Arterial fraction of blood volume in percent. 18 measurements were performed in 9 subjects. Paired measurements are connected by a line. Mean f_{art} was 28%. There is a large range of observed values (6-24%). Variability between subjects is larger than variability within single subjects by a factor of 4.8 (One-way ANOVA for repeated measurements).

4. DISCUSSION

The arterial fraction of cranial vascular volume f_{art} is estimated by a non-invasive technique using fast scanning NIRS and intravenous injection of ICG. Knowledge of f_{art} is important because the concentration of the principal chromophores investigated by NIRS (oxy- and deoxy-Hb) change along the longitudinal axis of the vessels due to desaturation. With unchanged oxygenation of hemoglobin, a change of optode position to an optical field with higher f_{art} will give higher contributions of oxy-Hb and lower contributions of deoxy-Hb.

The method is based on the decomposition in pulsatile and non-pulsatile components of NIRS absorption data collected before and during the passage of ICG through the vascular bed under the NIRS-detector. The method relies on the following assumptions.

1) The concentration of ICG becomes equal in all vascular sections in the optical field after many circulation times. This assumption is certainly reasonable, but it remains to be defined what exactly 'many circulation times' means. A strength of the present method is that it provides an indication of the time after which thorough mixing of ICG has occurred. The point of the parametric plot after which both pulsatile and non-pulsatile absorptions decrease towards zero at the same rate indicates that the outflow of ICG from both compartments is identical during any instant of time. Since the compartments share the same ICG influx, this statement is identical with

saying that the concentration of ICG is identical in the compartments. In the 18 measurements, identical concentrations in the compartments were reached approximately 2 minutes after injection of ICG.

2) The concentration in the non-pulsatile part of the vessels is zero when the indicator starts to arrive in the optical field. This assumption is true only, if the vascular system inside the optical field is pictured by a single vessel, in which sequential filling of arterial and venous compartments is guaranteed. The assumption may be in error in biological tissues, in which the optical field is perfused by a large number of vessels. Some of these vessels might have their pulsatile components outside the optical field, therefore not contributing to Ap, but their non-pulsatile components inside the optical field, thereby contributing to A_{np}. This anatomical characteristic of an unknown portion of V_{tot} per se does not influence the accuracy of the measurement. However, if this portion of vessels happens to be filled sooner after injection of ICG than the remaining vessels, a systematic error of measurement results. More precisely, the initial slope (r_1) of Figure 2 becomes false low because initial non-pulsatile absorption results not only from the core volumes V_{art} of the arterial vessels but also from the early-filling venous vessels that do not contribute to Ap. Thus, overestimation of the arterial volume fraction r_2/r_1 is the result. This effect might lead to serious errors when a large portion of V_{tot} is made up from vessels that have a large percentage of their pulsatile parts outside the optical field and if these vessels are filled early, for example in arterio-venous shunting. The error of measurement introduced by this effect in physiological situations cannot be estimated at the moment, but we think that it is small because both the anatomic peculiarity of the vessel and early filling have to coincide in order to affect the measurement.

3) The concentration of ICG is equal in V_p and $\overline{V}_{art}$. This statement is equivalent to the assumption of complete radial mixing in the arterial vessels. Uniform concentration of the indicator at any point of the cross-section of a vessel is established faster than mixing along the longitudinal axis because the distances the indicator has to travel are shorter. In the present case, due to intravenous injection of ICG, the time elapsed until ICG appears in the optical field is on the order of 10-15 seconds. During this time ICG has to cross the capillary bed of pulmonary circulation. Therefore, thorough radial mixing seems to be a safe assumption in this case.

A potential source of error inherent to this method is the determination of the initial slope r_1 of the parametric plot. While r_2 can be estimated accurately due to the large number of data points collected during elimination of ICG over several minutes, the value of r_1 relies on a few data points only (Figure 2). The reason for this is that ICG does not appear in the optical field as a step concentration increase. The leading edge of the ICG front approaching, made up of those ICG particles with the shortest transit times from the site of injection to the optical field has a lower ICG concentration than the bulk of indicator arriving a little later. Therefore, during the initial phase of the parametric plot used to measure r_l, the concentration of ICG is low in the arterial vessels. This could be improved on by a closer site of injection for ICG (e.g. central venous catheter) leading to a more compact ICG bolus. Furthermore, the error introduced by this effect is not systematic, because r_l can be equally over- and underestimated. Therefore, averaging more than one measurement of r_l should improve precision.

REFERENCES

1. Oppenheim AV and Schafer RW. Discrete-Time Signal Processing. In: Prentice Hall Signal Processing Series 1st ed. Englewood Cliffs: Prentice Hall, 1989.

Chapter 7

ABSOLUTE FREQUENCY-DOMAIN PULSE OXIMETRY OF THE BRAIN: METHODOLOGY AND MEASUREMENTS

Martin Wolf[1], Maria A. Franceschini[2], Lelia A. Paunescu, Vlad Toronov, Antonios Michalos, Ursula Wolf, Enrico Gratton & Sergio Fantini[2]

Laboratory for Fluorescence Dynamics, Department of Physics, University of Illinois at Urbana-Champaign, 1110 W. Green St., Urbana, IL 61801-3080, USA Present Addresses: 1Clinic for Neonatology, University Hospital, Frauenklinikstr. 10, 8091 Zurich, Switzerland email: martin.wolf@alumni.ethz.ch 2Electro-Optics Technology Center, Department of Electrical Engineering and Computer Science, Tufts University, 4 Colby Street, Medford, MA 02155, USA

Abstract: A new method to non-invasively measure the absolute tissue oxygen saturation (SO_2) and arterial oxygen saturation ($fdSaO_2$) by frequency-domain spectroscopy is described. This method is based on the quantitative measurement of the tissue absorption spectrum, which is used to determine global SO_2. From the amplitude of absorption changes caused by arterial pulsation oscillations, in the range of 633-841 nm, the $fdSaO_2$ can be calculated. During deoxygenation (air / N_2 mixture) experiments, we measured the $fdSaO_2$ and SO_2 on the forehead of three healthy volunteers and compared them to the arterial oxygen saturation measured by conventional pulse oximetry ($poSaO_2$) on the finger. $fdSaO_2$ and $poSaO_2$ agree very well (mean difference: $-1.2\pm2.6\%$). Changes in SO_2 were systematically smaller than in $fdSaO_2$ or $poSaO_2$ probably due to autoregulation. The measurements with 4 and 8 wavelengths had comparable quality.

Key words: arterial oxygen saturation, frequency domain spectroscopy, near infrared spectroscopy, tissue oxygen saturation, nitrogen inhalation

1. INTRODUCTION

In the near infrared region (wavelength range of 600-900 nm) the penetration depth of the light in tissues is relatively large and absolute optical coefficients and hemoglobin oxygen saturation (SO_2 in %) can be measured non-invasively [1,2]. Tissue oximetry is mostly sensitive to the blood in the capillaries [3] where the oxygen exchange with the tissue occurs. Optical spectroscopy also allows measurements of oxygen saturation of the arterial blood (SaO_2 in %). The arterial blood is pulsating with the systolic/diastolic pressure, which induces oscillations also in the hemoglobin concentration [4]. These can be observed as oscillations in the optical signal, which are attributed to the arterial blood. This is the basis of pulse oximetry. Conventional pulse oximetry gives a SaO_2 reading based on a preliminary calibration on reference subjects [5], which leads to errors in SaO_2 reading at low (<80%) and high values (>97%) [6].

This paper presents a new approach of absolute oximetry using frequency-domain spectroscopy, which does not require preliminary calibration. This method provides non-invasive, real-time, simultaneous and absolute measurements of the local tissue parameters: frequency-domain hemoglobin oxygen saturation (SO_2) and frequency-domain arterial oxygen saturation ($fdSaO_2$). We have studied the change in those parameters in the brain of healthy volunteers during cycles of reduced oxygen inhalation.

2. METHODS

Instrument. The frequency-domain tissue spectrometer used in this study is similar to the one previously presented by Franceschini [7] and is a modified version of the commercial two-wavelength frequency-domain tissue oximeter [8] (Mod. 96208, ISS, Inc., Champaign, IL). The frequency synthesizer modulates the intensity of the laser diodes at a frequency of 110 MHz and the second dynode of the two photomultiplier tubes (pmt a and b) at a frequency of 110.005 MHz.

The measurements were performed using two configurations of the instrument: a) 8 discrete wavelength-system, and b) 4 discrete wavelength-system.

a) The system uses two sets of 8 wavelengths (633, 670, 751, 776, 786, 814, 830, and 841 nm) to implement the multi-distance protocol [7]. The 16 light sources are electronically multiplexed at a rate of 71.4 Hz, so that each light source is on for 14 ms. The total acquisition time for a full cycle over the 16 light sources therefore is 224 ms. The source-detector distances for this arrangement are 3.0 and 3.6 cm.

The system uses two sets of 4 wavelengths (670, 758, 785, and 830 nm) again in a multi-distance arrangement. The 8 light sources are electronically multiplexed at a rate of 99.7 Hz, so that each light source is on for 10 ms. The total acquisition time for a full cycle over the 8 light sources therefore is 161 ms. The source-detector separations for this arrangement are 3.0 and 4.2 cm shown in figure 1.

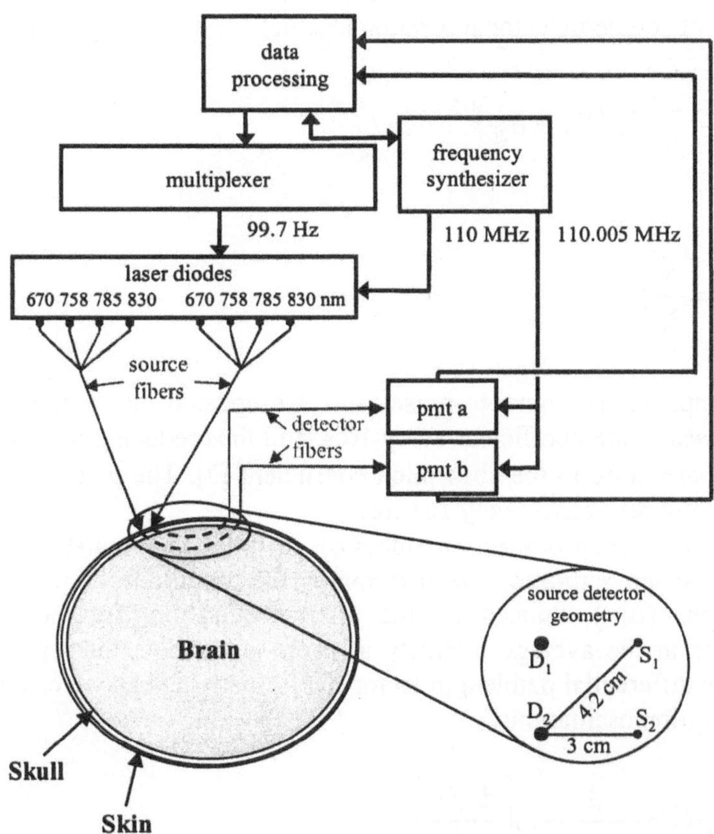

Figure 1. Scheme of the frequency-domain tissue spectrometer (4 wavelength-system) and the source-detector forehead display (inset).

Although the geometrical set-up of the configurations varies, it probes the same volume of brain tissue. Both provide quantitative spectroscopy independent of source, detector and optical coupling terms, without requiring any sort of instrumental calibration [9].

The optical signal at the tissue surface is sent to the two parallel detector channels of the instrument. The source and detector fibers are arranged in such a way that they are on the same side of the forehead, symmetrically.

Tissue Saturation (SO_2). In order to calculate the absolute optical coefficients of the investigated tissue, we have used the multi-distance frequency-domain method assuming a semi-infinite geometry and homogeneous tissue [10]. The absolute absorption (μ_a) and reduced scattering (μ_s') coefficients of the tissue are calculated using the slopes of the average light intensity (dc) and phase of the light intensity wave (Φ) as a function of source-detector separation (S_{dc} and S_Φ, respectively) [10]:

$$\mu_s' = \frac{2v}{3\omega} S_\Phi \left(S_\Phi^2 + S_{dc}^2\right)^{1/2} \tag{1}$$

$$\mu_a = \frac{S_{dc}^2}{3\mu_s'} \tag{2}$$

To improve the signal-to-noise ratio, we updated the calculation of the reduced scattering coefficient every 10 s, and thus reducing the contribution of the phase noise to the absorption coefficient [7]. The μ_a acquisition time was still 224 ms, respectively 161 ms.

Arterial Oxygen Saturation ($fdSaO_2$). In order to calculate the absolute arterial oxygen saturation we determined the amplitude of the absorption oscillations ($\Delta\mu_a$) induced by the arterial pulsation. By measuring the variations in the average intensity at each source-detector separation and using the differential pathlength factor (DPF) method [11], we can determine the absorption oscillations:

$$\Delta\mu_a(t) = \frac{1}{r \cdot \mathrm{DPF}} \ln\left[\frac{\mathrm{dc}(0)}{\mathrm{dc}(t)}\right] \tag{3}$$

where r is the source-detector distance, while dc(0) and dc(t) are the average intensities measured at time 0 and time t.

Since the DPF method assumes the knowledge of the average photon pathlength in tissue and its wavelength dependence, this leads to errors in the $fdSaO_2$ if one considers an incorrect DPF versus wavelength dependence. In order to avoid such errors, we directly measured the DPF spectrum during each measurement using the frequency-domain method in the same manner as reported by Franceschini [7]. Arterial pulsation-induced absorption variations were calculated by taking the amplitudes of the fast Fourier transform (FFT) of $\Delta\mu_a$ over the heartbeat band. In our cases, the FFT was calculated over 64 points. The relative concentrations of oxy-hemoglobin (O_2Hb in μmol/l) and deoxy-hemoglobin (HHb in μmol/l) were determined by performing a least square fit [12]. With O_2Hb and HHb obtained, one can calculate the arterial oxygen saturation in percent:

$$fdSaO_2=100\ O_2Hb/(O_2Hb+HHb) \qquad (4)$$

Deoxygenation experiment. By inspirations of different oxygen concentration (FiO_2 in %) one can induce variations in the $fdSaO_2$ and cerebral SO_2. The breathing protocol was the following: the subjects were breathing 21% of oxygen (air) for at least 1 minute and then the FiO_2 was reduced to 10% for 1 to 4 minutes depending on the SaO_2 decrease. For all the subjects, we set up 85% as the lower limit of the decrease in SaO_2, as given by pulse oximeter ($poSaO_2$). One of the subjects was also asked to inspire pure nitrogen for 0.5 min. This protocol was approved by the Institutional Review Board of the University of Illinois at Urbana-Champaign. During this protocol, we measured the $fdSaO_2$ on the forehead, with the frequency-domain spectrometer, and $poSaO_2$ at the finger of the subject, with a commercial pulse oximeter (Nellcor N-200). Given the source-detector distances used, all the $fdSaO_2$ and SO_2 refer to SaO_2 variations occurring at the surface of the brain [13].

Subjects: This study was done on three healthy male volunteers of 43, 34 and 53 years old, numbered consecutively 1 to 3. Written informed consent was obtained from all subjects.

3. RESULTS

Figure 2 shows a sample of the arterial-pulsation-induced absorption variations versus time obtained by the DPF method in the case of subject 3, using the 8 discrete wavelength-system. The fast periodic changes in light absorption ($\Delta\mu_a$ in the order of 0.0004, at 633 nm – Fig. 2) are caused by the pulsations of the arterial blood with systolic/diastolic blood pressure changes. We assume that these oscillations refer only to the arterial blood. The amplitudes of these oscillations at different wavelengths can be used to obtain a spectrum of the arterial blood. The large slow changes in the time interval of 3 to 4 min ($\Delta\mu_a$ in the order of 0.006, at 633 nm – Fig. 2) are caused by a deoxygenation of the tissue, which was induced by breathing 100% nitrogen for 30 s. These large changes reflect the global tissue oxygen saturation.

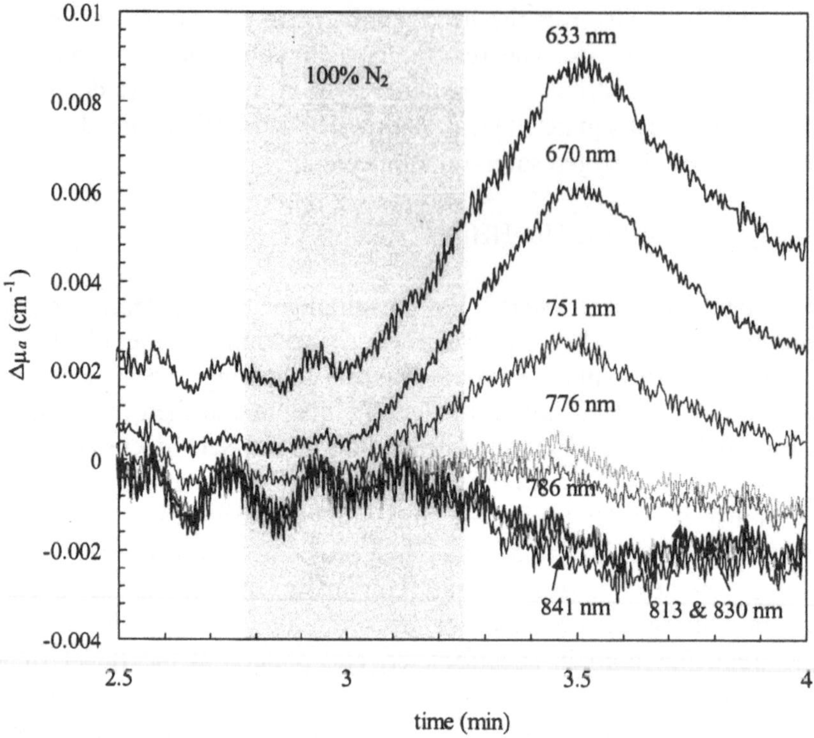

Figure 2. Pulse waves versus time for subject 3 using 8 wavelength-system. The shaded area represents the hypoxia (100% N2). The time scale agrees with the one in figure 4.

The fast Fourier transforms were obtained for each of the used wavelengths. An example of the FFT at 633 nm is shown in figure 3. From the FFT spectrum, the peak at the frequency of the heart rate (1.4 Hz – Fig. 3) can easily be seen. Taking the amplitude of the heart rate for each wavelength, we generated the spectrum of the arterial blood. From this spectrum $fdSaO_2$ was calculated.

Figures 4 (a and b) represents the time traces of $fdSaO_2$, $poSaO_2$, FiO_2, SO_2, O_2Hb, HHb and total hemoglobin concentration (tHb in μmol/l) during the deoxygenation experiment on subject 3. The inspired oxygen fraction of subject 3 was set to zero (FiO_2=0%) by giving 100% nitrogen twice for 30 s. It was followed by 2 admissions of a mixture of 50% air and 50% nitrogen (FiO_2=10.5%) for 30 s and 60 s, respectively. In all the measurements done on all three subjects we observed a time delay between $fdSaO_2$ and $poSaO_2$ in the order of 10-15 s.

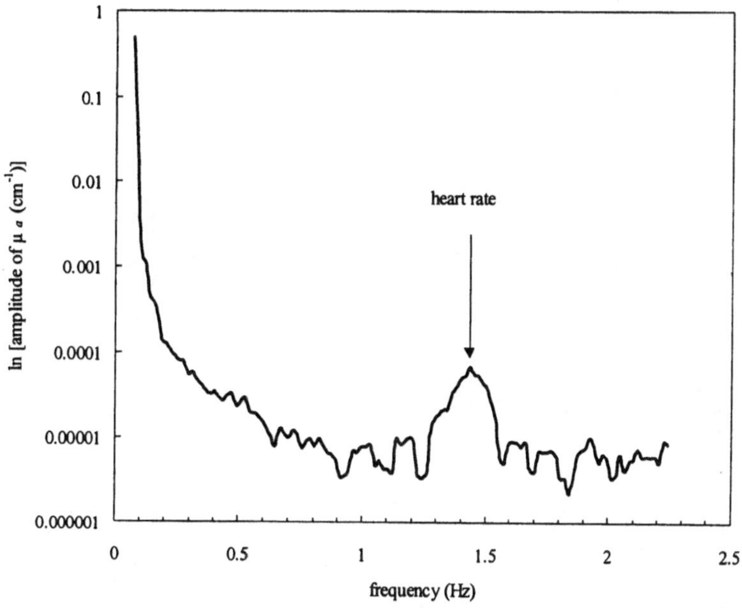

Figure 3. Power spectrum of subject 3 at 633 nm (8 wavelength-system).

The summary of the oxygen saturation values during normoxia and hypoxia for the deoxygenation experiment, for all the trials on the three subjects, are given in the table I. All the experiments on subjects 1 and 2 were done with the 4 wavelength-system and the ones done on subject 3 with the 8 wavelength-system. The baseline condition at room air, (FiO_2=21%) was determined by an average over 60 s ± standard deviation (SD) of the oxygen saturation values. The minimum saturation refers to the smallest average over 5 s ± SD, when FiO_2 was decreased to the value indicated for the specified duration. Only desaturation experiments with a minimal drop of at least 3% in $poSaO_2$ are reported in this table. One of the trials (no. 3) on subject 3 was not considered because the oxygenation changed only by 2%, as is also shown in figure 4a.

In order to compare $fdSaO_2$ and $poSaO_2$, which were obtained on all the subjects (over all the experiments), we plotted the difference between the two values against their mean value [14]. This type of plot was used because it effectively reveals systematic differences between two methods of measurement. The comparison is represented in figure 5 and it is seen that the two measurements show good agreement. The mean difference was –1.2±2.6%. We represented each of the three subjects with different symbols in order to show inter-subject differences.

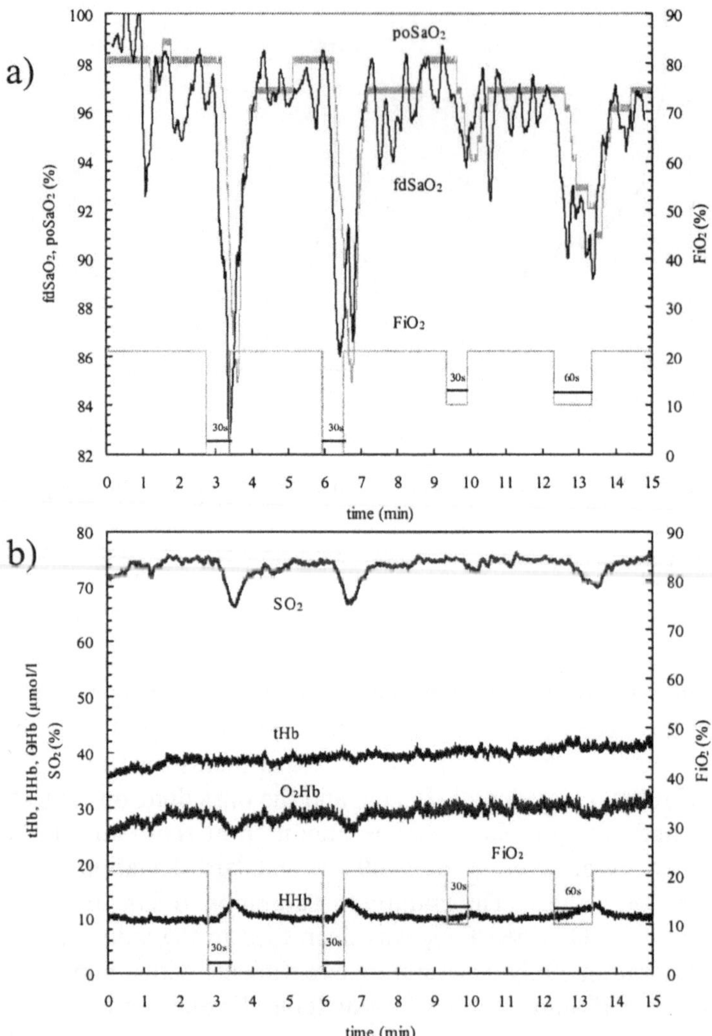

Figure 4. Time dependence of a) $fdSaO_2$ (forehead), $poSaO_2$ (finger), FiO_2, and b) cerebral O_2Hb, HHb, tHb concentration, cerebral SO_2, and FiO_2 on subject 3 using the 8 wavelength-system. The time delay between $fdSaO_2$ and $poSaO_2$ was 15 s in this particular subject.

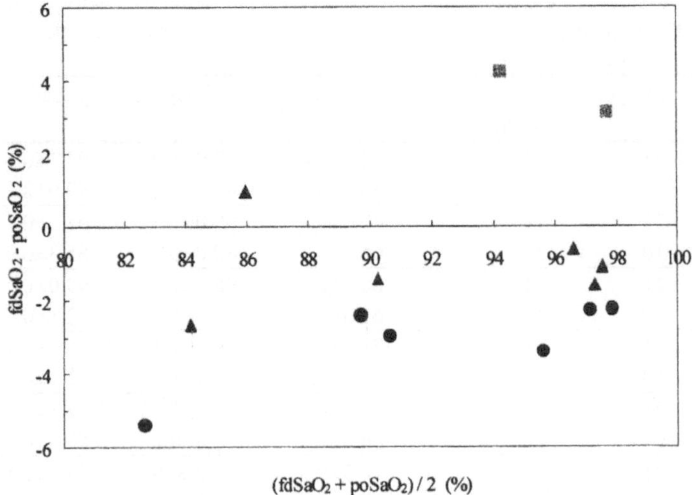

Figure 5. Difference between $fdSaO_2$ and $poSaO_2$ against their mean value, obtained over all the experiments. The symbols are: subject 1 (●), subject 2 (■), and subject 3 (▲).

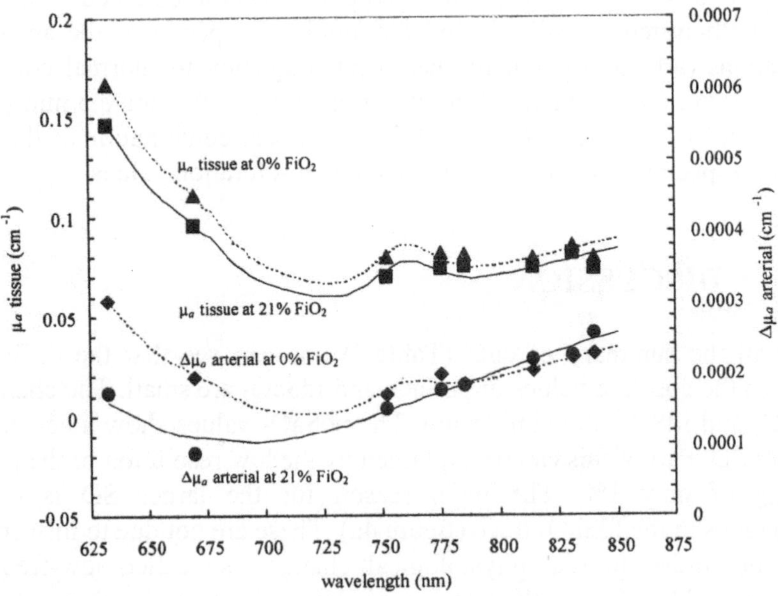

Figure 6. Tissue absorption spectrum (tissue μ_a), obtained with the 8 wavelength-system, at 0% FiO_2 (▲), at 21% FiO_2 (■), and spectrum of the amplitude of the oscillating absorption component at the heartbeat frequency (arterial $\Delta\mu_a$) at 0% FiO_2 (♦), at 21% FiO_2 (●). The symbols are the experimental data and the lines the best fits.

Table 1. Baseline (21% FiO_2) and minimum oxygen saturation values during deoxygenation experiment, on subjects 1 and 2 with 4 wavelength-system and subject 3 with 8 wavelength-system. SO_2 and $fdSaO_2$ are the tissue and arterial oxygen saturation, respectively, measured in the forehead by frequency-domain spectroscopy. $poSaO_2$ is the arterial oxygen saturation measured on the finger by a commercial pulse oximeter.

Subject	Trial	FiO_2 (%)	Duration (min)	SO_2 (%)	$fdSaO_2$ (%)	$poSaO_2$ (%)
1	1	21	>1	60.2±3.0	93.9±1.0	97.3±0.5
		10	3	58.9±3.1	88.5±2.4	90.9±2.2
	2	21	>1	65.8±3.4	96.0±1.3	98.2±0.1
		10	3	65.7±3.1	80.0±0.7	85.3±0.7
	3	21	>1	66.1±2.9	96.7±1.6	99.0±0.3
		10	2	64.6±2.9	89.1±1.3	92.1±0.8
2	1	21	>1	71.4±1.5	99.2±1.3	96.1±0.1
		10	4	70.3±1.5	96.4±1.4	92.1±0.3
3	1	21	>1	74.8±0.4	96.5±1.1	98.1±0.1
		0	0.5	66.7±0.1	82.8±0.2	85.5±0.4
	2	21	>1	74.3±0.3	97.0±0.8	98.1±0.1
		0	0.5	67.5±0.2	86.5±0.2	85.5±0.4
	4	21	>1	74.9±0.5	96.3±0.6	96.9±0.1
		10	1	70.1±0.1	89.5±0.3	91.0±0.1

Figure 6 shows the absolute absorption spectra obtained with the 8 wavelength-system (tissue μ_a) and the amplitude spectra of the absorption oscillations (arterial $\Delta\mu_a$) at the heartbeat frequency for normal conditions ($FiO_2 = 21\%$), and at maximal desaturation ($FiO_2 = 0\%$) for 0.5 min (trial 1 on subject 3). The lines are the best fits of a linear combination of the O_2Hb and HHb spectra [12] at the respective oxygen saturation values.

4. DISCUSSION

From the summary of data (Table I) one can see that the differences between the absolute values of $poSaO_2$ and $fdSaO_2$ are small. The changes in $poSaO_2$ and $fdSaO_2$ are similar too. The $poSaO_2$ values show a consistently smaller SD. Part of this can be explained by the low resolution of the $poSaO_2$ reading of only 1%. The main reason for the larger SD is cyclical fluctuations in the $fdSaO_2$ trace (figure 4a). These are not due to instrumental noise but rather to real physiological changes, so called low-frequency fluctuations [15]. These fluctuations are probably more pronounced in the brain than in the finger and therefore not detectable by the pulse oximeter. The one exceptionally high SD, of 2.2%, in the $poSaO_2$ reading is due to unusually large cyclical fluctuations in that particular experiment, clearly visible in the $poSaO_2$ trace as well.

During the desaturation experiment, the drop in SO_2 is consistently smaller than the drop in $fdSaO_2$ or $poSaO_2$. One can find two possible explanations for this:

1. The desaturation causes the cerebral blood flow to increase in order to maintain adequate oxygenation of the brain. This effect is well known as a mean of the cerebral protection [15].

2. The instrument does not quantify SO_2 accurately. A possible reason for this maybe that the algorithm assumes a semi-infinite, flat, homogenous tissue geometry. This is not strictly fulfilled by the geometry of the head. Although we do not know the exact size of this effect, from our experience, we believe that its contribution is small.

The frequency-domain data are crucial in our approach to obtain absolute tissue parameters. We have used intensity and phase data to obtain μ_a and μ_s' of the tissue to determine SO_2, and the DPF spectrum, which was used to calculate the $fdSaO_2$. The frequency-domain approach has several advantages:

1. It does not rely on any calibration. The pulse oximeter relies on such a calibration on a standard population. However, the mean calibration may be inaccurate for certain individuals, because it can not take into consideration the individual differences in μ_s'. In figure 5 a small but systematical inter-subject variability between $fdSaO_2$ and $poSaO_2$ can be seen. In subject 1, with the lowest μ_s'=5.6 cm^{-1}, $poSaO_2$ is always higher than $fdSaO_2$. In subject 2, with the highest μ_s'=7.6 cm^{-1}, $poSaO_2$ is always lower than $fdSaO_2$. In subject 3, μ_s'=6.9 cm^{-1}, $poSaO_2$ and $fdSaO_2$ are similar. Thus, we suggest that the inter-subject variability is due to the different μ_s' among the subjects, which causes small errors in the $poSaO_2$ values. In contrast, the frequency-domain method measures the necessary parameters for each subject individually.

2. The frequency-domain method is expected to be accurate for the full range of values from 0% to 100%, because the optical parameters μ_a and μ_s' are actually measured. Furthermore, the fit in figure 6 does not show evidence of unaccounted spectral features. The $poSaO_2$ is only accurate for values between 85% and 97%, because it was not calibrated for other values[5]. In clinical practice low values of SaO_2 appear in intensive care medicine or during apneic episodes.

3. The frequency-domain method measures the SaO_2 directly at the tissue of interest, e.g. the brain. Although the SaO_2 is a systemic parameter, there are circumstances under which the SaO_2 is different among body parts, such as an open ductus arteriosus in neonates, where peripheral measurements may give too low SaO_2 values. During shock situations the blood centralizes and it may be difficult to get accurate SaO_2 readings peripherally. This is one of the reasons for clinicians to insert an invasive arterial line. The brain is last affected by the centralization of blood during

shock and therefore is expected to give a much more reliable $fdSaO_2$ reading under such critical circumstances. Furthermore, the frequency-domain method will yield SO_2 values, which will even be available when there is no pulse.

4. For the frequency-domain method all the parameters ($fdSaO_2$, SO_2, O_2Hb, HHb, and tHb) are measured at the same location. The interpretation of the changes in these parameters is more precise since they are synchronous. Also, there is usually a delay between the $poSaO_2$ measured at the finger and the other parameters measured at the brain (the blood travels longer to reach the finger).

We have used two instrumental set-ups: with a) 8 wavelengths and b) 4 wavelengths. The measurements turned to be of similar quality. Therefore, the more inexpensive 4 wavelength-system is favorable.

5. CONCLUSION

Our paper presents a new, non-invasive, frequency-domain approach to measure in real time, simultaneously absolute values of the local SO_2 and SaO_2. This method provides SaO_2 directly, at the location of interest, and can be applied on different locations and over a full range of SaO_2 values. The 4 and 8 wavelength-system work equally well.

ACKNOWLEDGMENTS

This research is supported by the US National Institutes of Health (NIH) Grant No. CA57032, and by Whitaker-NIH Grant No. RR10966. We also thank John Maier for medical assistance and Dennis Hueber for technical support with the instrumentation.

REFERENCES

1. Fantini S, Franceschini-Fantini MA, Maier JS, Walker SA, Barbieri B, Gratton E. Frequency-domain multichannel optical detector for noninvasive tissue spectroscopy and oximetry. Opt Eng 1995;34:32-42.
2. Miwa M, Ueda Y, Chance B. Development of time-resolved spectroscopy system for quantitative noninvasive tissue measurement. SPIE 1995;2389:142-149.
3. Liu H, Chance B, Hielscher AH, Jacques SL, Tittel FK. Influence of blood vessels on the measurement of hemoglobin oxygenation as determined by time-resolved reflectance spectroscopy. Med Phys 1995;22:1209-1217.
4. Kohl M, Nolte C, Heekeren HR, Horst S, Scholz U, Obrig H, Villringer A. Changes in cytochrome-oxidase oxidation in the occipital cortex during visual simulation: improvement in sensitivity by the determination of the wavelength dependence of the differential pathlength. SPIE 1998;3194:18-27.

5. Mendelson Y. Pulse oximetry: theory and applications for noninvasive monitoring. Clin Chem 1992;38:1601-1607.
6. Webb RK, Ralston AC, Runciman WB. Potential errors in pulse oximetry. II. Effects of changes in saturation and signal quality. Anaesthesia 1991;46:207-212.
7. Franceschini MA, Gratton E, Fantini S. Noninvasive optical method of measuring tissue and arterial saturation: an application to absolute pulse oximetry of the brain. Opt Lett 1999;24:1-3.
8. Franceschini MA, Wallace D, Barbieri B, Fantini S, Mantulin WW, Pratesi S, Donzelli GP, Gratton E. Optical study of the skeletal muscle during exercise with a second generation frequency-domain tissue oximeter. SPIE 1997;2979:807-814.
9. Hueber DM, Fantini S, Cerussi AE, Barbieri B. New optical probe design for absolute (self-calibrating) NIR tissue hemoglobin measurements. SPIE 1999;3597:618-631.
10. Fantini S, Franceschini MA, Gratton E. Semi-infinite-geometry boundary problem for light migration in highly scattering media: a frequency-domain study in the diffusion approximation. J Opt Soc Am B 1994;11:2128-2138.
11. Delpy DT, Cope M, van der Zee P, Arridge S, Wray S, Wyatt J. Estimation of optical pathlength through tissue from direct time of flight measurement. Phys Med Biol 1988;33:1433-1442.
12. Wray S, Cope M, Delpy DT, Wyatt JS, Reynolds EO. Characterization of the near infrared absorption spectra of cytochrome aa3 and haemoglobin for the non-invasive monitoring of cerebral oxygenation. Biochim Biophys Acta 1988;933:184-192.
13. Villringer K, Minoshima S, Hock C, Obrig H, Ziegler S, Dirnagl U, Schwaiger M, Villringer A. Assessment of local brain activation. Adv Exp Med Biol 1997;413:149-153.
14. Bland JM, Altman DG. Statistical methods for assessing agreement between two methods of clinical measurement. Lancet 1986;1(8476):307-10.
15. Hudetz AG. Regulation of oxygen supply in the cerebral circulation. Adv Exp Med Biol 1997;428:513-520.

Chapter **8**

THE INFLUENCE OF A CLEAR LAYER ON NEAR INFRARED SPECTROPHOTOMETRY: COMPARISON OF MEASUREMENTS IN A LIQUID NEONATAL HEAD PHANTOM TO INFANTS *IN VIVO*

Martin Wolf, Matthias Keel, Vera Dietz, Kurt von Siebenthal, Jan Teller, Hans-Ulrich Bucher & Oskar Baenziger
Clinic for Neonatology, University Hospital, Frauenklinikstr. 10, 8091 Zurich, Switzerland

Abstract: Near infrared spectrophotometric (NIRS) algorithms to determine the tissue oxygen saturation (TOI) assume a semi-infinite, homogenous tissue geometry. At the head, the clear cerebrospinal fluid (CSF) layer may violate this assumption. The aim was to estimate the error in the TOI values caused by the CSF layer in vitro and to confirm the results *in vivo*. The liquid phantom mimicking the neonatal head, consisted of a spherical shell of silicone filled with a liquid solution (1% Intralipid®, 60 µmol/l haemoglobin, yeast) and a clear layer imitating CSF. The solution was oxygenated and deoxygenated, while measuring its TOI and pO_2. Without clear layer the mean TOI was $90.9 \pm 0.5\%$ at $pO_2 > 18$ kPa and decreased to $26.0 \pm 1.3\%$ at $pO_2 = 0$ kPa. With a clear layer the TOI *increased* from $27.8 \pm 0.8\%$ at $pO_2 > 18$ kPa to $68.0 \pm 0.8\%$ at $pO_2 = 0$ kPa. The clear layer caused a large error in the TOI. In ten mechanically ventilated infants (postnatal age 0.03 to 8 months) the TOI (at the head) and arterial oxygen saturation (SaO_2) were measured while the inspired oxygen fraction was altered. The TOI was always positively correlated with the SaO_2 (mean slope linear regression=0.89, r^2=0.62). Thus an adverse effect of the CSF layer on TOI measurements can be excluded for infants. The CSF layer is not modelled correctly in the phantom.

Key words: Cerebrospinal fluid, head of infant, neonatal head phantom, near infrared spectrophotometry, tissue oxygen saturation

1. INTRODUCTION

Instruments to measure the tissue oxygen saturation (TOI in %, is the abbreviation used by the manufacturer of our instrument) based on near-infrared spectrophotometry (NIRS) are commercially available. The TOI may be of great clinical importance to determine the oxygen supply of various organs, especially in intensive care medicine, where patients may have insufficient oxygenation. In our clinic for neonatology, the main focus of research is oxygenation of the brain, because this organ is very sensitive to maloxygenation, which can lead to brain lesions. Brain lesions may produce serious permanent handicaps.

In principle, NIRS is able to quantify TOI. The algorithms for the calculation of haemoglobin concentration and TOI from the raw optical data are based on geometrical assumptions, concerning the tissue of interest, which is usually supposed to be semi-infinite, flat and homogeneous.

The brain is surrounded by skin, skull and cerebrospinal fluid (CSF). Thus the anatomical structure is neither homogeneous nor semi-infinite nor flat. Especially the inhomogeneity caused by the CSF layer, which has a low scattering coefficient (μ_s') and absorption coefficient (μ_a), may produce considerable errors [1,2]. The focus of this study was to investigate the influence of the CSF layer.

It is difficult to validate NIRS instruments directly in infants, because there is no gold standard method available which would conform to clinical ethical standards. Animal experiments are not comparable due to the great geometrical differences of anatomical structures, especially of the head. Thus NIRS has already been shown to give erroneous values for cerebral blood flow in dogs [3], while two studies proved that NIRS is a valid method to determine the cerebral blood flow of neonates [4,5].

Therefore as our first aim, we decided to use a phantom model with geometrical characteristics similar to a neonatal head to test the influence of the clear layer *in vitro*.

How does the TOI reading behave *in vivo*, when a CSF layer is present? In infants, complete oxygenation or deoxygenation, as in the phantom experiments, is not possible, but small changes within the physiological range can be induced. Although the absolute value of TOI cannot be tested, changes in TOI can be compared quantitatively to the changes in arterial oxygen saturation (SaO_2 in %) measured by pulse oximetry. This was our second aim.

The third aim of the study was to compare the effects of a clear layer in vitro to those found *in vivo* in infants.

2. METHODS

2.1 NIRS instrument

For our study we used the NIRO 300 (Hamamatsu, Japan), which is based on spatially resolved spectroscopy. It uses four pulsed laser diodes with wavelengths at 775 nm, 810 nm, 847 nm and 913 nm. A fiber optic guides the light to the region of interest. Three silicon photodiode detectors are positioned at different distances from the source fiber. The slope of the light intensity over the three detectors is proportional to the effective light attenuation. Under certain assumptions, the shape of the light absorption of the tissue can be determined at the four mentioned wavelengths. This enables to calculate TOI. These assumptions and the methodology of this instrument are described in detail in the literature [6,7].

2.2 Phantom for the *in vitro* test

The purpose of the phantom was to investigate the influence of a CSF layer on NIRS measurements of the oxygenation of the brain. The phantom consists of three components: A solid layer mimicking skin and skull, a clear layer of polypropylene imitating CSF, and a liquid solution with variable scattering properties and haemoglobin concentrations, which represents the brain (Figure 1).

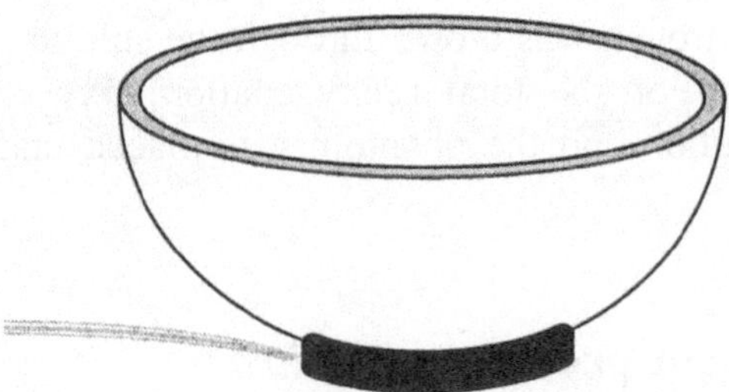

Figure 1. The silicone phantom is a spherically-shaped shell filled with a liquid Intralipid solution and hemoglobin imitating the optical properties of brain tissue. The sensor of the spectrophotometer is placed at the bottom on the outside of the shell. Above the sensor a clear layer of polypropylene is attached inside the shell.

The solid layer has the form of a spherical shell. Its diameter of 11cm corresponds to the size of the head of a newborn infant of 40 weeks gestational age [8]. The thickness of the layer is 3.5mm, which corresponds approximately to the thickness of skin and skull. It consists of transparent silicone rubber with a scatterer (TiO_2) and an absorber (Zeneca™, Manchester, UK) added: $\mu_s' \approx 1.7$ mm^{-1}, $\mu_a \approx 0.05$ mm^{-1}.

We used polypropylene to imitate the CSF layer, because its absorption and scattering coefficients proved to be equally low as in CSF during a spectrometric analysis. A 0.5mm thick layer of polypropylene was attached inside the shell directly above the sensor, which was attached on the outside of the shell. The clear layer exceeded the sensor's limits by 1.5cm in any direction.

The shell was filled with an aqueous liquid solution containing:

- 1% Intralipid® mimicking scattering of the brain.
- 0.9% sodium chloride.
- 0.5% yeast to consume the oxygen in the solution.
- 0.15% glucose was maximally given to activate the yeast. The amount was increased in small steps of 0.05% until the yeast consumed enough oxygen to produce total deoxygenation within 30 min.
- 60 µmol/l haemoglobin from packed red cell concentrate whose haemoglobin concentration was determined by a Corning 2500 co-oximeter.
- The pH was adjusted to a physiological level by adding a H_2PO_4 buffer.

The volume of the solution was approximately 200 ml. The liquid solution was temperature regulated to 37° C by a small heater and constantly stirred by a mixer. Its pO_2 was continuously measured by a Hellige Servomed pO_2 Monitor, whose probe was adjusted to 37° C and calibrated in water of this temperature.

To totally deoxygenate the solution, the phantom was put under a nitrogen atmosphere, nitrogen was blown through the solution and the yeast consumed the oxygen. For the total reoxygenation, oxygen was slowly blown through the solution and the phantom was placed under an oxygen atmosphere.

2.3 Measurement procedure *in vitro*

The complete liquid solution with an Intralipid® concentration of 1% was filled into the phantom without the clear layer. First the solution was totally deoxygenated until the pO_2 was 0.0 kPa and the TOI reading did not decrease anymore. Subsequently, it was reoxygenated until the pO_2 was >18 kPa. All optical data and calculated concentrations were recorded simultaneously with the pO_2. The same procedure was repeated with the clear polypropylene attached inside the shell.

2.4 Data analysis *in vitro*

Two states were defined for both phantom experiments:

1. Total oxygenation, which was defined as a pO_2 >18 kPa. In this state, almost all of the haemoglobin was oxygenated and hence the TOI should be approximately 100%.
2. Total deoxygenation, which was defined as a pO_2 = 0.00 kPa. In this state the TOI should be nearly 0%.

During the period of total oxygenation and the period of total deoxygenation, the mean TOI was calculated.

2.5 Measurement procedure *in vivo*

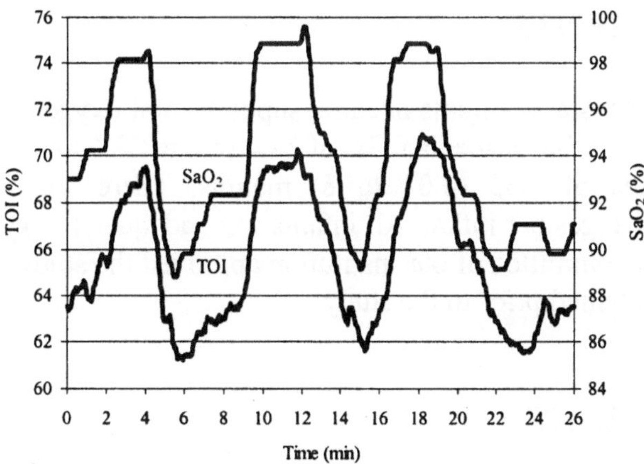

Figure 2. An example of a series of oxygenation changes induced by altering the inspired oxygen fraction in mechanically ventilated infants, which was reflected in the arterial oxygen saturation (SaO_2) and the tissue oxygen saturation (TOI) determined by NIRS.

The NIRS sensor was attached to the temple of the head of the infant, avoiding the area of the sagittal sinus. The pulse oximetry sensor was fixed to the right hand. Slow oxygenation changes were achieved by altering the inspired oxygen fraction (FiO_2 in %) in small steps every 20 s in mechanically ventilated infants. An oxygenation change was defined to be any change in SaO_2 of more than 3% over a period of more than 1.2 min, which was shown to be a period of time, where the oxygenation change is slow enough that the cerebral oxygenation remains in equilibrium [9]. The amount of alteration in FiO_2 in order to achieve a change of at least 3% in SaO_2 depended very much on the condition of the lungs of the infant and varied widely. In each infant at least 6 oxygenation changes were carried out consecutively (Figure 2). The SaO_2 was kept between 99% and 85%.

2.6 Data analysis *in vivo*

The data of the NIRS instrument and the pulse oximeter were collected simultaneously and stored in a PC. All data used in this study had sufficient technical quality, i.e. the NIRS instrument did not indicate poor signal quality. Oxygenation changes containing clear movement artifacts of the pulse oximeter, i.e. a sudden drop of SaO_2 by more than 20%, were removed.

The sample time of 2 s of the data was converted into 10 s by averaging. All oxygenation changes with a change in SaO_2 of more than 3% over a period of more than 1.2 min were included in the analysis. A regression line of TOI versus SaO_2 was calculated. For each infant the slope, intercept and r^2 of this line were determined.

3. PATIENTS

Ten clinically stable infants needing supplemental oxygen were included in this study. They had a median GA of 37.4 (range 29.4 to 41.6) weeks and a postnatal age of 0.35 (0.03 to 8) months. There were at least six oxygenation changes per infant. All infants were adequately oxygenated.

The ethical committee of our institution approved the study and informed consent was obtained prior to the study.

4. RESULTS

4.1 *In vitro* phantom study

The mean TOI during total oxygenation and deoxygenation is shown in Table 1 for the two phantom set-ups: once without a clear layer mimicking CSF and once with this layer. The time series of the experiment without and with a clear layer can be seen in Figure 3 and Figure 4, respectively. A dramatic error in the TOI reading is caused by the clear layer: the TOI increases when oxygen is withdrawn and vice versa, i.e. the TOI reacts paradoxically in this situation.

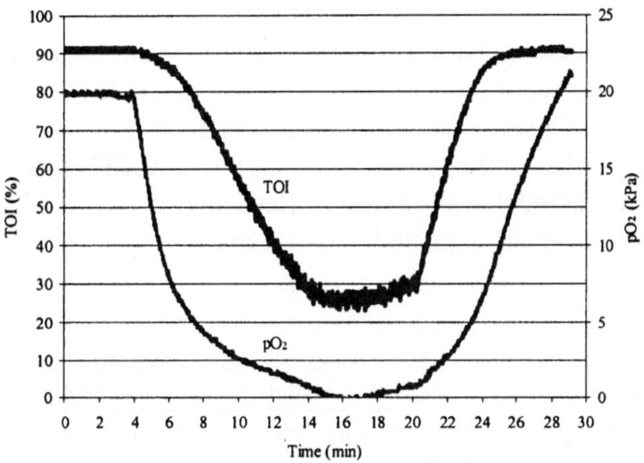

Figure 3. The phantom experiment: In this phantom there was no clear layer present. The TOI changes parallel with the pO_2 as oxygen is withdrawn and re-entered.

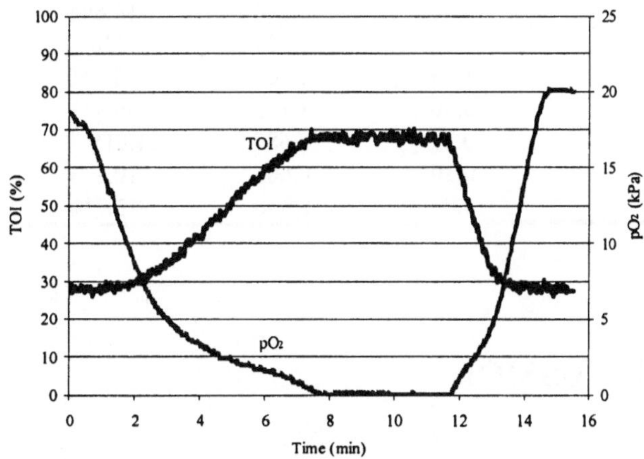

Figure 4. The phantom experiment: In this phantom there was a clear layer present. The TOI shows a dramatic paradox reaction: It increases as oxygen is removed and decreases as the liquid solution is reoxygenated.

Table 1. Results of the phantom experiment (*tissue oxygen saturation in mean ± standard deviation %).

State	pO_2	TOI* without clear layer	TOI* with clear layer	TOI* expected
total oxygenation	>18.0	90.9±0.5	27.8±0.8	100
total deoxygenation	= 0.0	26.0±1.3	68.0±0.8	0

4.2 *In vivo* study on the head of infants

The slopes, intercepts and r^2 of the regression line for each infant for the small induced oxygenation changes are shown in Table 2. Figure 2 depicts a typical example of a series of oxygenation changes.

Across all infants the slope was 0.894 ± 0.232 (mean ± standard deviation), the intercept was -18.50 ± 20.95 and r^2 was 0.620 ± 0.135. The mean slope was near the expected value of one.

Table 2. Results from the oxygenation changes in infants (regression lines for each infant for the small induced oxygenation changes: TOI = slope * SaO_2 + intercept).

Infant	gestational age weeks	postnatal age months	Slope	Intercept	r^2
1	41 4/7	8.00	0.847	-14.612	0.711
2	37 3/7	0.40	0.886	-18.061	0.802
3	36 0/7	0.10	0.818	-8.577	0.395
4	29 3/7	5.00	0.790	-14.604	0.579
5	37 2/7	0.03	0.727	1.520	0.708
6	39 0/7	1.25	1.408	-68.121	0.520
7	29 5/7	3.00	1.175	-39.931	0.820
8	39 0/7	0.27	0.687	1.915	0.599
9	38 4/7	0.30	0.665	-10.813	0.542
10	33 0/7	0.03	0.940	-13.711	0.528

5. DISCUSSION

5.1 Phantom experiments *in vitro*

Both phantom set-ups were exactly the same except for the clear layer. Thus the enormous error in the TOI reading could clearly be attributed to the clear layer. Most likely the effect was caused by light channelling along the clear layer instead of passing through the silicone shell or liquid solution, which represent the tissue [1]. In the phantom, which had no clear layer, the light intensity decreased with distance of the detector to the source, which is precisely what we expected. When the clear layer was inserted the highest light intensity was actually found at the detector, which was farthest from the light source, which is contrary to what we expected and may be explained by the mentioned light channelling effect.

5.2 Measurements on the infants' heads *in vivo*

The slope of the TOI reading in vivo versus the SaO_2 reading of the pulse oximeter was close to unity, i.e. the TOI reading was proportional to the oxygen concentration in all infants. This result, which is favourable for a clinical application of the instrument, shows that the phantom with a clear layer did not represent the situation *in vivo* accurately. However, the phantom without a clear layer represented the situation in vivo precisely. Therefore, we conclude that the clear layer we used in the phantom experiment was an erroneous model for the CSF layer of the neonatal head.

There are several possible reasons for this: *In vivo* the clear CSF layer is not evenly thick and of a spherical shape as the one we used in the phantom. The CSF layer follows the structure of the brain with its gyri and valleys. Furthermore there are blood vessels and arachnoid villi penetrating the CSF layer. All of these properties can be expected to reduce light channelling.

5.3 Reduced range of the TOI reading

In the phantom experiment without a clear layer the TOI reading showed 26% instead 0% when no oxygen was present, and 90.9% instead of 100% when almost all of the haemoglobin was oxygenated. Thus the range of the TOI reading, which was expected to be 100%, was decreased to 64.9%. In the *in vivo* experiment, we found that the mean slope is 0.894 instead of unity. Again the range of the TOI was decreased to 89.4%.

Suzuki et al. [7] found a slope corresponding to unity, when using a phantom model with a thin surface layer and a flat sensor geometry. Thus the decreased range of the TOI in our phantom model without a clear layer can be attributed to the curvature and/or the inert layer, which represented skin and skull. The inert layer did not contain any haemoglobin. Its optical properties were inert to changes in oxygen concentration of the liquid solution.

In vivo the skin and skull contain haemoglobin and change oxygenation like the brain tissue. This was the major difference compared with the phantom model without a clear layer. This may be why the range of the TOI reading was less decreased than in the phantom model. The remaining decreased range was most likely due to the curvature of the neonatal head.

5.4 Variability among infants

The intercept value depended on the oxygen extraction of the head. It is reasonable that it varied among infants.

The slope should not have varied among infants. It is beyond the scope of this study to identify the reasons for this variation. Possible reasons are the variation of the curvature of the head and changes in blood circulation. Clearly the slopes, which were most different from unity, were associated with a low r^2. This issue needs further investigation.

6. CONCLUSION

In the phantom model without a clear layer, the TOI reading was proportional to oxygen concentration as expected. When a clear layer mimicking the CSF was inserted, the TOI reading was the opposite of that expected: it decreased when oxygen was added and vice versa. In measurements on the infant's head, the TOI reacted similar to the phantom model without a clear layer. We therefore conclude that the CSF layer does not cause major errors in NIRS measurements of the infant's head and that it cannot be represented by a simple evenly thick layer in phantom models.

REFERENCES

1. Okada E, Delpy DT. The effect of overlying tissue on NIR light propagation in neonatal brain. Adv Optical Imag Photon Migration 1996;2:338-343.
2. Wolf M, Keel M, Dietz V, von Siebenthal K, Bucher HU, Baenziger O. The influence of a clear layer on near infrared spectrophotometry measurements using a liquid neonatal head phantom. Phys Med Biol 1999;44:1743-1753.
3. Newton CRJC, Wilson DA, Gunnoe B, Wagner B, Cope M, Traystman RJ. Measurement of cerebral blood flow in dogs with near infrared spectroscopy in the reflectance mode is invalid. J Cereb Blood Flow Metab 1997;17:695-703.
4. Skov L, Pryds O, Greisen G. Estimating cerebral blood flow in newborn infants: comparison of near infrared spectroscopy and Xe clearance. Pediatr Res 1991;30:570-573.
5. Bucher HU, Edwards AD, Lipp AE, Duc G. Comparison between near infrared spectroscopy and 133Xenon clearance for estimation of cerebral blood flow in critically ill preterm infants. Pediatr Res 1993;33:56-60.
6. Matcher SJ, Kirkpatrick P, Nahid K, Cope M, Delpy DT. Absolute quantification methods in tissue near infrared spectroscopy. SPIE 1993;2389:486-495.
7. Suzuki S, Takasaki S, Ozaki T, Kobayashi Y. A tissue oxygenation monitor using NIR spatially resolved spectroscopy. SPIE 1999, 3597:582-592.
8. Hall JG, Froster-Iskenius UG, Allanson JE. Handbook of normal physical measurements, Oxford University Press 1989.
9. Wolf M, Bucher HU, Dietz V, Keel M, von Siebenthal K, Duc G. How to evaluate slow oxygenation changes to estimate absolute cerebral haemoglobin concentration by near infrared spectrophotometry. Adv Exp Med Biol 1997;411:495-501.

Chapter 9

STRATEGIES FOR ABSOLUTE CALIBRATION OF NEAR INFRARED TOMOGRAPHIC TISSUE IMAGING

Troy O. McBride, Brian W. Pogue, Ulf L. Österberg, & Keith D. Paulsen
Thayer School of Engineering, 8000 Cummings Hall, Dartmouth College, Hanover, New Hampshire 03755

Abstract: Quantitative near infrared (NIR) imaging of tissue requires the use of a diffusion model-based reconstruction algorithm, which solves for the absorption and scattering coefficients of a tissue volume by matching transmission measurements of light to the predictive diffusion equation solution. Calibration problems as well as other practical considerations arise for an imaging system when using a model-based method for a real system. For example, systematic noise in the data acquisition hardware and source/detector fibers must be removed to prevent spurious results in the reconstructed image. Practical considerations for a NIR diffuse tomographic imaging system include: (1) calibration with a homogeneous phantom, (2) use of a homogeneous fitting algorithm to arrive at an initial optical property estimate for image reconstruction of a heterogeneous medium, and (3) correction for fluctuations in source strength and initial phase offset during data acquisition. These practical considerations, which rely on an accurate homogeneous fitting algorithm are described. They have allowed demonstration of a prototype imaging system that has the ability to quantitatively reconstruct heterogeneous images of hemoglobin concentrations within a highly scattering medium with no *a priori* information.

Key words: blood, calibration, hemoglobin, photon migration, reconstruction, tomography

Oxygen Transport to Tissue XXIV, edited by
Dunn and Swartz, Kluwer Academic/Plenum Publishers, 2003

1. INTRODUCTION

Tissue is highly scattering at visible and near-infrared (NIR) wavelengths, thus a simplistic back-projection image is highly blurred and difficult to interpret as the effects of scattering and absorption can not be distinguished. However, by using model-based imaging and computational methods in combination with frequency domain measurements, moderate resolution, quantitatively accurate images of absorption and scattering can be obtained within highly scattering media, such as tissue [1,2]. This method may lead to a potentially powerful medical imaging modality for non-invasive hemoglobin tomography.

Practical calibration issues arise when using a model-based image reconstruction approach which may not be present in modalities such as x-ray tomography due to the necessity of fitting the data to a partial differential equation solution. In addition, while many system calibrations can be ignored in qualitative imaging, more care is required for quantitative imaging of hemoglobin concentrations on an absolute scale. Issues encountered in the development of the NIR imaging system addressed in this paper include: (1) elimination of system-based offsets, (2) choice of heterogeneous starting value, and (3) offsets due to long-term drift. These issues are addressed through a calibration protocol which involves a homogeneous phantom and the use of a precise homogeneous fitting algorithm that is resistant to measurement noise.

A 4:1 increase in hemoglobin concentration [3] and a 1.4 - 4.4 times lower oxygen pressure [4] have been observed in breast cancer. By using NIR frequency domain measurements which have been properly calibrated at multiple wavelengths, quantitative absolute images of hemoglobin related parameters can be obtained [5]. Once calibrated imaging is implemented in practice it should be possible to determine whether NIR hemoglobin concentration and oxygen saturation information is useful in the diagnosis of breast tumors.

2. METHODS

2.1 Imaging system

The frequency-domain near-infrared imaging system consists of three main components: (1) data acquisition system, (2) image reconstruction

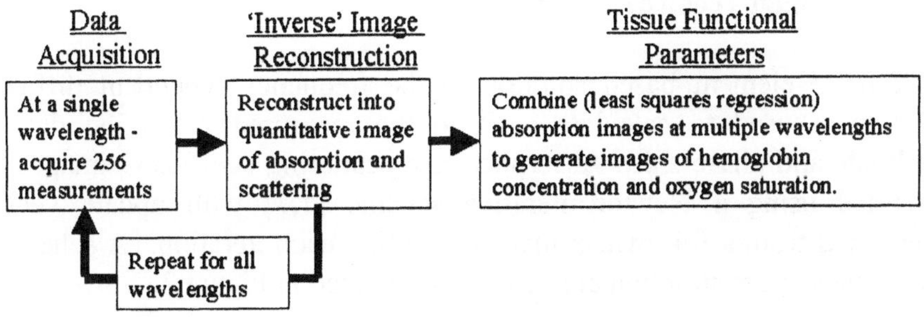

Figure 1. Schematic of near-infrared tomographic imaging system

algorithm, and (3) spectroscopic determination of functional properties. A schematic of the imaging process is shown in Figure 1.

2.1.1 Data acquisition system

The data acquisition system (shown in Figure 2) uses an amplitude-modulated (100 MHz) wavelength-tunable (700-850 nm) TiSapphire laser as the light source. Sixteen source and sixteen detector optical fibers are arranged in a circular geometry to analyze a single tomographic plane of the measured tissue or tissue-simulating phantom. The detector for the system is a photomultiplier tube with built-in heterodyning circuitry. The heterodyne signal (1 kHz) amplitude and phase shift are read into a computer using a commercial-grade data acquisition board. The source and detector are multiplexed to acquire the 256 data points using linear translation stages. [5,6]

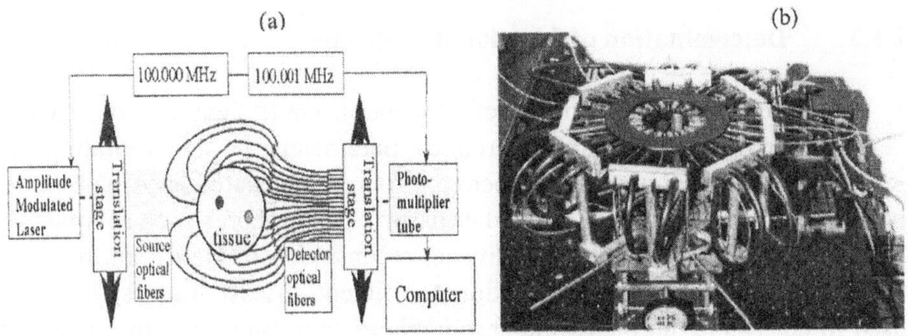

Figure 2. (a) Diagram and (b) photograph of the data acquisition system.

2.1.2 Image reconstruction

A finite element based solution to the frequency-dependent diffusion equation is used to calculate a fit to the measured projection data of amplitude and phase shift. Absorption and scattering coefficient images are generated using a Newton iterative scheme [1,7] with update vectors determined from a full matrix inversion during each iteration. A schematic of the image reconstruction algorithm is illustrated in Figure 3.

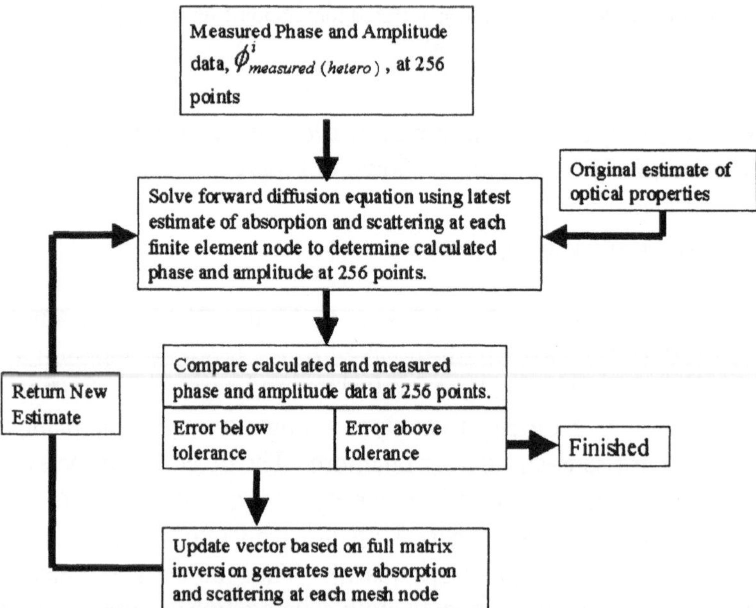

Figure 3. Schematic of the finite element based reconstruction algorithm based on the non-linear estimation problem of inferring optical property maps using the frequency-domain diffusion equation.

2.1.3 Determination of functional properties

Multiple wavelength images of absorption coefficient can be used to determine maps of hemoglobin related parameters by incorporating the previously measured absorption spectra of pure oxygenated hemoglobin, de-oxygenated hemoglobin [8], and other tissue chromophores such as lipids [9] and water [10]. Because the image reconstruction algorithm recovers quantitative values for the absorption coefficient, a least squares fit can be used to determine the metabolic chromophore concentrations from that data as shown schematically in Figure 4.

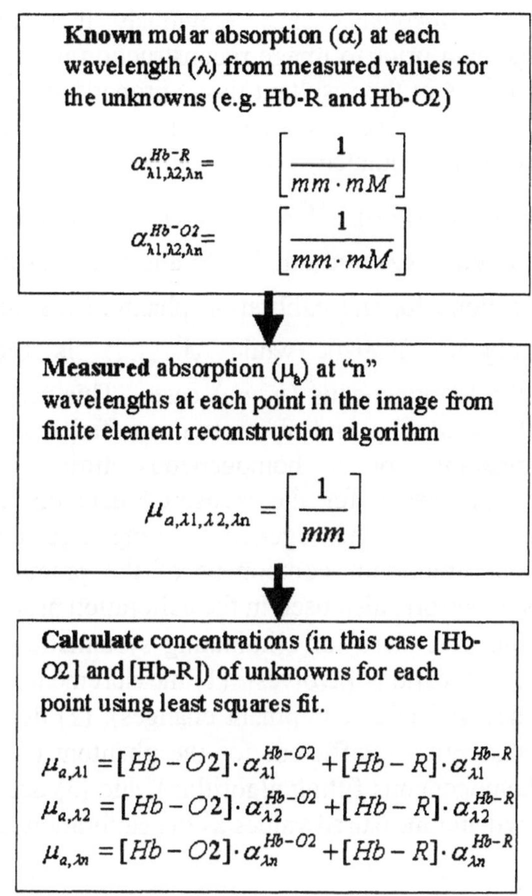

Figure 4. Schematic of methodology for determining hemoglobin related parameters from multiple wavelength images of absorption coefficient.

2.2 Practical considerations

Calibration and other practical considerations arise when using a model-based image reconstruction method with data acquired from a real imaging system. In the NIR diffuse tomography context these include: (1) system-based offsets, (2) initial optical property estimate for a heterogeneous medium, and (3) long term fluctuations of source strength and initial phase offset.

2.2.1 System-based offsets

Systematic offset in the measured data occurs at various source and detector locations due to optical fiber differences, multiplexing imprecision, and other systematic inconsistencies. Obviously, these offsets need to be removed from the data prior to image reconstruction in order to prevent the appearance of artifacts in the resultant tissue property profiles. The method adopted here measures a homogeneous phantom and subtracts differences between measured and calculated values:

$$\phi^{i}_{calibrated(hetero)} = \phi^{i}_{measured(hetero)} - (\phi^{i}_{measured(homo)} - \phi^{i}_{calculated(homo)}) \quad (1)$$

where $\phi^{i}_{measured}$ is the measured amplitude and phase at each i of the 256 measurement locations for the calibration phantom (*homo*) and the actual heterogeneous object (*hetero*), while $\phi^{i}_{calculated}$ is the corresponding calculation for the homogeneous (*homo*) case. This requires either exact knowledge of the scattering and absorption coefficients for the measured homogeneous phantom or a homogeneous fitting algorithm which determines these properties for the measured data on the homogeneous calibration phantom. The homogeneous fitting algorithm (described in Section 2.3) is an important component of the practical reconstruction algorithm and is, therefore, also used in the calibration procedure.

The calibration procedure for eliminating systematic offset at different source and detector locations involves: (1) measurement of a homogeneous phantom (each day and after equipment changes), (2) determination of the scattering and absorption coefficient for the phantom (from the measured data using the homogeneous fitting algorithm), and (3) use of the difference between calculated and measured values as the calibration factor.

2.2.2 Heterogeneous starting values

The image reconstruction algorithm starts from an estimated set of optical properties and then calculates an update vector based on χ^2 (the squared difference between the calculated and measured values). If the initially estimated set of optical properties is far from the actual optical properties of the heterogeneous object being imaged, this inaccurate starting point can slow convergence and even lead to erroneous answers. In order to determine the initial optical property estimate for patient data or a heterogeneous phantom, the measurements are averaged for each of the sixteen sources and the homogeneous fitting algorithm is used to determine a "homogeneous" estimate of the properties that are consistent with a diffusion equation solution which matches the averaged data. In this manner, the initial estimate is determined solely from the heterogeneous phantom or patient measurements, yet, is close enough to the unknown actual heterogeneous parameter distribution that the algorithm will converge.

2.2.3 Long term drift

Differences between the source strength and initial phase at the time of the measurement of the homogeneous calibration phantom and the actual source strength and initial phase observed at the time of the heterogeneous measurement can lead to an overall offset (mainly due to long term drift) between the calibrated measured and calculated data. This difference in overall offset needs to be removed in order for the reconstruction algorithm, which fits to the absolute amplitude and phase data, to be effective. To account for long term drift, the offset for both the homogeneous phantom measurements and the actual heterogeneous measurement are calculated based on the homogeneous fitting algorithm. (The homogeneous fitting algorithm responds to the slope of the data and is therefore independent of initial source strength and phase shift.) The offset is estimated as the average difference between the measured and calculated data.

$$\bar{\phi}_{offset(homo)} = \frac{\sum_{i=1}^{N}\left(\phi^{i}_{measured(homo)} - \phi^{i}_{calculated(homo)}\right)}{N} \tag{2}$$

$$\bar{\phi}_{offset(hetero)} = \frac{\sum_{i=1}^{N}\left(\phi^{i}_{measured(hetero)} - \phi^{i}_{calculated(hetero)}\right)}{N} \tag{3}$$

$$\bar{\phi}_{offset(net)} = \bar{\phi}_{offset(hetero)} - \bar{\phi}_{offset(homo)} \tag{4}$$

where $\bar{\phi}_{offset(net)}$ is the long term drift correction for initial phase and source strength.

2.3 Homogeneous fitting algorithm

An important part of practical imaging is the homogeneous fitting algorithm. The homogeneous fitting algorithm is critical for both calibration of the system and for providing the initial optical property estimate for the image reconstruction process. Assuming a homogeneous medium, only one source location is needed to determine the absorption and scattering coefficient of the material from measurements around its periphery. The absorption and scattering coefficients which result in the best fit to the measured data can be determined by using a Newton-Raphson iterative scheme applied to the finite element forward solution of the diffusion equation for the relevant geometry. To reduce the effect of noise on the fitting algorithm, the data acquired is averaged together for all sources based on detector distance from the source location. The Newton-Raphson solution is simplified by reducing the fit to two parameters: slope of the phase with respect to distance from the source location and slope of the log of intensity times distance with respect to distance. These two parameters were chosen because they are nearly constant and can be obtained from the data through linear regression. They are constant for the analytic infinite medium diffusion equation solution [11]. This method is insensitive to noise due to the large amount of averaging which occurs (256 measurements are used to find two parameters).

Figure 5 contrasts two versions of the homogeneous fitting algorithm. Method A performs a Newton-Raphson iterative scheme based on an estimate of the first derivative for the averaged data (number of detectors minus one (15) parameters), while Method B uses the slope of the phase with respect to distance from the source location and slope of the log of intensity times distance with respect to distance determined from linear regression (2 parameters).

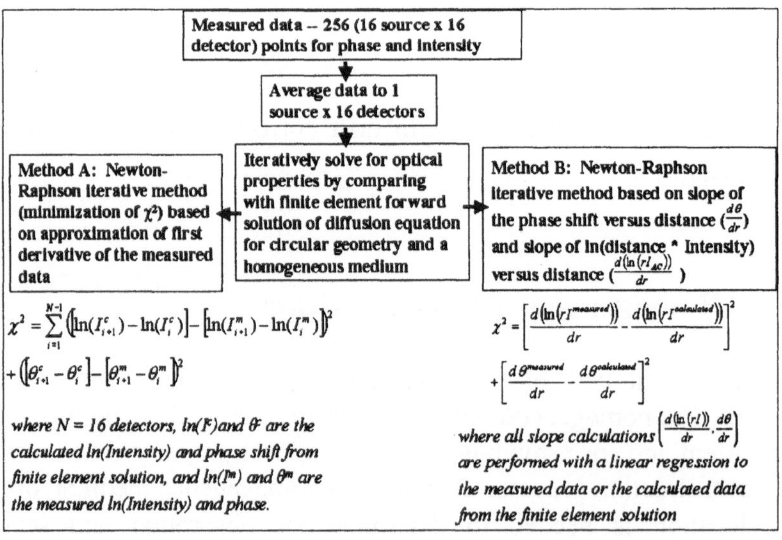

Figure 5. Schematic describing two homogeneous fitting algorithms.

3. RESULTS

3.1 Practical considerations

3.1.1 Homogeneous calibration

Figure 6 shows systematic offset from a slight difference in alignment of two optical fibers with the laser source fiber during multiplexing. This offset is consistent over a period of weeks and only changes during system modifications/maintenance such as re-alignment of the laser. The resultant difference data (Fig.6.c.) between the measured (Fig.6.a.) and the calculated data (Fig.6.b.) is used as a calibration factor in the reconstruction algorithm.

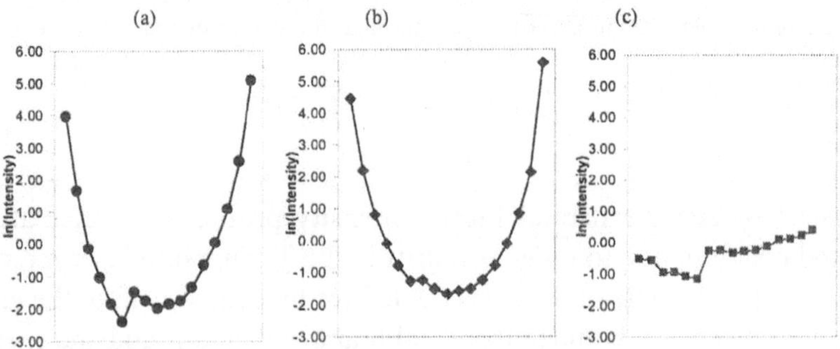

Figure 6. Plot of (a) ln(Intensity) data at 16 detector sites showing offset due to alignment differences, (b) calculated data from finite element solution, and (c) difference between (a) and (b). Difference data (c) is used as the 'calibration' factor.

3.1.2 Heterogeneous starting value

Simulations were performed to demonstrate the effect of different starting values during reconstruction of a heterogeneous object. Images for three different initial optical property estimates (0%, 10%, and 50% error from the actual average optical properties) are compared in Figure 7 after one iteration. The algorithm converges to the correct image (2:1 absorbing object with Gaussian profile) during the first iteration for an initial estimate which is equal to the average optical properties of the heterogeneous test object (Fig.7.a,d). When the initial estimate is significantly removed from the actual average optical properties, the algorithm takes longer to converge. This slowdown is evident in the other images in Figure 7.

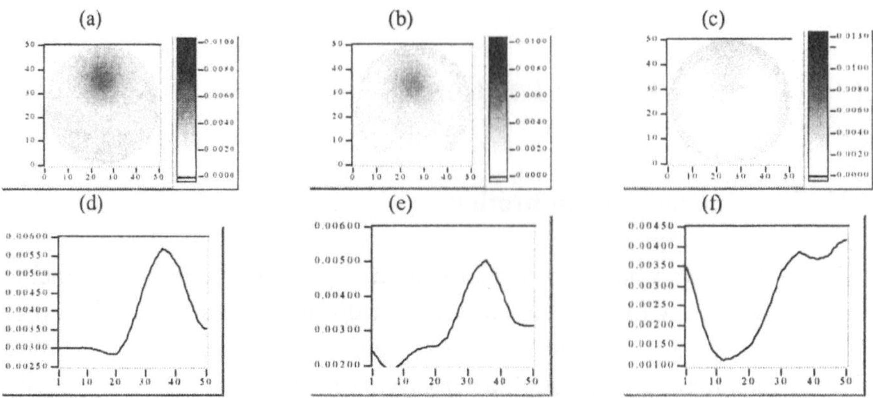

Figure 7. Image after first iteration of reconstruction of absorbing object with Gaussian profile with (a) initial estimate equal to average optical properties, (b) 10% error in initial estimate, and (c) 50% error in initial estimate. The vertical axis units are in mm^{-1}. (d), (e), and (f) are profile plots along a vertical line through the center of images (a), (b), and (c) respectively.

3.1.3 Offset

Offset between the natural log of intensity profile of a calculated and measured phantom due to long term drift in the Ti:Sapphire laser intensity is shown in Figure 8. The offset is accounted for by comparison of the changes in offset between the homogeneous calibration phantom and the measured heterogeneous object. In the extreme case shown in Figure 8, the offset in intensity is almost two natural log units.

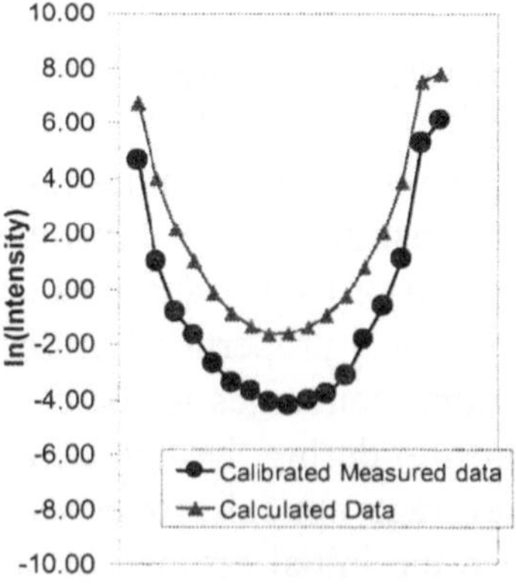

Figure 8. Plot demonstrating measured data with overall offset due to long term drift of laser intensity. The average difference is used to correct this offset.

3.2 Testing of homogeneous fitting algorithm

Using repeated measurements of the same phantom, the results of the homogeneous fitting algorithm (Method B in Figure 5) have been shown to have an average deviation of 0.5% even in the presence of 5% measurement noise. This compares very favorably with our previous approach (Method A in Figure 5), which did not take full advantage of the data averaging, where 5% measurement noise translated to 5% average deviation in the homogeneous fit. This decrease in deviation is demonstrated in Figure 9 which shows data for a phantom study involving triplicate measurement of objects with increasing amounts of blood. The absorption coefficient increases while the scattering coefficient remains constant. The study is performed at three wavelengths and the slope of the line of increasing absorption with added blood is compared with expected values of the molar extinction coefficient from Wray et al. [8] in Table 1.

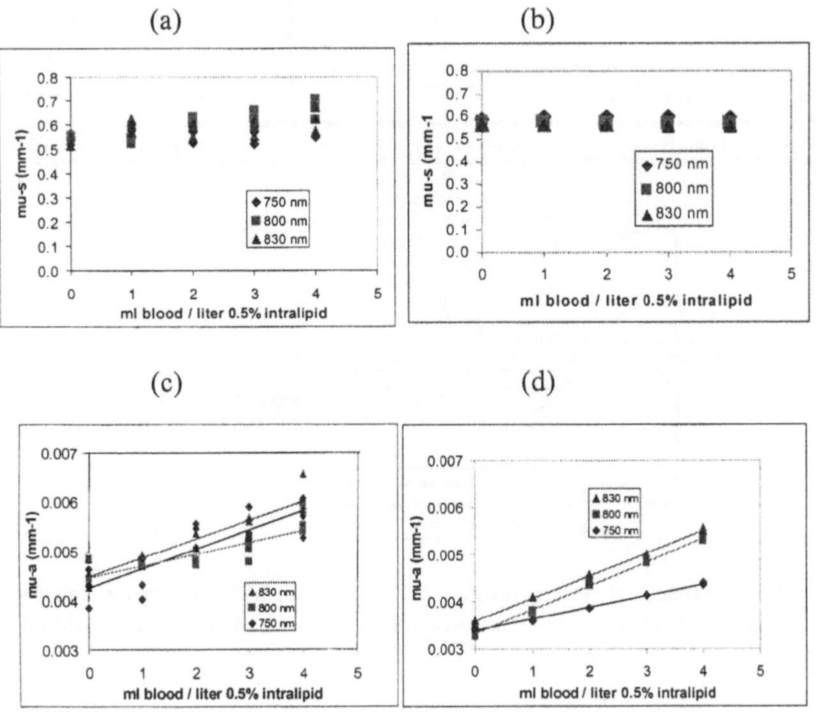

Figure 9. Homogeneous fits to measured data for the two methods described in Figure 5: (a) Method A scattering coefficient data, (b) Method B scattering coefficient data, (c) Method A absorption coefficient data, (d) Method B absorption coefficient data. Each concentration was measured three times at each of three wavelengths.

	Molar Extinction Coefficient (mm^{-1} mM^{-1})		
Wavelength	Wray et al. data	Method B	Percent difference
750 nm	3.10E-04	2.49E-04	-19.68%
800 nm	4.50E-04	5.06E-04	12.44%
830 nm	5.30E-04	4.79E-04	-9.62%

Table 1. Comparison of molar extinction coefficient for oxygenated hemoglobin measured by Wray et al. [8] with results from use of homogeneous fitting algorithm (Method B.) based on data from phantoms increasing hemoglobin concentration (Figure 9.d).

3.3 Modified imaging algorithm

The modified image reconstruction algorithm is shown schematically in Figure 10. This modified algorithm uses the three practical imaging considerations discussed in this paper. These modifications are denoted in the figure by the thicker border lines.

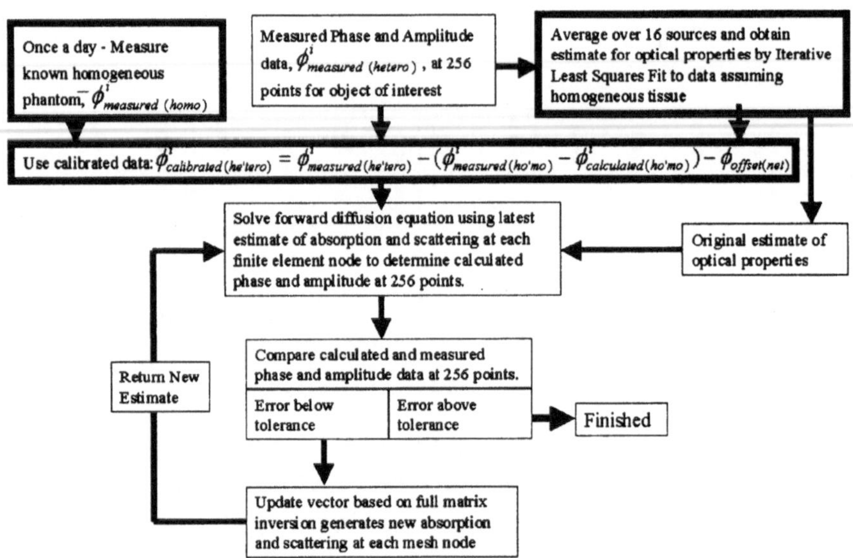

Figure 10. Schematic of revised image reconstruction algorithm (compare with Figure 3) which includes calibration and other practical imaging considerations added in **bold** boxes.

3.4 Phantom imaging results

The final calibrated system was used to measure a tissue-simulating phantom. A representative image is shown in Figure 11.

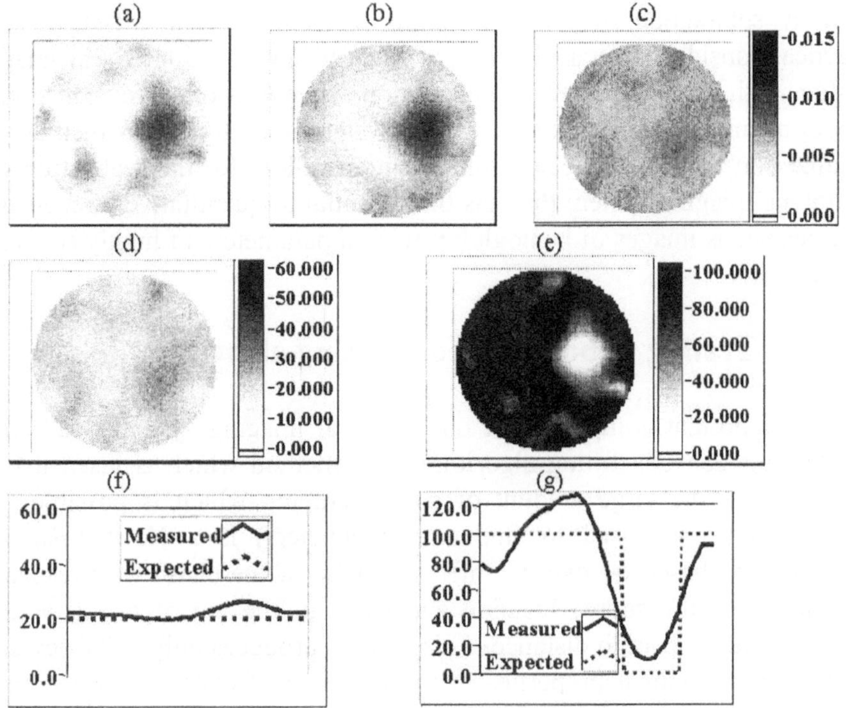

Figure 11. Representative phantom images from the calibrated imaging system. The 90 mm diameter phantom consists of a mixture of 0.5% Intralipid and 20 microMolar hemoglobin concentration in water. The hemoglobin in the phantom is fully oxygenated except for a 24 mm diameter object to the right of center which was de-oxygenated by bubbling with Nitrogen gas. Absorption coefficient images (scale is mm^{-1}) of the phantom are shown (a) at 720 nm, (b) at 750 nm, and (c) 800 nm. These three images were combined to form images of (d) hemoglobin concentration (units are microMolar) and (e) hemoglobin oxygen saturation (in percent oxygenation). (f) and (g) are horizontal profile plots through the center of (d) and (e) respectively.

4. DISCUSSION

4.1 Practical considerations

Practical considerations for a NIR diffuse imaging system include: (1) system calibration with a homogeneous phantom, (2) use of a homogeneous fitting algorithm to arrive at an initial optical property estimate for image reconstruction of a heterogeneous medium, and (3) correction for long term drift by subtraction of source strength and initial phase offsets. These practical considerations are necessary when using data from a real imaging system with a model-based reconstruction scheme in order to reduce data-model mismatch errors associated with system imperfections which are not idealized in the model. These modifications allow for the realization of a prototype imaging system that has the potential to quantitatively reconstruct heterogeneous images of hemoglobin-related parameters of highly scattering tissue *in vivo*.

4.2 Testing of homogeneous fitting algorithm

An important component of our practical imaging system is a robust, stable homogeneous fitting algorithm. One method which is insensitive to noise due to the large amount of data averaging which is possible (256 measurements are used to find two parameters) is a Newton-Raphson minimization based on two parameters: slope of the phase with respect to distance from the source location and slope of the log of intensity times distance with respect to distance. This method produces only 0.5% deviation in determined optical properties in the presence of over 5% measurement noise in repeated phantom measurements.

5. CONCLUSION

Quantitatively accurate images of hemoglobin related parameters recovered on an absolute scale have been demonstrated with no *a priori* information. Production of these quantitative images of hemoglobin concentration and oxygen saturation in a tissue-simulating phantom were critically dependent on a prototype imaging system implementation which includes the practical considerations of source/detector calibration, an accurate first estimate of the optical property distribution, and a correction for long term source/sensor drift. Once properly calibrated, this near-infrared tomographic imaging system has the potential to provide *in vivo* images of hemoglobin related parameters which may make it useful in breast cancer imaging and therapy monitoring.

ACKNOWLEDGEMENTS

This work has been supported by the NIH through grants #R01CA69544 and PO1CA80139 awarded by the National Cancer Institute. Authors gratefully acknowledge previous development work by Huabei Jiang and David Rinehart.

REFERENCES

1. Jiang H, Paulsen KD, Österberg UL, Pogue BW, Patterson MS. Optical image reconstruction using frequency-domain data: simulations and experiments. J Opt Soc Am A 1996;13:253-266.
2. Arridge SR, Schweiger M. Image reconstruction in optical tomography. Phil Trans R. Soc Lond B 1997;352:717-726.
3. Profio AF, Navarro GA. Scientific basis of breast diaphanography. Med Phys 1989; 16:60-65.
4. Vaupel P, Kallinowski F, Okunieff P. Blood flow, oxygen and nutrient supply, and metabolic microenvironment of human tumors: a review. Cancer Research 1989;49: 6449-6465.
5. McBride TO, Pogue BW, Gerety ED, Poplack SB, Österberg UL, Paulsen KD. Spectroscopic diffuse optical tomography for quantitatively assessing hemoglobin concentration and oxygenation in breast tissue. Appl Opt 1999;38:5480-5490.
6. Pogue BW, Testorf M, McBride T, Österberg U, Paulsen K. Instrumentation and design of a frequency-domain diffuse optical tomography imager for breast cancer detection. Optics Express 1997;1:391-403.
7. Pogue BW, McBride TO, Österberg UL, Paulsen KD. Spatially variant regularization improves diffuse optical tomography. Appl Opt 1999;38:2950-2961.
8. Wray S, Cope M, Delpy DT, Wyatt JS, Reynolds EOR. Characterization of the near infrared absorption spectra of cytochrome aa_3 and haemoglobin for the non-invasive monitoring of cerebral oxygenation. Biochim Biophys Acta 1988;933:184-192.
9. Quaresima V, Matcher SJ, Ferrari M. Identification and quantification of intrinsic optical contrast for near-infrared mammography. Photochem Photobiol 1998;67:4-14.
10. Hale GM, Querry MR, Optical constants of water in the 200-nm to 200- μm wavelength region. Appl Opt 1973;12:555-563.
11. Pogue BW, Patterson MS, Frequency-domain optical absorption spectroscopy of finite element tissue volumes using diffusion theory. Phys Med Bio 1994;39:1157-1180

Chapter **10**

BRAIN TISSUE AND SAGITTAL SINUS pO_2 MEASUREMENTS USING THE LIFETIMES OF OXYGEN-QUENCHED LUMINESCENCE OF A RUTHENIUM COMPOUND

Casmiar I. Nwaigwe, Marcie A. Roche, Oleg Grinberg*, Jeff F. Dunn
Biomedical NMR Laboratory, and EPR Research Center, Department of Radiology, Dartmouth Medical School, Hanover, NH 03755

Abstract: The study was done to assess the performance of a system that measures the partial pressures of oxygen (pO_2) from the lifetimes of oxygen-quenched luminescence of ruthenium compounds immobilized at the tip of fiber-optic optodes (Oxylite system). The system was used to measure the pO_2 in brain tissue (thalamus and hypothalamus) and in the sagittal sinus of isoflurane-anesthetized rats at different FiO_2's. The pO_2 recorded in the hypothalamus ($HPtO_2$) was consistently higher than the pO_2 in the thalamus ($TPtO_2$) at all FiO_2. $HPtO_2$ was closely related to PvO_2 during normoxia but not during hypoxia. The equilibrium time of Oxylite system was found to be rapid compared to *in vivo* tissue response to changes in FiO_2

Key words: brain, hypothalamus, oxygenation, ruthenium, sagittal sinus, thalamus

Abbreviations:

PtO_2: partial pressure of oxygen in brain tissue
PvO_2: partial pressure of oxygen in the sagittal sinus
$HPtO_2$: partial pressure of oxygen in the hypothalamus
$TPtO_2$: partial pressure of oxygen in the thalamus
FiO_2: fraction of oxygen in inspired gas expressed as percentage

1. INTRODUCTION

The present study was designed to measure cerebral tissue and venous partial pressure of oxygen (pO_2) using the Oxylite system. Tissue pO_2 can be measured using the luminescent lifetimes of ruthenium compounds fixed to the end of fiber-optic optodes (2). The ruthenium molecules absorb photons of incident radiation making them to move from the electronic ground state to excited singlet state. The excited singlet states lose their vibrational energy through collision with oxygen molecules, and when they return to ground state they emit a photon. Their fluorescence lifetimes decrease with increasing oxygen tensions. A modified Stern-Volmer relation governs the decrease of fluorescence lifetimes of the ruthenium compounds with increasing oxygen concentration (17). However, this relationship is not linear, and its sensitivity is highest at low pO_2 values.

The results of many studies indicate that there is a wide variability in the values of cerebral tissue pO_2 measured by different techniques (5, 6, 10, 11). The differences in the pO_2 obtained may also be as a result of differences in the region of the brain measured (5), the method of anesthesia used (7) or the ventilatory parameter (16).

Tenney (13) suggested that under normal resting conditions there is a remarkably close agreement between venous pO_2 and tissue PO_2. However, this close association between venous and tissue oxygenation has never been demonstrated in the brain. Nonetheless, many studies have tried to monitor cerebral oxygen saturation (3, 4, 14) and pO_2 (9) as an index of brain tissue oxygenation with varying results. We decided to measure brain tissue and sagittal sinus pO_2 at different percentage inspired oxygen (FiO_2) using the Oxylite method. To assess regional heterogeneity in the brain we measured pO_2 within the thalamus and the hypothalamus.

2. MATERIALS AND METHODS

2.1 Equipment

The Oxylite machine (Oxford Optronics, UK) comes with fiber-optic probes (optodes). Each optode has a terminal filament coated with a bulb of ruthenium compound (tris (4,7-dyphenyl-1, 10-phenanthroline) ruthenium (II) chloride). This end is about 0.25 mm in diameter and is the sensitive part of the optode. In addition there is a thermocouple probe which monitors the temperature and provides temperature compensation for the pO_2 measured by the optodes. The Oxylite machine then measures the pO_2 by calculating the luminescent lifetimes of the ruthenium compound following incident radiation via the fiber-optic fiber. The machine has multiple channels for simultaneous pO_2 and temperature measurements. Thus it was possible to monitor pO_2 in the thalamus ($TPtO_2$), hypothalamus ($HPtO_2$) and sagittal sinus (PvO_2) at the same time. The data is digitized to a computer with the aid of the software, PowerLab (AD Instruments Ltd., Australia).

2.2 Calibrations

Before use, each Oxylite probe (optode) was calibrated against known concentrations of oxygen in saline and gas with standardized pO_2's. Saline calibration was done by exposing the optode to saline solutions bubbled with gases of different concentrations of oxygen. The pO_2 of the solution was plotted against the pO_2 recorded by the optode. Saline pO_2 (SPO_2) was derived from the equation:

SPO_2 = $\%O_2$ in gas mixture. (barometric pressure – 47) (1)

where 47 is the saturated water vapor pressure.

Dry gas calibration was done similarly by exposing the optodes to standardized commercial gases of different O_2 concentrations flowing through a Varian temperature controller (Varian Associates, CA) and warmed to 37°C. The pO_2 measured by the optode was plotted against the pO_2 of the commercial gas calculated from:

Gas pO_2 = $\%0_2$. barometric pressure .... (2)

2.3 Equilibration time

The response time was determined by quickly transferring the optode between two solutions of different pO_2's (0.0 and 34.7 mmHg). The time taken for the optode to register the new oxygen tension from time of insertion was recorded as its response time.

2.4 Effect of temperature on recorded pO_2

With the optode in a saline solution of fixed pO_2 of 34.7 mmHg and temperature of 37°C, the thermocouple was immersed in solutions of different temperatures from 22.5 to 45°C. The pO_2 measured by the optode was measured and plotted against the temperatures recorded by the thermocouple.

2.5 *In vivo* studies

Male Sprague-Dawley rats (250 - 350g) were used for the study. They were purchased from Charles River Laboratories, Wilmington MA, were fed normal rat chow *ad libitum*, and kept on a 12hr diurnal cycle.

Anesthesia was induced with 3% isoflurane in a plastic box. After induction of anesthesia, established by the disappearance of the corneal reflex, the rats were taped to a dissecting board and the dose of isoflurane was reduced to 2% with the rats breathing through a nose cone. Tracheotomy was performed, and the femoral artery and vein were cannulated with PE50 tubing filled with heparinized saline. The level of isoflurane was then reduced to 1%. The rats were paralyzed by intravenous injection of pancuronium (0.1mg/kg/hr) and ventilated with a small animal ventilator (Model SAR-830/P, CWE Inc, Ardmore PA). The inspired and expired gases were connected to a calibrated gas analyzer (Datex Capnomac II, Helsinki, Finland). The ventilatory flow was maintained at a minute ventilation of 100 mL/100 g body weight. Blood pressure (BP) was monitored by connecting the arterial cannula through a pressure transducer to a data acquisition system (Model MP100WS, Biopac Systems, Goleta, CA).

Then with the animal lying prone on a stereotaxic board equipped with water bath, the back of the head was shaved and a midline incision made through the skin. The skull was then exposed by blunt dissection. Two burr holes were made through the skull. The landmarks were 2 mm posterior to the bregma and 3 mm lateral to the sagittal suture on either side. An optode was implanted through the left burr hole to a depth of 8 mm from the skull surface (corresponding to the hypothalamus) and in the right hole to a depth of 6mm, (corresponding to the thalamus). The tissue optodes were then pulled back about 1 mm, to reduce pressure on the bulb (1), and then held in position by clamps. Another hole was made near the caudal end of the sagittal suture and an optode used to pierce into the sagittal sinus. The optode was held in place by taping it to the back of the animal. Core temperature was maintained between 36.5 and 37.5°C by the circulating water bath. The experimental techniques and protocol were approved by the Dartmouth College Institutional Animal Care and Use Committee and conformed to the *Guide for the Care and Use of Laboratory Animals* (Institute for Laboratory Animal Resources, National Academy Press, 1996).

2.6 Protocol

After inserting all the optodes 15 – 30 min was allowed for equilibration. Animals with significant bleeding from the scalp or skull were sacrificed and excluded from the study at this time. The animals were then subjected to different levels of inspired oxygen (FiO_2): 25, 21, 13 and 10%. Each animal was maintained on the FiO_2 until a stable recording was obtained; usually 10 – 30 min. BP, HR, core temperature, brain temperature, and brain tissue pO_2 (PtO_2) were monitored. Arterial blood gas analysis was done for each FiO_2 at the end of data collection. This was done by collecting 0.1mL of blood from the femoral artery into a capillary tube immediately sealed at both ends and quickly taken to a Ciba-Corning 288 blood gas analyzer (Ciba-Corning Diagnostic Corp, MA).

The PvO_2 and PtO_2 results obtained were adjusted using the saline calibration curves for the optode used in the experiment. BP is presented as mean arterial pressure. Results are presented as mean ± standard error of mean (SE) except where otherwise stated, and $p < 0.05$ was considered significant.

3. RESULTS

All the optodes used in the experiment underestimated the pO_2 of the reference media both in dry gas and saline calibration with the slope varying from 0.6 to 0.9. Fig. 1 shows the calibration curves of an optode with the reference pO_2 terminating at 3 different points. As the reference pO_2 increased the slope diminished. At reference pO_2 of 38 mm Hg, the slope and intercept were 0.755 and –0.058 respectively; at 52 mm Hg, the slope and intercept were 0.705 and +2.054 while at 76 mm Hg they were 0.634 and 1.378 respectively. Notice that the points begin to deviate between reference pO_2 of 38 and 52 mm Hg.

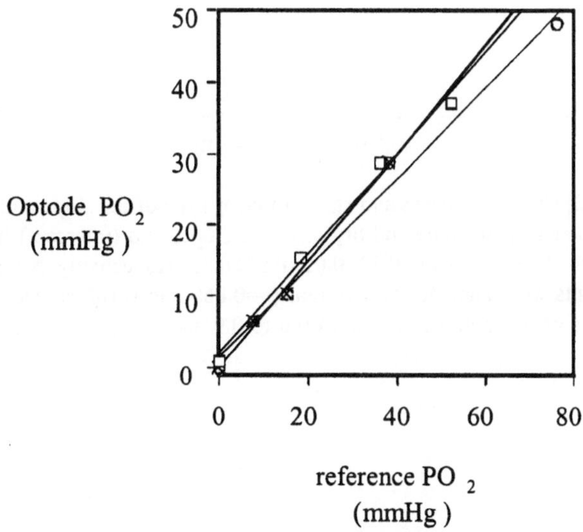

Figure 1. The calibration curves of an optode when the maximum reference pO_2 was 38 mm Hg (X), 52 mm Hg (□), and 76 mm Hg (O). Note that the 3 plots were similar below 52 mm Hg.

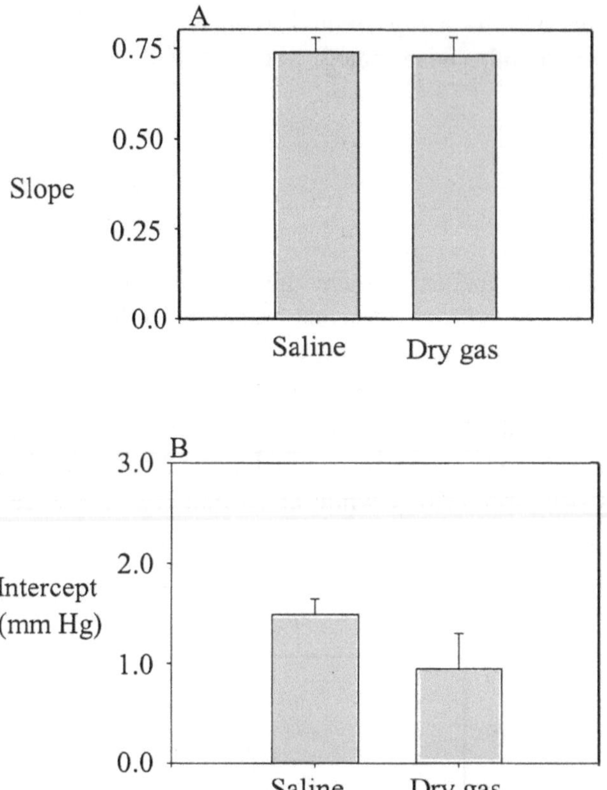

Figure 2 (A and B) shows the mean and standard error (SE) of the slope and intercept of the optodes used in the study in saline and dry gas. The slopes and (intercept) in saline and dry gas were 0.74±0.04 (1.49±0.15) and 0.73±0.05 (0.95±0.35) respectively. Note that neither the slopes nor intercepts were statistically different (p=0.869 and 0.186 respectively, n=6). The range of the slopes of our optodes was from 0.6 to 0.9 and the intercepts from –0.5 to +2.5 mm Hg.

Figure 3 shows the equilibration time for an optode when transferred from saline solutions bubbled with 0% O_2 to that with5% O_2 (pO_2 of 34.7 mm Hg) and then back. The rise was exponential lasting 8.0 sec while the decay curve appeared delayed with duration of 22 sec. When these data were plotted on a program for log transformation, the rise was exponential with a t½ of 5.24sec, while the decay was not. Both the rise and decay curves were similar for all the optodes assessed. When compared to the equilibrium time in brain tissue and sagittal sinus, the rise was about 40X faster than *in vivo* response.

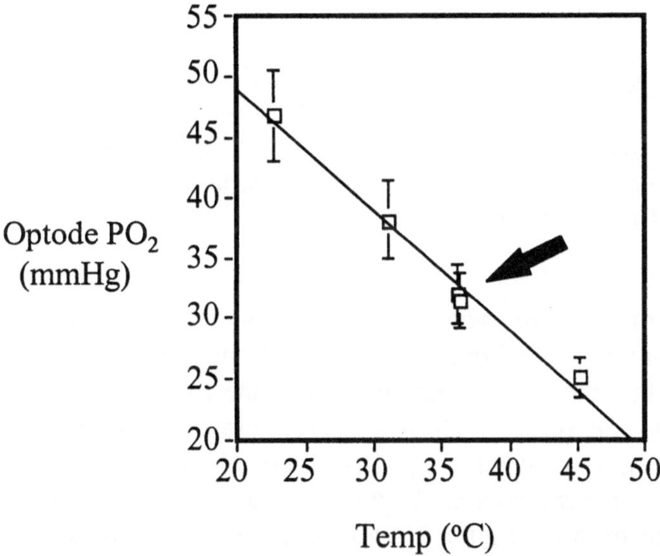

Figure 3. The temperature coefficient of the optodes. With the pO_2 and temperature of the saline solution maintained at 35.7mm Hg and 37°C, the temperature monitored by the thermocouple was varied and plotted against the PO_2 recorded by the Oxylite system (optode PO_2). The data shows the mean±S.E from 3 optodes. The arrow indicates the two data points recorded in the saline solution at the correct temperature. The pO_2's were highly temperature dependent ($r = 0.993$, $p = 0.0007$). The temperature coefficient was 0.995mm Hg/°C.

The pO_2 measured by the optodes had a strong temperature coefficient (Figure 4). When the optodes were equilibrated in saline solution at 37°C and bubbled with 5% oxygen, the pO_2 measured by the optodes was strongly correlated to the temperature sensed by the thermocouple ($r = 0.993$, $p = 0.0007$). The temperature coefficient of the relationship was 1.0 mm Hg/°C at 37°C in the 3 optodes we assessed.

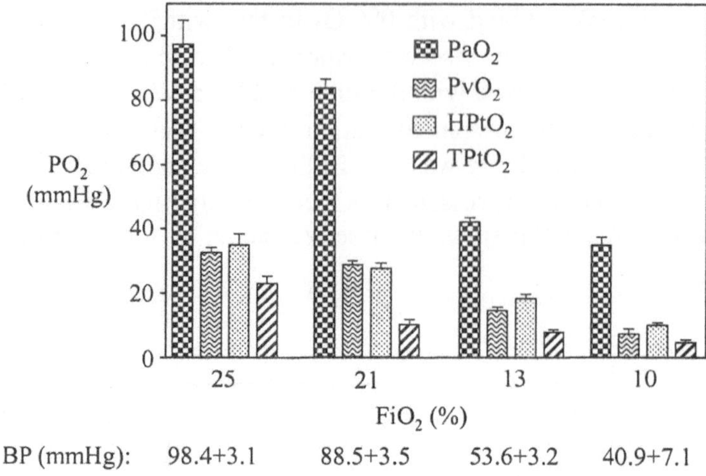

Figure 4. The physiological data from the subjects (mean $\pm$ SE,) at different FiO_2's (n = 5). The blood pressure was not controlled and the corresponding mean blood pressures (BP) are shown below each FiO_2. For details see text.

The accuracy of the *in vivo* measurements was assessed by macroscopically examining the brain of the animals post-mortem and by histology to ensure that the probes were located in the right region of the brain. Figure 5 shows the mean ($\pm$SE) of the partial pressures of oxygen in arterial blood (p_aO_2), sagittal sinus (pvO_2), hypothalamus ($HPtO_2$) and thalamus ($TPtO_2$) at different FiO_2's. $HPtO_2$ was consistently higher than $TPtO_2$ as can be seen from the figure. Even though the change in p_aO_2 and PvO_2 from 25 to 21% FiO_2 were not significant, the drop in $HPtO_2$ and $TPtO_2$ were significant ($p<0.05$) indicating that 25% FiO_2 gives better cerebral oxygenation than 21%. PvO_2 decreased significantly more than PtO_2 (hypothalamus and thalamus) when FiO_2 was changed from 21 to 13% ($p<0.05$). Our data also indicate that PvO_2 was similar to $HPtO_2$ during normoxia ($FiO_2 = 25\%$) but not during hypoxia ($FiO_2 = 13$ and 10%).

4. DISCUSSION

We measured pO_2 from brain tissue and sagittal sinus of rats using a technology that depends on the luminescent lifetimes of ruthenium. This lifetime changes with oxygen concentration according to the Stern-Volmer equation (17) with maximum sensitivity and linearity at low pO_2 values. Our calibration curves were linear in both dry gas and wet media at low pO_2's, below 52 mm Hg. With the example shown in Fig. 1, calibrating the same optode up to 76 mm Hg will result in about 20% over-estimation of its pO_2 reading at 38 mm Hg. Therefore, the technology is adequate for pre-calibrated pO_2 measurement from 0 to about 50 mm Hg.

We did not observe any difference between the saline (wet) and dry gas calibration of our optodes once the water vapor pressure was taken into account in calculating the reference pO_2 in wet calibration. This is an obvious advantage since it indicates that hydration state of the ruthenium compound does not effect its luminescent lifetimes, and the system can thus be used to assess the pO_2 in dry and aqueous media. The equilibrium time of the optodes is rapid when compared to *in vivo* response time. Therefore, the equilibration time of the optodes does not appear to affect the tissue response time obtained in the *in vivo* experiments. When compared to the equilibrium time in brain tissue and sagittal sinus, the rise was about 40X faster than *in vivo* response.

The pO_2 measured by the optodes had a strong temperature coefficient (Figure 4). The linearity of this curve makes ideal for use to manually correct for pO_2 measurement when the instrument was not set to the temperature of the environment in which the pO_2 was monitored.

Even though the change in p_aO_2 and PvO_2 from 25 to 21% FiO_2 were not significant, the drop in $HPtO_2$ and $TPtO_2$ were significant ($p<0.05$) indicating that 25% FiO_2 gives better cerebral oxygenation than 21%. PvO_2 decreased significantly more than PtO_2 (hypothalamus and thalamus) when FiO_2 was changed from 21 to 13% ($p<0.05$). This is in agreement with the reported increased oxygen extraction by brain tissue during hypoxia (12, 15). Our data also indicate that PvO_2 was similar to $HPtO_2$ during normoxia (FiO_2=25%) but not during hypoxia (13 and 10%). This supports the model that suggests that under normal resting conditions venous pO_2 may approximate tissue pO_2 (13). However, this is not applicable to the entire brain tissue as in the thalamus indicating that there is regional heterogeneity in brain tissue oxygenation, also observed in other studies (5, 8, 11).

PtO_2 values from the present study are similar to the values reported in the sub-cortical regions of the brain in rats by other methods (5, 7, 11). The tip of the Oxylite optode is large (250 μm), compared to microelectrodes (1μm); therefore, the values of pO_2 obtained may only be comparable to the mean of microelectrode or electron paramagnetic resonance (EPR) oximetry values, since the optode integrates mean pO_2 over larger areas. These differences in PtO_2 may account, to some extent, for the differences in the values of brain pO_2 reported by various other studies.

5. CONCLUSION

The present study has been able to measure cerebral venous and tissue pO_2 using the Oxylite system at various levels of inspired oxygen. The system appears stable and sensitive to rapid changes in pO_2 *in vivo*. However, to ensure the accuracy of the recordings, it was necessary to re-calibrate our optodes before use. Our data indicate that there is regional heterogeneity in brain pO_2 and the relationship between PvO_2 and PtO_2 varies in different regions of the brain.

REFERENCES

1. Baumgartl H., W Zimelka and D W Lubbers. Effects of puncturing on the measurement of local oxygen pressure using polarographic microelectrodes. Adv Exp Med Biol 1997;411:527-541.
2. Collingridge D R., W K Young, B Vojnovic, P Wardman, E M Lynch, S A Hill, and D J Chaplin. Masurement of tumor oxygenation: a comparison between polarographic needle electrodes and a time-resolved luminescence-based optical sensor. Radiat Res 1997;147:329-334.
3. Croughwell, N D, M F Newman, J A Blumenthal, W D White, J B Lewis, P E Frasco, L R Smith, E A Thyrum, B J Hurwitz, B J Leone, and et al. Jugular bulb saturation and cognitive dysfunction after cardiopulmonary bypass [see comments]. Ann Thorac Surg 1994;58:1702-1708.
4. Dexter F and B J Hindman. Theoretical analysis of cerebral venous blood hemoglobin oxygen saturation as an index of cerebral oxygenation during hypothermic cardiopulmonary bypass. A counterproposal to the "luxury perfusion" hypothesis [see comments]. Anesthesiology 1995;83:405-412.
5. Dunn, J F, E S Rhodes and T Panz. Heterogeneity of brain oxidative metabolism and hypoxia response. Mammalian systems and nature's solutions. Adv Exp Med Biol 1997; 428:425-432.
6. Gaab, M R, B Poch and V Heller. Oxygen tension, oxygen metabolism, and microcirculation in vasogenic brain edema. Adv Neurol 1990; 52:247-256.
7. Liu, K J, P J.Hoopes, E.L Rolett, BJ Beerle, A Azzawi, F Goda, J F Dunn, and H M Swartz. Effect of anesthesia on cerebral tissue oxygen and cardiopulmonary parameters in rats. Adv Exp Med Biol 1997;411:33-39.
8. Lubbers, D W and H Baumgartl. Heterogeneities and profiles of oxygen pressure in brain and kidney as examples of the pO_2 distribution in the living tissue. Kidney Int 1997;51:372-380.
9. MacMillan, V and B K Siesjo. Brain energy metabolism in hypoxemia. Scand J Clin Lab Invest 1972;30:127-136.
10. Metzger, H. S Heuber-Metzger, A Steinacker, J Struber. Staining pO_2 measurement sites in the rat brain cortex and quantitative morphometry of the surrounding capillaries. Pflugers Arch 1980;388:21-27.
11. Seyde, W C and D E Longnecker. Cerebral oxygen tension in rats during deliberate hypotension with sodium nitroprusside, 2-chloroadenosine, or deep isoflurane anesthesia. Anesthesiology 1986;64:480-485.

12. Sokolova, I B, and E P Vovenko. Oxygen tension in the brain cortex arterioles during spontaneous respiration with the hypoxic gas mixture in rats. Ross Fiziol Zh Im I M Sechenova 1998;84:527-535.
13. Tenney, S M. A theoretical analysis of the relationship between venous blood and mean tissue oxygen pressures. Respir Physiol 1974;20:283-96.
14. von Helden, A, G H Schneider, A Unterberg, W R Lanksch. Monitoring of jugular venous oxygen saturation in comatose patients with subarachnoid haemorrhage and intracerebral haematomas. Acta Neurochir Suppl (Wien) 1993;59:102-106.
15. Vovenko, E P and I B Sokolova. [Oxygen tension in the brain cortex arterioles during spontaneous respiration with the hypoxic gas mixture in rats]. Ross Fiziol Zh Im I M Sechenova 1998;84:527-535.
16. Wilson, D F, A Pastuszko, J E DiGiacomo, M Pawlowski, R Schneiderman, M Delivoria-Papadopoulos. Effect of hyperventilation on oxygenation of the brain cortex of newborn piglets. J Appl Physiol 1991;70:2691-2696.
17. Xu, M, J M Demas, B A DeGraff. Highly luminescent transition metal complexes as sensors 1994; vol. 2131. Society of optical instrumentation engineers, Los Angeles.

Chapter **11**

PRELIMINARY STUDIES ON THE PHOTON PATH IN BREAST TISSUE MODEL BY NIR-TRS

*Arshia L. Honar, #Cory Ricks, and Kyung A. Kang

*Department of Chemical Engineering, University of Louisville, Louisville, KY 40292; *Department of Chemical and Biochemical Engineering, #Department of Mathematics and Statistics, University of Maryland Baltimore County, 1000 Hilltop Circle, Baltimore, MD 21250*

Abstract: A typical tumor releases angiogenesis factor that induces the capillary growth around the tumor and, therefore, a greater amount of blood is present around the tumor. A tumor usually, therefore, becomes a local absorber due to its higher hemoglobin concentration. The ultimate goal of this project is to localize the position of a tumor in a breast tissue at an early stage using near infrared spectroscopy.

Computer simulations were performed to obtain TRS spectra at various locations in a system with optical properties of human breast. The time domain output pulses, then, were transferred to the frequency domain and the data were analyzed at various modulated frequencies. The changes in the photon path with respect to the frequency were systematically studied for absorber localization in three-dimensions.

Two different source-detector configurations, transmittance and reflectance were studied to explore the best TRS spectra acquisition procedure for tumor localization. Our previous study results have shown that in reflectance measurements the photon penetration depth is dependent on both the source-detector (S-D) separation distance and the modulation frequency. More specifically, the photon penetration depth decreases at smaller source and detector separations and higher frequencies. In this study, the effects of these two parameters on the photon path were studied in both transmittance and reflectance. This study results may be used effectively in determining the position.

Key words: Optical Mammogram, Breast cancer detection, Frequency response analysis, Near Infrared, Time resolved spectroscopy

Abbreviations: Near infrared time resolved spectroscopy (NIR-TRS), Source and detector (S-D), Magnitude ratio (MR), phase (ϕ)

1. INTRODUCTION

Mammography utilizes X-ray for the breast cancer screening. Due to its ionizing nature, the patient cannot use it frequently. Also, because of the dense nature of their breast tissue, X-ray does not detect cancer for younger women's breast well. Tumors release tumor angiogenesis factor that stimulates the proliferation of blood vessels for an adequate blood supply. As the tumor grows, the blood vessels within the tumor become compressed and necrosis occurs. Shortly after necrosis formation, the tumor tissue becomes deprived of oxygen causing an increase in deoxygenated hemoglobin concentration [1]. This difference in hemoglobin and oxygen concentration creates an optical heterogeneity in the breast tissue. Optical bio-heterogeneity screening is not an easy task since mammalian tissues are highly scattering, causing the resolution and contrast to suffer.

Near infrared time resolved spectroscopy (NIR-TRS) has been studied as a potential optical mammogram screening technique. NIR-TRS uses a sharp pulse as a source. The detector is either placed on the same plane as the source (reflectance), or across the sample (transmittance). The usual way of analyzing a TRS spectrum is by the number of photons arriving at the detector during a certain amount of time or the mean transient time of photons between the source and the detector [2]. One can also fit the TRS spectra to the theoretical solutions to construct actual images. Here, for a more efficient NIR-TRS spectra analysis, frequency response analysis that is frequently used in the engineering system identification is applied for TRS spectra analysis. This method transfers TRS spectra from the time domain to the frequency domain [3]. Light intensity (magnitude ratio; MR) and time delay (phase) of the detected photons were obtained over a wide range of modulation frequencies.

Our previous study results indicated that in the reflectance measurement, at a constant S-D separation, multi-frequency phase and magnitude ratio values may provide the absorber depth information [4]. Therefore, the objective of this study is to explore the optimum NIR-TRS source/detector configuration for breast cancer detection.

2. METHODS

A study was designed to obtain the photon paths at various modulated frequencies in a homogeneous model. Then the change in the light intensity (ΔMR) and phase (ϕ) values at multiple frequencies were correlated with the absorber position.

The mathematical model configuration was chosen as 20 x 10 x 6 cm and the effective scattering and absorption coefficients were 8.0 and 0.01 cm^{-1}, respectively. The geometry was selected as the similar shape of the breast between two mammogram plates and optical properties of the model were set to match those of the human breast. Computer simulations were performed in both reflectance and transmittance.

Figure 1 is the schematic description of the first case. A 0.5 x 0.5 x 0.5 cm cubic black absorber was moved both vertically and horizontally in the x-z plane (y = 5.0). This is to observe the effect of the absorber presence at a particular position, while the positions of the source and the detect are fixed. TRS output out spectra were computed and transferred to the frequency domain through a data reduction code developed by Bruley [3]. Magnitude ratio and phase values were obtained over a range of the modulated frequency. The accumulation of the effects registered at the detector for the various absorber locations was interpreted as the photon path in the plane that the absorber was placed, in the homogenous media: the reflected (a) and transmitted (b).

The main difference between the real photon path between a source and the detector and the photon path that we use here is that our photon path is

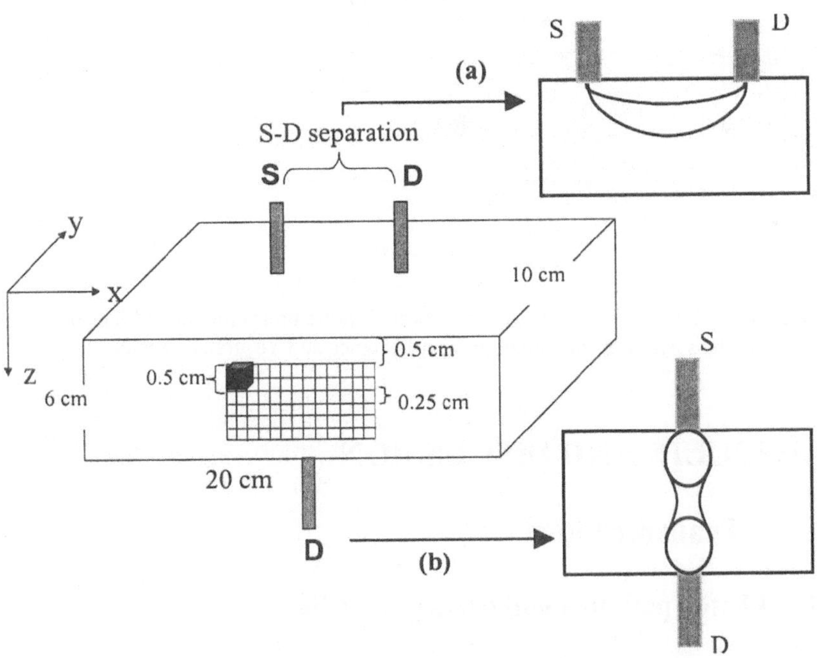

Figure 1. Breast tissue model used to obtain photon paths. A small cubic absorber was moved both vertically and horizontally at a 0.25 cm interval in x-z plane (y = 5.0) to obtain the photon path for (a) reflectance and (b) transmittance.

based on a specific nature of a bioheterogeneity and, therefore, it can change with the optical property and the size of the bio-heterogeneity. Also, since the computer simulation is very difficult to perform where the absorber is located immediately beneath the source or the detector, the path between z = 1.0 and z = 5.0 will be shown.

The second case involves moving the source and detector position in the x-y plane, (z = 0 for the reflectance and z = 6 for the transmittance) while the absorber was located at a specific location. This will be a possible case of the actual clinical measurements for tumor localization. Two different absorber depths (z) of 1.25 and 3.0 cm were studied (Figure 2). The relative changes in multi-frequency magnitude ratio and phase values were analyzed. Note that the depth of the absorber is specified as the distance between the surface of the breast model to the center of the absorber.

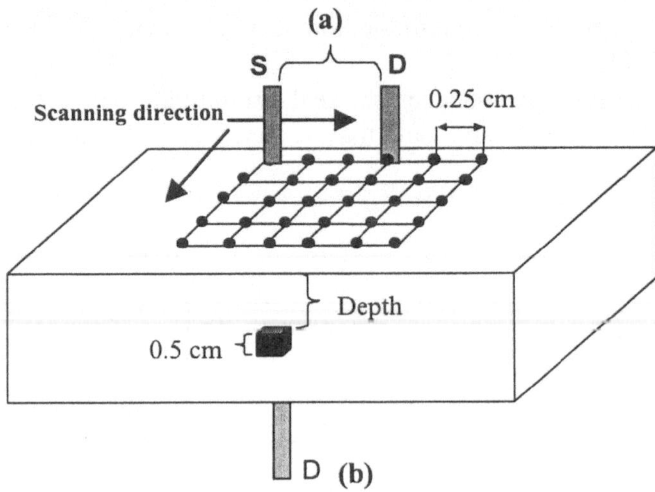

Figure 2. A 0.5 x 0.5 x 0.5 cm absorber is located in the breast model and the source and the detector scan the surface of the model in (a) reflectance and (b) transmittance.

3. RESULTS AND DISCUSSION

3.1 Transmittance

3.1.1 Photon path in a homogeneous media

The photon path (z = 1 ~ 5 cm) for transmittance was obtained at a low (0.1 GHz) and a high (1 GHz) modulation frequencies (Figure 3). The area near to the source and the detector contains a dense photon path and the area far from them (in the middle of the breast model) photons are scattered more. At the low frequency [3(a)], photons are spread more throughout the medium, and at the high frequency [3(b)] photons propagated in a more coherent path from the source to the detector.

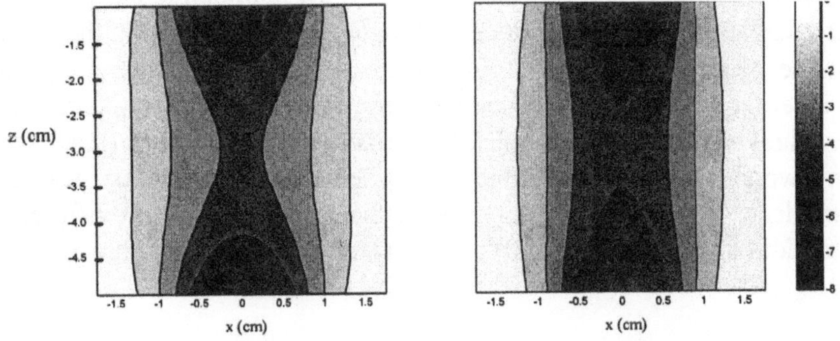

Figure 3. Photon paths in transmittance at modulation frequencies of (a) 0.1 and (b) 1 GHz.

3.1.2. Absorber localization and modulation frequency

Figure 4 shows the relative light intensity (ΔMR) values for modulation frequencies of 0.1 and 1 GHz, while the absorber is located at the depth of 1.25 cm. The absorber was successfully localized in two dimensions in both frequency. For both frequencies, the maximum decrease in MR of 8 db is observed. However, the degree of ΔMR decrease around the absorber is greater for the high frequency result, showing the more coherent nature of photon path at higher frequency.

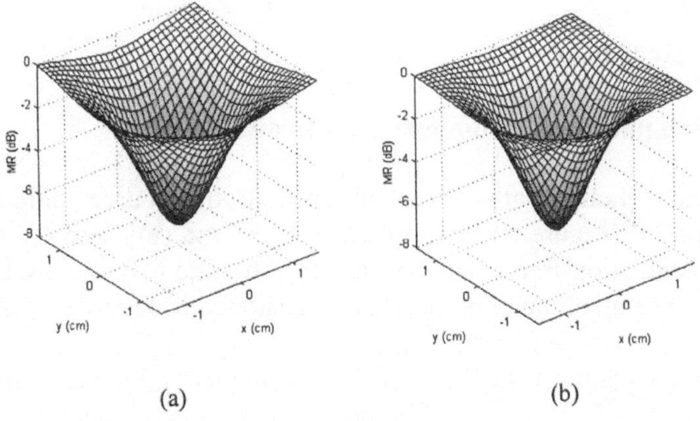

Figure 4. ΔMR values at frequencies of (a) 0.1 and (b) 1 GHz when the absorber is located at 1.25 cm deep from the surface of the breast model.

The change in the phase was also studied but, for transmittance, they showed little contrast at either frequency (results not shown).

When the absorber is located at 3 cm deep, the modulation frequency has a greater impact on the ΔMR values [Figure 5(a) and 5(b)] than in the former case. The ΔMR reduction at 1 GHz is greater since the higher frequency forms a denser photon path between the source and the detector. Normalized phase values showed an increase of approximately 4 degrees (results not shown), which is a greater change than when the absorber was located at 1.25 cm deep. It may be probably due to the greater pathlength differences in the middle of the model (z=3.0) between two frequencies (Figure 3).

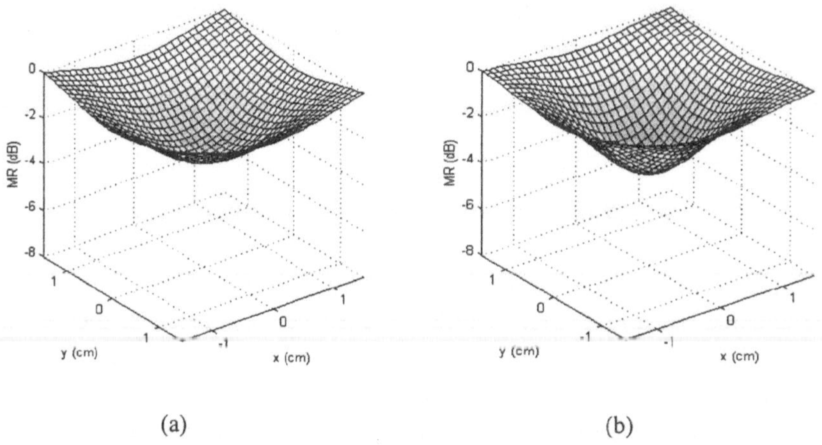

Figure 5. ΔMR when the absorber is at 3.0 cm deep (middle of the breast model) from the surface at the frequency of (a) 0.1 and (b) 1 GHz.

3.2 Reflectance

3.2.1 Photon path in the homogeneous medium

Reflectance photon paths were obtained at the source and detector separations of 2.0 [Figure 6 (a) and (d)], 3.0 [(b) and (e)], and 4.0 cm [(c) and (f)], at modulation frequencies of 0.1 [(a), (b), and (c)] and 1 GHz [(d), (e), and (f)]. As stated, the photon path demonstrated in Figure 3 is from z = 1 ~ 5.0 cm.

At 2.0 cm separation, as the modulation frequency increases from 0.1 GHz [6 (a)] to 1 GHz [6 (b)] the photon penetration depth decreases. This effect also occurs for the 3.0 cm separation [6 (b) and (e)]. At the 4.0 cm separation [6 (c) and (f)], a different trend occurred. When the S-D separation is long then photons travel a longer distance before they get to the detector, causing more photons being absorbed in the media. However, at the high frequency, photons tend to move coherently and their loss between the source and the detector is less. Therefore, they remain in the system longer than the case with low frequencies.

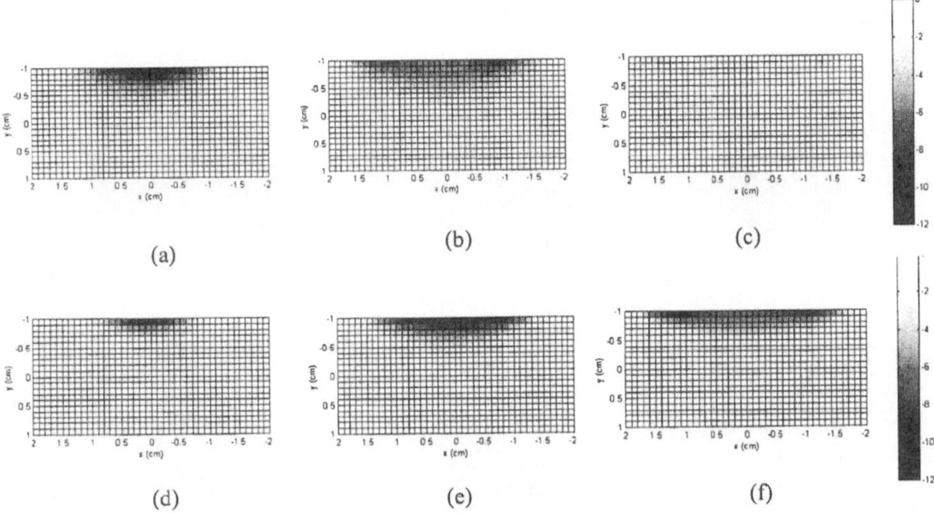

Figure 6. Photon paths for the reflectance measurements in x-z plane for 0.1 (a, b, c) and (a, d), 3 (b, e), and 4 (c, f) cm.

3.2.2. Absorber localization at various source/detector separations and modulation frequencies

As in the transmittance studies, a 0.5 x 0.5 x 0.5 cm black absorber was located at the depth of 1.25 cm in the breast model. Figure 7 illustrates ΔMR at frequencies of 0.1 and 1 GHz for S-D separations of 2.0, 3.0, and 4.0 cm. The absorber position is marked as a white dashed square in the figures. For the S-D separation of 2.0 cm, at 0.1 GHz [7(a)], a monotonic decrease of 7 db around the absorber occurred. At 1.0 GHz [7(d)] there is less ΔMR decrease due to the less photon penetration depth. At 1 GHz, for the S-D separations of 3.0 cm [7(b) and (e)] and 4.0 cm [7(c) and (f)], a different trend is obtained. Two absorbing regions are observed, while the position of the absorber is in the middle of these two regions. Also this phenomenon is more obvious at the lower modulation frequencies.

For the S-D separations of 2 and 3 cm, at a constant modulation frequency, as the S-D separation increases ΔMR reduction increases because the photon penetration depth increases, which was also shown in the shape of the photon path in Figure 6. At a constant S-D separation, a greater ΔMR reduction is obtained at lower modulation frequencies. It indicates that the photons have penetrated deeper in the media, shown by the photon paths obtained at the S-D separations of 2 and 3.0 cm S-D (Figure 6).

When the absorber was located 3.0 cm deep, no significant changes in the parameter were obtained (results not shown). This shows that, in reflectance measurements, the photons do not penetrate as deep as in the transmission.

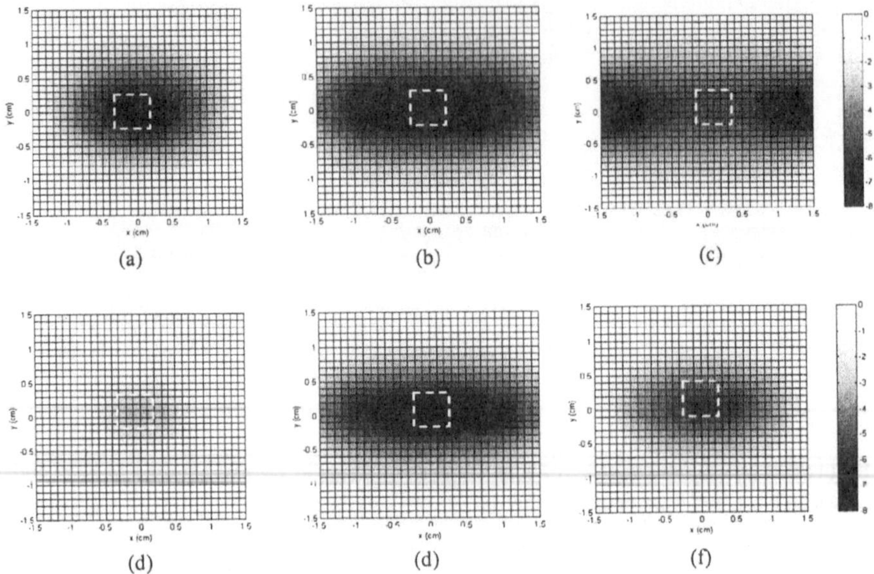

Figure 7. ΔMR values in the breast model with an absorber located at the center of x-y plane 1.25 cm deep, at the modulation frequency of 0.1 GHz for the S-D separations of (a) 2.0, (b) 3.0, and (c) 4.0 cm, and at 1 GHz for S-D separations of (d) 2.0, (e) 3.0, and (f) 4.0 cm.

4. CONCLUSIONS

Absorbers can be better localized in two-dimensions by transmittance at high modulation frequencies because the photon path in high frequency transmittance measurements is more coherent between source and the detector. The photon path in reflectance measurements changes more with the change in modulation frequency than that in transmittance. It has been generally accepted that at the greater S-D separation photons penetrate deeper. However, in a system with optical properties of human breast, the results indicate that the increase in the photon penetration depth by increasing the S-D separation has a limitation due to the photon loss in the media. Therefore, it appears that 4.0 cm of the S-D separation is not an ideal separation for breast cancer localization.

Our preliminary results indicate that the tumors that are positioned at a shallow depth (e.g., 1.25 cm) can be successfully localized in three-dimensions by conducting transmittance measurements followed by multi-frequency, multi-S-D separation reflectance measurements. High modulation frequency transmittance measurements showed better sensitivity for the deep (3.0 cm) tumor. However, reflectance measurements were not able to detect the small tumors located deep.

ACKNOWLEDGEMENT

The authors would like to thank the National Science Foundation for their funding for the NIR-TRS machine, also the authors would like to acknowledge Mark Hemelt for his technical assistance.

REFERENCES

1. Creasey WA. Tumor angiogenesis. The growth of tumors:Growth of tumor masses. New York: Oxford, 1981; 60-62.
2. Hemelt MW, Barnett JT, Bruley DF, Kang KA. Two and three dimensional localization of deep vein thrombosis using near infrared time resolved spectroscopy. Proc 1997 south biomed eng conf. 1997;449-454.
3. Kang KA, Bruley DF, Londono JM, Chance B. Highly scattering optical system identification via frequency response analysis of optical NIR-TRS spectra. Ann Biomed eng 1994;22:240-252.
4. Hemelt MW, Barnett JT, Bruley DF, Kang KA. The application of NIR-TRS and frequency response analysis for deep vein thrombosis detection. Biotechnol Prog 1997; 13: 640-648.

Chapter 12

EPR SPECTROSCOPY AND IMAGING OF OXYGEN: APPLICATIONS TO THE GASTROINTESTINAL TRACT

Jay L. Zweier, Guanglong He, Alexandre Samouilov, and Periannan Kuppusamy
Center For Biomedical EPR Spectroscopy and Imaging, Davis Heart and Lung Research Institute, The Ohio State University College of Medicine, Columbus, Ohio

Abstract: EPR imaging has emerged as an important tool for noninvasive three-dimensional (3D) spatial mapping of free radicals in biological tissues. Spectral-spatial EPR imaging enables mapping of the spectral information at each spatial position, and, from the observed linewidth, the localized tissue oxygenation can be mapped. We report the application of EPR imaging techniques enabling 3D spatial and spectral-spatial EPR imaging of small animals. This instrumentation, along with the use of a biocompatible charcoal oximetry-probe suspension, enabled 3D spatial imaging of the gastrointestinal (GI) tract, along with mapping of oxygenation in living mice. By using this technique, the oxygen tension was mapped at different levels of the GI tract from the stomach to the rectum. The results clearly show the presence of a marked oxygen gradient from the proximal to the distal GI tract, which decreases after respiratory arrest. This technique for *in vivo* mapping of oxygenation is a promising method, enabling the noninvasive imaging of oxygen within the normal GI tract. This method should be useful in determining the alterations in oxygenation associated with disease.

Key Words: activated charcoal, free radical, oximetry, EPRI

1. INTRODUCTION

Tissue oxygenation is an important regulator of physiological function, and alterations in oxygenation are of critical importance in the pathogenesis of disease. Oxygen concentrations vary greatly in different tissues and are modulated by alterations in vascular flow and metabolic work. Therefore there has been a great need for noninvasive techniques suitable for measurement and spatial mapping of oxygen in models of disease and in humans.

In the gastrointestinal (GI) tract, tissue oxygen tension has been used to assess intestinal function and viability [1, 2]. Oxygen tension varies in bowel ischemia where there is limited perfusion and in inflammatory bowel disease [3]. Past techniques for measurement of oxygen tension in the GI tract have been based primarily on modified Clark electrodes with surgical laparotomy or endoscopy [1, 4, 5]. The invasive nature of these methods has limited greatly the ability to perform measurements of oxygen in the clinical diagnosis and treatment of disease, as well as in animal models of the pathogenesis of disease [6-9]. In addition, there are concerns that these invasive methods with external cannulation could perturb the values of oxygen that are measured. Thus, there is a great need for a noninvasive technique suitable for *in vivo* measurement and spatial mapping of oxygen within the GI tract.

EPR imaging techniques have enabled the spatial mapping of paramagnetic species in isolated organs and other *ex vivo* or *in vitro* biological systems [10, 11, 12, 13, 14]. With the use of suitable oximetry probes this technique can measure tissue oxygenation or oxygen tension. Molecular oxygen is paramagnetic; it produces linewidth broadening caused by spin-spin interaction of oxygen with the probe, and the magnitude of this broadening is a function of the concentration of oxygen. Although spatial EPR imaging provides a map of spin distribution in the sample, it does not provide information related to spectral line shape [14-16]. However, spectral-spatial imaging techniques have been developed to enable spatial imaging with determination of the EPR line shape at each spatial point [12, 17-19]. Thus, by using these spectral-spatial imaging techniques, oxygen concentration or tension can be imaged.

We report the mapping of oxygen tension within the lumen of the GI tract by using a charcoal oximetry probe that is suitable for clinical use. Spatial EPR 3D imaging enabled visualization of the spatial structure and anatomy of the GI tract, whereas spectral-spatial EPR imaging enabled measurement of the oxygen tension at different locations. This methodology enables noninvasive mapping of oxygenation in living animals and has potential for clinical applications in humans.

2. MATERIALS AND METHODS

2.1 Spin probe

Activated charcoal (American Norit, Milwaukee, WI) was used as the EPR oximetry probe. The charcoal was mixed with glucose in a 1:1 (wt/wt) ratio and stirred thoroughly with water to a semisolid form. This mixture was used to feed the mice before the EPR imaging experiments.

2.2 Mouse preparation

C3H mice weighing ≈30 g were used [20]. Before feeding the mice with charcoal, solid-food intake was stopped for 12 h, after which the mice were fed a mixture of charcoal and glucose for 24 h. The mice then were anesthetized with 20 mg/kg pentobarbital by i.p. injection. The tail vein was cannulated to allow administration of additional anesthesia as required. Animals survived the feeding and imaging, unless purposely killed in the experimental protocol. For experiments in which respiratory arrest was induced, the mice were given an overdose of pentobarbital by rapid bolus intravenous infusion of 20 mg of the drug. A total of 12 mice were studied for the spatial and spectral-spatial imaging protocols described.

2.3 EPR imaging instrumentation

EPR spectroscopy and EPR imaging experiments were performed using instrumentation described in [20].

2.4 Oximetry calibration and measurement

Measurements of the line width of activated charcoal suspension after equilibration with a series of oxygen and nitrogen gas mixtures were performed with the 750 MHz spectrometer. Calibration was performed over the oxygen-concentration range (0–21%) with oxygen/nitrogen gas mixtures. In these experiments, the time response of the observed linewidth changes was measured, and equilibration for a given oxygen tension occurred in less than 5 min. It was observed that the oxygen dependent linewidth of these charcoal suspensions was stable and unchanged even 1 week after preparation. No significant alterations in oxygen dependent broadening were seen over the temperature range from 24°C to 37°C. The oxygen sensitivity also was unaffected after oral ingestion in mice after periods of 3 days or more. The O_2 tension values from the spectral-spatial image data were measured from the linewidth observed in five pixels at the specific region of GI tract, and values are represented as mean (in torr) ± SEM. (Note that 1 torr = 133 Pa.).

3. RESULTS AND DISCUSSION

The EPR linewidth of activated charcoal is sensitive to oxygen because of spin-spin coupling of paramagnetic oxygen molecules with the unpaired electron of activated charcoal. Activated charcoal has been used previously to measure oxygen tension in mammalian cells and tissues [13, 21]. In the present study, we used activated charcoal as spin label to measure oxygen concentration inside the GI tract of mice. The activated charcoal was mixed with glucose to make it nutritious and to enhance its spontaneous oral intake by the mice. The mice were allowed to feed on the charcoal-sugar diet for periods of 24-72 h before measurements were performed.

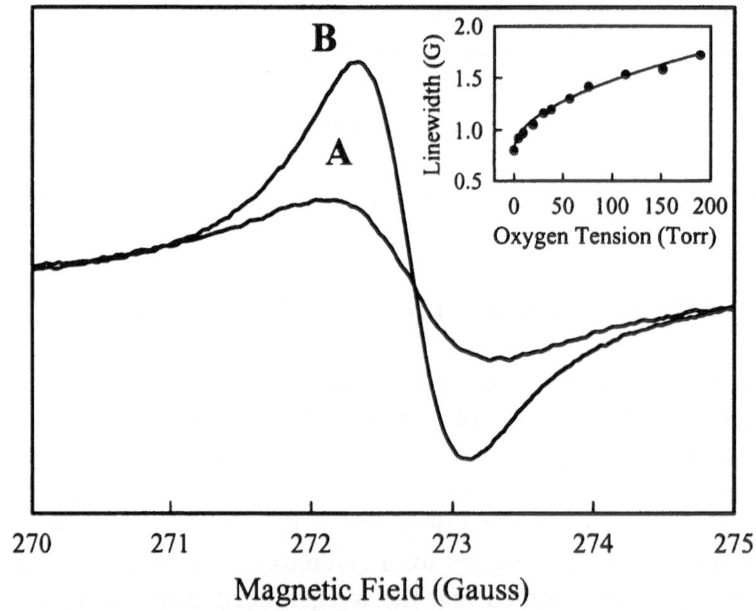

Figure 1. EPR spectra of charcoal suspension equilibrated with room air (21% oxygen, line A) and 0% oxygen (line B). The *inset* shows the variation of line width that occurs as a function of oxygen tension. Measurements were performed with a microwave frequency of 746 MHz, a modulation amplitude of 0.4 G, a field modulation of 100 kHz, and a microwave power of 60 mW.

The EPR linewidth of a charcoal and water suspension was calibrated as a function of oxygen tension. The linewidth was 1.50 G in room air, and under anaerobic conditions, the linewidth was 0.80 G (Figure 1). Over the physiological oxygen concentration range (0 - 21%, 0-160 torr), the linewidth varied in a nonlinear manner as shown in the inset of Figure 1. The experimental calibration curve was fitted with a power function $L = 0.80 + 0.06 \times [O_2]^{0.51}$, where L is the peak-to-peak linewidth of the activated charcoal and $[O_2]$ is the oxygen concentration.

In preliminary experiments with mice fed charcoal for periods of 24-72 h, it was observed on laparotomy that 24 h of feeding was sufficient to fill the GI tract from the stomach to the distal colon and rectum. *In vivo* EPR measurements performed on these mice also showed that a maximum EPR signal was observed with 24 h of charcoal feeding, and no further increase occurred with longer periods of feeding. Therefore, EPR imaging experiments were performed on mice after 24 h charcoal feeding. The mice were anesthetized and placed inside the resonator facing down, and then 2D spectral-spatial imaging and 3D spatial imaging measurements were performed. Fig. 2A shows a photograph of the mouse with demarcation of the active area imaged, whereas Fig. 2B shows the corresponding surface rendered 3D spatial EPR image. The 3D spatial EPR image data enable visualization of the location of the charcoal probe in the GI tract from the stomach, to the duodenum, mid and distal small intestine, as well as the ascending, transverse, and descending sigmoid colon and rectum. The spatial image data seen in Fig. 2 corresponded to the anatomic structure and charcoal distribution found on postmortem laparotomy. In Fig. 2C, the corresponding 2D spectral-spatial image along the longitudinal axis of the mouse from the proximal to the distal GI tract is shown. The linewidth was measured along the longitudinal axis and was found to decrease from the proximal to the distal GI tract.

To confirm the decrease in linewidth and oxygenation from the proximal to the distal GI tract, experiments were performed with administration or feeding of the animals for discrete times, after which EPR measurements of the linewidth were taken and the animals were sacrificed to verify anatomic location. With acute infusion, after 15-30 min, the linewidth values and calculated oxygen tensions were in the range of 50-60 torr, in agreement with the values obtained from the imaging at the level of the stomach (58 ± 15 torr). Both 3D spatial EPR imaging and laparotomy confirmed that the charcoal was present primarily in the stomach. After acute feeding followed by 2 days, the charcoal oximetry probe was found on autopsy in the distal colon and rectum; the oximetry values obtained before the mice were sacrificed were similar to those from the imaging experiments (3 ± 1 torr).

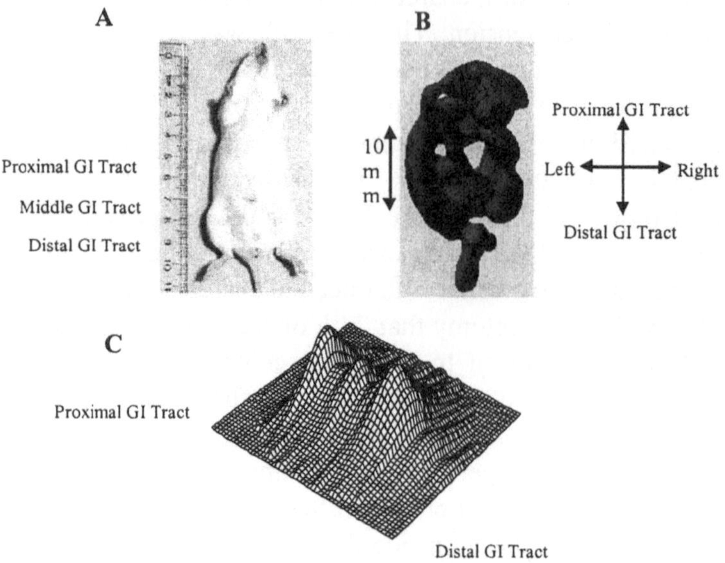

Figure 2. Photograph of the mouse studied with demarcation of the region imaged and corresponding 3D spatial and 2D spectral-spatial image data. **A** photograph of the mouse with a ruler for scale. The area imaged is shown between black lines. The mouse was fed with the charcoal probe for 1 day. **B** Spatial EPR 3D image visualizing the location of the charcoal probe in the GI tract. **C** Spectral-spatial 2D image data along the longitudinal axis from the proximal to the distal GI tract. The parameters for 3D spatial imaging were: microwave power 60 mW, microwave frequency 746 MHz, projections 1024, field gradient 15 G/cm, projection acquisition time 5 s, modulation amplitude 0.4 G, spatial window 40 mm. The parameters for 2D spectral-spatial image were: microwave power 60 mW, microwave frequency 746 MHz, spectral window 5 G, spatial window 40 mm, projection acquisition time 5 s.

After respiratory arrest, systemic oxygen concentrations fall, and it would be expected that oxygenation within the gut also would decrease; however, the time course of this process at different levels of the GI tract is not known. To determine the effect and time course of alterations in oxygenation within the GI tract after respiratory arrest, mice were administered an overdose of pentobarbital by i.v. infusion. The variation of oxygen concentration was obtained as a function of time with a series of 2D spectral-spatial imaging measurements. At the end of the spectral-spatial measurements, 3D spatial imaging was performed to obtain 3D mapping of the distribution of the charcoal spin label. This 3D mapping enabled coregistration of the spectral-spatial image with the 3D spatial image, which in turn facilitated assignment of the spectral data and calculated oxygen values within the GI tract.

The time course of the change in oxygen within the GI tract after respiratory arrest was followed by spectral-spatial imaging for 2 h. The oxygen tension was calculated from the observed linewidth data. Although, before arrest, the oxygen tension at the level of mid stomach (Fig. 3, line **a**) was 58 ± 15 torr, it decreased rapidly within the first 5 minutes after arrest to a value of 22 ± 4 torr. At the level of mid duodenum (Fig. 3, line **b**), the oxygen tension was 32 ± 8 torr before ischemia, and it decreased to 14 ± 3 torr only 5 min after arrest. At the level of mid small intestine and mid colon (Fig. 3, line **c**), the oxygen tension was 11 ± 3 torr before arrest and decreased gradually over 30 min to 5 ± 1 torr. At the level of distal sigmoid colon-rectal junction (Fig. 3, line **d**), the basal oxygen tension was 3 ± 1 torr, and it gradually decreased to 0.5 ± 0.3 torr over 30 min.

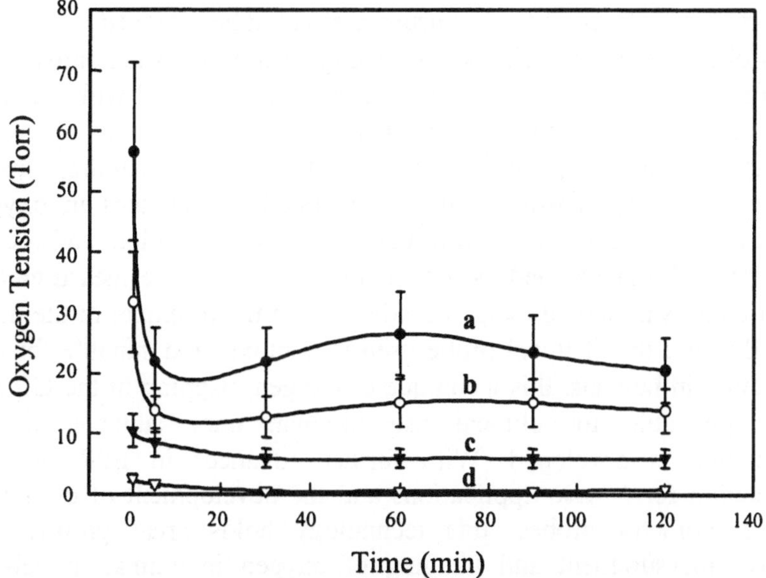

Figure 3. Time course of the change in oxygen tension at different levels of the GI tract before and after respiratory arrest. Values were calculated from the oxygen dependent linewidth broadening by using the calibration data shown in Fig. 1. The levels are as defined by the planar cross sections. Line **a**, the level of mid stomach; line **b**, the level of mid duodenum; line **c**, the level of mid colon and mid small intestine; and line **d**, level of sigmoid colon-rectal junction.

The observed oxygen gradient seen along the GI tract can be explained by a combination of processes. When food is swallowed, it would be initially equilibrated with the oxygen tension of room air. On passage to the stomach and later the small intestine, the oxygen levels would fall as oxygen diffused across the mucosal membrane. A gradual process of equilibration with the capillary levels of oxygen (i.e., 5-10 torr; ref. [9]) would occur. On passage to the colon, with its heavy bacterial colonization, further decreases in oxygenation would be expected. The presence of marked hypoxia within the lumen of the distal colon is consistent with the known abundance of anaerobic bacteria at this site.

Activated charcoal is commonly used in humans for the clinical treatment of a variety of types of oral poisoning or drug overdose. Activated charcoal also is used in oral medications for the treatment of indigestion. As such, it is known to be well tolerated in humans. Although the material used in the present study was a commonly available charcoal, a number of laboratories have shown that with special controlled formulations or types of chars, both the paramagnetic content and the oxygen dependent linewidth broadening can be enhanced [22]. With future development and application of optimized formulations, it would be expected that further enhancement of the image quality and sensitivity of oxygen measurement could be achieved.

It was observed that the charcoal oximetry spin probe was nontoxic and well tolerated with feeding for up to 3 days. In animals fed with this spin probe, spatial EPRI enabled clear 3D spatial visualization of the entire GI tract distal to the esophagus. With spectral-spatial imaging, spatial differences in oxygen tension could be mapped. Differences in oxygen tension at levels from the stomach to the small intestine, colon, and rectum were determined and mapped as a function of time. These measurements show that there is a marked oxygen gradient from the proximal to the distal GI tract. Because the charcoal probe used is nontoxic and suitable for oral administration in humans, this technique of oxygen mapping in the GI tract may be applicable for clinical use in man once suitable clinical instrumentation is developed. With further advances in EPR imaging instrumentation for *in vivo* applications and the development of optimized solid state oximetry probes, this technology holds great promise for noninvasive measurement and imaging of oxygen in animal models of disease and as well as eventual clinical use in man.

REFERENCE

1. Sheridan WG, Lowndes RH, Young HL. Intraoperative tissue oximetry in the human gastrointestinal tract. Am J Surg 1990;159:314-319.
2. Knudson MM, Bermudez KM, Doyle CA, Mackersie RC, Hopf HP, Morabito D. Use of tissue oxygen tension measurements during resuscitation from hemorrhagic shock. J Trauma 1997;42:608-616.
3. Hauser CJ, Locke RR, Kao HW, Patterson J, Zipser RD. Visceral surface oxygen-tension in experimental colitis in the rabbit. J Lab Clin Med 1988;112:68-71.
4. Cooper GJ, Sherry KM, Thorpe JA. Changes in gastric tissue oxygenation during mobilization for esophageal replacement. Eur J Cardiothorac Surg 1995;9:158-160.
5. Larsen PN, Moesgaard F, Naver L, Rosenberg J, Gottrup F, Kirkegaard P, Helledie N. Gastric and colonic oxygen-tension measured with a vacuum-fixed oxygen-electrode. Scand J Gastroenterol 1991;26:409-418.

6. Landow L, Phillips DA, Heard SO, Prevost D, Vandersalm TJ, Fink MP. Gastric tonometry and venous oximetry in cardiac-surgery patients. Crit Care Med 1991;19:1226-1233.
7. Kram HB, Appel PL, Fleming AW, Shoemaker WC. Assessment of intestinal and renal perfusion using surface oximetry. Crit Care Med 1986;14:707-713.
8. Zabel DD, Hopf HW, Hunt TK. The role of nitric oxide in subcutaneous and transmural gut tissue oxygenation. Shock 1996;5:341-343.
9. Uribe N, Garcia-Granero E, Belda J, Calvete J, Alos R, Marti F, Gallen T, Lledo S. Evaluation of residual vascularization in esophageal substitution gastroplasty by surface oximetry-capnography and photoplethysmography. Eur J Surg 1995;161:569-573.
10. Berliner LJ, Fujii H. Magnetic-resonance imaging of biological specimens by electron-paramagnetic resonance of nitroxide spin labels. Science 1985;227:517-519.
11. Halpern HJ, Bowman MK, Spencer DP, Polen JV, Dowey EM, Massoth RJ, Nelson AC, Teicher BA. Imaging radio-frequency electron-spin-resonance spectrometer with high-resolution and sensitivity for invivo measurements. Rev Sci Instrum 1989;60:1040-1050.
12. Maltempo MM. Differentiation of spectral and spatial components in Electron-Paramagnetic-Res imaging using 2-D image-reconstruction algorithms. J Magn Reson 1986;69:156-161.
13. Goda F, Liu KJ, Walczak T, O'Hara JA, Jiang J, Swartz HM. In-vivo oximetry using EPR and India ink. Magn Reson Med 1995;33:237-245.
14. Woods RK, Bacic GC, Lauterbur PC, Swartz HM. 3-dimensional electron-spin resonance imaging. J Magn Reson 1989;84:247-254.
15. Alecci M, Colacicchi S, Indovina PL, Momo F, Pavone P, Sotigiu A. 3-dimensional invivo ESR imaging in rats. Magn Reson Imaging 1990;8:59-63.
16. Kuppusumy P, Chzhan M, Zweier JL. Development and optimization of 3-dimensional spatial EPR imaging for biological organs and tissues. J Magn Reson Ser B 1995;106:122-130.
17. Ewert U, Herrling T. Spectrally resolved Electron-Paramagnetic-Res tomography with stationary gradient. Chem Phys Lett 1986;129:516-520.
18. Woods RK, Dobrucki JW, Glockner JF, Morse PD, Swartz HM. Spectral spatial ESR imaging as a method of noninvasive biological oximetry. J Magn Reson 1989;85:50-59.
19. Kuppusumy P, Chzhan M, Vij K, Shteynbuk M, Gianella E, Lefer DJ, Zweier JL. 3-dimensional spectral spatial EPR imaging of free-radicals in the heart - a technique for imaging tissue metabolism and oxygenation. Proc Natl Acad Sci USA 1994;91:3388-3392.
20. He G, Shankar RA, Chzhan M, Samouilov A, Kuppusamy P, Zweier J L. Noninvasive measurement of anatomic structure and intraluminal oxygenation in the gastrointestinal tract of living mice with spatial and spectral EPR imaging. Proc Natl Acad Sci USA 1999;96:4586-4591.
21. Zweier JL, Chzhan M, Ewert U, Schneider G, Kuppusamy P. Development of a highly sensitive probe for measuring oxygen in biological tissues. J Magn Reson Ser B 1994;105:52-57.
22. Swartz HM, Clarkson RB. The measurement of oxygen in vivo using EPR techniques. Phys Med Biol 1998;43:1957-1975.

Chapter 13

EFFECTS OF BLOCKING BUFFERS AND PLASMA PROTEINS ON THE PROTEIN C BIOSENSOR PERFORMANCE

*Heath I. Balcer, *James O. Spiker, Kyung A. Kang

*Department of Chemical Engineering, University of Louisville, Louisville, KY 40292; *Department of Biological Sciences, University of Maryland Baltimore County, Baltimore, Maryland 21250, USA*

Abstract: Protein C (PC) deficiency can lead to abnormal thrombus formation in blood vessels, obstructing oxygen and nutrient transport to various organs or tissues. Quantifying PC amount in blood plasma usually takes long time, is difficult and expensive due to its low concentration and other homologous proteins in it. A fiber-optic immunosensor has been under development for several years for the PC real-time assay. In this study, blocking buffers have been examined to reduce the noise caused by non-specific adsorption. Ethanolamine was found to reduce the noise significantly. The effect of various PC homologues (factors II, VII, IX, and X) on the biosensor performance was investigated. The sensor showed little cross reactivity with these homologues. Human serum albumin (HSA) in the sample decreased the signal intensity. However, PC could be still quantified in samples with HSA at the physiological level.

Key words: Protein C, Immunosensor, Protein C deficiency, real-time detection

Abbreviations: American Red Cross (ARC), Bovine Serum Albumin (BSA), dimethyl sulfoxide (DMSO), enzyme-linked immunosorbent assay (ELISA), Human Serum Albumin (HSA), monoclonal antibody (mAb), phosphate buffered solution (PBS), phosphate buffered solution with 0.1% Tween (PBST), picoamperes (pA), Protein C (PC)

1. INTRODUCTION

1.1 Protein C

PC, produced in the liver, circulates in blood as a zymogen and is activated upon contact with thrombin-thrombomodulin complex. Activated PC serves as an anticoagulant by digesting activated clotting factors V and VIII. PC exhibits antithrombotic activity by inhibiting tissue plasminogen activation inhibitor [1].

In normal individual's plasma, PC concentration is approximately 4 μg/ml. Heterozygous deficiency is defined to be 40-60% of the normal level. Those with heterozygous deficiencies do not show thrombo-embolic symptoms until the event causing high consumption of PC in the body occurs. These events include surgery, pregnancy, and severe physical activities. A failure to receive immediate treatment for the events can lead to major thrombo-embolic complications, such as, lung embolism, strokes, heart attack, etc. [2].

1.2 Protein C immuno-optical biosensor

Conventional assay for PC is ELISA and it has a number of limitations, such as a long assay time (6 hours), a need for a skilled technician, and high assay cost. The PC biosensor consists of a 12.5 cm long optical fiber with a diameter of 600 μm, whose tip is tapered down to 100 μm for efficient photon transfer. A type of anti-PC monoclonal antibodies (1° mAb) is immobilized on the surface of the fiber by means of an avidin-biotin reaction [3]. The fiber sensor is then inserted in a sample chamber and is connected to the fluorometer. Samples are applied to the sample chamber, where PC binds the 1° mAb on the fiber. The fiber is washed with washing buffer to remove all non-reacting molecules in the sample. Then, another type of monoclonal antibody (2° mAb) tagged with a fluorophore (Cy5) probes the PC/1° mAb complex. The sensor is washed again and the excitation laser beam passes through the fiber. The light emitted by the fluorophores is relayed back through the fiber to the detector, allowing the signal to be correlated with the PC concentration in the sample [4].

1.3 Homologous proteins

Five coagulant proteins (factors II, VII, VIII, IX, and X) have a high degree of structural homology with PC. Their structural similarities raised concerns over cross reactivity with the antibodies against PC and the sensor performance with their presence was studied.

1.4 Blocking buffers

The purpose of a blocking buffer is to prevent antigens or 2° mAb from nonspecifically binding to the fiber surface. Factors affecting the performance of blocking reagents include their size, charge, and polarity. A study has been conducted to compare the performance of the PC biosensor using BSA and ethanolamine. BSA was initially chosen because it is a commonly used blocking reagent for ELISA. Ethanolamine was also tested because it is a commonly used blocking molecule in affinity chromatography [5]. BSA is a large (68 kDa), polar protein, while ethanolamine is a small (62 Da) molecule with only a slight charge.

1.5 Human serum albumin

Human serum albumin (HSA) is a key plasma protein, responsible for the primary transport of fatty acids, along with ions, proteins, and enzymes. HSA is also the most abundant protein in plasma (50 mg/ml), constituting approximately 55% of the entire protein content. Our previous study results showed that increasing concentrations of HSA in the sample reduced the signal intensity [6]. It is to understand the characteristics of HSA on the sensor performance so that PC can be quantified in plasma.

2. MATERIALS AND METHODS

2.1 Antibody preparation

The 1° mAb (MAB7D7) and the 2° mAb (MAB8861) are murine originated, specific to two different epitopes of PC. They are donated by the American Red Cross (Rockville, MD). The 1° mAb is linked to biotin (MW=575.7; Sigma; St. Louis, MO) by mixing antibody in sodium bicarbonate buffer with biotin (dissolved in DMSO) at a 1:5 molar ratio for thirty minutes. The biotinylated antibody is then separated from free biotin molecules using gel chromatography. P-10 gel (Bio-Rad Laboratories; Hercules, CA) is used for the purification, with PBS (Sigma) as the equilibrium and elution buffer. 2° mAb labeling with Cy5 (Amersham-Pharmacia; Sweden) and its purification are performed according to the manufacturer's instructions.

2.2 Chemical treatment of fibers

Quartz optical fibers (Research International; Woodinville, WA) are tapered and constructed according to the methods outlined by the manufacturer [4]. They are chemically treated according to the method developed by Bhatia, *et al.* [7] to couple avidin to the fiber. Avidin-biotin binding is used to couple the antibody to the fiber as shown by Spiker et al. [3]. 1° mAb immobilization using avidin-biotin binding minimizes the distortion of the three-dimensional structure of the antibody. Consequently, the antibody does not lose affinity to PC as much as when the antibody is directly coupled to the fiber [8]. After twenty-four hours, the antibody solution is washed out, and the blocking buffer is allowed to incubate for thirty minutes. The blocking buffer is then washed out with PBST.

2.3 Protein C assay protocol

After the background fluorescence had been established, a sample was applied to the chamber and incubated for 10 minutes. The sensor was then washed with 1 ml of PBST, and the signal intensity was measured (intensity A). 2° mAb (5 μg/ml) solution was then applied to the chamber and incubated for five minutes. 2° mAb was then removed, and the flow cell was washed out with 1 ml of PBST and the signal intensity was measured (intensity B). The difference between intensity B and A represents the signal intensity by the PC amount in the sample.

3. RESULTS AND DISCUSSION

3.1 Homologous proteins

The effect of four most homologous proteins (factors II, VII, IX, and X) on the biosensor is studied to determine if they would react with the PC antibodies. Fibers were blocked with 2 mg/ml BSA prior to use. Physiological concentrations (100, 0.7, 4.0, and 7.5 μg/ml, respectively) of the homologous proteins were added individually and also collectively to samples with PC (0, 1.25, and 2.5 μg/ml). The signals from the samples with PC and the homologous proteins show very little difference (Figure 1). This result confirms that these proteins are not cross-reactive to the 1° mAb or the 2° mAb to form a sandwich complex.

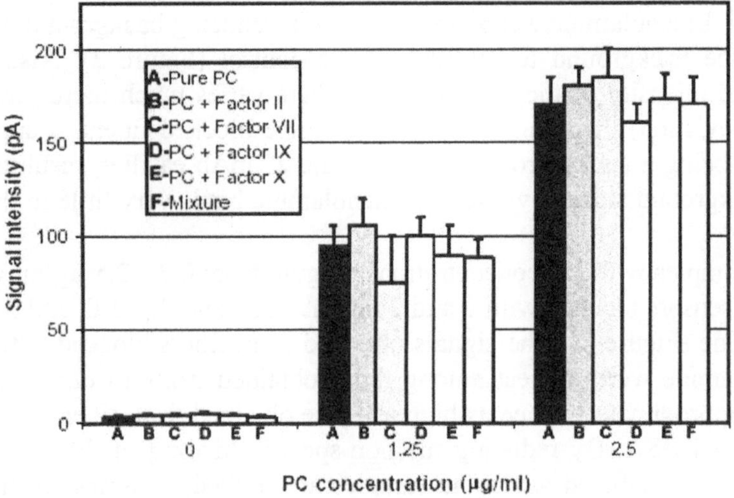

Figure 1. Cross-reactivity of homologous proteins with anti-PC antibodies.

3.2 Blocking buffers

Fibers were blocked with 1 and 2 mg/ml BSA, and 0.1, 1.0 , and 5.0 M ethanolamine. BSA concentrations were based on protocols for ELISA, and ethanolamine concentrations based on protocols for immunoaffinity chromatography [5]. Unblocked fibers were used as controls for this experiment. Then the background fluorescence was examined to determine the effectiveness of BSA and ethanolamine as blocking buffers. Fibers blocked with BSA gave 85-90% of the background fluorescence intensity of

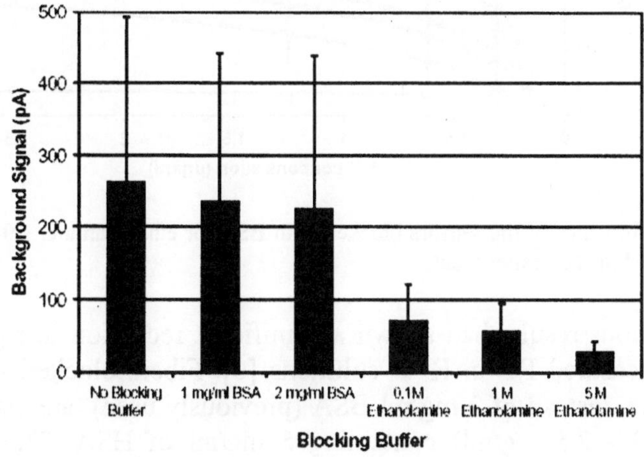

Figure 2. Background intensities of fibers blocked with BSA or ethanolamine at various concentrations.

the control. Ethanolamine was more effective in reducing background noise, reducing the background to 10-20% of the control (Figure 2). Also, the background intensity of the BSA blocked fibers varies much more (greater standard deviation) than that of the ethanolamine blocked fibers. It appears that BSA, being a sticky protein, adsorbes the 2° mAb easilier, resulting in higher background signals, while the ethanolamine binds very little to the 2° mAb.

Also, samples with PC concentrations ranging from 0.3 - 2.5 μg/ml were applied to sensors blocked with 1 and 2 mg/ml BSA, and 0.1, 1.0, and 5.0 M ethanolamine (Figure 3). The signals obtained from fibers blocked with 1.0 M ethanolamine were highest among those obtained from all other fibers, and were approximately twice as high as those obtained from fibers blocked with 2 mg/ml BSA. By reducing the non-specific binding of the 2° mAb, more PC can be labeled when the samples are probed, resulting in higher signal intensities. The fibers tested with 5.0 M ethanolamine gave very little signal intensity over the whole range of PC tested. High concentration of ethanolamine may have denatured the 1° mAb. In general, ethanolamine is proven to be more effective than BSA in preventing non-specific binding of 2° mAb to the fiber.

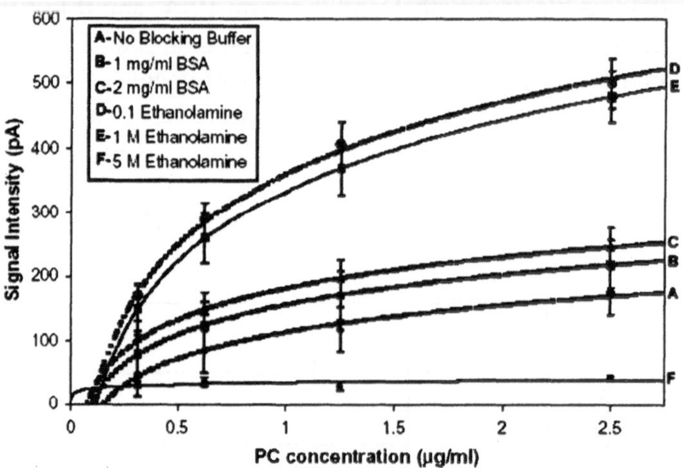

Figure 3. Performance of the sensors blocked with BSA, or ethanolamine. 1.0 mg/ml of PC sample was used in each experiment.

Our previous results had shown a significant reduction in signals when biosensor measured PC in HAS solutions [6]. Fibers blocked with 0.1 M ethanolamine, instead of 2 mg/ml BSA (previously used), are tested for PC samples (0.3 - 2.5 g/ml) containing 5 mg/ml of HSA. The change in blocking buffer provided an increase of 49%-97% over the previous results (Figure 4).

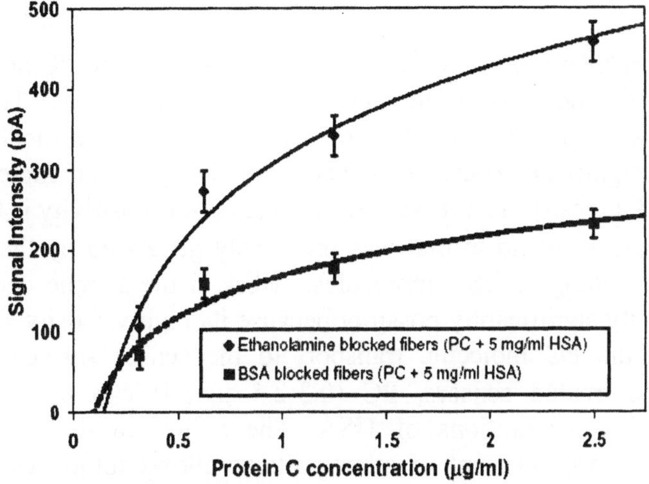

Figure 4. The effect of blocking buffers on PC samples with 5mg/ml HSA.

3.3 Human serum albumin

Experiments were then conducted with PC samples (0.3-2.5 g/ml) containing 5 - 50 mg/ml of HSA, with the ethanolamine treated sensors. As HSA concentration increases the signal intensity decreased more. The signal intensity reduction by 25%-65% was observed when the sample contains HSA at the physiological level (50 mg/ml; Figure 5). Nevertheless, the signals were strong enough to clearly discriminate PC concentrations. The cause of the signal reduction in the samples with HSA is currently under investigation and it appears to be related to the increased sample viscosity caused by the high concentration of HSA.

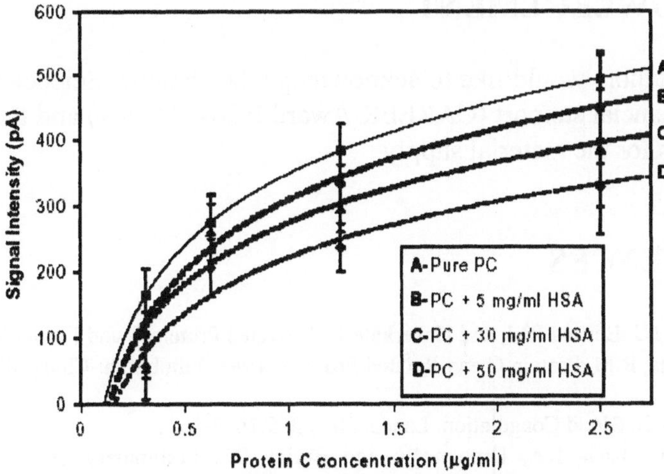

Figure 5. PC sensor performance for samples containing HSA.

4. CONCLUSIONS

The experiments with the four proteins homologous to PC demonstrated that these do not affect the PC biosensor selectivity. The use of ethanolamine, at the concentration of 0.1 and 1.0 M, as a blocking buffer provided a significant decrease in background signal and increase in the actual signal intensity. The reason for this increased sensitivity is believed to be the depression of noise that was previously generated by the 2° mAb's nonspecific binding to BSA molecules. HSA in the sample decreases the signal intensity significantly possibly because its high viscosity slows down the rate of the PC molecule transport to the sensor surface. By using ethanolamine treated sensors, PC (0.3-2.5 μg/ml) can be measured in physiological concentrations of HSA. The ability to detect PC in the presence of a large amount of other proteins allows future testing with a variety of samples, including plasma, transgenic milk containing PC, and animal cell culture broths.

5. FUTURE WORK

These include optimizing 1° and 2° mAb concentrations, minimizing incubation times for samples while allowing sufficient signal generation, testing of actual physiological samples, incorporating of a circulation pump to increase PC transport to the fiber surface, and fully automating the PC biosensor.

ACKNOWLEGEMENT

The authors would like to acknowledge the National Science Foundation for the financial support (CAREER Award BES-9733207) and the American Red Cross for the material supply.

REFERENCES

1. de Fouw NJ, Bertina RM, and Haverkate F. Activated Protein C and Fibrinolysis. In editor Bertina, R.M. Protein C and Related Proteins. 1988; Edinburgh: Churchill Livingstone; 21-54.
2. Dahlbäck B. Blood Coagulation. Lancet 2000;355:1627-1632.
3. Spiker JO, Kang KA, Drohan W, and Bruley DF. Preliminary Study of Biosensor Optimization for the Detection of Protein C. Adv Exp Med and Biol 1998;454:681-688.

4. Research International Inc. Analyte 2000 Fiber Optic Multianalyte Sensor System Instruction Manual. 1995; Woodinville, Washington.
5. Kang, KA, Ryu, D, Drohan, WN, and Orthner, CL. Effect of Matrices on the Immunoaffinity Purification of Protein C. Biotechnology and Bioengineering. 1992; 39:1086-1098.
6. Balcer HI, Spiker JO, and Kang KA. Sensitivity of Protein C Immunosensor With and Without Human Serum Albumin. Adv Exp and Biol 1999;471:605-611.
7. Bhatia S, Shriver-Lake L, Prior K, Geoger J, Calvert J, Bredehorst R, and Ligler F. Use of Thiol-Terminal and Hetrobifunctinoal Crosslinkers for Immobilization of Antibodies on Silica Surfaces. Anal Biochem 1989;178: 408-413.
8. Narnag U, Anderson G, Ligler F, and Buans J. Fiber optic-based biosensor for Ricin. Biosen and Bioelec 1997;12: 937-945.

Chapter **14**

PROTEIN C SEPARATION FROM HUMAN BLOOD PLASMA DERIVATIVES USING LOW COST CHROMATOGRAPHY

Huiping Wu & Duane F. Bruley*

Department of Chemical and Biochemical Engineering, University of Maryland, Baltimore County, 1000 Hilltop Circle, Baltimore, Maryland 21250, USA

Abstract: Protein C (PC) deficiency can cause thrombosis, inhibiting oxygen transport to tissue thus resulting in many complications, including death. Present treatment can cause catastrophic bleeding and other major medical problems. PC treatment has no bleeding or skin necrosis problems because it circulates in the blood as a zymogen and is only activated when and where it is needed.

The vitamin K dependent (VKD) proteins are homologous proteins, making the separation of PC from plasma extremely difficult. Immobilized metal affinity chromatography (IMAC) is investigated to separate the VKD proteins to replace immunoaffinity chromatography, because of the high cost of monoclonal antibodies.

An IDA-Cu column was found effective for the separation of PC from prothrombin, the most harmful contaminant. For Cohn fraction IV-1 separation, a DEAE column was found an efficient initial step, with about 25-fold PC purity increase. Following this step, an IDA-Cu column could remove many contaminants including prothrombin. The combination of DEAE and IDA-Cu resulted in PC purity increase of about 100-fold.

Key words: IMAC, protein C, prothrombin, separation, VKD proteins

1. INTRODUCTION

Protein C (PC) is an important anticoagulant and antithrombotic found in the human blood coagulation cascade. PC is a member of the vitamin K dependent (VKD) family, also consisting of coagulation factors VII, IX, X, protein S, Z, and prothrombin. PC circulates in the blood primarily as a two chain inactive zymogen. When it is activated by thrombin-thrombomodulin, it deactivates factors Va and VIIIa [1], proteins which are crucial to the function of two important coagulation proteases: factors IXa and Xa (subscript "a" denotes the activated form of the protein).

PC circulates in the human blood at a concentration of 4 μg/ml. PC deficient patients are at a greater risk for developing thrombosis, which prevents blood from flowing normally and inhibits oxygen transport to tissue. This in turn can cause pulmonary embolism, heart attack and stroke [2].

The currently prescribed anticoagulation therapy for PC deficiency is heparin and coumarin. This treatment can create many dangerous side effects. The most serious risks are hemorrhage in any tissue or organ, and necrosis and gangrene of skin and other tissue, resulting in oxygen-deprived dead tissue. Hemorrhage and necrosis have been reported to result in amputation or death.

PC is a natural therapeutic protein. It functions only when and where the human body needs it. PC has no known side effect, and it can be used at larger than blood concentration without bleeding complications. Therefore, it is desirable to produce PC as a safe anticoagulant for hereditary and acquired PC deficiency treatment (and other medical indications, such as sepsis, etc.).

Human blood plasma is a natural source of PC. PC is presented along with other VKD proteins in Cohn fraction IV-1 paste. The VKD proteins are homologous proteins, and it is extremely difficult to separate them using traditional chromatography, such as ion exchange. Immunoaffinity chromatography columns are used to separate them [3]. Because of the high cost of monoclonal antibodies, this technology is very expensive. Also the immunogenic substance that leach from the adsorbent are harmful to the human body and harsh elution conditions cause protein denaturation and yield losses.

Immobilized metal affinity chromatography (IMAC) utilizes the protein's metal binding affinity for their separation. The surface exposed electron-donating amino acid residues, such as histidine, cysteine, and tryptophan, have affinity for metal ions, such as Cu2+, Ni2+, Fe3+, which are immobilized on the chelator, such as iminodiacetic acid (IDA). The literature reports that IMAC could result in high biospecific separation [4]. PC has twelve surface accessible histidines, which is the major metal binding unit. Therefore, it is worthwhile to study PC separation using

IMAC. Since the initial investigations of IMAC for the separation of PC from Cohn fraction IV-1, there have been significant progresses [5, 6, 7].

2. METHODS

2.1 Protein separation using IMAC column

A 5 ml HiTrap chelating Sepharose column was washed with water and charged with metal ions by pumping metal ion solution through it. The unbound metal ions were washed out of the column with water. The column was then charged with imidazole by pumping 15 mM imidazole buffer through it and then equilibrated with equilibration buffer containing 2 mM imidazole. Sample was loaded onto the IMAC column at 20 ml/h. The column was washed with wash buffer, the same buffer as the equilibration buffer. Then PC was eluted using an elution buffer containing 15 mM imidazole.

2.2 Enzyme linked immunosorbent assay (ELISA)

ELISA was performed to measure the concentration of the antigen. The principle of the assay is that PC is caught in a sandwich between immobilized anti-PC IgG and immunopurified anti-PC IgG coupled to peroxidase, after which the peroxidase activity is measured with a suitable substrate.

2.3 Amidolytic activity assay (AAA)

PC activity was measured using protac activator and S-2366 chromogenic substrate by adapting the method of Odegaard et al.. This activator rapidly and specifically catalyzed the conversion of PC into an active form. Chromogenic substrate was used to measure the activated PC.

2.4 SDS-PAGE

The fraction were electrophoresed under reduced (5% β-mercaptoethanol) conditions using Novex 8-16% tris-glycine gels. SDS-PAGE was performed using a Novex electrophoresis system under the method developed by Laemmli.

3. RESULTS

3.1 PC separation from the mixture of PC and prothrombin using an IDA-Cu column

Prothrombin is a coagulant, which has an opposite activity as PC. Prothrombin is present in the blood plasma at about 30 times the concentration of PC with 15 times the in vivo half-life of PC. If prothrombin exists in the final PC product, it upsets pro- and anticoagulation, which can be life threatening. Since these two proteins are homologous proteins, immunoaffinity chromatography is presently used for their separation, and the cost is very high.

An IDA-Cu IMAC column was selected to study PC separation from prothrombin. Since PC concentration was relatively high in Cohn fraction IV-1, the ratio of prothrombin to PC was lower than in normal plasma. The mixture of PC and prothrombin, which contained 328 µg of PC (AAA) and 586 µg of prothrombin, was loaded onto a 5 ml IDA-Cu column at room temperature. The column had been pre-equilibrated with 2 mM imidazole, 20 mM sodium phosphate, 0.5 M sodium chloride at pH 7.0 prior to loading. Since the column capacity was high, there was no protein in the fall through fraction. After 15 minutes binding time, the column was washed with 30 ml of the same buffer to remove prothrombin. The bound PC was then eluted with 15 ml of 15 mM imidazole, 20 mM sodium phosphate, 0.5 M sodium chloride at pH 7.0, containing other contaminants from the starting material, which bound to the column more tightly than prothrombin. PC recovery was 95% (AAA). Prothrombin could be removed from PC very well as illustrated on the SDS-PAGE gel (Figure 1). If the wash buffer amount was increased, prothrombin could be removed totally.

3.2 Study of Cohn fraction IV-1 separation using a DEAE column

Because the majority of contaminant proteins, such as α-1 antitrypsin and human albumin, can be dissolved in sodium citrate buffer, ion exchange can be used as a pre-step to remove these contaminants.

Cohn fraction IV-1 was dissolved into 30 mM sodium citrate buffer at pH around 6.8 with magnetic stirring at 4 °C, then centrifuged at 8000 rpm for 10 minutes. The supernatant was poured into a container, and pH was

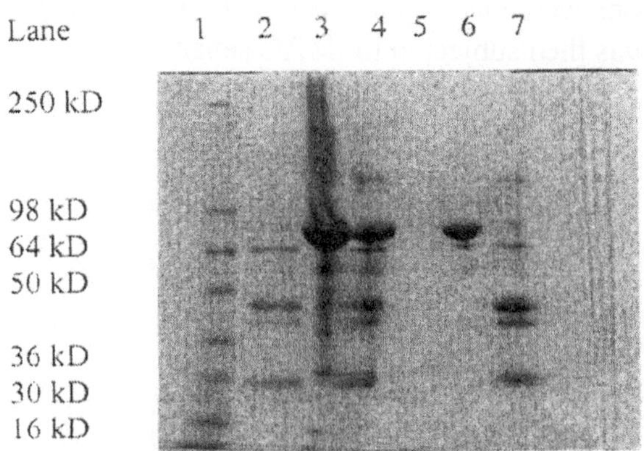

Figure 1. SDS-PAGE analysis of IMAC separation of PC and prothrombin mixture using an IDA-Cu column. Samples were reduced with 5% β-mercaptoethanol and stained with Coomassie blue. Lane 1: SeeBlue protein standards myosin (250 kD), BSA (98 kD), glutamic dehydrogenase (64 kD), alcohol dehygrogenase (50 kD), carbonic anhydrase (36 kD), myoglobin (30 kD) and lysozyme (16 kD); lane 2: human PC separated from PC-albumin mixture by ion exchange; lane 3: prothrombin; lane 4: the mixture of PC and prothrombin; lane 5: fall through; lane 6: wash with 2 mM imidazole buffer; lane 7: eluate with 15 mM imidazole.

adjusted to 6.0. It was then loaded onto a 16 ml DEAE (diethylamino ethyl) column at 4 °C, which had been preequilibrated with 30 mM sodium citrate at pH 6.0 buffer, at flow rate 80 ml/h. The column was washed with about 50 column volumes of the same buffer to remove some major contaminant proteins, and then washed with about 20 to 30 column volumes of 170 mM sodium chloride, 20 mM Mes at pH 6.0 buffer to remove lightly bound contaminants. There was no PC in these wash fractions according to ELISA and AAA. Finally, the bound VKD proteins were eluted by three column volumes of 370 mM sodium chloride, 20 mM Mes at pH 6.0 elution buffer with about 80% PC recovery (AAA) and about 25-fold PC purity increase. The collected fractions were electrophoresed as shown on the gel (Figure 2). This DEAE eluate was then subjected to IMAC separation.

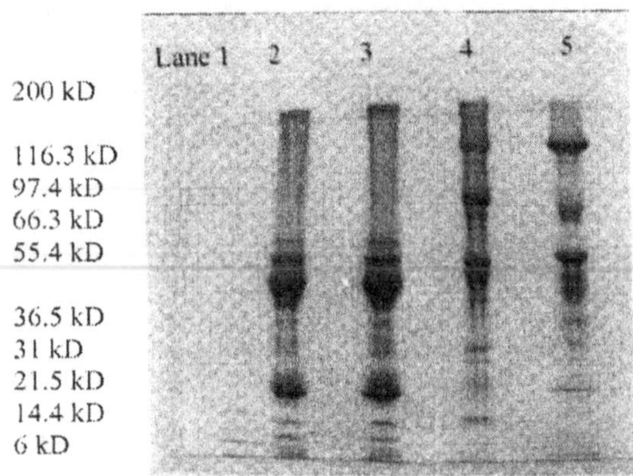

Figure 2. SDS-PAGE analysis of Cohn fraction IV-1 separation using a DEAE column. Lane 1: Mark12TM protein standard myosin (200 kD), β galactosidase (116.3 kD), phosphorylase b (97.4 kD), Bovine serum albumin (66.3 kD), glutamic dehydrogenase (55.4 kD), lasozyme (14.4 kD); lane 2: Cohn fraction IV-1; lane 3: wash with 30 mM sodium citrate at pH 6.0; lane 4: wash with 170 mM sodium chloride; lane 5: eluate with 370 mM sodium chloride.

3.3 Study of DEAE eluate separation using an IDA-Cu column

DEAE eluate was added to 2 mM imidazole, and pH was adjusted to 7.0. It was then loaded onto the 5 ml IDA-Cu column at 4 °C, which had been preequilibrated with the buffer containing 2 mM imidazole. The column was washed with 2 mM imidazole buffer to remove the lightly bound contaminants. The bound PC was eluted by 20 ml of 15 mM imidazole

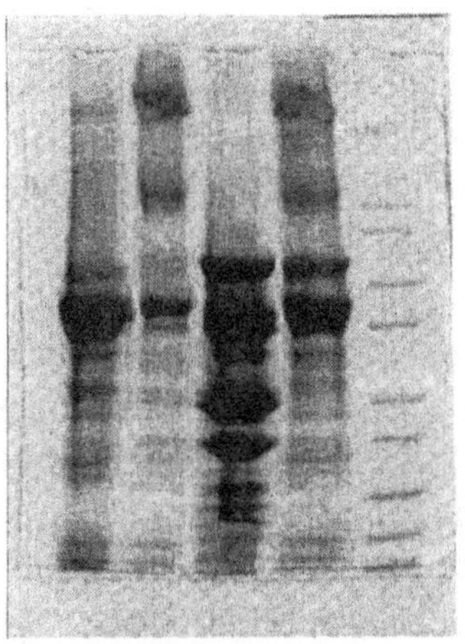

Figure 3. SDS-PAGE analysis of IDA-Cu separation of DEAE eluate. Lane 1: Mark12TM protein standard; lane 2: DEAE eluate; lane 3: wash with 2 mM imidazole; lane 4: first four column volume eluate with 15 mM imidazole; lane 5: later eluate.

buffer. PC activity recovery was 85% (AAA). This separation could increase PC purity about 4-fold. SDS-PAGE was performed to analyze these fractions (Figure 3). The low molecular weight contaminants were removed in the wash fraction, and tightly bound contaminants remained on the column, after PC recovery.

4. DISCUSSION

An IDA-Cu column has been found effective for the separation of PC from prothrombin/PC mixture. This is a very significant result, because the molecules are homologous and have opposite biological activity. Prothrombin can be washed out using a buffer containing 2 mM imidazole, and PC can be recovered using a buffer containing 15 mM imidazole. Because of the mild elution condition, the recovered PC remains totally active.

Because Cohn fraction IV-1 contains many contaminants, and most of them have similar metal binding affinity, it is hard to use IMAC directly for its separation. Ion exchange chromatography has high protein capacity, and is an efficient initial step for protein separation. After different buffer concentrations were studied, the optimal conditions were found. This step increases PC purity about 25-fold. It removes most contaminant proteins. Therefore DEAE is a very efficient step prior to IMAC separation.

IDA-Cu is a very popular IMAC column. It can be used to separate prothrombin from PC and remove many low molecule weight contaminants. It increases PC purity about 4-fold. Therefore it is an important step in the separation process.

The separation using combination columns of DEAE and IDA-Cu results in PC purity increase of about 100-fold. This process can remove the most harmful contaminant, prothrombin and many other contaminant proteins. It can produce a PC cocktail with PC activity of about 5%. Therefore, we feel confident that a process with IMAC elements could help to achieve a low cost, high purity separation of PC.

5. CONCLUSION

Since other IMAC columns, such as IDA-Fe and IDA-Ni, have different binding affinity and separation selectivity, they can be studied for this separation. Also, these columns need to be fit into the separation process to remove some other contaminants and to increase PC purity.

ACKNOWLEDGEMENTS

The authors would like to express thanks to Dr. William N. Drohan and Dr. Annemarie H. Ralston for their technical assistance and for the supply of material from the American Red Cross. Also, we would like to express our appreciation to the National Science Foundation for Grant No. CTS-9904465. Thanks also to The Whitaker Foundation for their financial support through a Special Opportunities Award in Biomedical Engineering.

REFERENCES

1. Esmon CT. The roles of protein C and thrombomodulin in the regulation of blood coagulation. J Bio Chem 1989;264:4743.
2. Bruley DF, Drohan WN. Protein C and related anticoagulants. Advances in applied biotechnology series, Volume 11, Gulf Publishing Company (1990).
3. Velander WH, Orthner CL, Tharakan JP, Madurawe RD, Ralston AH, Strickland DK, Drohan WN. Process implications for metal-dependent immunoaffinity interactions. Biotechnol Prog 1989;5:119-125.
4. Mantovaara T, Pertoft H, Porath J. Purification of factor VIII:c coagulant activity from rat liver nonparenchymal cell culture medium by immobilized metal ion affinity chromatography. Biotechnol Appl Biochem 1991;13:120-126.

5. Wu H, Bruley DF. Homologous human blood protein separation using immobilized metal affinity chromatography-protein C separation from prothrombin with application to the separation of factor IX and prothrombin. Biotechnol Prog 1999;15:928-931.
6. Wu H, Bruley DF, Kang KA. Protein C separation from human blood plasma Cohn fraction IV-1 using immobilized metal affinity chromatography. In:Hudetz AG, Bruley DF, editors. Advances in Experimental Medicine and Biology. Plenum:New York and London, 1998;454:697-704.
7. Tadepalli SS, Bruley DF, Kang KA, Drohan WN. Immobilized metal affinity chromatography process identification and scale-up for protein C production. In:Harrison DK, Delpy DT, editors. Advances in Experimental Medicine and Biology Plenum:New York and London, 1997;428:31-43.

Chapter **15**

GLUTATHIONE DEPLETION OR RADIATION TREATMENT ALTERS RESPIRATION AND INDUCES APOPTOSIS IN R3230AC MAMMARY CARCINOMA

John E. Biaglow[1], Intae Lee[1], Jerry Donahue[1], Kathy Held[2] , John Mieyal[3], Mark Dewhirst[4] & Steve Tuttle[1]

[1]Departments of Biochemistry and Radiation Oncology, University of Pennsylvania Medical School, Philadelphia Pa; [2]Department of Radiation oncology, Mass General Hospital, Harvard University, Boston Mass; [3]Department of Pharmacology, Case Western Reserve University, Cleveland, Ohio; [4]Department of Radiation Oncology, Duke University, Durham North Carolina.

Abstract: Glutathione depletion by L-buthionine sulfoximine inhibits the growth of Ehrlich mouse mammary carcinoma, R3230Ac rat mammary carcinoma and the PC3 human prostrate carcinoma cells, *in vitro*. Inhibition of growth occurs within the first 24 hours after exposure to the drug. The cell density does not increase over the initial cell density over 7 days. A549 human lung carcinoma and the DU145 human prostrate carcinoma cells show no inhibition of growth under the same treatment conditions. A comparative study of the R323OAc and A549 cells demonstrated a marked increase in apoptosis following L-BSO treatment in R3230Ac, which was dependent on L-BSO concentration and incubation time. L-BSO did not induce apoptosis in A549 cells at any of the concentrations tested. The incidence of apoptosis for R323OAc cells following exposure to 0.1 mM L-BSO was similar to the incidence of radiation-induced apoptosis observed after exposure to 10 Gy. Treatment with L-BSO or radiation alone inhibited O_2 utilization in of R323Oac, while no effect on O_2 utilization was observed in A549 cells. LBSO altered the bioreductive capacity of both the R323OAc and A549 cells. These results suggest that the ability of L-BSO to block mitochondrial O_2 utilization may be involved in the apoptotic response in R3230Ac cells.

Key words: apoptosis, bioreduction, carcinoma cells, glutathione, oxygen uptake, radiation

1. INTRODUCTION

Previously we demonstrated a role for glutathione (GSH) in the cellular radiation response [1,2]. Glutathione protects against the oxidative damage induced by ionizing radiation by three distinct mechanisms: First, GSH reacts chemically with reactive oxygen species (ROS) [1,2]. Second, GSH is a substrate for selenium-linked GSH peroxidases [1-3], which protect cells against radiation toxicity by reducing peroxide and hydroperoxides [1-3]. Glutathione disulfide (GSSG) produced in both of the above reactions is reduced *via* GSSG reductase with NADPH donated by the oxidative limb of the pentose phosphate cycle [4]. Finally, GSH-S-transferases conjugate GSH with the breakdown products of lipid peroxidation namely, malondialdehyde and 4-hydroxynonenal [5].

GSH has been shown to protect cells against radiation-induced apoptosis [6,7]. Moreover, the loss of cellular and mitochondrial GSH content is an early event in the generalized cellular apoptotic response [7-10]. While inhibiting the loss of GSH has been shown to block apoptosis [11-14], the direct inhibition of GSH synthesis by L-BSO can induce apoptosis in a number of cell lines [15-18]. It has been suggested that depletion of mitochondrial GSH precedes the loss of cytochrome c in the apoptotic response [13,14]. Loss of cellular GSH, during apoptosis, may increase oxidative damage via inhibition of antioxidant defenses [19]. While the loss of cytochrome c would inhibit mitochondrial energy production and cellular O_2 consumption [20].

Today there is continued interest in the role of glutathione (GSH) in apoptosis [cf. 6-20]. The exact mechanism for its involvement in the apoptotic response is not known. Cellular GSH is an important biochemical involved in the growth of many cells [14-18,21]. Some cells like the human A549 lung carcinoma cell and GSH^- mutants are exceptions and can grow without GSH [1-3,22]. For the cells undergoing apoptosis GSH depletion inhibits protein, and DNA synthesis. as well as alters protein redox status. Glutathione is a substrate for γ-glutamyl-transpeptidase catalyzed amino acid transport and subsequent protein synthesis. Deoxyribonucleotide and DNA synthesis also in part involves GSH as coenzyme in the thioltransferase catalyzed reduction of ribonucleotide reductase [21]. GSH maintains thioltransferase in a reduced state necessary for reduction of nucleotide precursors to deoxyribonucleotides [23]. Thioltransferase's enzyme activity is also necessary for maintaining intracellular proteins in their correct oxidation status [1-3,23].

Based on the above information we decided to study the importance of GSH depletion in apoptosis. In this study, we examined the apoptotic response in two cell lines, R3230Ac mammary carcinoma and A549 human lung carcinoma. Growth of the R3230 Ac cells was inhibited by L-BSO, while A549 cell growth was unaffected by L-BSO treatment. L-BSO treatment and exposure to ionizing radiation each induced apoptosis in R3230Ac cells, but not in A549. We examine O_2 consumption and disulfide reduction capacity for both cell lines treated with L-BSO, demonstrating that the ability of L-BSO to inhibit O_2 consumption correlates with the ability of L-BSO to induce apoptosis.

2. METHODS

The incidence of apoptosis was determined by fluorescent microscopy as previously described [23]. Briefly, the cells in monolayer are harvested by trypsinization, after resuspending cells spent media to inhibit the trypsin; the cells were pelleted by spinning at 1200 rpm for 5 minutes. The cells were then resuspended in 0.2 ml of ice cold McCoy's 5a with 5% FCS and 20 mM Hepes, pH = 7.0. Twenty λ of the cell suspension was mixed with 5λ of .04% propidium iodide staining solution containing, 0.1% NP-40, 5 mM spermine hydrochloride, 3.4 mM citrate and 0.5mM Tris (pH=7.6). The incidence of apoptosis was assayed by examining nuclear morphology at 200x magnification, using a Nikon epiflourescence microscope. Fragmented nuclei (Figure 2a) were scored as apoptotic while intact and mitotic nuclei were scored as normal [24]. Three fields of 100 cells were counted for each variable within an individual experiment. Data points represent the mean values obtained from at least three separate experiments.

Oxygen consumption measurements were determined with the Clark electrode as previously described [25]. Cells are harvested by scrapping, centrifuged at 1200 rpm for 5 minutes, washed twice and resuspended in DPBS, with one mM Ca^{2+}, one mM Mg^{2+}, pH 7.4, 37 in 3 cc. Proteins were measured on the cell suspension following the measurement of O_2 uptake using the Biorad protein assay [26, 27].

The thiol-linked reduction of hydroxyethyldisulfide is described in detail elsewhere [27]. All cells are routinely grown in McCoy's 5a supplemented with 10% calf serum and buffered with 20 mM Hepes buffer pH 7.2. Cell number is determined with the aid of a Coulter counter. All chemicals were purchased from the Sigma chemical company.

3. RESULTS

3.1 Effect of GSH depletion on cell growth

In our early studies, we inhibited growth of the 66 and 67 murine carcinomas *in vitro* [28,29]. We were interested in extending these studies to other tumor cell models. Figure 1 shows the effects of continuous LBSO treatment on the growth of monolayer cultures of various tumor cell lines. The Ehrlich cells (EC), and the R323OAc, both mammary carcinoma cell lines, exhibited the highest growth rate in the absence of LBSO, both with a doubling time of approximately 12 hours. EC cells reach plateau phase at a density of 3 x 10^7 cells/25cm^2 flask. R323OAc mammary tumor cells reach plateau phase at a density of 6 x 10^6 cells /25 cm^2. LBSO treatment depletes GSH to less than 1 % of the untreated control by 24 hours and there was no detectable GSH by 48 hrs in both of these cell lines (data not shown). The continuous inclusion of 0.1 mM LBSO in the growth medium totally blocked growth of both EC and R323OAc cells by 24 hours.

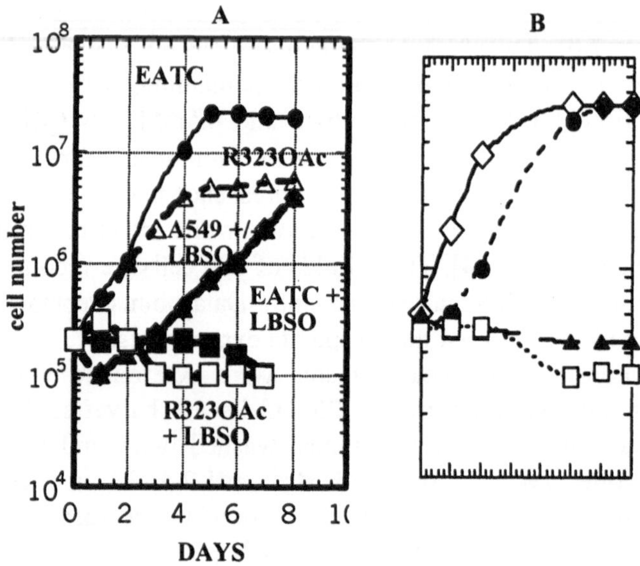

Figure 1A and 1B. The inhibition of cell growth caused by continuous treatment with LBSO. The cells are trypsinized and counted daily in duplicate flasks. The inclusion of 0.1 mM in McCoy's 5a with 10% calf serum inhibits the growth of Ehrlich and R323OAc. There is no effect on A5449 growth. Growth of the human prostate PC3 and DU 245 is blocked by 0.1 mM LBSO.

A549 cells grow at a much lower rate than the murine mammary carcinoma lines, having doubling time of 22-24 hours (Figure 1A). A549 entered plateau phase by 9-10 days at a density 5-6 x 10^6 cells/25 cm^2 flask. GSH is depleted to non-detectable levels in A549 cells after 72 hours in the presence of 0.1 mM L-BSO. There was no effect of continuous exposure to 0.1 mM LBSO on the growth of A549 cells (Figure 1A). We also tested for the effects of 0.1-mM LBSO on the growth of the human prostrate cell lines Du145, PC3 (Figure 1B) and LnCap. L-BSO inhibited the growth of PC3 and DU145 cells (Figure 1B). Doubling times of 20-22 hours were observed for PC3 and DU145 in the absence of LBSO and plateau phase was reached at a density of 5 x10^6 cells and 6 x 10^5-cells/25cm^2 flask, respectively.

3.2 Effect of LBSO on apoptosis

We tested the hypothesis that L-BSO-induced inhibition of growth in R3230Ac was due to L-BSO-induced apoptosis. L-BSO treatment induces apoptosis in a number of cell types [14-18]. The effect of L-BSO concentration on apoptosis as a function of time is shown in Figure 2. The

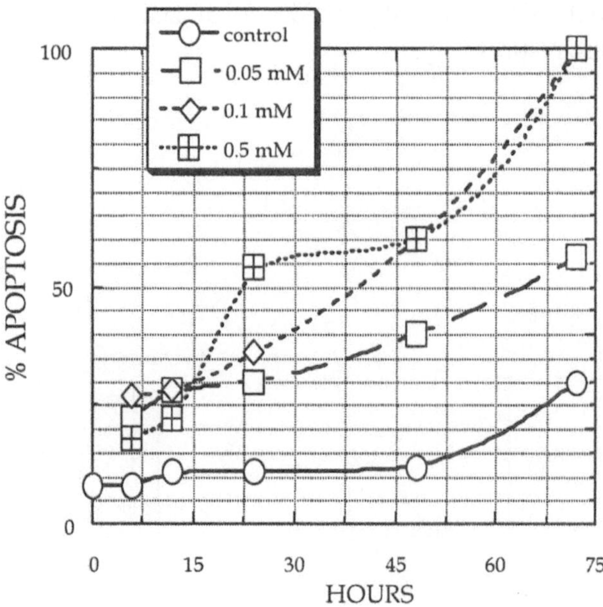

Figure 2. The effect of LBSO concentration on the apoptotic fraction for R3230Ac cells. Apoptosis is measurement as previously describe (24).

R3230Ac cells have a relatively high background rate of apoptosis, with a mean value of approximately 8 % in log phase cultures. The background incidence of apoptosis increases as the cells enter the late log phase and plateau phases of growth, (cf. Lower curve Figure 2). LBSO induced apoptosis was visible by 6 hours at all concentrations examined. By 24 hours 0.5 mM L-BSO induced apoptosis in 50% of the cells, while 0.1 mM LBSO induced 30% of the cells to undergo apoptosis. Exposure to LBSO concentrations > 0.1 mM resulted in 100% of the cells undergoing apoptotic cell death by 72 hours. The apoptotic fraction continues to increase with time In the presence of 0.05 mM LBSO, approaching 50% by 72 hours.

3.3 Effect of radiation on apoptosisγ

Our results demonstrated that the inhibition of cell growth observed in L-BSO treated R3230Ac cells is due to apoptosis. Ionizing radiation can also induce apoptosis [6,7]. Radiation-induced apoptosis is linked to a rapid efflux of cellular GSH [8-10]. Therefore, we compared the incidence of apoptosis after exposure to either ionizing radiation or L-BSO. Neither L-BSO treatment or exposure to 10 Gy of γ-radiation induced apoptosis in A549 cells (Figure 3). R3230Ac cells exposed to 10 Gy exhibited levels of apoptosis at 24 hours similar to those observed after a 24 hour exposure to 0.1 mM LBSO, 55% and 64% incidence of apoptosis for radiation and LBSO exposure respectively.

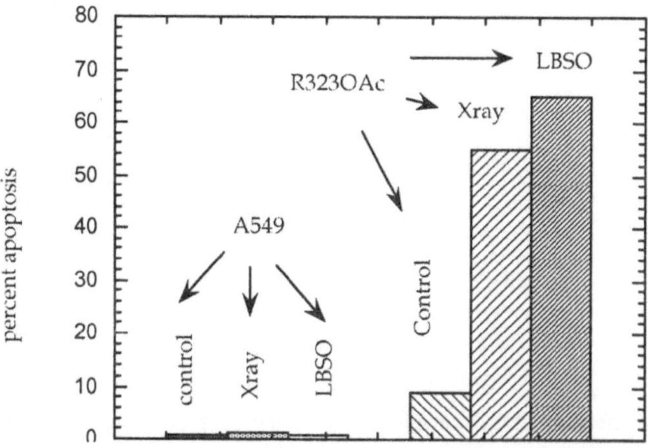

Figure 3. Radiation induced apoptosis. Ten Gray of radiation produces maximal apoptosis with the R323OAc and is similar in magnitude for that produced by LBSO treatment.

3.4 Effect of radiation and L-BSO treatment on O2 uptake

Inhibition of cellular O_2 consumption is observed early during apoptosis, at the time that cytochrome c is observed leaking across the mitochondrial membrane [13,14]. Therefore, we were interested in testing whether radiation or L-BSO treatment could inhibition O_2 cellular utilization. Figure 4 ,shows the effect of exposure to 10 Gy radiation or L-BSO on O_2 consumption in R323OAc. Radiations effect on O_2 consumption occurred within 2 hours. The radiation-induced inhibition of O_2 consumption continued to gradually increase with time out to 48 hours. The rate of O_2 consumption dropped to 45% of the rate in untreated controls by 48 hrs. Measurement of clonogenic death indicates that > 97% of R3230Ac cells ultimately die following exposure to 10 Gy. Oxygen consumption following the addition of 0.1mM LBSO decreases to < 20% of the control value by 24 hours. We didn't observe immediate changes in O_2 consumption for A549 cells treated with radiation or for cells depleted of their GSH by treatment with 0.1 mM L-BSO (upper curve).

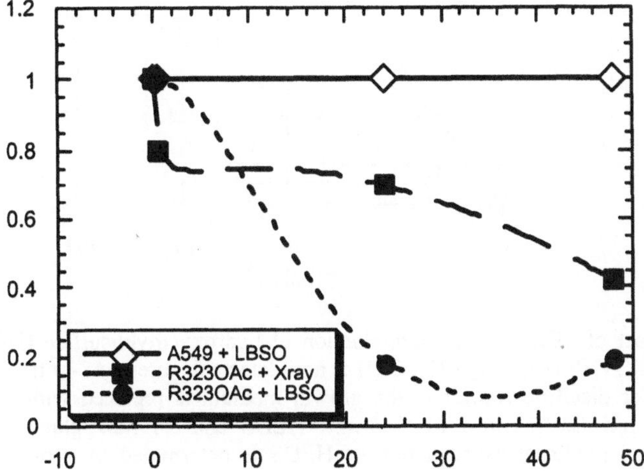

Figure 4. Radiation and LBSO treated monolayer show inhibition of O_2 consumption. Minelayers of cells were treated with either single 10 Gy radiation or continuously with LBSO for the indicated times, trypsinized and assayed for O2 consumption with the aid of a Clark electrode system as previously described (25). Cells in the chamber are saved for protein determinations for each O_2 measurement.

3.5 Effect of LBSO on bioreduction

It is assumed that radiation induced loss of GSH will enhance oxidative damage via inhibition of essential antioxidant defenses (14,19). GSH depletion by L-BSO also affects other processes dependent on GSH, i.e., amino acid transport, DNA synthesis and bioreduction. We found that reduction of hydroxyethyldisulfide (HEDS), catalysed in part by GSH dependent thiol transferase [23,27], is a good indicator of the overall cytosolic bioreductive capacity [27]. Cells with low bioreductive capacity have been shown to be more sensitive to radiation induced apoptosis [30]. Therefore, we examined how LBSO pretreatment alters the capacity of R3230 and A549 cells to reduce HEDS. Figure 5 shows that exposing

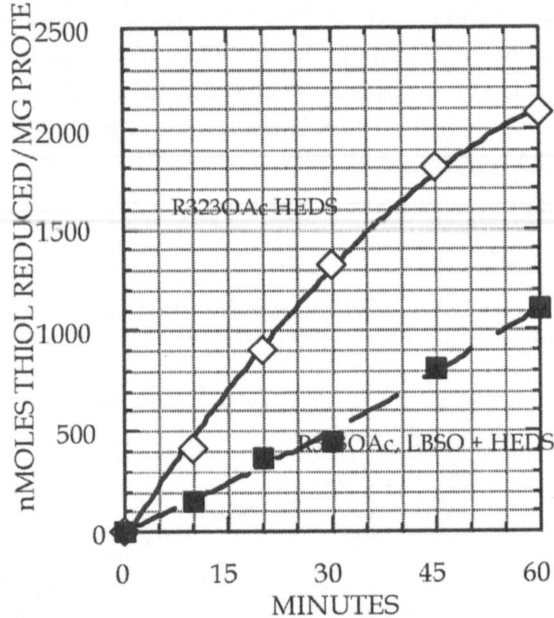

Figure 5. Effect of LBSO on the bioreduction of hydroxyethyldisulfide (HEDS). Thiol is determined via DTNB reactivity (27). GSH is non-detectable after a 48-hr treatment with 0.1-mM LBSO. For disulfide reduction, the growth medium is replaced with 5 ml of DPBS, containing 1 mM Ca^{2+}, 1-mM Mg^{2+} 5-mM HEDS and 20 mM glucose, and pH 7.4. Mercaptoethanol produced by reduction of HEDS is determined by removing aliquots of supernatant from the monolayer cultures and measuring color development at 420 nm (27).

R3230Ac cells to 0.1-mM LBSO for 24 hours inhibits the production of mercaptoethanol as measured by formation of DTNB reactive thiol [27]. The initial rate of HEDS reduction is three times faster in untreated R3230Ac cells. However, the reduction rate decreased in the untreated cells over the 60 minute measurement period, which may be related to HEDS induced decrease in GSH [30]. Radiation had no immediate effect on the reduction of HEDS (data not shown). 0.1 mM L-BSO treatment blocks the reduction of HEDS in A549 by 33% without apoptosis. These data indicate that loss of bioreductive capacity is related to the induction of apoptosis by L-BSO treated R323Oac cells.

4. DISCUSSION

We previously demonstrated that LBSO inhibits the growth of the 66 and 67 mouse mammary carcinoma cells *in vitro* [28,29]. In this paper we describe similar results in the Ehrlich mammary carcinoma and R3230Ac mammary carcinoma cell lines (cf. Figure 1A and B). L-BSO also inhibits the growth of the prostate cancer lines PC3, DU145and LnCap (data not shown). However, growth of A549 lung carcinoma is not affected by exposure to 0.1 mM LBSO (Figure 1A and B). Inhibition of growth by L-BSO may be due to altered amino acid uptake, DNA synthesis , bioreductive capacity or mitochondrial function [31]. Depletion of GSH by L-BSO occurs in both the cytoplasm and mitochondria [14,15]. Indeed, removal of GSH in the mitochondria by substrates of GSH-S-transferases produces rapid onset of apoptosis [14]. It was demonstrated that the initial decrease in mitochondrial GSH occurs before change in the transmembrane potential and the leakage of cytochrome c [14]. Leakage of cytochrome c inturn would block mitochondrial oxygen consumption. Our data is the first to show a correlation between L-BSO induced thiol depletion, inhibition of O_2 consumption, and apoptosis

Our data (Figure 1A) has shown that continuous LBSO treatment does not influence the growth of A549 cells. GSH depletion in A549 cells by L-BSO treatment, does not result in apoptosis or inhibition of O_2 uptake. A549 cells also do not exhibit an apoptotic response to ionizing radiation. However, other laboratories measured inhibition of growth with concentrations of 5-1O mM LBSO [31,32]. The uses of such high concentrations of LBSO suggest that there may be additional nonspecific effects on cell biochemistry. We have found that prolonged treatment, >3 days, with 0.1 mM LBSO can alter the protein thiols in A549 cells [1-3]. GSH depletion would prevent the maintenance of a proper reduced protein thiol status [23,30]. Protein thiols are critical for many enzyme activities [34]. GSH is involved in the cellular maintenance of PSH in the reduced form via thioltransferase [23,35]. Our data (Figure 5) shows that L-BSO treatment inhibits GSH dependent thioltransferase activity in R3230Ac cells. The inability to totally block the reduction of HEDS suggests that part of the reduction is due to thioredoxin reductase in agreement with our previous results [27,30]. Depletion of GSH only partially inhibited the reduction of HEDS by A549 cells [27].

GSH depletion blocks O_2 consumption in R3230Ac cells, but has no effect on A549 cell O_2 consumption, even when GSH is depleted to non-detectable levels (Figure 4). This result suggests that inhibition of O_2 consumption by L-BSO directly correlate with the capacity of a cell line to undergo L-BSO induced apoptosis. The variance between the A549 and R323OAc response, to L-BSO treatment, suggest different sensitivities of important mitochondrial protein thiols to GSH depletion.

The effects of GSH depletion and radiation on cellular O_2 utilization may have additional implications for the treatment of tumors *in vivo*. There is the possibility for L-BSO to induce reoxygenation of tumors. Inhibition of O_2 consumption would show a time dependency for reoxygenation of previously hypoxic cells *in vivo* and increase the radiation response. L-BSO induced reoxygenation could therefore increase the therapeutic potential of subsequent treatment with radiation or chemotherapy. Milas et al. has demonstrated that apoptosis induced by taxol results in reoxygenation of tumors [36]. However, his laboratory did not consider the possibility that inhibition of O_2 utilization is involved in the apoptotic response produced by taxol.

5. CONCLUSIONS

Thiol depletion inhibits the growth of EC and R3230Ac mammary carcinomas as well as the PC3 and DU145 human prostrate cells *in vitro*. The inhibition of growth correlates with increased LBSO-induced apoptotic response for R323OAc cells. There is no inhibition of growth or apoptosis with A549 human lung carcinoma upon continuous exposure to LBSO. Radiation induced apoptosis occurs with the R323OAc and not with the A549 cells. Thiol depletion and radiation both induce a time dependent inhibition of R323Ac O_2 utilization. Suggesting that sensitivity to LBSO induced decrease in oxygen consumption and apoptosis are directly related. Clinically, treatment planning resulting in improved tumor response might be accomplished by tumor reoxygenation that would occur due to LBSO-induced apoptosis.

ACKNOWLEDGEMENTS

This research was supported by grant number CA 44982 to JEB and NIA(NIH) program project AG15885 and a VA Merit Review Grant to JJM

REFERENCES

1. Biaglow, J.E., and Varnes, M. The role of thiols in cellular response to radiation and drugs. Radiation Res 1983; 95:437-455.
2. Biaglow JE, Mitchell JB, Held K. The importance of peroxide and superoxide in the X-ray response. Int J Oncology Biol Phys 1992;22:665-669.
3. Biaglow, J.E and Varnes. Cellular protection against damage by hydroperoxides: Role of Glutathione. Basic Life Sci 1988;49:567-573.
4. Tuttle, S.E. and Biaglow, J.E. Sensitivity to chemical oxidants and radiation in CHO lines deficient in oxidative pentose cycle activity. Int. J. Radiat; Oncol Biol. Phys. 1992, 22:671-675.
5. Puertas Fj, Diaz-Llopis, Chiupoint E, Roma A, Romero FJ. Glutathione system of human retina: enzymatic conjugation of lipid peroxidation products. Free Radic Biol Ned 1993;14:549-551.
6. Vlachaki MT, Meyn RE. ASTRO research fellowship: the role of BCL-2 and glutathione in an antioxidant pathway to prevent radiation-induced apoptosis. Int J Radiat Oncol Biol Phys 1998;42:185-190.
7. Mirkovic N, Voehringer DW, Story MD, Mcconkey DJ, Mcdonnell TJ, Meyn RE. Resistance to radiation-induced apoptosis in Bcl-2-expressing cells is reversed by depleting cellular thiols. Oncogene 1997;19:1461-1469.
8. Chibelli L, Fanelli C, Rotilio G, Lafavia E, Coppola S, Colussi C, Civitareeale P, Ciriolo MKR. Rescue of cells from apoptosis by inhibition of active GSH extrusion. FASEB J 1988:12:479-486.
9. Slater AF, Stefan C, Nobel I, van den Dobbelsteen DJ, Orrenius S. Signaling mechanisms and oxidative stress in apoptosis. Toxicol Lett 1995;82-83:149-153
10. Celli A, Que FG, Gores GJ, Larusso NF. Glutathione depletion is associated with decreased Bcl-2 expression and increased apoptosis in cholagiocytes. Am J Physiol 1998;274:G749-757.
11. Bojes HK, Datta K, Xu J, Chin A, Simonian P, Nunez G, Kehrer JP. Bcl-xL over expression attenuates glutathione depletion in Fl5.12 cells following interleukin3 withdrawal. Biochem J 1997;325:315-319.
12. Van den Dobbelsteen DJ, Nobel CSI, Schlegel J, Cotgreave IA, Orrenius S, Slater AF. Rapid and specific efflux of reduced glutathione during apoptosis induced by anti-Fas/APO-1 antibody. J Biol Chem 1996;26:15420-15427.
13. Kurihara M, Kurihara H, Tsukuda M. Elevation of cytoplasmic cytochrome c in radiation-induced apoptosis. Jpn J Cancer Res 1998;10:1082-1086.
14. Wullner U, Seyfried J, Groscurth P, Beinroth S, Winter S, Gleichmann M, Heneka M, Leschmann P, Schulz JB, Weller M, Klockgetherr T. Glutathione depletion and neuronal cell death: the role of reactive oxygen intermediates and mitochondrial function. Brain Res 1999;24: 150-160.
15. Wan XS, St Clair DK. Differential cytotoxicity of buthionine sulfoximine to "normal" and transformed human lung fibroblasts. Cancer Chemother Pharmacol 1993;33:210-220.
16. Nicole A, Santiard-Baron D, Ceballos-Picot, I. Direct evidence for glutathione as mediator of apoptosis in neuronal cells. Biomed Pharmacother 1988;52:349-355.
17. Asohiba K, Yasui S, Nishimura K, Nagai A. Thiol depletion induces apoptosis in cultured lung fibroblasts. Am J Respir Cell Mol Biol 1999;21:54-60.
18. Boggs SE, McCormick TS, Lapetina EG. Glutathione levels determine apoptosis in macrophages. Biochem Biophys Res Commun 1988;247:229-233.

19. Lizard G, Gueldry S, Sordet O, Monier S, Athias A, Miguet C, Bessede G, Lemaire S, Solary E, Gambert P. Glutathione is implied in the control of 7-ketocholesterol-induced apoptosis, which is associated with radical oxygen species. FASEB J 1998;15:1651-1663.
20. Varnes ME, Chiu SM, Xue LY, Oleinick NL. Photodynamic therapy-induced apoptosis in lymphoma cells: translocation of cytochrome c causes inhibition of respiration as well as caspase activation. Biochem Biophys Res Commun 1999;255:673-679.
21. Voet, D. and Voet J.G. Biochemistry (1995). John Wiley & Sons, Inc. New York. Pp. 617-623.
22. Biaglow, JE. Radiation effects in mammalian cells. In: Radiation Chemistry: Principles and Applications, M. A. J. Rogers and Dr. Farhataziz, Eds., 1987 VGH Publishers, 527-564.
23. Starke, DW, Chen, YC, Bapna, CP, Lesnefsky, EJ, Mieyal, JJ. Sensitivity of protein sufhydryl repair enzymes to oxidative stress. Free Rad Biol Med 1997; 23:373-384.
24. Muschel RJ, Bernhard EJ, Graze L, McKenna GW, Koch CJ. Cancer Research, 1995;55:995-998.
25. Biaglow JB, Tuttle S, Evans S. MIBG inhibits respiration: potential for radio and hypertermic sensitization. Int J Radiat Biol Phys 1998;42:871-876.
26. Biorad protein assay purchased from Cal Biochem Co.
27. Biaglow J, Donahue J, Tuttle S, Held K, Mieyal J. A method for measuring disulfide reduction by cultured mammalian cells: role of GSH dependent and GSH independent mechanisms. Analytical Biochemistry 2000, 281.71-85
28. Dethlefson LA, Biaglow JE, Peck VM, Ridinger DN. Toxic effects if extended glutathione depletion by buthionine sulfoximine on murine mammary carcinoma cells. Int J Radiat Oncol Biol Physics 1986;12:1157-1160.
29. Dethleson LA, Lehaman CM, Biaglow JE, Peck VM. Toxic effects of acute glutathione depletion by buthionine sulfoximine and dimethylfumarate on murine mammary carcinoma cells. Radiation Res 1988;114:215-224.
30. Biaglow JE, Iramundi I, Koch, CJ, Donahue J, Mieyal, JJ, Stomato T, Tuttle S Raadiation response of cells during altered protein redox. 2003 Radiation Research in press.
31. Kang VJ, Emery D., Enger MD. Buthionine sulfoximine induced growth inhibition in human lung carcinoma cells does not correlate with glutathione depletion. Cell Biol Toxicol 1991;7:249-261.
32. Kang YJ, Enger MD. Glutathione content and growth in A549 human lung carcinoma cells. Exp Cell Res 1990;187:177-179.
33. Baker MA, Hagner,B. Diamide induced shift in protein and glutathione thiol: disulfide status delays DNA rejoining X-irradiation of human cancer cells. Biochem Biophys Acta 1990;1037: 39-47.
34. Webb JL. (1966) Enzyme and Metabolic Inhibitors, Vol. 3, pp. 602-656, Academic Press, New York
35. Powlis G, Gasdaska JR, Baker A. Redox signaling and the control of cell growth and death. (1997) Academic Press, N.Y., 329-359.
36. Milas L, Saito Y, Hunter N, Milross CG, Mason KA. Therapeutic potential of placitaxel-radiation of a murine ovarian carcinoma. Radiother Oncol 1996;40:163-170.

Chapter 16

OXYGENATION AND VASCULAR PERFUSION IN SPONTANEOUS AND TRANSPLANTED TUMOR MODELS

Bruce M. Fenton and Scott F. Paoni
Department of Radiation Oncology, University of Rochester School of Medicine, Rochester, NY 14642 USA

Abstract: Since quantitative measurements of tumor vascular function cannot be obtained in human tumors, appropriate animal tumor models must be utilized. The current studies were undertaken to compare transplantable, murine KHT tumors with primary and 1st generation transplants of spontaneous mammary carcinomas. To evaluate changes in tumor vascular structure and function, immunostaining of total and perfused vascular spacing, and cryospectrophotometric measurement of intravascular HbO_2 saturations were utilized. KHT tumors demonstrated a distinct pattern of decreasing oxygenation with increasing distance from the tumor surface, while spontaneous tumors exhibited striking intertumor heterogeneities and a reduced dependence of oxygenation on distance from tumor surface. Anatomical/perfused vessel distributions and functional response were similar between the primary and transplanted tumor models, as was tissue histological appearance, but were quite different from KHT tumors. These results indicate that spontaneous tumor vascular configuration and function tend to be preserved in 1st generation trochar transplanted tumors.

Key words: hypoxia, oxygenation, perfusion, spontaneous, tumor, vasculature

1. INTRODUCTION

Over the past few years, the importance of tumor vascular development has become much more widely appreciated, particularly in relation to the interdependence between tumor growth and angiogenesis [1]. Clinical trials utilizing strategies to manipulate tumor oxygenation and blood flow are underway, despite the fact that the underlying physiological mechanisms remain poorly understood. Since detailed quantitative measurements of tumor vascular structure and function cannot be obtained in human tumors, appropriate animal tumor models must instead be utilized. A key question is whether or not transplantable tumor models adequately mimic the vascular structure and functional response of more slowly growing primary tumors [2]. The current studies were undertaken primarily to compare transplantable, fast-growing, KHT murine tumors with primary and 1st generation transplanted spontaneous murine tumors, using immunohistochemical staining and cryospectrophotometric procedures to quantitate both tumor vascular structure and function.

2. METHODS

2.1 Mice, tumor models, and tumor freezing

KHT tumor cells (2×10^5) were implanted *i.m.* into the legs of 6-8 week old C3H/HeJ mice and grown to a volume of ~500 mm^3. C3H mammary tumors arose spontaneously in retired C3H breeder mice (at 12-24 months of age). From these primary tumors, 1st generation tumors were trochar transplanted *s.c.* into the flanks of additional C3H mice and grown to ~500 mm^3 in 6-8 weeks. Guidelines for the humane treatment of animals were followed as approved by the University Committee on Animal Resources.

2.2 Injection of fluorescent stains and hydralazine

To visualize blood vessels open to flow, $DiOC_7(3)$, was injected *i.v.* one minute prior to tumor freezing at a concentration of 1.0 mg/kg (dissolved in 75% dimethyl sulphoxide in phosphate-buffered saline) [3]. To minimize any possible vasoactive effects of the $DiOC_7(3)$ on the HbO_2 saturations, separate groups of tumors were used for the HbO_2 analyses and for immunohistochemistry. Hydralazine (HYD) was administered *i.p.* (5 mg/kg) at a time of either 30 or 60 min prior to freezing, as a gauge of tumor vascular response.

2.3 Tumor freezing and determination of intravascular HbO_2 saturations

To accelerate the tumor freezing procedure, tumors were shaved, and a depilatory agent was applied. The mice were cervically dislocated and the tumors immediately quick-frozen using a liquid N_2-cooled copper block and stored in cryotanks for later cryospectrophotometric analysis. For the HbO_2 determinations, four cross-sections of the tumor were analyzed on a liquid nitrogen-cooled microscope stage [4]. Approximately 95 blood vessels were quantified per tumor, systematically selecting vessels of diameter ≥ 6 µm. Spatial positions of the blood vessels were recorded using stage micrometers, and intravascular HbO_2 saturations were determined cryospectrophotometrically [5]. For this technique, HbO_2 saturations were calibrated as a function of reflected light intensity at three discrete wavelengths, based on the spectral differences between oxy- and deoxy-hemoglobin.

2.4 Immunohistochemistry and image analysis

Immediately following cryostat sectioning, tumor slices were imaged for $DiOC_7(3)$ perfusion staining using an epi-fluorescence equipped microscope, digitized, background-corrected, and image-analyzed using Image Pro software and a Pentium computer [6]. Color images from adjacent microscope fields were automatically acquired and digitally combined to form 4×4 montages (20× objective) of the tumor cross-section. After the immunohistochemical staining procedures were completed [6], the same tumor section was returned to the microscope stage and automatically rescanned using the same coordinates. Using transmitted light, matching brownish-red montages of the PECAM-1/CD31 staining were then acquired. The PECAM-1 image (total blood vessels) was enhanced, and a binary image of the selected colors was created. For automated counting of the vascular structures, the binary image was first subjected to an 11×11 binary closing (a morphological filter used to omit narrow gaps between objects), vascular structures of area less than 10 μm^2 were removed, and an area of interest was outlined to omit artifacts and normal tissue.

The corresponding $DiOC_7(3)$ stained vessels (perfused blood vessels) were color segmented using a predefined range of colors and filtered twice, again resulting in a binary image. To quantify vessel distributions, a rectangular matrix of sampling points was computer superimposed over the selected tumor montage, and the distances from each sampling point to the nearest total anatomical or perfused blood vessel were then determined [7].

2.5 Statistical methods

Differences between groups were evaluated using Student's unpaired t-test, and a p value of ≤ 0.05 was considered statistically significant.

3. RESULTS

To compare oxygen availability among the three tumor models, intravascular HbO_2 saturations were first determined cryospectrophotometrically. Although HbO_2 saturations varied substantially among individual KHT tumors, a distinct pattern of decreasing oxygenation with increasing distance from the tumor surface was invariably observed. Spontaneous tumors, in contrast, exhibited a reduced dependence of oxygenation on distance from the tumor surface. Figure 1 shows the percentage of blood vessels containing >25% HbO_2 levels as a function of distance from the tumor surface. Although HbO_2 levels in Zone 1 (0-1.3 mm from the tumor surface) were somewhat similar for KHT and spontaneous

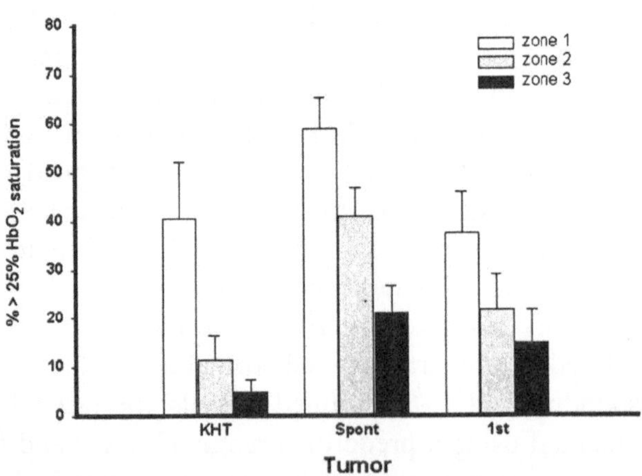

Figure 1. Percentage of blood vessels containing >25% HbO_2 saturations (mean ± s.e.) as a function of distance from tumor surface. Zone 1 = 0-1.3 mm, zone 2 = 1.3-2.6 mm, and zone 3 = 2.6-3.9 mm. KHT: N=6, mean tumor volume=660 mm^3; spontaneous: N=6, volume=650 mm^3; 1st generation spontaneous: N=6, volume=430 mm^3

tumors ($p = 0.18$), the spontaneous tumors demonstrated statistically higher HbO_2 levels in both Zones 2 and 3 ($p = 0.003$ and 0.029, respectively). 1st generation transplants were somewhere in between and not significantly different from either KHT or spontaneous tumors.

Total and perfused vascular configurations were quantified by measuring the distribution of distances between tumor cells and the nearest blood vessel. Total blood vessels were recognized using the PECAM-1/CD31 endothelial cell marker and perfused vessels, using the *i.v.* injected $DiOC_7(3)$ perfusion marker. Fig. 2 illustrates changes in median distances to both total (anatomical) and perfused blood vessels for sequential sections through two spontaneous tumors. Note that this parameter is inversely related to vascular density. For tumor A, median distance to the nearest total vessel (open bars) remained constant at about 30 μm among successive sections. However, median distance to the nearest perfused vessel (cross-hatched bars) increased markedly, from 49 μm near the tumor center (Section A) to 164 μm near the tumor periphery (Section E). In tumor B, distances to total vessels were again fairly constant throughout the tumor. In this tumor, however, perfused vessel distances varied less substantially and tended to decreased towards the tumor surface, rather than increasing as in A. For each of the remaining figures, tumors were quantified based on a single tumor section (which included four 4×4 image montages), taken at a central location similar to that of Section A

Figure 3 presents total and perfused distances for five KHT, five spontaneous, and five 1st generation spontaneous tumors. For KHT tumors (Figure 3A), distances to total vessels remained fairly constant among tumors, while distances to perfused vessels varied markedly, indicating substantial inter-tumor differences in perfusion. For spontaneous tumors (Figure 3B) and 1st generation tumors (Figure 3C), in contrast, distances to total vessels were more variable. In addition, more of the vessels were perfused (as indicated by reduced differences between total and perfused bars). Figure 4A summarizes these results, highlighting the disparities between median total and perfused distances in these tumor models. To better compare relative vascular function among these tumors, Figure 4B replots the data in terms of the mean ratio of median perfused distance to total distance. Thus, a ratio of 2.5 indicates that the median distance to a perfused vessel is more than twice the distance to an anatomical vessel. Although equivalent for spontaneous and 1st generation tumors ($p = 0.36$), this ratio was significantly increased for the KHT tumors ($p = .0001$ and $p = .00001$, compared to spontaneous and 1st generation, respectively). These results indicate that the spontaneous tumors were better perfused than the KHT to begin with, and were able to maintain this improved perfusion in the 1st generation transplants.

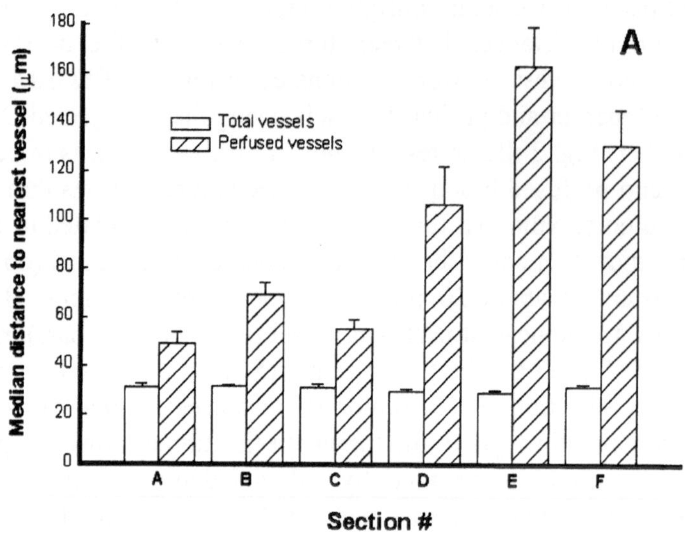

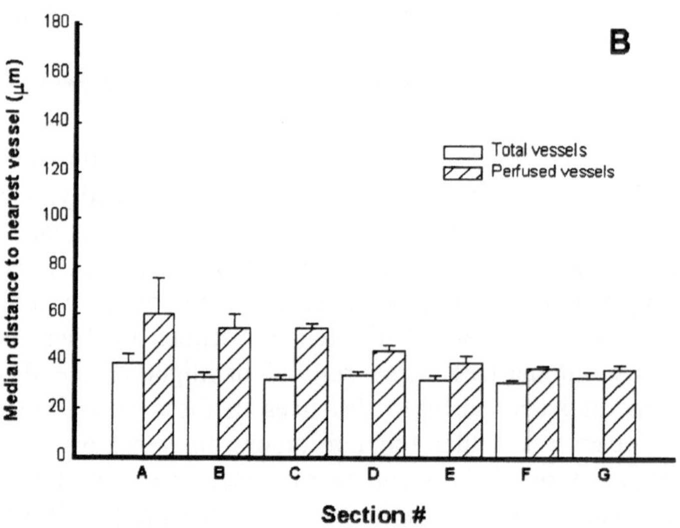

Figure 2. Median distance to nearest blood vessel (mean ± s.e.) across multiple sections of two spontaneous tumors. Open bars correspond to total anatomical vessels and cross-hatched bars to perfused vessels. Four 16 image montages were taken for each tumor section. A) 500 mm^3 tumor, 700 μm between sections, section F is nearest the tumor periphery, section A is near the tumor center; B) 450 mm^3 tumor, 1400 μm between sections, section G is nearest the tumor periphery.

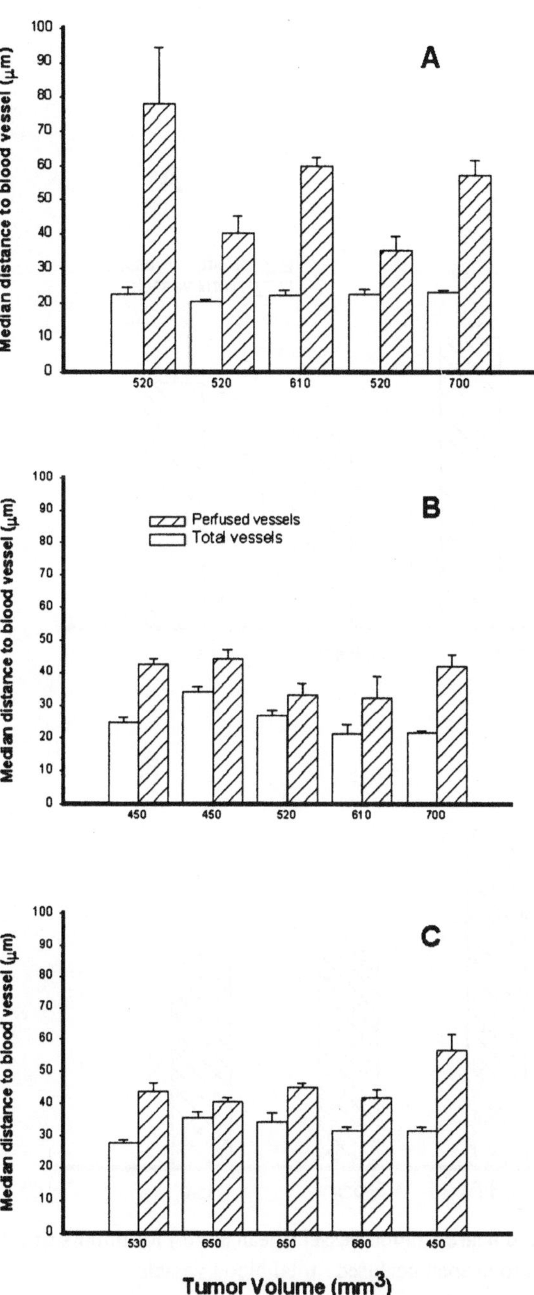

Figure 3. Median distance to nearest blood vessel for individual tumors. Four 16 image montages were taken for each tumor. A) KHT tumors (n=5, volume=565 mm^3), B) spontaneous tumors (n=5, volume=545mm^3), C) 1st generation spontaneous tumors (n=5, volume=590 mm^3).

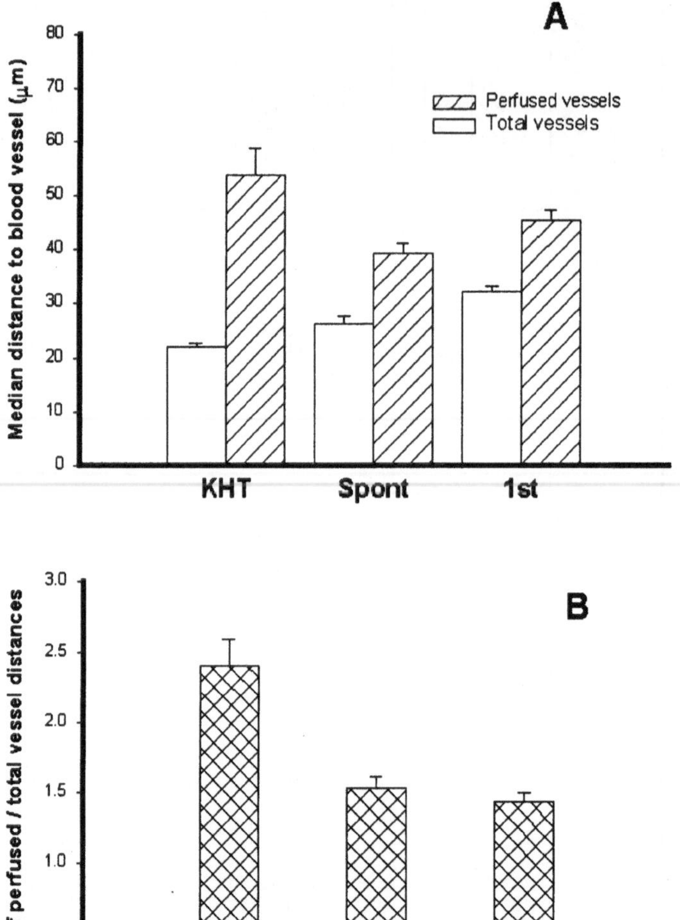

Figure 4. A) Median distance to nearest blood vessel (mean ± s.e.) for tumors of figure 3; B) Mean ratio of median distances to nearest perfused / total blood vessels.

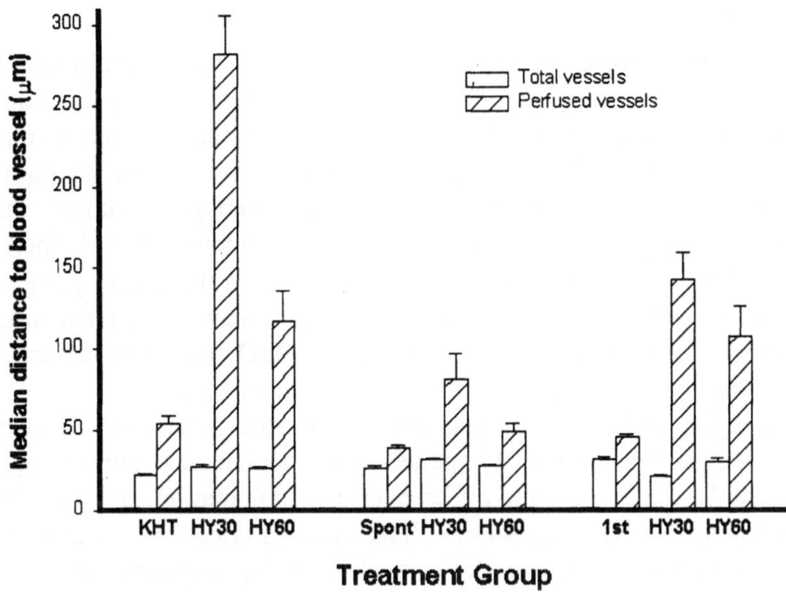

Figure 5. Median distance to nearest blood vessel (mean ± s.e.) for control KHT (N=5,volume=550 mm^3, 30 min post-HYD KHT (N=5,volume=500), 60 min post-HYD KHT (N=8,volume=500), control spontaneous (N=5,volume=640), 30 min post-HYD spontaneous (N=5,volume=390), 60 min post-HYD spontaneous (N=3,volume=610), control 1st generation (N=5,volume=680), 30 min post-HYD 1st (N=7,volume=430), and 60 min post-HYD 1st (N=7,volume=440).

Figure 5 illustrates changes in total and perfused distances at 30 and 60 min following hydralazine administration (a vasodilator that generally results in decreased tumor blood flow [2]). For the KHT tumors, HYD produced a marked decrease in the number of perfused blood vessels at 30 min (as indicated by the substantial increase in the perfused vessel distances in comparison to controls, $p<0.0001$). By 60 min post-HYD, this effect was somewhat less evident, but still significantly different from controls (p = 0.013). For spontaneous tumors, effects on perfusion were much less striking at 30 min ($p = 0.012$), and disappeared entirely by 60 min ($p = 0.55$). Finally, although 1st generation perfused distances were less than for KHT at 30 min ($p < 0.0001$) and equivalent by 60 min ($p = 0.72$), distances were greater than for spontaneous ($p < 0.001$ and $p = 0.004$, respectively). These data indicate that the 1st generation response to the drug was somewhere in between the responses of the spontaneous and KHT tumors.

4. DISCUSSION

The primary objective of this study was to compare transplanted and spontaneous tumor models in terms of vascular structure and function. As has been previously reported [8], tumor intravascular oxygen availability decreases with increasing distance from the tumor surface in both fast-growing, transplanted, murine and human xenograft tumor models. Furthermore, this trend becomes even more pronounced with increasing tumor volume. In spontaneous tumors, this trend is much less pronounced, and oxygen availability in the tumor center is somewhat better maintained. In addition, heterogeneities in intravascular HbO_2 distributions are much more striking among individual primary spontaneous tumors.

Although HbO_2 levels were not significantly different between the subcutaneously implanted 1st generation spontaneous tumors and the intramuscular KHT tumors, differences in implantation site may be the primary factor. Previous studies [9] have demonstrated that *s.c.* implantation results in significantly lower HbO_2 levels than *i.m.* implantation for similar volume KHT tumors, possibly due to a decreased number of preexisting host vessels available at the *s.c.* site. Thus, if the 1st generation spontaneous tumors are compared to similarly implanted *s.c.* KHT tumors, significantly higher HbO_2 levels are found in the 1st generation tumors.

Based on fluorescent and immunohistochemical staining, total anatomical blood vessel spacing remains fairly uniform throughout spontaneous mammary tumors, while perfused vessel spacing can vary markedly with distance from the tumor surface both within and among individual tumors. On average, distances to the nearest total blood vessel are significantly reduced for the KHT tumors in comparison to either the spontaneous (p = 0.009) or 1st generation tumors ($p < 0.0001$), indicating an increase in total vasculature in the KHT tumors. Despite this increase in available vessels, perfused vessel densities are much lower in the KHT (as evidenced by the increased perfused vessel distances). Thus, although the spontaneous tumors have fewer vessels to begin with, the proportion of perfused vessels is increased. This improvement in perfusion is similarly maintained in the 1st generation transplants, where the ratio of perfused/total vessel distances are equivalent to that of primary tumors.

Previously reported comparisons of the response to HYD administration between spontaneous and transplanted tumor models have been somewhat contradictory. Although this drug has generally been shown to substantially decrease tumor blood flow and oxygenation in transplanted animal tumors [2,10,11], in spontaneous tumors the effect is less clear. Some ^{31}P magnetic resonance spectroscopy studies have reported that spontaneous tumors were unresponsive [12,13], while more recent studies have reported that HYD can indeed decrease both oxygenation and ^{31}P-MRS energetics in spontaneous tumors [2,14]. Although the current study demonstrates a significant decrease in numbers of perfused vessels in spontaneous tumors following HYD, the effect is not nearly as striking nor prolonged as that observed in the KHT tumors. In the transplanted spontaneous tumors, the effects of HYD were somewhere in between the other two tumor models: significantly increased in comparison to the spontaneous tumors, but not as pronounced as in the KHT.

In summary, distances between tumor cells and the nearest anatomical or perfused blood vessel were similar between the primary spontaneous and 1st generation transplanted spontaneous tumor models, as was overall vascular morphology and tissue histological appearance. Although functional response to vasoactive agents was not identical in the two tumor models, 1st generation tumors were more closely related to parent primary tumors than to the fast-growing KHT tumors. In addition, oxygen availability was not significantly different between spontaneous and 1st generation, as evidenced by HbO_2 saturation distributions. Taken together, these results indicate that spontaneous tumor vascular configuration and function are preserved in the 1st generation transplanted tumors and support the validity of using transplanted spontaneous tumors as a model for primary tumor vascular structure and function. Although these types of tumor models are more expensive and time-consuming to acquire and propagate, vascular structure is clearly dependent on specific choice of model. Thus, fast-growing, transplantable tumors may result in quite different and possibly incorrect conclusions with regards to primary tumor functional response.

Ongoing studies are comparing later generations of the transplanted spontaneous tumors to determine whether functional characteristics can be further maintained. If this is indeed the case, such transplants will permit larger scale experimental studies to be performed, including growth delay assays or multi-timepoint studies that require larger tumor numbers.

ACKNOWLEDGMENTS

Financial support was provided by NIH grant CA52586.

REFERENCES

1. Folkman J. Mendelsohn J, Howley PM, Israel MA, Liotta LA, editors. The Molecular Basis of Cancer. Philadelphia: WB Saunders; 1995; Tumor angiogenesis. p. 206-232.

2. Nordsmark M, Maxwell RJ, Wood PJ, Stratford IJ, Adams GE, Overgaard J, Horsman MR. Effect of hydralazine in spontaneous tumours assessed by oxygen electrodes and P-31-magnetic resonance spectroscopy. Br.J.Cancer 1996;74:S 232-S 235
3. Trotter MJ, Chaplin DJ, Olive PL. Use of a carbocyanine dye as a marker of functional vasculature in murine tumours. Br.J.Cancer 1989;59:706-709.
4. Fenton BM, Boyce DJ. Micro-regional mapping of HbO_2 saturations and blood flow following nicotinamide administration. Int.J.Radiat.Oncol.Biol.Phys. 1994;29:459-462.
5. Fenton BM, Gayeski TEJ. Determination of microvascular oxyhemoglobin saturations using cryospectrophotometry. Am.J.Physiol. 1990;259:H1912-H1920
6. Fenton BM, Paoni SF, Lee J, Koch CJ, Lord EM. Quantification of tumor vascular development and hypoxia by immunohistochemical staining and HbO_2 saturation measurements. Br.J.Cancer 1999;79:472-477.
7. Fenton BM, Way BW. Vascular morphometry of KHT and RIF-1 murine sarcomas. Radiother.Oncol. 1993;28:57-62.
8. Fenton BM, Rofstad EK, Degner FL, Sutherland RM. Cryospectrophotometric determination of tumor intravascular oxyhemoglobin saturations: Dependence on vascular geometry and tumor growth. J.Natl.Cancer Inst. 1988;80:1612-1619.
9. Fenton BM. The effects of carbogen and nicotinamide on intravascular HbO_2 saturations in SCCVII and KHT murine tumours. Br.J.Cancer 1995;71:945-949.
10. Dewhirst MW, Madwed D, Meyer RE, et al. Reduction in tumour blood flow in skin flap tumour after hydralazine is not due to a vascular steal phenomenon. Radiat.Oncol.Invest. 1994;1:270-278.
11. Chaplin DJ, Acker B. The effect of hydralazine on the tumor cytotoxicity of the hypoxic cell cytotoxin RSU-1069. Evidence for therapeutic gain. Int.J.Radiat.Oncol.Biol.Phys. 1987;13:579-585.
12. Field SB, Burney IA, Needham S, Maxwell RJ, Coggle J, Griffiths JR. Are transplanted tumours suitable as models for studies on vasculature? Int.J.Radiat.Biol. 1991;60:255-260.
13. Wood PJ, Stratford IJ, Sansom JM, Cattanach BM, Quinney RM, Adams GE. The response of spontaneous and transplantable murine tumors to vasoactive agents measured by 31P magnetic resonance spectroscopy. Int.J.Radiat.Oncol.Biol.Phys. 1992;22:473-476.
14. Horsman MR, Nordsmark M, Hoyer M, Overgaard J. Direct evidence that hydralazine can induce hypoxia in both transplanted and spontaneous murine tumours. Br.J.Cancer 1995;72:1474-1478.

Chapter **17**

EFFECT OF MILD HYPERGLYCEMIA ± META-IODO-BENZYLGUANIDINE ON THE RADIATION RESPONSE OF R3230 AC TUMORS

Intae Lee[1], Jerry D. Glickson[2], Mark W. Dewhirst[3], Dennis B. Leeper[4], Randy Burd[4], Harish Poptani[2], Lydie Nadal[2], W. Gillies McKenna[1] & John E. Biaglow[1]

[1]Department of Radiation Oncology and [2]Department of Radiology, University of Pennsylvania School of Medicine, Philadelphia, PA 19104; [3]Department of Radiation Oncology, Duke University Medical Center, Durham, NC 27710; [4]Department of Radiation Oncology, Thomas Jefferson University, Philadelphia, PA 19107.

Abstract: The effects of glucose or meta-iodo-benzylguanidine (MIBG) on oxygen utilization (QO_2) of several tumor cell lines were studied using a Clark-type electrode chamber. For *in vivo* studies, rats bearing R3230 Ac rat mammary adenocarcinomas were utilized. To evaluate changes in tumor oxygenation induced by glucose or MIBG, intratumoral pO_2 and skeletal muscle pO_2 were measured using Eppendorf Histography. To find the effect of mild hyperglycemia (i.p., 1 g/kg) ± MIBG (i.p., 20 mg/kg) on the radiation response, a growth delay assay was used. Glucose alone produced a ~20% inhibition of QO_2 in several tumor cells we tested except Q7 tumor cells. MIBG inhibited QO_2 in R3230 Ac tumors. The median tumor pO_2 for glucose + MIBG was increased from 5.3 mm Hg to 13.8 mm Hg. We hypothesized that combined treatment with glucose + MIBG significantly enhanced radiation-induced tumoricidal effects on R3230 Ac tumors, mainly due to reduction in QO_2 and increase in tumor pO_2.

Key words: glucose, meta-iodo-benzylguanidine, QO_2, pO_2, radiation response

Abbreviations: meta-iodo-benzylguanidine (MIBG)

1. INTRODUCTION

Tumor cells become hypoxic when the rate of O_2 usage is greater than the rate of delivery. It is well-documented that several factors determine O_2 utilization (QO_2) such as pO_2, blood flow, the availability of oxidizable substrates (i.e., glucose, glutamine), and O_2 consumption. Our present interest was to demonstrate that transient changes in metabolic flux by the addition of glucose can decrease tumor cell oxygen uptake. Thereby, enhancing radiation response, due to the diffusion of O_2 to previously hypoxic tumor cells. Biaglow et al. was the first to demonstrate an increase in radiation sensitivity for cultured tumor cells resulting from inhibition of QO_2 with glucose [1]. However, as early as 1929, Crabtree reported the inhibition of respiration induced by addition of glucose to cells [2]. This inhibition is called either the reversed Pasteur effect or Crabtree effect. The Crabtree effect is due to competition between glycolysis and respiration for adenosine diphosphate [3]. Hyperglycemia induced by the addition of glucose significantly reduces tumor blood flow. Increased glycolysis results in tumor acidification [4]. Our intent is to improve the radiation response without significant alterations in tumor physiological parameters. We hypothesized that mild hyperglycemia, defined as "not inducing significant decreases in TBF by a glucose administration"; would enhance glycolysis and inhibit cellular respiration, thereby inducing an O_2 sparing effect enhancing the radiation response.

To further the reduction in QO_2, glucose was combined with a functional analogue of the neurotransmitter norepinephrine, meta-iodo-benzylguanidine (MIBG). MIBG is known to reduce QO_2 as well as intracellular ATP levels by impairment of mitochondrial respiration [5]. The guanidine moiety of the MIBG was responsible for its cytotoxic and anti-tumoral effects [6]. To test the effect of either glucose or MIBG on the inhibition of QO_2, nine tumor cell lines, (R3230 Ac, DU-145, PC-3, LNCaP, RIF-1, 9L, Q7, MCF-7, A549) were studied using a Clark-type electrode chamber.

Inhibition of O_2 consumption *in vitro* by glucose suggested similar results would occur *in vivo*. Therefore, we evaluated changes in tumor oxygenation induced by glucose, MIBG, glucose + MIBG, and intratumoral pO_2 in R3230 Ac rat mammary adenocarcinomas which were measured using an Eppendorf Histograph. Additionally, we measured pO_2 in skeletal muscles and the kidney. Based on both QO_2 and pO_2 studies, we examined the effect of mild hyperglycemia ± MIBG on radiation-induced responses in *vivo* of R3230 Ac tumors using a growth delay assay

2. MATERIALS & METHODS

2.1 Measurements in oxygen consumption rate using a Clark-type electrode chamber

The nine tumor cell lines for QO_2 were R3230 Ac rat mammary adenocarcinoma, three human prostate adenocarcinomas (DU145, PC-3, LNCaP), RIF-1 mouse fibrosarcoma, 9L rat glioma, Q7 rat hepatoma, MCF-7 human breast adenocarcinoma, and A549 human lung adenocarcinoma. The cells were in the confluent (plateau growth-phase) levels, and the preparation of single cell suspensions of tumor cells was done with a cell scraper without adding a trypsin solution. QO_2 measurements were made at 37 °C with a Clark oxygen electrode, using a constant temperature water bath and an amplifier (Yellow Springs Instruments Co., Yellow Springs, OH). The procedure for the QO_2 measurements has been detailed previously [1].

2.2 Animals and tumor implantation

Female Fisher rats ~100 gr. bearing R3230 Ac rat mammary adenocarcinomas were used. Prior to tumor inoculation, rats were exposed to 2 Gy irradiation using a cesium irradiator. The institution's guidelines for the care and use of laboratory animals were followed. Tumors were grown by inoculating, subcutaneously, a suspension of 10^6 cells in 100 μl of 0.9% NaCl solution on the hind leg of rats. The tumors were selected for use, except when otherwise stated, at a mean diameter of 9~10 mm. Tumor volumes were calculated using the formula $V=0.4 \times ab^2$ as the longer and shorter caliper diameters of the tumor, respectively [7].

2.3 Anesthesia

Rats were anesthetized i.m. with ketamine (90 mg/kg) and acepromazine (9 mg/kg). Rats were placed on a heating pad to keep the body temperature ~37.5 C maintained by monitoring their rectal temperature using a Type-T thermocouple and BAT-10 thermometer (Physitemp Inc., Clifton, NJ).

2.4 X-irradiation procedure

Each lightly anesthetized rat bearing R3230 Ac tumors was put inside a plexiglass irradiation jig, which was placed on an irradiation stage. The stage was situated inside of a 250 keV X-irradiator (RT250, Philips) with a source-surface distance of 30 cm. Normal tissues cover shielded. A single dose of various radiation exposure (dose rate: 3.4 Gy per min) was delivered.

2.5 Radiation-induced growth delay assay after treatment with glucose ± MIBG

Prior to x-irradiation, glucose (1 g/kg) or/ and MIBG (20 mg/kg) was administered via an intraperitoneal route (i.p.). After x-irradiation, tumor dimensions were measured every other day with calipers. The end point was defined as the time for the tumor volume to exceed four times the initial tumor volume.

2.6 Measurements of pO_2 in skeletal muscles, kidneys, and tumors

Oxygen tension in tumors, skeletal muscle, and kidney was measured. The needle-type electrodes (commercially available with Eppendorf Histography, Hamburg, Germany) were calibrated following at least 3 cycles of 2 min of air (20.9% of O_2) and 5 min of nitrogen (0% of O_2) bubbling. The steplength and the overstroke were set at 0.4 and 0.3 mm, respectively. Thus, the needle automatically moved 0.7 mm forward and then 0.3 mm backward every 1.4 second, when making a pO_2 determination. After shaving, an electrocardiogram patch was attached to the abdominal wall as an anode. A 23-gauge needle was used to make a small hole on the surface of the tumors, the skin of hind legs, or the capsule of left-side kidneys, and the tip of the needle-electrode was placed in the hole. This procedure was detailed previously [8].

2.7 Statistical evaluation

All measured values were shown as mean ± standard error of each group, and t-test was applied for statistical evaluations. To analyze the oxygen status in tissues, all pO_2 values obtained from several tumors of experimental group were grouped, and were evaluated using Mann-Whitney U-test. P values smaller than 0.05 were considered as significant.

3. RESULTS

Figure 1 shows relative oxygen utilization of various tumor cell lines after treatment with glucose at 16.7 mM. We have observed large variability in QO_2 (nmoles O_2/ min/ mg protein) for tumor cells grown *in vitro*. In

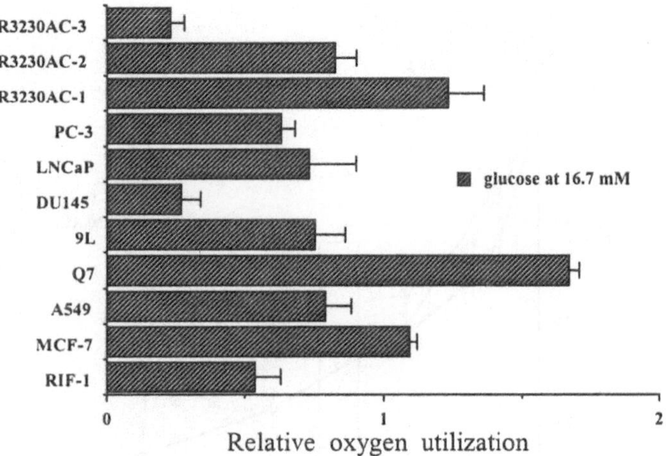

Figure 1. Effect of glucose at 16.7 mM on oxygen consumption during the plateau growth phase of nine tumors *in vitro*. The reaction medium was DPBS, and contained 1mM Ca^{++}, 1 mM Mg^{++}, 20 mM Pi, pH 7.3 at 37 °C. Each cell line was the average of at least 5 independent samples before and after glucose administration into the sample chamber. In the case of R3230 Ac tumors, 1= DPBS-pH 7.3, 2= Hepes-pH 7.3, 3= Hepes-pH 6.7.

general, glucose alone produced a ~20% inhibition of QO_2 in several tumor cells incubated in Dulbecco's phosphate buffered saline (DPBS) except Q7 and R3230 Ac tumor cells. Glucose concentration (16.7 mM) significantly inhibited the respiration of all human prostate tumor cell lines (i.e., PC-3, LNCaP, DU145). The rodent tumors RIF-1 and 9L as well as the human A549 lung carcinoma also showed the Crabtree effect. In contrast, Q7 and R3230 Ac showed the stimulation of respiration by an elevated glucose concentration. Interestingly, glucose inhibition of cellular respiration rates in R3230 Ac tumor cells was dependent on the test media. When R3230 Ac tumor cells were incubated in Hepes buffered saline at pH 6.3 or 7.3, respiration was significantly inhibited by glucose. Glucose did not inhibit O_2 uptake in DPBS. The major difference between two media was the level of phosphate.

As shown in Figure 2-A, MIBG also significantly inhibited QO_2 in RIF-1, R3230 Ac, and 9L tumors. The decrease in QO_2 with MIBG was concentration dependent for all tumor cell lines we tested (data not shown).

In 9L tumors, the QO_2 values significantly decreased with increased exposure time to 0.15 mM MIBG. However, in RIF-1 and R3230 Ac tumors, the QO_2 values decreased with MIBG exposure up to 10 min, then stabilized between 10 and 20 min. Figure 2-B shows the inhibition of QO_2 in several tumor cell lines induced by MIBG at 0.15 mM after 15 min

A

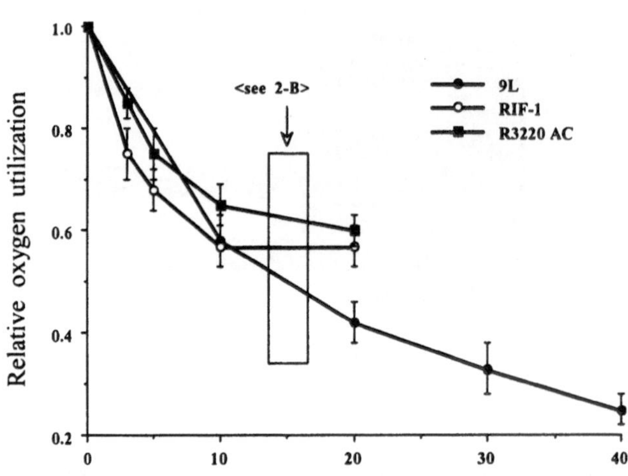

B

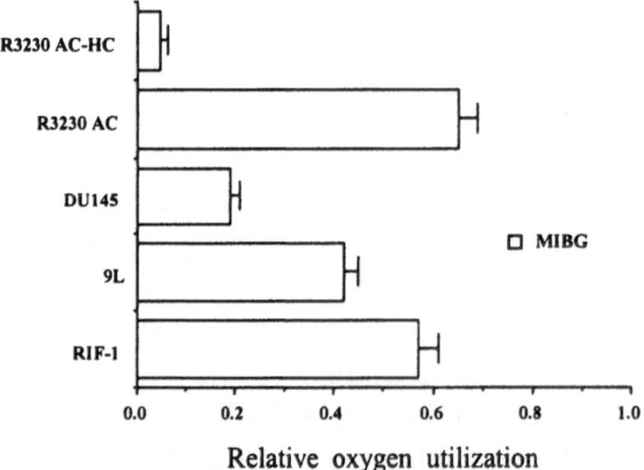

Figure 2. The reaction medium was DPBS, and contained 1mM Ca^{++}, 1 mM Mg^{++}, 20 mM Pi, pH 7.3 at 37 °C. Each cell line was the average of at least 5 independent samples before and after MIBG administration into the sample chamber.

A. Effect of MIBG at 0.15 mM on oxygen consumption during the plateau growth phase of three tumors *in vitro*, as a function of MIBG exposure time.

B. Effect of MIBG on oxygen consumption during the plateau growth phase of four tumors *in vitro*. The QO_2 values were determined at 15 min after exposure to MIBG. In the case of R3230 Ac tumors, HC= the higher concentration of MIBG (0.3 mM).

exposure. Treatment with MIBG at 0.15 mM was the most effective on DU145. In R3230 Ac tumors 0.15 mM of MIBG reduced QO_2 by 40%. The QO_2 was completely inhibited when MIBG was increased to 0.3 mM

Based on our *in vitro* QO_2 studies, we expanded to *in vivo* studies using rats bearing R3230 Ac tumors. Table 1 shows the results of tissue oxygenation of tumors, skeletal muscle, and kidney after an administration of glucose, MIBG, and glucose + MIBG. An i.p. administration of glucose at a dose of 1 g/kg increased median pO_2 in tumors from 3.7 to 11.2 mm Hg. Median pO_2 values in skeletal muscle or kidney were not significantly increased by hyperglycemia. An i.p. administration of MIBG at 20 mg/kg improved tumor oxygenation from 3.3 mm Hg to 14.3 mm Hg without changes in pO_2 of skeletal muscle. The median tumor pO_2 for the control group was 3.7 mm Hg, while the combined treatment with glucose + MIBG increased to 13.8 mm Hg.

Table 1. Oxygen tension in tissues before and after various treatments*

	tumors	skeletal muscles	kidneys
Glucose <before>	3.7	30.4	17.2
Glucose <after>	11.2 [b]	33.6 [a]	17.8 [a]
MIBG <before>	3.3	22.8	n.d.
MIBG <after>	14.3 [b]	22.7 [a]	n.d.
Glucose + MIBG <before>	5.3	25.5	n.d.
Glucose + MIBG <after>	13.8 [b]	19.2 [a]	n.d.

*: unit= mm Hg, median value, N= 4 - 5 rats per group. n.d.: not determined.
Treated groups were compared to the control (=before treatment) for their own treatment protocols (i.e., glucose, MIBG, or glucose + MIBG). For each comparison Mann-Whitney U-test was applied. [a] vs. before treatment, not significant; [b] vs. before treatment, $p<0.05$.

A single i.p. injection of either glucose or MIBG in rats 24 hr food-deprived prior to x-irradiation did not show the retardation of tumor growth compared with the radiation alone (Figure 3). However, combined treatment with glucose + MIBG significantly inhibited tumor growth (i.e., ~8 days growth delay compared with 5 Gy alone). Thus, the treatment with RT and glucose + MIBG was more than additive.

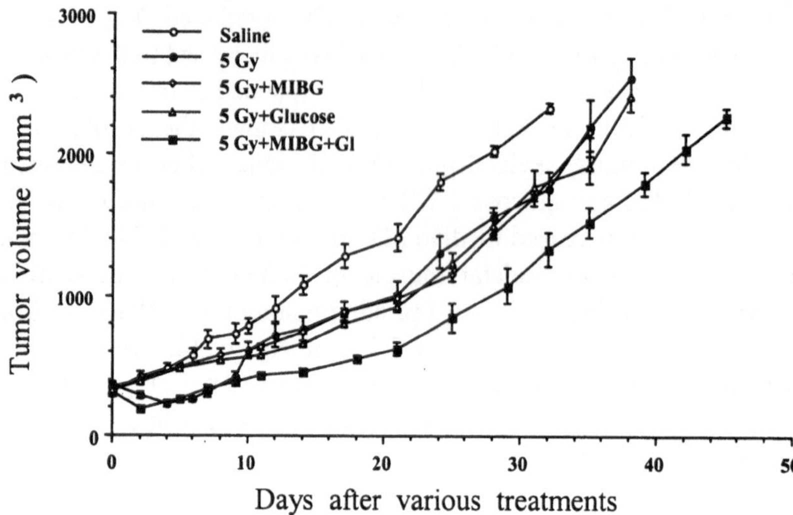

Figure 3. Radiation-induced tumor growth curves after various treatments (N= 6 rats per group).

4. DISCUSSION

The first goal of this study was to evaluate changes in QO_2 rate induced by the addition of glucose during the plateau growth phase of various tumors *in vitro*. Using the temporary metabolic alteration through the use of the Crabtree effect (i.e., the inhibition of respiration by the addition of glucose), it was possible to decrease the rate of QO_2 in various tumor cell lines in our present study. In the case of R3230 Ac tumors, the addition of phosphate influenced the effect of glucose on the rate of respiration of the cells, as shown in Figure 1. The Crabtree effect was observed when R3230 Ac tumor cells were in Hepes buffered saline, but not in DPBS. It was in agreement with other investigators who have shown a partial reversal of glucose inhibition by the addition of phosphate, even though the effect of phosphate appears variable [9]. Much of the stimulatory effect which phosphate had been reported to cause was due to the effect of phosphate ion on glycolysis, but not to a direct effect on respiration. Therefore, the influence of a phosphate effect with R3230 Ac tumor cells suggested that the phosphate concentration may be limited in this system. More detailed experiments regarding the phosphate effect on QO_2 are in progress.

Our second goal was to examine MIBG effects on the QO_2 for plateau phase cultures of various tumors *in vitro*. As shown in Figures 2-A & -B, the effect of MIBG on QO_2 was time- and dosage- dependent. QO_2, in all four cell lines was effectively inhibited. Biaglow *et al.* [10] reported that MIBG inhibited O_2 consumption resulting in the stimulation of glycolysis. Increasing acidification occurred due to increases in lactic acid production. Also, MIBG primarily affected mitochondrial ATP production. Increased glycolysis was necessary to maintain ATP levels. Therefore, decreased energy production as well as acidification may have altered the response of tumor cells to ionizing radiation.

Higher doses of glucose have resulted in significant alterations in glucose metabolism and respiration in tumor cells. Hyperglycemia induced by an i.p. glucose loading increased hypovolemic hemoconcentration, due to a significant osmotic water shift from the intravascular space into the abdominal cavity [4]. Consequently, venous return was impaired, and it caused a significant rise in water content in tumors, which led to an increase in tumor interstitial fluid pressure (TIFP) [11]. While mild hyperglycemia may have benefited from changes in glucose metabolism and the Crabtree effect, it did not cause any significant changes in water content. Furthermore, glucose-altered radiation sensitivity may be exploited in animal tumor systems where the glucose concentration is known to be low [12]. The tumor oxygenation was significantly improved by glucose alone, MIBG alone, and combination of glucose and MIBG, as shown in Table 1. However, tissue oxygen tensions in normal organs, such as skeletal muscle or kidney, were not influenced by any of the treatments. From our preliminary results, changes in tumor physiological parameters induced by glucose or MIBG were not significant. MIBG at 20 mg/kg slightly increased TBF with high variability, and decreased TIFP from 14 mm Hg to 11.5 mmHg within 1 hr's observation (Lee I. *et al.*, unpublished observations, 1999). Thus, we concluded that improved tumor oxygenation by MIBG was mainly due to a decrease in cellular QO_2, rather than alterations in availability of O_2.

Our final goal in this investigation was to find whether an application of glucose, MIBG, and combination of glucose + MIBG, without significant alterations in tumor physiological parameters, improved radiation responses. Mild hyperglycemia enhanced glycolysis and inhibited cellular respiration, thereby inducing an O_2 sparing effect. Using Eppendorf Histography, we observed the improvement in tumor oxygenation by glucose, MIBG, and combination with glucose + MIBG. Median pO_2 in the combination of glucose + MIBG was the same as that in either of the treatments alone (Table 1). In contrast, a single i.p. injection of either glucose or MIBG in rats 24 hr food-deprived prior to x-irradiation did not show the retardation of tumor growth compared with the radiation alone (Figure 3). However, we observed that multiple fractions of x-irradiation with multiple i.p. loading of glucose inhibited tumor growth (further studies using various dosages of glucose or MIBG are in progress). We concluded that combined treatment with glucose + MIBG significantly inhibited radiation-induced tumor growth, mainly due to reduction in QO_2 and increase in tumor pO_2. The improvement in the radiation response may have been related to altered glycolysis and decreased tumor pH, resulting in increased radiation-induced apoptosis (Biaglow JE, *et al.*, unpublished observations, 1999) or altered in PLDR. We are pursuing additional research in this area.

REFERENCES

1. Biaglow JE, Lavik P, Ferencz N. Modification of radiation response through glucose-controlled respiration. Radiat Res 1969;39:623-633.
2. Crabtree HG. Observation on the carbohydrate metabolism of tumours. Biochem J 1929; 23:536-545.
3. Chance B. Phosphorylation efficiency of the intact cell. J Biol Chem 1959;234:3036-3043.
4. Vaupel PW, Okunieff PG. Role of hypovolemic hemoconcentration in dose-dependent flow decline observed in murine tumors after intraperitoneal administration of glucose or mannitol. Cancer Res 1988;48:7102-7106.
5. Loesberg C, Van Rooij H, Nooijen WJ, Meiher AJ, Smets LA. Impaired mitochondrial respiration and stimulated glycolysis by m-Idobenzylbuanidine (MIBG). Int J Cancer 1990; 46:276-281.
6. Smets LA, Bout B, Wisse J. Cytotoxic and antitumor effects of the norepinephrine analogue meta-iodo-benzylguanidine (MIBG). Cancer Chemother Pharmacol 1988;21:9-13.
7. Lee I, Boucher Y, Jain RK. Nicotinamide can lower tumor interstitial fluid pressure: mechanistic and therapeutic implications. Cancer Res 1992;52:3237-3240.
8. Nozue M, Lee I, Manning JM, Manning LR, Jain RK. Oxygenation in tumors by modified hemoglobin. J Surg Oncol 1996;62:109-114.
9. Ibsen KH. The Crabtree effect: a review. Cancer Res 1961;21:829-841.
10. Biaglow JE, Manevich Y, Leeper D, Chance B, Dewhirst MW, Jenkins WT, Tuttle SW, Wroblewski K, Glickson JD, Stevens C, Evans SM. MIBG inhibits respiration: potential for radio- and hyperthermic sensitization. Int J Radiat Oncol Biol Phys 1998;42:871-876.
11. Lee I, Lee YH. The effect of various therapeutic solutions including colloidal chromic ^{32}P via an intratumoral injection on the tumor pathophysiological parameters of AsPC-1 human pancreatic tumor xenografts in nude mice. Clinical Cancer Res 1999; 5:3139S-3124S.
12. Gullino PM, Yi PN, Grantham FH. Relationship between temperature and blood supply or consumption of oxygen and glucose by rat mammary carcinomas. J Natl Cancer Inst 1966;60:835-847.

Chapter **18**

EFFECT OF ONCONASE ± TAMOXIFEN ON ASPC-1 HUMAN PANCREATIC TUMORS IN NUDE MICE

Intae Lee[1,2], Young H. Lee[2], Stanislaw M. Mikulski[3] & Kuslima Shogen[3]
[1]Department of Radiation Oncology, University of Pennsylvania School of Medicine, Philadelphia, PA 19104; [2]Division of Radiation Research, UMDNJ-Robert Wood Johnson Medical School, Camden, NJ 08103; [3]Alfacell Corporation, Bloomfield, NJ 07003

Abstract: To evaluate changes in tumor physiological parameters after treatment with ONCONASE (ONC), we used laser Doppler flowmetery for tumor blood flow (TBF) in legs of nude mice bearing AsPC-1 human pancreatic carcinoma. Tumor interstitial fluid pressure (TIFP) was measured using the wick-in-needle technique, while intratumoral pO_2 utilized O_2 sensitive needle electrodes. TBF was significantly increased by an i.p. injection of ONC, then returned to the untreated levels at 150 min post-treatment. Single and multiple intraperitoneal injections of ONC significantly reduced TIFP (post-treatment at 1-7 days). ONC significantly improved tumor oxygenation evidenced by an increase in median pO_2 from 4.2 mm Hg to 8.2 mm Hg. Since ONC had a synergistic interaction with tamoxifen (TAMX) on the cytotoxic clonogenicity of AsPC-1 tumor cells *in vitro*, we evaluated the antitumoral effects *in vivo* by ONC ± TAMX, using nude mice bearing AsPC-1 pancreatic tumor cells of ascite (intraperitoneally implanted) or solid (subcutaneously implanted to legs) tumors. ONC alone effectively retarded the tumor growth, but TAMX alone did not. ONC + TAMX was synergistically effective in inhibiting tumor growth.

Key words: ONCONASE, tamoxifen, tumor blood flow, tumor interstitial fluid pressure, intratumoral pO_2

Abbreviations: ONCONASE® (ONC), tamoxifen (TAMX), tumor blood flow (TBF), tumor interstitial fluid pressure (TIFP).

1. INTRODUCTION

Malignant tumor cells die by apoptosis or necrosis in response to cytotoxic treatments with drugs and radiation depending on the severity of the treatment. The failure of anticancer treatment is due in part to the intrinsic resistance of malignant tumor cells to radiation and chemotherapy. Various tumor microenvironments can also alter tumor cells and response to treatment. The results from preclinical and clinical trials have not been encouraging and indicate that the delivery of chemotherapeutic treatments to solid tumors for their effectiveness are hindered due to the physiological barriers [1]. One main problem is that all systemic anti-neoplastic agents must cross several barriers between the bloodstream and the tumor cells. First, the agent must distribute within the vascular compartment, and then cross the vascular wall and distribute throughout the interstitial space. Then, the agent must be transported across the cell membrane. However, an increase in the tumor interstitial fluid pressure (TIFP) hinders both transvascular and interstitial transport. It is well-documented that the elevated TIFP is a typical characteristic of most solid human and transplantable tumors [1-3]. However, no data are available for the pancreatic tumors in the literature.

An amphibian RNase-like substance, ONCONASE (ONC), isolated from *Rana pipiens* eggs and early embryos [4], has the capability of controlling tumor cell growth. This implicates some embryonic regulatory mechanisms, affecting tumor cell growth and differentiation. One potential mechanism of action leading to cell death is known to be the inhibition of protein synthesis by the inactivation of cellular tRNA [5]. ONC, then, exerts its effect on the proliferative cycle, resulting in the accumulation of cells in G_1 phase with concomitant decreases of the S and M + G_2 fractions [6].

Recently, we demonstrated with increasing tumor volume, an inverse correlation between pO_2 and TIFP in FSaII tumors [2]. Therefore, we hypothesized that if ONC could decrease resistance to blood flow, it could also lower tumor hypertension. Therefore, ONC was used as a TIFP-lowering agent, and TIFP were measured with the wick-in-needles. We also measured changes in tumor blood flow (TBF) using a laser Doppler flowmetry, and changes in intratumoral pO_2 by utilizing the O_2 sensitive needle electrodes. Thus, we could provide some insight and physiological mechanisms regarding the potential benefit of ONC for the clinical trials as a new therapeutic agent in advanced pancreatic carcinomas.

From our *in vitro* observations, AsPC-1 tumor cells were sensitive to ONC, and insensitive to the anti-estrogen agent tamoxifen (TAMX). Ascites and subcutaneous: two models of AsPC-1 tumor xenografts in nude mice were used for evaluation of the anti-tumoral effect of ONC *in vivo*. Mikulski *et al.* [7] reported that ONC had a synergistic interaction with several cytotoxic and/or cytostatic agents including TAMX on anti-proliferative and cytotoxic clonogenic responses of AsPC-1 tumors *in vitro*. The use of the two agents in combination was based on the rationale that the anti-estrogen activity coincide with the early G_1 phase inhibitory effect of protein kinase inhibitors, and that TAMX could act synergistically with the cell cycle inhibitory effect of ONC. However, no studies had been performed *in vivo* regarding the inhibitory effect of ONC in combination with TAMX on the growth of solid tumor models. Therefore, a study for the tumoricidal effects of treatment with ONC ± TAMX *in vivo* was warranted prior to further clinical evaluations.

2. MATERIALS AND METHODS

2.1 Animals and tumors

The institution's guidelines for the care and use of laboratory animals were followed. Eight to ten week-old, female nude mice (purchased from the Cox-animal facility, Massachusetts General Hospital, Boston, MA) bearing human tumor xenografts (AsPC-1 human pancreatic carcinoma cells) were utilized. Frozen AsPC-1 tumor cells (purchased from the American Type Culture Collection, Rockville, MD) were thawed, cultured, and grown *in vitro*. Single cell suspensions were prepared using 0.25% trypsin solution. ~1 x 10^6 viable cells suspended in 50 µl of dMEM medium were injected either intraperitoneally (i.p.) or subcutaneously (s.c.) into the right thighs of mice. Experiments were carried out when the tumor volume was between 100 and 2000 mm^3. Tumor volumes were calculated using the formula $V = 0.4 \times ab^2$, with a and b as the longer and shorter diameters of the tumor, respectively [2].

2.2 ONC treatment *in vivo*

ONC was dissolved in sterile 0.9% NaCl solution before the experiments. The mice were given either a single injection (i.p. or intratumorally: i.t.) or multiple injections (an i.p. injection per week for 4 consecutive weeks) of ONC (0-100 µg/ mouse) at a volume of 0.2 ml/ 20 g of body weight. Experiments were carried out when the tumor volume reached ~250 mm^3.

2.3 TAMX treatment *in vivo*

The stock solution of TAMX was dissolved in dimethyl sulfoxide (1 mg TAMX/ ml), and kept in the freezer. Prior to the experiments, the aliquots of TAMX were diluted with sesame oil. TAMX at 100 μg per mouse was given subcutaneously (s.c.) and daily for 30 days at a volume of 0.1 ml/ 20 g of body weight. Experiments were carried out when the tumor volume reached ~250 mm^3.

2.4 Anesthesia

Mice were anesthetized with ketamine (90 mg/kg) and xylazine (9 mg/kg) via i.m. route. Mice were placed on a heating pad to keep the body temperature ~37.5 C managed by monitoring their rectal temperature using a Type-T thermocouple and BAT-10 thermometer (Physitemp Inc., Clifton, NJ).

2.5 Measurements of systemic pressure and TIFP

After the right carotid artery was cannulated with a PE-10 catheter using a binocular microscope, the tubing was connected with a pressure transducer [7]. TIFP was measured with the wick-in-needle technique using 23 gauge needles overall with a side hole 2 mm from the tip. Then, all TIFPs were measured at 2 hr prior to Day 0 and Days 1, 4, 7 after the treatment. In separate experiments, TIFP and MABP were measured during 1 hr observation after treatment with ONC. Measurements were made by introducing wick-in-needles into the central regions of the tumors using a Mac Lab/ 4 analog digital system (ADInstruments, Milford, MA) linked to a Macintosh computer [3].

2.6 Measurements of TBF

Relative TBF was measured using the Laserflow Blood Perfusion Monitor 403A (Vasamedics, St. Paul, MN) with a 0.8 mm-diameter laser Doppler needle probe. Briefly, a small hole was made in the tumor using a 23 gauge needle and a needle probe was inserted into the tumor center, then slightly withdrawn to ensure that there was no compression of the tumor under the probe tip. The electrical signals of flow, volume, and velocity from the laser Doppler system were digitally processed using a Mac Lab/ 4 analog digital system linked to a Macintosh computer with output voltage ranging from 0 to 2.5 V. At the end of the experiments the zero-flow signal was measured by sacrificing the animals with an overdose of anesthesia [3].

2.7 Measurements of intratumor pO_2

Tumor pO_2 values were measured using needle-type electrodes commercially available from Diamond-General (Ann Arbor, MI), detailed previously [3]. In brief, a 27 gauge needle with an insulated membrane-recessed cathode (25 μm) was used as a sensor electrode. The electrode was calibrated by immersing it in isotonic saline solution saturated with four different known oxygen levels (0, 1, 5 and 10% oxygen). Movement of the needle was accomplished manually using a micromanipulator.

2.8 Statistical evaluation

Relative changes were determined individually for each mouse based on pre-treatment values. In Table 1, the individual median value was derived from independent 5-7 samples per time (i.e., 1 min, 10 min, 20 min, etc.). For each comparison Wilcoxon's two-sample test was applied. To analyze the oxygen status in tissue, all pO_2 values obtained from several tumors of experimental group were grouped together. Then, the accumulated frequency (%) of pO_2 distribution with a class width of 2.5 mm Hg was constructed. To analyze the oxygen status in tumors, all pO_2 values obtained from several tumors were grouped, and were evaluated using Mann-Whitney U-test. P values smaller than 0.05 were considered as significant.

Table 1. Changes in tumor interstitial fluid pressure (TIFP) and tumor blood flow (TBF) of AsPC-1 tumors after an intratumoral or intraperitoneal infusion of Onconase®

	0 min	1 min	5 min	10 min	20 min	30 min	1 hr	<24hr>
1. Intratumoral infusion*: (infusion volume, 100 μl of 0.9% NaCl or 100 μg/μl of ONC)								
1-A. TIFP (unit: mmHg).								
no substance**	21.0^{a}	$21.0^{b,c}$	$21.0^{a,c}$	$21.0^{a,c}$	$21.5^{a,c}$	$21.5^{a,c}$	$21.5^{a,c}$	$22.0^{a,c}$
0.9% NaCl	23.0	30.0^{d}	25.2^{c}	23.5^{c}	21.8^{c}	22.5^{c}	23.5^{c}	23.5^{c}
Onconase	20.0^{a}	$7.0^{b,d}$	$8.5^{b,d}$	$18.5^{b,d}$	$20.5^{a,c}$	$20.5^{a,c}$	$20.5^{a,c}$	$10.5^{b,d}$
1-B. TBF.								
no substance**	1.00^{a}	$1.00^{b,c}$	$1.00^{a,c}$	$1.00^{a,c}$	$1.03^{a,c}$	$1.03^{a,c}$	$1.03^{a,c}$	
0.9% NaCl	1.00	0.83^{d}	0.97^{c}	0.95^{c}	1.02^{c}	1.05^{c}	1.08^{c}	
Onconase	1.00^{a}	$1.40^{b,d}$	$1.20^{b,d}$	$1.10^{a,c}$	$1.02^{a,c}$	$1.02^{a,c}$	$1.05^{a,c}$	
2. Intraperitoneal infusion*: (infusion volume, 100 μl of 0.9% NaCl or 100 μg/μl of ONC)								
2-A. TIFP (unit: mmHg).								
0.9% NaCl	22.5	22.5^{c}	22.5^{c}	22.5^{c}	22.5^{c}	22.5^{c}	22.5^{c}	24.0^{c}
Onconase	23.5^{a}	$23.5^{a,c}$	$23.5^{a,c}$	$23.5^{a,c}$	$23.5^{a,c}$	$23.5^{a,c}$	$23.5^{a,c}$	$15.5^{b,d}$
2-B. TBF.								<150 mir
0.9% NaCl	1.00	1.00^{c}	1.05^{c}	1.03^{c}	1.05^{c}	1.05^{c}	1.05^{c}	1.05^{c}
Onconase	1.00^{a}	$1.05^{a,c}$	$1.10^{a,c}$	$1.25^{a,d}$	$1.40^{b,d}$	$1.52^{b,d}$	$1.28^{b,d}$	$1.08^{a,c}$

*Median value, N= 5 - 7 mice per group; ** TIFP measurements without an i.t. infusion of substances.
Treated tumors were compared to the control (0.9% NaCl solution) at each time point, and the values obtained at various time points were compared to their initial values (0 min). For each comparision Wilcoxon's two-tailed test was applied. [a] vs. 0.9% NaCl solution, not significant; [b] vs. 0.9% NaCl solution, $p<0.05$; [c] vs. 0 min, not significant; [d] vs. 0 min, $p<0.05$.

3. RESULTS

We observed that AsPC-1 tumors in legs had an elevated TIFP (median= 21.3 mm Hg <range: 8.1, 34.1>, N=10, median tumor volume= 220 mm^3) compared to the surrounding normal tissues, skeletal muscles in legs (median= -1.5 mm Hg <range: -4.5, 0>, N=10). The effects of various routes of ONC administration on tumor physiological parameters are shown in Table 1. After an intratumoral injection of ONC, TIFP was reduced for 10 min before returning to the untreated level. In contrast, the effect of ONC on TBF was in the opposite direction as TIFP values. We did not observe such trends in the intratumorally 0.9% NaCl solution-injected group. After a single i.p. injection of ONC, neither systemic pressure nor TIFP dropped during 2 hr. of observations. Interestingly, TBF significantly increased in AsPC-1 tumors after treatment with ONC. Tumor pO_2 was also monitored before and after an i.p. injection of ONC. Although AsPC-1 tumors grown in legs were hypovascular and marble-like in hardness, the median pO_2 value for 0.9% NaCl-control group (N= 8 tumors) was 4.1 mm Hg, but that for 5 mg/kg ONC-treated group (N= 7 tumors) was 8.2 mm Hg. Thus, tumor oxygenation was observed ($p<0.05$) when tumor pO_2 was measured between 40 to 60 min post-injection of ONC (Figure 1). Additionally, multiple i.p. injections of ONC significantly reduced TIFP from 22 mm Hg to 5 mm Hg (N= 10 tumors, $p<0.01$) at day 2 post-treatment (data not shown). Finally, TIFP remained at that level, and it did not return to pretreatment control levels within 7 days.

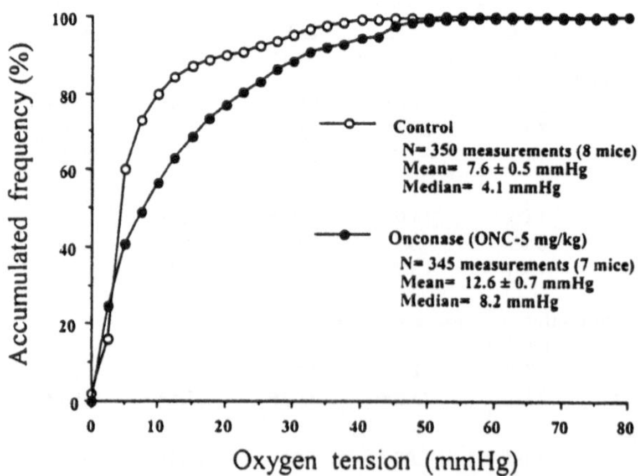

Figure 1. The accumulated frequency (%) of intratumoral pO_2 distribution before and after an i.p. injection of ONC at 5 mg/ kg.

Table 2 shows the effect of ONC on ascites of AsPC-1 human pancreatic tumors. When AsPC-1 tumors were i.p. implanted, the development of ascites was noticeable within 30 days post implantation. However, when ONC at 100 μg/ mouse (equivalent to 5 mg/ kg) was i.p. injected at day 7 post-implantation, the ascites did not develop within the 30 day period (data not shown). Thus, day 30 post-implantation was chosen as the starting date for ONC treatment. Nodule development was followed for 32 days after various treatments (i.e., all mice were sacrificed at day 62 post-implantation). Nine to 10 nodules of 6.8 ± 0.13 mm in average diameter developed per animal in the 0.9% NaCl-treated control group. A significant reduction in the number (2.6 ± 0.18 nodules per animal) and size (1.6 ± 0.16 mm in an average diameter) of nodules was observed in the 50 μg ONC-treated group. The anti-tumor effect of dosages between 50 and 100 μg of ONC was not separable. When combined with ONC at 5 mg/kg, TAMX totally blocked tumor growth, and no nodules were observed in a total of 8 mice.

Table 2. Number & size of nodules developed as ascite tumor models*

	0.9% NaCl[1]	ONC-50[2]	ONC-100[2]	TAMX[3]	TAMX + ONC-50[4]
# of nodules per mouse:	9.6±0.18	2.6±0.18	2.4±0.18	7.9±0.44	0
size of nodules (mm):	6.8±0.13	1.6±0.16	1.4±0.14	5.9±0.11	0
N (mice per group):	8	8	8	8	8

*Mean value ± standard error; mice bearing ascite tumors were sacrificed at 62 days after tumor implantation. All treatments began at day 30.
[1]0.9% NaCl solution at 10 ml/ kg, injected s.c., daily for 30 days, as a control group.
[2]Onconase at 50 μg or 100 μg / 20 g of mouse, injected i.p., once per week for 4 consecutive weeks.
[3]Tamoxifen at 100 μg/ mouse, injected s.c., daily for 30 days.
[4]When ONC and TAMX were given at the same date, ONC was given 1 hr prior to TAMX.

The effect of ONC at 100 μg/ mouse (weekly for 4 consecutive weeks) ± TAMX at 100 μg/ mouse (daily for 4 consecutive weeks) on AsPC-1 human pancreatic tumors implanted s.c. is shown in Figure 2. ONC alone effectively retarded tumor growth. In contrast to ascite tumor models, dosage-dependence was found. ONC at 100 μg/ mouse was more effective than treatment with ONC at either 25 or 50 μg/ mouse (data not shown). However, TAMX alone minimally inhibited the tumor growth. When combined with ONC and TAMX, the combination significantly inhibited tumor growth. However, for the first two weeks, we did not observe a difference in inhibiting tumor growth between ONC treatments in the presence and absence of TAMX.

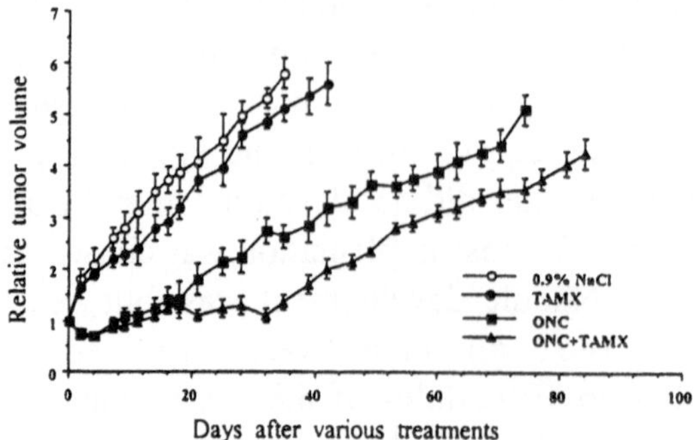

Figure 2. The effect of ONC ± TAMX on AsPC-1 human pancreatic tumors implanted s.c. in the hind legs of nude mice (N= 10 mice per group: mean ± standard error). TAMX alone at 100 µg for 4 weeks (closed circles) slightly inhibited the tumor growth. Multiple injections of ONC alone (closed squares) significantly retarded tumor growth. The combined treatment with ONC and TAMX (closed triangles) synergistically retarded the tumor growth.

4. DISCUSSION

The first goal of this study was to test the hypothesis that ONC could lower TIFP. It was the first data concerning the elevation of TIFP in human pancreatic tumors. The median TIFP in AsPC-1 tumors was significantly elevated to 21.3 mm Hg. Then, TIFP in AsPC-1 tumors significantly increased as a function of tumor size between 100 and 1000 mm^3 (unpublished observations). The positive correlation between TIFP and tumor size in AsPC-1 tumors was characteristic of various human tumor xenografts and murine tumor isografts in small rodents as well as human cancers in patients [1,2]. Furthermore, the increase in TIFP during tumor growth disappeared after the treatment with ONC. Between day 1 and day 7, after the treatment with either single or multiple injections of ONC, TIFP in AsPC-1 tumors was significantly reduced.

As shown in Table 1, this investigation provides insight into the ONC-induced TBF modification. Our results suggest that ONC may be useful in improving the delivery of novel therapeutic agents to tumors. Hyperthermia can lower the TIFP but at the cost of lowering the tumor blood flow rate [8]. On the other hand, ONC can improve the blood supply and also lower TIFP. Vasandani *et al.* reported that ONC can move into tissues quickly, due to the fact that muscle capillary permeability of ONC is 15-fold and 60-fold higher than that of diphtheria toxin and monoclonal antibodies, respectively [9]. That may be in part a cause of the increase in TBF after the intratumoral injection of ONC.

The second goal of this study was to evaluate the effect of ONC on AsPC-1 ascite tumor models. As shown in Table 2, ONC significantly reduced the number and size of nodules of these intraperitoneal ascite tumors. We found that the anti-tumoral effect of ONC was enhanced when TAMX was added. It agreed with the results from Mikulski et al. [7], who showed that the combined treatment of ONC and TAMX was effective *in vitro*. It was in part due to the fact that the ascite tumor models are similar to the *in vitro* situations. ONC significantly inhibited QO_2 in various tumor cells *in vitro* and induced apoptosis (unpublished observations). More detailed experiments regarding the inhibition of QO_2 is in progress (whether ONC causes the stimulation of glycolysis). However, we speculated that the anti-tumoral effect of ONC was enhanced, in part due to the acceleration of apoptosis by TAMX.

The third goal of this investigation was to determine whether TAMX could potentiate the effectiveness of ONC on AsPC-1 pancreatic tumors, implanted s.c. in the hind legs of nude mice. AsPC-1 tumors *in vitro* were relatively resistant to TAMX by a factor of ~100 compared to MCaIV murine mammary adenocarcinoma. Although the growth of MCaIV tumors *in vivo* was significantly retarded by TAMX alone (unpublished observations), TAMX did not inhibit the growth of AsPC-1 tumors *in vivo*. As shown in Figure 2, the first two weeks of ONC treatment in the presence and absence of TAMX showed no difference in the inhibition of tumor growth, but the third injection of ONC in the absence of TAMX developed ONC resistance. Interestingly, TAMX abolished resistance to ONC, hence the combined treatment with ONC and TAMX became synergistic. TAMX is known to interfere with the ability of tumor cells to develop resistance to cisplatin [10]. This mechanism while unclear, was not due to accumulation of cisplatin, intracellular levels of glutathione, or on the repair of adduct between cisplatin and DNA. Therefore, we postulated that TAMX resistance may diminish synergy, but resistance to ONC can be overcome by increasing the concentration of TAMX in the system (i.e., 3 weeks post-treatment of TAMX).

Our recent results using C3H mice bearing MCaIV tumors (sensitive to TAMX) showed that TIFP decreased significantly within 30 min after treatment with TAMX at 100 μg per mouse, reaching a minimum of ~40% of the control value by 2 hr (unpublished observations, Covone KC & Lee I, 1998). Then, it returned to the original level ~7 to ~10 days post-treatment. Although AsPC-1 tumors were resistant to TAMX, we found similar results on tumor physiology (unpublished observations). TBF and TIFP were simultaneously measured during 2 hr after treatment with TAMX. Within 15 min after a subcutaneous injection of TAMX at 100 μg/ 20 g of mice, TBF declined. Then, a decline of TIFP followed at 30 min post-injection. Therefore, the effect of TAMX on tumor physiological parameters, based on sensitivity to TAMX, is worthwhile to pursue in future investigations

In conclusion, TIFP was elevated in AsPC-1 human pancreatic tumor xenografts in nude mice compared to skeletal muscles in legs. Single and multiple i.p. injections of ONC significantly reduced TIFP. ONC alone, effectively retarded the tumor growth, but ONC + TAMX were synergistically effective in inhibiting tumor growth. Therefore, our preliminary results have supported ONC ± TAMX for the clinical trials as a new therapeutic agent in the treatment of advanced pancreatic carcinomas. In addition, study of changes in physiological parameters after ONC + TAMX are in progress to clarify the synergism from the use of two agents in combination *in vivo*.

ACKNOWLEDGEMENT

I. Lee sincerely thanks Dr. W.Gilles McKenna and Dr. John E. Biaglow, Dept. of Radiation Oncology, University of Pennsylvania School of Medicine for their financial support.

REFERENCES

1. Jain RK. Barriers to drug delivery in solid tumors. Scientific Amer 1994;271:58-65.
2. Lee I, Boucher Y, Jain RK. Nicotinamide can lower tumor interstitial hypertension: mechanistic and therapeutic implications. Cancer Res 1992;52:3237-3240.
3. Lee I, Demhartner TJ, Boucher Y, Jain RK, Intaglietta M. Effects of hemodilution and resuscitation on tumor interstitial fluid pressure, blood flow, and oxygenation. Microvasc Res 1994;48:1-12.
4. Wu YN, Mikulski SM, Ardelt W, Rybak SM, Ypole RJ. A cytotoxic ribonuclease, study of the mechanism of onconase cytotoxicity. J Biol Chem 1993;268:10686-10693.
5. Lin JJ, Newton DL, Mikulski SM, Kung HF, Youle RJ, Rybak SM. Characterization of the mechanism of cellular and cell free protein synthesis inhibition by an anti-tumor ribonuclease. Biochem Biophysical Res Comm 1994;204:156-162.
6. Mikulski SM, Viera A, Ardelt W, Menduke H, Shogen K. Tamoxifen and trifluoroperazine (Stelazine) potentiate cytostatic/ cytotoxic effects of P-30 protein, a novel protein possessing anti-tumor activity. Cell Tissue Kinet 1990;23:237-246.
7. Mikulski SM, Viera A, Shogen K. In vitro synergism between a novel amphibian oocytic ribonuclease (Onconase®) and tamoxifen, lovastatin and cisplatin, in human OVAR-3 ovarian carcinoma cell line. Int J Oncol 1992;1:779-785.
8. Leunig M, Goetz AE, Dellian M, Zetterer G, Gamarra F, Jain RK, Messmer K. Interstitial fluid pressure in solid tumors following hyperthermia: possible correlation with therapeutic response. Cancer Res 1992;52:487-490.
9. Vasandani VM, Wu Y, Mikulski SM, Youle RJ, Sung C. Molecular determinants in the plasma clearance and tissue distribution of ribonucleases of the ribonuclease A superfamily. Cancer Res 1996;56:4180-4186.
10. McClay EM, Winski PJ, Jones JA, Jennertte II J, Gattoni-Celli S. Δ^{12}-prostaglandin-J_2 is cytotoxic in human malignancies and synergizes with both cisplatin and radiation. Cancer Res 1996;56:3866-3869.

Chapter 19

OXYGENATION IN A HUMAN TUMOR XENOGRAFT: MANIPULATION THROUGH RESPIRATORY CHALLENGE AND ANTIBODY-DIRECTED INFARCTION

Ralph P. Mason*, Anca Constantinescu, Sophia Ran+ & Philip E. Thorpe+

Departments of Radiology and +Pharmacology, U.T. Southwestern Medical Center, Dallas, TX 75390, USA

Abstract: We recently demonstrated the use of ^{19}F NMR relaxometry of hexafluorobenzene to monitor regional tumor oxygen tension dynamics in rats. We have now extended the application to human tumors implanted in immunocompromised (SCID) mice. This has allowed us both to investigate dynamic response to respiratory challenge (carbogen) and to probe the mechanisms of a new anti-vascular therapy designed to produce tumor-specific infarction.

Key words: magnetic resonance imaging, hexafluorobenzene, oxygen, tumor, tissue factor, antibody targeting

Abbreviations: ARDVARC (Alternated Relaxation Delays with Variable Acquisitions to Reduce Clearance effects); EPI (echo planar imaging); FREDOM (Fluorocarbon Relaxometry using Echo planar imaging for Dynamic Oxygen Mapping); HFB (hexafluorobenzene); i.t (intra tumoral)

1. INTRODUCTION

Tumor response to therapy is strongly influenced by physiological parameters. Quantitative measurements of tumor oxygen tension (pO_2) could provide insight into tumor physiology and mechanisms of novel therapeutic approaches. We recently demonstrated a new tumor oximetry protocol (TOP), which we have called FREDOM (Fluorocarbon Relaxometry using Echo planar imaging for Dynamic Oxygen Mapping). Using this approach based on ^{19}F NMR relaxometry of hexafluorobenzene (HFB) we were able to monitor regional tumor oxygen tension in rats [1, 2]. Hexafluorobenzene is used as the interrogating molecule since the spin-lattice relaxation rate, R1, is exceptionally sensitive to changes in pO_2, relatively insensitive to temperature and the material is readily available, non-toxic, and has a single ^{19}F resonance [1, 2].

It is widely recognized that hypoxic tumors are relatively resistant to irradiation [3,4], and photodynamic therapy [5], while being good candidates for hypoxia selective chemotherapy [6]. Thus, the ability to measure tumor hypoxia may allow therapy to be tailored to the individual characteristics of a tumor. Indeed, recent clinical trials have shown strong prognostic value for oxygen tension measurements made using the Eppendorf electrode system [7, 8]. Equally important may be the ability to monitor changes in tumor oxygenation in response to interventions designed to modulate tumor oxygenation for therapeutic enhancement, *e.g.*, carbogen inhalation [9]. In this case, it is desirable to follow the fate of multiple individual tumor regions simultaneously and non-invasively.

There is increasing emphasis in oncology on developing novel therapeutic approaches that act on the vasculature, *e.g.*, anti angiogenesis [10], anti vascular [11] and infarcting agents [12]. Historically, the efficacy of such agents has been determined from histological analyses of excised tissue at various time points, or by observing relative tumor growth/regression. An *in vivo* approach can provide insight into the physiological response and point to the optimal times for sacrifice and detailed histology. We have now applied the ^{19}F NMR approach to investigating the novel therapeutic approach of anti-body directed infarcts in human tumors.

2. METHODS

Human L540 Hodgkin's tumors were implanted subcutaneously in the flanks of male CB17 SCID mice (Charles River) and allowed to grow to ~1.5 cm diameter (3-6 mm thick). Mice were anesthetized using ketamine/xylazine. HFB (~40 µl, Lancaster) was injected directly in both central and peripheral regions of the disc-like tumors (i.t.) using a Hamilton syringe (Reno, NV) with a custom made fine sharp needle (32 G). Each mouse was placed within a 2 cm single turn solenoid coil in an Omega CSI 4.7 Tesla horizontal bore magnet system with actively shielded gradients. Proton and ^{19}F MR images were obtained to confirm the distribution of HFB. Tumor oxygenation was estimated on a regional basis using ^{19}F echo planar imaging (EPI) relaxometry of the HFB, in a similar fashion to that described previously for rat tumors [2]. Briefly, following a pulse burst saturation recovery (PBSR) preparation sequence with a variable recovery time (τ), a single spin echo planar image was acquired. However, by adding the ARDVARC protocol (Alternated Relaxation Delays with Variable Acquisitions to Reduce Clearance effects) [13] the quality of the relaxation data was enhanced, and typically a precision of 2-5 torr at each of 20 - 50 voxels was achieved within a tumor in 8 minutes. The spin-lattice relaxation rate (R1 = 1/T1) was calculated for each voxel using a three-parameter fit of signal intensity $y_i = A\ (1 - (1+W)\ \exp(-R1*\tau_i))$ by the Levenberg-Marquardt least-squares algorithm. Selection criteria were applied to the data: data were accepted provided that the relative standard error of T1 (σT1/T1) < 25% and 0 < σT1 < 2.0 s. Oxygen tension was estimated using the relationship pO_2 (torr) = 7.6 x (R1 - 0.077 + 0.00009T) / (0.018-0.00017T), where T is rectal temperature in ^oC and R1 is the spin lattice relaxation rate in s^{-1} [2].

The inhaled gas could be manipulated via a nose cone from air to carbogen (95%O_2/5%CO_2; 1 dm^3/min). In other cases, the air breathing mice were injected *i.v.* with anti-VCAM1-tTF (truncated tissue factor) coaguligand (200 μl≡40 μg protein) to produce tumor-specific infarct [12]. The coaguligand was constructed by conjugating the MK2.7 monoclonal antibody directed against mouse VCAM-1 to the extracellular domain of human tissue factor (TF), as described previously [14].

Statistical significance of changes in oxygenation was assessed using analysis of variance (ANOVA) on the basis of Fisher PLSD using Statview II (Abacus Concepts, Berkeley CA) and data are quoted as mean ± standard error of the mean. Experiments were approved by the Institutional Animal Care and Advisory Committee conducted in accordance with National Laws.

3. RESULTS

Typical baseline oxygenation ranged from hypoxia to 50 torr with mean 9.7 ± 1.4 (s.e) torr and median 7.2 torr (Fig. 1): no significant changes occurred during 3 repeat measurements over a period of 24 mins. Within 8 mins of altering the inspired gas to carbogen there was a change in oxygenation with elevation in mean ($p < 0.01$) and median pO_2, and decreased hypoxic fraction (Figs. 1 and 2). Oxygen tension continued to rise generally reaching a plateau after 16 mins with mean pO_2 significantly above baseline ($p < 0.00001$). Upon returning the inhaled gas to air, tumor pO_2 gradually declined, but after 24 mins was still significantly elevated compared with baseline ($p < 0.01$).

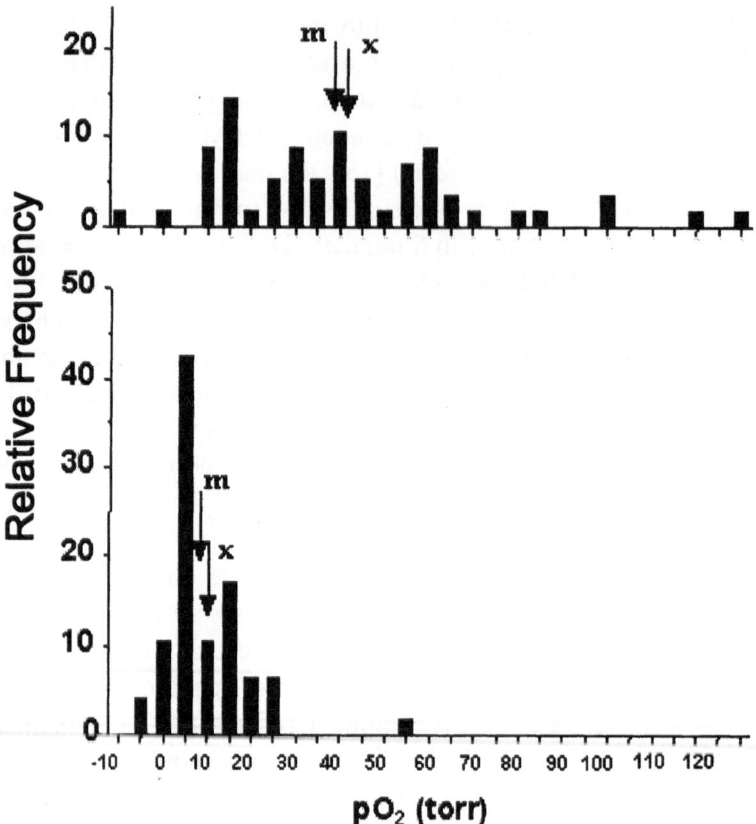

Figure 1. Histograms of tumor oxygenation in response to respiratory challenge for human Hodgkin's tumors growing subcutaneously in mice: in each case data have been pooled from 3 repeat determinations.

Lower: Baseline with mouse breathing air. Mean (x) pO_2 = 9.7 + 1.4 (s.e.) torr, median pO_2 (m) = 7 torr.

Upper: Mouse inhaling carbogen (95%O_2/5%CO_2): mean pO_2 = 41 + 3.9 torr, median 39 torr.

Administration of the anti-VCAM1-tTF coaguligand caused a rapid reduction in pO_2 in well oxygenated tumor regions (initial pO_2 > 10 torr) with mean declining from 15.5 $\pm$ 2.7 torr to 4.2 $\pm$ 2.4 torr ($p \leq 0.01$) after 16 mins and to -1.5 $\pm$ 2.4 torr ($p < 0.001$) after 70 mins. Little change was observed in regions, which were initially poorly oxygenated (initial pO_2 < 10 torr) (Fig. 3).

4. DISCUSSION

While the importance of tumor oxygenation with respect to therapy has been appreciated for many years, most clinical trials designed to improve outcome based on manipulating tumor oxygenation, *e.g.*, hyperbaric chambers, have been disappointing. It has been suggested that failure may

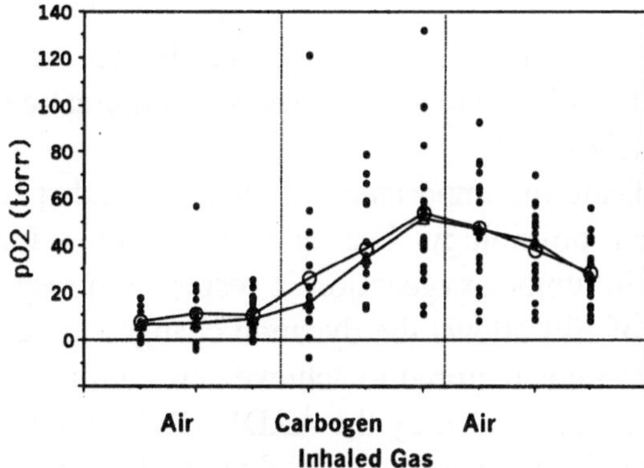

Figure 2. Oxygenation of individual voxels from respective pO_2 maps. The general trend shows increased oxygenation with carbogen and a decline, albeit slower, upon return to breathing air. This emphasizes need for rapid time resolution and danger of pooling data. Variation in mean O and median Δ pO_2 with respect to respiratory challenge is overlaid. Each measurement required 8 mins.

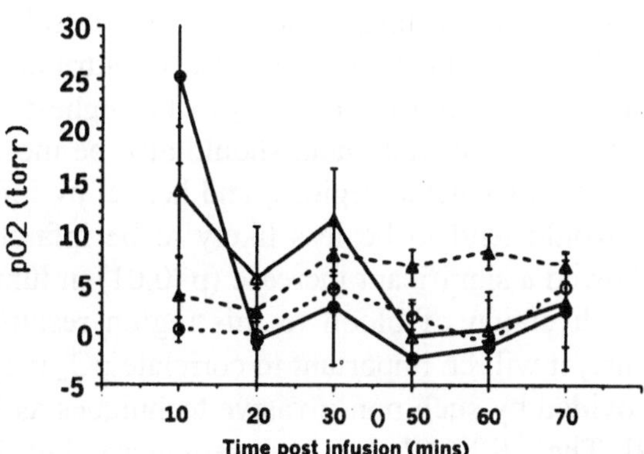

Figure 3. Administration of the coaguligand caused a rapid reduction in pO_2 in well oxygenated tumor regions (initial $pO_2 > 10$ torr) with mean declining from 15.5 ± 2.7 torr to 4.2 ± 2.4 torr ($p< 0.01$) after 16 minutes and to -1.5 ± 2.4 torr ($p < 0.001$) after 70 mins, *e.g.*, two tumor regions (voxels) ● and Δ. Little change was observed in regions, which were initially poorly oxygenated (initial $pO_2 < 10$ torr), *e.g.*, O, although in some regions pO_2 was found to rise significantly ▲.

have resulted from the inability to differentiate those tumors, which would derive benefit, from others, which were unaffected [15]. Thus, the ability to measure tumor oxygenation could have a significant impact on the development of novel therapeutic approaches and the implementation of clinical trials. A number of methods are becoming available to reliably measure tumor pO_2 [16] and this coincides with a dramatic increase in novel therapeutic approaches [17] .

The data indicate the importance of temporal and spatial resolution in investigating tumor physiology. During a period of 8 mins considerable changes occurred in tumor oxygenation in response to respiratory challenge (Fig. 2). The use of EPI allows the dynamic changes to be followed during the typical 16 to 24 mins required to achieve new equilibrium (plateau). The 8 min. time resolution provided by the ARDVARC approach thus represents considerable improvement over the initial imaging technique, which required 20 mins per map [2, 18]. The mapping capability is critical, since it reveals distinct heterogeneity in tumor oxygenation and differential response to intervention. In particular, the initially well oxygenated regions of the tumor showed a significant decline in pO_2 (hypoxiation) following administration of the coaguligand, whereas the less well oxygenated regions showed minimal response (Fig. 3).

The selective decline in pO_2 of well oxygenated tumor regions following administration of the targeted infarcting agent may be related to the absence of the VCAM-1 target molecule on vessels in the hypoxic central region of the tumor [14]. It could also be attributed to differential vascular efficiency. One would expect pO_2 to be highest in well perfused regions and delivery of the coaguligand should also be most efficient there. Meanwhile, less well oxygenated regions, and hence, by inference less well perfused regions would tend to be less likely to be infarcted. Indeed, one region (Fig. 3) showed a significant increase ($p<0.01$) in tumor oxygenation. This could imply diversion of blood to this region resulting from infarct elsewhere. In future, it will be important to correlate pO_2 measurements with blood flow as provided by such non-invasive techniques as ^{1}H MRI arterial spin tagging [19]. The ^{19}F NMR data are preliminary, but do coincide with histological investigations. Specifically, many, but not all tumor capillaries are found to be occluded.

With the increasing emphasis on speed of development for new therapeutic approaches, non-invasive techniques become increasingly important. As combinatorial approaches gain application, many more agents are available, but typically in small quantities. As an example, prior assessment of the efficacy of novel coaguligands has relied on histological endpoints (destructive) or gross anatomical evaluation (slow). The ^{19}F NMR procedure, described here, allows individual locations within a tumor to be followed for a period of hours. These procedures can be used to screen rapidly new agents and suggest optimal time points for sacrifice and histological evaluation. This will both spare animals and, perhaps more significantly, reduce the amount of drug required. The non-destructive imaging technique will accelerate the discovery process for targets, efficacious conjugates and successful development of drugs.

The FREDOM approach provides quantitative measurement of regional tumor oxygenation, but does require introduction of a reporter molecule, here HFB. Administration of a sensing molecule raises issues of potential toxicity. In the case of HFB there is extensive literature indicating a remarkable tolerance and lack of toxicity: specifically, HFB exhibited no mutagenicity [20], no teratogenicity or fetotoxicity [21] and had an $LD_{50} >$ 10 g/kg orally or i.p. in the rat [22]. The manufacturer's material data safety sheet [23] indicates $LD_{50} > 25$ g/kg (oral- rat) and LC_{50} 95 g/m^3/2 hours (inhalation-mouse).

The precision of the measurements, together with the ability to simultaneously examine dynamic changes in multiple specified regions will provide a useful technique for investigating tumor hypoxia with respect to interventions. Successful observation of tumors implanted in mice now provides the opportunity to examine a variety of human tumors.

ACKNOWLEDGMENTS

This work was supported in part by The American Cancer Society (RPG-97-116-010CCE) and The National Cancer Institute (RO1 79515, CA74951 and CA54168): NMR experiments were performed at the Mary Nell & Ralph B. Rogers MR Center, an NIH BRTP Facility #5-P41-RR02584.

REFERENCES

1. Mason RP, Rodbumrung W, Antich PP. Hexafluorobenzene: a sensitive ^{19}F NMR indicator of tumor oxygenation. NMR in Biomed. 1996;9:125-134.
2. Le D, Mason RP, Hunjan S, Constantinescu A, Barker BR, Antich PP. Regional tumor oxygen dynamics: ^{19}F PBSR EPI of hexafluorobenzene. Magn Reson Imaging, 1997;15:971-981.
3. Denekamp J. Physiological hypoxia and its influence on radiotherapy. In: Steel GG, Adams GE, Horwich A, eds. The Biological Basis of Radiotherapy, 2nd ed., Amsterdam: Elsevier, 1989;115-143.
4. Wouters BG, Brown JM. Cells at intermediate oxygen levels can be more important than the "hypoxic fraction" in determining tumor response to fractionated radiotherapy. Radiat Res 1997;147:514-550.

5. Chapman JD, Stobbe CC, Arnfield MR, Santus R, Lee J, McPhee MS. Oxygen Dependency of Tumor Cell Killing *In Vitro* by Light Activated Photofrin II. Radiat Res 1991;126:73-79.
6. Coleman CN. Chemical sensitizers and protectors. Int J Radiat Oncol Biol Phys 1998;42:781-783.
7. Höckel M, Schlenger K, Aral B, Mitze M, Schäffer U, Vaupel P. Association between tumor hypoxia and malignant progression in advanced cancer of the uterine cervix. Cancer Res. 1996;56:4509-4515.
8. Fyles AW, Milosevic M, Wong R, et al. Oxygenation predicts radiation response and survival in patients with cervix cancer. Radiother. Oncol. 1998;48:149-56.
9. Denekamp J, Fowler JF. ARCON-current status: summary of a workshop on preclinical and clinical studies. Acta Oncol 1997;36:517-525.
10. Boehm T, Folkman J, Browder T, O'Reilly MS. Antiangiogenic therapy of experimental cancer does not induce acquired drug resistance. Nature 1997;390(6658):404-407.
11. Dark GG, Hill SA, Prise VE, Tozer GM, Pettit GR, Chaplin DJ. Combretastatin A-4, an agent that displays potent and selective toxicity toward tumor vasculature. Cancer Res. 1997;57:1829-1834\.
12. Huang X, Molema G, King S, Watkins L, Edgington TS, Thorpe PE. Tumor infarction in mice by antibody directed targeting of tissue factor to tumor vasculature. Science 1997;275:547-550.
13. Hunjan S, Mason RP, Constantinescu A, Peschke P, Hahn EW, Antich PP. Regional tumor oximetry: ^{19}F NMR spectroscopy of hexafluorobenzene. Int J Radiat Oncol Biol Phys 1998;40:161-71.
14. Ran S, Gao B, Duffy S, Watkins L, Rote N, Thorpe PE. Infarction of solid Hodgkin's tumors in mice by antibody-directed targeting of tissue factor to tumor vasculature. Cancer Res 1998;58:4646-4653.
15. Siemann DW, Johansen IM, Horsman MR. Radiobiological hypoxia in the KHT sarcoma: predictions using the Eppendorf Histograph. Int J Radiat Oncol Biol Phys 1998;40:1171-1176.
16. Stone HB, Brown JM, Phillips T, Sutherland RM. Oxygen in human tumors: correlations between methods of measurement and response to therapy. Radiat Res 1993;136:422-434.
17. Rockwell S. Oxygen delivery: implications for the biology and therapy of solid tumors. Oncol Res 1997;9:383-90.
18. Mason RP, Constantinescu A, Hunjan S, et al. Regional tumor oxygenation and measurement of dynamic changes. Radiat Res 1999;152:239.
19. Brown SL, Ewing JR, Kolozsvary A, Butt S, Cao Y, Kim JH. Magnetic Resonance Imaging of perfusion in rat cerebral 9L tumor after nicotinamide administration. Int J Radiat Oncol Biol Phys 1999;43:627-33.
20. Mortelmanns KM. "*In vitro*" microbiological mutagenicity assays of eight fluorocarbon taggant samples. Gov. Rep. Announce Index (US) 1981;81:2555.
21. Courtney KD, Andrews JE. Teratogenic evaluation and fetal deposition of hexabromobenzene (HBB) and hexafluorobenzene (HFB) in CD-1 mice. J Environ Sci Health B 1984;19:83-94.
22. Grosman YS, Kapitonenko TA. Pharmacology and toxicology of hexafluorobenzene. Izv Estestvennonauchu Inst Pevinsk 1973;15:155-63.
23. Lancaster. Material Safety Data Sheet. In: Lancaster Synthesis Inc., 1998.

Chapter **20**

TUMOR pO_2 ASSESSMENTS IN HUMAN XENOGRAFT TUMORS MEASURED BY EPR OXIMETRY: LOCATION OF PARAMAGNETIC MATERIALS

Julia A. O'Hara, ‡Rosalyn D. Blumenthal, Oleg Y. Grinberg, Stalina Grinberg, Carmen Wilmot, ‡David M. Goldenberg & Harold M. Swartz
Department of Diagnostic Radiology, Dartmouth Medical School, Hanover NH, ‡Garden State Cancer Center, Belleville NJ, USA

Abstract: Radioantibody immunotherapy (RAIT) is a promising treatment modality but the effectiveness of this targeted low dose radiation varies from tumor to tumor. Since RAIT is an oxygen dependent treatment, baseline pO_2 or growth-induced changes in the microenvironment may alter treatment response. In this pilot work we monitored tumor pO_2 in untreated human xenograft tumors growing s.c. in nude mice. These data will be used to plan a study of the relationship between the effectiveness of RAIT and tumor pO_2. Growth or treatment-induced changes in the microenvironment may alter the tumor pO_2 and thus affect the response to therapy but may also affect location and microenvironment of the particulate oxygen sensor. We monitored tumor pO_2 during growth and also examined the tumor histological structure overall and in the region of the paramagnetic material in the tumor at the time of necropsy.

Key words: Electron paramagnetic resonance, histology, human xenografts, Radioimmunotherapy (RAIT), tumor pO_2

1. INTRODUCTION

Control of tumor growth has been reported for several human tumor xenograft models [1, 2] and for a few clinical trials [3, 4] using radioactivity delivered to tumors by linking a radionuclide to an antibody that is targeted to a tumor-associated antigen (e.g., CEA, EGP-1). These radioactive antibodies accumulate and are retained in tumor tissue for many days, thereby delivering a continuous low dose rate of radiation in a cancer treatment called radioimmunotherapy (RAIT) [5].

Although the trials have been promising there is a need to understand the effects of RAIT on tumor physiology, in order to improve results and to determine why some tumors fail to be controlled by the therapy [6]. The radio-antibodies localize in the tumor near the blood vessels and cause changes in vascular function that may influence intratumoral pH and pO_2 and intratumoral interstitial pressure [7]. Because low LET radiation, such as is delivered by RAIT, is more effective in oxygenated than in hypoxic tissue, the baseline tumor pO_2 as well as the vascular changes post-treatment may be related to the response of human xenografts to RAIT. The net effect of the vascular changes may not only impact on the oxygen-dependent response to radiation but may also affect accumulation of additional doses of radioantibody in a multiple cycle scheme or the uptake of other anti-tumor agents (e.g. drugs, or biological response modifiers) [8, 9].

EPR oximetry has been used successfully to monitor murine tumor pO_2 during normal growth and after single doses of ionizing radiation [10, 11]. Contrast agent uptake measured by dynamic magnetic resonance imaging, a parameter that indicates tumor perfusion, correlated with decreased tumor pO_2 after irradiation of RIF-1 tumors [12]. Using RIF-1 tumors the EPR pO_2 measurements served to identify appropriate times to schedule split dose radiation in order to increase effectiveness of the split dose regimen [11].

In this work we used EPR oximetry to monitor tumor pO_2 in several human tumor xenografts growing s.c. in nude mice, to assess whether different human xenografts had inherently different pO_2 that could be associated with histological structures of the tumors and their reported responses to RAIT. Growth-induced changes in the microenvironment may alter the tumor pO_2 and tumor heterogeneity. Therefore we monitored tumor pO_2 in the same tumors multiple times. After all pO_2 measurements we examined tissues to determine tumor histology and location of the paramagnetic material with regard to tumor margins and vasculature.

2. METHODS

All protocols involving animal use were approved by the Institutional Animal Care and Use Committee of Dartmouth College. Animals were NIH/nude mice (T-cell deficient) bearing one of six different human tumor xenografts grown s.c on the flank. These tumors were human colonic carcinomas, GW-39, LS174T, HT-29 or LoVo; human pancreatic cancer, CAPAN-1; or non-small cell lung adenocarcinoma, CALU- 3. Subcutaneous tumors were initiated at the Garden State Cancer Center by Dr. Blumenthal

(0.2 ml of a 10% suspension of serially transplanted GW-39 or 1 x 10^7 cells from culture of the other five lines) and then shipped 3 days later to Dartmouth for pO_2 measurements. Tumor volume was measured daily beginning on day 8 after initiation of tumor growth. Mice were allowed to acclimate for 7 days before injection of the paramagnetic material into the tumor, at which time tumor volumes are <0.2 cm^3. All colonic tumors are CEA+. CAPAN-1 is both CEA+ and MUC-1+ and CALU-3 is EGP-1+ and CEA-. All tumors in this study were untreated. Statistical analysis was done using Student's t-test. Significance was taken at $p<0.05$.

2.1 Tumor pO_2 measurements

Tumor pO_2 was monitored using EPR oximetry, which utilizes an oxygen sensitive material that is placed into the area of interest [10, 11, 13]. The linewidth, a parameter of the EPR spectrum of this material, varies with oxygen level and allows measuring the local pO_2. This method can monitor pO_2 non-invasively (after the paramagnetic material is invasively placed into the tumor) at one point in the tumor at multiple time points.

In this study wood charcoal CX0670-1, (EM Science Inc.) and char bubinga (University of IL) were used as the oxygen-sensitive materials. A small amount of the sterilized char, about 0.5 mm^3, was placed directly into tumors via a 22-gauge needle, 10--17 days after tumor initiation. Baseline tumor pO_2 measurements were made at least one day after insertion of the paramagnetic material. Mice were anesthetized with 1.3% isoflurane plus 26% oxygen with spontaneous breathing or ketamine/xylazine (90:9mg/kg). Temperature was monitored with a rectal probe and maintained using a circulating warm water pad.

2.2 *In vivo* EPR oximetry

The EPR spectra of chars in tissues were obtained using a custom-made low frequency L-band (1.2 GHz) EPR spectrometer that utilizes a homemade microwave bridge and external loop resonator specifically designed for *in vivo* experiments as previously described [14, 15]. The EPR spectra were recorded, with care to avoid power saturation and over modulation, with scan time varying from 5 to 20s. EPR spectra were averaged over 5 min. Total time for acquisition of EPR spectra was about 20 minutes/mouse for each time point. Spectral fitting by the EWVOIGT program (Scientific software Inc.) was used to derive EPR linewidths. The linewidths were transferred to pO_2 using the calibration curve of the material, which was performed at five different concentrations of oxygen.

2.3 Histology

Mice were sacrificed 24 to 31 days after tumor initiation after 3-4 measurements of pO_2 in each tumor. Each tumor was cut in the plane of the paramagnetic material. Tissues were preserved in 10% phosphate buffered formalin, embedded in paraffin and 4 μm sections were stained with hematoxylin & eosin for examination of tissue structure and location of char material. The char is visible as black particulate deposits in a localized region of the tissue. Eosinophilic regions were scored as necrotic and viable tissue was stained with hematoxylin.

3. RESULTS

The baseline pO_2 in the four xenograft tumors examined initially (GW-39, LS174T, HT-29 and CAPAN-1) was low (1--6 mm Hg) compared to normal tissue levels (23 ± 3 mm Hg for mouse muscle [16]) and differed among the types of tumors (Figure 1 and Table 1). There was no correlation with rate of growth. (Figure 2). In most cases, tumor pO_2 just prior to necropsy was lower than the initial values shown in Table 1. (The final pO_2 averaged about 1 mm Hg in each tumor type). To investigate further the effect of tumor or individual differences on tumor pO_2 estimates, we examined pO_2 by EPR oximetry in 6 tumors of two tumor types (n=12); LoVo and CALU-3. These tumors differ histologically and grew slower than the other four xenografts (Figure 1). CALU-3 tumor baseline pO_2s were significantly higher (average 17 ± 1.5 mm Hg) than LoVo tumors (7 ± 1.2 mm Hg) and CALU-3 remained higher than LoVo during the period of measurement (Figure 3).

Table 1. Baseline pO_2 of 6 Human Tumor Xenografts

TUMOR	# Measures	#Mice	Baseline pO_2± SEM	Tumor Volume mm^3 ± SEM
CALU-3	143	20	17.2 ± 1.5	177 ± 15
LOVO	79	6	6.9 ± 1.2	184 ± 20
CAPAN-1	4	2	6.1 ± 2.1	60 ± 17
LS17-4T	4	2	3.2 ± 0.2	84 ± 21
GW-39	4	2	0.6 ± 0.1	252 ± 21
HT-29	6	3	0.8 ± 0.1	159 ± 24

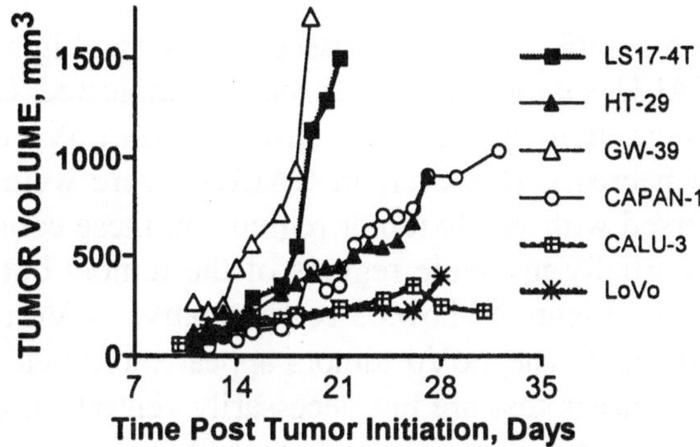

Figure 1. Growth rates of each of six human xenografts. pO_2 data for these mice are shown in Table 1 and Figure 3. Points are Mean ± SEM of tumor volume. (LoVo and CALU-3 tumors, n=10 each type: others (n=2-3)).

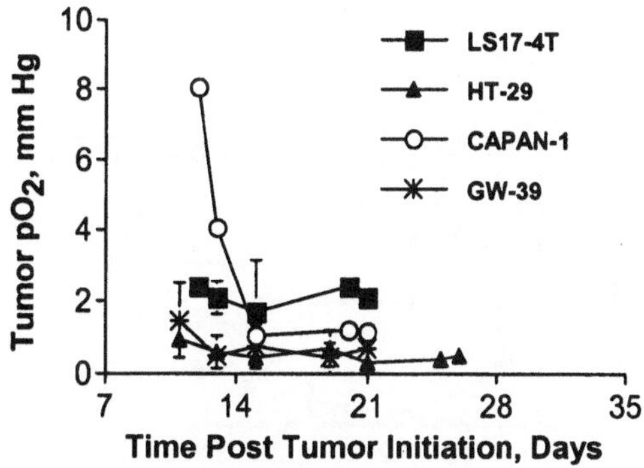

Figure 2. Mean pO_2 of four types of tumors that are also shown in Figure 1 and Table 1. Note that at later time points tumor pO_2 is almost uniformly hypoxic. Day 0 for pO_2 measurements was the 11th or 12th day after tumor implantation.

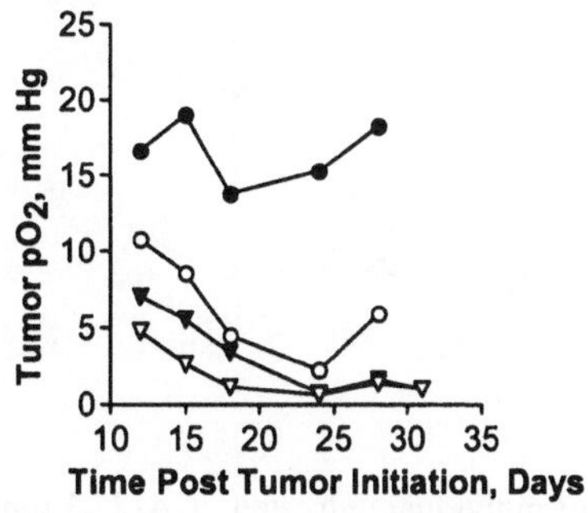

Figure 3. Tumor pO_2 in CALU-3 (circles) and LoVo (triangles) human xenograft tumors measured from the 10th to the 31st day of growth. Open symbols are means and closed symbols the median values for 6 mice in each group.

The location of the char with regard to tumor architecture was determined using H&E stains of tissue sections. Figure 4 shows 3 representative CALU-3 tumors with char location indicated. CALU-3 tumors were in general more homogeneous and vascular than LoVo tumors, without large regions of necrosis. However, in CALU-3 there were focal areas of necrosis interspersed with viable tumor regions. In these cases, the char was located at least partially in viable regions of the tumors but in some cases adjacent to necrosis. Figure 5 shows 3 representative LoVo tumors with the char location indicated. The LoVo tumors appear more heterogeneous with large regions of necrosis that are not necessarily central. In these cases, the char was usually mostly in necrotic areas and in some tumors was not observed possibly due to the loss of material from acellular tumor regions. Table 2 summarizes the pO_2 in individual tumors and observations about char location.

A

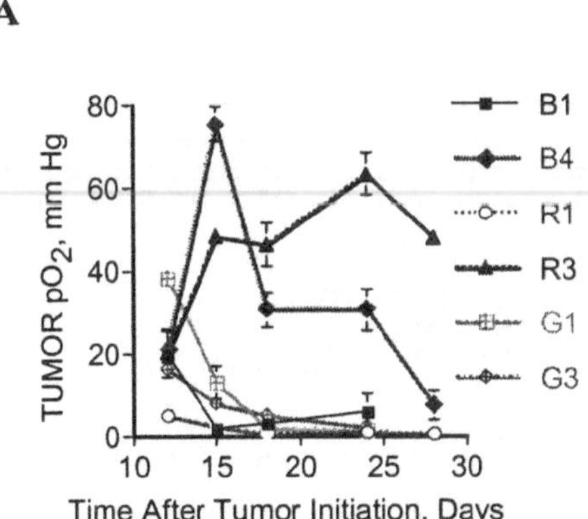

B

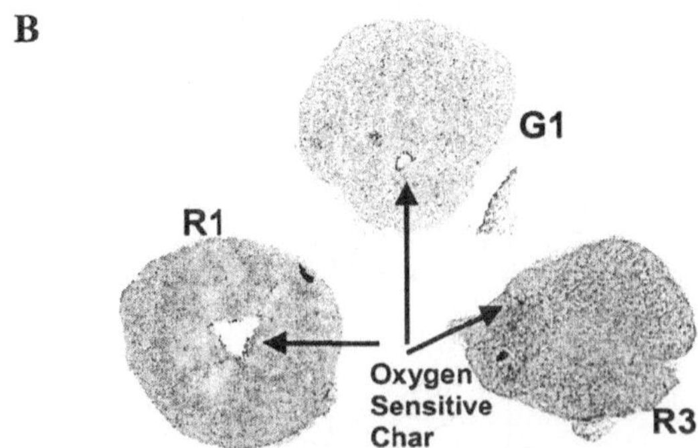

Figure 4. A. Tumor pO_2 of individual untreated CALU-3 tumors. B. Typical histology of 3 untreated Calu-3 tumors (from A) showing the location of EMS char material. Generally the char was near focal necrotic regions except in R3 where most of the tumor was viable.

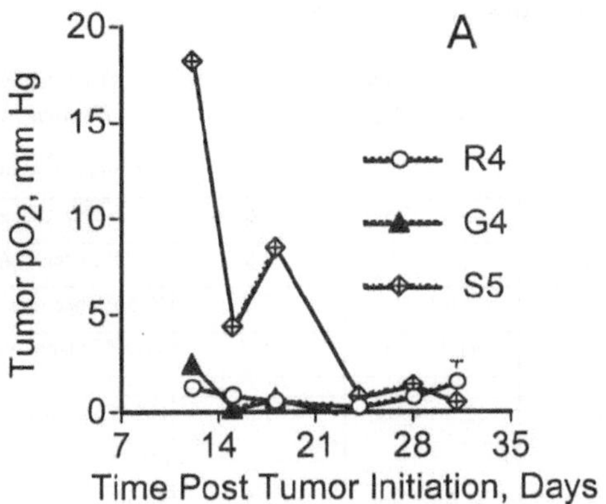

B

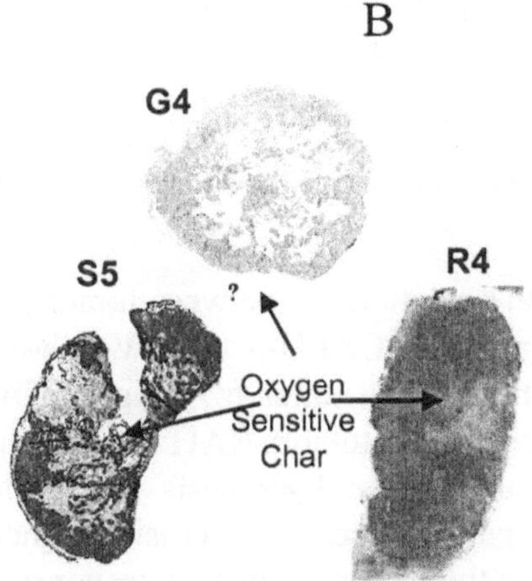

Figure 5. A. Tumor pO_2 of individual untreated LoVo tumors. B. Typical histology of 3 LoVo tumors with the position of the char indicated. (none was seen in G4).

4. DISCUSSION

Baseline pO_2 measurements in the first four xenografts allowed us to rank them from highest to lowest tumor pO_2 as: CAPAN-1 > LS17-4T > GW-39 = HT-29. This was interesting in that the relative sensitivity to RAIT was reported to be roughly in the same order of the pO_2. In terms of sensitivity to RAIT the rankings were: LoVo > GW-39 >HT-29 [6]. Although the trend is intriguing, the numbers of tumors in our study needed

Table 2: EPR assessments of Tumor pO_2 compared with a histological assessment of the tumor and the position of the char material.

Mouse	Tumor	EPR measurements			‡Histological Assessment of tumor / area near char
		Baseline pO_2	*Mean pO_2	Final pO_2	
B4	CALU	13.6	31.3 ± 5.2	5.9	90% viable/vascular
G1	CALU	27	14.1 ± 3.5	1.6	90% viable/vascular
G3	CALU	7.0	7.1 ± 1.3	2.8	65-70% necrotic/poorly vascular
R1	CALU	3.0	2.0 ± 0.5	2.7	40-50% necrotic/ poorly vascularized
R3	CALU	36.0	41.5 ± 2.8	49.4	80-90% viable /vascular
R2	LoVo	6.8	3.3 ± 0.6	1.6	viable periphery, necrotic core/char at periphery
R4	LoVo	1.3	1.0 ± 0.1	1.5	80-90% necrotic / poorly vascular
S5	LoVo	18.1	5.9 ± 1.3	0.4	90% necrotic
B5	LoVo	11.4	14.0 ± 3.5	3.4	80-90% necrotic / poorly vascular
G4	LoVo	2.4	1.1 ± 0.2	0	char not visible/ 70% viable tumor many necrotic areas

*± SEM

‡necrotic = acellular, eosinophilic regions; viable = intact tumors cells and may be well or poorly vascularized.

to be increased to establish a relationship between tumor pO_2 and a treatment outcome. Our results comparing CALU-3 to LoVo showed that CALU-3 tumor pO_2 was greater than LoVo tumor pO_2. Again CALU-3 tumors are more sensitive to RAIT than LoVo tumors RAIT [6]. Our histological results with these two tumors support the hypothesis that higher tumor pO_2 is associated with a lower percent area of necrosis, higher vascularity and greater homogeneity across the tumors as well as response to RAIT [6]. In a related study we showed that there were transient changes in tumor pO_2 during the response to a single dose of radioimmunotherapy and that the response was related to tumor pO_2 at the time of treatment [22].

There was a great deal of inter-individual heterogeneity of tumor pO_2, which may be related to the variation in structural heterogeneity and vascularity of the tumors. The char location in relation to the tumor margins or the tumor vasculature may affect the reported pO_2, since some tumors typically have a more developed vasculature in the perimeter, compared with the core of the tissue [17, 18]. We attempted to implant the char into the outer third of the tumor volume. In most cases the char was found in that region even after the period of tumor growth but this was not true in all tumors. Other possible explanations are differences in vascular density, interstitial pressure, and vascular organization among individual tumors. To investigate these questions we will need to examine implanted materials earlier in the growth process of the tumors. Human colonic tumors are mucinogenic and heterogeneous so that accurate pO_2 measurements of viable tissue may only be possible early in tumor development or by measuring multiple sites within each tumor. This will be further investigated since it may be important to measure long term tumor pO_2 as well as initial tumor pO_2.

The variations between measurements in individual tumors of the same type on different days exceeded that expected from simple measurement errors. We speculate that this may be due to changing microenvironment in the tumor region containing the char. Although in this experiment we have no data to support this hypothesis, others have shown that there are long and short-term fluctuations in blood flow, nutrient supply and tumor pO_2 occur [19, 20].

Further work will compare our EPR results with other indications of hypoxia. These include immunohistochemistry to a) identify hypoxic regions that bind the hypoxia specific marker, pimonidazole, b) localize a hypoxia-induced protein: VEGF, and c) stain for proliferation markers [21]. Previously, Blumenthal and coworkers have shown that single doses of RAIT result in differential responses in different tumor lines and decreases in vascular permeability in human colonic xenografts [6, 8, 9]. Further work will investigate the relationship between tumor pO_2 and outcome after RAIT treatment, and whether the observed changes in tumor pO_2 after RAIT [22] are associated with previously reported vascular changes and the sensitivity to a multi-dose cycle of RAIT.

REFERENCES

1. Sharkey RM, Pykett MJ, Siegel JA, Alger EA, Primus FJ, Goldenberg DM. Radioimmunotherapy of the GW-39 human colonic tumor xenograft with 131I-labeled murine monoclonal antibody to carcinoembryonic antigen. Cancer Res 1987;47:5672-5677.
2. Ceriani RL, Blank EW. Experimental therapy of human breast tumors with 131I-labeled monoclonal antibodies prepared against the human milk fat globule [published erratum appears in Cancer Res 1989 Sep 15;49:5236]. Cancer Res 1988;48:4664-4672.
3. DeNardo SJ, Warhoe KA, O'Grady LF, Hellstrom I, Hellstrom KE, Mills SL, Macey DJ, Goodnight JE, DeNardo GL. Radioimmunotherapy for breast cancer: treatment of a patient with I-131 L6 chimeric monoclonal antibody. Intern Jour of Biological Markers 1991;6: 221-230.
4. Hird V, Maraveyas A, Snook D, Dhokia B, Soutter WP, Meares C, Stewart JS, Mason P, Lambert HE, Epenetos AA. Adjuvant therapy of ovarian cancer with radioactive monoclonal antibody. Brit Jour Canc 1993; 68:403-406.
5. Goldenberg DM. Future role of radiolabeled monoclonal antibodies in oncological diagnosis and therapy. Semin In Nucl Med 1989;19:332-339.
6. Blumenthal RD, Sharkey RM, Natale AM, Kashi R, Wong G, Goldenberg DM. Comparison of equitoxic radioimmunotherapy and chemotherapy in the treatment of human colonic cancer xenografts. Cancer Res 1994;54: 142-151.

7. Blumenthal RD, Sharkey RM, Kashi R, Goldenberg DM. Suppression of tumor vascular activity by radioantibody therapy: implications for multiple cycle treatments. Sel Cancer Ther 1991;7:9-16.
8. Blumenthal RD, Kashi R, Sharkey RM, Goldenberg DM. Quantitative and qualitative effects of experimental radioimmunotherapy on tumor vascular permeability. Int J Cancer 1995;61:557-566.
9. Blumenthal RD, Sharkey RM, Kashi R, Sides K, Stein R, Goldenberg DM. Changes in tumor vascular permeability in response to experimental radioimmunotherapy: a comparative study of 11 xenografts. Tumour Biol 1997;18:367-377.
10. O'Hara JA, Goda F, Liu KJ, Bacic G, Hoopes PJ, Swartz HM. The pO_2 in a murine tumor after irradiation: an *in vivo* electron paramagnetic resonance oximetry study. Radiat Res 1995;144:222-229.
11. O'Hara JA, Goda F, Demidenko E, Swartz HM. Effect on regrowth delay in a murine tumor of scheduling split dose radiation based on direct pO_2 measurements by EPR oximetry. Radiat Res 1998;150:549-556.
12. Goda F, Bacic G, O'Hara JA, Gallez B, Swartz HM, Dunn JF. The relationship between partial pressure of oxygen and perfusion in two murine tumors after x-ray irradiation: a combined gadopentetate dimeglumine dynamic magnetic resonance imaging and *in vivo* electron paramagnetic resonance study. Cancer Res 1996;56:3344-3349.
13. Goda F, O'Hara JA, Rhodes ES, Liu KJ, Dunn JF, Bacic G, Swartz HM. Changes of oxygen tension in experimental tumors after a single dose of X-ray irradiation. Cancer Res 1995;55:2249-2252.
14. Nilges M, Walczak T, Swartz H. 1 GHz *in vivo* ESR spectrometer operating with a surface probe. Physica Medica 1990;5:195-201.
15. Smirnov AI, Norby SW, Walczak T, Liu KJ, Swartz HM. Physical and instrumental considerations in the use of lithium phthalocyanine for measurements of the concentration of the oxygen. J Magn Reson B 1994;103:95-102.
16. Glockner JF, Swartz HM. *In vivo* EPR oximetry using two novel probes: fusinite and lithium phthalocyanine. In: Erdmann W, Bruley DF, ed. Oxygen Transport to Tissue. New York: Plenum Publishing Corp, 1992:229-245. vol XIV).
17. Baish JW, Gazit Y, Berk DA, Nozue M, Baxter LT, Jain RK. Role of tumor vascular architecture in nutrient and drug delivery: an invasion percolation-based network model. Microvasc Res 1996;51:327-346.
18. Harris AL, Zhang H, Moghaddam A, Fox S, Scott P, Pattison A, Gatter K, Stratford I, Bicknell R. Breast cancer angiogenesis--new approaches to therapy via antiangiogenesis, hypoxic activated drugs, and vascular targeting. Breast Canc Res Treat 1996;38:97-108.
19. Dewhirst MW, Kimura H, Rehmus SW, Braun RD, Papahadjopoulos D, Hong K, Secomb TW. Microvascular studies on the origins of perfusion-limited hypoxia. Brit Jour Canc-Suppl 1996;27:S247-S251.
20. Braun RD, Lanzen JL, Dewhirst MW. Fourier analysis of fluctuations of oxygen tension and blood flow in R3230Ac tumors and muscle in rats. Amer Jour Physiol 1999;277(2 Pt 2):H551-H568.
21. Kennedy AS, Raleigh JA, Perez GM, Calkins DP, Thrall DE, Novotny DB, Varia MA. Proliferation and hypoxia in human squamous cell carcinoma of the cervix: first report of combined immunohistochemical assays. International Journal of Radiation Oncology, Biology, Physics 1997;37: 897-905.
22. O'Hara JA, Blumenthal RD, Demidenko E, Grinberg OY, Grinberg S, Wilmot CM, Taylor AM, Goldenberg DM, Swartz HM. Response to radioimmunotherapy correlates with tumor pO_2 measured by EPR oximetry in human tumor xenografts. Radiat Res 2001;155:466-473.

Chapter **21**

HEMOGLOBIN IMAGING WITH HYBRID MAGNETIC RESONANCE AND NEAR-INFRARED DIFFUSE TOMOGRAPHY

Brian W. Pogue[1]*, Haoqin Zhu[2], Casmair Nwaigwe[2], Troy O. McBride[1], Ulf L. Osterberg[1], Keith D. Paulsen[1] & Jeffery F. Dunn[2]

[1]Thayer School of Engineering, Dartmouth College, [2]Department of Radiology, Dartmouth Medical School, Lebanon, NH 03756

Abstract: This study examines the methodology of combining high-resolution information from magnetic resonance imaging into the reconstruction of near-infrared images of hemoglobin concentration and oxygen saturation. This type of hybrid imaging modality has the potential to provide non-invasive maps of hemoglobin concentration and oxygen saturation with relatively high spatial resolution with a fast time response. The study uses (i) tissue-simulating phantoms, as well as (ii) a rat cranial model, to test the method in two well-controlled situations. The phantom test demonstrates that better reconstruction accuracy can be achieved with the use of MRI-generated spatial regions in near-infrared reconstruction. The rat functional testing reveals that the technique can be applied to *in vivo* physiology, even in situations where the tissue is quite heterogeneous. It also shows that the application of *a priori* structure in the finite element mesh as well as spatial constraints in the near-infrared image reconstruction, can significantly improve the quality of the resulting hemoglobin images.

Key words: tomography, reconstruction, photon migration, blood, hemoglobin

1. INTRODUCTION

Imaging blood dynamics has been a key factor in assessing physiologic changes as well as tumor pathophysiolgy. While significant advances have come along in contrast-agent based imaging with computed tomography (CT) and magnetic resonance imaging (MRI), there are no reliable methods to image hemoglobin concentration and oxygen saturation non-invasively. Blood oxygen level dependent (BOLD) MRI has produced some success at imaging changes in deoxy-hemoglobin levels in the brain, however the ability to discriminate between total hemoglobin changes or saturation changes has remained elusive [1]. Alternatively, near-infrared spectroscopy (NIRS) has been used to discriminate hemoglobin oxygenation from concentration changes *in vivo* [2], yet imaging with near-infrared light suffers from very low resolution, due to the diffuse path of travel within tissue. By combining the spatial information of MRI with the spectral information of NIRS, it is possible to create a new modality [3, 4], which can be used to non-invasively image hemoglobin-dynamics with high spatial resolution. This paper investigates some of the methods needed to create this hybrid image modality.

2. METHODS

2.1 Magnetic resonance imaging system

A 7-T horizontal bore NMR system was used to acquire a coronal image of rat cranial tissue. A spin-echo weighted image of one rat was generated in the planar region being imaged with NIRS, using a 40 mm field of view and 156 x 156 x 1000 micron voxel resolution.

2.2 Near-infrared imaging system

A frequency-modulated imaging system was developed, using a Ti:sapphire laser operating at 750, 802, and 833 nm, and detection with an R928 photomultiplier tube (Hamamatsu, Japan) contained within a heterodyne driven housing (ISS Instruments, Champaign-Urbana IL). The light source is intensity-modulated at 100MHz by a Pockel's cell in combination with a polarizer, which is driven by current from a frequency generator which is synchronized to an identical generator driving the detector dynode chain. The difference frequency between the two is 1kHz, allowing detection of this signal through direct sampling by a data acquisition board in the computer at 10kHz. The tissue to be imaged was contained within an array of 8 source fibers and 8 detection fibers, as shown in Figure 1, with the source and detector multiplexed sequentially into each fiber by linear translation stages. The fibers are all 2 mm diameter plastic, and are coupled to the tissue/phantom surface by direct placement in intimate contact with the boundary. An initial pre-fitting routine is used to homogeneous estimate of the medium optical properties, as a rough initial guess, before the reconstruction program begins.

2.3 Tissue-phantoms

Tissue simulating phantoms were created to test the ability for precise reconstruction of hemoglobin concentrations (reduced, oxygenated and total) in a well-controlled medium, which approximates the optical properties of mammalian tissue. Ceramic phantoms, made of Alraldite (D. H. Litter, Elmsford NY), were created containing TiO_2 powder (1.4g per 400 ml) and India ink (0.025 ml/L), which was mixed and allowed to harden. One phantom was machined to a smooth 40 mm diameter cylinder, with two 5 mm diameter holes at the center and edge, as shown in Figure 2. The bulk scattering and absorption coefficients of this phantom at 802 nm were measured as 1.08 mm^{-1} and 0.0050 mm^{-1}, respectively, using our tomographic system before drilling the interior holes. A mixture of 1.08% Intralipid® (Pharmacia and Upjohn Inc.) was used with varying amounts of whole blood to create contrasting regions within the holes of this phantom, to test the imaging system. Intralipid® is an aqueous suspension of lipids, which is commonly used to simulate the optical scattering properties of mammalian tissues.

2.4 Animals

All procedures in this study were approved by our institutional committee for the use and care of animals in research. Rats were used in this study to image cranial hemoglobin dynamics in response to changes in the inspired oxygen fraction (FiO_2). Animals were anesthetized with 1.5% inhaled isofluorane in a continuous flow of oxygen/nitrogen mixture. Once placed in the NIR imaging system the animal was cycled for 5 minutes from FiO_2 = 0.3 to 0.13, then from 0.30 to 0.08, and then was sacrificed in the NIR imaging system by i.v. injection of potassium chloride. The geometry of the imaging system is shown in Figure 1.

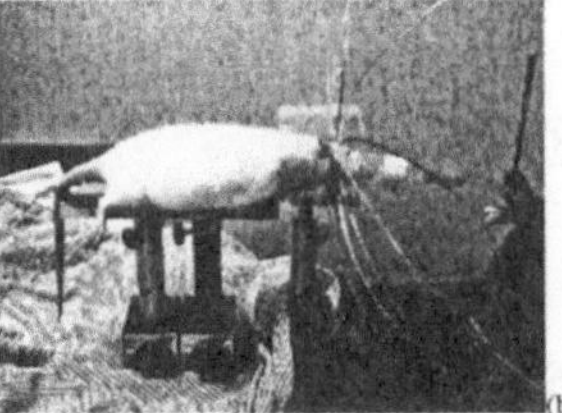

Figure 1. Photographs of the imaging system used for imaging phantoms and tissues are shown (a) and the positioning of the rat in this array is shown (b).

2.5 Near-infrared image reconstruction

A finite element based solution to the frequency-dependent diffusion equation was used to calculate a fit to the measured projection data of amplitude and phase shift. A Newton-type fit [5, 6] was used requiring a full matrix inversion to update the image at each iteration, creating both absorption and scattering coefficient images. To incorporate the MRI information, a MATLAB program was created to import an image file from the MRI and threshold the image based upon user input gray scale values. The external boundaries of the tissue were used to create the mesh of nodes, and then the initial image was used again to tag nodes with representative material properties. These material properties could then be used to either (i) allow a reduced number of free parameters in the fit, as was tested with the solid phantom, or (ii) to input optical properties into the reconstruction to improve the spatial resolution, as was tested in the rat cranial imaging. The full details of the optical properties used and implementation is described in a previous study [4].

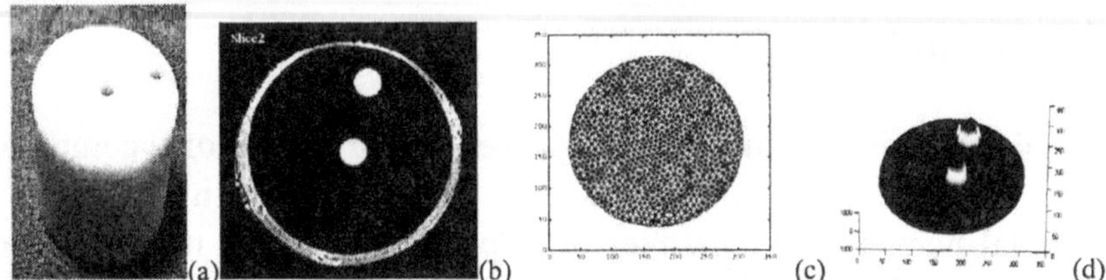

Figure 2. (a) Photograph of the tissue simulating phantom used for imaging incremental changes in hemoglobin concentration within the two holes shown. Outer diameter was 40mm, with holes of 5mm diameter at the center and 5mm from the edge. (b) Magnetic resonance image of the tissue-phantom across the mid section, which was used to create the finite element mesh, shown in the following figure. This image was thresholded to descriminate the exterior and internal boundaries between material regions. (c) The finite element mesh created within MATLAB using the external boundary of the thresholded image. (d) Display of the material properties for the finite element mesh, where the two holes have automatically been given different material numbers than the surrounding mesh.

3. RESULTS

3.1 Tissue phantom imaging

The tissue-simulating phantom shown in Figure 2 was imaged with blood concentrations of 0.3, 0.6, 0.8 and 1.6% into an aqueous solution of 1%

Intralipid. The resulting data was reconstructed with two methods. The first method used our standard Newton-type reconstruction, which allowed all nodes in the image to vary freely, with the resulting images of edge hole and center hole filling shown in the upper rows of Figures 4 and 5, respectively. In this set of images, one hole had varied blood concentration while the other hole had a constant 1% Intralipid concentration to match the rest of the phantoms scattering coefficient background (i.e. making the hole somewhat invisible to the optical tomography by matching the background optical properties). In the second row of these two figures, the ability to reconstruct larger fixed regions was tested by specifying all nodes with the same material number together, and simplifying the reconstruction into a 2-region fitting process. The resulting images are shown in the lower rows of Figures 4 and 5, and are best fits to the data based upon homogeneous regions. The peak absorption coefficient values of these regions were then plotted in Figure 6 as a function of the blood concentration. The theoretical prediction based upon the known spectrum of hemoglobin and our own measurement of the hemoglobin concentration in blood 156 g/L, as measured by a clinical co-oximetry system) were used to make this prediction.

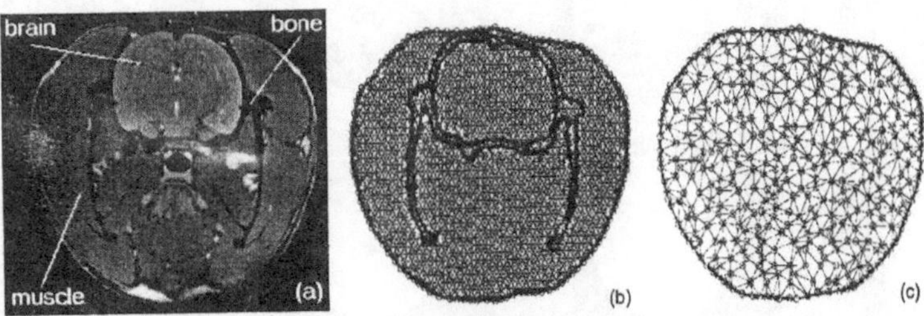

Figure 3. A rat cranial MR image in the plane being imaged here is shown (a), and the MATLAB generated finite element mesh from this image is shown containing the relevant tissue boundaries from separating bone, brain and muscle (b). A lower resolution mesh was also generated to reduce the computational burdon of the reconstruction (c).

3.2 Rat cranial imaging

The rat cranial imaging was completed to provide a test of the system in a more realisticmodel, where there was higher structural complexity as well as the problems associated with blood flow. The rat images were acquired at three different wavelengths and absorption coefficient maps were recovered for each, and then the data at each pixel was used in a multi-wavelength fit to recover maps of hemoglobin concentration and oxygen saturation.

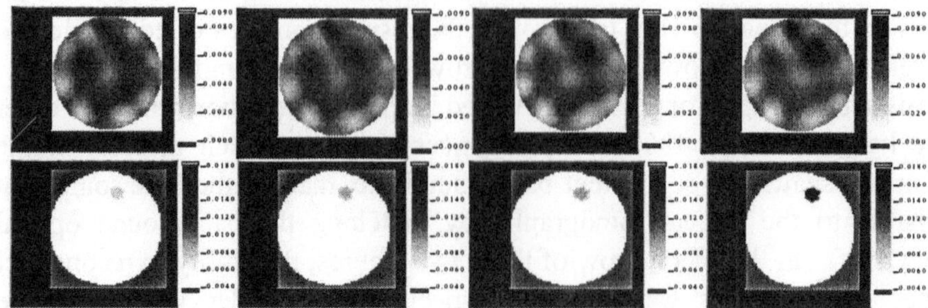

Figure 4. (top-row) Reconstructed images of the phantom in Fig. 2, with increasing blood concentration in the outer hole, mixed into 1.08% Intralipid to match the scattering of the solid phantom. The left scale bar in each image shows the absorption coefficient value in units of mm^{-1}. The blood concentrations were 0.3, 0.6, 0.8 and 1.6%, going from left to right, and the reconstruction was based upon a full iterative Newton-type solution to the diffusion equation minimizing the square-difference between measurement data and forward diffusion calculations. (bottom row) The same data as in the top row was used to reconstruct images using a simpler 2-region fit, driven by a least-squares fit of the predicted diffusion solution to the measured data.

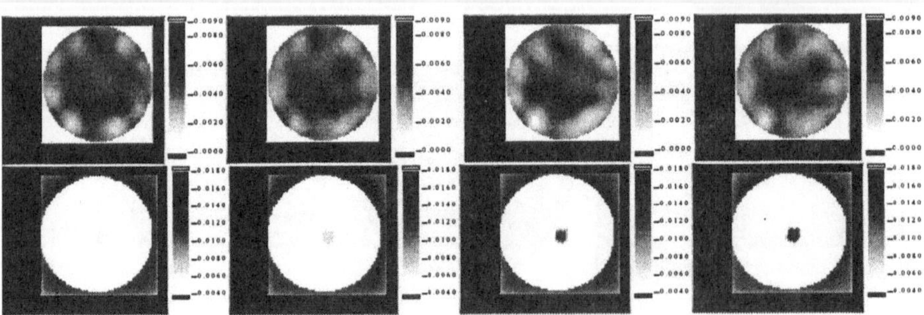

Figure 5. The same tests as shown in Fig. 4 were completed, but with the same series of blood concentrations as in Fig. 4 put into only the center hole, with the edge hole kept fixed to the background optical properties. The reconstructed images are analogous to those in Fig. 4.

The details of this algorithm have been presented previously [4, 7]. The resulting images at FiO_2 of 13%, 8% and after sacrifice are displayed in Figure 7. The left set of images demonstrates the imaging quality when no apriori internal information is used in the near-infrared reconstruction. The right set of images shows the higher precision and improved accuracy which can be achieved when the approximate absorption and scattering coefficients are input into the initial guess of the distributions, based upon the MRI structure map and literature values for brain, bone and muscle tissues.

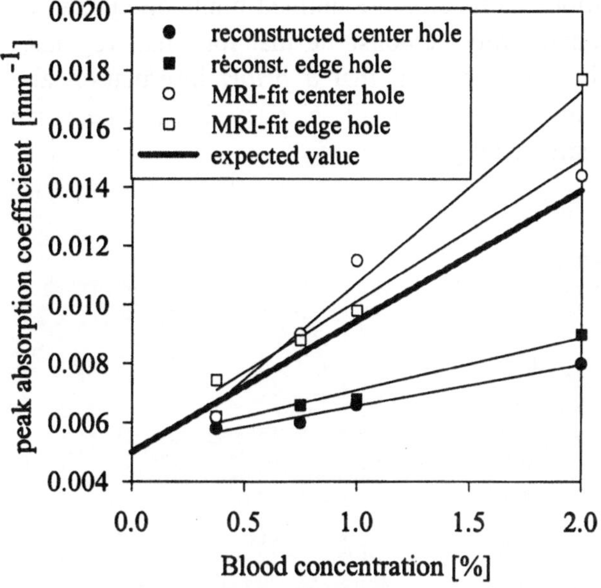

Figure 6. For each image in Figures 4 and 5, the absorption coefficient at the location of the object was calculated, and a plot of the peak absorption coefficient in the reconstructed region versus blood concentration was created. The thick solid line on the figure demonstrates the expected value, based upon the known absorption spectrum of hemoglobin.

4. DISCUSSION

4.1 Phantom imaging

In this data comparison, the 2-parameter fit to the images produced a more accurate fit than did the full Newton-type reconstruction, as shown in Figure 6. This result was expected since the number of parameters to reconstruct is significantly reduced, however the scatter in the data also suggests that the simple 2-region fitting does not suffice for reconstruction accuracies better than 20%. It is likely, that a Newton-type reconstruction must be used where the regions are lumped together, while preserving the ability of all nodes to update somewhat freely outside of specified sub-zones. This type of loose-clumping of pixels may be achieved either through (i) spatially varying the regularization parameter [8] as we used in the rat study, or (ii) using Bayesian-statisical methods to pre-define acceptable regions of clumping, or (iii) using a progressive zoning method which cycles through varying spatial resolutions with some pre-defined limits on update level [9]. This analysis does conclusively demonstrate that for small regions, which change in hemoglobin concentration, that *a priori* information about the location and size can improve the overall accuracy, but not necessarily the precision, of the reconstructed regions.

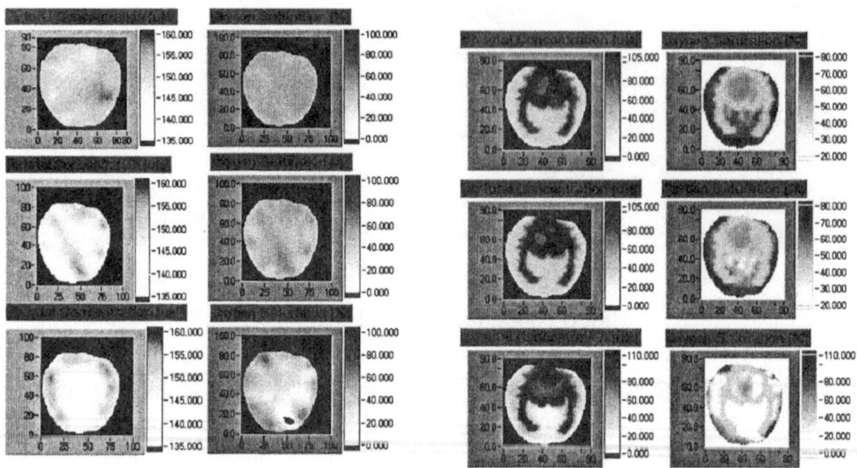

Figure 7. The reconstructed images of total hemoglobin concentration and oxygen saturation are shown for three different physiologic conditions (shown in three rows), and for two different reconstruction methods (a) and (b). The top row shows the image for a reduction from FiO2 of 30% to 13%, the second row shows the images for a reduction from FiO2 from 30% to 8%, and the third row shows the image after sacrifice by i.v. injection of potassium chloride. The first type of reconstruction (a) was done with knowledge of the exterior profile of the rat head, but with no other information, which results in low resolution images. The second type of reconstruction (b) utilized knowledge about the approximate tissue optical properties in the initial guess and constrained the skull bone regions not to fluctuate in hemoglobin concentration. This latter series of images produces high resolution images which are much closer to what would be expected within the tissue.

4.2 Rat cranial imaging

The rat cranial images provide an illustration of the benefit of using the MRI structure as an *a priori* guess of the initial structure and optical properties. In the NIR images without this structure, shown in Figure 7 (a), there is very poor spatial resolution, and it is difficult to resolve changes in brain from changes in surrounding muscle tissues. Because of the diffuse path that light travels in tissue, it is unlikely that optical tomography can progress beyond blurry reconstruction of the tissue interior without further information about the tissue. However in Figure 7 (b), the higher resolution structure of the brain and muscle tissues are input into the initial guess of the distribution, and the skull regions have been constrained not to change during the reconstruction process. This allows a potentially more accurate reconstruction of the hemoglobin changes in response to changes in FiO_2. As well, upon examining the changes after sacrifice in Fig. 7(a) the muscle tissue only decreases to approximately 20% oxygen saturation with the blurred image, whereas in Fig. 7(b) the muscle saturation shows a full decrease to 0%, as would be expected after death. This suggests that the additional information of the tissue optical properties has aided the fit to oxy- and deoxy-hemoglobin concentrations and thereby improve the accuracy of oxygen saturation images. Also in Fig. 7(a) there are artifactual regions near the tissue surface which show high and low regions of blood concentration, which are not present when more accurate initial guesses are provided for the tissue optical properties (Fig 7(b).

Because the precise changes in hemoglobin *in vivo* are not exactly known, under these particular conditions, we do not propose using the rat model for validating our reconstruction method. Rather, reconstruction of the tissue phantoms must be used to validate the reconstruction program, and then testing in the rat model must be carried out with an online secondary check from tissue pO_2 change. However, the use of imaging animals hemoglobin dynamic in response to metabolic changes such as FiO_2 alteration can provide a fundamentally new modality to examine tissue physiology and pathophysiology. Further *in vivo* studies will compose be the next phase of this study.

ACKNOWLEDGEMENTS

This work has been supported by the NIH through contracts R01CA69544 (KDP, BWP, TOM, ULO) and R01CA67431 (JFD). Authors gratefully acknowledge assistance from Rendy Strawbridge and Jack Hoopes in animal handling, and also Mike Miga, David Kung and Claire Wilscher for assistance with finite element mesh generation and data acquisition.

REFERENCES

1. Dunn JF, Zaim Wadghiri Y, Pogue B, Kida I, BOLD vs. NIR spectroscopy: will the best technique come forward? Adv Exp Med Biol, 1998;454: 103-113.
2. Delpy DT, Cope M, Quantification in tissue near-infrared spectroscopy. Phil. Trans. R. Soc. Lond. B, 1997; 352:649-659.

3. Ntziachristos V, Ma XH, Chance B.. Time-correlated single photon counting imager for simultaneous magnetic resonance and near-infrared mammography. Review of Scientific Instruments, 1998;69:4221-4233.
4. Pogue BW, McBride TO, Nwaigwe C, Osterberg UL, Dunn JF, Paulsen KD. Near-infrared diffuse tomography with a priori MRI structural information: testing a hybrid image reconstruction methodology with functional imaging of the rat cranium. Proc. SPIE, 1999;3597: 484-492.
5. Arridge, SR, Schweiger M Inverse methods for optical tomography. in Information Processing in Medical Imaging. 1993. Flagstaff, AZ: Springer-Verlag.
6. Paulsen KD, Jiang H. Enhanced frequency-domain optical image reconstruction in tissues through total-variation minimization. Appl Opt 1996;35:3447-3458.
7. McBride TO, Pogue BW, Gerety E, Poplack S, Osterberg UL, Paulsen KD. Spectroscopic diffuse optical tomography for quantitatively assessing hemoglobin concentration and oxygenation in tissue. Appl Opt 1999; (in press).
8. Pogue BW, McBride T O, Prewitt J, Osterberg U L, Paulsen K D. Spatially variant regularization improves diffuse optical tomography. Appl Opt 1999;38: 2950-2961.
9. Eppstein MJ, Dougherty DE, Troy TL, Sevick-Muraca EM. Biomedical optical tomography using dynamic parameterization and Bayesian conditioning on photon migration measurements. Applied Optics 1999;38: 2138-2150.

Chapter 22

TUMOR OXYGEN DYNAMICS: COMPARISON OF ^{19}F MR EPI AND FREQUENCY DOMAIN NIR SPECTROSCOPY

Yulin Song[1,2], Kate L. Worden[1], Xin Jiang[1], Dawen Zhao[2], Anca Constantinescu[2], Hanli Liu[1], and Ralph P. Mason[2]

[1]Joint Graduate Program in Biomedical Engineering; [2]Department of Radiology, UT Southwestern Medical Center, Dallas, TX 75390, USA

Abstract: Oxygen plays a key role in tumor therapy and may be related to tumor development: *e.g.*, angiogenesis and metastasis. Using noninvasive techniques to accurately measure tumor oxygenation could assist in developing novel therapies. Here, we have used the FREDOM (Fluorocarbon Relaxometry using Echo planar imaging for Dynamic Oxygen Mapping) approach based on hexafluorobenzene (HFB) to monitor tissue oxygen tension (pO_2) of rat breast and prostate tumors and compared the results with changes in tumor vascular hemoglobin saturation (sO_2) and concentration observed using a new dual wavelength homodyne near-infrared (NIR) system. The dynamic changes in pO_2 and sO_2 were assessed while rats were breathing various gases. NIR showed significant changes in vascular oxygenation accompanying respiratory interventions. ^{19}F MR-EPI also showed significant changes in tissue pO_2 and revealed considerable regional heterogeneity in both absolute values and rate of change accompanying interventions. Generally, changes in vascular sO_2 preceded tissue pO_2, particularly for smaller tumors.

Key words: Oxygen tension, Echo planar imaging, MRI, tumor, NIR spectroscopy

Abbreviations: NIR (Near infrared); EPI (Echo Planar Imaging); HFB (Hexafluorobenzene); FREDOM (Fluorocarbon Relaxometry using Echo planar imaging for Dynamic Oxygen Mapping)

1. INTRODUCTION

The growth and development of tumors are greatly influenced by oxygen tension (pO_2), e.g., tumor hypoxia leads to increased expression of vascular endothelial growth factor (VEGF), and thus, angiogenesis [1].

Hypoxia reduces radiosensitivity [2], but chemotherapeutic approaches have been proposed to exploit the tumor hypoxia based on selective cytotoxicity of bioreductive drugs [3, 4]. In addition, increasing evidence from clinical trials has revealed that poorly oxygenated tumors have poor prognosis for patients [5, 6]. Therefore, accurate measurements of oxygenation could enhance cancer treatment planning. Here, we present two methods of measuring tumor oxygenation: the FREDOM approach to measure tumor tissue oxygen tension (pO_2) and NIR spectroscopy to measure changes in tumor vascular hemoglobin saturation (sO_2) and concentration [Hb]. By comparing these two techniques, we can examine the relationship between tumor tissue pO_2 and vascular sO_2.

The FREDOM approach is based on ^{19}F PBSR-EPI of hexafluorobenzene (HFB). It has been shown that the spin-lattice relaxation rate, *R*1 (1/T1), is linearly proportional to dissolved oxygen concentration [7]. HFB offers exceptional sensitivity to changes in pO_2 while having little response to temperature [8]. Because of structural symmetry, HFB has a single resonance and thus, is free from chemical shift artifact, providing an optimal signal-to-noise ratio (SNR). Maps of the tumor tissue pO_2 were obtained in 8 minutes with millimeter resolution, allowing the fate of individual voxels to be traced. NIR spectroscopy can be used to measure tumor vascular sO_2 because the absorption coefficients of deoxy-hemoglobin differ from those of oxy-hemoglobin at the wavelengths selected (758 nm and 782 nm) [9]. The system has many attractive features, including being completely non-invasive, inexpensive, portable, and amenable to real-time measurements.

2. METHODS

2.1 Tumor model

NF 13762 breast and Dunning prostate R3327-AT1 adenocarcinomas were implanted in skin pedicles on the forebacks of adult female Fischer and male Copenhagen rats (~250 g), respectively, as described previously [10]. Once the tumors reached ~1 cm diameter, the rats were anesthetized with 200 μl ketamine hydrochloride (100 mg/ml) and maintained under general gaseous anesthesia (33% O_2, 66% N_2O and 0.5%) through a mask placed over the mouth and nose. Body temperature was maintained at 37 °C by a thermal blanket. A fiber optic pulse oximeter was placed on the hind foot to monitor arterial hemoglobin saturation and heart rate, and a fiber optic probe was inserted rectally to monitor core temperature. NIR spectroscopy and ^{19}F PBSR-EPI measurements were then performed sequentially, while inhaled gas was alternated between 33% O_2, carbogen (95% O_2 + 5% CO_2), and 100% O_2.

2.2 NIR spectroscopy

The tumor vascular sO_2 was assessed by NIR spectroscopy using a new dual wavelength, homodyne system (wavelengths 758 nm and 782 nm). These wavelengths were selected because they not only allow the calculation of sO_2, but also fall into the range of wavelengths compatible with the low cost photo multiplier tube (PMT). The system uses only one RF source to determine amplitude and phase changes of light. Figure 1 shows a schematic diagram of the system. An RF source modulates the light from two laser diodes at 140 MHz. The light passes through fiber optic cables, is transmitted through the tumor tissue, and is collected by a second fiber bundle. The light is then detected, amplified, filtered, and demodulated into I and Q components. Amplitude and phase changes caused by the tumor are related to changes in hemoglobin concentration [Hb] and hemoglobin saturation [HbO_2], i.e., sO_2.

To obtain the absorption coefficients of deoxy-hemoglobin and oxy-hemoglobin, we have assumed background absorbance to be negligible and estimated the absorption coefficients by multiplying the extinction coefficients for deoxy-hemoglobin and oxy-hemoglobin with their respective concentrations.

$$\mu_a^{758} = \varepsilon_{Hb}^{758}[Hb] + \varepsilon_{HbO2}^{758}[HbO_2] \quad (1)$$

$$\mu_a^{782} = \varepsilon_{Hb}^{782}[Hb] + \varepsilon_{HbO2}^{782}[HbO_2] \quad (2)$$

where μ_a^{758} and μ_a^{782} are the absorption coefficients, ε_{Hb}^{758} and ε_{Hb}^{782} the extinction coefficients for deoxy-hemoglobin, ε_{HbO2}^{758} and ε_{HbO2}^{782} the extinction coefficients for oxy-hemoglobin at the wavelengths 758 nm and 782 nm, respectively, and [Hb] and [HbO_2] are the deoxy-hemoglobin and oxy-hemoglobin concentrations, respectively.

Although the IQ system does give both phase and amplitude values, given the tumors' small sizes and our fiber configuration, conventional diffusion theory doesn't hold. To overcome this difficulty, we modified Beer-Lambert's law and used the amplitude values to find trends in the changing absorption coefficients.

$$\mu_{aC} - \mu_{aI} = (1/L)*\log(A_I/A_C) \quad (3)$$

where A_I is the initial amplitude (amplitude of baseline), A_C the current amplitude, and L the optical pathlength between source and detector.

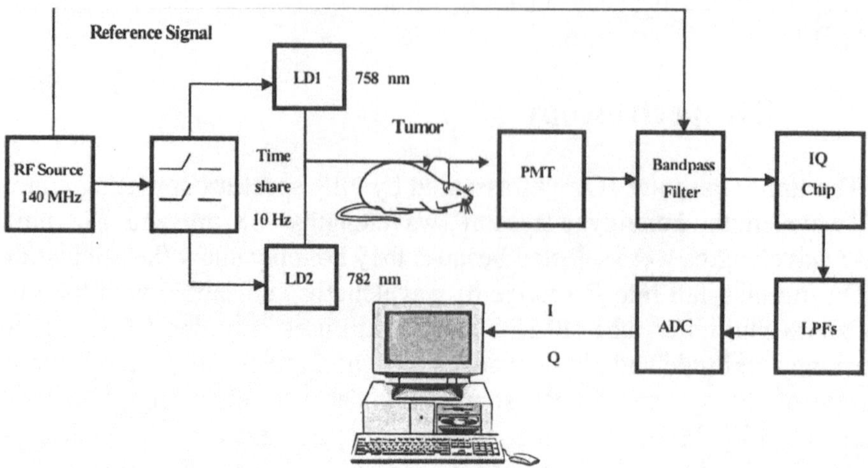

Figure 1. A schematic diagram of the NIR IQ system.

By manipulating equations 1-3, we can compute changes in blood volume and saturation between the initial state and the intervention state from the transmitted amplitude of the light through the tumor.

$$\Delta[Hb]_{total} = -(3.63/L)^* \log (A_I/A_C)^{758} + (8.68/L)^* \log (A_I/A_C)^{782} \quad (4)$$

$$\Delta[HbO_2] - \Delta[Hb] = -(18.49/L)^* \log (A_I/A_C)^{758} + (21.20/L) * \log (A_I/A_C)^{782} \quad (5)$$

where $\Delta[]$ represents change in concentration. The constants were computed with extinction coefficients for oxy- and deoxy- hemoglobin at the two wavelengths used.

Once stable baseline measurements were achieved, the inhaled gas was altered to pure oxygen or carbogen and dynamic changes were observed over a period of two hours. Both the magnitude and rate of change of sO_2 were examined. Following the NIR experiments, the MRI experiments were performed.

2.3 ^{19}F MR-EPI

All MRI experiments were performed on an Omega CSI 4.7 T 40 cm system with actively shielded gradients. A tunable 2 cm ^{1}H/^{19}F single turn solenoid coil was placed around the tumor and 40 µl HFB were injected directly into both central and peripheral regions of the tumor using a 32 G needle. Shimming was performed on the ^{1}H signal of the tissue water to a typical linewidth of 50 Hz. 3D spin-echo (SE) ^{1}H images were acquired for anatomical reference and corresponding ^{19}F images were then obtained to show the distribution of HFB in the tumor. Regional tumor pO_2 maps were generated using ^{19}F PBSR-EPI based on the relationship: $pO_2(\text{torr}) = [R1 - 0.074]/0.0016$, where $R1$ is the spin lattice relaxation rate of HFB, as described in detail previously [11]. Twenty-three pO_2 maps were produced in 3 hours with respect to respiratory challenge.

3. RESULTS

3.1 NIR results

Figure 2 shows the time course of changes in tumor vascular hemoglobin saturation and concentration accompanying alterations in inhaled gases for a breast tumor and Figure 3 shows the result for a prostate tumor. Hemoglobin saturation and concentration are presented as relative millimolar changes. Both tumors show significant changes in vascular oxygenation accompanying respiratory interventions. Hemoglobin saturation increased almost immediately after a gas switch from baseline (33% O_2) to either carbogen or 100% O_2 and increased steadily for several minutes, and then gradually returned to baseline after the gas was switched back to baseline. In contrast, total hemoglobin change is insignificant, indicating relatively constant blood volume in the tumor.

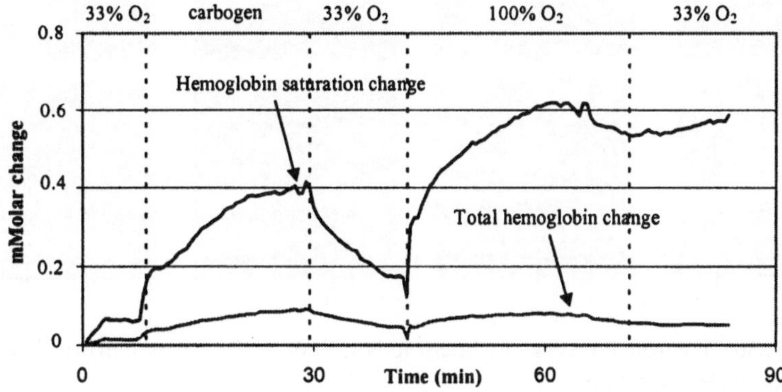

Figure 2. Hemoglobin saturation and concentration change in a 4.0 cm³ breast tumor.

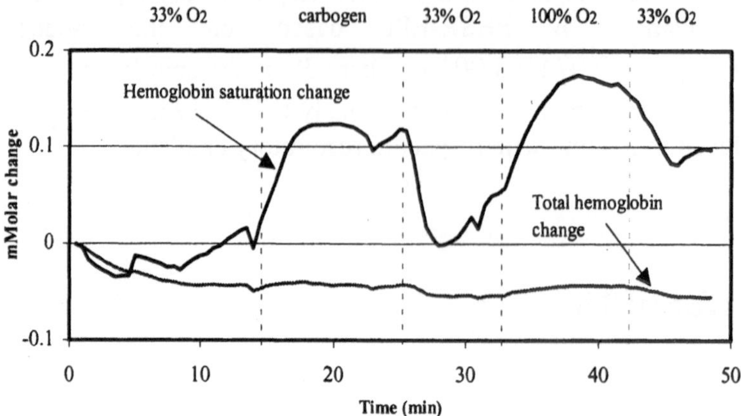

Figure 3. Hemoglobin saturation and concentration change in a 4.0 cm^3 prostate tumor.

3.2 MRI results

Figure 4 shows conventional spin echo (SE) ^{1}H images (upper) and corresponding ^{19}F images of a representative breast tumor. Figure 5a shows a ^{19}F PBSR-EPI projection image obtained from the tumor shown in Figure 4 in a single acquisition (τ = 90 s) and Figure 5b shows corresponding pO_2 map (expanded).

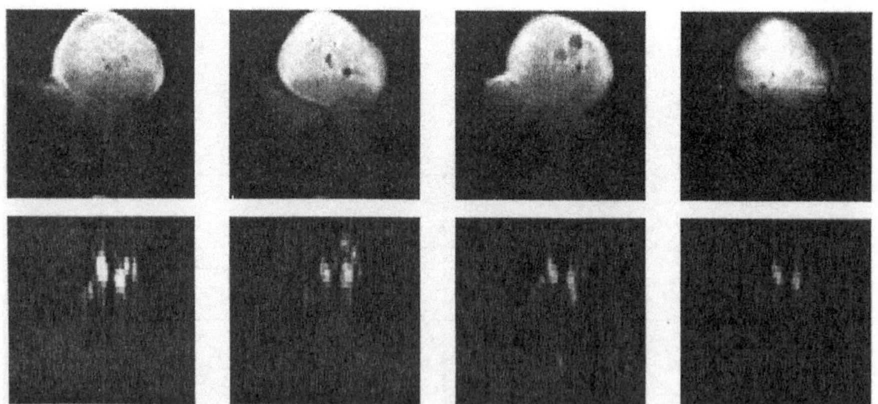

Figure 4. ^{1}H and ^{19}F coronal images of a breast tumor. FOV = 48 x 48 mm, matrix size = 128 x 64, and slice thickness = 4 mm.

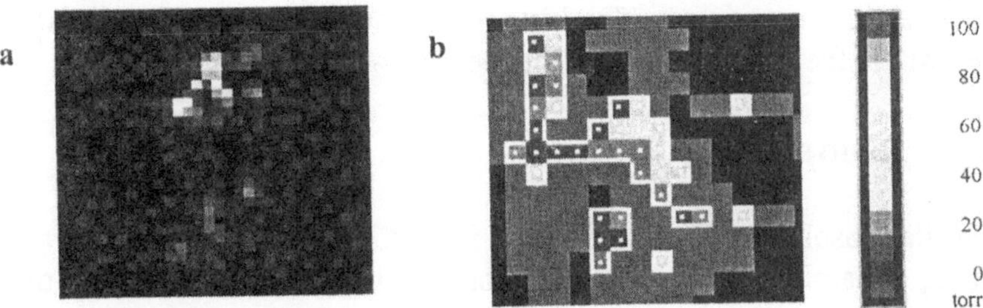

Figure 5. (**a**) A ^{19}F PBSR-EPI projection image obtained from the tumor shown in Figure 4 in a single acquisition (τ = 90 s). Fourteen images were acquired with variable relaxation delays (τ) ranging from 200 ms to 90 sec and 1.5 mm in plane resolution. (**b**) Corresponding pO_2 map (expanded).

The ^{19}F MR-EPI oximetry of tumor has the distinct advantage over other techniques that subsequent measurements are completely non-invasive. The greatest strength of this method is the ability to trace the fate of individual voxels (regions) with respect to therapeutic interventions. Figure 6 shows dynamic changes in pO_2 of six specific voxels of a breast tumor with respect to different inhaled gases. It is noteworthy that voxels with high baseline pO_2 had significantly different response characteristics from those with initially low pO_2, which showed small changes.

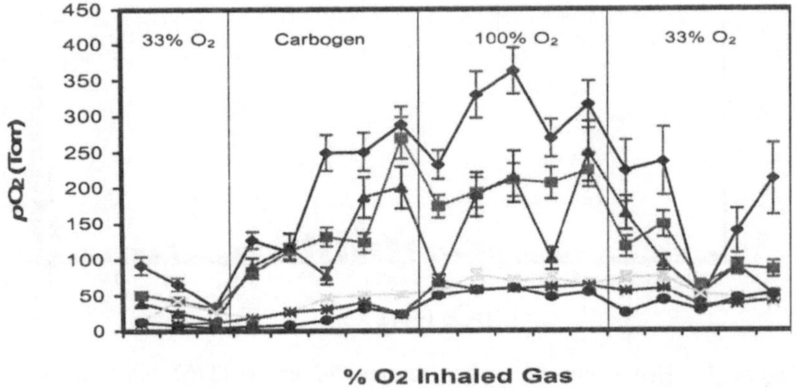

Figure 6. Dynamic changes in pO_2 of six specific voxels of a breast tumor.

Figure 7 shows pO_2 histograms obtained by FREDOM of HFB for a representative breast tumor. The histograms show the heterogeneity of pO_2 values within the tumor as well as the mean pO_2 values.

3.3 Comparison

While absolute pO_2 values are important for investigating tumor hypoxia, dynamic changes may be more valuable for investigating tumor

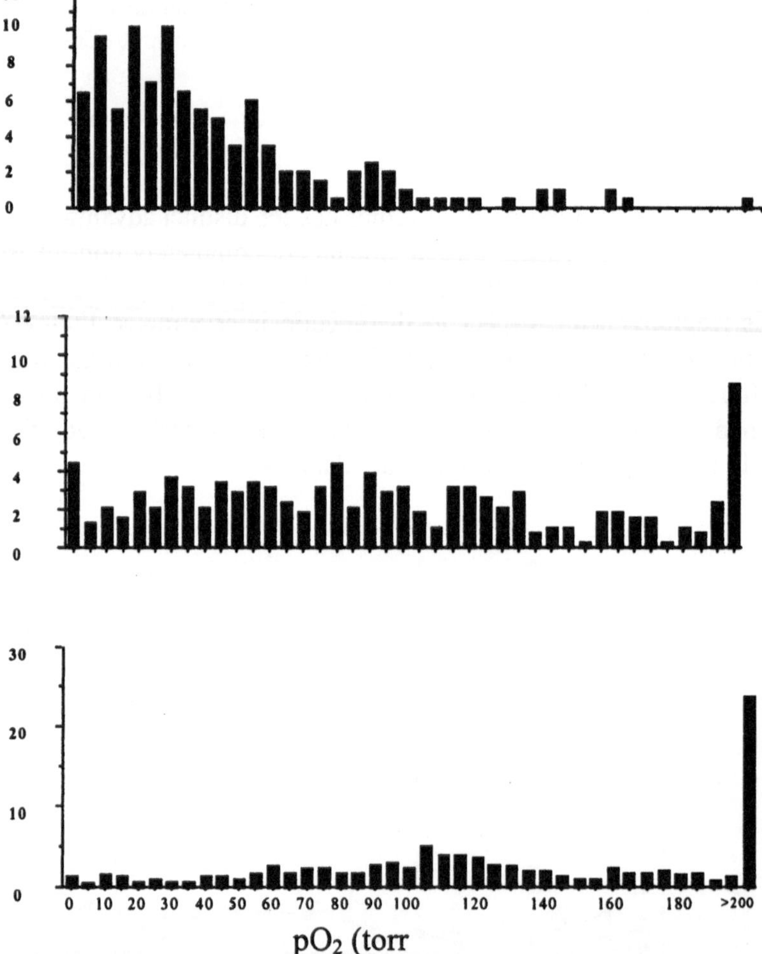

Figure 7. Histograms of oxygen tension for an NF13762 breast tumor determined by FREDOM of HFB. **(Top)** pO_2 distribution while the rat breathed 33% O_2. Mean $pO_2 = 44 \pm 3$ torr. **(Middle)** pO_2 distribution while the rat breathed carbogen (95% O_2 + 5% CO_2). **(Bottom)** Mean $pO_2 = 99 \pm 4$ torr. pO_2 distribution while the rat breathed 100% O_2. Mean $pO_2 = 145 \pm 4$ torr.

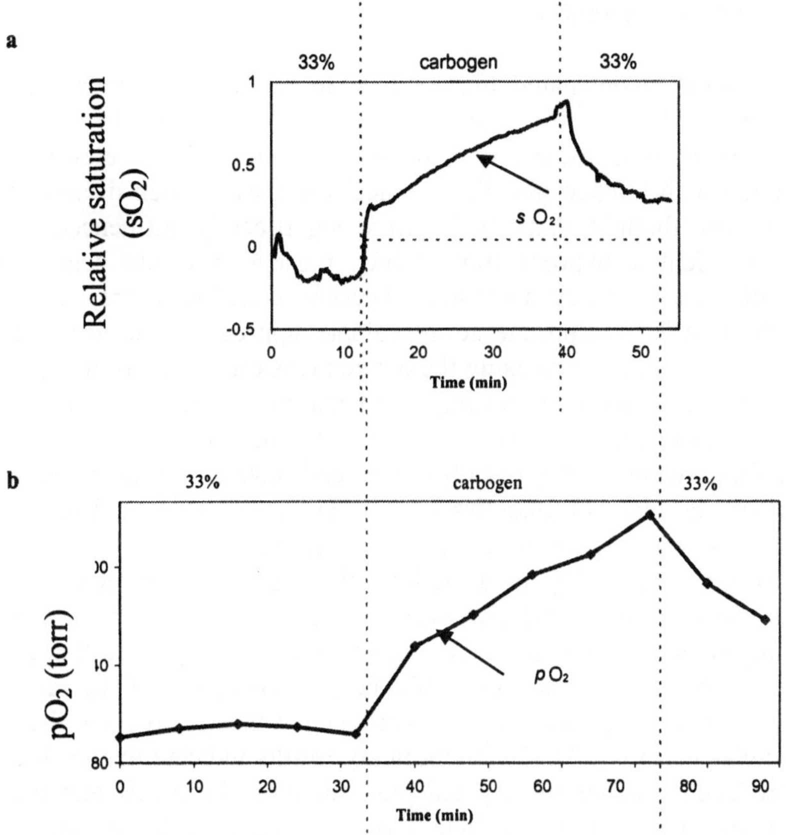

Figure 8. Comparison of pO_2 and hemoglobin saturation (sO_2) of a breast tumor. (**a**) sO_2 determined by NIR spectroscopy. (**b**) pO_2 determined by FREDOM. Both pO_2 and sO_2 increase with a transition from 33% to carbogen and begin to decrease when switched back to 33% O_2. Changes in vascular sO_2 precede tissue pO_2. The tumor size = 1.9 x 2.2 x 2.0 cm.

response to therapeutic interventions. Figure 8 shows comparisons between the dynamic changes in pO_2 and sO_2. Since the IQ system provides a global sO_2 value, each pO_2 data point is represented as an average over all voxels of each pO_2 map.

To further investigate the response, we modeled the temporal response in pO_2 and sO_2 using exponential equations:

1) $y = a + b \cdot (1 - e^{-t/\tau})$, for increasing trend

2) $y = a + b \cdot e^{-t/\tau}$, for decreasing trend

where y is pO_2 or sO_2, a and b are two constants, t is time, and τ is the time constant. Changes in sO_2 were faster than pO_2, especially in the case of the larger tumor, which was found to be less well oxygenated and presumably less well perfused.

4. DISCUSSION

Since poorly oxygenated tumors tend to resist conventional therapy, there have been many efforts to re-oxygenate tumors prior to therapy. A simple intervention is respiratory challenge, i.e., attempting to elevate tumor oxygenation with inhaled gas. Past clinical trials were often disappointing, but it is now thought that results were significantly influenced by the inability to identify hypoxic tumors (i.e., those that would benefit from manipulation), a priori. As techniques become available to measure tumor oxygenation, it is appropriate to reevaluate approaches to manipulating tumor oxygenation. By increasing the oxygen tension of the inspired gas, the arterial sO_2 should increase, leading to increased hemoglobin saturation of the tumor vasculature, and hence, increased tumor tissue pO_2. Our data indicate that breathing elevated O_2 did indeed have a significant effect on both tumor vascular sO_2 and tissue pO_2. There has been a debate as to whether carbogen is more effective at modulating tumor oxygenation than 100% oxygen since CO_2 is a peripheral vasoconstrictor. Recent work suggests that the relative effectiveness can depend on tumor type and also site of implantation. Thus, for example, Brizel *et al.* [12] found that neither normobaric oxygen nor carbogen influenced oxygenation of the rat breast tumor R3230Ac implanted s.c., but carbogen was effective for the same tumor implanted i.m. [13]. Our preliminary results indicate that each of the gases produced changes in sO_2 and pO_2, for both the breast and prostate tumors implanted s.c. in the pedicle. However, the variable behavior found in many tumors emphasizes the value of a rapid and totally non-invasive approach to investigating response to interventions. Indeed, NIR was recently used by van der Sanden *et al.* [14] to show that carbogen, but not oxygen altered vascular oxygenation in a human glioma xenograft.
We recently showed that the FREDOM approach indicates similar pO_2 values in tumors to electrodes [15]. By contrast, NIR investigates vascular oxygenation. As expected, changes in vascular oxygenation (sO_2) were found to be more rapid than tumor pO_2, with greater differences in large tumors. This probably reflects the extensive perfusion of the small tumors with lesser perfusion of large tumors as reflected by lower mean pO_2 and larger hypoxic fraction [15]. Such observations are also consistent with the microscopic analyses of pO_2 distribution in window chamber models [16].

Regional tumor tissue pO_2 and blood sO_2 are important physiological parameters. The capability to measure them will provide insight into progressive physiological changes in a tumor accompanying interventions. NIR has the advantages of being entirely noninvasive, inexpensive, portable, and real-time. But the MRI approach clearly reveals detailed oxygenation heterogeneity. The correlation of NIR and MR technologies will provide insight into issues of oxygen delivery and consumption. We believe that application of multiple approaches to tumor oxygenation can lead to better understanding of tumor physiology and probably optimized tumor therapy.

ACKNOWLEDGMENTS

This work was supported in part by grants from The American Cancer Society (RPG-97-116-010CCE; RPM), The Whitaker Foundation (RPM and HL), the DOD Breast Cancer Initiative (DAMD17-97-1-7261;YS), and NIH NCI RO1 CA79515 and BRTP Facility P41-RR02584.

REFERENCES

1. Pötgens AJG, Westphal HR, de Waal RMW, Ruiter DJ. The Role of Vascular Permeability Factor and Basic Fibroblast Growth Factor in Tumor Angiogenesis. Biol Chem Hopper-Seyler, 1995; 376:57-70.
2. Gray LH, Conger AD, Ebert M, Hornsey S, Scott OCA. The Concentration of Oxygen Dissolved in Tissues at the Time of Irradiation as a Factor in Radiotherapy. Br J Radiol 1953;26:638-648.
3. Brown J.M, Giaccia AJ. Tumor Hypoxia: the Picture has Changed in the 1990s. Int J Radiat Biol 1994; 65:95-102.
4. Stratford I, Adams G, Bremmer J, Cole S, et al. Manipulation and Exploitation of the Tumor Environment for Therapeutic Benefit. Int J Radiat Biol 1994; 85-94.
5. Höckel M, Schlenger K, Aral B, Mitze M, Schäffer U, Vaupel P. Association between Tumor Hypoxia and Malignant Progression in Advanced Cancer of the Uterine Cervix. Cancer Res 1996;56:4509-4515.
6. Fyles AW, Milosevic M, Wong R, Kavanagh MC, Pintile M, Sun A, Chapman W, Levin W, Manchul L, Keane TJ, Hill RP. Oxygenation Predicts Radiation Response and Survival in Patients with Cervix Cancer. Radiother Oncol 1998;48:149-56.
7. Mason R.P. Non-invasive Physiology: ^{19}F NMR of Perfluorocarbons. Artif Cells Blood Sub. Immob Biotechnol. 1994;22:1141-1153.
8. Mason RP, Rodbumrung W, Antich PP. Hexafluorobenzene: a Sensitive ^{19}F NMR Indicator of Tumor Oxygenation. NMR in Biomed 1996; 9:125-134.
9. Yunsong Y, Liu H, Li X, Chance B. Low-Cost Frequency-Domain Photon Migration Instrument for Tissue Spectroscopy, Oximetry, and Imaging. Opt Eng 1997;36:1562-1569.
10. Hahn EW, Peschke P, Mason RP, Babcock EE, Antich PP. Isolated Tumor Growth in a Surgically Formed Skin Pedicle in the Rat: A New Tumor Model for NMR Studies. Magn Reson Imaging. 1993;11:1007-1017.
11. Le D, Mason RP, Hunjan S, Constantinescu A, Barker BR, Antich PP. Regional Tumor Oxygen Dynamics: ^{19}F PBSR EPI of Hexafluorobenzene. Magn Reson Imaging 1997;15:971-981.
12. Brizel DM, Lin S, Johnson JL, Brooks J, Dewhirst MW, Piantadosi CA. The Mechanisms by Which Hyperbaric Oxygen and Carbogen Improve Tumour Oxygenation. Br J of Cancer 1995;72:1120-1124.

13. Lanzen JL, Braun RD, Ong AL., Dewhirst MW. Variability in Blood Flow and pO_2 in Tumors in Response to Carbogen Breathing. Int J Radiat Oncol Biol Phys 1998; 42:855-859.
14. van der Sanden BPJ, Heerschap A, Simonetti AW, Rijken PFJW, Peters HPW, Stüben G, van der Kogel AJ. Characterization and Validation of Noninvasive Oxygen Tension Measurements in Human Glioma Xenografts by ^{19}F-MR Relaxometry. Int J Radiat Oncol Biol Phys 1999;44:649-658.
15. Mason RP, Constantinescu A, Hunjan S, Le D, Hahn EW, Blum C, Antich PP, Peschke P. Regional Tumor Oxygenation and Measurement of Dynamic Changes. Radiat Res 1999;152:239-249.
16. Dewhirst MW, Ong ET, Braun RD, Smith B, Klitzman B, Evans SM Wilson D, Quantification of Longitudinal Tissue pO_2 Gradients in Window Chamber Tumours: Impact on Tumour Hypoxia. Br J of Cancer 1999;79:1717-1722.

Chapter **23**

MICROCIRCULATORY FUNCTION, TISSUE OXYGENATION, MICROREGIONAL REDOX STATUS AND ATP DISTRIBUTION IN TUMORS UPON LOCALIZED INFRARED-A-HYPERTHERMIA AT 42°C

Oliver Thews*, Yanping Li**, Debra K. Kelleher*, Britton Chance** & Peter Vaupel*

(*) Institute of Physiology and Pathophysiology, University of Mainz, Duesbrgweg 6, 55099 Mainz, Germany; (**) Department of Biochemistry and Biophysics, University of Pennsylvania, Philadelphia, PA 19104, USA

Abstract: Since local hyperthermia (HT) affects microenvironmental parameters, the aim of the study was to analyze the impact of 42°C-HT on microcirculatory function, tumor pO_2, microregional redox status and ATP distribution in experimental rat tumors. Subcutaneously growing DS-sarcomas were treated with localized HT using infrared-A radiation resulting in a mean tumor temperature of 42°C. The relative red blood cell (RBC) flux in the tumor was assessed using the laser Doppler technique and the mean tumor pO_2 measured continuously using O_2-sensitive catheter electrodes. In a second series of experiments, the microregional distribution of the mitochondrial redox status and ATP concentration was measured. Although the average RBC flux increased by 63%, tumor pO_2 rose only by approx. 6%. No distinct changes were seen in the mitochondrial redox status. The microregional distribution of the redox status as well as of the ATP concentration showed considerable heterogeneity. In conclusion, although 42°C-HT leads to a distinct improvement in tumor perfusion, there is practically no change in the oxygenation status. The latter finding can be explained by an equivalent increase in the oxygen consumption rate of the cells which increases by approx. 58% at 42°C compared to normothermia.

Key words: ATP concentration, infrared-A radiation, localized hyperthermia, mitochondrial redox status, tumor oxygenation, tumor perfusion

Oxygen Transport to Tissue XXIV, edited by
Dunn and Swartz, Kluwer Academic/Plenum Publishers, 2003

1. INTRODUCTION

Local hyperthermia (HT) with tissue temperatures above 43°C appears to have a direct damaging effect on tumor cells. However, below this temperature, not only the direct cell killing effect but also alterations of several microenvironmental parameters might be responsible for the effectiveness of hyperthermia as a supportive treatment in clinical oncology. The tumor microenvironment can be affected either directly or indirectly by HT. For instance, increasing the tissue temperature was found to increase (at least temporarily) tumor perfusion and thus improve the nutrient supply (oxygen and glucose) to the tissue as well as the transport of anticancer agents to the tumor. At the same time, HT seems to affect the oxidative metabolism of the tumor and thus might influence the bioenergetic status or the pH homeostasis (for a review see [1]).

Even so, the effects of localized HT on parameters of the tumor micromilieu are non-uniform and somewhat contradictory. For instance, an increase of tumor perfusion has been described upon HT [1] as well as a pronounced worsening of tumor blood flow [1-3]. In some studies, the ATP concentration was almost unaffected by HT treatment [1,4], in others a significant decrease in bioenergetic parameters was found [1,2]. Parameters predominantly responsible for this variable behavior of tissue upon HT seem to be the temperature reached in the tissue and the duration of HT treatment [1]. Since many microenvironmental parameters are closely related and can influence one another, the result of hyperthermia on the tumor micromilieu is difficult to predict.

The aim of the study was to analyze the impact of localized 42°C-HT on microcirculatory function, tumor pO_2, microregional redox status and ATP distribution in experimental rat tumors. The study should particularly address the question of how HT treatment at this temperature influences tumor perfusion and whether these changes in tumor blood flow are also reflected in equivalent alterations of the oxygen and bioenergetic status. In addition, the distribution of these changes will be examined to see whether they are equally distributed within the tumor tissue or if regional differences were seen (e.g., between tumor periphery and more central regions).

2. METHODS

2.1 Animals and tumors

Male Sprague-Dawley rats (Charles River Wiga, Sulzfeld, Germany) housed in our animal care facility were used for all experiments in this study.

All experimentation had previously been approved by the regional animal ethics committee and was conducted according to German federal law and the UKCCCR guidelines [5].

DS-sarcomas were implanted on the hind food dorsum by s.c. injection of 0.4 ml DS-ascites (approx. 10^4 tumor cells/μl). Tumors grew as flat, spherical segments and replaced the subcutis and corium completely. Tumor volumes were determined by measuring the three orthogonal diameters of the tumor and using an ellipsoid approximation with the formula: $V = d_1 \times d_2 \times d_3 \times \pi/6$. Tumors reached the target volume of 0.5 - 1.0 ml approx. 6 to 8 days after inoculation.

2.2 Hyperthermia treatment

Localized hyperthermia was induced using of a novel infrared-A radiation technique [6]. Here, an infrared light source was fitted with a water-containing cuvette which absorbs almost all radiation ≥ 1400 nm and within distinct wavelength bands at 940 and 1180 nm. This water absorption of infrared radiation prevents superficial water-containing skin and tissue layers from painful overheating and exsiccosis resulting in a higher energy deposition in deeper seated tissue regions. The infrared-A radiation source was coupled to a thermocontrol feedback system measuring the temperature in the center of the tumor using 250 μm needle-type thermocouples (type 2ABAc, Philips, Kassel, Germany). During the first 20 min of hyperthermia treatment, heating was performed at a heat-up rate of 0.4°C/min until a final steady state temperature of 43°C in the tumor center was achieved, which was maintained over 60 min. This heating protocol resulted in a mean tumor temperature of 42°C over the whole treatment period.

2.3 Laser Doppler flowmetry

As a measure of tumor perfusion red blood cell (RBC) flux was determined using the laser Doppler technique (2 mW He-Ne laser, wavelength 632.8 nm, cut off frequency 12 kHz, Periflux PF 2B, Perimed, Stockholm, Sweden). The measured signal predominantly represents the RBC flux which is defined as the product of the mean local RBC velocity and the volume fraction of erythrocytes within the illuminated volume. This technique is able to monitor the microcirculatory function in small, discrete tissue areas [7]. For insertion of the needle probe (type PF 302, o.d. 0.45 mm) a small incision was made with a 24-gauge needle in the skin covering the tumor such that bleeding from the wound was kept to a minimum. Total backscatter light was recorded continuously in order to optimize probe positioning and ensure a constant probe location. In cases where flux artifacts, due to alteration of probe position (e.g., as a result of movement) occurred (as indicated by a sudden change in total backscattered light) the whole measurement of this tumor was excluded from further analysis. At the end of each experiment, the probes were left in place and the animal was sacrificed by an overdose of pentobarbital-Na to obtain the "biological zero" laser Doppler signal (which was in every case < 10% of the RBC flux at t = 0 min). After subtracting the "biological zero" value, data were expressed as relative RBC flux and represent percentage values related to the flux measurement immediately before hyperthermia treatment.

2.4 Tumor oxygenation

For assessing the temporal changes in tissue oxygenation, mean tumor pO_2 was measured continuously using polarographic catheter electrodes of the Clark type (Licox, GMS, Kiel, Germany). The O_2-sensitive cathode has a length of 5 mm and is placed in a O_2-permeable flexible catheter with an outer diameter of approx. 350 µm. For positioning of the electrode an intravenous catheter was inserted into the tumor along its long axis. After placing the electrode in the trocar the catheter was redrawn leaving the electrode in the tumor center. Measured pO_2 values were averaged over a period of 60 s.

2.5 Determination of the microregional distribution of the redox status

In a second series of experiments, tumors were removed at the end of HT treatment and rapidly frozen in liquid nitrogen. In cryosections of these tissues, the microregional distribution of oxidation-reduction related fluorescence signals of oxidized flavoprotein (FP) and reduced pyridine nucleotides (PN) was measured in order to assess the mitochondrial redox status. For this, the tumor specimen was mounted on a holder constantly cooled with liquid nitrogen. The light of a mercury arc source was filtered (366 nm for excitation of the fluorescence of reduced PN and 436 nm for excitation of oxidized FP) and then directed through one arm of a bifurcated light guide to the surface of the tissue specimen. The fluorescence emitted from the chromophores was then conducted through the other arm of the light guide to a photomultiplier. The fluorescent light was also conducted through two alternative emission filters (450 nm and 540 nm) so that the tissue under the light guide was studied with four pairs of wavelengths (excitation and emission). For analyzing the spatial distribution of PN and FP as well as of the total reflectance the light guide (with a diameter of 20 µm) was moved in steps of 40 µm in two dimensions over the surface of the sample (for a more detailed description of the technique see [8,9]). In each tumor 2 to 4 tissue layers with a vertical distance of approx. 1.5 mm were analyzed. For quantification of the redox status the ratio FP/(FP+PN) was calculated.

2.6 Determination of the microregional and global ATP distribution

From frozen tumor sections obtained at the end of HT treatment, global ATP concentrations were analyzed by HPLC. The microregional ATP distribution within the tissue was measured using single photon counting and quantitative bioluminescence as described previously [10,11]. In brief, tumors were rapidly frozen in liquid N_2. 20 μm sections were cut from at -20°C, heat inactivated at 100°C and subsequently stored at -80°C. For imaging of the ATP distribution, a cover glass with a frozen section adhering to its lower side was placed upon a glass slide with a rectangular casting-mould containing a cooled (4°C) mixture of enzymes and coenzymes to link ATP to a luciferase reaction. The whole preparation was then warmed to 20°C and the spatial distribution of the luminescence intensity within the tissue section directly measured using a microscope (Axiophot, Zeiss, Oberkochen, Germany) and an imaging photon counting system (Argus 100, Hamamatsu, Herrsching, Germany). The spatial resolution gained by this method was approximately 50 μm. The intensity of the bioluminescence signal was calibrated using heat-inactivated tissue homogenates with defined ATP levels. The bioluminescence intensities of the homogenates are linearly related to the metabolic concentrations [10,11].

2.7 Statistical analysis

Results are expressed as means ± standard error of the mean (SEM). Differences between various groups were assessed by two-tailed Wilcoxon-test for unpaired samples. The significance level was set at $\alpha = 5\%$. In order to quantify the spatial distribution of the redox and ATP-images histograms of each slice were calculated and pooled together for each treatment group.

3. RESULTS

Infrared-A HT at an average tumor temperature of 42°C increased tumor perfusion substantially. The RBC flux improved continuously during the whole heating period reaching maximum values towards the end of the treatment (Fig. 1A). The average increase in laser Doppler flux (considering the whole hyperthermia treatment) was approx. 63%. Especially during the last 30 min of HT, large inter-tumor variability of changes in the RBC flux signals was found. The improvement in tumor perfusion was also seen in a lowering of the total reflectance signal at 558 nm in the fluorimetric measurements of the redox status. This signal reflects the total hemoglobin content in the tissue. A decrease in the reflectance (as seen in tumors treated with HT; Fig. 2) corresponds to an increase in the microregional hemoglobin concentration. From these data, it becomes obvious that the HT-induced increase in laser Doppler flux (defined as the product of the mean local RBC velocity and the volume fraction of erythrocytes within the illuminated volume) was at least partially the result of a higher volume fraction of erythrocytes in the tumor tissue.

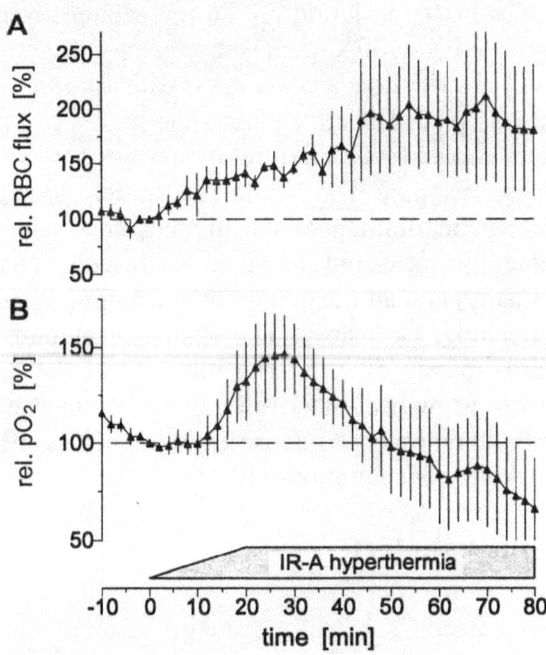

Figure 1. (A) Red blood cell (RBC) flux and (B) mean tumor pO_2 during localized hyperthermia at 42°C. The gray bar indicates the heating time. Each data point represents the mean ± SEM of at least 3 tumors. All measured data are normalized to value immediately before commencement of HT treatment.

During localized hyperthermia the tumor pO_2 showed a biphasic change rising over the first 30 min of treatment before continuously decreasing down to approx. 75% of the baseline pO_2 value prior to heating (Fig. 1B). When averaged over the whole treatment period, only a slight increase of the mean tumor pO_2 of approx. 6% resulted. This negligible improvement in tumor oxygenation corresponded well with the fluorimetric measurements of the mitochondrial redox status. The redox ratio FP/(FP+PN) showed almost no difference between HT-treated and control tumors (Fig. 3). However, the microregional distribution of this ratio exhibited considerable heterogeneity within the tumor with regions showing a ratio of almost 1 adjacent to areas with a ratio close to zero. Nevertheless, the spatial distribution either in control or in HT-treated tumors revealed no regular pattern within the tissue.

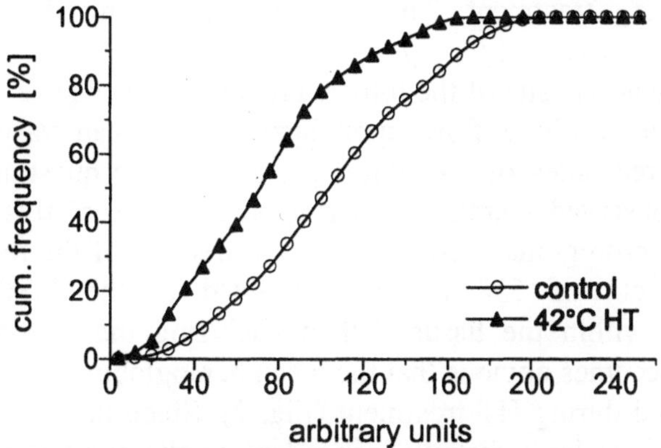

Figure 2. Cumulative frequency distribution of total reflectance at 558 nm during microregional fluorimetric analysis as a measure of local hemoglobin concentration upon 42°C-hyperthermia and under control conditions. Each data point represents the mean of at least 5 tissue slices.

Although the tumor oxygenation did not change during HT and tumor perfusion even increased, the global ATP decreased from 0.99 ± 0.24 µmol/g ($n = 9$) under control conditions down to 0.56 ± 0.10 µmol/g ($n = 10$) at the end of HT treatment. Even though a two-fold change in ATP content could be considered rather substantial, this increase was not statistically significant due to large inter-tumor variability of the ATP concentrations. As was the case for the redox status, the spatial distribution of the ATP concentrations also showed pronounced heterogeneity under control conditions as well as after HT without any regular pattern.

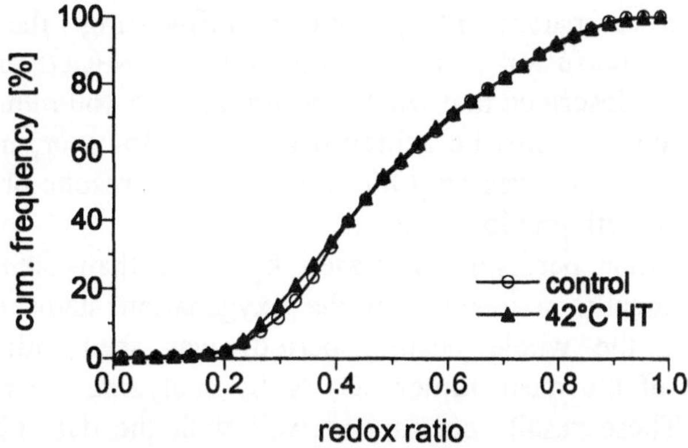

Figure 3. Cumulative frequency distribution of the redox ratio as defined by FP/(FP+PN) (with FP being the concentration of oxidized flavoproteins and PN the concentration of reduced pyridine nucleotides) during microregional fluorimetric measurement upon 42°C hyperthermia and under control conditions. Each data point represents the mean of at least 5 tissue slices.

4. DISCUSSION

Localized 42°C-hyperthermia led to a distinct improvement in tumor perfusion during treatment. The measurements using the laser Doppler technique were performed only with a single light guide thus assessing perfusion only at one site of the tumor. Previous studies have shown that the changes in tumor blood flow upon hyperthermia can vary substantially between different sites of the tumor [2] so that the question arises as to whether the observed increase in perfusion is representative of the whole tumor. The microregional fluorimetric measurement of the total reflectance at a wavelength of 558 nm corresponded to the local hemoglobin concentration within the tissue. When analyzing the distribution of this parameter it becomes obvious that the mean hemoglobin concentration in the tissue increased during HT treatment (Fig. 2). Since the laser Doppler flux measurement is at least partially dependent on the number of erythrocytes within the illuminated tissue volume, the reflectance measurements are in good accordance with the RBC flux changes indicating an increase in tumor perfusion during 42°C hyperthermia. The analysis of the spatial distribution of the reflectance showed no regular pattern in either HT-treated or control tumors. In some sarcomas, more peripheral regions of the tumor exhibited a higher hemoglobin content whereas in others the reflectance signal was more pronounced in the tumor center. These results indicated that the changes in tumor perfusion upon 42°C HT were randomly distributed over the whole tumor tissue.

The response of tumor blood flow upon localized HT showed pronounced variability in a series of studies (for a review see [1]). In animal studies, a decrease in tumor perfusion was mostly seen with tissue temperatures $\geq$ 43°C [1-3,12]. Below this "critical" temperature, the blood flow (at least temporarily) increased or remained at the pre-treatment level [1]. The second paramount parameter influencing the impact of hyperthermia on perfusion is the heating time. REINHOLD and VAN DEN BERG-BLOK [13] described that with exposure times > 160 min, decreases in tumor blood flow can also be achieved at 42°C. However, in the present study total heating time was only 80 min so that the results obtained are in good accordance with previous studies.

Although tumor perfusion increased by more than 60% during HT treatment, practically no change in the oxygenation status of the tissue (averaged over the whole heating period) was seen either by direct measurements of the mean tumor pO_2 or by analyzing the mitochondrial redox status. These results correspond well with the data obtained upon hyperthermia at 43 - 44°C [2] showing no change in tumor pO_2 distribution at the end of HT treatment. However, the question arises as to why tumor oxygenation showed no change even though the oxygen supply to the tissue was substantially increased by an improvement in tumor perfusion. The lack of distinct changes in tumor oxygenation upon 42°C-HT (despite a 63% increase in RBC flux) might be explained by a concomitant increase in the oxygen consumption rate of the cancer cells which has been found to be approx. 58% higher at 42°C compared to normothermia (≈ 35°C) [1].

Analyzing the time course of oxygenation and perfusion shows that significant changes in tumor pO_2 were seen during the first 20 min of heating. During this period the pO_2 increased parallel to the changes in tumor perfusion indicating that at the beginning of the HT treatment the oxygenation status improved as a result of a better O_2 supply. In order to explain the increase in tumor pO_2, it can be concluded that during this early period the oxygen consumption remained almost constant. Later during HT the oxygenation status worsened although tumor perfusion remained improved (Fig. 1). From these results it can be concluded that the HT-induced increase in O_2 consumption occurs at least 20 min after the onset of heating indicating that the switch to a higher metabolic rate (at raised tissue temperature) takes several minutes. To our knowledge there are no *in-vitro* data analyzing the kinetic effects of temperature on oxygen consumption which might be comparable to the results obtained in the present study.

At the end of the HT treatment tumor pO_2 was lower than the pretreatment baseline value although tumor perfusion remained increased almost by a factor of two. The results indicate a pronounced increase in O_2 consumption during this period. Theoretical analyses using mathematical simulation models indicate that changes in oxygen consumption of even 50% would be sufficient to explain the worsening of oxygenation under these conditions [14].

Upon HT treatment the ATP level was markedly reduced by approx. 40% which is in good accordance with previous studies [2,15,16]. In general, a decrease in energy level could be the result of either a reduced ATP formation or an increase in ATP turnover rate. Since in the present study at 42°C the metabolic rate of oxidative reactions was obviously increased (as indicated by a higher O_2 consumption rate of the tumor cells), it seems to be more likely that the reduction in ATP levels were caused by a higher ATP turnover rate rather than from a reduced ATP formation which might take place at higher temperatures (i.e., ≥ 43.5°C) as a consequence of a switch from oxidative to glycolytic glucose degradation.

The tissue samples for determining ATP and the mitochondrial redox potential distributions were taken at the end of the HT treatment. For this reason, the data of these parameters reflect the situation in the tumor after 80 min at elevated temperature. Since the continuous measurement of the tumor pO_2 showed pronounced changes during the heating period (Fig. 1B) it might be expected that the redox status as well as the ATP levels may also vary during the treatment. Presumably, the redox status as an indicator of the cellular oxygenation might be improved during the initial heating period and becomes worsened with ongoing hyperthermia treatment (parallel to the tumor pO_2 measured). However, these assumptions need to be verified in further experiments. The temporal changes in ATP concentration during HT have been measured indicating continuous decrease of this parameter with ongoing HT [16].

5. CONCLUSION

Although 42°C-HT leads to a distinct improvement in tumor perfusion during treatment, there is no significant change in the oxygenation or redox status considering the whole heating period. The lack of distinct changes in tumor oxygenation upon HT can be explained by an increase in the oxygen consumption rate of the tumor. The changes in microenvironmental parameters observed at a tumor temperature of 42°C (improvement of tumor perfusion, reduced ATP levels) might be relevant for the effectiveness of other non-surgical treatment modalities combined with HT (e.g., chemotherapy) or for the direct cell killing effect of HT.

ACKNOWLEDGEMENTS

Supported in part by the Dr.med.h.c. Erwin Braun Foundation (Basel, Switzerland).

REFERENCES

1. Vaupel P. Pathophysiological mechanisms of hyperthermia in cancer therapy. In: Gautherie M, editor. Biological basis of oncologic thermotherapy. Berlin: Springer, 1990; 73-134.
2. Kelleher DK, Engel T, Vaupel PW. Changes in microregional perfusion, oxygenation, ATP and lactate distribution in subcutaneous rat tumours upon water-filtered IR-A hyperthermia. Int J Hyperthermia 1995;11:241-255.
3. Lin JC, Song CW. Influence of vascular thermotolerance on the heat-induced changes in blood flow, pO_2, and cell survival in tumors. Cancer Res. 1993;53:2076-2080.
4. Krüger W, Mayer WK, Schaefer C, Stohrer M, Vaupel P. Acute changes of systemic parameters in tumour-bearing rats, and of tumour glucose, lactate, and ATP levels upon local hyperthermia and/or hyperglycaemia. J Cancer Res Clin Oncol 1991;117:409-415.
5. UKCCCR. United Kingdom Co-ordinating Committee on Cancer Research (UKCCCR) guidelines for the welfare of animals in experimental neoplasia (second edition). Br J Cancer 1998;77:1-10.
6. Vaupel P, Kelleher DK, Krüger W. Water-filtered infrared-A radiation: A novel technique to heat superficial tumors. Strahlenther Onkol 1992;11:633-639.
7. Smits GJ, Roman RJ, Lombard JH. Evaluation of laser-Doppler flowmetry as a measure of tissue blood flow. J Appl Physiol 1986;61:666-672.

8. Chance B, Schoener B, Oshino R, Itshak F, Nakase Y. Oxidation-reduction ratio studies of mitochondria in freeze-trapped samples. NADH and flavoprotein fluorescence signals. J Biol Chem 1979; 254: 4764-4771.
9. Olgin J, Connett RJ, Chance B. Mitochondrial redox changes during rest-work transition in dog gracilis muscle. Adv Exp Med Biol 1986; 200: 545-554.
10. Müller-Klieser W, Walenta S, Paschen W, Kallinowski F, Vaupel PW. Metabolic imaging in microregions of tumours and normal tissues with bioluminescence and photon counting. J Natl Cancer Inst 1988; 80: 842-848.
11. Walenta S, Dötsch J, Müller-Klieser W. ATP concentrations in multicellular spheroids assessed by single photon imaging and quantitative bioluminescence. Eur J Cell Biol 1990; 52: 389-393.
12. Mayer WK, Stohrer M, Krüger W, Vaupel P. Laser Doppler flux and tissue oxygenation of experimental tumours upon local hyperthermia and/or hyperglycaemia. J Cancer Res Clin Oncol 1992; 118: 523-528.
13. Reinhold HS, van den Berg-Blok A. Enhancement of thermal damage to the microcirculation of "sandwich" tumours by additional treatment. Eur J Cancer Clin Oncol 1981; 17: 781-795.
14. Secomb TW, Hsu R, Ong ET, Gross JF, Dewhirst MW. Analysis of the effects of oxygen supply and demand on hypoxic fraction in tumors. Acta Oncol 1995; 34: 313-316.
15. Streffer C. Aspects of metabolic change after hyperthermia. Recent Results Cancer Res 1988; 107: 7-16.
16. Vaupel P, Okunieff P, Neuringer LJ. *In vivo* ^{31}P-NMR spectroscopy of murine tumours before and after localized hyperthermia. Int J Hyperthermia 1990; 6: 15-31.

Chapter **24**

ALLOSTERIC MODIFICATION OF HEMOGLOBIN BY RSR13 AS A THERAPEUTIC STRATEGY

Robert P. Steffen, Jean-Francois Liard, Michael J. Gerber, & Stephen J. Hoffman

Allos Therapeutics, Inc., 11080 CirclePoint Road, Suite 200, Westminster, CO 80020

Abstract: RSR13 binds to hemoglobin (Hb), reduces oxygen (O_2) binding affinity, and enhances O_2 unloading from Hb to hypoxic tissue. Tissue hypoxia is common to cancer, surgery, myocardial ischemia, and stroke. RSR13 increases tumor pO_2, reduces tumor hypoxic fraction and because O_2 is necessary to maximize the effectiveness of radiation therapy, RSR13 enhances the efficacy of radiation therapy. Patients with brain metastases or glioblastoma multiforme receiving RSR13 and radiation therapy have improved median survival, compared to matched historical controls. Myocardial and cerebral hypoxia can be complications to cardiopulmonary bypass (CPB) surgery. RSR13 improves myocardial oxidative metabolism and contractile function in models of myocardial ischemia, including CPB. In CPB patients, RSR13 improved cardiac contractile function and reduced blood product use. In animals, RSR13 increased brain pO_2 and reduced neuronal cell death following cerebral ischemia, alone or in combination with excitotoxic neurotransmitter inhibition. Allosteric modification of Hb by RSR13 represents a unique therapeutic strategy

Key words: Allosteric modifier, Cancer radiotherapy, Cerebral ischemia, Hemoglobin oxygen-binding affinity, Myocardial ischemia, Radioenhancing agent

1. INTRODUCTION

RSR13, 2-[4-[[(3,5-Dimethylanilino)carbonyl]methyl]phenoxy]-2-methyl propionic acid sodium salt (MW 363) (Figure 1), is a synthetic allosteric modifier of hemoglobin (Hb). RSR13 is a small molecule that noncovalently binds in the central water cavity of the hemoglobin tetramer, reduces hemoglobin-oxygen affinity [1], described by an increase in p50 (pO_2 for 50% Hb saturation), and enhances the diffusion of oxygen from the blood to the tissues [2, 3, 4]. RSR13 emulates the function of natural allosteric modifiers such as hydrogen ions, carbon dioxide, and 2,3-diphosphoglycerate. This therapeutic strategy of enhancing oxygen unloading from hemoglobin to tissue emulates and amplifies physiological tissue oxygenation. This approach has potential application in clinical conditions characterized by tissue hypoxia due to: 1) inadequate blood flow (regional or global), 2) insufficient oxygen carrying capacity (e.g., hemorrhage or dilutional anemia), 3) increased tissue oxygen demand unmatched by supply (e.g., myocardial ischemia), and/or 4) insufficient oxygen loading/unloading capacity of hemoglobin (e.g., hypothermia).

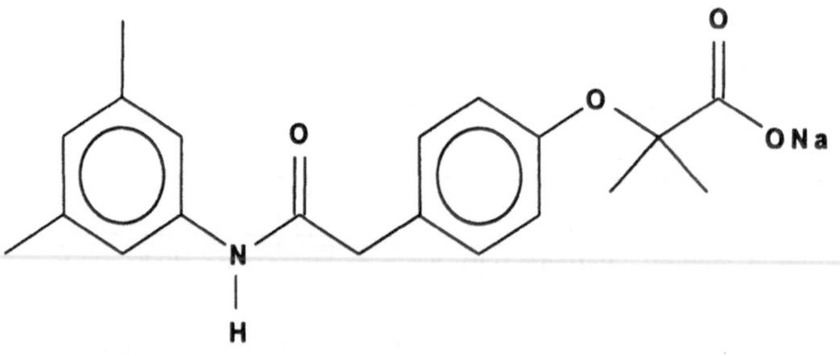

Figure 1. Chemical Structure of RSR13 Sodium.

RSR13 has completed and is currently undergoing a number of animal studies and human clinical trials designed to evaluate the potential clinical utility of this agent in several clinical indications characterized by tissue hypoxia, including radioenhancement in oncology, ischemic coronary artery disease, and surgery/critical care.

All studies reported using animals were approved by the respective institutions Animal Care and Use Committee and all clinical trials reported using humans were approved by the respective institutions Investigational Review Board and the Food and Drug Administration.

2. RADIOENHANCEMENT

Radiation therapy is the principle non-surgical means to achieve local control of cancer, with tumor oxygenation playing an important role in the efficacy of radiation therapy. Hypoxic tumor cells are an important cause of radiation treatment failure because of their relative resistance to cell damage by radiation [5]. Tumor hypoxia adversely affects the clinical prognosis of radiation therapy [6, 7, 8, 9]. Oxygen measurements in human tumors have confirmed tumor hypoxia in squamous cell carcinomas of the uterine cervix [10] and head and neck [11], glioblastoma multiforme [12], breast carcinoma [13], and brain metastases [12, 14]. Because hypoxic cells are substantially more resistant to radiation than oxygenated cells, even small hypoxic fractions in a tumor may affect overall response of the tumor to radiation by increasing the probability that some tumor cells will survive radiation treatment.

Animal pharmacology studies have shown that RSR13 dose-dependently increases blood p50 [3, 15, 16, 17, 18], increases tissue pO_2 [3, 19, 20], and increases oxygen diffusive transport [4] in normal tissue. In animals bearing mammary tumors, RSR13 dose-dependently decreased tumor hypoxic fraction (Figure 2) and increases tumor oxygenation [2, 21]. As shown in Figure 2, RSR13 more effectively reduced tumor hypoxic fraction, under either air or carbogen breathing, than the experimental blood substitutes, perflubron or bovine hemoglobin. In animals bearing FSaII fibrosarcomas or squamous cell carcinomas [17], lung carcinoma [2], or mammary tumors [22], RSR13 increased the efficacy of fractionated radiation. The radioenhancement of RSR13 was shown to be oxygen dependent, with no direct cytotoxic effect on the tumor, bone marrow [2, 22], or skin [22].

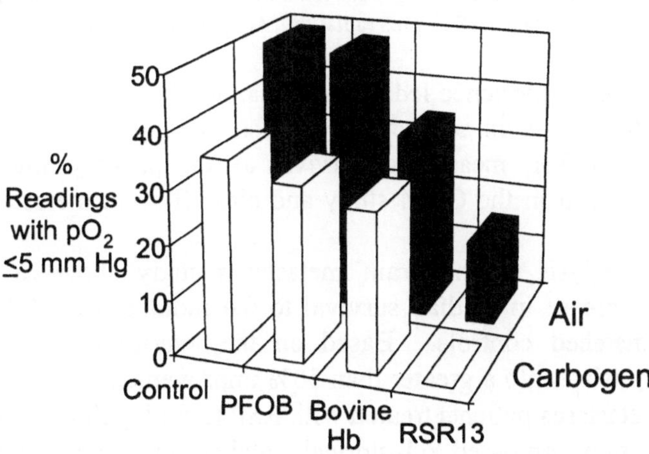

Figure 2. Effect of saline (control), perflubron emulsion (PFOB), bovine hemoglobin, and RSR13 (150mg/kg, iv) on rat mammary carcinoma hypoxic fraction under air breathing (black bars) or carbogen breathing (shaded bars). Adapted from [2, 21].

Clinical trials were initiated with RSR13 based on its mechanism of action, its ability to enhance the efficacy of radiotherapy in animal models, and its preclinical safety profile. RSR13 has successfully completed a Phase

Ia clinical trial in healthy volunteers [23], Phase Ib dose-escalation safety studies in patients with brain metastases requiring palliative radiation therapy [24] or primary brain cancer (glioblastoma multiforme, GBM) [25] undergoing radiation therapy for their cancers. Results have demonstrated encouraging response data and improved survival trends in both patient populations. Illustrated in Figure 3 is the Kaplan-Meier survival for GBM patients receiving RSR13 plus radiation for 6 weeks with a median survival of 13.7 months.

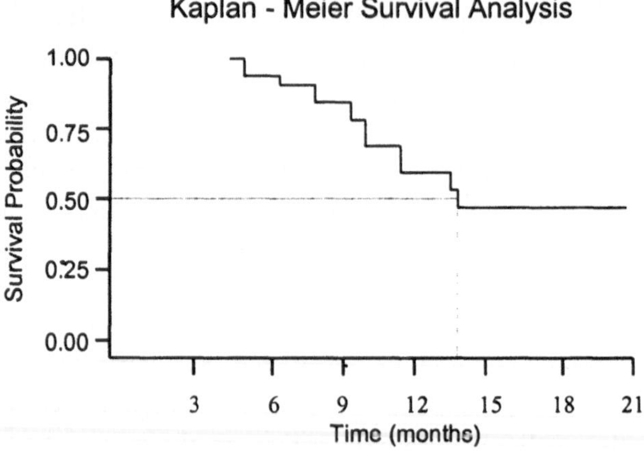

Figure 3. NABTT newly diagnosed glioblastoma multiforme (GBM) Kaplan-Meier survival analysis of patient receiving RSR13 plus radiation. Adapted from [25].

Early clinical experience led to two separate Phase II safety and efficacy trials with RSR13 as an adjunct to radiation therapy in patients with brain metastasis or GBM, measuring survival as the primary study endpoint. Patient enrollment in the GBM study and class II brain metastases study is complete.

Interim analysis in the brain metastases study shows a statistically significant increase in median survival to 6.4 months from 4.2 months for historical matched controls. Based on the interim data, Kaplan-Meier estimates translate into a greater than 50% improvement in mean survival in the brain metastases patients treated with RSR13 with palliative, whole-brain radiation therapy compared to historical matched controls receiving radiation therapy alone.

In collaboration with the NCI's New Approaches to Brain Tumor Therapy (NABTT) CNS Consortium, a 50-patient Phase II clinical trial with RSR13 in conjunction with radiotherapy in patients with GBM was completed. Preliminary analysis of the Kaplan-Meier survival-curve analysis of the Phase I GBM-patients combined with interim Phase II GBM patients treated with RSR13 in conjunction with standard radiation therapy shows that RSR13 treatment significantly increases median survival compared to the NABTT historical control database [26].

3. ISCHEMIC CORONARY ARTERY DISEASE AND SURGICAL HYPOXIA

Myocardial hypoxia occurs as the result of an imbalance of oxygen supply to demand, predominantly due to ischemia. Conventional treatment for ischemia-mediated myocardial hypoxia is to increase coronary blood flow or decrease myocardial oxygen demand. By increasing the release of oxygen from hemoglobin, RSR13 may provide and alternate means of reducing myocardial hypoxia. Previous studies assessed the effect of RSR13 on cardiac function and metabolism in animal models of low coronary perfusion pressure and low blood flow myocardial ischemia. Using a rat isolated-heart model under normothermic conditions, Woods [27] reported that RSR13 attenuated the ischemia-induced decrease in high-energy phosphates, adenosine triphosphate, and myocardial creatine phosphate. In an open-chest canine model of myocardial ischemia, Weiss [28] showed that myocardial high-energy phosphates, pH, and fractional shortening were preserved with RSR13, given prior to or following ischemia (Figure 4). Pagel [16] published data demonstrating that the RSR13 dose-dependently enhanced the recovery of ischemic bed left ventricular segment shortening throughout reperfusion compared to vehicle-treated animals in a model of coronary artery occlusion and reperfusion induced-induced myocardial stunning without affecting function of the non-ischemic circumflex perfused region. In the RSR13-treated animals, the improvement in myocardial contractile function was significantly and positively correlated to the increase in p50.

Significant myocardial and cerebral ischemia may occur in the setting of cardiopulmonary bypass (CPB) surgery. Avoiding ischemic injury during cardiac surgery is dependent on supplying sufficient energy to meet metabolic needs [29, 30]. Although standard clinical cardioplegia and hypothermia provide myocardial protection during surgically-induced ischemia for cardiac surgery, perioperative infarction, stunning, and poor postoperative ventricular function remain significant problems in CPB surgery, especially in high risk patients. Cold cardioplegia has been shown to reduce myocardial energy demand by 95%. However, energy-dependent dependent maintenance of basal cellular metabolism, ionic equilibrium, and membrane integrity is required. Hypothermia not only impairs glycolysis and energy utilization [31], it reduces oxygen delivery by increasing oxygen binding affinity of hemoglobin [32]. In human whole blood, RSR13 reverses hypothermia-mediated increase in hemoglobin oxygen binding affinity [33]. In a dog model of hypothermic blood cardioplegia following normothermic cardiac ischemia, RSR13 supplementation in the cardioplegia significantly improved myocardial mechanical (Figure 5) and oxidative metabolic functions (Figure 6) and improved myocardial and endothelium morphology compared to blood cardioplegia alone [34].

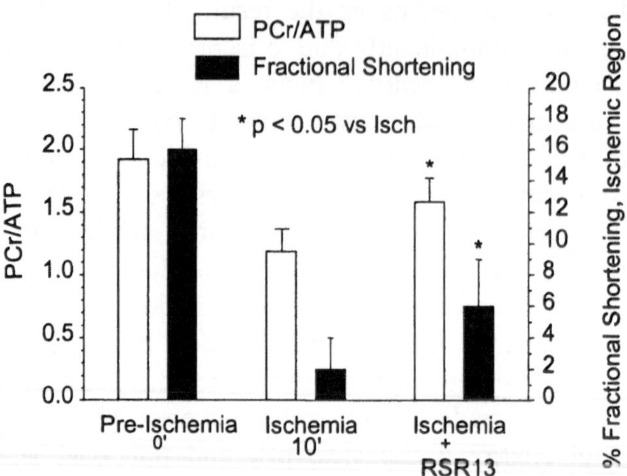

Figure 4. Myocardial PCr/ATP ratios (open bars) and % fractional shortening (solid bars) in the left ventricular ischemic zone prior to ischemia, during ischemia, and during ischemia with RSR13 in anesthetized open-chest dogs. Adapted from [28].

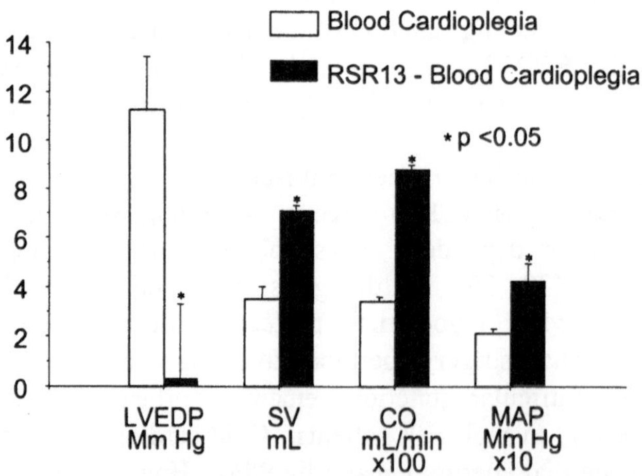

Figure 5. Effect of blood cardioplegia (open bars) or RSR13 (1.75mM) supplemented blood cardioplegia (solid bars) on left ventricular end diastolic pressure (LVEDP), stroke volume (SV), cardiac output (CO), and mean arterial blood pressure (MAP) in anesthetized open-chest dogs, following 30 minute reperfusion at 37°C after 75 minutes of hypothermic (13°C) cardioplegia. Adapted from [34].

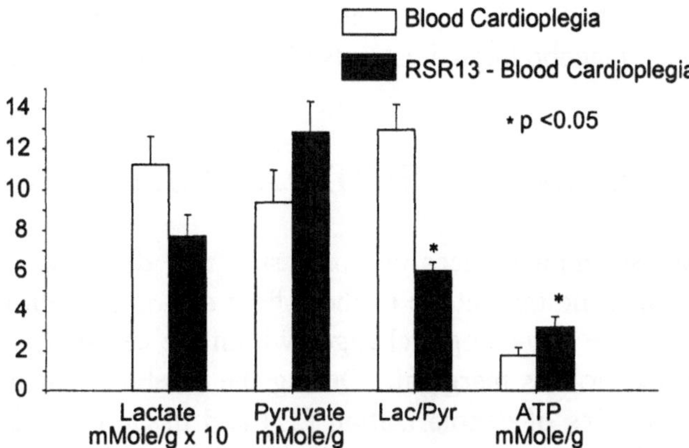

Figure 6. Effect of blood cardioplegia (open bars) or RSR13 (1.75mM) supplemented blood cardioplegia (solid bars) on left ventricular lactate, pyruvate, lactate/pyruvate ratio (Lac/Pyr), and ATP in anesthetized open-chest dogs, following 30 minute reperfusion at 37°C after 75 minutes of hypothermic (13°C) cardioplegia. Adapted from [34].

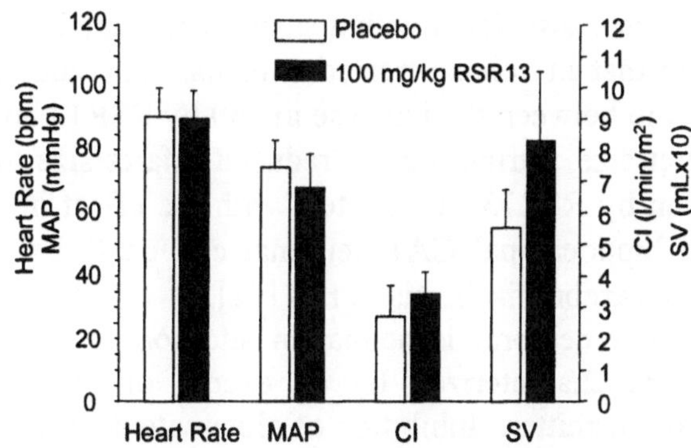

Figure 7. Effect of placebo (open bars, n=14) or RSR13 at 100 mg/kg iv, (closed bars, n=8) on heart rate, mean arterial blood pressure (MAP), cardiac index (CI), and stroke volume (SV) 30 minutes following aortic cross-clamp removal in low-risk patients undergoing elective first time coronary artery bypass graft surgery.

Based on the improvement in myocardial function, oxidative metabolism, and evidence of myocyte and endothelial preservation in models of myocardial ischemia, a 30-patient clinical trial with RSR13 in low-risk patients undergoing CPB for first time coronary artery bypass graft surgery was successfully completed. This randomized double-blind, placebo-controlled study demonstrated that RSR13 could be safely administered and provided evidence that patients receiving RSR13 had improved post-surgical outcomes. Cardiac function tended to be improved in the RSR13-treated patients, supported by improved stroke volume and cardiac output (Figure 7). There was a general trend in the RSR13-treated group for less perioperative packed red blood cell use in the RSR13 treated group compared to the placebo treated patients (413 ml vs 1167 ml, respectively).

4. CEREBRAL ISCHEMIA AND STROKE

Cerebral ischemia reduces neuronal substrate delivery. During focal cerebral ischemia, the magnitude of the infarct size is thought to depend, at least in part, on the physiologic changes within the cerebral penumbra [35, 36] where blood flow is marginal. During the cerebral ischemia and early reperfusion, episodes of depolarization [37] and tissue hypoxia [38] occur, contribution to infarct expansion [39, 40]. By increasing oxygen availability by increasing release of oxygen from hemoglobin, RSR13 may attenuate neuronal damage secondary to cerebral ischemia.

The effect of RSR13 to limit cerebral infarct size under normothermic conditions has been reported. Ischemia-induced cerebral hypoxia was achieved in a cat model by permanent middle cerebral artery occlusion [41]. RSR13 treatment was associated with a significantly smaller cerebral infarct size compared to that in vehicle-treated animals. The authors observed an inverse relationship between the increase in p50 by RSR13 and the reduction in cerebral infarct size, attributing the reduced infarct size to the improved brain oxygenation by RSR13. Consistent with the effect in focal ischemia, RSR13 reduced hippocampal CA1 neuronal cell death by 28% following incomplete global cerebral ischemia in rats [42].

During and after cerebral ischemia, in addition to being hypoxic, the penumbral area is characterized by an excess of glutamate release, a neurotoxic neurotransmitter. Inhibition of glutamate neurotoxicity by the N-methyl-D-aspartate receptor antagonist dizocilpine has been shown to reduce brain damage following focal cerebral ischemia [43, 44]. RSR13 was evaluated in the setting of focal cerebral artery occlusion and reperfusion in combination with dizocilpine. The combination treatment was shown to decrease cerebral infarct size better than dizocilpine alone, when given prior to ischemic insult [45] or after the ischemic insult at the time of reperfusion [46]. In fact, post-ischemic treatment with RSR13 and dizocilpine also was shown not only to reduce cerebral infarct size, but also to significantly improve neurological outcome, compared to dizocilpine alone [46].

5. SUMMARY

RSR13 is a small molecule that noncovalently binds to hemoglobin, reducing hemoglobin-oxygen (Hb-O_2) affinity, and enhances oxygen unloading from hemoglobin, leading to an enhancement of oxygen diffusion from the blood to hypoxic tissue. By enhancing the unloading of oxygen from hemoglobin, RSR13 represents a new therapeutic strategy to affect multiple clinical conditions in which tissue hypoxia plays a central role. RSR13 may improve the outcome of conditions caused by tissue hypoxia due to: 1) inadequate blood flow (regional or global), 2) insufficient oxygen carrying capacity (e.g., hemorrhage or dilutional anemia), 3) increased tissue oxygen demand unmatched by supply (e.g., myocardial ischemia), and/or 4) insufficient oxygen loading/unloading capacity of hemoglobin (e.g., hypothermia). Additional phase III clinical trials will evaluate the efficacy of RSR13 in these clinical indications.

REFERENCES

1. Abraham DJ, Wireko FC, Randad RS, Poyart C, Kister J, Bohn B, Liard JF, Kunart MP. Allosteric modifiers of hemoglobin: 2-[4-[[(3,5-disubstituted anilino)carbonyl]methyl]phenoxy]- 2-methylpropionic acid derivatives that lower the oxygen affinity of hemoglobin in red cell suspensions, in whole blood, and *in vivo* in rats. Biochemistry 1992;31:9141-9149.
2. Teicher BA, Ara G, Emi Y, Kakeji Y, Ikebe M, Maehara Y, Buxton D. RSR13: effects on tumor oxygenation and response to therapy. Drug Development Research 1996; 38:1-11.
3. Kunert MP, Liard JF, Abraham, DJ, Lombard JH. Low affinity hemoglobin increases tissue pO_2 and decreases arteriolar diameter and flow in the rat cremaster muscle. Microvascular Research 1996;52:58-68.
4. Richardson RS, Jordan MC, Haseler LJ, Tagore KS, Wagner PD. Increased VO2max with right shifted Hb-O2 dissociation curve at a constant O2 delivery in dog muscle in situ. J Appl Physiol 1998; 84:995-1002.
5. Hall EJ. The oxygen effect and reoxygenation in radiobiology for the radiologist. 3rd edition, Philadelphia: Lipincott, 1988; 137-160.
6. Brizel DM, Sibley GS, Prosnitz, Scher RL, Dewhirst MW. Tumor hypoxia adversely affects the prognosis of carcinoma of the head and neck. Int J Oncol Biol Phys 1997;38:285-289.
7. Stadler P, Becker A, Feldmann HJ, Hansgen G, Dunst J, Wurschmidt F, Molls M. Influence of the hypoxic subvolume on the survival of patients with head and neck cancer. Int J Radiat Oncol Biol Phys 1999; 44:749-754.
8. Nordsmark M, Overgaard M, Overgaard J. Pretreatment oxygenation predicts radiation response in advanced squanous cell carcinoma of the head and neck. Radiother Oncol 1996; 41:31-39.
9. Fyles AW, Milosevic M, Wong R, Kavanagh MC, Pintilie M, Sun A, Chapman W, Levin W, Manchul L, Keane TJ, Hill RP. Oxygenation predicts radiation response and survival in patients with cervical cancer. Radiother Oncol 1998;48:149-156.
10. Hockel M, Schlenger K, Knoop C. Oxygenation of carcinomas of the uterine cervix: evaluation by computerized O_2 tension measurements. Cancer Res 1991;51:6098-6102.
11. Gatenby RA, Kessler HB, Rosenblum JS, Coia LR, Moldofsky PJ, Hartz WH, Broder GJ. Oxygen distribution in squamous cell carcinoma metastases and its relationship to outcome of radiation Therapy. Int J Radiat Oncol Biol Phys 1988;14:831-838.
12. Ramping R, Cruickshank G, Lewis AD, Fitzsimmons SA, Workman P. Direct measurement of pO_2 distribution and bioreductive enzymes in human malignant brain tumors. Int J Radiat Oncol Biol Phys 1994;29:427-431.

13. Vaupel P, Schlenger K, Knoop C, Hockel M. Oxygenation of human tumors: evaluation of tissue oxygen distribution in breast cancers by computerized O_2 tension measurements. Cancer Res 1991;51:3316-3322.
14. De Santis M, Balducci M, Basilico L, Marucci L, Mattiucci GC, Niespolo RM, Smaniotto D. Radiotherapy, local control and survival in brain tumors. Rays 1998;23:543-548.
15. Kunert MP, Liard JF, Abraham, DJ. RSR13, an allosteric effector of hemoglobin, increases systemic and iliac vascular resistance in rats. Am J Physiol 1996 271(Heart Circ Physiol 40); H602-H613.
16. Pagel PS, Hettrick DA, Montomery MW, Kersten JR, Steffen RP, Warltier DC. 1997, RSR13, a synthetic modifier of hemoglobin-oxygen affinity, enhances recovery of stunned myocardium in dogs. J Pharm Exp Ther 1998;285:1-8.
17. Khandelwal SR, Lin PS, Hall CE, Truong QT, Lu J, Laurent JJ, Joshi GS, Abraham DJ, Schmidt-Ullrich RK. Increased radiation response of FSAII fibrosarcomas in C3H mice following administration of an allosteric effector of hemoglobin-oxygen affinity. Radiat Oncol Invest 1996;4:51-59.
18. Khandelwal SR, Kavanagh BD, Lin PS, Truong QT, Lu J, Abraham DJ, Schmidt-Ullrich RK. RSR13, an allosteric effector of haemoglobin, and carbogen radiosensitize FSAII and SCCVII tumors in C3H mice. Br J Cancer 1998;79:814-820.
19. Khandelwal SR, Randad RS, Lin PS, Schmidt-Ullrich R, Meng H, Pittman RN, Kontos HA, Abraham DJ. Enhanced oxygenation *in vivo* by allosteric inhibitors of hemoglobin saturation. Am. J. Physiol. 1993; 265(Heart Circ 34): H1450-H1453.
20. Grinberg OY, Miyake M, Hou H, Steffen R, Swartz H. The dose-dependent effect of RSR13, a synthetic allosteric modifier of hemoglobin, on physiological parameters and brain tissue oxygenation in rats. Oxygen Transport to Tissue XXII Edited by H Swartz New York, Plenum Press, 1999.
21. Dupuis NP, Kusumoto T, Robinson MF, Liu F, Teicher BA. Restoration of tumor oxygenation after cytotoxic therapy by perflubron emulsion/carbogen breathing. Artif Cells Blood Substit Immobil Biotechnol 1995;23:423-429.
22. Rockwell S, Kelley M. RSR13, a synthetic allosteric modifier of hemoglobin, as an adjunct to radiotherapy: preliminary studies with EMT6 cells and tumors and normal tissues in mice. Radiat Oncol Invest 1998;6:199-208.
23. Venitz J, Gerber M, Abraham D. Pharmacological effects of escalating IV doses of an allosteric hemoglobin (Hb) modifier, RSR13, in healthy volunteers. Pharm Res 1996;13:S-115.
24. Kavanagh BD, Pearlman AD, Schmidt-Ullrich RK, Shaw EG, Dusenbery KE, Gerber MJ. Clinical experience with RSR13, a radiation-enhancing hemoglobin modifier, for recurrent and metastatic brain tumors. Tenth International Conference on Chemical Modifiers of Cancer Treatment 1998.
25. Kleinberg L, Grossman SA, Pantadosi S, Pearlman J, Engelman H, Lesser G, Ruffer J, Gerber M. Phase I trial to determin the safety and pharmacodynamics of RSR13, a novel radioenhancer, in newly diagnosed glioblastoma multiforme. J Clin Oncol 1999;17:2593-2603.
26. Kleinberg L, Grossman SA, Piantadosi S, Pearlman J, Engelhard H, Lesser G, Ruffer J, Priet R, Gerber M. Dose-Frequency, Pharmacokinetics, and Pharmacodynamics (PD) of RSR13 in Patients with Newly Diagnosed Glioblastoma Multiforme (GBM). Proceedings of ASCO 1999;18: 542.
27. Woods JA, Storey CJ, Babcock EE, Malloy CR. Right-shifting the oxyhemoglobin dissociation curve with RSR13: effects on high-energy phosphates and myocardial recovery after low-flow ischemia. 1998; J Cardiovasc Pharmacol 31:359-363.
28. Weiss RG, Mejia MA, Kass DA, DiPaula AF, Becker LC, Gerstenblith G, Chacko VP. Preservation of canine myocardial high-energy phosphates during low-flow ischemia with modification of hemoglobin-oxygen affinity. J Clin Invest 1999;103:739-746.
29. Allard MF, Henning SL, Wambolt RB, Gransleese SR, English DR, Lopaschuk GD. Glycogen metabolism in the aerobic hypertrophied rat heart. Circulation 1997;96:676-682.

30. Ning XH, Childs KF, Bolling SF. Glucose level and myocardial recovery after warm arrest. Ann Thorac Surg 1996;62:1825-1829.
31. Yau TM, Ikonimidis JS, Weisel RD, Mickle DAG, Ivanov J, Moharabeer MK, Tumiati L, Carson S, Liu P. Ventricular function after normothermic versus hypothermic cardioplegia. J Thorac Cardiovasc Surg 1993;105:833-844.
32. Vinten-Johansen J, Julian JS, Yokoyama H. Efficacy of myocardial protection with hypothermic blood cardioplegia depends on oxygen. Ann Thorac Surg 1991; 52:939-948.
33. Steffen RP. Effect of RSR13 on temperature dependent changes in hemoglobin oxygen affinity of human whole blood. Advances in Experimental Medicine and Biology, Vol. 454, Oxygen Transport to Tissue XX, Edited by Hudetz and Bruley, New York: Plenum Press, 1998;653-661.
34. Kilgore KS, Shwartz CF, Gallagher MA, Steffen, RP, Mosca RS, Bolling SF. RSR13, a synthetic allosteric modifier of hemoglobin, improves myocardial recovery following hypothermic cardiopulmonary bypass. Circulation 1999;100:351-356.
35. Memezawa H, Minamisawa H, Smith M, Siesjo B. Ischemic penumbra in a model of reversable middle cerebral artery occlusion in the rat. Exp Brain Res 1992;89:67-78.
36. Nedergaard M, Gjedde A, Diemer N. Focal ischemia of the rat brain: autoradiographic determination of cerebral glucose utilization, glucose content, and blood flow. J Cereb Blood Flow Metab 1986;6:414-424.
37. Back T, Kohno K, Hossman K-A. Cortical negative dc deflections following middle cerebral artery occlusion and KCL-induced spreading depression: effect on blood flow, tissue oxygenation, and electroencephalogram. J Cereb Blood Flow Metab 1994;14:12-19.
38. Back T, Ginsberg M, Dietrich W, Watson B. Induction of spreading depression in the ischemic hemisphere following experimental middle cerebral artery occlusion: effect on infarct morphology. J Cereb Blod Flow Metab 1996;16:202-213.
39. Heiss WD, Grond M, Thiel A, von Stockhausen HM, Rudolf J, Ghaemi M, Lottgen J, Stenzel C, Pawlik G. Tissue at risk if infarction rescued by early reperfusion: a positron emmision tomography study in systemic recombinant tissue plasminogen activator thrombolysis of acute stroke. J Cereb Blood Flow Metab 1998;18:1298-1307.
40. Belayev L, Zhao W, Busto R, Ginsberg MD. Transient middle cerebral artery occlusion by intraluminal suture: I. Three dimensional autoradiographic image analysis of local cerebral glucose metabolism-blood flow relationships during ischemia and early recirculation. J Cereb Blood Flow Metab 1997;17:1266-1280.
41. Watson JC, Doppenberg EMR, Bullock MR, Zauner A, Rice MR, Abraham D, Young, HF. Effects of the allosteric modification of hemoglobin on brain oxygen and infarct size in a feline model of stroke. Stroke 1997;28:1625-1631.
42. Grocott HP, Bart RD, Sheng HX, Miura Y, Steffen RP, Pealstein RD, Warner DS. Effect of a synthetic allosteric modifier of hemoglobin oxygen affinity on outcome from global cerebral ischemia in the rat. Stroke 1998;29:1650-1655.
43. Hatfield RH, Gill R, Brazell C. The dose-response relationship and therapeuticwindow for dizocilpine (MK-801) in a rat focal ischaemia model. Eur J Pharmacol 1991;216:1-7.
44. Iijaima T, Mies G, Hossman K-A. Repeated negative DC deflections in rat cortex following middle cerebral artery occlusion are abolished by MK-801 – effect on volume of ischemic injury. J Cereb Blood Flow Metabol 1992;12:727-733.
45. Sarraf-Yazdi S, Sheng H, Grocott HP, Bart RD, Pearlstein RD, Steffen RP, Warner DS. Effects of RSR13, a synthetic allosteric modifier of hemoglobin, alone and in combination with dizocilpine, on outcome from transient focal cerebral ischemia in the rat. Brain Research 1999;826:172-180.
46. Mackensen GB, Nellgård B, Sarraf-Yazdi S, Dexter SF, Steffen RP, Grocott HP, Warner DS. Post-ischemic RSR13 amplifies the effect of dizocilpine on outcome from focal cerebral ischemia in the rat. Anesthe Analg 1999;88:SCA94.

Chapter **25**

THE PHARMACOLOGY OF TISSUE OXYGENATION BY BIOPURE'S HEMOGLOBIN-BASED OXYGEN CARRIER, HEMOPURE® (HBOC-201)

L. Bruce Pearce and Maria S. Gawryl

Biopure Corporation, 11 Hurley Street, Cambridge Massachusetts 02141

Abstract: Biopure's hemoglobin-based oxygen carrier, HBOC-201 (Hemopure®), enhances oxygen transport by promoting both the convective and diffusive components of transport in the microcirculation. Convective transport is modified by HBOC-201 in three ways; (i) volume expansion promotes organ and tissue perfusion, (ii) the low viscosity of HBOC-201 improves flow to tissues, and (iii) oxygen delivery by HBOC Hb in the plasma is relatively insensitive to mechanisms regulating RBC distribution in the microcirculation. Diffusive oxygen transport is increased by the higher P_{50} compared with native RBC Hb which increases the off-loading of oxygen to tissues. Oxygen transport is also increased by reducing the diffusional barrier to oxygen transport associated with the plasma, in which oxygen is sparingly soluble. Biopure's HBOC solutions have been shown *in vitro* and *in vivo* to take up and off-load oxygen more efficiently than RBC Hb, and when added to blood can increase the efficiency of RBC oxygen transport.

Key words: Blood substitutes, hemoglobin, oxygen carrier, stroma-free

Abbreviations: Hb: hemoglobin
HBOC: hemoglobin-based oxygen carrier
HBOC-201: Hemopure®
HBOC-301: Oxyglobin®
pO_2: partial pressure of oxygen

Oxygen Transport to Tissue XXIV, edited by
Dunn and Swartz, Kluwer Academic/Plenum Publishers, 2003

1. INTRODUCTION

The pharmacology of hemoglobin-based oxygen carrying solutions is defined by a complex combination of effects of these solutions on the rheology of blood and the efficiency of oxygen transport. The primary pharmacodynamic effects of HBOC-201, however, are associated with oxygen transport from the lung to the tissues. HBOC-201 enhances oxygenation by enhancing both convective and diffusive oxygen transport (Figure 1). Convective transport refers to the movement of oxygen-laden Hb in blood which, in the case of HBOC-201, can be modified in three ways. First, HBOC-201 acts as a volume expander, promoting tissue and organ perfusion. Second, the microvascular rheology of HBOC-201 contributes to increased flow to tissues by reducing viscosity during hemodilution (viscosity of HBOC-201 = 1.3 centipoise at 37°C); Hemodilution to hematocrits as low as 15% have been suggested to be optimal in the cerebral microcirculation[1]. Third, the distribution of HBOC-201 is not limited by the normal mechanisms that restrict/redirect delivery of oxygen to tissues by RBCs. While mammals have evolved elegant mechanisms for conserving and controlling oxygen delivery to the body's tissues, the perfusion of capillaries by HBOC solutions is relatively insensitive to these mechanisms. Accordingly, infusion of even a small amount of HBOC-201 can produce an increase in the amount of oxygen carrying Hb that enters all capillary beds. The latter may be important in disease and tissue injury where capillaries are not perfused adequately by red blood cells.

The diffusive component of oxygen transport, specifically the flux of oxygen into and out of blood, is significantly altered by HBOC-201. There are two mechanisms by which HBOC-201 modifies diffusive oxygen transport. The higher P_{50} of HBOC-201 increases the tendency to off-load oxygen to tissues compared to native RBC Hb. The most significant effect on the diffusive component of oxygen transport results from the presence of HBOC-201 in the plasma. Oxygen is sparingly soluble in plasma and thus the plasma acts as a barrier limiting the transfer of oxygen from RBC Hb to the tissues. The distribution of HBOC-201 in the plasma phase decreases the magnitude of this resistance by facilitating the transfer of oxygen from RBC Hb to the tissues.

2. THE EFFECTS OF HBOC FORMULATIONS ON OXYGEN DIFFUSIVITY

The resistance to movement of oxygen from RBC Hb in the circulation to the sites of utilization at mitochondria in tissues can be divided into several

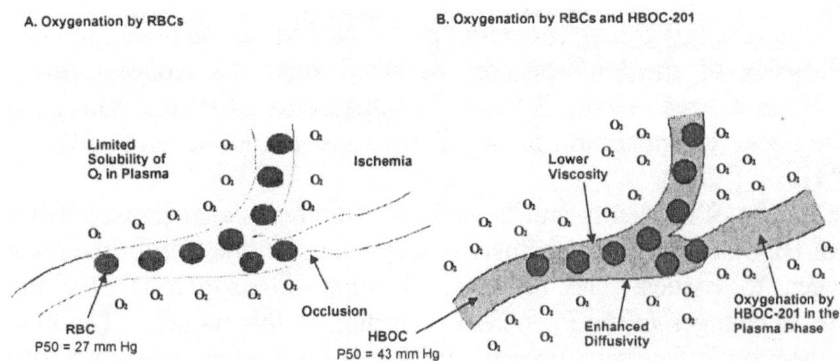

Figure 1. Oxygenation by RBCs and HBOC-201. Panel A depicts normal tissue oxygenation in areas adequately perfused by RBCs (upper portion of the figure) and restricted RBC flow into an occluded vessel resulting in tissue ischemia (lower portion of figure). Panel B depicts oxygenation of tissue by RBCs and HBOC-201 showing the enhanced oxygenation due to improved flow due to decreased viscosity of blood and increased diffusivity resulting in improved off-loading of oxygen. Oxygenation distal to the occlusion via HBOC-201 carried in the plasma is depicted in the branching vessel to the right.

components. One of the components of this resistance is contributed by the plasma, which has been estimated to account for approximately 50% of the total resistance[2]. Theoretically, it should be possible to increase the diffusivity of oxygen by reducing the resistance to oxygen flux contributed by the plasma. This question has been directly examined by Page et al.[3,4,5] who have studied the oxygen transport properties of various hemoglobin solutions *in vitro*. They directly examined oxygen flux in a 25-100 μm diameter artificial capillary model system. This model system was validated in a study comparing experimental measurements with predictive mathematical models of oxygen transport[5]. These investigators developed a mathematical model that incorporated erythrocyte and acellular Hb solution phases, radial Hct and velocity gradients, axial convection, radial diffusion of oxygen and oxyHb in arteriolar-sized artificial capillaries. Using this validated microcapillary system, oxygen flux measurements were made for hemoglobin solutions, red blood cell suspensions, and RBC/Hb solution mixtures. The model very closely predicted the observed oxygen fluxes in RBC, Hb, and RBC/Hb mixtures measured *in vitro* in the artificial capillaries.

Biopure's HBOC solutions (polymerized and native) were found to be substantially more efficient than RBC suspensions in both the uptake and release of oxygen. Increasing the extracellular hemoglobin content of mixtures of RBCs and hemoglobin solutions, resulted in an increase in oxygen transport efficiency when the total hemoglobin content remained constant at 10g/dL. When the fraction of HBOC Hb content reached 50%

(5g/dL), the mixture had the equivalent oxygen transport efficiency of pure HBOC solutions (10 g/dL). Modeling predicted that the maximal increase in the efficiency of oxygen transport occurred when the concentration of HBOC Hb in plasma reached 3-5 g/dL in the absence of RBCs. This agreed with the observed maximum achieved with the 1:1 mixture of RBCs and HBOC Hb.

The results of modeling and *in vitro* measurements also showed that the effect of HBOCs on oxygen diffusivity appears to be greatest for the release of oxygen to tissues. The release/off-loading of oxygen is also more sensitive to changes in the P_{50} of Hb according to this model. The higher the P_{50} the more effectively oxygen is off-loaded to tissues. Thus, we would predict that the higher P_{50} of HBOC-201, 37 mmHg, compared with native human RBC Hb, would further enhance off-loading. The lesser effect of HBOCs on the uptake of O_2 is probably related to the decreased O_2 gradient during uptake.

More recently, Page et al.[6] have examined oxygen fluxes in a capillary model system with artificial capillaries having a 10 μm diameter. One of the most significant observations to come out of this study is that these plasma soluble Hb formulations have a larger effect on the efficiency of oxygen fluxes when the diameter of the capillaries is smaller. Thus, the results obtained with 25-100 μm diameter capillaries underestimated the magnitude of the effect of HBOCs on oxygen transport in the microcirculation comprising capillary beds.

Accordingly, modeling of oxygen fluxes and the observed fluxes *in vitro* suggest that Biopure's HBOC solutions more efficiently transport oxygen than red blood cells, and when added to RBCs can also increase the efficiency of red blood cell oxygen transport. This predicted enhancement of the effectiveness of oxygenation by HBOC formulations is supported by the results of *in vivo* studies described below.

3. HBOC-201 IN ARTERIAL STENOSIS AND OCCLUSION

The presence of HBOC-201 in the plasma can affect the efficiency with which oxygen is delivered to tissues, and this is best illustrated in an animal model in which vessels have been partially occluded so as to limit red blood cell flow to tissues distal to the restriction. Horn et al.[7] investigated the effects of infusion of Oxyglobin® (HBOC-301) in a dog model of arterial stenosis. Animals underwent isovolemic hemodilution with lactated Ringer's solution to a hematocrit of 25%, then 95% stenosis of the popliteal artery was established to limit blood flow to the gastrocnemius muscle. Animals were randomized to treatment with either Oxyglobin® or 6% hetastarch. Hemodynamic parameters, blood gases, temperature, and tissue oxygen levels were measured. Skeletal muscle tissue oxygen tension was directly measured in the gastrocnemius muscle using an Eppendorff needle probe. In both groups oxygen delivery and oxygen consumption decreased in parallel with decreasing blood flow due to the arterial stenosis. During stenosis the 50th percentile tissue pO_2 (tpO_2) was lower (7.9±3.0 and 8.6±2.2 mm Hg) than baseline measurements taken following isovolemeic hemodilution (27.2±7.8 and 22.9±5.5 mm Hg) in both groups. Following treatment the tpO_2 remained low in the hetastarch group (9.9±2.9 mm Hg), but returned (29.4±7.6 mm Hg) to pre-occlusion levels following infusion of Oxyglobin® as shown in Figure 2. These data suggest that HBOC carried in the plasma phase of blood and flowing past such severe occlusions (as depicted in Figure 1) can oxygenate tissues distal to vascular blockages that would normally result in tissue ischemia and possibly tissue injury.

The study by Horn et al.[7] illustrates the importance of having Hb-mediated oxygen carrying capacity in the plasma and that HBOC-201 can be used to take advantage of the plasma flow through areas that are poorly perfused with RBCs. This approach to enhancing tissue oxygenation may be important in coronary or cerebrovascular disease, where significant narrowing of these vessels contributes to ischemia, compromised function and ultimately infarction.

4. HBOC-201 IN ANEMIA DUE TO ISOVOLEMIC HEMODILUTION

The ability of HBOC-201 to enhance oxygen transport can be illustrated *in vivo* by directly monitoring the effects of infusing HBOC-201 in animals made anemic by isovolemic hemodilution. Standl et al.[8] compared the effects of treatment with Biopure's HBOC solutions, stored RBCs, and freshly donated blood on hemodynamics, oxygen transport capacity, and tissue oxygenation after acute and almost complete isovolemic hemodilution in a canine model. Animals were exposed to stepwise hemodilution down to hematocrit values of 20%, 15%, and 10% before receiving stepwise isovolemic augmentation of 1g Hb per deciliter of blood. The added Hb was in the form of HBOC-201 or RBC Hb. The Eppendorf microelectrode technique was used to directly measure the muscle tissue oxygenation under these experimental conditions. In all groups, initial hemodilution resulted in increased heart rate and cardiac index and decreased vascular resistance in

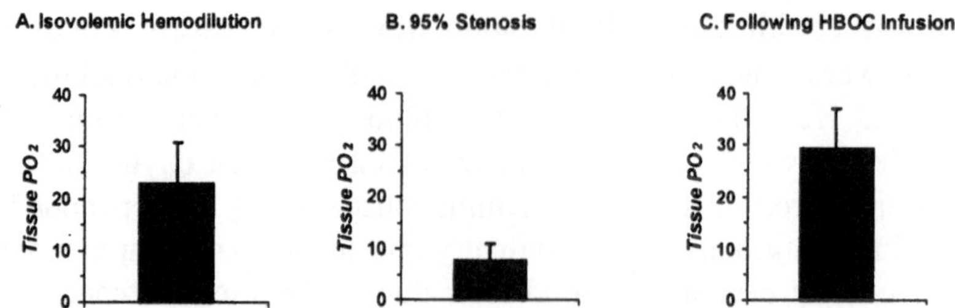

Figure 2. Tissue Oxygenation in a Canine Model of Severe Arterial Stenosis. Horn et al. investigated the effects of infusion of Oxyglobin® in a dog model of arterial stenosis. Animals underwent isovolemic hemodilution with lactated Ringer's solution to a hematocrit of 25%, then 95% stenosis of the popliteal artery was established to limit blood flow to the gastrocnemius muscle. Muscle 50th percentile tissue pO_2 values are shown in A following isovolemic hemodilution (27.2±7.8 mm Hg), in B following 95% occlusion (7.9±3.0 mm Hg), and in C following administration of 100 ml (13g) Oxyglobin (29.4±7.6 mm Hg).

response to the decrease in total Hb. Tissue oxygen tensions were reduced in response to hemodilution and restored with 0.7g/dL HBOC, 2.7g/dL stored RBCs, and 2.1g/dL freshly donated RBCs. These data indicate that the HBOC-201 was approximately three times more potent than stored or fresh RBCs at restoration of baseline tissue oxygenation following severe acute anemia. The restoration of tissue oxygen tensions was also associated with a higher extraction ratio in the HBOC-201 group (59±8%, $p<0.01$) compared with either RBC group. That is, a larger fraction of the oxygen carried by the circulation was delivered to the tissues in the presence of HBOC-201. These results are consistent with and would be predicted by the *in vitro* observations of Page et al.[4,6] The increased extraction fraction observed with HBOC is predicted by the effect of the elevated P_{50} , the enhanced oxygen off-loading from HBOC-201 in plasma, and the facilitation of oxygen off-loading from RBC Hb.

5. HBOC-201 ENHANCES OXYGENATION AND PROTECTS AGAINST REPERFUSION INJURY

Further evidence for enhanced oxygenation comes from the results of studies conducted in a canine model of myocardial infarction. Cole et al.[9] recently investigated the cardioprotective effects of HBOC-201 infusion in a canine model of complete occlusion of the left anterior descending coronary artery. HBOC-201 (30g) was infused just prior to complete occlusion of the coronary artery for 90 minutes whereupon reperfusion was established for

4.5 hours. Histological analysis showed a >55% reduction in infarct size due to infusion of HBOC-201 compared to control (saline). These data show that the enhanced oxygenation associated with having an oxygen carrier in the plasma can also mitigate injury due to complete occlusion. The mechanism(s) by which this occurs has not been delineated, but these results clearly suggest that HBOC-201 can alter the dynamics of tissue oxygenation in such a way as to "buffer" against the reperfusion injury, characteristic of tissue ischemia. Furthermore, Hayward and Lefer[10] have recently reported that administration of HBOC-201 (10% of estimated blood volume) to rats following whole body trauma resulted in stabilization of hemodynamics, protection of endothelial function, and increased survival.

6. HBOC-201 IN OXYGENATION OF TUMORS

Enhanced oxygenation of tumors has been of interest to radiologists for some time because of the observed relationship between tumor tissue hypoxia and resistance to the cell killing effects of irradiation. The ability of HBOC-201 to enhance oxygen off-loading to tissues and to access all capillaries via plasma flow, gives it the potential to reduce tumor hypoxia and perhaps enhance the efficacy of radiation therapy. Teicher et al.[11,12,13] investigated the ability of Biopure's HBOC formulations to increase tumor pO_2 levels in 9L gliosarcoma or 13672 mammary adenocarcinoma tumors in rats. Tumors were induced by hind limb or intracranial implantation of tumor cells. Tumor pO_2 levels were measured directly using a polarographic needle microelectrode while rats were breathing either normal air or carbogen ($95\%0_2/5\%CO_2$). Tumor growth delay was measured following chemotherapy, radiation treatment, or the combination of chemotherapy and fractionated radiation treatment. The results of these studies show that infusion of Biopure's HBOC formulations increased tumor pO_2 levels and decreased the hypoxic fraction of tumors. In almost all treatment paradigms investigated, the increased tumor oxygenation corresponded with an increase in tumor growth delay.

One of the factors that may contribute to the resistance to radiation therapy of malignant tumors is the reported heterogeneity of oxygen distribution in these tumors.[14] This heterogeneity is seen even with the administration of treatments to increase tumor oxygenation such as carbogen breathing. Theoretically, the widespread distribution of HBOC Hb carried by plasma may decrease the net heterogeneity of oxygen delivery and further reduce the fraction of tumor tissue that is resistant to radiotherapy. Although not directly demonstrated in the studies described by Teicher and colleagues, it is possible that the observed net decrease in the hypoxic fraction of the tumors was due, at least in part, to a reduction in the heterogeneity of oxygen distribution throughout the tumor. HBOCs may decrease the heterogeneity of oxygen distribution in tumors because the distribution of HBOC Hb via plasma flow into all capillaries of tumors, coupled with the enhanced off-loading of O_2 to tissues, may result in more uniform delivery of oxygen to the total tumor.

7. CLINICAL EVIDENCE OF ENHANCED OXYGENATION

The earliest Phase I clinical trials performed with HBOC-201, investigated the effects of this HBOC in normal volunteers. In these studies, the pharmacokinetics[15] of HBOC-201 and the effects of this oxygen carrier on exercise performance were examined[16] Hughes et al.[16] studied the effect of autologous blood transfusion compared to transfusion with HBOC-201 on exercise performance in a bicycle exercise stress test. Performance was evaluated at the anaerobic threshold, and diffusing capacity of the lung was estimated using the single breath carbon monoxide technique. Normal volunteers were first phlebotomized, 15% of their estimated blood volume was removed and replaced with lactated Ringer's solution, after which their performance on the bicycle stress tested was evaluated. Following autologous blood transfusion or infusion of HBOC-201, their performance was evaluated again. The results of these studies showed that infusion of 45g Hb in the form of HBOC-201 resulted in the equivalent exercise performance and diffusing capacity as observed with 150g Hb added as RBCs. This 1:3 ratio is similar to the difference in potency reported by Standl et al.[8] (see above) in dogs where tissue oxygenation was measured directly.

Although the relative potency of the HBOC-201 and RBCs was not specifically compared in more recent clinical studies, the results of these studies show that in postoperative and perioperative settings, including cardiopulmonary bypass, abdominal aortic reconstruction, and orthopedic surgery, that HBOC-201 can completely eliminate the need for allogenic blood transfusion in 34, 27, and 43% of subjects, respectively. These three studies in combination with the 16 other clinical studies (total of 19 completed studies) involving 420 subjects exposed to HBOC-201, also demonstrate that this oxygen therapeutic is well tolerated under a variety of circumstances.

8. CONCLUSION

The results of *in vitro* oxygen flux studies have provided the foundation for understanding the basic mechanisms underlying the enhanced oxygenation observed in hemodilution studies, ischemic skeletal muscle, implanted tumors, and exercise tolerance in human subjects as well as the effectiveness in limiting allogeneic blood transfusion in surgery. Further work is needed with model systems to help in understanding the mechanisms that underlie the protective effects of HBOC-201 in reperfusion injury. Investigation of the impact of enhanced oxygenation by Biopure's HBOC formulations on the outcome and management of medical conditions involving ischemia is an important area of concentration as we expand our human clinical studies of additional indications for the use of HBOC-201. Clearly, optimal medical use of HBOC-201 will be dependent upon establishing a clear understanding of how these oxygen carriers enhance oxygen delivery to tissues in both health and disease. With the knowledge derived from *in vitro* model systems, we may be better able to identify the optimal conditions for the clinical use of these oxygen carriers as well as the scope of their potential application.

REFERENCES

1 Hudetz AG, Wood JD, Biswal BB, Krolo I, Kampine, JP. Effect of hemodilution on RBC velocity, supply rate, and hematocrit in the cerebral capillary network. J. Appl Physiol 1999; 87(2):505-509.

2 Hellums, JD The resistance to oxygen transport in the capillaries relative to that in the surrounding tissue. Microvas Res 1977;13: 131-136.

3 Page, TC, McKay, CB, Light, WR, Hellums, JD. Experimental simulation of oxygen transport in microvessels. In RM Winslow, KD Vandegriff, and M Intaglietta, editors. Blood Substitutes: New Challenges. Boston: Birkhauser, 1996, 132-145.

4 Page, TC, Light, WR, McKay, CB, and Hellums, JD. Oxygen transport by erythrocyte/hemoglobin solution mixtures in an in vitro capillary as a model of hemoglobin-based oxygen carrier performance. Microvas Res 1998; 55:54-64.

5 Page, TC, Light, WR, and Hellums, JD. Prediction of microcirulatory oxygen transport by erythrocytes/hemoglobin solution mixtures. Microvas Res 1998; 56:113-126.

6 Page TC, Light, WR, and Hellums, JD. Oxygen transport in 10 µm artificial capillaries In Oxygen Transport to Tissue XXI . New York: Kluwer Academic/Plenum Publishers, 1999, 715-722.

7 Horn, E-P, Standl, T, Wilhelm, S, Jacobs, EE, Freitag, U, Freitag, M. Bovine hemoglobin increases skeletal muscle oxygenation during 95% artificial arterial stenosis. Surgery 1997; 121:411-418.

8 Standl, T, Horn, P, Wilhelm, S, Greim, C, Freitag, M, Freitag, U. Bovine haemoglobin is more potent than autologous red blood cells in restoring muscular tissue oxygenation after profound isovolaemic haemodilution in dogs. Can. J. Anesth. 1996; 43: 714-723.

9 Cole, P.A., Caswell, JE, Jones, SP, Lefer, DJ. A novel blood substitute protects against myocardial reperfusion injury. ISOTT, 1999;Abstr 53.

10 Hayward, R, Lefer, AM. Administration of polymerized bovine hemoglobin improves survival in a rat model of traumatic shock. Methods Find Exp Clin Pharmacol 1999; 21(6):427-433.

11 Teicher, BA, Schwartz, GN, Sotomayor, EA, et al. Oxygenation of tumors by a hemoglobin solution. J Cancer Res Clin Oncol 1993; 120:85-90.

12 Teicher, BA, Holden, SA, Menon, K, Hopkins, RE, Gawryl, MS. Effect of hemoglobin solution on the response of intracranial and subcutaneous 9L tumors to antitumor alkylating agents. Cancer Chemother Pharmacol 1993; 33:57-62.

13 Teicher, BA Dupus, NP, Emi, Y, et al. Increased efficacy of chemo- and radio-therapy by a hemoglobin solution in the 9L gliosarcoma. in vivo 1995; 9: 11-18.

14 Al-HallaQ, HA, River, JN, Zamora, M, Oikawa, H, Karczmar, GS. Correlation of magnetic resonance and oxygen microelectode measurements of carbogen-induced changes in tumor oxygenation. Int. J. Rad Oncol Biol Phys 1998; 41(1):151-159.

15 Hughes, G.S, Antal, EJ, Locker, PK, Francom, SF, Adams, WJ, Jacobs, EE .Physiology and pharmacokinetics of a novel hemoglobin-based oxygen carrier in humans. Crit Care Med. 1996; 24:756-764.

16 Hughes, GS, Yancey, E.P, Albrecht, R, Locker, PK, Francom, SF, Orringer, EP, Antal, EJ, Jacobs, EE. Hemoglobin-based oxygen carrier preserves submaximal exercise capacity in humans. Clin Pharmacol Ther 1995; 58434-443.

Chapter **26**

THE CONCEPT OF HEMOGLOBIN EQUIVALENCY OF PERFLUOROCHEMICAL EMULSIONS

N. Simon Faithfull
Alliance Pharmaceutical Corp., San Diego, California

Abstract: Perfluorochemical (PFC) emulsions have been in development as intravenous oxygen carriers for a number of years and many publications have dealt with their oxygen transport characteristics in both experimental models and in clinical trials. Though it has been stressed on numerous occasions that PFCs deliver oxygen to the tissues in very different ways to those by which Hemoglobin (Hb) releases oxygen (O_2), no serious attempts have been made to correlate the oxygen delivery capacity of PFCs to those of Hb. This paper presents theoretical ways in which this can be done and demonstrates that a 2.7 g/kg dose of PFC is approximately equivalent to 4 g/dL [Hb]. Clinical trial planning is discussed.

1. OXYGEN TRANSPORT BY PERFLUOROCHEMICALS

The relationship between the partial pressure of oxygen (PO_2) and the Oxyhemoglobin saturation (SO_2) follows the well-known S-shaped curve. This curve can be shifted to the right or left depending on the concentration of the allosteric modifier, 2,3 diphosphoglycerate (2,3 DPG), in the red cells - less 2,3 DPG shifts the curve to the left and vice versa. The curve is also shifted and the PO_2 for half-saturation value (P50), increased by ambient conditions such as increase in body temperature, decrease in pH or increase in carbon dioxide tension (PCO_2). Decreases in temperature and PCO_2 or increases in pH have an opposite effect and the oxyhemoglobin dissociation curve shifts to the left. The amount of O_2 that can be bound by one gram of Hb is variously quoted in the literature as varying from 1.306 ml [1], 1.34 ml or 1.39 ml based on the molecular weight of Hb [2]. The value used in calculations in this paper was 1.34 ml per gm of Hb as this has been shown to give good results at normal acid base values over a wide range of arteriovenous O_2 content differences [3].

Oxygen content of the plasma, in contrast to that of hemoglobin, is directly related to the PO_2 . The absolute O_2 content is slightly effected by changes in temperature as the oxygen solubility coefficients are decreased slightly as temperature rises; for all practical purposes the oxygen solubility in plasma can be taken to be 0.3 ml per dL of plasma per 100 mm Hg PO_2. PFC emulsions transport O_2 in precisely the same way as does plasma, ie., as a direct function of their solubility coefficient and the PO_2. The difference, of course lies in their considerably greater solubility for O_2 - in the case of perflubron (perfluoro-octyl bromide), the major PFC component in Oxygent^{TM} (Alliance Pharmaceutical Corp., San Diego, CA) this solubility amounts to about 52.7 ml of O_2 per 100 ml of Perflubron. All calculations in this paper have been based on this number.

Oxygen transport characteristics of Hemoglobin and PFCs are shown in Fig.1, which demonstrates the relationship between O_2 content and PO2 in blood with a normal [Hb] of about 14 g/dL and whose red cell Hb, when fully saturated will contain about 20 ml of O_2 per dL. At a normal ambient PO_2 at sea level of about 150 mm Hg, the patient with normal pulmonary function will be able to achieve a PaO_2 of approx. 100 mm Hg, sufficient to ensure about 98 percent oxyhemoglobin saturation. Plasma will contain about 0.3 ml/dL and the PFC emulsion *Oxygent* (60 percent w/v Perflubron) will contain about 2.2 ml/dL. As PO_2 is decreased to about 40 mm Hg, about the level of the mixed venous oxygen tension ($P\bar{v}O_2$) approximately 20 to 25% of the oxygen bound to hemoglobin (amounting to about 4 ml/dL) will be released to the tissues. In contrast, due to their straight line "dissociation curve" plasma and *Oxygent* will release 60 percent of their dissolved oxygen, amounting to about 0.18 and 1.3 ml of O_2 respectively.

When the patient breathes an inspired O_2 fraction (FiO_2) of 1.0, PaO_2 rises and under favorable circumstances can easily reach 500 mm Hg. This results in marginal increase in O_2 content of the already almost saturated Hemoglobin, but a considerable increase in O_2 content in both the plasma and the PFC to about 1.5 and 10.8 ml/dL respectively. When the PO_2 is lowered under these circumstances to $P\bar{v}O_2$ values, the PFC can now release 92 percent of its contained O_2, or 9.75 ml/dL. This amount is about twice that released by hemoglobin under normal circumstances and is more than enough to sustain life. This fact was first dramatically demonstrated many years age when the blood of oxygen breathing rats was almost entirely replaced by PFC emulsion [4].

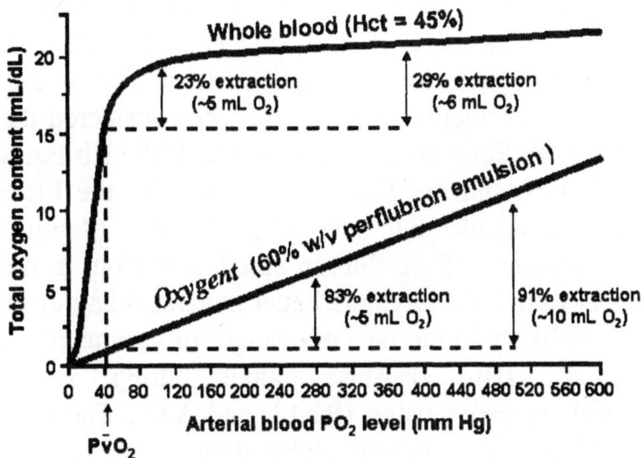

Figure 1. Oxygen binding and delivery characteristics of blood and perflubron emulsion

2. HEMOGLOBIN EQUIVALENCY

When a clinical dose of 2.7 g/kg of a PFC emulsion, such as *Oxygent* (60 percent wt/vol perflubron, Alliance Pharmaceutical Corp., San Diego CA) is given to a patient who is breathing 100 percent O_2 and having a PaO_2 of 500 mm Hg, the oxygen dissolved in PFC amounts to about 1.04 ml/dL. This calculation assumes that the patient weighs 70 kg and has a blood volume of 5 liters. Assuming that the PFC would be given at a stage that blood transfusion might be necessary, ie. at a [Hb] of about 8 g/dl, oxygen transported bound to Hemoglobin would be 10.7 ml/dL. Hence, in terms of O_2 transported by PFC we could say that PFC was equivalent to (1.04/10.7)*8 = 0.78 g/dL [Hb].

Clearly the above calculation "equivalency" of perflubron of 0.78 g/dL [Hb] does not take into account the much greater extraction of Oxygen from PFC (over 90 percent) than occurs from Hemoglobin (20 to 25 percent) as blood passes through the tissues.

The primary role of hemoglobin in the body is to take up and bind O_2 in the lungs and to release it in the tissues - in other words it provides O_2 for oxygen consumption ($\dot{V}O_2$). An oxygen carrier should also provide O_2 for $\dot{V}O_2$ and it would seem reasonable to use its contribution to $\dot{V}O_2$ as a measure of its effectiveness. A means of describing the added O_2-delivery capacity provided by the PFC as a "Hemoglobin Equivalent" (Hb

Equivalent) value has been developed. The percent contribution of the PFC-carried O_2 to the total $\dot{V}O_2$ is expressed in terms of the percent contribution of 1 g/dL red blood cell (RBC) Hb to the total $\dot{V}O_2$. For instance, if a 1 g/kg dose of perflubron provides 10% of $\dot{V}O_2$ and each 1 g/dL of RBC Hb provides 5% of the $\dot{V}O_2$, then the "Hb Equivalent" is 10/5 = 2 g/dL. The "Effective Hemoglobin" (Effective Hb) can then be considered the sum of the RBC [Hb] and the Hb Equivalent. For example, if the Hb Equivalent of a dose of PFC is 2 g/dL and the RBC [Hb] is 8 g/dL, the Effective Hb concentration should be considered to be 8 g/dL + 2 g/dL = 10 g/dL.

During the development of perflubron-based emulsion, a number of physiological models were built to gain a better understanding of the effects of perflubron on O_2 delivery in the face of changes of various physiological parameters, such as, [Hb], cardiac output during normovolemic hemodilution, arterial oxygen tension (PaO_2), and $\dot{V}O_2$. The models take into consideration shifts of the oxyhemoglobin dissociation curve under the influence of arterial pH (pHa), arterial carbon dioxide tension ($PaCO_2$), and body temperature using accepted equations [5,6]. The model determines O_2 content in three compartments of the blood, namely, bound to Hb, in solution in the plasma, and dissolved in the added perflubron. Using an iterative procedure, $P\bar{v}O_2$ can be calculated using both experimental and clinical data; this value has been shown to have a reasonable correlation to measured $P\bar{v}O_2$ [7]. This model has been able to predict increases in $P\bar{v}O_2$ when PFCs are introduced into the circulation, and these higher $P\bar{v}O_2$ values have been demonstrated to reflect both increases in the PO_2 of venous blood draining an organ and increases in directly measured tissue PO_2 [8,9]. In addition, the model is able to predict changes in $P\bar{v}O_2$ that occur with reductions in [Hb] and perflubron concentration ([perflubron]) due to blood loss according the equation developed by Weiskopf [9] and to account for the loss of perflubron due to the normal clearance mechanism via uptake into the reticuloendothelial system (RES).

2.1 Calculation of hemoglobin equivalents

Knowing the cardiac output, arterial and mixed venous blood gases, body temperature, [Hb], and the [perflubron] in the blood at any given time point, it is possible to calculate the contribution of both Hb and perflubron to total body $\dot{V}O_2$. This is done using the following accepted physiologic formulae. For all abbreviations used please refer to the Glossary at the end of this paper.

CaO_2 = $CaO_2Hb + CaO_2Pl + CaO_2PFC$
CaO_2Hb = $(1.34*[Hb]*SaO_2/100)$
CaO_2Pl = $(0.003*p_aO_2)$
CaO_2PFC = $(0.527*1/1.92*[PFB]*p_aO_2/760)$
$C\bar{v}O_2$ = $C\bar{v}O_2Hb + C\bar{v}O_2Pl + C\bar{v}O_2PFC$
CvO_2Hb = $(1.34*[Hb]*SvO_2/100)$
$C\bar{v}O_2Pl$ = $(0.003*P\bar{v}O_2)$
$C\bar{v}O_2PFC$ = $(0.527*1/1.92*[PFB]*P\bar{v}O_2/760)$

It is further possible to calculate the contribution of hemoglobin to $\dot{V}O_2$ and, knowing the Hemoglobin concentration of the blood, the percentage contribution of each g/dL of [Hb] to total $\dot{V}O_2$ can be determined. Hence, the [Hb] that contributes the same percentage of $\dot{V}O_2$ as the PFC can be calculated and is referred to as the "Hemoglobin Equivalent" (see above).

$\dot{V}O_2$ = $(CaO_2 - C\bar{v}O_2) * CO * 10$
$\dot{V}O_2$ from PFB(%) = $(((CaO_2PFB - C\bar{v}O_2PFB) * CO * 10)/ \dot{V}O_2) * 100$
$\dot{V}O_2$ from Hb (%) = $(((CaO_2Hb - C\bar{v}O_2Hb) * CO * 10)/ \dot{V}O_2) * 100$
Hemce: Hb Equivalent = ($\dot{V}O_2$ from PFB (%)/ $\dot{V}O_2$ from Hb (%)) * [Hb}

2.2 Phase 2 trials and determination of hemoglobin equivalent

In the Phase 2 trials of *Oxygent,* the drug was administered when any one of several predefined non-hemoglobin physiological "transfusion triggers" were reached during surgery. After the dose had been administered (during ongoing surgical blood loss), a set of hemodynamic and oxygenation measurements were taken to assess trigger reversal. These data have been used retrospectively to calculate the Hb Equivalents for the subjects in these studies.

The method of calculating Hb Equivalents is complex, but can be simplified to the steps outlined below. Example sets of measured and calculated data are provided in Table 1; cell references (column and row; e.g., B12) given below, refer to cells in this table. A list of measured variables and their abbreviations as well as a list of constants used are provided in the Glossary. The procedure for calculation of Hb Equivalent was as follows:

1) From the subject's weight, height and sex (Cells B13 to B15, respectively), blood volume (Cell B16) was calculated according to the equations of Nadler [10]. This calculated blood volume value was then increased by 10% to account for the increase in blood volume that occurs when a patient is anesthetized due to the vasodilatory effects of anesthesia. This dilutional increase was confirmed in the Phase 2 data by comparing the pre- and post-induction [Hb] values, which showed an approximately 10% decrease in [Hb] following induction.

2) The postdosing data set used in the calculations included [Hb], arterial blood gases (PvO_2, $PvCO_2$, pHa, body temperature, mixed venous blood gases ($P\bar{v}O_2$, $P\bar{v}CO_2$, $pH\bar{v}$) and cardiac output (Cells B18 to B26, respectively). [Hb] in each unit of RBCs, which was assumed to contain 50 g of Hb (Cell B28), and measured [perflubron] (Cell B30) were also used.

3) For each complete data set, the "Standard PaO_2" (i.e., PaO_2 corrected for $PaCO_2$, pHa and body temperature) was calculated using measured PaO_2, $PaCO_2$, pHa and body temperature. This was inserted in Cell B48 and then used together with the Kelman Constants [5] (Cells B50 to B56) to calculate the arterial oxyhemoglobin saturation (SaO_2) (Cell B45). The SaO_2 was then used together with the measured [Hb] to calculate the O_2 content of the Hb compartment (CaO_2Hb) (Cell B36).

4) Identical steps were used to calculate the O_2 content of the mixed venous Hb compartment ($C\bar{v}O_2Hb$ - B41) using "Standard $P\bar{v}O_2$" (B49) and $S\bar{v}O_2$ (Cell B46).

5) The plasma and PFC (perflubron) compartment contributions to CaO_2 (Cells B37 and B38, respectively) and $C\bar{v}O_2$ (Cells B42 and B43, respectively) were calculated from PaO_2 and $P\bar{v}O_2$, using the O_2 solubility coefficients for plasma and perflubron respectively.

Total CaO_2 (Cell B35) and total $C\bar{v}O_2$ (Cell B40) were obtained by summing the O_2 contents of the Hb, plasma, and perflubron compartments. Using the subject's cardiac output value (Cell B26), the relative contributions of the compartments to total body $\dot{V}O_2$ (displayed in Cell B32) were calculated (Cells B58 to B60, respectively)

The Hb Equivalent was then calculated and displayed in Cell B7. By summing this value and the RBC [Hb] from Cell B18, the Effective Hb was determined and is displayed in Cell B8.

Using the program outlined above, the prediction of Hb Equivalents underestimates the potential Hb Equivalents because, at the time the blood samples for [perflubron] were obtained, circulating drug levels had already been reduced from the immediate postdosing levels due to blood loss and clearance of the PFC emulsion particles by the reticuloendothelial system. Since dosing of perflubron in its proposed clinical use will be initiated before surgical blood loss occurs, a better simulation of this condition can be made by using the [perflubron] immediately postdosing (prior to loss due to bleeding and clearance). While no perflubron blood levels were obtained at this point during the Phase 2 studies, a reasonable estimate of this value can be determined for each subject by dividing the total dose delivered by the subject's blood volume, which can be calculated using the Nadler equations [10].

Thus, a more relevant estimate of the initial Hb Equivalent can be obtained by using the calculated [perflubron] and the measured physiologic data in the computer model as described above. The individual subject values from these calculations were then used as described above to predict the Hb Equivalent of a 2.7-g/kg dose of *Oxygent* in the completed European Phase 3 transfusion-reduction study [11]. In this case, the mean predicted Hb Equivalent (Mean ± Standard Deviation) was 3.99 ± 2.71 g/dL.

The variability of the available data appears to be due, in part, to the apparent inclusion of subjects in whom PaO_2 was substantially below the level that should have been obtained on an FiO_2 of 1.0. While this suggests possible protocol deviations, the data were included in the estimates of Hb Equivalency; exclusion of these subjects from the calculation would have resulted in slightly higher Hb Equivalent values with less variability of the means. A similar high degree of variability was also obtained when the efficacy (i.e., O_2 contribution to $\dot{V}O_2$) of a 50-g "Unit" of RBC Hb in the Phase 2 studies was assessed; the contribution of this [Hb] is also dependant upon the physiological state of the patient.

3. LOSSES OF PERFLUOROCHEMICAL DURING SURGERY

As mentioned above, to understand the changes in Hb Equivalency that will occur following perflubron dosing during surgical bleeding, the dosing regimen needs to be considered and the estimates need to account for the expected losses of perflubron from the circulation (due to blood loss and RES clearance).

It has been shown in preclinical and clinical Phase 1 studies that PFC removal from the circulation by the RES is dose-dependent. The larger the dose of perflubron, the longer the half-life. Phase 1 studies demonstrated that

Spreadsheet Example

	A	B	C	D	E	F	G
1					Mean	SD	
2				HE	1.8	0.5	
3				BI Unit Eq	6.6	2.0	
4				%VO2PFC	10.4	1.7	
5	Perflubron Dosing (g/kg)	1.8					
6							
7	Hemoglobin Equivalent (g/dL)	2.0	1.1	1.6	2.4	1.5	2.2
8	Effective Hemoglobin (g/dL)	8.0	7.4	8.1	10.4	7.9	9.0
9							
10	BI Unit Equivalence of PFC	1.8	0.9	2.1	2.3	1.1	1.9
11							
12	Patient ID	1	2	3	4	5	6
13	Weight (Kg)	75	58	97	74	56	65
14	Height (Meters)	1.65	1.6	1.82	1.7	1.54	1.65
15	Sex (MALE/FEMALE)	Female	Female	Male	Female	Female	Female
16	Estimated Blood Volume	4690	3917	6532	4819	3670	4326
17							
18	Hb (g/kg)	6	6.3	6.5	8	6.4	6.8
19	PaO2 (mm Hg)	486	519	359	491	461	483
20	PaCO2 (mm Hg)	42.3	38.7	39.9	33.5	39.3	35.3
21	pHa	7.35	7.36	7.38	7.4	7.4	7.37
22	Temperature (°C)	35.4	34.6	36.5	34.7	34.5	35.1
23	PvO2 (mm Hg)	55.0	51.2	63.2	55.7	44.0	58.0
24	PvCO2 (mm Hg)	47.3	44.5	45.9	39.0	45.9	39.2
25	pHv	7.31	7.32	7.33	7.36	7.35	7.34
26	Cardiac Output (L/min)	7.4	4.1	8.5	4.5	3.0	6.7
27							
28	Hb in Blood Unit (gm)	50	50	50	50	50	50
29							
30	PFC Conc (mic/100mL)	192.7	107.5	176.4	170.2	228.6	161.5
31							
32	VO2 (mL/min)	186	107	155	110	89	153
33	VO2/kg	2.48	1.85	1.60	1.49	1.59	2.36
34							
35	CaO2 (mL/dL)	9.83	10.20	10.00	12.49	10.33	10.83
36	CaO2Hb (mL/dL)	8.03	8.44	8.69	10.71	8.57	9.10
37	CaO2Pl (mL/dL)	1.46	1.56	1.08	1.47	1.38	1.45
38	CaO2PFC (mL/dL)	0.34	0.20	0.23	0.30	0.38	0.28
39							
40	CvO2 (mL/dL)	7.31	7.60	8.18	10.01	7.31	8.54
41	CvO2Hb (mL/dL)	7.11	7.43	7.95	9.81	7.14	8.33
42	CvO2Pl (mL/dL)	0.17	0.15	0.19	0.17	0.13	0.17
43	CvO2PFC (mL/dL)	0.04	0.02	0.04	0.03	0.04	0.03
44							
45	SaO2 (percent/100)	1.00	1.00	1.00	1.00	1.00	1.00
46	SvO2 (percent/100)	0.88	0.88	0.91	0.91	0.83	0.91
47							
48	Standard PaO2 (mm Hg)	505.15	572.87	362.81	563.09	529.48	525.45
49	Standard PvO2 (mm Hg)	54.75	53.96	60.42	61.06	47.85	61.03
50	Ka1	-8532.2289					
51	Ka2	2121.401					
52	Ka3	-67.073989					
53	Ka4	935960.87					
54	Ka5	-31346.258					
55	Ka6	2396.1674					
56	Ka7	-67.104406					
57	VO2 from Hb (%)	36.6	38.9	41.0	36.5	47.2	33.6
58	VO2 from Pl (%)	51.4	54.1	48.7	52.7	41.4	55.6
59	VO2 from PFC (%)	11.9	7.0	10.3	10.8	11.4	10.8

the intravascular half-life of perflubron in the blood is related to the absolute dose of PFC administered and can be predicted by the following formula:

Perflubron Half-life (hours) = -1.24 + 0.1047*Dose - 0.00017*Dose^2 Eq. 1

The Hb Equivalent (HE) after uptake of perflubron by the RES can be found by the following equation:

$HE_{\text{half-life corrected}}$ = Initial HE*(0.5) $^{((1/\text{half-life})*\text{time elapsed})}$ Eq. 2

Perflubron is also removed from the circulation as bleeding occurs. By adapting the formula given by Weiskopf [9], the amount remaining after bleeding is determined as follows:

$HE_{\text{blood-loss corrected}}$ = Initial HE/exp(Blood Loss/Blood Volume) Eq. 3

Blood volume (BV) can be calculated from height (Ht) and weight (Wt) by using formulas from Nadler [10]

For men: BV = 0.3669*H^3 +0.03219*W + 0.6041 Eq. 4

For women: BV = 0.3561*H^3 + 0.03308*W + 0.1833 Eq. 5

If the initial and final Hb concentrations are unknown, Eq. 3 can also be expressed as:

$HE_{\text{blood-loss corrected}}$ = Initial HE*(Final [Hb]/Initial [Hb]) Eq. 6

By using the above equations, it is possible to estimate the Hb Equivalent adjusted for RES clearance and surgical blood loss at any point during the surgical procedure.

In the surgical setting, a number of factors, such as, patient size and gender, or type and duration of surgery, will influence the "Effective Hb" (sum of RBC Hb and PFC Hb Equivalent). Therefore, a computer program was used to assess the Hb Equivalents and, hence, to predict the Effective Hb concentrations for a possible clinical scenario in which 1.35 g/kg *Oxygent* would be given (in response to the need for more oxygen delivery)at a [Hb] of 8 g/dl, followed by a similar dose when bleeding had reduced [Hb] a level of 7 g/dL. This theoretical situation is shown in Figure 3 for two average rates of bleeding - 500 ml/hour and 1000 ml/hour. The faster rate of bleeding, (where for the same blood, loss less perflubron is taken up by the RES), allows more bleeding before the Effective [Hb] (RBC [Hb] plus Hb Equivalent) is reduced to 8 g/dL and where blood transfusion might be indicated. This extra permissible blood loss is relatively small and amounts to only about 120 ml of blood loss in the approximately 2000 ml of permissible loss in this patient. At this time red cell [Hb] will be approximately 5.4 g/dL.

The computer model, while a valuable tool for predicting Effective Hb, has a number of limitations. The model primarily calculates Hb Equivalent based solely on blood loss and time elapsed. Inherent in this model is the assumption that the subject's physiological parameters do not change during normovolemic bleeding. This, of course, may not always be the case, as it is well known that cardiac output increases as [Hb] is reduced, due to decreases

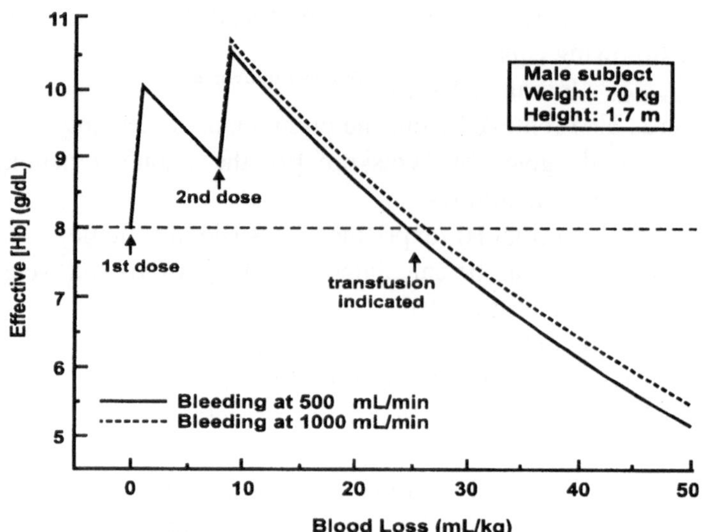

Figure 2. The effects of surgical blood loss on Hemoglobin Equivalency.

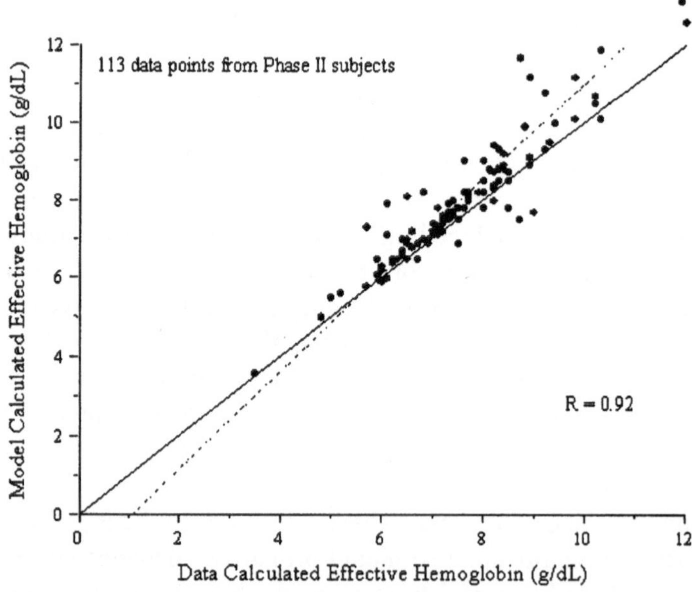

Figure 3. Comparison of model-calculated and data-calculated Hemoglobin Equivalency.

in viscosity and consequent reduction of the vascular resistance [12]. The model does not account for this potential change in cardiac output; however, the consequence of this normal compensatory mechanism is for the model to underestimate Hb Equivalents. Hence, the model may calculate the worst-case scenario for the individual patient, because if cardiac output increases, the extraction of total O_2 from perflubron will increase, yielding a greater Hb Equivalent.

4. MODEL VERIFICATION

Verification of the model calculations has been accomplished by comparison to clinical data from Phase 2 studies. In the studies, physiologic transfusion triggers were used to determine the point at which the dose of *Oxygent* was to be administered. Subsequently, when any of the physiologic transfusion triggers were reached a second time, a blood transfusion was indicated. The computer model used Hb Equivalent at the postdosing time point, together with blood loss and time elapsed to reach the second transfusion trigger, to predict the Hb Equivalent at the second transfusion trigger. Additionally a number of other complete data sets ([Hb], arterial and mixed venous blood cases, cardiac output and body temperature) were available at timepoints after the first transfusion trigger allowing further points at which prediction of Hb Equivalents could be made. Effective [Hb] calculations (sum of RBC [Hb] and Hb Equivalent) were made using both data calculated and model calculated Hb Equivalents.

A total of 113 data calculated Effective [Hb] were obtained that could be compared against their corresponding model calculated values. The respective values are plotted against each other in Figure 3. It can be seen that the majority of points lie in close proximity to the line of identity. Also drawn on the graph as a dotted line is the regression line (R = 0.9), which closely approaches the line of identity in the region where Effective [Hb] values might be expected to cause concern [4]. Effective [Hb]s between 6 and 9 g/dL. This indicates that the model is able to predict Effective [Hb] with a fair degree of accuracy over this range. Indeed more than 75 percent of the model predicted values lie within 0.5 g/dL of the measured values.

REFERENCES

1. Gregory IC. The oxygen and carbon monoxide capacities of foetal and adult blood. J Physiol 1974;326:625-634.
2. Braunitzer G. Molekulaire struktur von haemoglobine. Nova Acta Acad Caesar Leop Caral 1963;26:471.

3. Prys-Roberts C, Foex P, Hahn CEW. Calculation of Blood PO2. Anesthesiology 1971;34:581-582.
4. Geyer RP, Monroe RG, Taylor K. Survival of rats totally perfused with a fluorocarbon-detergent preparation. In: Normal JC Ed. Organ perfusion and preservation. New York:Appleton Century Crofts 1986;85-96.
5. Kelman GR. Digital subroutine for the conversion of oxygen tension into saturation. J Appl Physiol 1966;21:1375-76.
6. Faithfull NS, Rhoades GE, Keipert PE, et al. A program to calculate mixed venous oxygen tension - A guide to transfusion? Adv Exp Med Biol 1994;361:41-9.
7. Habler OP, Keen MS, Hutter JW, et al. Hemodilution and intravenous perflubron emulsion as an alternative to blood transfusion: Effects on tissue oxygenation during profound hemodilution in anesthetized dogs. Transfusion 1998;38:145-155.
8. Batra S, Peters BP, Symonds JD, et al. Perflubron emulsion increases tissue oxygenation in a canine model of hemodilution. Artif Cells, Blood Subs and Immob Technol 1996;24:307.
9. Weiskopf RB. Mathematical analysis of isovolemic hemodilution indicates that it can decrease the need for allogeneic blood transfusion. *Transfusion* 1995;35:37-41.
10. Nadler SB, Hidalgo JU, Bloch T. Prediction of blood volume in normal human adults. Surgery 1962;51:224-232.
11. Spahn DR, Waschke K, Standl T, et al. Use of perflubron emulsion to decrease allogeneic blood transfusion in high-blood-loss non-cardiac surgery: results of a European Phase 3 study. Anesthes 2002;97. In Press.
12. Tuman KJ. Tissue oxygen delivery. Anesthesiology Clinics of North America 1990;8:451-69.

Glossary

List of Measured and Calculated Values, and Constants and Definitions of Terminology Used in the Computer Models

Measured Variable -Definition	**Abbreviation**	**Units**
Hemoglobin Equivalent - VO_2 provided from PFC expressed as the [Hb] needed to provide the same VO_2 under the same conditions	HE	g/dL
Effective Hemoglobin - Sum of the RBC [Hb] and the Hemoglobin Equivalent	---	g/dL
Blood Unit Equivalence of PFC - Number of 50-g Units of Hb necessary to provide same VO_2 as the PFC dose		
Body Weight	Wt	kg
Height	Ht	M
Estimated Blood Volume	EBV	mL
Hemoglobin	Hb	g/dL
Arterial Oxygen Partial Pressure	PaO_2	mm Hg
Arterial Carbon Dioxide Partial Pressure	$PaCO_2$	mm Hg
Arterial pH	pHa	
Body Temperature	Temp	°C
Mixed Venous Oxygen Partial Pressure	$P\bar{v}O_2$	mm Hg
Mixed Venous Carbon Dioxide Partial Pressure	$P\bar{v}CO_2$	mm Hg
Mixed Venous pH	$pH\bar{v}$	
Cardiac Output	CO	L/min
Hemoglobin Content in a Unit of Blood	Hb in Blood Unit	g
Perfluorochemical Concentration	PFC conc	µg/100 mL
Oxygen Consumption	VO_2	mL/min

Glossary - *continued*

Measured Variable -Definition	**Abbreviation**	**Units**
Arterial Oxygen Content	CaO_2	mL/dL
Arterial Oxygen Content in Hemoglobin	CaO_2Hb	mL/dL
Arterial Oxygen Content in Plasma	CaO_2Pl	ml/dL
Arterial Oxygen Content in PFC	CaO_2PFC	mL/dL
Mixed Venous Oxygen Content	$C\bar{v}O_2$	mL/dL
Mixed Venous Oxygen Content in Hemoglobin	$C\bar{v}O_2Hb$	mL/dL
Mixed Venous Oxygen Content in Plasma	$C\bar{v}O_2Pl$	mL/dL
Mixed Venous Oxygen Content in PFC	$C\bar{v}O_2PFC$	mL/dL
Arterial Oxyhemoglobin Saturation	SaO_2	%
Mixed Venous Oxyhemoglobin Saturation	$S\bar{v}O_2$	%
Standard PaO_2 - PaO_2 corrected for pH, PCO_2 and Temp.		mmHg
Standard $P\bar{v}O_2$ - $P\bar{v}O_2$ corrected for pH, PCO_2 and Temp.		mmHg
Percent Oxygen Consumption from Hemoglobin	$\dot{V}O_2$ from Hb	%
Percent Oxygen Consumption from Plasma	$\dot{V}O_2$ from Pl	%
Percent Oxygen Consumption from PFC	$\dot{V}O_2$ from PFC	%
Percent $\dot{V}O_2$ provided by from a Unit of Blood	$\dot{V}O_2$ from Unit	%

O_2 solubility of Plasma — 2.28 mL/dL at 760 mm Hg

O_2 Solubility in Perflubron — 52.7 mL/dL at 760 mm Hg

Specific Gravity of Perflubron — 1.92 g/mL

Constants Used		**Value**
Kelman Constants -used for deriving the oxyhemoglobin dissociation curve		
	Ka1	- 8532.2289
	Ka2	2121.401
	Ka3	-67.073989
	Ka4	935960.87
	Ka5	-31346.258
	Ka6	2396.1674
	Ka7	-67.104406

Potential O_2 transport by 1 g of fully saturated Hb — 1.34 mL/g

Equations used in the Hemoglobin Equivalence Model

The equations used in calculation of oxygen content and consumption in the three compartments of the model are as follows:

Oxygen Content (in mL/dL whole blood):

$$CaO_2 = CaO_2Hb + CaO_2Pl + CaO_2PFC$$

$$CaO_2Hb = (1.34*[Hb]*SaO_2/100)$$

$$CaO_2Pl = (0.003*PaO_2)$$

$$CaO_2PFC = (0.527*1/1.92*[PFB]*PaO_2/760)$$

$$C\bar{V}O_2 = C\bar{V}O_2Hb + C\bar{V}O_2Pl + C\bar{V}O_2PFC$$

$$CvO_2Hb = (1.34*[Hb]*SvO_2/100)$$

$$C\bar{V}O_2Pl = (0.003*P\bar{V}O_2)$$

$$C\bar{V}O_2PFC = (0.527*1/1.92*[PFB]*P\bar{V}O_2/760)$$

Oxygen Consumption (in mL/min):

$$\dot{V}O_2 = (CaO_2 - C\bar{V}O_2) * CO * 10$$

Percent of Oxygen Consumption Due to Perflubron:

$$\dot{V}O_2 \text{ from PFB}(\%) = (((CaO_2PFB - C\bar{V}O_2PFB) * CO * 10)/\ \dot{V}O_2) * 100$$

Percent of Oxygen Consumption Due One Gram of Hemoglobin:

$$\dot{V}O_2 \text{ from Hb}(\%) = (((CaO_2Hb - C\bar{V}O_2Hb) * CO * 10)/\ \dot{V}O_2) * 100$$

Equations to be used in calculating blood volume (Nadler *et al*,1962):

For men: $BV = 0.3669*H^3 + 0.03219*W + 0.6041$.
For women: $BV = 0.3561*H^3 + 0.03308*W + 0.1833$.

Equation to be used to correct for decreasing [PFC] and [Hb] due to blood loss (modified from Weiskopf, 1995):

$$HE_{\text{blood loss corrected}} = \text{Initial HE/exp(Blood Loss/Blood Volume)}.$$

Equation to be used to calculate perflubron half-life:

$$\text{Perflubron Half-life (hours)} = -1.24 + 0.1047*\text{Dose} - 0.00017*\text{Dose}^2.$$

Chapter 27

THE DOSE-DEPENDENT EFFECT OF RSR13, A SYNTHETIC ALLOSTERIC MODIFIER OF HEMOGLOBIN, ON PHYSIOLOGICAL PARAMETERS AND BRAIN TISSUE OXYGENATION IN RATS

Oleg Y. Grinberg, Minoru Miyake, Huagang Hou, Robert P. Steffen* & Harold M. Swartz

*EPR Center for the Study of Viable Systems, Dept. of Radiology, Dartmouth Medical School, 7785 Vail, Hanover, NH 03755 USA; *Allos Therapeutics, Inc., 11080 CirclePoint Road, Suite 200, Westminster, CO 80020 USA*

Abstract: RSR13 is a synthetic allosteric modifier of hemoglobin that decreases the oxygen-binding affinity of hemoglobin, increasing the P50. As a result, tissue oxygen tension is expected to increase. Using the capabilities of *in vivo* EPR, we directly examined the effect of RSR13 on brain pO_2 in rats and the relationship between any change in brain oxygenation and changes in physiological parameters, including blood gases. The brain pO_2 and arterial blood p_aO_2 were increased significantly ($p<0.005$) following RSR13 administration. The peak increase of brain tissue pO_2 was 8.8±1.2 mm Hg in the animals receiving 150 mg/kg RSR13 and 13±3 mm Hg in the animals receiving 300 mg/kg RSR13. There was no difference among groups in MBP, heart rate, p_aCO_2, pH, or HCO_3. These data indicate that in anesthetized rats, RSR13 dose-dependently increases brain pO_2 without affecting other physiologic parameters. This capability is likely to be very useful in circumstances where the pO_2 of the brain is compromised.

Key words: Allosteric modifier, RSR13, Cerebral pO_2, EPR oximetry

1. INTRODUCTION

RSR13 (2-[4-[[(3,5-dimethylanilino) carbonyl] methyl]-phenoxyl]-2-methylpropionic acid sodium salt) is a synthetic allosteric modifier of hemoglobin that produces an allosteric effect on hemoglobin, similar to that of 2,3-diphosphoglycerate (2,3-DPG). However, this effect is achieved by a mechanism and at a site of action different from 2,3-DPG [1,2,3]. *In vitro* screening by measuring the shift in the oxygen-binding affinity of hemoglobin revealed that RSR13 was the most efficient modifier of hemoglobin among over 120 analogues tested [1]. It is well established that RSR13 decreases the oxygen-binding affinity of red blood cells suspensions [3] and of whole blood *in vivo* [4] by its allosteric effect. It has been shown that this decrease of hemoglobin oxygen-binding affinity in turn increases oxygenation of normal tissue [5], hypoxic tumors [6], ischemic myocardium [7,8], and ischemic brain [9,10]. These results suggest that RSR13 may have potentially important clinical applications.

Prior studies have shown that EPR (Electron Paramagnetic Resonance) oximetry has some advantages for repeatedly measuring the partial pressure of oxygen (pO_2) *in vivo* [11,12]. Using the capabilities of *in vivo* EPR oximetry in this study we examined how RSR13 affects brain tissue oxygenation (pO_2) in rats and how brain oxygenation relates to physiological parameters, including blood gases.

2. METHODS

2.1 Animals and materials

Male Sprague-Dawley rats weighing 280-330g were supplied by Charles River Lab, Wilmington, MA. RSR13 (batch CTM-612, 2g RSR13 acid in 100mL: 20mg/ml solution) was provided by Allos Therapeutics, Inc. The lithium phthalocyanine (LiPc) used in this study was produced in our lab.

2.2 Cerebral pO_2 measurements

EPR oximetry technique for this study utilized an oxygen sensitive material (LiPc, a neutral p-radical crystalline substance) that was placed in the area of interest. The EPR line width of LiPc is a linear function of pO_2 and independent of local metabolic processes, the presence of other paramagnetic species, and pH [13]. LiPc was implanted into the cerebral cortex of rats (under ketamine/xylazine 80 mg/8 mg/kg i.m. anesthesia) 1-week before the experiment. The LiPc (about 0.05mg) was placed using stereotactic techniques via 25 gauge needles to a depth of 2 mm from the surface of the skull. The implantation site was drilled with a 23 gauge needle 3.0 mm off the midline and 1 mm behind the bregma. An L-band (1.2 GHz) EPR spectrometer which utilizes a microwave bridge and external loop resonator specially designed for *in vivo* experiments was used. The EPR spectra of LiPc were recorded with careful regard to avoid power saturation and over modulation, with scan times varying from 5 to 20 sec. The spectrum for each time point was accumulated during 3-4 minutes to increase the signal to noise ratio and obtain better accuracy in the pO_2 measurement. The mean of these pO_2 values over 30 minute period was considered as the cerebral pO_2 for each time point.

Spectral fitting by the EWVoigt program (Scientific Software Inc.) was used to derive the EPR line width. The line width was transformed to cerebral pO_2 using a calibration of the LiPc that was performed at five different partial pressures of oxygen.

2.3 Experimental protocol and monitoring of animal physiology

Rats were anesthetized using 26% oxygen with 2.5-3% isofluorane. Then, to provide better control of respiration during the experimental procedure, endotracheal intubation was carried out nonsurgically using a laryngoscope and an "over-the-needle" 14 gauge catheter. During the experiment, rats were anesthetized and were maintained with 1.2-1.5% isofluorane under controlled respiration (FiO_2 25-28%). Respiratory support (tidal volume ~1.1 ml/100 g body weight; rate ~65-75 strokes/min) was provided by a volume cycled respirator (Harvard Apparatus) after endotracheal intubation. FiO_2, the concentration of isofluorane, and end-tidal pCO_2 were monitored by a Capnomac II Datex (Model AMG-103-27-00) throughout the experiment. Polyethylene catheters (PE-50, Clay-Adams, NJ) were placed in the left femoral artery and vein for continuous monitoring of arterial blood pressure (systolic, diastolic, mean BP), heart rate, periodic blood gas measurements, and for administration of RSR13, respectively. All parameters and core body temperature (by a rectal probe) were taken as average values over 30 sec at the time points, using a Biopac system. Arterial blood (100μl) was drawn into a glass capillary tube and blood gas (pO_2, pCO_2, pH, and HCO_3) measurements were obtained immediately using a Ciba-Corning 238 Blood Gas System. After each sample was withdrawn, 0.2ml of heparinized saline (2 U/ml) was infused into the catheters to prevent clotting within the catheter. Two to three arterial-blood samples were taken before the experiment to adjust the optimal setting for ventilation. The last of these blood gas measurements was used for baseline comparisons. During the experiment, the animal was kept on a water pad that was maintained at 37 to 38° C. During 20 minutes before the start of RSR13 administration (time 0) all parameters were taken at the time points -20 and -10 minutes. The mean brain pO_2 over this interval was considered as the baseline brain pO_2.

At time 0 the animals received either 150 mg/kg (6 rats) or 300 mg/kg (6 rats) RSR13 intravenously over 15 min, using an infusion pump (KD Scientific model 100). Control animals (6 rats) received the same volume of saline intravenously, over 15 minutes. All physiological parameters were monitored for 300 min after treatment and taken every 3 min during the first 15 min after administration of RSR13 or saline and every 15 min thereafter. Blood samples were taken at 15 min, 120 min, 210 min, and 300 min post RSR13 or saline administration. At the end of the experiment some of the brains were perfused with a 10% formalin solution and stained for subsequent histologic study.

After the measurements were completed, the animals were euthanized by the method approved by Animal Resource Center, Dartmouth College, NH, USA.

3. RESULTS

3.1 Cerebral pO_2

Figure 1 presents the effect of RSR13 on cerebral pO_2 as the difference (mean ±SE) between the values at each time point and the baseline. There were similar increases in brain pO_2 for both RSR13-treated groups during the first 45 minutes after RSR13 administration. The 300mg/kg dose group had a 14 mm Hg increase in cerebral pO_2, peaking at 135 minutes post infusion. At the 150 mg/kg dose, the peak increase was 8 mm Hg at 45 minutes post infusion. The controls had a decrease from the baseline value during the time period 115-275 min.

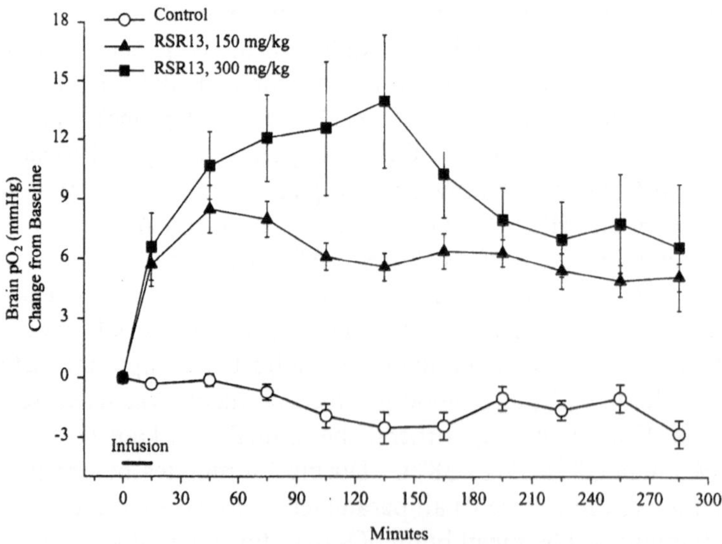

Figure 1. Effect of RSR13 on cerebral pO_2, mean±SE.

3.2 p_aO_2

Figure 2 presents the effect of RSR13 on p_aO_2 as the difference (mean ±SE) between the values at each time point and the baseline. There was a similar significant increase of p_aO_2 for low and high-dose groups immediately after RSR13 administration. At the subsequent time points the high-dose group had a larger and more prolonged effect than the low-dose group. In the 300 mg/kg -dose group the peak increase in p_aO_2 (about 36 mm Hg) occurred at 120 min post infusion while the peak for the low-dose group was at 15 min.

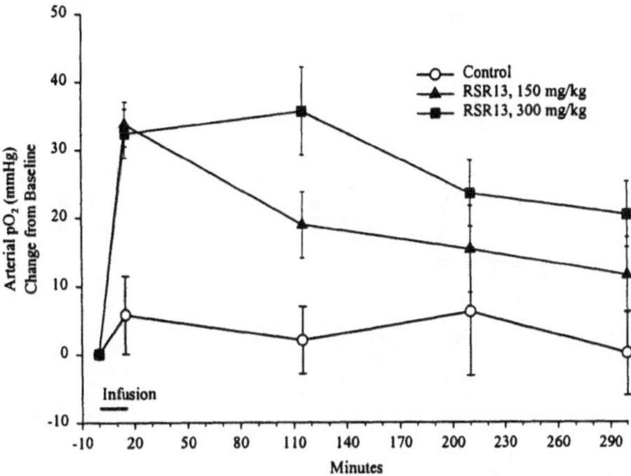

Figure 2. Effect of RSR13 on arterial p_aO_2, mean±SE.

3.3 Cardiovascular function

3.3.1 Mean arterial blood pressure (MBP)

The MBP in the control and experimental groups was stable throughout the entire 300 minute post-infusion period (Figure 3a). However, in the experimental groups a transient MBP drop was observed (in 3 min after the start of RSR13 injection, Figure 3b, the decrease of 17±4 mm Hg, $p<0.05$ lasted only about 6 minutes), then MBP returned to baseline in 30 min. There was no significant difference among the experimental groups for this transient drop in MBP.

3.3.2 Heart rate (HR)

The HR was quite stable in the control group. In the high-dose RSR13-treated group the HR gradually decreased during the first 120 min, about 75

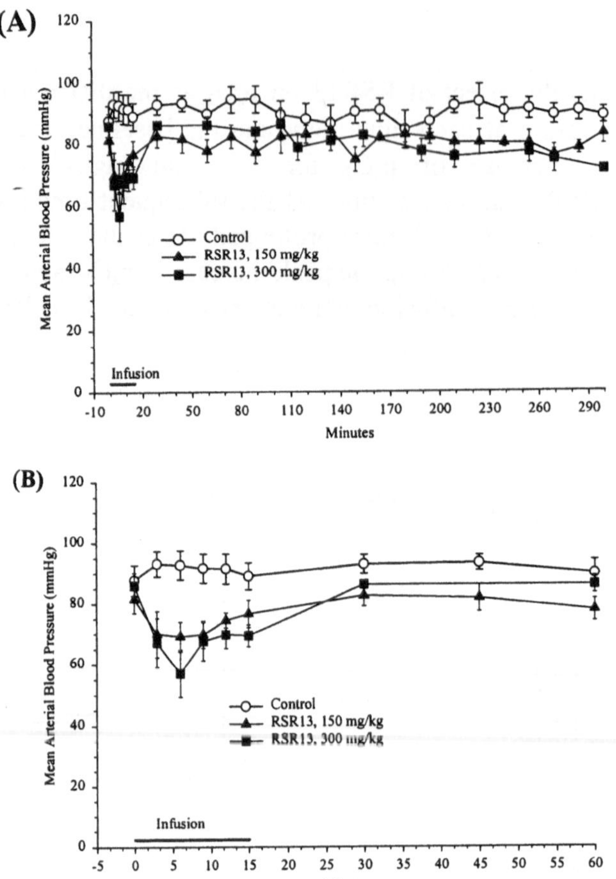

Figure 3. Mean Arterial Blood Pressure (mean±SE) during whole experiment (a) and during first 60 min after RSR13 injection (b).

beats/min. below the value at 0 time point and remained lower until 300 min. For the low-dose group the HR at 120 min after RSR13 administration gradually increased to 70 beats/min. These data indicate a dose-dependent effect of RSR13 on the HR. However, this observation cannot be considered as strong evidence of this effect because the HR value at 0 time point was different among groups (Figure 4).

3.4 Other physiological parameters

There was no difference among groups in pCO_2, HCO_3, or pH, Table 1. The rectal temperature (RT) was stable in the controls and decreased modestly in the experimental groups (Figure 5), reaching statistical significance (about 3^oC, $p<0.05$) for the high-dose RSR13 group at 120 min.

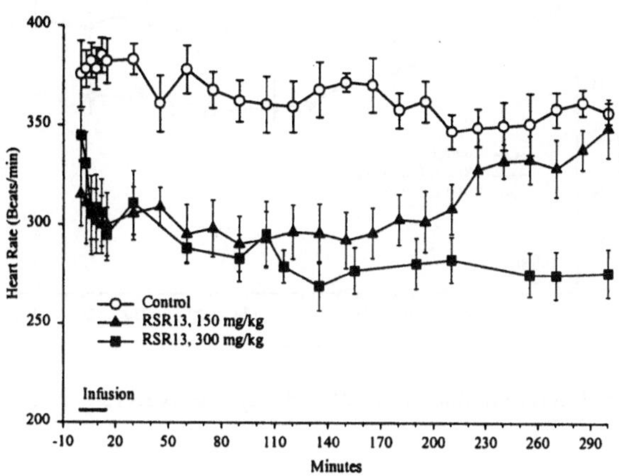

Figure 4. Heart rate (mean±SE) after RSR13 administration.

Table 1. Effect of RSR13 administration on the blood parameters (mean ± SE)

Parameters	0min	15 min	120 min	210 min	300 min
pCO_2, 0 mg	38.6±2.1	36.9±2.5	35.2±1.6	35.3±2.8	34.8±4.1
pCO_2 150 mg	35.0±5.0	36.3±2.3	37.6±3.6	37.3±3.7	36.6±3.8
pCO_2 300 mg	39.7±4.2	35.1±6.3	38.4±6.3	37.8±6.6	36.0±4.9
HCO_3 0 mg	24.7±0.7	24.7±2.1	23.7±1.5	24.3±1.2	22.9±1.5
HCO_3 150 mg	26.0±1.0	24.8±0.9	26.1±2.1	26.0±0.8	25.6±2.2
HCO_3 300 mg	26.1±2.1	23.2±1.5	24.2±1.8	22.8±2.2	22.5±3.3
pH 0 mg	7.43± 0.02	7.43± 0.02	7.44± 0.02	7.44± 0.02	7.44± 0.02
pH 150 mg	7.47 ± 0.02	7.44± 0.01	7.45± 0.03	7.45± 0.03	7.45± 0.03
pH 300 mg	7.42 ± 0.03	7.41± 0.03	7.406± 0.04	7.390± 0.05	7.399± 0.05

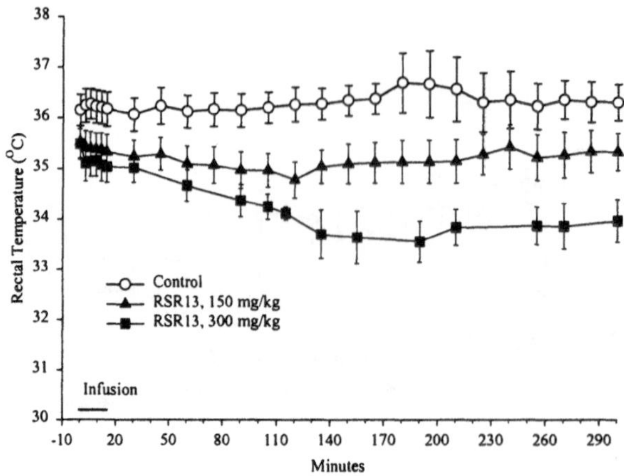

Figure 5. Rectum temperature (mean±SE) after RSR13 administration.

3.5 Histological analysis

There were no signs of brain damage caused by the implantation of the LiPc. There also were no indications of any neurobehavioral abnormalities due to the presence of LiPc in the cortex.

4. DISCUSSION

The results clearly indicate that the administration of RSR13 increased the brain pO_2 and p_aO_2 (Figures 1 and 2, respectively). RSR13 had little or no effect on other physiological parameters in rats during five hours following administration of RSR13, (Figures 3-5) and Table 1. The demonstration of an increase in the pO_2 in the brain is a new finding that, while not surprising, also was not considered inevitable. The effects of RSR13 on brain pO_2 appear to be dose dependent but the range of doses was too small to determine the quantitative nature of the dose-dependency. The increases in arterial p_aO_2 and cerebral pO_2 observed in this study are not predicted from simple considerations of the physiological parameters that were measured. To determine if the p_aO_2 could be predicted quantitatively from changes in hemoglobin oxygen-binding affinity, we carried out the following, admittedly over-simplified calculation, based on the changes of the p_aO_2 using commonly accepted quantitative methods, numerical values, and assumptions. We used as normal values in rats: p_aO_2=90 mm Hg, venous pO_2=40 mm Hg. Using the Hill equation with parameters for rats n=2.6 and P50=38 mm Hg [4], $S=(pO_2/P50)^n/[1+(pO_2/P50)^n]$, the arterial Sa=0.904 and venous Sv=0.533. It is known that RSR13 shifts p50 to the right. Kunert et al., 1996, observed a shift of P50 from 38 mm Hg to 58 mm Hg in rats (1 hour after RSR13 infusion of 200 mg/kg). Assuming that administration of RSR13 changes only P50, using the Hill equation the arterial blood oxygenation after RSR13 administration should be p_aO_2=137 mm Hg. Please note that a large assumption was made that RSR13 administration does not change any other physiological parameters such as arterial and venous blood saturation, cerebral metabolic rate (CMR), and cerebral blood flow (CBF). Unfortunately, there is very little experimental evidence for this assumption. The results of others [7] and the absence of an effect of RSR13 on the MBP in our experiments might support only one of them, namely, unchanged CBF. It also is true, however, that the predicted increase in p_aO_2 of 47 mm Hg is only a modest amount more than the observed change of 35 mm Hg.

To determine if the increase of the brain pO_2 could be predicted quantitatively from the observed change in the p_aO_2, we also assumed that there would be no effect of RSR13 on CBF and CMR. Using the Hill equation, the change in venous blood oxygenation after RSR13 administration was expected to be 26 mm Hg. One could expect the increase of the brain pO_2 should be between the changes in venous pO_2 and arterial p_aO_2, i.e. between 26 and 35 mm Hg. However, the observed change in brain pO_2 (between 8 and 14 mm Hg) is much smaller than the predicted change. While this result is not surprising, it does confirm that until we have more sophisticated models that take into account all of the complexities of real physiological systems with feedback and multiple sources of control, direct measurements will be needed.

These results suggest that in view of RSR13's ability to increase brain pO_2, RSR13 is likely to have therapeutic benefit in conditions associated with cerebral hypoxia.

5. CONCLUSION

This study suggests that:

1. The pO_2 of the rat brain can be repeatedly measured directly in vivo using EPR oximetry.
2. RSR13 increased p_aO_2.
3. RSR13 dose-dependently increased the pO_2 of the brain.
4. Except for the changes in pO_2 and p_aO_2, RSR13 had little or no effect on other physiological parameters.
5. The effects of RSR13 on the brain pO_2 could not be directly and simply related to conventional parameters.
6. RSR13 might have clinical utility for increasing brain pO_2 in circumstances where the pO_2 of the brain is potentially compromised.

ACKNOWLEDGEMENTS

This study was supported by NIH Grant PO1 GM51630 and used the facilities of the EPR Center for the Study of Viable Biological Studies, supported by NCRR and NIH Grant P41 RR11602. Allos Therapeutics, Inc also provided financial support and RSR13 for this study.

REFERENCES

1. Randad RS, Mahran MA, Mehanna AS, and Abraham DJ.Allosteric modifiers of hemoglobin. 1. Design, synthesis, testing, and structure-allosteric activity relationship of novel hemoglobin oxygen affinity decreasing agents. J Med Chem 1991;34:752-757
2. Wireko FC, Glen E. Kellogg GE, and Abraham DJ. Allosteric modifiers of hemoglobin. 2. Crystallographically determined binding sites and hydrophobic binding/interaction. Analysis of novel hemoglobin oxygen effectors. J Med Chem 1991;34:758-767.
3. Abraham DJ, Wireko FC, and Randad RS. Allosteric modifiers of hemoglobin: 2-[4-[[(3,5-Disubstituted anilino) carbonyl]methyl]phenoxy]-2-methylpropionic acid derivatives that lower the oxygen affinity of hemoglobin in red cell suspensions, in whole blood, and in vivo in rats Biochemistry 1992;31:9141-9149.
4. Kunert MP, Liard JF, and Abraham DJ. RSR-13, an allosteric effector of hemoglobin, increases systemic and iliac vascular resistance in rats. Am J Physiol 1996;271:H602-613.
5. Khandelwal SR, Randad RS, Lin PS, Meng H, Pittman RN, Kontos HA, Choi SC, Abraham DJ, and RupertSchmidt-Ullrich. Enhanced oxygenation in vivo by allosteric inhibitors of hemoglobin saturation. Am J. Physiol 1993;265:H1450-H1453.
6. Teicher BA, Ara G, Emi Y, et al. RSR:effects on tumor oxygenation and response to therapy. Drug Dev Res 1996;38:1-11.
7. Pagel PS, Hettrick DA, Montgomery MW, Kersten JR, Steffen RP and Warltier DC. RSR 13 , a synthetic modifier of hemoglobin-oxygen affinity, enhances the recovery of stunned myocardium in anesthetized dogs. J Pharmacology and Experimental Therapeutics 1998;2851-2858.
8. Jason A. Woods, Charles J. Storey, Evelyn E. Babcock and Craig R. Malloy.Right-shifting the oxyhemoglobin dissociation curve with rsr13:effects on high-energy phosphates and myocardial recovery after low-flow ischemia J Cardiovascular Pharmacology 1998;31:359-364.
9. Wei EP, Randad RS, Levasseur JE, Abraham DJ, and Kontos HA. Effect of local change in O2 saturation of hemoglobin on cerebral vasodilution from hypoxia and hypotension. Am J Physiol 1993;265:H1439-H1443.
10. Grocott HP, Bart RD, Sheng H, Miura Y, Steffen R, Pearlstein RD, and Warner DC. Effects of a synthetic allosteric modifier of hemoglobin oxygen affinity on outcome from global cerebral ischemia in the rat Stroke.1998;29:1650-1655.
11. Swartz HM, Boyer S, Gast P, Glockner JF, Hu H, Liu KJ, Moussavi M, Norby SW, Walczak T, Vahidi N, Wu M, and Clarkson RB. Measurements of pertinent concentrations of oxygen in vivo. Magn Reson Med. 1991;20:333-339.
12. Swartz HM and Clarkson RB. The measurement of oxygen in vivo using EPR techniques. Phys Med Biol 1998;43:1957-1975.
13. Liu KJ, Gast P, Moussavi M, Norby SW, Vahidi N, Walczak T, Wu M, and Swartz HM. Lithium phthalocyanine:A probe for EPR oximetry in viable biological systems. Proc Natl Acad Sci 1993;90:5438-5442.

Chapter **28**

COMPUTER MODELING OF RELATIONSHIP BETWEEN CRITICAL PVO_2, VO_2MAX AND BLOOD SUPPLY OF SKELETAL MUSCLE AT WORKING WITH A RIGHT-SHIFTED BLOOD O_2 DISSOCIATION CURVE

Katherine G. Lyabakh & Irina N. Mankovskaya

Department of Information Technology in Medicine, Institute of Cybernetics, Kiev 01022, Ukraine; Department of Hypoxia, Bogomolets Institute of Physiology, Kiev 01024, Ukraine

Abstract: Investigations were performed on a computer model of O_2 delivery and O_2 consumption in the one working muscle. At working with increasing power and achieving the critical value of VO_2 (VO_2crit), the muscle VO_2 began to lag behind the oxygen demand qO_2 .The model permits to find critical pO_2 in effluent venous blood Pvcrit at VO2crit as well as to calculate VO2max and PvO_2 at VO_2max under exercise with changing muscle blood flow F and blood pH.Pvcrit was computed from the condition VO_2crit=0.9qO_2, and VO_2max– from the condition (dVO_2/dqO_2) = 0.1. VO_2max, Pv at VO_2max , Pvcrit, and VO_2crit were calculated for: $40 \leq F \leq 120$ ml/min/100g; $6.8 \leq pH \leq 7.4$. It was shown that the faster is F and the lesser is blood pH, the greater were the Pvcrit and VO_2max values. With decreasing F and blood pH, the influence of F on Pvcrit and VO_2max increases, whereas the influence of blood pH on these values decreases. With increasing F and, hence, an increasing VO_2max, the blood supply efficiency decreases due to the important limiting factor – tissue oxygen diffusion.

Key Words: critical pO_2, exercise, modeling, skeletal muscle, VO_2max

1. INTRODUCTION

Maximal energy demand of working muscles is unequivocally linked with the tissue respiration capacity and evaluated by maximal oxygen uptake VO_2max. There exist two theories as regard its limitation. According to the first theory, it is limited by the oxidative capacity of muscle [1]. In accordance with the second theory, the main limiting factor for VO_2max is the oxygen delivery to working muscle fibers. The literature data indicate that preference has been given to the second theory, as it has been proved experimentally. It has been shown that, depending on a change in the parameters of oxygen transport by blood, the VO_2max is proportional to the oxygen delivery to tissues [1-5].There is an agreement between physiologists that the respiratory links of the oxygen transport chain (ventilation and diffusion) determine limitation of muscle aerobic capacity, while such a function of circulatory link is not so clear [6]. Among limitations within the circulatory link, the muscle blood flow F is probably of the primary importance [2,7]. Exercise blood flow patterns within and among muscles were found to limit VO_2max [7]. Energy production can be accomplished by a purely aerobic way to a point when the definite oxidative capacity is reached. Thereafter, the rate of oxygen consumption VO_2 lags behind the oxygen demand qO_2. This consumption rate is taken as critical VO_2 (VO_2crit) and a respective end-capillary pO_2 value as critical PvO_2 (Pvcrit). We consider that beginning from the VO_2crit, the oxygen transport to muscle mitochondria gets limited by diffusion. For VO_2crit there is a reserve to rise, if the load still increases, until it has reached a plateau, when the limit of aerobic capacity is practically reached and, correspondingly, the maximal oxygen consumption rate attained. The corresponding value of PvO_2 at VO_2max (Pv at VO_2max) may be of interest as an informative indicator of the oxygen transport efficiency to working muscle. From reaching the Pvcrit and on, with an increasing load there occurs an increase in lactate and other acid product levels in the blood, altering the blood pH and shifting a blood O_2 dissociation curve (ODC) to the right (Bohr's effect). The ODC shift changes the blood-tissue pO_2 gradient and affects the oxygen delivery to the muscles influencing the threshold of anaerobic metabolism VO_2crit and VO_2max. Thus, sustaining the hypothesis which holds that the main limiting factor of the anaerobic metabolism threshold and VO_2max is the oxygen transport to mitochondria of active muscles, we may formulate the goals of the present computer investigation.

The goals of this study were (a) to calculate the VO_2max values with corresponding Pv at VO_2max as well as Pvcrit with corresponding VO_2crit, and (b) to investigate the influence of F and an ODC position on VO_2 max, VO_2crit, Pv at VO2max, and Pvcrit by means of computer modeling.

2. METHODS

Investigations were performed on a computer model of O_2 delivery and O_2 consumption in contracting muscle [8]. A computer model of oxygen delivery to and consumption in a contracting muscle was used to describe a steady state of capillary blood flow and three-dimensional oxygen diffusion-reaction in the microcirculatory unit (MCU) with homogenous distribution of mitochondria. The configuration of MCU was taken from [9]. The pO_2 and VO_2 in MCU were computed. We used a biochemical criterion of hypoxia - the state when oxygen consumption rate $VO_2(x,y,z)$ becomes less than oxygen demand qO_2 because of the lack of oxygen at (x,y,z). That provides an opportunity to get a graphical image of tissue hypoxia. The calculated value of ratio $S(x,y,z) = VO_2(x,y,z)/qO_2 x100(\%)$ describes the degree of hypoxia at (x,y,z). The surfaces Si(x,y,z) = const set up a field of $VO_{2.}$ This field represents an image of hypoxia in MCU and shows an area where oxygen debt is being produced., We shall show the MCU symmetry cross-section, which passes through the axis and a capillary, thus obtaining a flat picture of hypoxia zone. We rendered the simplest case: con-current capillary blood flow with equal values of red blood cells velocity. This provides a symmetry distribution of S-lines. For this we show only a half of a picture (Figure 1). The model served as a basis of the information technology for the computation of Pvcrit, VO_2max and corresponding Pv at VO_2max and VO_2 crit. In order to calculate Pvcrit and corresponding VO_2crit, we proceed from the condition $VO_2crit=0.9qO_2$. In order to calculate VO_2max and Pv at VO_2max, we used the condition $(dVO_2/dqO_2) = 0.1.VO_2max$, Pv at VO_2max , Pvcrit, and VO_2crit were calculated from the assumptions: $40 \leq F \leq 120$ ml/min/100g; $6.8 \leq pH \leq 7.4$; intercapillary distance $d=40x10^{-6}m$; oxygen concentration in arterial blood 190 ml/l; oxygen blood capacity 200 ml/l. The morphological and physical values of a model were chosen as: an area of capillary cross-section $16\times10^{-8}cm^2$, length of capillaries $L=900\times10^{-6}$ m , O_2 diffusivity in tissue 1.3×10^{-5} cm^2s^{-1}, O_2 solubility in blood and tissue 2.8×10^{-5} mm Hg^{-1}, apparent value of Michaelis constant K_m=3 mm Hg.

In order to explore the diffusion transport limitations in an active muscle fiber, we have estimated the image and location of hypoxia zone in that fiber working at VO_2max and supplied with F=40 and 120 ml/min/100g.

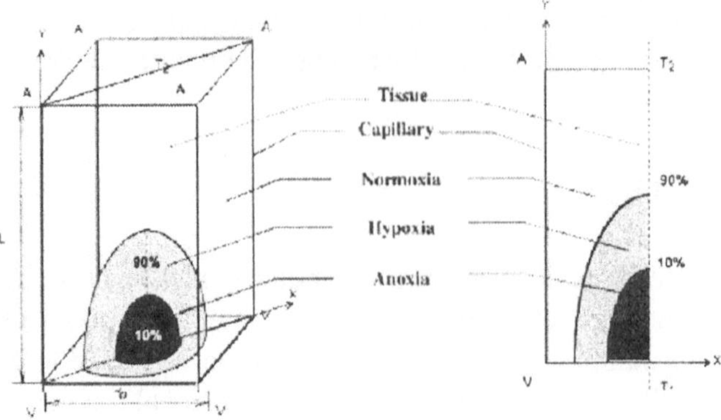

Figure 1. Microcirculatory unit (MCU) with hypoxia and anoxia zones inside. On the left: the three-dimensional image of MCU, on the right: a half of MCU cross-section . The ends of capillaries are designated as A (arterial) and V (venous). The center of MCU is designated as T_1T_2. % values represent (VO_2/ qO_2) x100

3. RESULTS

3.1 The effect of F and pH values on the critical pO_2 in muscle end- capillary blood (Pvcrit) and on the corresponding oxygen consumption rate (VO_2crit)

The diffusion limitations begin to manifest themselves when the oxygen consumption rate reaches a level, at which part of the working muscle energy is produced by an anaerobic way. The diffusion oxygen transport limitations do not permit to meet an oxygen demand to the fullest. This limitation follows from the expression that we used in the model [8]. It shows the connection between VO_2 , qO_2, and oxygen partial pressure P at any (x,y,z) of tissue: $VO_2= qO_2 \times P^2 / (P^2+ Km^2)$ [10]. The diffusion-limited decrease of tissue pO_2 below the critical level leads to the appearance of an area where the oxygen debt is produced - hypoxia zone. Pvcrit indirectly reflects a diffusion limitation and represents an informative indicator associated with the anaerobic metabolism threshold. Our investigations have shown that the calculated Pvcrit depends not only on the intercapillary distance in working muscle, but also on the other parameters of O_2 transport, in particular, blood pH in muscle capillaries. According to estimates, there is an almost linear relationship between the Pvcrit and blood pH: the more an ODC is shifted to the right, the more is the Pvcrit (Figure 2). This finding has been confirmed experimentally [11]. Naturally, this relationship is most

influenced by the blood flow within the muscle. The slope of the curve Pvcrit (pH) is determined by parameter F.
The greater the blood flow, the greater the pH values influence Pvcrit. The Pvcrit varies over some mm Hg from minimal to maximal values (Figure 2). If F is less than 60 ml/min/100g, the change in pH may only insignificantly influence the Pvcrit. Thus increasing pH from 6.8 to 7.4 decreases the Pvcrit from 25 to 22.5 mm Hg. At F=40 ml/min/100g, the Pvcrit level does not depend on pH changes at all. The highest value of Pvcrit=38 mm Hg is observed at the highest blood flow level of all levels considered - 120 ml/min/100g at pH=6.8 (Figure 2). The value VO_2crit versus Pvcrit depends, to a greater degree, on blood pH, but it repeats the run of the curve Pvcrit (pH, F) (Table 1). The mean value of tissue pO2 (Pt) decreases with an increasing blood flow. As seen from the estimates, the change in blood pH from 6.8 to 7.4 diminishes the value of Pt by approximately 6 mm Hg. VO_2 crit changes from 11.2 to 6.1 ml/min/100g at pH=6.8 and from 8.5 to 5 ml/min/100g at pH=7.4. We may conclude that the sensitivity of Pt and VO_2 crit to the changes of F is greater at pH=6.8 than at pH=7.4.

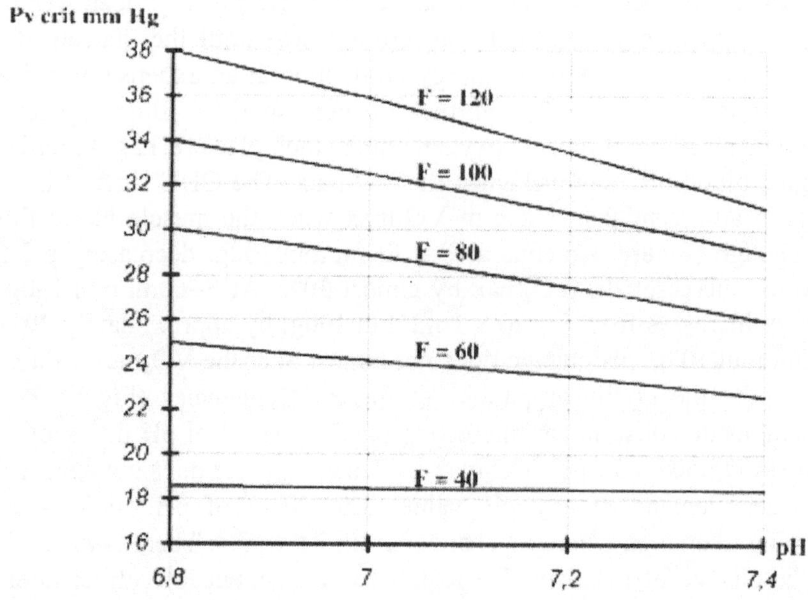

Figure 2. The effect of muscle capillary blood pH on Pvcrit at different values of muscle blood flow F (ml/min/100g)

Table 1. The influence of muscle blood flow F and blood pH on VO_2 crit and mean oxygen pressure in tissue Pt.

F ml/min/100g	pH	VO_2 crit ml/min/100g	Pt mm Hg
120	6.8	11.2	22.0
	7.4	8.5	16.0
100	6.8	10.2	22.5
	7.4	7.8	16.4
80	6.8	9.2	23.0
	7.4	7.2	17.0
60	6.8	7.9	24.0
	7.4	6.2	17.7
40	6.8	6.1	26.0
	7.4	5.0	18.3

3.2 The effect of F and pH values on the VO_2max and Pv at VO_2max

It was established that the dependence $VO_2(qO_2)$ is nonlinear. Beginning from the critical mode (partially anaerobic), and until the plateau of VO_2 (qO_2) is attained, the share of energy production at an expense of oxidative processes decreases gradually. Further, an increase of VO_2 did not practically occur, and a ceiling of muscle oxygen uptake took place [8,12]. According to our data, blood pH may influence the VO_2max. The ODC shift to the right results in an essential increase in VO_2max when the muscle blood flow is high enough (Figure 3). Thus at F=120 ml/min/100g, decreasing pH from 7.4 to 6.8 increases the VO_2max by almost 30%. At F=60ml/min/100g, the VO_2max increases from 7.5 to 9.1 ml/min/100g, by approximately 20%. At F=40ml/min/100g, pH change does not almost alter the VO_2max value. One should note the nonlinear pattern of these dependencies (Figure 3). Our calculations demonstrate an interesting result: varying of pH does not effect on ratio VO_2max/ qO_2, i.e., VO_2max in skeletal muscle does not exceed 70% of oxygen demand at any pH value. The effect of pH on VO_2max is beginning to emerge when pH changes from 7.0 to 7.4. There is no evidence that the change of pH from 7.0 to 6.8 has any noticeable effect upon the maximal aerobic capacity of muscles. This range of pH changes, according to our estimates, has appeared to be unsuitable for an adjustment of oxygen transport to tissue.

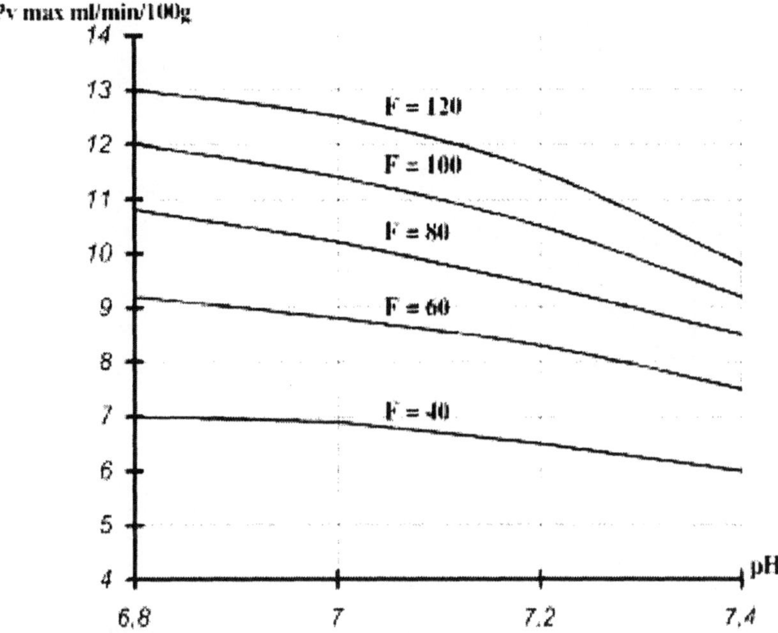

Figure 3. The effect of muscle capillary blood pH on VO_2max at different levels of muscle blood flow F (ml/min/100g).

Our calculations revealed that both VO_2 and Pv change in a similar manner under two investigated modes - during exercise at VO_2max and at critical VO_2 With increasing blood flow, both VO_2max and the lactate threshold rise (Figure 4). At the relatively low blood flow values, the beginning of VO_2 limitation (critical mode) to a state of maximal aerobic capacity mode are close in theVO_2 values: 6 and 7 ml/min/100g respectively. With increasing blood flow, the VO_2 values show a greater difference between themselves. By contrast, the Pv values become closer, as the blood flow increases (Figure 5). As seen from Figures 4, 5, and Table 2, there exists the relationship between VO_2max, Pv at VO_2max and F. With decreasing F, both VO_2max and pO_2 in venous blood, draining the exercising muscles, fell. These results are compatible with that of [6,13,14]. According to [6], because of the lower VO_2max the pO_2 would be reduced more in venous blood if the tissue diffusing capacity remains the same, then tissue capillary pO_2 will fall as the VO_2max falls. It is evident that decreasing Pv at VO_2max actually means greater tissue extraction.

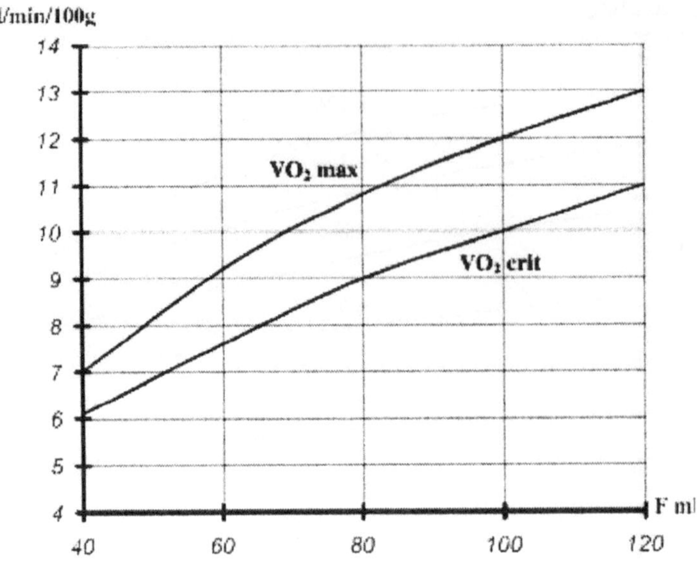

Figure 4. The effect of muscle blood flow F on the VO_2max and VO_2crit.

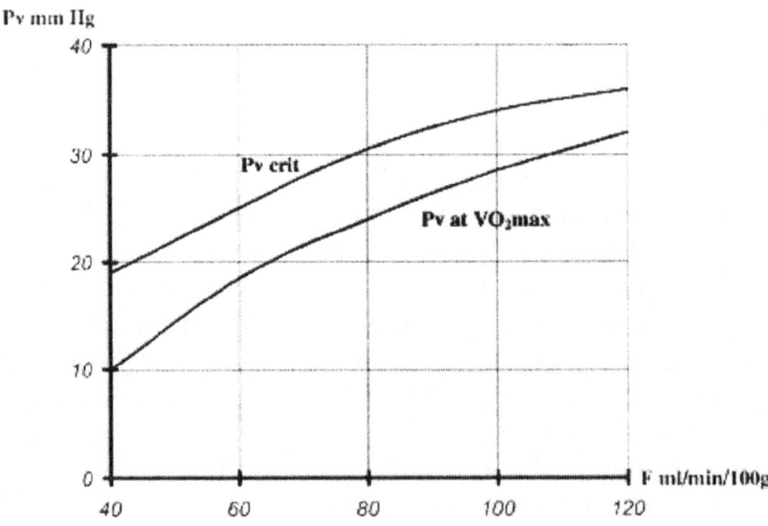

Figure 5. The effect of muscle blood flow F on the Pvcrit and Pv at VO_2max.

Table 2. Exercise muscle blood flow F patterns and corresponding values PvO_2 and VO_2 at the critical mode and VO_2max mode *

F	1.0	1.5	2.0	2.5	3.0
VO_2crit	1.0	1.3	1.5	1.7	1.9
P_V crit	1.0	1.4	1.7	1.8	2.0
VO_2max	1.0	1.3	1.6	1.7	1.9
P_V at VO_2max	1.0	1.8	2.4	2.8	3.2

* Data were given in dimensionless forms as regard to the initial state designated as 1.0: F=40ml/min/100g, VO_2crit= 6 ml/min/100g, Pvcrit=10 mm Hg, VO_2max=7ml/min/100g, Pv at VO_2max=18.7 mm Hg

The above results are explained by the diffusive and convective resistance ratios. With increasing blood flow, the resistance to the convective oxygen flow decreases, while that to the diffusive oxygen flow remains unaltered. Accordingly, the oxygen flows were re-distributed. We shall compare two examples of the oxygen supply to skeletal muscle working at VO_2max. In the first case, the blood flow is equal to 40 ml/min/100g, in the second case it is 3 times greater (120 ml/min/100g). In according to our calculations, the arterio-venous difference for O_2 is 180 ml/l for F=40 ml/min/100g and 120 ml/l at F 3 times greater. In the former case, the diffusive flow is equal to 18/19 oxygen extracted from one liter of the blood. In other words, the ratio of the oxygen arterio-venous difference to the arterial blood oxygen content is about 90%, as the diffusive versus convective resistance is low. In the second case, these are comparable and, hence, the diffusive and convective oxygen flows to tissue are related as 12/19 to 7/19 respectively. In this case, 60% of whole oxygen is extracted from one liter of the blood. Certainly, an absolute value of the oxygen flow to muscle mitochondria in the second case is much higher. It is nearly 2 times greater and the blood flow is 3 times greater compared to the values, which have been computed in the first case.

This points to the reduced mode efficiency in the second case. So, an increase in the blood supply results in the rise of VO_2max, though the blood flow efficiency is diminished.

3.3 The hypoxia image within working muscle fiber at VO_2max

Figure 6 presents the hypoxia image within a muscle fiber, working at VO_2max and being supplied with blood flow F=40 ml/min/100g (top) and F=120 ml/min/100g (bottom). The convective oxygen flow is directed from A to V along the length of capillaries. The three-dimensional diffusion oxygen flow from the capillaries is directed to the central fiber axis T_1T_2 (see Figure1). The oxygen pass length from the end-capillary to worst supply point – a lethal corner in MCU – is equal to 28×10^{-6}m. From the right of the cross-section, percentage values from 10% to 90% designate the boundaries of the areas for a reduced oxygen consumption rate (share of oxygen demand, %). Below each of these lines, the oxygen consumption does not exceed the level expressed in percentage. The white field of the section (VO_2 is more than 90% qO_2) represents an area in which oxygen demand is fully met (normoxia). The black field (VO_2 is less than 10% qO_2) represents an anoxia zone, and the grey fields represent the hypoxia zone. We suppose that these pictures provide understanding of complicated interconnections between Pv, VO_2, qO_2, and F at VO_2max. The hypoxia images vary in the two cases under consideration. At F=40 ml/min/100g, there is an anoxia at the worst supply point with the value of Pv of up to 10 mm Hg, that is evidence of great pO_2 gradients along the capillary. The pO_2- and VO_2-gradients in tissue are also high. Since the oxygen extraction from the blood is very intensive, there takes place venous hypoxemia. One gets convinced of its existence from the hypoxia image, even not knowing the pO_2 value of the blood - the hypoxia zone involves the venous end of the capillary. Enlargement of the grey area means that the greater part of the fiber is under hypoxia. At F=120 ml/min/100g, the end-capillary oxygen pressure is much greater (Pv=32 mm Hg) compared to that observed in the previous case. The hypoxia picture confirms this finding: the area of full meeting of oxygen demand surrounds a capillary and the hypoxia zone appears to be concentrated in the central part of the fiber. The diffusive oxygen flow is very intensive and the oxygen pressure gradients are high. In addition, the field of VO_2 (and pO_2) shows its specificity with respect to the tissue hypoxia caused by an intensive work. The areas with a very high oxygen utilization rate are neighboring with the areas, where oxygen consumption rate is noticeably reduced. In this respect, the points of an area adjacent to the venous capillary end are found to be most typical. So oxygen consumption rate at a lethal corner is about 10% qO_2 , and pO_2=1.5 mm Hg.We suppose that lactate production within the hypoxia zone is very considerable thus inducing lactic acidosis.

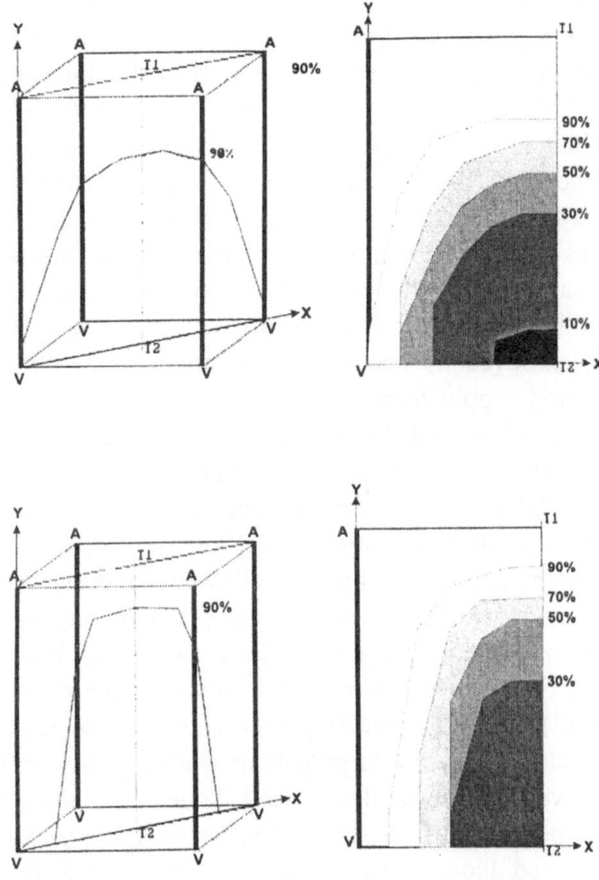

Figure 6. The comparison of hypoxia images in muscle MCU at F=40 ml/min/100g (top) and F=120 ml/min/100g (bottom). On the left: the three dimensional image of MCU, on the right: a half of MCU cross-section which passes through the center of MCU T_1T_2 and a capillary AV.

4. DISCUSSION

To investigate quantitatively the effect of the muscle blood flow velocity and blood pH on the oxygen consumption rate and the oxygen pressure in venous blood, we introduced the values F and pH, being independent from each other, in the model. This permitted us to calculate the oxygen pressure in blood and muscular tissue as well as an average oxygen consumption rate. The model deals with the average capillary blood pH values, being constant along capillary length. The use of mathematical modeling had certain advantages, namely: to estimate, along with natural combinations of parameters, some artificial cases of O_2 transport and utilization in working muscle. Thus we have assumed that the blood pH in capillaries may undergo changes, irrespective of blood supply and level of the muscle loading under consideration. It should be noted that we shall consider the VO_2max and the blood flow values in one separate muscle. Hence, these values may exceed considerably those of the VO_2max values obtained from estimating the oxygen consumption per kg of the total body mass or per kg of the total mass of muscles as is usually done in experimental measurements. It should be also noted that the amount of maximally activated muscle is unknown but may be 50% of the total mass of muscles [6].

While VO_2 achieved its maximal value under highest levels of F, the efficiency of blood supply regarding the tissue oxygenation diminished. It may be explained by the fact that the great convective O_2 flow required the corresponding diffusion O_2 flux to the muscle tissue but the latter is diffusionally limited and creates the hypoxia zone in MCU. An evaluation of the oxygen transport and utilization in an intensively working muscle by means of mathematical modeling has allowed to describe the hypoxia characteristics, on a whole. These include: (1) high oxygen demand and oxygen consumption rate values, a powerful gas exchange between the blood and tissue; (2) oxygen diffusion in muscle tissue as the main O_2 transport-limiting factor; (3) high pO_2- and VO_2- gradients in tissue as typical signs of hypoxia zone, extended over a large part of working muscle fiber; (4) high red blood cells velocity in capillaries and, hence, the potential functional oxygen shunt appearance; (5) absence of an extracapillary oxygen transport due to the high red blood cells velocity; (6) a possible development of venous hypoxemia accompanying an intensive blood oxygen utilization; (7) large oxygen debt formation, an accumulation of the acid products in the blood, and the ODC shift to the right, and (8) low average oxygen tension in tissue and potential anoxia in the worst supply areas. The above-listed characteristics, which are well-known from physiology and complemented with the mathematical modeling results, give grounds to classify the working muscle hypoxia as a special type of work hypoxia in its own right. Earlier the term "hypoxia under loading" was proposed by A.Z. Kolchinskaya et al [15] to define a hypoxic state of the whole organism working intensively.

The above investigations have allowed us to make the following proposal. The main factor promoting a VO_2max rise is increasing muscle blood flow. Nevertheless, the rise of blood supply to the working muscle is limited not only by cardiac output. There is an economical limit for the rise in blood supply and VO_2max - a decline of the F efficiency, which makes an enhancement of the blood flow of little sense. It is widely known that under extreme blood perfusion functional shunting emerges and deteriorates O_2 transport to tissue. Furthermore, muscle blood flow may rise without apparent limit when the mass of active muscle is too small to overwhelm the heart [16]. If so, with the maximal effort by a large muscle mass some central control will be needed to limit F [14]. It is available not only for large muscle mass. Muscle blood flow F is supposed to play a key role in the limitation of VO_2max and VO_2crit. The less F, the less VO_2max and VO_2crit. It is clear that heterogeneous F distribution within the muscle under neural control may provide F/ VO_2 mismatching and, as a result, a decrease of VO_2max [6,14].Thus, upon attaining the certain high level in blood flow rate within working muscle, the blood pH value begins to rise owing to washing out the acid products. But according to our estimates, this should lead to a VO_2max reduction. And finally at high oxygen demand values, despite a high blood flow rate, O_2 diffusion in tissue limits VOcrit and VO_2max and reduces the blood flow efficiency.

We may conclude:

(1) The faster is the muscle blood flow and the lesser is the blood pH, the greater are the Pvcrit and VO_2max values.

(2) With decreasing F and blood pH, the influence of F on Pvcrit and VO_2max increases whereas the influence of blood pH on these values decreases.

(3) With increasing muscle blood flow velocity and, hence, an increasing VO_2max, the blood supply efficiency decreases due to the important limiting factor – tissue oxygen diffusion.

REFERENCES

1.Holloszy J. Adaptation of skeletal muscle to endurance exercise and their metabolic consequences. J Appl Physiol 1984;56:831-838.

2.Astrand P, Saltin B. Maximal oxygen uptake and heart rate in various types of muscular exercise. J Appl Physiol 1961;16:977-981.

3. Pirnay F, Dujardin J, Deroanne R, Petit J. Muscular exercise during intoxication by carbon monoxide. J Appl Physiol 1971; 31:573-575.

4. Ekblom B, Goldbarg A, Gullbring B. Response to exercise after blood loss and reinfusion. J Appl Physiol 1972;33:175-180.

5. Saltin B, Rowell L, Gostil L. Functional adaptations to physical activity and inactivity. Fed Proc 1980;39:1506-1513.

6. Rowell L, Wagner P. Debate. Limitations to exercise in extreme hypoxia. Why does the exercise cardiac output fall during altitude residence and is it important? 11th International Hypoxia Symp. Jasper Park Lodge. Spec Progr Suppl 1999;1-11.

7. Armstrong R, Laughlin M. Exercise blood flow patterns within and among rat muscles after training. Amer J Physiol 1984;246:H59-H68.

8. Lyabakh K. Mathematical modeling of oxygen transport in skeletal muscle during exercise: hypoxia and VO_2 max . In:Eke A, Delpy DT, editors. Oxygen Transport to Tissue XX1, Adv Exper Med Biol New-York-London: Plenum Press,1999;471:585-593.

9. Grunewald W, Sowa W. Capillary structures and O_2 supply to tissue. Rev Physiol. Biochem Pharmacol 1977;77:149-209.

10. Bagdasarova T, Haskin V The effect of adaptation to hypoxia on the dependence of tissue breathing of temperature and oxygen pressure in vitro. Sechenov Physiol J 1977; 63:1598-1609.

11. Wasserman K. Critical capillary pO_2, and the role of lactate production in oxyhemoglobin dissociation during exercise. In: Eke A, Delpy DT, editors. Oxygen Transport to Tissue XX1, Adv Exper. Med. Biol. New-York-London: Plenum Press,1999; 471:321-333.

12. Lyabakh K. Quantitative analysis of determinants of aerobic metabolism threshold and maximal oxygen consumption in working muscle. In: Boloban AI, Kochur BM, editors. The proceeding of the modern Olympic sport. International scientific congress. Kyiv: University of Physical Education and Sport, 1997;146-147.

13. Wagner P. Theoretical analysis of factors determining VO_2 max at sea level and altitude. Respir Physiol 1996;196:329-343.

14. Rowell L. Neural control of muscle blood flow: importance during dynamic exercise. Clin Exp Pharmacol Physiol 1997;24:117-125.

15. Kolchinskaya A, Lyabakh K, Philippov M. General description of hypoxia under loading: genesis and compensation. In: Kolchinskaya AZ, ed. Secondary tissue hypoxia. Kiev: Naukova Dumka, 1983;183-190.

16. Reeves J, Groves B, Sutton J, Wagner P, Cymerman A, Malconian M, Rock P young, Houston C. Operation Everest II: preservation of cardiac function at extreme altitude. J Appl Physiol 1987;63:531-539.

Chapter 29

SMALL-VOLUME RESUSCITATION WITH THE HEMOGLOBIN SUBSTITUTE HBOC-201: EFFECT ON BRAIN TISSUE OXYGENATION

Geoffrey T. Manley[1,2], J. Claude Hemphill[1,3], Diane Morabito[1,4], Vanessa Erickson[1], John J Holcroft[1], Nakita Derugin[1,2], and M. Margaret Knudson[1,4]
School of Medicine, University of California San Francisco, San Francisco Injury Center,[1] and the Departments of Neurological Surgery,[2] Neurology,[3] and Surgery[4]

Abstract: **OBJECTIVES:** To investigate the effects of small-volume resuscitation with a hemoglobin based oxygen carrier on brain tissue oxygen tension ($PbrO_2$) in hemorrhaged swine. **METHODS:** Clark-type electrodes were inserted into the brain tissue of 6 swine to measure $PbrO_2$ directly. Swine were hemorrhaged to a MAP of 40 mm Hg for 20 minutes. Resuscitation was performed with a bolus infusion of HBOC-201 (6 cc/kg; Biopure Corp.) and high-flow oxygen (100%). Swine were observed for an additional 2 hours. **RESULTS:** $PbrO_2$ prior to hemorrhage was 48.7 ± 4.7 mm Hg with 100% inspired oxygen. PbrO2 rapidly declined to 7.6 ± 5.3 mm Hg in response to hemorrhage. Small-volume resuscitation with HBOC-201 and high-flow oxygen resulted in a significant increase ($p<0.001$) in PbrO2 to 44.6 ± 8.1 mm Hg. MAP was also significantly increased to 84% of baseline. These elevations were sustained during the observation period. **CONCLUSIONS:** Resuscitation with HBOC-201 can restore and sustain cerebral oxygenation and MAP. These results suggest that a small-volume bolus of HBOC-201 may provide adequate oxygen and pressure support during the initial management of hemorrhage.

Key words: Hemorrhage, cerebral resuscitation, cerebral oxygenation, blood substitutes

1. INTRODUCTION

Hemorrhagic shock and head injury are the principle causes of traumatic morbidity and mortality [1]. Post mortem and experimental studies have demonstrated that cerebral ischemia plays a significant role in the pathogenesis of severe hypotension and traumatic brain injury [2-4]. Thus, rapid early cerebral resuscitation provides the foundation for the treatment of shock and head injury. However, controversy exists surrounding the optimal method of cerebral and systemic resuscitation, particularly in the pre-hospital setting.

Currently, large-volume crystalloid fluid resuscitation is used to restore hemodynamic parameters until O^- or cross-matched blood is available. This approach has recently been called into question by a clinical study that demonstrated a worse outcome in non-head injured trauma patients aggressively resuscitated with crystalloid, suggesting that delayed resuscitation may be warranted [5]. While restoration of cerebral perfusion and tissue oxygenation is critical, thereby arguing against delayed resuscitation, there is experimental evidence suggesting that crystalloid fluid resuscitation may lead to increased cerebral edema in patients with head injuries [6]. Thus, it appears that a resuscitation solution should ideally restore cerebral oxygenation with limited volume.

To satisfy the requirements of adequate cerebral resuscitation with minimal fluid administration, we have investigated the effects of small-volume resuscitation with a hemoglobin based oxygen carrier (HBOC-201) on brain tissue oxygen tension ($PbrO_2$) and mean arterial pressure (MAP) in hemorrhaged swine.

2. MATERIALS AND METHODS

The experimental protocol was reviewed and approved by the Committee on Animal Research at the University of California San Francisco (UCSF). All experiments were performed in the UCSF Experimental Surgery Laboratory.

2.1 Animal preparation

Six male Yorkshire swine weighing from 40 kg to 46 kg were studied. Following intubation, anesthesia was maintained with a fentanyl infusion (1 mcg/kg/hr) and inhaled isoflurane (0.5% to 2%). The swine were paralyzed initially with 0.05 mg/kg of pancuronium and received supplemental bolus injections as needed. Cerebral and peripheral venous and arterial access was established. Body temperature of 38.5C+/-1.0C was maintained by using a forced air warmer (BAIR Hugger, Augustine Medical, Eden Prairie, MN). Burr holes were made for the placement of intracranial monitors as previously described [7].

2.2 Oxygen monitoring

$PbrO_2$ was measured directly with the Licox Clark-type polarographic pO_2 probe (GMS mbH, Kiel, Germany). $PbrO_2$ measurements represented the mean values of 2 probes. At the completion of each experiment, probes were removed, placed in their respective calibration barrels, and observed for return to the calibrated oxygen levels recorded before the experiment began.

2.3 Hemorrhage and resuscitation protocol

The time-line and treatment interventions for the experimental protocol were designed to approximate a pre-hospital clinical situation. Following instrumentation, the swine were allowed to stabilize for 1 hour. Maintenance intravenous fluid (Lactated Ringer's) was adjusted to maintain blood pressure, heart rate, and urine output. The *in vivo* response of the brain oxygen probes was then evaluated by increasing the FiO_2 to 1.0. Once the 100% oxygen response was verified, the FiO_2 was decreased to 0.30 and the swine were allowed to stabilize for another 30 minutes. Maintenance fluids were discontinued; the swine were then hemorrhaged through an arterial catheter over 30Êminutes until a MAP of 40 mm Hg was achieved. This pressure was maintained for 20 minutes by the withdrawal of additional blood. The volume of blood withdrawn from the swine using this modified Wigger's model, typically resulted in a Grade IV hemorrhage. After completion of the baseline hemorrhage measurements, the FiO_2 was increased to 100% to simulate the institution of high flow 100% oxygen in the field. After 10 minutes, the swine were resuscitated with HBOC-201 (4 cc/kg; Biopure Corp.) through a peripheral venous line over 5 minutes and the maintenance fluids were resumed. Swine were then observed for 2 hours to evaluate the durability of these interventions. $PbrO_2$ and other physiological variables were monitored continuously.

2.4 Numerical and statistical analysis

All physiologic characteristics were recorded at 1-min intervals throughout each experiment unless otherwise noted. All values were depicted as the mean ± the standard deviation if not specified otherwise. Repeated measures analysis of variance was used to test the differences between various experimental conditions. Tukey's method was used for the post-hoc tests. A p value of less than 0.05 was considered significant.

3. RESULTS

The experimental time intervals for this model were based on clinically relevant prehospital data reflecting average response times, length of initial assessment, and time to first treatment. The effect of small-volume resuscitation with HBOC-201 following hemorrhage is shown for a representative swine in Figure 1. After stabilization, an increase in the FiO_2 from 0.5 to 1.0 resulted in an increase in $PbrO_2$ (not shown). Decreasing the FiO_2 to 0.30 resulted in a predictable decrease in $PbrO_2$. There was no significant change in the MAP during these alterations in FiO_2. To approximate the clinical setting, the FiO_2 was maintained at 0.3 to target a p_aO_2 of 100 mm Hg to 130 mm Hg prior to hemorrhage. Hemorrhage resulted in a rapid decrease in $PbrO_2$ and MAP (Figure 1). Following cessation of the hemorrhage, the MAP was maintained at approximately 40 mm Hg with intermittent withdrawal of blood for 20 minutes. The total volume of blood removed was equivalent to a severe Grade IV hemorrhage. Initial resuscitation with FiO_2 of 1.0 resulted in a small increase in $PbrO_2$ over 10 minutes with no change in MAP. Small-volume resuscitation with HBOC-201 (4 cc/kg) led to a significant increase in PbrO2 that exceeded the pre-hemorrhage value with an FiO_2 of 1.0. This slowly decreased to a post-hemorrhage baseline value that was greater than the pre-hemorrhage value obtained with a FiO_2 of 0.3. The MAP also increased to 75% of the pre-hemorrhage baseline.

The data for all six swine are summarized in Table 1. The mean $PbrO_2$ prior to hemorrhage with an FiO_2 of 1.0 was 48.7 ± 4.7 mm Hg. Reduction of the FiO_2 to 0.30 led to a decrease in $PbrO_2$ to 25.7 ± 1.4 mm Hg. In response to hemorrhage, $PbrO_2$ rapidly declined to 7.6 ± 5.3 mm Hg and MAP decreased to 40.8 ± 0.8 mm Hg. Institution of high-flow oxygen resuscitation (FiO_2= 1.0) led to a small increase in $PbrO_2$. Small-volume resuscitation with HBOC-201 resulted in a significant increase ($p<0.001$) in $PbrO_2$ to 44.6 ± 8.1 mm Hg. MAP also was significantly increased to 84% of baseline. These elevations in $PbrO_2$ and MAP were sustained over the 2 hour observation period.

4. DISCUSSION

The ideal resuscitation solution for traumatic hemorrhage should stabilize systemic hemodynamics and support tissue metabolism. Although blood is generally considered to be the "gold standard" for resuscitation, it is not

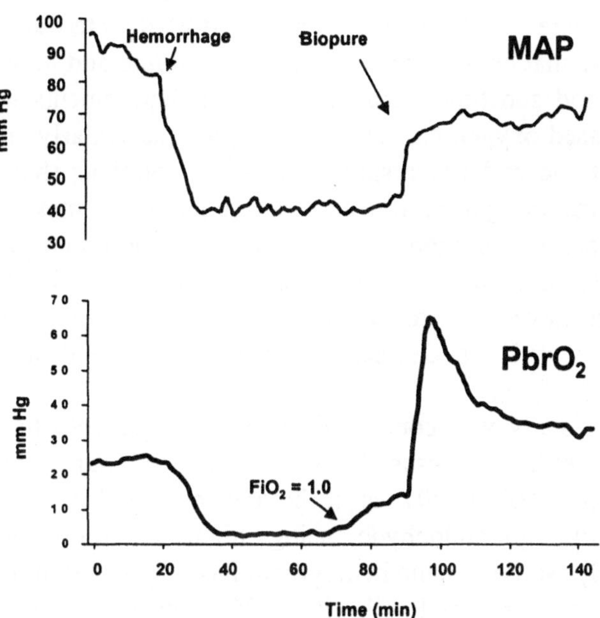

Figure 1. Characteristic effects of small-volume HBOC-201 resuscitation on oxygen concentrations in brain tissue ($PbrO_2$) and mean arterial pressure (MAP) from a representative swine following hemorrhage.

TABLE I. Effects of Hemorrhage and Resuscitation on Cerebral Variables (n = 6)

	PbrO2	MAP	HR	CO
Baseline (100% O_2)	48.7 ± 4.7	88.4 ± 5.3	82 ± 2.9	4.1 ± 0.4
Baseline (30% O_2)	25.7 ± 1.4	84.6 ± 5	87 ± 6	3.9 ± 0.4
End Hemorrhage	7.6 ± 5.3	40.8 ± 0.8	153 ± 8.4	2.3 ± 0.2
Resuscitation with 100% O_2	14.8 ± 6.4	38.3 ± 1.4	138 ± 11	2.14 ± .2
Resuscitation with HBOC-201	44.6 ± 8.1	71.0 ± 4.7	122 ± 11.3	3.2 ± 0.3

Mean values of concentrations in brain tissue oxygen (PbrO2), mean arterial pressure (MAP), heart rate (HR), and cardiac output (CO) in 6 swine undergoing hemorrhage and resuscitation with 100% oxygen and HBOC-201 (4 cc/kg).

typically available in the prehospital setting. Furthermore, there are significant issues of compatibility, disease transmission, and storage limitations. With the added recognition of the limitations of crystalloid fluid resuscitation, there has been considerable effort directed at developing a blood substitute that could expand the plasma volume and transport oxygen. A class of blood substitutes that meets these requirements known as the hemoglobin-based oxygen carriers (HBOCs) are particularly appealing. The HBOCs are sterile and universally compatible solutions that can be used without cross-matching or viral testing. These properties make the HBOCs a potentially important solution for resuscitation in the prehospital arena and the emergency room. However, a number of concerns have been raised about HBOCs. There have been reports of significant systemic and pulmonary hypertension as well as decreased cardiac output and impaired tissue perfusion [8, 9].

The current study was conducted to evaluate the potentially beneficial and deleterious effects of resuscitation with an HBOC know as HBOC-201 (Biopure Corp.). HBOC-201 is a bovine hemoglobin product that is crosslinked with glutaraldehyde using highly reproducible automated procedures [10]. Native bovine hemoglobin has a P_{50} of 30 mm Hg, which is similar to that of human red cells (P_{50} = 26.5 mm Hg). In its polymerized form, HBOC-201 has a P50 of 38 mm Hg. The HBOC-201 product has an osmolarity of 300 mOsm/kg with a colloid osmotic pressure of 18 mm Hg. The half-life in humans is 24 hours. The most attractive feature of this product is its stability. In contrast to the other HBOCs currently in clinical trials, HBOC-201 does not require refrigeration [11]. HBOC-201 has a shelf life of 2 years at room temperature thereby making it an ideal resuscitation fluid for the prehospital arena.

In the present study we observed a significant and reproducible increase in brain tissue oxygenation following administration of a small volume of HBOC-201. While two previous reports describing studies in humans [12] and dogs [13] have suggested that a reduction in cardiac output could impair oxygen delivery, our data obtained from a direct measurement of tissue oxygen partial pressure ($PbrO_2$) does not support this hypothesis. In contrast to these prior studies where there was isovolemic hemodilution with HBOC-201, we employed a small-volume resuscitation in a severe hypovolemic state. We observed a significant improvement in both MAP and cardiac output that resulted in enhanced tissue oxygenation. These finding suggest that a small-volume strategy for resuscitation of hemorrhagic shock maximizes the beneficial effects of HBOC-201 while minimizing any of the potential deleterious effects.

In summary, direct measurement of $PbrO_2$ demonstrated that resuscitation with HBOC-201 can restore and sustain cerebral oxygenation and MAP. These results suggest that HBOC-201 is an ideal cerebral resuscitation fluid, particularly in the pre-hospital arena where a small-volume bolus may provide an oxygen and pressure support "bridge" during transport and initial hospital management.

ACKNOWLEDGEMENTS

These studies were supported by Grant R49/CCR903697-07. The HBOC-201 solution was provided by Biopure Corporation. The investigators have no financial interests or consulting agreements with the Biopure Corporation.

REFERENCES

1. Trunkey DD. Trauma. Sci Am 1983;249:28-35.
2. Graham DI, Adams JH. Ischaemic brain damage in fatal head injuries. Lancet 1971;1:265-266.
3. Adams JH. Head injuries. J Clin Pathol Suppl (R Coll Pathol) 1970;4:176-177.
4. Adams JH, Brierley JB. The effects of systemic hypotension upon the human brain. Clinical and pathological observations in 11 cases. Head injuries. Brain 1966;89:235-268.
5. Bickell WH, Wall MJ, Pepe PE, et al. Immediate versus delayed fluid resuscitation for hypotensive patients with penetrating torso injuries. N Engl J Med. 1994;331:1105-1109.
6. Walsh JC, Zhuang J, Shackford SR. A comparison of hypertonic to isotonic fluid in the resuscitation of brain injury and hemorrhagic shock. J Surg Res. 1991;50:284.
7. Manley GT, Pitts LH, Morabito D, Manley GT, Doyle CA, Gibson J, Hopf HW, Knudson MM. Brain tissue oxygenation during hemorrhagic shock, resuscitation, and alterations in ventilation. J of Trauma 1999; 46:261-267.
8. Hess JR, MacDonald VW, Brinkley WW. Systemic and pulmonary hypertension after resuscitation with cell-free hemoglobin. J Appl Physiol. 1993;74:1769-1778.
9. de Figueiredo LFP, Mathru M, Solanki D, MacDonald VW, Hess J, Kramer GC. Pulmonary hypertension and systemic vasoconstriction may offset the benefits of acellular hemoglobin blood substitutes. J Trauma 1997;42:487-856 .
10. Pearce LB, Gawryl MS. Overview of preclinical and clinical efficacy of Biopure's HBOC's. In: Chang TMS editor. Blood Substitutes: Principles, Methods, Products, and Clinical Trials. New York: Karger Landes Systems, 1998; 82-100.
11. Ketcham EM, Cairns CB. Hemoglobin-based oxygen carriers: development and clinical potential. Ann Emerg Med 1999;33:326-337.
12. Kasper SM, Walter M, Grune F, et al. Effect of a hemoglobin-based oxygen carrier (HBOC-201) on hemodynamics and oxygen transport in patients undergoing preoperative hemodilution for elective abdominal aortic surgery. Anesth Analg 1996;83:921-927.
13. Krieter H, Hagen G, Waschke KF et al. Isovolemic hemodilution with a bovine hemoglobin-based oxygen carrier: Effects on hemodynamics and oxygen transport in comparison with a nonoxygen-carrying volume substitute. J Cardiothorac Vasc Anesth 1997;11:3-9.

Chapter **30**

THE EFFECT OF RSR13, A SYNTHETIC ALLOSTERIC MODIFIER OF HEMOGLOBIN, ON BRAIN TISSUE pO_2 (MEASURED BY EPR OXIMETRY) FOLLOWING SEVERE HEMORRHAGIC SHOCK IN RATS

Minoru Miyake, Oleg Y. Grinberg, Huagang Hou, *Robert P. Steffen, **Hisham Elkadi, Harold M. Swartz

*EPR Center for the Study of Viable Systems, Department of Radiology, 7785 Vail Dartmouth Medical School, Hanover, NH USA; *Allos Therapeutics, Inc. 11080 CirclePoint Road, Suite 200, Westminster, CO US; **Department of Anesthesiology, Boston University Hospital, Boston, MA USA*

Abstract: RSR13 is a synthetic allosteric modifier of hemoglobin that decreases the oxygen binding affinity of hemoglobin, potentially increasing oxygen availability to hypoxic tissues. Using *in vivo* EPR to directly measure cortical pO_2, we examined whether RSR13 would improve brain tissue pO_2 following severe hemorrhagic shock in rats. Hemorrhagic shock was induced by withdrawing blood (2.7-2.8 mL/100 g/15 min). Following a 30 min shock period, resuscitation was performed by infusion with Ringer lactate plus RSR13 (150mg/kg) or saline (control). Following hemorrhage, brain pO_2 decreased by about 14 mm Hg in both groups. Following crystalloid resuscitation brain pO_2 remained depressed in the control group but returned to the pre-hemorrhage values in the rats that received RSR13. RSR13 immediately increased and maintained the p_aO_2 while controls had a very gradual increase towards pre-hemorrhage values. There was no difference in the blood pressure or heart rate between groups. RSR13 may have useful applications to decrease the effects of acute hemorrhagic hypoxemia by increasing brain oxygenation.

Key words: Allosteric modifier RSR13, cerebral pO_2, EPR oximetry, hemorrhagic shock

1. INTRODUCTION

Severe blood loss sufficient to impair oxygen transport leads to hemorrhagic shock, which is characterized by the failure of compensatory mechanisms and therefore can result in loss of homeostasis and death. Hemorrhagic shock is one of the major causes of morbidity and mortality in trauma patients [1]. The effects of hemorrhagic shock are especially manifested in the brain because of its dependence on oxidative metabolism. A variety of approaches to augment oxygen delivery have been investigated in laboratory models of hemorrhagic shock, including manipulations of systemic perfusion pressure, blood flow, and the oxygen-carrying capacity of blood. An alternative approach is to increase the release of the available oxygen from hemoglobin. RSR13 (2-[4-[[(3,5dimethylanilino) carbonyl] methyl] phenoxyl]-2-methylproprionic acid) binds to deoxyhemoglobin, shifting the Hb-O_2 dissociation curve to the right and thus increases the release of oxygen in low oxygen environments [2-10]. The ability of RSR13 to enhance the release of the available oxygen bound to hemoglobin potentially may increase the local pO_2 of the brain in hemorrhagic shock.

It has been difficult to relate the effects of pathophysiological conditions and their treatments directly to the pO_2 of the brain because of a lack of a means to measure the brain pO_2 directly and repeatedly during acute events. Electron paramagnetic resonance (EPR) oximetry appears to have the capability to make such measurements [11-17]. EPR oximetry is based on the fact molecular oxygen will proportionately change the EPR spectra of some paramagnetic materials. After implantation of the paramagnetic material, *in vivo* EPR oximetry can provide repeated non-invasive measurements from the same site with high sensitivity to the low pO_2 that may occur with shock. This technique already has been productively employed for the measurement of pO_2 in the brain [12,14, 17-19]. In this study we have used *in vivo* EPR oximetry and simultaneous physiological monitoring to investigate the effect of hemorrhagic shock on the pO_2 of brain tissue during and after severe hemorrhagic shock and to measure the effect on brain pO_2 of crystalloid resuscitation with or without RSR13.

2. MATERIALS AND METHODS

This study used oxygen sensitive material lithium phthalocyanine (LiPc) that was synthesized in our laboratory. The line width of LiPc is a linear function of pO_2 and independent of local metabolism, the presence of other paramagnetic species, and pH [12]. The measurements of pO_2 in the brain were made after the stereotactic implantation of LiPc (about 0.05mg). The RSR13 (batch CTM-612, 2g RSR13 Acid in 100mL: 20 mg/ml solution) was provided by Allos Therapeutics, Inc.

The studies were carried out using 15 male Sprague-Dawley rats weighing 280-350 g (Charles River Lab., Wilmington, MA). LiPc was implanted under anesthesia (ketamine/xylazine 80mg/8mg/kg i.m.) one week before the experimental procedures. Using stereotactic techniques the LiPc was placed into the frontal cerebral cortex 3.0 mm from the midline and 1 mm in front of the bregma at a depth of 2 mm from the surface of the skull (Figure 1).

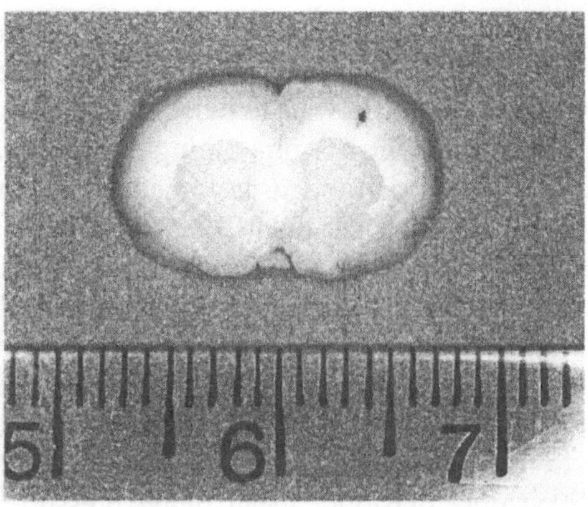

Figure 1. Brain slice showing LiPc implanted in the cerebral cortex in the right hemisphere.

The rats were randomly assigned to one of two groups and received either RSR13 (150 mg/kg i.v.; n=9), or normal saline (n=6) at the time of crystalloid resuscitation. Anesthesia was induced with isofluorane administered via a snout cone, and then endotracheal intubation was established with a 14 gauge catheter. Respiratory support (tidal volume ~1.1 ml/100 g body weight; rate ~60-65 strokes/min) was maintained with a volume cycled respirator (Harvard Apparatus) after endotracheal intubation. The FiO_2, concentration of inhalation anesthetics, and end-tidal pCO_2 were monitored throughout the experiment (Capnomac II Datex, Model AMG-103-27-00). Polyethylene catheters (PE-50, Clay-Adams, NJ) were placed in the left and right femoral arteries and left femoral vein for withdrawing the blood to induce hemorrhagic shock, continuous monitoring of blood pressure and periodic blood gas measurements, and administration of resuscitations fluid and RSR13 or saline. Core temperature of the rat was monitored by a rectal probe and the animal was kept on a warm water pad that was maintained at 37 to 38° C. FIO_2 was maintained at 25-27% (isofluorane 1.2%) during the entire experiment.

The oxygen-dependent spectra of LiPc were obtained using a low frequency (1.2 GHz, L-band) EPR spectrometer constructed in our laboratory. An extended loop resonator was placed over the site of interest and the position was adjusted to obtain the maximum signal from the LiPc in the cerebral cortex. Typical settings for the spectrometer were: incident microwave power, 10 mW; magnetic field center, 425 gauss; scan range, 1 gauss; modulation frequency, 27 kHz. The modulation amplitude was set at less than one-third of the EPR line width. The scan time was 5 sec. Twenty scans were averaged to achieve a satisfactory signal to noise ratio for the spectra.

After setting the ventilation, baseline measurements of the pO_2 were obtained, and then 50 μL of blood was taken for measurement of the hematocrit. Hemorrhagic shock was induced by withdrawing 2.7-2.9 ml blood/100 g body weight within 15 min, through the femoral artery catheter. Following a 30 min shock period, resuscitation was started with Ringer's lactated (RL) solution, to maintain the MBP within 20% of the value of the pre-hemorrhage value, by infusion of 2 to 3 times the blood loss volume. At the same time as the start of the RL resuscitation, either RSR13 or saline was infused over 15 min, using an infusion pump (KD Scientific model 100). After resuscitation, another 50 μL of blood was taken to measure the hematocrit. Measurements of brain pO_2 were obtained throughout the experiment, up to 120 min after the end of infusion of RSR13 or saline. Blood pressure (systolic, diastolic, mean blood pressure (MBP) from the femoral artery), heart rate (HR), and core body temperature were monitored via a Biopac system interfaced to a personal computer and were recorded every 3 min during the 15 min of administration of RSR13 or saline and then every 15 min over the next 120 min.

Arterial blood (150 μL) was drawn into a glass capillary tube and blood gases (p_aO_2, p_aCO_2, pH, HCO_3, and base excess) were measured immediately on a Ciba-Corning 238 Blood Gas System. Two measurements were made initially to determine the optimal settings for the ventilation, and the blood was sampled at 30, 60, 90, and 120 min after the end infusion of RSR13 or saline. After each sample was withdrawn, 0.2 ml of heparinized saline (5U/mL) was infused into the catheters to prevent clotting and maintain the blood volume. At the conclusion of the measurements, the animals were euthanized by a method approved by the Animal Resource Center, Dartmouth College, NH, USA.

3. RESULTS

The withdrawal of 2.7-2.9 per 100 g body weight within 15 min resulted in an approximate 50% loss of the initial circulating blood volume. Due to the dilution of the red blood cells, the hematocrit decreased by about 50% from the base line (Figure 2). The MBP and HR pre- and post hemorrhage is summarized in Table 1. The MBP decreased to less than 20 mm Hg at the end of the hemorrhage period and a transient decrease in heart rate occurred. During the subsequent "shock period" of 30 min, the MBP and HR improved modestly. During the 30 min period when the resuscitation fluid was administered the MBP and HR immediately recovered to pre-hemorrhage values. In the saline-treated controls the MBP remained within 80% of the baseline value to the end of the measurements (120 min). There was no significant difference in the MBP and HR between the RSR13-treated group and the saline-treated group throughout the experiment. This suggests that RSR13 (150mg/kg) does not have a direct effect on the cardiovascular system during hemorrhage and resuscitation.

Table 1. Mean blood pressure and hear rate (mean±SE) during HS and experimental period.

	baseline	after blood withdrawing	before resuscitation	after resuscitation	Average BP during maintained
Control MBP, mm Hg	88.1±3.3	19.2±1.4	34.1±2.9	83.8±5.5	70.0±5.3
Control HR, beats/min	352±10	261±10	272±7	333±9	341±13
RSR13 MBP, mm Hg	88.0±2.3	18.9±0.9	32.8±2.3	78.1±5.3	70.1±4.3
RSR13 HR, beats/min	348±12	240±13	252±10	313±8	302±20

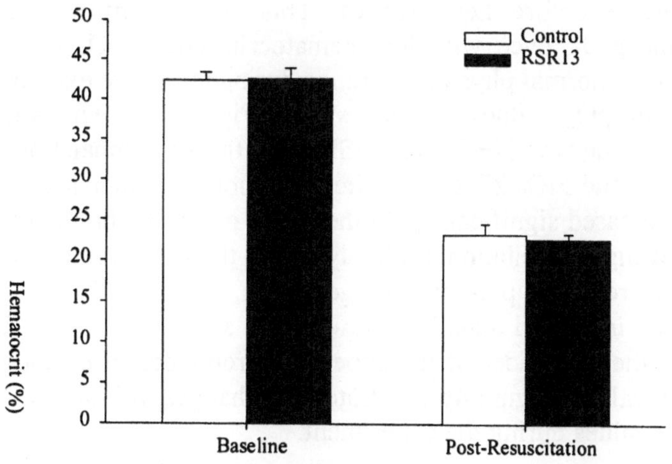

Figure 2. Hematocrit (mean±SE) after resuscitation with Ringers Lactate.

Table 2 summarizes the pH, bicarbonate (HCO_3^-), and base excess throughout the experiment. There were no significant differences between the saline- and RSR13-treated groups. The arterial pH remained within the normal range. HCO_3^- and base excess decreased modestly from the pre-hemorrhage during the course of the experiment.

Table 2. Hematological data (Mean ±SE) after hemorrhagic shock

	baseline	after resuscitation	30min	60min	90min	120min
pH control	7.42±0.01	7.40±0.01	7.42±0.02	7.42±0.02	7.43±0.02	7.43±0.03
pH RSR13	7.42±0.02	7.38±0.02	7.41±0.03	7.42±0.03	7.41±0.03	7.39±0.04
HCO_3^- control	25.9±0.7	23.4±0.6	22.3±0.8	22.1±0.9	22.3±0.7	21.7±1.2
HCO_3^- RSR13	26.6±0.6	21.5±1.3	19.4±2.5	21.8±1.2	21.3±1.2	21.6±1.7
Base Excess, mmole/L Control	2.0±0.8	-0.7±0.6	-1.4±0.6	-1.7±1.1	-1.3±0.9	0.7±0.6
Base Excess, mmole/L RSR13	2.4±0.6	-2.8±1.6	-2.1±1.6	-1.9±1.3	-2.5±1.4	-0.4±1.9

The p_aO_2 and p_aCO_2 are shown in Figure 3. The ventilation for each animal initially was set to keep these within the normal physiologic range for rats (p_aO_2:85-95 mm Hg, p_aCO_2:35-45 mm Hg) and then was not further adjusted throughout the experiment. After severe hemorrhage, the p_aO_2 was maintained within physiological normal range in saline group. The animals that received RSR13 had a significant increase in p_aO_2 (about 40-35% more than the value before hemorrhage). This result confirms that RSR13 increases the p_aO_2, even with low hematocrit. The p_aCO_2 in both groups remained in the normal physiologic range during the entire experiment.

The brain pO_2 values are shown in Figure 4a. The value before hemorrhage value was 28±7(mean ±SE) mmHg under inhalation anesthesia (1.2% Isofluorane FiO_2 27-28%). After initiation of hemorrhage, the pO_2 of the brain decreased significantly. In the saline group the brain pO_2 increased modestly during resuscitation, probably due to the improvement in the MBP, but it did not return to pre-hemorrhage values. In the RSR13-treated group, the brain pO_2 increased quickly following the administration of RSR13 and then during the remainder of the experiment remained at or above the pre-hemorrhage values. Figure 4b illustrates the changes in brain pO_2 from pre-hemorrhage values during the experiment.

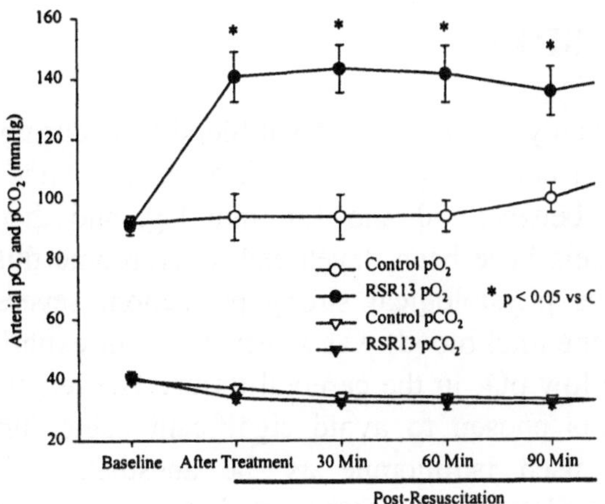

Figure 3. The p_aO_2 and p_aCO_2 (mean±SE) during hemorrhagic shock and resuscitation.

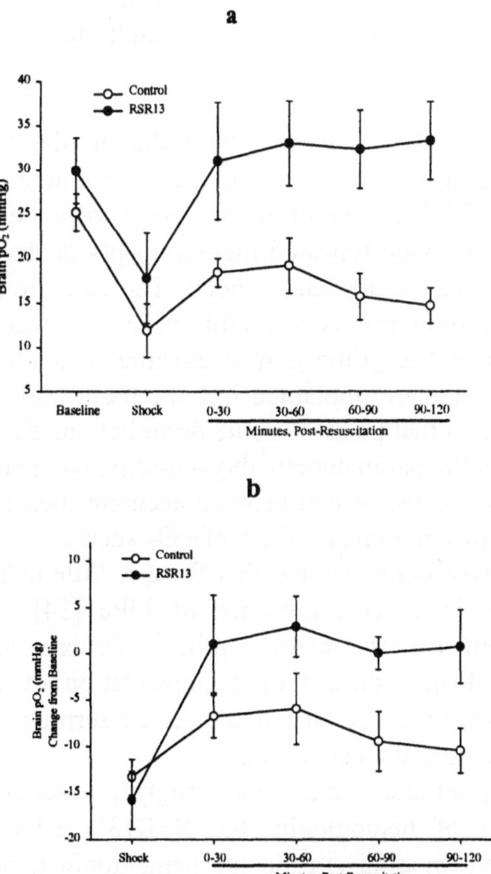

Figure 4. Cerebral pO2 (mean±SE) in HS and control groups: the absolute values mean±SE (a) and the changes in brain pO2 from pre-hemorrhage values during the experiment (b).

4. DISCUSSION

The brain remarkably conserves cerebral blood flow during hypertension and hypotension. Constant blood flow to the brain is maintained over a range of blood pressures between 50 and 150 mm Hg, and therefore several different shock models have been developed to overcome this control [20-28]. Because of this physiological strong protection, severe hemorrhage (removal of half of the total blood) was selected for our experimental model in order to produce low pO_2 in the cerebral tissue. The length of the shock period (30 min) was chosen to avoid significant acute functional brain damage [22]. We used isofluorane as the anesthetic with controlled respiration with a ventilator for the present study to avoid potential affects of anesthesia on brain tissue pO_2 during the course of the experiments [14,18,19]. Our results indicate that we were able to keep p_aO_2 and brain pO_2 near baseline values throughout hemorrhage and resuscitation. The baseline brain pO_2 in our experiment appeared to be similar to the value in conscious rats [14,18].

Brain damage during hemorrhagic shock is considered to arise from insufficient oxygen delivery to the brain due to the low arterial blood pressure and low blood flow [29]. In addition, oxidative damage can occur after reperfusion [30-33]. Therefore, it seems very desirable to have a technique that can provide repeated measurements of the actual pO_2 in the brain during and after hemorrhagic shock. The data from this experiment indicate that EPR oximetry is a useful means of measuring tissue pO_2 (continuously and in 'real time'). It is essential that the oxygen-sensitive paramagnetic materials are implanted into the tissue well in advance of the acute experiments, so that possible acute perturbations from the implant can be avoided. LiPc is the paramagnetic oxygen-sensitive material of choice for these experiments because it can provide accurate measurement of pO_2 at moderate to high physiological values of pO_2 such as in the brain [12,17]. Previous studies have demonstrated that there is little or no reaction of the CNS to even the long term presence of LiPc [34]. In the present experiment, we confirmed the inertness of the LiPc through both observation of the behavior of the animals after implantation of the LiPc and by histological assessment at the site of the cortex surrounding the implanted LiPc in brain slices after the experiment.

RSR13 acts similar to 2,3-diphosphoglyceric acid (DPG), the natural allosteric modifier of hemoglobin, but RSR13's effect is achieved by affecting a different binding site on the hemoglobin tetramer [4,6,9]. The advantages of RSR13 over 2,3DPG principally relate to pharmacological stability and dose-dependence of the increase in P50 [9]. Our findings demonstrated that RSR13 increased the p_aO_2 immediately and that within 10 min the brain pO_2 increased. It is likely that the effect of RSR13 on the brain pO_2 was mediated directly through increased availability of oxygen via more facile release from hemoglobin. While our results cannot rule out other possible mechanisms such as an increase in regional blood flow or a decrease in oxygen consumption, these seem less likely. Our data and other reports [7,8] indicate that the administration of RSR13 (150-200 mg/kg) does not alter systemic homodynamic such as cardiac output and systemic vascular resistance, and therefore, administration of RSR13 would not be expected to change cerebral blood flow. There is no obvious reason to expect RSR to decrease oxygen consumption.

5. CONCLUSION

EPR oximetry provides a useful way to measure the brain pO_2 throughout the period of experimental hemorrhagic shock in rats. RSR13 can significantly increase the partial pressure of oxygen in the cerebral cortex following hemorrhagic shock. This effect of RSR13 potentially may reduce neuronal damage resulting from an insufficient supply of oxygen to the brain induced by severe hemorrhagic shock. It also seems likely that RSR13 could be helpful in other situations where the brain potentially is subjected to short-lived acute hypoxia, such as reversible acute lung pathology and carbon monoxide poisoning.

ACKNOWLEDGEMENTS

This study was supported by NIH Grant PO1 GM51630 and used the facilities of the EPR Center for Viable Biological Systems, supported by NCRR and NIH grant P41 RR11602. Allos Therapeutics, Inc. also provided financial support and RSR13 for this study.

REFERENCES

1. Heckbert RS, Vedder NB, Hoffman W, Winn RK, Hudson LD, Jurkovich GJ, Copass MK, Harlan JM, Rice CL, Maier RV. Outcome after hemorrhagic shock in trauma patients. J. Trauma 1998 Sep;45:545-549
2. Randad RS, MA, Mahran MA, Mehanna AS, Abraham DJ. Allosteric modifiers of hemoglobin. 1. Design, synthesis, testing, and structure-allosteric activity relationship of novel hemoglobin oxygen affinity decreasing agents. J Med Chem 1991;34:752-757.
3. Wireko FC, Kellogg GE, Abraham DJ. Allosteric modifiers of hemoglobin. 2. crystallographically determined binding sites and hydrophobic binding/interaction analysis of novel hemoglobin oxygen effectors. J Med Chem 1991;34:758-767.

4. Abraham DJ, Wireko FC, Randad RS. Allosteric modifiers of hemoglobin: 2-[4-[[(3,5-disubstituted anilino) carbonyl]methyl]phenoxy]-2-methylpropionic acid derivatives that lower the oxygen affinity of hemoglobin in red cell suspensions, in whole blood, and *in vivo* in rats. Biochemistry 1991;31:9141-9149.
5. Wei EP, Randad RS, Levasseur JE, Abraham DJ, Kontos HA. Effect of local change in O2 saturation of hemoglobin on cerebral vasodilution from hypoxia and hypotension. Am J Physiol 1993;265:H1439-H1443.
6. Khandelwal SR, Randad RS, Lin P-S, Meng H, Pittman RN, Kontos HA, Choi SC, Abraham DJ, Schmidt-Ullrich R. Enhanced oxygenation in vivo by allosteric inhibitors of hemoglobin saturation. Am J Physiol 1993;265:H1450-H1453.
7. Kunert MP, Liard JF, Abraham DJ. RSR-13, an allosteric effector of hemoglobin, increases systemic and iliac vascular resistance in rats. Am J Physiol 1996;271:H602-613.
8. Pagel PS, Hettrick DA, Montgomery MW, Kersten JR, Steffen RP, Warltier DC. RSR13, a synthetic modifier of hemoglobin-oxygen affinity, enhances the recovery of stunned myocardium in anesthetized dogs. J Pharmacology and Experimental Therapeutics, 1998;285:1-8.
9. Grocott HP, Bart RD, Sheng, H, Miura Y, Steffen, RP, Pearlstein RD, Warner DC. Effects of a synthetic allosteric modifier of hemoglobin oxygen affinity on outcome from global cerebral ischemia in the rat. Stroke 1998;29:1650-1655.
10. Woods JA, Storey CJ, Babcock EE, Malloy CR. Right-shifting the oxyhemoglobin dissociation curve with RSR13: effects on high-energy phosphates and myocardial recovery after low-flow ischemia. J Cardiovascular Pharmacology 1998;31:359-364.
11. Swartz HM, Boyer S, Gast P, Glockner JF, Hu H, Liu KJ, Moussavi M, Norby SW, Walczak T, Vahidi N, Wu M, Clarkson RB. Measurements of pertinent concentrations of oxygen in vivo. Magn Reson Med. 1991;20:333-339.
12. Liu KJ, Gast P, Moussavi M, Norby SW, Vahidi N, Walczak T, Wu M, Swartz HM. Lithium phthalocyanine: a probe for EPR oximetry in viable biological systems. Proc Natl Acad Sci 1993;90:5438-5442.
13. Bacic G, Liu KJ, O'Hara JA, Harris RD, Szybinski K, Goda F, Swartz HM. Oxygen tension in a murine tumor: a combined EPR and MRI study. Magn Reson Med. 1993;30:568-572.
14. Liu KJ, Bacic G, Hoopes PJ, Jiang J, Du H, Ou LC, Dunn JF, Swartz HM. Assessment of cerebral pO2 by EPR oximetry in rodents: effects of anesthesia, ischemia, and breathing gas. Brain Res 1995;685:91-98.
15. Goda F., Bacic G, O'Hara JA, Gallez B, Swartz HM, Dunn JF. The relationship between pO2 and perfusion in two murine tumors after X-ray Irradiation: a combined Gd-DTPA dynamic MRI and in vivo EPR oximetry study. Cancer Res 1996;56:3344-3349.
16. James P, Grinberg O, Goda F, Panz T, O'Hara J, Swartz HM. Gloxy: an oxygen-sensitive coal for accurate measurement of low oxygen tensions in biological systems. Magn Reson Med 19971 38:48-58.
17. Swartz HM, Clarkson RB. The measurement of oxygen *in vivo* using EPR techniques. Phys Med Biol 1998;43:957-1975.
18. Liu KJ, Hoopes PJ, Rolett EL, Beerle BJ, Azzawi A, Goda F, Dunn JF, Swartz HM. Effect of anesthesia on cerebral tissue oxygen and cardiopulmonary parameters in rats. Adv Exp Med Biol 1997;428:33-39.
19. Taie S, Leichtweis SB, Liu KJ, Miyake M, Grinberg O, Demidenko E, Swartz HM. The effect of ketamine/xylazine and pentobarbital anesthesia on cerebral tissue oxygen tension, blood pressure, and arterial blood gas in rats. Adv Exp Med Biol 1999;471:189-198.

20. Rhee P, Waxman K, Clark L, Tominaga G, Soliman MH. Superoxide dismutase polyethylene glycol improves survival in hemorrhagic shock. The American Surgeon 1991;57:747-750.
21. Ismail H, Ulus B, Arslan Y, Savic V, Kiran K. Restoration of blood pressure by chroline treatment in rats made hypotensive by hemorrhage. British Journal of Pharmacology 1995;116:1911-1917.
22. Carillo P, Takatsu A, Safar P, Tisherman S, Stezoski W, Stolz G, Dixon CE, Radovsky A. Prolonged severe hemorrhagic shock and resuscitation in rats does not cause subtle brain damage. Journal of Trauma 1998;2:239-249.
23. Munchitsh E-M, Auer W, Pichler L. Effects of a1 acid glycoprotein in different rodent models of shock. Fundam Clin Pharmacol 1998;12:173-181.
24. Sazabo C. Potential role of the peroxynitrite-poly(ADP-RIBOSE) synthetase pathway in a rat model of severe hemorrhagic shock. Shock 1998;9:341-344.

Chapter **31**

EXPRESSION OF MYOGLOBIN IN THE TRANSGENIC MOUSE BRAIN

Ross D. Shonat[1] and Alan P. Koretsky[2]

[1]Department of Biomedical Engineering, Worcester Polytechnic Institute, 100 Institute Road, Worcester, MA 01609; [2]In Vivo NMR Research Center, NINDS, National Institutes of Health, 10 Center Drive, Rm. B1D-69B, MSC 1060, 9000 Rockville Pike, Bethesda, MD 20892

Abstract: The main purpose of this study was to express human myoglobin in mouse brain neurons and investigate the effects of this expression on metabolism and blood flow using phosphorous (^{31}P) NMR spectroscopy and NMR perfusion imaging. Transgenic mice expressing brain myoglobin were created using a cDNA sequence for human myoglobin placed under the transcriptional control of either a human platelet-derived grown factor polypeptide B (*PDGF-B*) promoter sequence or a rat neuron-specific enolase (*NSE*) promoter sequence. The presence of myoglobin having a functional, reduced-state, heme group was demonstrated by protein analysis and immunocytochemistry. Expression levels were highest in the hippocampus, cerebellum, and cerebral cortex. No gross morphological adaptations of neural tissue resulting from the expression were observed and no statistically significant differences in the energetic state, as measured by ^{31}P NMR, or baseline cortical perfusion, as measured by an NMR perfusion imaging technique, were found.

Key words: blood flow, brain energetics, NMR perfusion imaging, NMR spectroscopy, oxygenation, transgenic mice

Abbreviations: NSE – neuron-specific enolase; PDGF – platelet-derived growth factor

1. INTRODUCTION

The brain, lacking a capacity to store energy, requires a constant delivery of oxygen and substrates to maintain metabolic function. While some authors have suggested that the "brain is regulated at the level of slight hypoxia" [1], phosphorous (^{31}P) NMR studies in brain indicate a high phosphorylation potential, with the ADP levels approximately 0.02 μmol per gram wet weight of tissue, suggesting that mitochondrial function may not be diminished [2]. Studies investigating the oxygen dependence of oxidative phosphorylation in skeletal muscle, a tissue expressing myoglobin, have generally revealed that oxygen is not rate limiting [3], although this is still a subject of some debate [4]. Because skeletal muscle must have the capacity to change its respiration over a wide dynamic range, it is thought that muscle myoglobin is functioning as an oxygen buffer and possibly a facilitator of oxygen delivery to maintain mitochondrial oxygen tension sufficiently high during dramatic increases in muscle activity [5]. However, with the recent description of mice exhibiting normal exercise behavior after removal of myoglobin by gene-knockout [6], a complete explanation for myoglobin's role in skeletal muscle metabolism remains unresolved.

Because there is controversy regarding the oxygenation of brain and the role of myoglobin in oxygen transport, we hypothesized that the introduction of myoglobin into the brain might affect the oxygenation state and, more profoundly, substantially alter the relationship between cerebral metabolism and blood flow. In addition, we hypothesized that this expression and any resulting changes in brain physiology would provide some clues about the role that myoglobin plays in tissues where it is expressed normally. To address this, we expressed human myoglobin in mouse brain neurons and investigated the effects of this expression on metabolism and blood flow using ^{31}P NMR spectroscopy and NMR perfusion imaging.

2. METHODS

2.1 Construction of the PDGF-myoglobin fusion gene

The plasmid pPWL504 (P. Laird, unpublished results) containing a mouse phosphoglycerate kinase (*PGK1*) promoter sequence [7], a heterologous intron splicing (*IVS*) sequence [8], and a *PGK1* polyadenylation (*polyA*) sequence [9] was digested with *BamHI* and *XhoI* and the larger 4.53 kb fragment isolated. The plasmid pMbO [10] containing a human myoglobin cDNA sequence (*Mb*) was digested with *PvuII* and *SmaI* (partial) and the 563 bp *Mb* fragment (including the myoglobin CAP site and stop codon) was subcloned into the *HincII* site of pBluescript SK+ (Stratagene, La Jolla, CA), then removed as a *BamHI-XhoI* fragment, and finally inserted into the opened pPWL504 fragment, generating pPGKMyo with an *IVS-Mb-polyA* cassette. The nucleotide sequence of the *IVS-Mb-polyA* cassette was confirmed by automated DNA sequencing (data not shown). The plasmid pSIS-1 [11] containing a human platelet-derived growth factor polypeptide B (*PDGF-B*) promoter sequence was digested with *XbaI* and *AvrII* and the 1.47 kb promoter fragment subcloned into the *XbaI* site of pBluescript SK+. The *IVS*-Mb-*polyA* cassette was then removed from pPGKMyo as a 1.64 kb *PvuII* fragment and ligated downstream of the *PDGF-B* promoter at the *SpeI* site (blunt-ended), generating pPDGFMyo. The 2.89 kb *PDGF*-myoglobin fusion gene fragment isolated from pPDGFMyo by *NotI* digestion was used to inject the pronuclei of fertilized oocytes.

2.2 Construction of the NSE-myoglobin fusion gene

The plasmid pNSElacZ [12] containing a rat neuron-specific enolase (*NSE*) promoter sequence was digested with *EcoRI* and *HindIII* and the 1.8 kb NSE promoter fragment subcloned into the *EcoRI-HindIII* sites of pBluescript SK+. The *IVS*-Mb-*polyA* cassette was then removed from pPGKMyo as a 1.64 kb *PvuII* fragment and ligated downstream of the *NSE* promoter at the *HindIII* site (blunt-ended), generating pNSEMyo. The 3.25 kb NSE-myoglobin fusion gene fragment isolated from pNSEMyo by *NotI* digestion was used to inject the pronuclei of fertilized oocytes.

2.3 Generation of transgenic mice

Mice were maintained under standard conditions and provided with food and water *ad libitum*. Animal use for all aspects of this work was in accord with the National Institutes of Health guidelines and approved by the Carnegie Mellon University Institutional Animal Care and Use Committee. Transgenic mice were produced by injecting the pronuclei of fertilized FVB/N oocytes using standard microinjection techniques [13]. The presence of either the *PDGF*- or *NSE*-driven myoglobin transgene was detected by PCR analysis (30 cycles: denature 94 °C for 1 min, anneal 55 °C for 2 min, extend 72 °C for 3 min with a 15 second/cycle increase) of tail DNA, using primers unique to the human myoglobin cDNA: the sense primer (5'-CTCATCAGGCTCTTTAAGGG-3') bridging the exon 1,2 junction and the antisense primer (5'-ATTCCGAGATGAACTCCAGG-3') bridging the exon 2,3 junction of the myoglobin cDNA. Random insertion of either transgene into the mouse genome generated a single 243 bp amplified PCR product. Standard breeding and crossing techniques were used to generate homozygous transgenic lines from the heterozygous founder lines.

2.4 Protein detection and quantification

Transgenic and control FVB/N mice were deeply anesthetized with an injection of 2.5% Avertin (0.60 ml, IP) containing 200 U/ml of heparin and the blood replaced with a perfusion solution (0.9% NaCl, 0.4% dextrose, 0.8% sucrose, 10 U/ml heparin, 0.25% dextran, and 10^{-7}M papaverin) by left-ventricular puncture and aortic cannulation through an opened chest. Soluble protein extracts were made by homogenizing the whole brain in approximately 6 volumes of ice-cold extraction solution (10mM potassium phosphate [pH 7.0], 5mM EGTA) and centrifuging at 23,000 x *g* for 30 minutes. These protein extracts were further concentrated by a centrifugal concentrator (Centricon-3, MW cut-off 3000 Daltons, Amicon, Beverly, MA) and the total protein concentration fixed at 10 mg/ml (dotMetric total protein assay kit, Geno Technology, St. Louis, MO). Proteins were separated (Mini-Protean II system, Bio-Rad, Hercules, CA) by SDS-PAGE on 15%T Tris-HCl gels (Ready Gel, Bio-Rad) or by native-PAGE on 7.5%T Tris-HCl gels (Ready Gel, Bio-Rad).

For Western blot analysis of human myoglobin, SDS-PAGE gels were transferred electrophoretically (Mini Trans-Blot system, Bio-Rad) to nitrocellulose (0.45 μm) in 25 mM Tris, 192 mM glycine, 20% methanol (pH 8.3) for 1 hour at 100 V and blocked in a Tris-buffered saline-Tween solution (500 mM NaCl, 100 mM Tris [pH 7.5], 0.1% Tween-20). The filters were then incubated with a mouse monoclonal anti-human myoglobin IgG (ICN Pharmaceuticals, Costa Mesa, CA) for 30 minutes at room temperature. Biotinylated anti-mouse IgG and avidin-biotinylated peroxidase (Vector Laboratories, Burlingame, CA) were used to detect the primary antibody and the peroxidase activity was visualized with a DAB substrate kit (Vector Laboratories). Human heart myoglobin (Sigma Chemical Co., St. Louis. MO) served as a positive control and brain protein extracts from non-transgenic FVB/N mice were used to assess the level of background staining.

For the detection of H_2O_2-peroxidase activity from the heme using 3,3',5,5'-tetramethylbenzidine (TMBZ) [14], native-PAGE gels were transferred electrophoretically to nitrocellulose in 25 mM Tris, 192 mM glycine and placed in a TMBZ staining solution (3 parts 6.3 mM TMBZ in methanol to 7 parts 0.25 M sodium acetate [pH 5.0]). After 2 hours, H_2O_2 was added to a final concentration of 30 mM and 30 minutes allowed for full color development. Filters were then digitized to preserve the transient staining pattern. Native conditions were used throughout the procedure to limit dissociation of the heme from myoglobin.

2.5 Immunocytochemistry

Transgenic and control FVB/N mice were deeply anesthetized with an injection of 2.5% Avertin (0.60 ml, IP) containing 200 U/ml of heparin and the blood replaced first with a perfusion solution (0.9% NaCl, 0.4% dextrose, 0.8% sucrose, 10 U/ml heparin, 0.25% dextran, and 10^{-7}M papaverin) and then with a fixative solution (4% paraformaldehyde) by left-ventricular puncture and aortic cannulation through an opened chest. After removal of the fixed brain, 14 μm sagittal, coronal, and transverse sections were cut on a cryostat (Zeiss) at –23°C and mounted on poly-L-lysine coated slides (Polysciences, Inc) for immunocytochemical staining of myoglobin.

Sections were blocked in a phosphate-buffered saline solution containing 4% goat serum (Vector Laboratories) and 0.1% Triton-X and then incubated overnight at 4°C with a rabbit polyclonal anti-human myoglobin IgG (1:200 dilution, Vector Laboratories). Biotinylated anti-rabbit IgG from goat (10 μg/ml) and avidin-biotinylated alkaline phosphatase (Vector Laboratories) were used to detect the primary antibody and the alkaline phosphatase activity was visualized with an AP substrate kit (Vector Laboratories). Both the primary and secondary antibodies were prepared in the blocking serum and pre-absorbed with FVB/N brain tissue to reduce background staining. Stained sections were photographed on a Zeiss upright microscope using 2X and 20X objectives.

2.6 Animal preparation for NMR experiments

Transgenic and control FVB/N mice were anesthetized with an injection of 2.5% Avertin (0.015 ml/g, IP) and tracheotomized. After a topical anesthetic (1% mepivacaine, 0.05 ml, sc) was injected under the scalp, a clamp was attached between the nasal and palantine bones to immobilize the skull in the magnet. Rectal temperature was regulated at 38°C using a water-filled heating blanket and anesthesia was maintained with intraperitoneal bolus injections of α-chloralose (0.04 mg/g) as needed. NMR experiments were performed on a 4.7T, 40-cm bore Bruker BIOSPEC instrument operating at a ^{1}H frequency of 200 MHz and equipped with a 15-cm gradient insert.

2.7 Phosphorous NMR spectroscopy

For phosphorous (^{31}P) spectroscopy measurements in the anesthetized mouse, a 14 mm, single-turn, surface coil tuned to the ^{31}P frequency of 81 MHz was placed on the dorsal surface of the exposed mouse skull. Individual phosphorous spectra were acquired using a hard-pulse width of 60 μs, a spectral width of 15 kHz, a repetition time of 10 sec, and 128 averages. Approximately 21 minutes were required to produce each spectra. To reduce the broad phosphorous component due to bone, power to the coil was adjusted to produce a 180° pulse inside the skull. To improve signal-to-noise, three spectra were separately processed with an exponential line broadening of 20 Hz and summed.

2.8 NMR perfusion imaging by arterial-spin tagging

Cortical blood flow in the anesthetized mouse was measured using a two-coil NMR perfusion imaging technique [15]. The two-coil system consisted of a small 3 mm diameter butterfly-shaped surface coil for labeling the arterial water spins in the neck region, and a 14 mm, single-turn, surface imaging coil placed on the dorsal surface of the exposed skull. The imaging coil was actively de-tuned with a PIN diode during the labeling interval to eliminate perturbation of the macromolecular spins and MTC effects in the brain [15].

For perfusion measurements, MR images were acquired using a fast gradient-echo imaging sequence using an echo time of 3.3 ms, a repetition time (TR) of 3.75 s, a matrix size of 64 X 64 (zero-filled from 34 linear phase-encoding steps), and a field-of-view of 1.8 cm. X 1.8 cm. Using four averages, a single image was acquired in 15 s. These images were oriented parallel to the imaging coil and approximately 0.5 mm below the surface of the brain with a slice thickness of 1.2 mm (see Fig 5A). Carotid and basilar arterial spin inversion by adiabatic-fast-passage was accomplished during the TR interval by applying 85 mG of off-resonance RF power to the labeling coil for 3.55 s in the presence of a 20 mT/m longitudinal gradient.

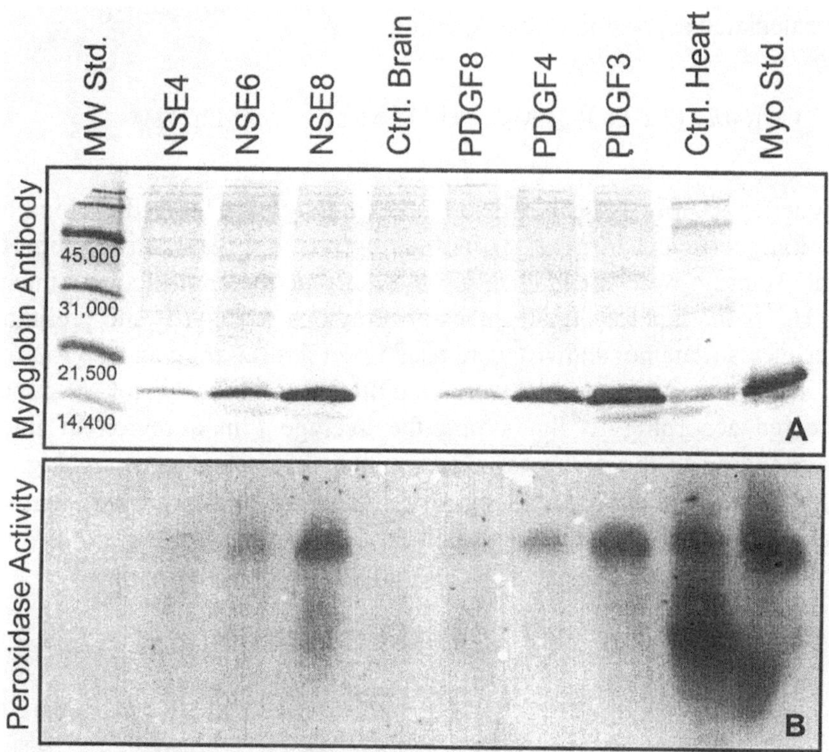

Figure 1. Immunoblot and peroxidase activity staining of protein extracts from the brains of six heterozygous transgenic mice and the heart and brain from one FVB/N control mouse. For both the *NSE*- and *PDGF*-promoter, three transgenic lines having increasing levels of human myoglobin expression are shown. No myoglobin expression was detected in the control brain extract and the low degree of staining in the control heart extract indicates the high degree of specificity of the monoclonal antibody for human myoglobin. Abbreviations are: **MW Std.**: biotinylated molecular weight standards (Bio-Rad), **NSE4, 6, and 8:** Brain extracts from *NSE*-promoter derived lines #4, #6, and #8, **Ctrl Brain:** Brain extract from an FVB/N control mouse, **PDGF8, 4, and 3:** Brain extracts from *PDGF*-promoter derived lines #8, #4, and #3, **Ctrl. Heart:** Heart extract from an FVB/N control mouse, **Myo Std.:** Human myoglobin standard (0.2 µg, Sigma).

Preliminary experiments confirmed that the degree of labeling, α, under these conditions was approximately 0.70 (data not shown). Control images without arterial spin inversion were acquired with the RF power applied at an off-resonance frequency far removed from the labeling plane. Perfusion was calculated according to the equation

$$\bar{f} = [60\lambda / (2\alpha\, T_{1b})]\ \{(M_b^C - M_b^L) / M_b^C\} \qquad \text{Eq. (1)}$$

where f is the tissue perfusion rate in ml/g/min, λ is the blood:brain partition coefficient in ml/g, T_{1b} is the observed spin-lattice relaxation time of mouse brain water magnetization in s, M_b^C is the control signal intensity and M_b^L is the labeled signal intensity. Previous work [16] and preliminary experiments (data not shown) were used to set $\lambda = 0.9$ ml/g and $T_{1b} = 1.6$ s.

Perfusion image pairs were acquired continuously for 5 minutes, processed according to Eq. 1 and the average f in a region of interest corresponding to the cortex was determined for each image (see Fig. 5B). The data were then grouped by sex and strain and averaged. Statistical analyses were based on the two-tailed Student's *t*-test.

3. RESULTS

3.1 Transgenic mice expressing brain myoglobin

The presence of either the *PDGF*- or *NSE*- driven myoglobin transgene was detected by PCR analysis using primers specific for human myoglobin cDNA and 3 stable heterozygous lines were established and maintained for each promoter. As shown in Fig. 1, western blot analysis of human myoglobin and native-PAGE staining for H_2O_2-peroxidase activity in whole brain protein extracts revealed the presence of myoglobin having a heme group in all 6 lines. The color of all protein extracts containing myoglobin was red, suggesting that the heme was functional and reduced. In the highest expressing PDGF-myoglobin line (PDGF3), expression levels were estimated to be between 25 – 50% of the levels in control mouse heart. Immunocytochemistry revealed that the expression was highest in the hippocampus, cerebellum, and cerebral cortex (Fig. 2). No gross morphological adaptation of neural tissue resulting from myoglobin expression was observed.

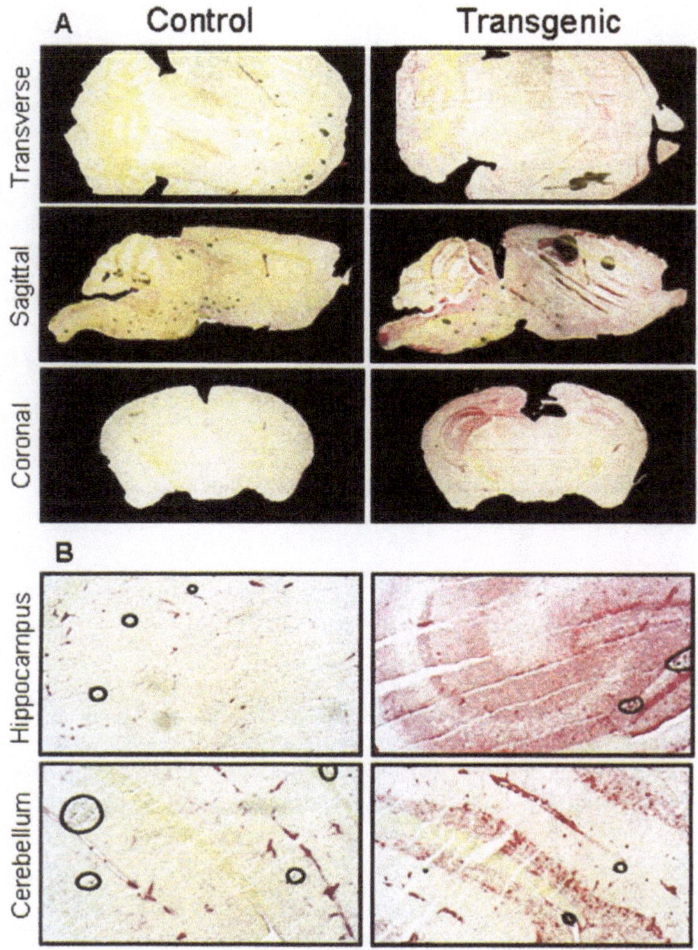

Figure 2. Immunocytochemical staining for human myoglobin in a FVB/N control mouse and a PDGF3 heterozygous transgenic mouse. Myoglobin staining above background is indicated by the darkened regions, particularly in the hippocampus and cerebellum. A: transverse, sagittal, and coronal sections photographed at 2X magnification. B: magnified images at 20X showing well defined neuronal staining for myoglobin in transgenic hippocampus and cerebellar regions.

FIGURE 2. [illegible] by the outlined regions, particularly in the [illegible] section, [illegible] photographed at 2X magnification. B, [illegible] showing well-defined neuronal staining [illegible] and [illegible] regions.

The results of independently breeding both the PDGF-myoglobin line #3 (PDGF3) and the NSE-myoglobin line #8 (NSE8) to homozygosity are shown in the two Western blots of Fig. 3. Using PDGF3 as a reference for each blot, myoglobin expression levels for both the PDGF3 and NSE8 homozygous lines appear approximately doubled, as expected, with the levels for the NSE8 homozygous line having increased to levels approaching the PDGF3 line. However, for unknown reasons, the PDGF3 homozygous mice did not breed very efficiently and were not available in sufficient numbers for inclusion in the phenotyping experiments described in this paper.

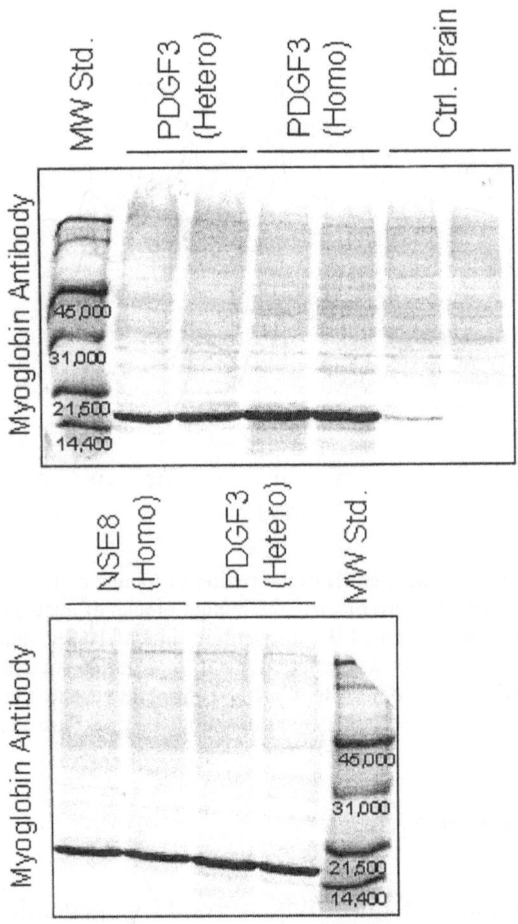

Figure 3. Immunoblot of protein extracts from the brains of control, heterozygous and homozygous transgenic mice. See Fig. 1 for additional details.

3.2 Energetic state in control and transgenic mice

Fig. 4 shows ^{31}P spectra for two female FVB/N control and two female PDGF3 transgenic mice. No statistically significant differences were found between spectra, either by direct subtraction or integration of peaks.

3.3 Cortical perfusion in control and transgenic mice

A gray scaled perfusion image from an FVB/N male mouse is shown in Fig. 5C and perfusion data for FVB/N control and NSE8 homozygous mice are shown in Table 1. While substantial gender differences were observed and are the subject of a separate investigation, no statistically significant differences were found between control and NSE8 homozygous lines.

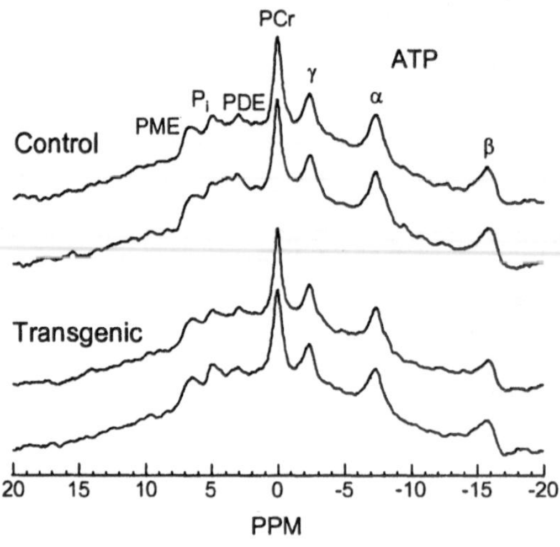

Figure 4. ^{31}P NMR spectra obtained from the region of mouse cortex illustrated in Fig. 5(A) and (B). Spectra represent the sum of three 21 minute blocks of acquisition, with each block acquired with a spectral width of 15 kHz, a repetition time of 10 sec, and 128 hard pulses. 20 Hz exponential line-broadening was applied prior to Fourier transformation. The upper spectra were acquired from two female FVB/N control mice and the lower spectra were acquired from two female PDGF3 transgenic mice. Spectral peaks are as labeled.

4. DISCUSSION

The principle aims of this study were: 1) to generate transgenic mice expressing human myoglobin in the brain and 2) to determine whether the presence of brain myoglobin had an effect on tissue metabolism and blood flow. At the current levels of myoglobin expression, we observed no obvious developmental or physiologic adaptation. No statistically significant differences in the energetic state, as measured by ^{31}P NMR, or baseline cortical perfusion, as measured by an NMR perfusion imaging technique, were found.

One of the original motivations for this work was a previous spectrophotometric study in the cerebral cortex of cats indicating that the terminal oxidase in the electron transport chain, cytochrome *a,a3*, was more reduced in the resting state than isolated mitochondria. [1]. Since this finding is generally consistent with a state of hypoxia, these authors suggested that the brain must be regulated at a level of slight hypoxia. [1] Because brain tissue lacks the capacity to store oxygen and has no mechanism to facilitate the diffusion of intracellular oxygen, we hypothesized that the presence of myoglobin might relieve this hypoxic condition and, more profoundly, substantially alter the relationship between cerebral metabolism and blood flow. However, at the current levels of myoglobin expression, perturbations in tissue metabolism and cortical perfusion were not observed. It remains to be seen whether higher levels of expression resulting from the cross-breeding of the NSE8 homozygous and PDGF3 homozygous lines will produce a phenotype and these experiments are continuing.

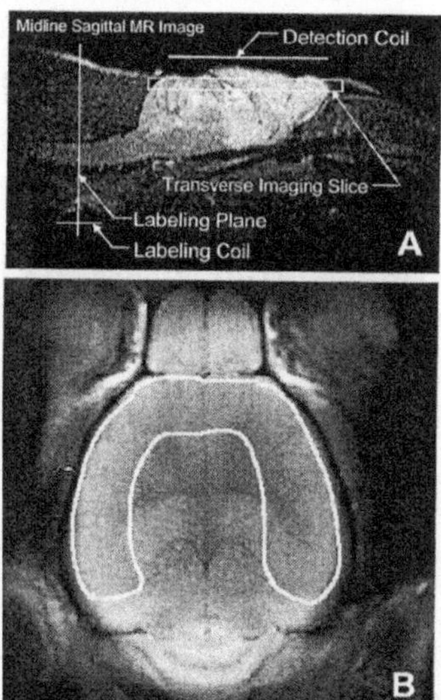

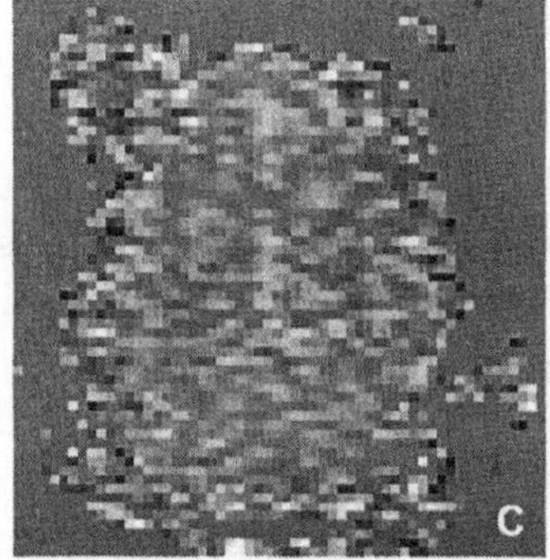

Figure 5. NMR perfusion imaging in the rat cortex using a two-coil system. A: midline sagittal section of the mouse head, showing the placement of the labeling and detection coils, the labeling plane, and the 1.2 mm thick transverse image slice. B: high resolution (256 X 256) spin-echo proton image of the transverse image slice, showing the outline of a cortical region used for quantifying perfusion. C: processed gray-scale image of the perfusion parameter *f* (Eq. 1) from a male FVB/N control mouse, with a full scale range of 0 – 1.5 ml/g/min. Pixels located in regions of very low signal-to-noise have been masked to an arbitrary gray-scale value.

Table 1. *Cortical perfusion data by strain and sex*

Strain (number)	Sex	Weight ± SD (g)	Age ±SD (days)	Perfusion ± SD (ml/mg/min)
FVB/N (8)	F	23 ± 2	63 ± 11	1.0 ± 0.2
NSE8 Homo (5)	F	25 ± 1	91 ± 20	1.0 ± 0.2
FVB/N (6)	M	29 ± 3	131 ± 66	0.7 ± 0.1
NSE8 Homo (5)	M	29 ± 3	72 ± 7	0.8 ± 0.1

The facilitated diffusion of oxygen by myoglobin is based on the premise that the binding of oxygen to deoxymyoglobin in the sarcoplasm enhances the oxygen flux to mitochondria [17]. If there are oxygen gradients between the capillary and cell, the presence of myoglobin should permit a higher oxygen flux in a shallower oxygen gradient. However, studies that refute this theory [18] have also been published and the issue of whether myoglobin acts to facilitate the delivery of oxygen to the mitochondria remains unresolved. Before beginning these experiments, we hypothesized that if the addition of brain myoglobin altered cellular energy metabolism and blood flow regulation, then it might be argued that it does play a role in the facilitation of oxygen in cells where it is normally expressed, namely heart and skeletal muscle. However, it was recently shown that gene-knockout mice lacking myoglobin functioned normally when challenged to exercise or exposed to hypoxic gas mixtures [6]. The demonstration that myoglobin may not be essential for normal cardiovascular and musculoskeletal function raises profound questions about oxygen transport and mitochondrial metabolism in all cell types. Perhaps the lack of a phenotype in our transgenic mice is consistent with this recent finding. It is, however, difficult to imagine why a muscle cell would express a functionless protein and clearly more work is needed.

A tight coupling between neuronal metabolic activity and blood flow in the normal brain was proposed as early as 1890 [19] and has remained the dominant hypothesis explaining blood flow regulation in the brain. However, this hypothesis does not account for the disproportionate increase in blood flow and oxygen delivery that has been postulated to occur in a number of circumstances. For example, it has been suggested that the regulation of cerebral blood flow during somatosensory stimulation is dependent on a neuronal or biochemical mechanism related to neuronal firing and not to the metabolic consumption of oxygen [20]. The implication is that task activation causes a focal hyperemia. If there is a hyperemia associated with task activation, then this is counter to the classically held belief that the delivery of oxygen to a tissue is matched (or coupled) to the utilization of that oxygen by the parenchymal cells. Unfortunately, it has been impossible to measure intracellular oxygen tension directly in the brain, making it difficult to address these issues. Given that the expression of myoglobin in our transgenic mice does not appear to affect blood flow or metabolism, the presence of myoglobin will allow for the monitoring of intracellular oxygen to determine the oxygen status during rest and if there are changes in cellular oxygen during neuronal activation. Myoglobin has been used as an intracellular oxygen probe in muscle, using either the changes in its optical absorption [21,22] or NMR spectrum [23,24], due to its alterations in the oxymyoglobin to deoxymyoglobin ratio. Future experiments are planned to address these issues.

5. CONCLUSION

In conclusion, we have demonstrated the successful expression of myoglobin in the brain of transgenic mice and found no developmental or physiologic adaptation resulting from this expression. Current work is directed at increasing the levels of brain myoglobin, by crossing the PDGF3 and NSE8 homozygous lines. Such a procedure is expected to boost myoglobin levels to near that found in heart. Spectral imaging techniques are also being developed to study the relationship between intracellular oxygen tension and metabolic state under normal and physiologic stress conditions in these transgenic mice.

ACKNOWLEDGEMENTS

The authors gratefully acknowledge Dr. Steven G. Boxer for providing the plasmid pMbO containing the human myoglobin cDNA, Dr. Peter W. Laird for providing the plasmid pPGK504 containing the intron and polyadenylation sequences, Dr. J. Gregor Sutcliffe for providing the plasmid pNSElacZ containing the NSE promoter and Dr. L. Ratner for providing the plasmid pSIS-1 containing the PDGF-B promoter. For the NMR experiments, the technical and intellectual input provided by Drs. Emmanuel L. Barbier, Afonso C. Silva, and Donald S. Williams is also gratefully acknowledged. We thank Kathy M. Sharer for her immeasurable skill in generating the transgenic mouse lines and maintaining the mouse facility. This research was supported by National Institutes of Health Grants HL-40354 and HL-02847 (to A.P.K.), HL-09418-02 (to R.D.S.), and RR-03631 (to the Pittsburgh NMR Center for Biomedical Research) and by a Fellowship from the W.M. Keck Center for Advanced Training in Computational Biology (to R.D.S.).

REFERENCES

1. Rosenthal MR, LaManna JC, Jöbsis FF, Levasseur JE, Kontos HA, and Patterson JL. Effects of respiratory gases on cytochrome a in intact cerebral cortex: Is there a critical pO_2? Brain Res 1976;108:143.

2. Detre JA, Williams DS, and Koretsky AP. Nuclear magnetic resonance determination of flow, lactate, and phosphate metabolites during amphetamine stimulation of the rat brain. NMR Biomed 1990;3:272.
3. Gayeski TEJ, Connett RJ, and Honig CR. Minimum intracellular pO_2 for maximum cytochrome turnover in red muscle in situ. Am J Physiol 1987;252:H906.
4. Jones DP. Intracellular diffusion gradients of O_2 and ATP. Am J Physiol 1986;250:C663.
5. Wittenberg BA and Wittenberg JB. Transport of oxygen in muscle. Annu Rev Physiol 1989;51:857.
6. Garry DJ, Ordway GA, Lorenz JN, Radford NB, Chin ER, Grange RW, Bassel-Duby R, and Williams RS. Mice without myoglobin. Nature 1999;395:905.
7. Adra CN, Boer PH, and McBurney MW. Cloning and expression of the mouse pgk-1 gene and the nucleotide sequence of its promoter. Gene 1987;60:65.
8. Choi T, Huang M, Gorman C, and Jaenisch R. A genetic intron increases gene expression in transgenic mice. Mol Cell Biol 1991;11:3070.
9. Potten H, Jendraschak E, Hauck S, Amar LC, Avner P, and Müllhofer G. Molecular cloning and sequencing of a murine pgk-1 pseudogene family. Gene 1988;71:461.
10. Varadarajan, Szabo RA, and Boxer SG. Cloning, expression in Escherichia coli, and reconstitution of human myoglobin. Proc Natl Acad Sci USA 1985;82:5681.
11. Ratner L, Thielan B, and Collins T. Sequences of the 5' portion of the human c-sis gene: Characterization of the transcriptional promoter and regulation of expression of the protein product by 5' untranslated mRNA sequences. Nucleic Acids Res 1987;15:6017.
12. Forss-Petter S, Danielson PE, Catsicas S, Battenberg E, Price J, Nerenberg M, and Sutcliffe JG. Transgenic mice expressing β-galactosidase in mature neurons under neuron-specific enolase promoter control. Neuron 1990;5:187.
13. Hogan B, Beddington R, Costantini F, and Lacy E. Manipulating the Mouse Embryo: A Laboratory Manual. Plainview, NY: Cold Spring Harbor Laboratory Press, 1994.
14. Thomas PE, Ryan D, and Levin W. An improved staining procedure for the detection of the peroxidase activity of cytochrome P-450 on sodium dodecyl sulfate polyacrylamide gels. Anal Biochem 1976;75:168.
15. Silva AC, Zhang W, Williams DS, and Koretsky AP. Multi-slice MRI of rat brain perfusion during amphetamine stimulation using arterial spin labeling. Magn Reson Med 1995;33:209.
16. Williams DS, Detre JA, Leigh JS, and Koretsky AP. Magnetic resonance imaging of perfusion using spin inversion of arterial water. Proc Natl Acad Sci USA 1992;89:212.
17. Cole RP, Sukanek PC, Wittenberg JB, and Wittenberg BA. Mitochondrial function in the presence of myoglobin. J Appl Physiol. 1982;53:1116.
18. Jones DP and Kennedy F. Intracellular oxygen gradients in cardiac myocytes: Lack of a role for myoglobin in facilitation of intracellular oxygen diffusion. Biochem Biophys Res Commun. 1982;105:419.
19. Roy CW and Sherrington CS. On the regulation of the blood supply of the brain. J Physiol Lond 1890;11:85.
20. Fox PT and Raichle ME. Focal physiological uncoupling of cerebral blood flow and oxidative metabolism during somatosensory stimulation in human subjects. Proc Natl Acad Sci USA 1986;83:1140.

21. Arakaki LSL, Kushmerick MJ, and Burns DH. Myoglobin oxygen saturation measured independently of hemoglobin in scattering media by optical reflectance spectroscopy. Appl Spectrosc 1996;50:697.
22. Tamura M, Oshino N, Chance B, and Silver IA. Optical measurements of intracellular oxygen concentration of rat heart in vitro. Arch Biochem Biophys 1978191:8.
23. Jue T and Anderson S. ^{1}H NMR observation of tissue myoglobin: An indicator of cellular oxygenation in vivo. Magn Reson Med 1990;13:524.
24. Wang Z, Noyszewski EA, and Leigh JS Jr. In vivo MRS measurement of deoxymyoglobin in human forearms. Magn Reson Med 199014:562.

Chapter 32

THE CEREBRAL MICROCIRCULATION IN ISCHEMIA AND HYPOXEMIA

The Arisztid G. B. Kovách Memorial Lecture

Antal G. Hudetz
Department of Anesthesiology, Medical College of Wisconsin, Milwaukee, WI 53226

1. INTRODUCTION

One of the most recognized aspects of Professor Kovách's work is his extensive study on the mechanisms of hemorrhagic shock, particularly the role of sympathetic activation and vascular dysfunction in producing irreversible injury to vital organs. He was interested in the tolerance and limits of physiological adaptation of the circulatory system and how these tolerances were compromised, in particular in the brain, in shock, hypoxia and ischemia.

Inspired by Professor Kovách's insight I began about a decade ago an investigation into the behavior cerebral microcirculation in hemorrhagic hypotension. I was interested in the dynamics of flow in the capillary bed, which with had been scarcely studied as compared to the microcirculation of other organs. We developed a cranial window and video-microscopic technique to directly visualize the movement of erythrocytes in subsurface capillaries of the cerebral cortex of anesthetized rats [20]. Fluorescent labeling of red blood cells was used to generate image contrast and aid the measurement erythrocyte velocity, flow rate, and capillary hematocrit in several capillaries simultaneously. With this technique, much information has been collected on the functional architecture and physiological behavior of the cerebral capillary circulation at rest and during challenges of cerebral ischemia, anemia, hypotension, hypoxemia, and hypercapnia.

Following the intended educational style of the original presentation, this paper consists of three parts. The first part briefly reviews the basic properties of cerebral blood flow autoregulation to reduced perfusion pressure. The second part focuses on the physiology of cerebral microcirculation and the third part discusses the cerebral microcirculation in ischemia.

2. AUTOREGULATION OF CEREBRAL BLOOD FLOW

The brain distinguishes itself from other organs with its extreme dependence on constant oxygen supply. To ensure the required milieu interior for its neurons, the brain developed several interdependent vascular regulatory mechanisms. Autoregulation aims to maintain cerebral blood flow (CBF) in the face of a change in systemic arterial pressure, intracranial pressure or blood viscosity. Autoregulation involves cerebral arteries of all sizes and is therefore an essentially global response. When arterial pressure is reduced, larger arteries dilate more than the smaller arteries. The more compensation is needed the smaller arteries get involved. When the intravascular pressure is reduced in a small territory, for example by compressing a pial artery, vessels respond locally which implies a local vasodilatory mechanism. Veins also participate in the autoregulatory response [11]. Certain brain regions, such as the brain stem, autoregulate more effectively than the rest of the brain or the cerebral cortex.

It is important to note that CBF autoregulation is not absolute. First, CBF is not constant but changes to a certain degree with pressure in the range of autoregulation (60-160 mm Hg). In rare cases, super-autoregulation can be observed (overcompensation of CBF). Second, autoregulation takes time; the time course of vasodilation is in the range of 10 seconds to minutes, following a rapid change in pressure. Autoregulation is more effective during slow changes in arterial pressure. Fast pressure changes (seconds) are directly reflected in CBF. Trauma, cerebral ischemia, hypercapnia, seizure, etc. disrupt autoregulation for an extended period (minutes to hours) after termination of the insult.

Recently, another aspect of CBF autoregulation has been realized. When cerebral perfusion pressure is reduced, either directly or by pharmacological means, CBF begins to fluctuate at a regular pattern in the 4-12 cycles per minute range. The amplitude of flow fluctuations reaches 40 percent of baseline and appears to depend on local intravascular pressure [21]. Cerebral vasoconstriction enhances, vasodilation suspends the fluctuations in flow. Fluctuations in tissue pO_2 and metabolic state [9] follow the fluctuations in cerebral blood volume and flow. Thus, mean CBF over time is autoregulated but instantaneous CBF is not. In fact, spontaneous fluctuations are absent when autoregulation is disrupted. While most of these observations come from anesthetized animals, recent studies with human fMRI suggest that similar fluctuations in CBF and BOLD are present in the normal awake human brain and are associated with functionally defined areas of the cerebral cortex [3]. These observations challenge our traditional concept of CBF autoregulation aimed at maintaining constant CBF.

The cellular mechanisms of cerebral autoregulation are in nearly as much doubt today as they were 30 years ago. There continue to be three major theories: myogenic, metabolic and neurogenic. The myogenic mechanism is responsible for vessel diameter changes to transmural pressure and shear stress in the absence of extravascular neuronal and hormonal influences as seen in isolated vascular segments. Its physiological significance is probably minor as neuronal and metabolic effects in vivo modify or override the myogenic response [7,32].

Metabolic theories of CBF autoregulation can be divided into two types. The first one is thought to adjust blood flow according to the current concentration or pressure of a metabolic substrate, principally oxygen. Despite its generally assumed role, oxygen dependent regulation of CBF remains unproven. Dóra and Kovách [8] concluded that some factor other than tissue hypoxia is responsible for the dilatation of cerebrocortical vessels during moderate arterial hypotension. Dóra and Urbanics [10] supported that the brain cortex does not become hypoxic at moderate arterial hypotension.

The second type of metabolic regulation would adjust CBF according to the accumulation of a metabolic product. There are many candidates, the most well known is adenosine. Adenosine could be a metabolic regulator in severe hypoxia but would not suffice for paradigms when CBF increases sufficiently to prevent tissue hypoxia such moderate hypoxemia or moderate hypotension. Kovách [27] proposed that a glycolytic intermediary metabolite may mediate the autoregulatory response to decreased arterial pressure.

Neurogenic theories postulate the involvement of perivascular nerves and various neurotransmitters (acetylcholine, NO, neuropeptides, etc.) in producing vasodilatation or vasoconstriction to a central stimulus. Some of these mechanisms may oppose autoregulation. For example, Kovách [26] showed that cerebral vasoconstriction due to sympathetic activation occurs in prolonged hemorrhagic hypotension. Recently, Harder et al [16] proposed that an arachidonic acid metabolite 20-HETE produced via the P450 pathway would mediate cerebral autoregulation to increased pressure. The mechanism of autoregulation to reduced pressure, however, remains uncertain.

3. BLOOD FLOW IN THE CEREBRAL MICROCIRCULATION

We define the realm of microcirculation as the movement of discrete cellular elements of blood in microvessels of comparable size and the factors that regulate or interact with the movement of blood components. With this definition, our current knowledge of cerebral microcirculation seems meager. This is evidently due to the greater technical difficulty at which microcirculation of the brain can be directly examined. Recently, significant advances have been made toward a better understanding of the dynamics of cerebral microcirculation, owing to novel fluorescence video-microscopic and image analysis techniques that are now applied to directly study the dynamics of flow of erythrocytes and leukocytes in intracortical microvessels *in vivo*.

The most striking characteristics of the cerebrocortical microcirculation that make it distinct from many other microvascular beds are the apparently irregular, tortuous course of capillaries and the high velocity of flow. Viewing from the surface of the brain, capillaries appear to have no preferential orientation or characteristic length. The cerebral capillary bed is traditionally viewed as a dense network of intercommunicating vessels that is continuous throughout the cerebral gray matter [31]. Recent observations challenge this view and suggest a rather dichotomous organization of microvessels [28]. The functional significance of capillary anastomoses, thought to endow the capillary network with protection against perfusion failure, has yet to be ascertained. Our own observations suggest an orderly branching pattern of cortical capillaries as opposed to a random topology [17].

Red cells travel in single file across several capillary segments as they traverse the microvascular network. The transit time of RBC through the capillary network at resting condition varies between 100 and 300 ms with corresponding flow path lengths between 150 and 500 μm (19). The transit times of plasma are about twice as long. The velocity of red cells varies from capillary to capillary and is mostly between 0.5 to 1 mm/s. It is higher in the proximal and distal segments than in the middle segment that we call terminal or "true" capillary. Capillaries with velocity greater than 1 mm/s are less numerous and may play the role of preferential channels (Figure 1).

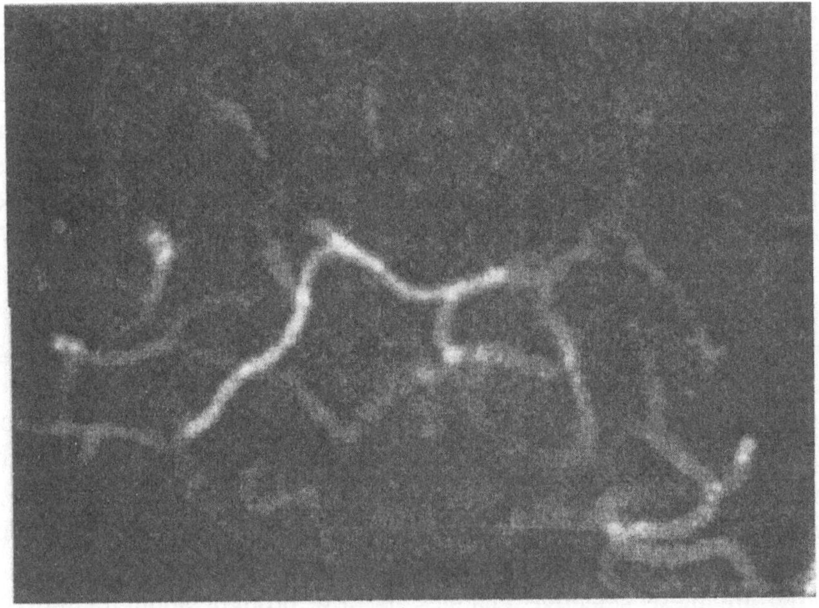

Figure 1. Reconstruction of a cortical capillary network from fluorescence video-microscopic images. A series of 3600 video images obtained at 60 Hz video rate were filtered and superimposed to yield the displayed image. Contrast is generated by moving red blood cells labeled with FITC. Brightness of the capillaries is proportional to the frequency of cell passage. The bright vessel in the middle is suspect for a postulated preferential channel. Image size is approximately 300 by 400 μm.

The velocity of flow in capillaries is not constant but fluctuates rapidly. In a small fraction of capillaries, RBC perfusion may stop for brief periods not longer than a few seconds. Flow cessation in capillaries occurs rarely and at random, probably associated with the passage of leukocytes and erythrocyte aggregates. On average, at least 90 percent of the capillaries are constantly perfused by red cells and probably all capillaries are constantly perfused by plasma. This leaves little room for capillary recruitment – a phenomenon observed in tissues other than the brain to describe an opening of new capillaries to increase blood flow.

The physiological questions of interest with respect to the capillary circulation are different from those that pertain to arterial flow regulation. Blood cells and plasma move according to different physical rules and transcapillary exchange is influenced by the configuration of cellular and plasma flow within each capillary. The capillaries can be considered as endothelial tubes with no capacity to actively dilate or constrict. However, the diameter of capillaries can change passively under the effect of distending pressure (see 17 for references). Furthermore, the rate of perfusion can change to a different degree among neighboring capillaries during physiological challenges. When cerebral perfusion pressure is reduced by controlled hemorrhage, erythrocyte flow is maintained in capillaries with low or normal resting velocity (<1 mm/s) but reduced in capillaries with high resting velocity (>1 mm/s). This suggests that RBC flow is maintained in the slowly perfused capillaries at the expense of the fast flow capillaries. The latter are thought to represent preferential or thoroughfare channels [18]. This observation suggests that classical autoregulation of CBF is incomplete and additional physiological adjustments of RBC perfusion in individual capillaries take place. Similar changes in RBC flow distribution occur in hypoxemia and hypercapnia; RBC flow increases more in capillaries with low resting flow than in those with high resting flow. Thus, in each of these hemodynamic responses, the pattern of flow distribution in the capillary network is altered and moves toward a greater homogeneity of perfusion.

How are these selective changes in flow brought about in individual capillaries devoid of vascular smooth muscle? In capillaries of organs other than brain, contractility has been ascribed to endothelial cells and pericytes. Pericytes are good candidates because they are abundant in brain capillaries particularly at the pre- and postcapillary sites [30]. Nakai [29] described ring-like impressions in cerebromicrovascular casts at the orifice of intracerebral arteriolar branches and attributed these to precapillary sphincters. To date, no functional evidence exists to support the flow regulatory role of pericytes in the brain *in vivo*. *In vitro*, retinal pericytes can contract and relax in response to a number of hormones and autocoids including angiotensin II, endothelin, serotonin, thromboxane A_2, NO and prostacyclin [7,15,25]. Capillary flow may also be influenced by rapid changes in volume of endothelial cells or perivascular astrocytes. The hypoxemic response of cerebral capillary flow depends critically on NO from neurons [22].

If blood flow in cerebral capillaries is physiologically regulated, then there must be a corresponding sensor for capillary flow, intracapillary pO_2 or some other parameter. Endothelial shear receptors, the glycocalyx, or adhesion receptors for leukocytes may play a role in capillary flow sensing. Capillary pO_2 may be sensed by endothelial heme containing enzymes or potassium channels. Signals representing blood flow and neuronal activity may be integrated by the endothelium [33] and communicated to the pericyte via gap junctions [3].

An amazing capacity for physiological compensation to maintain oxygen transport in the cerebral microcirculation is seen in acute anemia. During stepwise isovolemic hemodilution and maintained mean arterial pressure, RBC velocity and flux increase such that instantaneous capillary hematocrit is maintained at control levels in the face of a decrease in systemic hematocrit to as low as 15 percent [23]. Since arterial pO_2 is normal, oxygen delivery to cerebral tissue is maintained together with spontaneous cortical activity and cerebral energy state [2].

4. THE CEREBRAL MICROCIRCULATION IN ISCHEMIA

Kovách [26] demonstrated that vital functions of the brain are impaired during prolonged hypovolemic conditions. They described the development of patchy and circumscribed ischemic areas during hemorrhagic shock which persisted after re-infusion and implied a role for sympathetic activation that lead to the impairment of cerebral microcirculation (blood sludge formation). Increased tissue metabolism and the accumulation of metabolites combined with low flow would produce an imbalance between oxygen delivery and oxygen utilization during hemorrhagic shock.

Intravital microscopic studies reveal that when cerebral perfusion pressure is reduced, RBC perfusion in capillaries becomes intermittent [20]. Yamakawa described transient plugging of capillaries by leukocytes at the orifice of capillaries that increased in frequency and duration, as the arterial pressure was reduced [34]. Plasma flow continues around the trapped leukocytes in the same vessels. An example of intermittent RBC flow in cortical capillaries during hemorrhagic hypotension from one of our experiments is illustrated in Figure 2.

Whether microvascular perfusion defects persist following transient cerebral ischemia, the so-called "no-reflow" phenomenon [1], has been unclear. The early reperfusion phase after 1-2 hours of partial cerebral ischemia is characterized by a mismatch between blood flow and perfused capillary density [12]. However, this situation seems to normalize with longer periods of reperfusion. Dirnagl et al. [6] found no plugging leukocytes in the cortical capillary bed following brief, complete cerebral ischemia. Likewise, in our experiments, capillary plugging was not seen during reperfusion following one-hour partial forebrain ischemia in the rat [24].

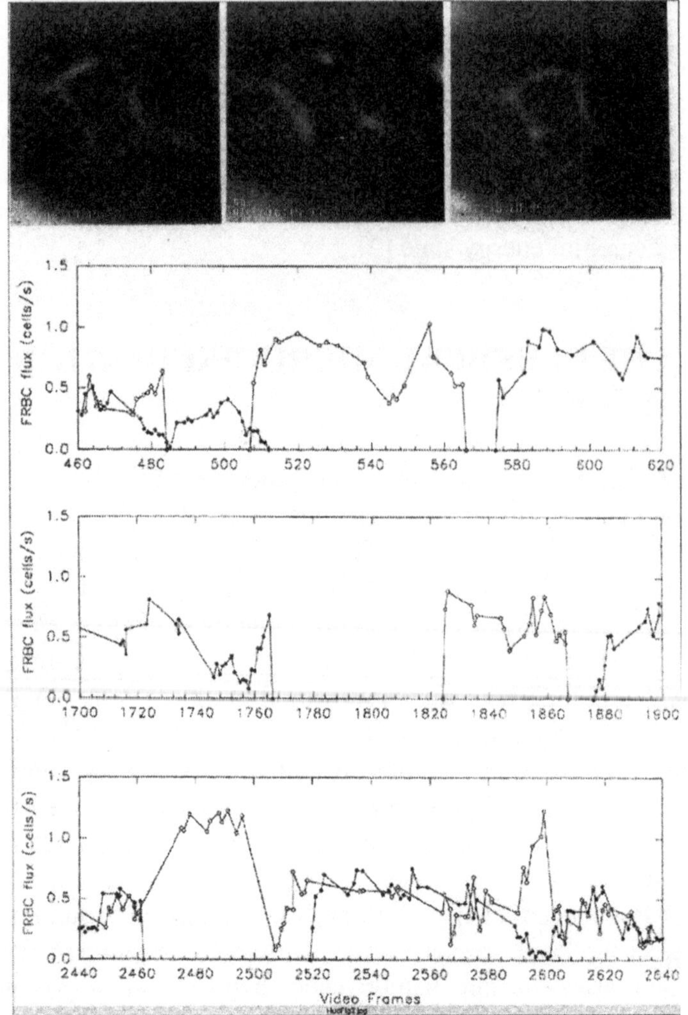

Figure 2. Example of intermittent RBC flow in two parallel capillaries in the cerebral cortex of the rat during prolonged hemorrhagic hypotension (40 mmHg mean arterial pressure). The top panels illustrate the passage of FITC-labeled red blood cells (FRBC) through the two arms of the capillary loop. Cells with elongated image are moving, cells with round image are stagnating. The three graphs below illustrated periods of on-off flow in the two capillaries (open and closed circles). FRBC flux, defined as the number of labeled cells passing each capillary per second, was calculated for 3 second periods. Video frames correspond to 16 ms time intervals. The repeated switching of flow between the two parallel capillaries cannot be explained by rheological factors alone and suggest that a hypoxic signal may be involved in the control of flow. A rapid fall in capillary pO_2 in transiently obstructed capillaries may produce pericyte relaxation, endothelial or astrocytic cell volume change, or leukocyte deactivation to jumpstart RBC flow in the respective capillary.

Severe, prolonged cerebral ischemia may present a different picture. In this case, capillary obstruction may be enhanced by endothelial swelling, formation of microvilli, platelet aggregation, and external compression of capillaries due to edema and perivascular astrocyte swelling [5]. Leukocytes may continue to accumulate in the brain if cerebral perfusion remains low for several hours [13] and may decrease capillary patency as seen in human stroke [14]. An inflammatory response may enhance leukocyte-endothelial interaction [5]. Thus, a combination of factors may be involved in capillary perfusion failure depending on the severity, regional distribution and duration of the ischemia and await further investigation.

5. SUMMARY AND CONCLUSIONS

The cerebral capillary circulation exhibits heterogeneous perfusion and undergoes characteristic changes in the distribution of RBC flow in response to systemic physiological stimuli. Hypoxemia, hypercapnia and hypotension increase the homogeneity of capillary perfusion, which is thought to preserve or enhance transcapillary exchange. Redistribution of capillary RBC flow between nutritive capillaries and preferential channels may contribute to this response. Selective changes in capillary flow may be brought about by non-smooth muscle-based contractile or blood-borne mechanisms. Isovolemic hemodilution anemia increases RBC velocity and supply rate with no decrease in capillary hematocrit. The effect of cerebral ischemia on microvascular patency depends on the severity and time course of the insult and whether the injury is global or focal. Capillary plugging is not observed following transient forebrain ischemia in the rat cerebral cortex but may contribute to tissue injury prior to reperfusion and during prolonged and severe ischemia.

In the future, a better understanding of the functional architecture of the cerebral capillary network and its significance in the adaptation to altered circulatory conditions will continue to be an important goal of research. More work will have to be done to (i) substantiate the postulated physiological regulation of cerebral capillary flow, (ii) determine the cellular mechanism of integration of flow-dependent and neuronal activity-dependent signals, and (iii) identify the principal mediators, their cellular sources and molecular targets. The final answer to these questions will in a large part depend on our ability to directly, i.e. microscopically, visualize microvascular, neuronal and molecular phenomena as they occur in the brain in a spatially and temporally distributed manner.

ACKNOWLEDGMENTS

This review was based in part on work supported by grants from the American Heart Association, GIA-95009340, the National Science Foundation, BES-9411631, and the National Institute of Health, GM-56398. Image processing to produce Figure 1 was performed by William O'Reilly. Figure 2 is the work of Derek E. Knuese. The experiments were performed by James D. Wood.

REFERENCES

1. Ames A, Wright RL, Kowada M, Thurston JM and Majno G. Cerebral ischemia II. The no-reflow phenomenon. Am J Pathol 1968;52:437-453.
2. Bauer R, Iijima T and Hossmann KA Influence of severe hemodilution on brain function and brain oxidative metabolism in the cat. Intensive Care Med 1996;22:47-51.
3. Biswal B. Hudetz AG. Yetkin FZ. Haughton VM. Hyde JS. Hypercapnia reversibly suppresses low-frequency fluctuations in the human motor cortex during rest using echo-planar MRI. J Cereb Blood Flow Metab 1997;17:301-308.
4. Cuevas P, Gutierrez-Diaz JA, Reimers D, Dujovny M, Diaz FG and Ausman JI. Pericyte endothelial gap junctions in human cerebral capillaries. Anat Embryol (Berlin) 1984;170:155-159.
5. del Zoppo GJ. Microvascular changes during cerebral ischemia and reperfusion. Cerebrovasc Brain Metab Rev 1994; 6:47-96.
6. Dirnagl U, Niwa K, Sixt G and Villringer A. Cortical hypoperfusion after global forebrain ischemia in rats is not caused by microvascular leukocyte plugging. Stroke 1994; 25:1028-1038.
7. Dodge AB, Hechtman HB and Shepro D. Microvascular endothelial-derived autacoids regulate pericyte contractility. Cell Motil Cytoskeleton 1991;18:180-188.
8. Dóra E., Kovách AG. Effect of acute arterial hypo- and hypertension on cerebrocortical NAD/NADH redox state and vascular volume. J Cereb Blood Flow Metabol 1982; 2:209-219.
9. Dóra E., and Kovách AG. Metabolic and vascular volume oscillations in the cat brain cortex. Acta Physiol. Acad. Sci Hung 1981; 57:261-75.
10. Dóra E., Urbanics R. Effect of surplus amount of oxygen on the cerebrocortical microcirculatory reactions associated to moderate arterial hypotension. Acta Physiol Hung 1986; 67:213-21.
11. Dóra E Effect of topically administered epinephrine, norepinephrine and acetylcholine on cerebralcortical vascular volume and NAD/NADH redox state. In: The Cerebral Veins (Auer LM and Loew F. Eds) Springer-Verlag Wien. New York. 1983;pp 193-199.
12. Ennis SR, Keep RF, Schielke GP and Betz AL. Decrease in perfusion of cerebral capillaries during incomplete ischemia and reperfusion. J Cereb Blood Flow Metab 1990;10:213-220.
13. Garcia JH, Liu KF, Yoshida Y, Lian J, Chen S and del Zoppo GJ. Influx of leukocytes and platelets in an evolving brain infarct (Wistar rat). Am J Pathol 1994; 144:188-199.
14. Gjedde A, Kuwabara H and Hakim AM. Reduction of functional capillary density in human brain after stroke. J Cereb Blood Flow Metab 1990;10:317-326.
15. Haefliger IO, Zschauer A and Anderson DA. Relaxation of retinal pericyte contractile tone through the nitric-oxide-cyclic guanosine monophosphate pathway. Invest Opthalamol Vis Sci 1994;35:991-997.
16. Harder DR, Narayanan J, Gebremedhin D and Roman RJ Transduction of physical force by the vascular wall. Trends Cardiovasc Med 1995; 5:7-14.

17. Hudetz AG Blood flow in the cerebral capillary network: A review emphasizing observations with intravital microscopy. Microcirculation 1997;4:233-252.
18. Hudetz AG, Fehér G and Kampine JP. Heterogeneous autoregulation of cerebrocortical capillary flow: Evidence for functional thoroughfare channels? Microvasc Res 1996; 51:131-136.
19. Hudetz AG, Fehér G, Knuese DE and Kampine JP. Erythrocyte flow heterogeneity in the cerebrocortical capillary network. Adv Exp Med Biol 1994; 345:633-642.
20. Hudetz AG, Fehér G, Weigle CGM, Knuese DE and Kampine JP. Video microscopy of cerebrocortical capillary flow: response to hypotension and intracranial hypertension. Am J Physiol 1995; 268:H2202-H2210.
21. Hudetz AG, Roman RJ and Harder DR. Spontaneous flow oscillations in the cerebral cortex during acute changes in mean arterial pressure. J Cereb Blood Flow Metab 1992;12:491-499.
22. Hudetz AG, Shen H and Kampine, JP. Nitric oxide from neuronal NOS plays critical role in cerebral capillary flow response to hypoxia. Am J Physiol 1998; 274:H982-989.
23. Hudetz AG, Wood JD, Biswal BB, Krolo I and Kampine JP Effect of hemodilution on RBC velocity, supply rate and hematocrit in the cerebral capillary network. J Appl Physiol 1999; 87:505-509.
24. Hudetz AG, Wood JD and Kampine JP (1999) Nitric oxide synthase inhibitor augments post-ischemic leukocyte adhesion in the cerebral microcirculation in vivo. Neurol. Res. 21:378-384.
25. Kelley C, D'Amore P, Hechtman HB and Shepro D. Vasoactive hormones and cAMP affect pericyte contraction and stress fibres in vitro. Journal of Muscle Research Cell Motil 1988; 9:184-194.
26. Kovách, A.G. Cerebral circulation in hypoxia and ischemia. Progr Clin Biol Res 1988; 264:147-158.
27. Kovach AG, Dora E, Hamar J, Eke A and Szabo L Transient metabolic and vascular volume changes following rapid blood pressure alterations which precede the autoregulatory vasodilation of cerebral cortical vessels. Adv Exp Med Biol 1977;94: 705-711.
28. Motti ED, Imhof HG and Yasargil MG. The terminal vascular bed in the superficial cortex of the rat. An SEM study of corrosion casts. J Neurosurg 1986; 65:834-846.
29. Nakai K, Imai H, Kamel I, Itakura T, Komari N, Kimura H, Nagai T and Meada T. Microangioarchitecture of rat parietal cortex with special reference to vascular sphincters. Stroke 1981;12:653-659.
30. Nehls V and Drenckhahn D. The versatility of microvascular pericytes: From mesenchyme to smooth muscle? Histochemistry 1993; 99:1-12.
31. Pfeifer RA. (1928) In: Die Angioarchitectonik der Grosshirnrinde Eds.) Springer. Berlin.
32. Wei EP and Kontos HA Responses of pial arterioles to increased venous pressure. J Cereb Blood Flow Metab 1981;1: S329.
33. Woolsey TA, Rovainen CM, Cox SB, Henegar MH, Liang GE, Liu D, Moskalenko YE, Sui J and Wei L. Neuronal units linked to microvascular modules in cerebral cortex: response elements for imaging the brain. Cereb Cortex 1996;6:647-660.
34. Yamakawa T, Yamaguchi S, Niimi H and Sugiyama I. White blood cell plugging and blood flow maldistribution in the capillary network of cat cerebral cortex in acute hemorrhagic hypotension: An intravital microscopic study. Circ Shock 1987;22:323-332.

Chapter **33**

INVESTIGATING THE ROLE OF NITRIC OXIDE IN REGULATING BLOOD FLOW AND OXYGEN DELIVERY FROM *IN VIVO* ELECTROCHEMICAL MEASUREMENTS IN EYE AND BRAIN

Donald G. Buerk[1,2,3], Dmitriy N. Atochin[3], and Charles E. Riva[4]

Departments of [1]Physiology and [2]Bioengineering, and [3]Institute for Environmental Medicine, University of Pennsylvania School of Medicine, Philadelphia, PA 19104 USA: and [4]Institut de Recherches en Ophthalmologie, Sion Switzerland

Abstract: We have previously shown from direct, *in vivo* measurements of NO in cats with recessed electrochemical microsensors that NO mediates increases in ONH blood flow during functional activation of the eye by flickering light. We have also reported that there are low frequency (< 15 cycles/min) spontaneous oscillations in NO that appear to be passively coupled to oscillations in blood flow at similar frequencies in the cat ONH. In this paper, we describe similarities between *in vivo* measurements of NO in the ONH of the cat eye and in the cortex of the rat brain. These data are consistent with a role for NO in the coupling of blood flow with increases in neuronal activity, autoregulation of blood flow, hyperemia, and vasodilation during hypoxia and hypercapnia.

Key words: autoregulation, blood flow, brain, electrochemical microsensor, functional activation, hypercapnia, nitric oxide, oxygen, optic nerve, somatosensory cortex.

Abbreviations:

EEG	electroencephalograph	IOP	intra-ocular pressure
K_m	Michaelis-Menten constant	LDF	laser Doppler flowmetry
L-NAME	N^{ω}-nitro-L-arginine methyl ester		
7-NI	7-nitroindazole	NO	nitric oxide, nitric monoxide
NOS	nitric oxide synthase (neuronal nNOS, endothelial eNOS, inducible iNOS)		
ONH	optic nerve head		

1. INTRODUCTION

There is no question that NO plays a vital role in regulating blood flow and O_2 delivery to most if not all tissues, primarily through its potent vasodilator function as an activator of guanylate cyclase in vascular smooth muscle cells. Recent reviews summarize evidence for multiple roles of NO in the eye [1] and its involvement in normal and pathological responses to hypoxia and ischemia in the brain [2].

There are numerous interactions between O_2 and NO which need to be incorporated into mathematical models of O_2 transport to tissues. First, all of the NOS isoforms (nNOS, eNOS, iNOS) require O_2 to synthesize NO from l-arginine. The enzymatic biosynthesis is O_2-dependent [3], and one group has found nNOS to be strongly dependent on O_2, with a very high K_m around 200 μM [4]. After it is formed, NO can react directly with O_2 to form nitrate and nitrite. Also, NO reversibly binds to cytochrome c oxidase, inhibiting O_2 consumption, glucose oxidation and ATP generation [5,6]. A newly discovered NO synthase isoform located in mitochondria may play an important role in regulating O_2 consumption [7]. There is also evidence that NO can alter oxyhemoglobin affinity, shifting the equilibrium curve to the right [8].

Clementi *et al.* [9] recently proposed that eNOS could serve as an O_2 sensor. They commented that "the classical paradigm about the independence of oxygen consumption along a wide range of oxygen concentrations should be corrected in terms of NO metabolism." We have shown that NO donors and inhibitors of NO biosynthesis can respectively decrease or increase O_2 consumption in the intact, perfused cat carotid body [10], and have presented evidence consistent with a role for NO as an O_2 sensor in this chemosensory organ [11].

Often the evidence for a specific biochemical role of NO in living systems is based on indirect observations of changes in key physiological parameters after inhibiting NO synthesis, or opposite changes with NO donors. Usually, there is no information about NO levels in tissue. We have previously shown from direct, *in vivo* measurements in cats with Nafion polymer coated recessed NO microsensors that NO mediates increases in ONH blood flow during functional activation of the eye by flickering light [12]. During the flickering light stimulus, the increase in NO was nearly 100 nM above basal levels. We have also reported that there are low frequency (< 15 cycles/min) spontaneous oscillations in NO in the cat ONH that appear to be passively coupled to oscillations in blood flow at similar frequencies [13]. There was no single dominant frequency, but several distinct modes for both NO and blood flow oscillations in this frequency range.

In addition to our NO studies in the cat ONH, we have recently made similar NO measurements in cortex of the rat brain for several different experimental conditions. Similarities and differences between these findings will be compared and their relevance to investigating the role of NO in the control of blood flow in the eye and brain will be discussed.

2. METHODS

2.1 Cat ONH studies

Experimental methods for studies in the cat ONH and calibration methods for Nafion polymer coated recessed NO microsensors have been published [12--14]. Measurements of relative blood flow changes in the cat ONH were made with a microscope mounted infrared laser Doppler instrument, with excellent spatial resolution from an illuminated region of tissue around 100 μm in diameter. In some studies, recessed NO microsensors with tip diameters between 5 to 15 were positioned into the region of ONH tissue illuminated by the laser. In other studies, a double barrel NO and PO_2 sensor was placed on the surface of the ONH immediately adjacent to the laser spot. In addition to flicker studies [12,14] and characterization of NO and blood flow oscillations [13], the effects of acute hypercapnia, hypoxia, hyperoxia, and increased IOP have been studied in the cat ONH. All of the experimental data were acquired by computer, usually at 2 Hz sampling rates.

2.2 Brain activation-blood flow coupling studies

NO was measured in the somatosensory cortex of 12 Sprague-Dawley rats anesthetized with α-choralose (60 mg/kg initially, 30 mg/kg/hr supplemental). Cortical sites where there was an increase in blood flow in response to electrical stimulation of the contralateral forepaw (1 mA current at 5 Hz) were identified by transcranial LDF measurements through the thinned skull. A small opening was made in the skull over one of the responding sites and a NO microsensor was inserted into the brain. The optical probe was placed on the section of thinned skull immediately adjacent to the opening for LDF measurements. Signal averaging methods, as described by Detre *et al.* [15], were used to collect NO and LDF data at a sampling rate of 10 Hz, synchronized with electrical stimulation of the contralateral forepaw. Each trial consisted of 10 consecutive stimuli over a 240 sec period, with either 4 or 8 sec of forepaw stimulation and either 20 or 16 sec of recovery between stimuli.

2.3 Brain hypoxia studies

NO was measured in the right parietal cortex of 6 male Wistar rats anesthetized with sodium pentobarbital (65 mg/kg *i.p.* initially, 15 mg/kg/hr *i.v.* supplemental) in a small pilot study. After recording at least 15 min of baseline tissue NO, LDF and EEG activity, the ventilator was stopped and the airway was occluded for 3 min. The rat was then resuscitated by reopening the airway and restarting the ventilator, giving sternal chest compression to restore cardiac function if necessary. Recovery was followed for 30 to 90 min. A second 3 min period of asphyxia was given in animals that had full recovery of EEG activity from the first challenge. In a few animals, a laser Doppler probe was available for transcranial LDF measurements. All data were acquired by computer for further analysis.

3. RESULTS

A typical flicker response from the cat ONH is shown in Fig. 1, during two min of flickering light at 30 Hz. In an earlier study, we found that the maximum blood flow increase usually occurred at this frequency [16]. The rise in NO (top panel) preceded the rise in the LDF signal (bottom panel) at the start of the flicker stimulus at t = 0 (dashed line). The fall in blood flow after stopping stimulation was delayed for several sec, while NO decreased immediately. In our earlier study [12], the average blood flow during flicker increased to 144% of control with NO increasing by 88 nM above baseline. After L-NAME, baseline NO decreased and both responses were significantly attenuated (117% and 14 nM).

As shown in Fig. 2, similar increases in NO and blood flow were seen in the rat somatosensory cortex during functional activation of this region of the brain by electrical stimulation of the contralateral forepaw. A frequency of 5 Hz was found to cause the greatest peak increase in cortical blood flow at 1 mA current (average 22% above control), while ipsilateral forepaw stimulation caused no changes in blood flow [16]. Greater increases in cortical blood flow could be evoked with higher forepaw stimulation currents, but were avoided since they could also cause changes in systemic pressure that would confound the results. The signal averaged NO and LDF responses are superimposed in Fig. 2 to clearly show that NO rises to a peak value of 145 nM above baseline before blood flow increases (onset marked by dashed line). The average NO was 82 nM above baseline during the period of stimulation. At the end of the stimulus, NO dropped rapidly, with a marked undershoot. The minimum NO was reached just as flow began to fall, after a delay of ~ 1 sec (dashed line). This temporal pattern, with the rise

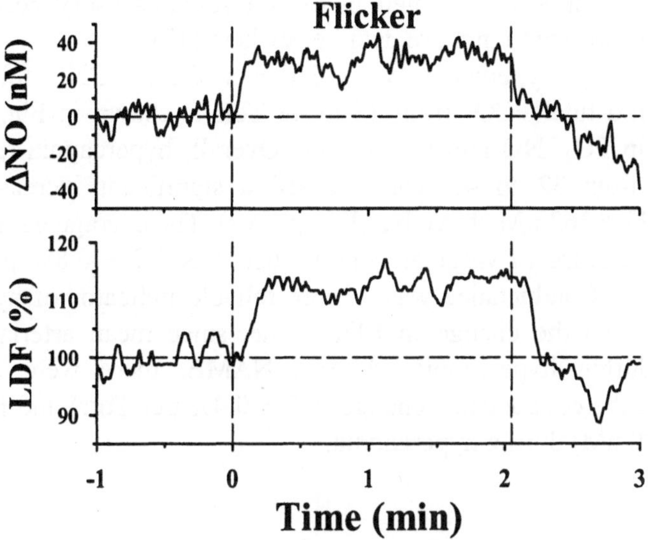

Figure 1. Simultaneous measurements of tissue NO with an electrochemical microsensor and blood flow by infrared LDF in the cat ONH during ~ 2 min period of stimulation of the cat eye by flickering light at 30 Hz. Tissue NO increased above baseline immediately after beginning the stimulus, preceding the increase in blood flow by several seconds. After stopping the stimulus, tissue NO dropped rapidly with a delayed fall in blood flow.

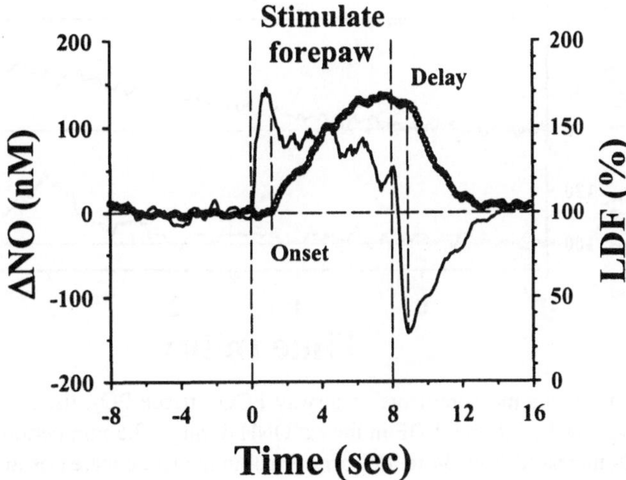

Figure 2. Measurements of brain tissue NO (solid line) with a recessed NO microsensor and blood flow by LDF (circles) in rat somatosensory cortex during electrical stimulation of the contralateral forepaw. A total of 20 repeated stimuli (8 sec on with 1 mA at 5 Hz, 16 sec off) were signal averaged. Tissue NO reached a peak value before flow increased (onset, dashed line), with an undershoot and delayed drop in blood flow after stopping the stimulus.

in NO preceding the rise in blood flow, was seen for the majority of measurements that were obtained in the rat somatosensory cortex. These results will be presented in more detail elsewhere [17].

An example of hypercapnic vasodilation in the cat ONH following an increase in end tidal PCO_2 from 34 to 48 Torr is shown in Fig. 3, with a typical rise in PO_2, NO and blood flow. Overall, hypercapnia (mean end-tidal PCO_2 from 32 to 45 Torr) caused a significant increase in NO, averaging 823 ± 362 nM above baseline (9 cats). The average gain factor for the relative increase in vascular conductance was 2.9 ± 0.8% per Torr of PCO_2 increase. Conductance was a more reliable indicator of hypercapnic vasodilation than the change in LDF alone, since mean arterial pressure often fell during hypercapnia. After L-NAME, there were significant reductions in the conductance change (0.7 ± 0.4% per Torr) and increase in NO (144 ± 90 nM) during hypercapnia.

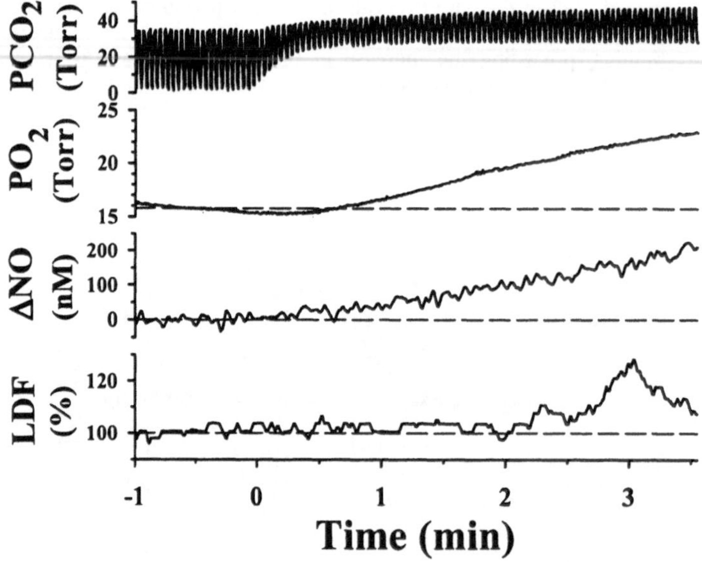

Figure 3. Simultaneous measurements of airway PCO_2, tissue PO_2, tissue NO and relative change in blood flow by infrared LDF in the cat ONH during ~ 3.5 min period of hypercapnia (end tidal PCO_2 increased from 34 to 48 Torr). Small to moderate increases in blood flow and tissue NO were seen with hypercapnia in the cat ONH.

An example of hypercapnic vasodilation in the rat somatosensory cortex is shown in Fig. 4. The gain factor for the blood flow increase was 23% per Torr of PCO_2 based on the measured change in arterial blood gases indicated in the Fig. NO increased by an average of 166 nM above baseline during the

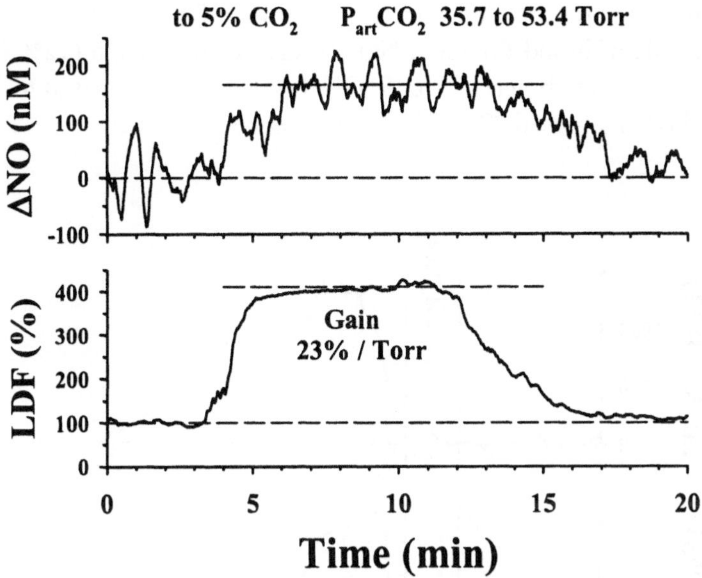

Figure 4. Measurements of brain tissue NO and relative change in blood flow by LDF in the rat somatosensory cortex during ~ 6 min of ventilation with 5% CO_2. The increase in arterial blood PCO_2 and relative gain factor for the hypercapnic vasodilation response are indicated. Large increases in blood flow during hypercapnia were typically seen in the rat brain.

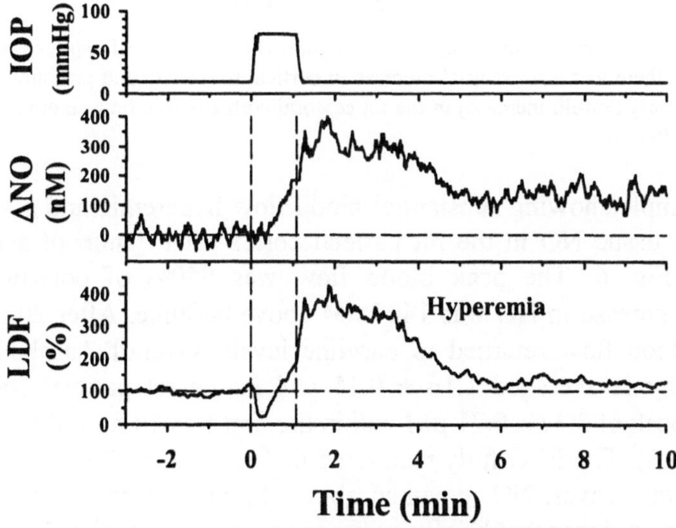

Figure 5. An acute increase in intraocular pressure (IOP) for ~ 1 min caused a temporary decrease in blood flow in the cat optic nerve head. Shortly after elevating IOP, there was an increase in tissue NO which preceded the rise in blood flow as it began to autoregulate. After restoring normal IOP, tissue NO remained high during several min of blood flow hyperemia.

period of hypercapnia. In general, the increase in blood flow with hypercapnia in the rat brain was much greater than for the cat ONH.

An example of blood flow and NO changes with hypoxia (12% O_2) from a cat ONH study are shown in Fig. 5. ONH blood flow increased to 155% of control and NO increased 278 nM above baseline in this cat.

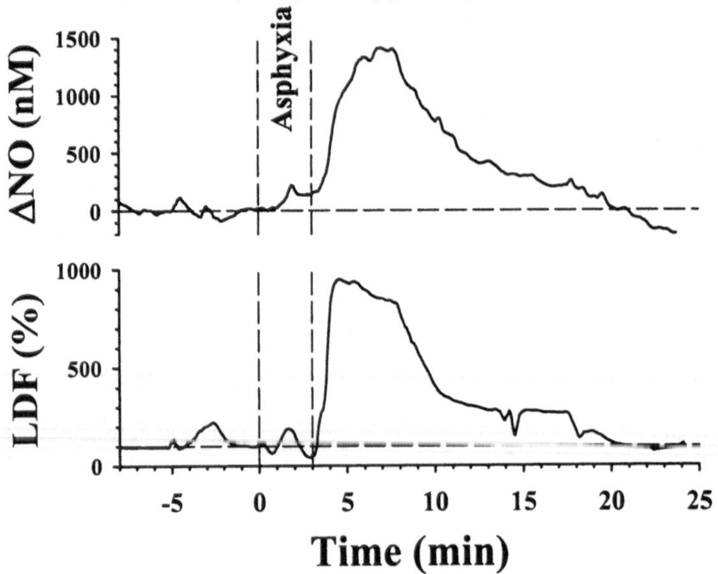

Figure 6. After restoring ventilation to the rat following ~ 3 min of asphyxia (marked by dashed lines), there was a substantial increase in cortical tissue NO and profound blood flow hyperemia (nearly tenfold increase) in the rat cerebral cortex before both returned to baseline levels within 30 min.

An example showing substantial blood flow hyperemia and a very large increase in tissue NO in the rat parietal cortex after 3 min of asphyxia is shown in Fig. 6. The peak blood flow was 950% of control and the maximum increase in NO was 1407 nM above baseline. After 20 min, both NO and blood flow returned to baseline levels. Overall, basal NO levels before asphyxia averaged 1.16 ± 0.15 μM in the rat parietal cortex, and reached a peak of 2.37 ± 0.71 μM within the first 5 to 10 min of the recovery period (6 rats). The EEG fully recovered in 4 rats. In the 2 rats in which the EEG did not recover, NO remained above the pre-asphyxia baseline for at least 90 min. A paper that briefly summarizes some of these results has been published [18], and a preliminary Fourier power spectral analysis showing that there were transient increases in the amplitude of NO oscillations during the recovery period was presented [19].

Responses to moderate hyperoxia were also obtained in the cat ONH, with variable results (data not shown). There was an increase in NO in 9 cats, but a decrease in 6 cats. Usually, there was a small decrease in ONH blood flow by 5 to 10% when tissue PO_2 was elevated. A few hyperoxia tests in the rat somatosensory cortex have also been done with slight reductions in blood flow with no change or very minor increases in NO (data not shown).

4. DISCUSSION

It has been known for some time that there is tight coupling between electrical activity, blood flow and metabolic activity in the brain, and modern optical neuroimaging methods such as functional magnetic resonance imaging, position-emission tomography and blood oxygenation level-dependent imaging are being refined for research and clinical use. Functional activation studies in the cat visual cortex have found an initial decrease in reduced oxyhemoglobin immediately after a stimulus in the short time period before blood flow starts to rise [20], suggesting that there is an immediate increase in O_2 consumption. Consistent with this hypothesis, we have measured transient decreases in tissue PO_2 on the surface of the cat ONH at the beginning of the flickering light stimulus followed by an increase after blood flow has increased [14]. Furthermore, we often saw a rapid drop in tissue PO_2 after stopping the stimulus, suggesting that O_2 consumption remained elevated for a short time during the recovery phase.

Our direct *in vivo* measurements of NO in the cat ONH and rat somatosensory cortex demonstrate that there is a rapid increase in NO that precedes the rise in blood flow. There is good evidence that NO produced from nNOS in the brain is important for functional activation [21], although one group has shown that chronic treatment with L-NAME that inhibited total brain NOS activity by 84% did not reduce increases in blood flow during whisker stimulation [22]. This group has also reported that cerebral autoregulation is intact after inhibiting NOS [23], although others have found that NO is important [24,25]. There is also good evidence that NO from nNOS is important for hypercapnic [26] and hypoxic vasodilation [27] in the brain. It has been suggested, based on the strong O_2-dependent kinetics of nNOS [4], that the production of NO might be limited by O_2 availability under normal physiological conditions. If that is the case, one might expect that increased tissue PO_2 in the brain could substantially increase tissue NO levels. We did not see any evidence for this response from a limited number of tests in the rat brain, although there were a substantial number of observations where this was true in the cat ONH. Further studies need to be directed towards quantifying the relative contribution of NO from nNOS, especially since it has been implicated in neurotoxicity and brain injury [2].

Studies of the ocular circulation in conscious rats using the inhibitor 7-NI have estimated that 29% of the ocular vascular resistance might be attributed to nNOS [28]. With the non–specific inhibitor L-NAME, there was a 130% increase in vascular resistance. Microsphere studies of blood flow in the retina and optic nerve in cats found a marked reduction in the flicker response in cats treated by L-NAME compared to control animals [29]. Studies in humans have also implicated NO in hypercapnic vasodilation of the choroidal circulation of the eye [30]. Further work to define the role of NO produced by nNOS in the eye needs to be done. We have investigated whether 7-NI impairs the hypercapnic vasodilation in the cat ONH, but have not seen any reduction in the response in studies completed to date.

While our experimental results provide direct evidence for the involvement of NO in the different physiological responses that we have investigated, it is clear that NO is not the only mediator since there are many other vasoactive factors that are important for the dynamic coupling of blood flow, O_2 metabolism and O_2 delivery [31]. It is possible that contradictory findings with regard to the role of NO are masked by compensatory changes in other pathways. We have demonstrated that the direct *in vivo* measurements of NO with electrochemical microsensors can be a valuable tool for investigating the role of NO in physiology.

ACKNOWLEDGMENTS

Cat ONH studies were supported by EY09269 from the National Eye Institute, NIH. Collaborative studies in the brain were conducted with Dr. Nitish V. Thakor, Department of Biomedical Engineering, Johns Hopkins University, Baltimore, MD and Dr. Joel H. Greenberg, Department of Neurology and Cerebrovascular Research Center, University of Pennsylvania, Philadelphia, PA.

REFERENCES

1. Goldstein IM, Ostwald P, Roth S. Nitric oxide: a review of its role in retinal function and disease. Vision Res 1996;36:2979-2994.
2. Bolanos JP, Almeida A. Roles of nitric oxide in brain hypoxia-ischemia. Biochim Biophys Acta 1999;1411:415-436.
3. Rengasamy A, Johns RA. Determination of K_m for oxygen of nitric oxide synthase isoforms. Am Soc Pharmacol Exp Therapeu 1996;276:30-33.

4. Abu-Soud HM, Rousseau DL, Stuehr DJ. Nitric oxide binding to the heme of neuronal nitric-oxide synthase links its activity to changes in oxygen tension. J. Biol. Chem. 1996;271: 32515-32518.
5. Brown GC, Cooper CE. Nanomolar concentrations of nitric oxide reversibly inhibit synaptosomal respiration by competing with oxygen at cytochrome oxidase. FEBS Lett 1994;356:295-298.
6. Cleeter MJW, Cooper JM, Darley-Usmar VM, Moncada S, Schapira AH. Reversible inhibition of cytochrome c oxidase, the terminal enzyme of the mitochondrial respiratory chain, by nitric oxide. FEBS Lett 1994;345:50-54.
7. Boveris A, Costa LE, Cadenas E, Poderoso JJ. Regulation of mitochondrial respiration by adenosine diphosphate, oxygen, and nitric oxide. Meth Enzymol 1999;301:188-198.
8. Kosaka H. Nitric oxide and hemoglobin interactions in the vasculature. Biochim Biophys Acta 1999;1411:370-377.
9. Clementi E, Brown GC, Foxwell N, Moncada S. On the mechanism by which vascular endothelial cells regulate their oxygen consumption. Proc Natl Acad Sci USA 1999;96:1559-1562.
10. Lahiri S, Buerk DG. Vascular and metabolic effects of nitric oxide synthase inhibition evaluated by tissue PO_2 measurements in carotid body. In: Hudetz A, Bruley D, editors. Oxygen Transport to Tissue XX, Advances in Experimental Medicine and Biology. New York: Plenum Press, 1998;455-460.
11. Buerk DG, Lahiri S. Evidence that nitric oxide plays a role in O_2 sensing from tissue NO and PO_2 measurements in cat carotid body. In: Lahiri S, Forster R, Prabhakar N, editors. Oxygen Sensing: Molecule to Man. New York: Plenum Press, in press.
12. Buerk DG, Riva CE, Cranstoun SD. Nitric oxide has a vasodilatory role in cat optic nerve head during flicker stimuli. Microvas Res 1996;52:13-26.
13. Buerk DG, Riva CE. Vasomotion and spontaneous low frequency oscillations in blood flow and nitric oxide in cat optic nerve head. Microvas Res 1998;55:103-112.
14. Buerk DG, Atochin DN, Riva CE. Simultaneous tissue PO_2, nitric oxide and laser Doppler blood flow measurements during neuronal activation of optic nerve. In: Hudetz A, Bruley D, editors. Oxygen Transport to Tissue XX, Advances in Experimental Medicine and Biology. New York: Plenum Press, 1998;159--164.
15. Detre JA, Ances BM, Takahashi K, Greenberg JH. Signal averaged laser Doppler measurements of activation-flow coupling in the rat forepaw somatosensory cortex. Brain Res 1998;796:91-98.
16. Buerk DG, Riva CE, Cranstoun SD. Frequency and luminance dependent blood flow and K^+ ion changes during flicker stimuli in cat optic nerve head. Invest Ophthalmol Vis Sci 1995;36:2216-2227.
17. Buerk DG, Ances BM, Detre JA, Greenberg JH. Increased tissue nitric oxide (NO) and cortical blood flow during functional activation of rat brain. Ann Biomed Eng. 1999;(abstract) in press.
18. Ghodadra R Buerk DG, Hao L, Mohadjer Y, Thakor NV. Nitric oxide changes in adult rat brain after transient global ischemia. Proc IEEE-EMB 1997;pp. 2152--2153.
19. Buerk DG, Ghodadra R, Hao L, Mohadjer Y, Thakor NV. Nitric oxide production and fluctuations in rat brain are transiently augmented after hypoxia. Ann Biomed Eng 1997;25(Suppl. 1):S58 (abstract).
20. Malonek D, Dirnagl U, Lindauer U, Yamada K, Kanno I, Grinvald A. Vascular imprints of neuronal activity: Relationships between the dynamics of cortical blood flow, oxygenation, and volume changes following sensory stimulation. Proc Natl Acad Sci 1997;94:14826-14831.

21. Cholet N, Seylez J, Lacombe P, Bonvento G. Local uncoupling of the cerebrovascular and metabolic responses to somatosensory stimulation after neuronal nitric oxide synthase inhibition. J Cereb Blood Flow Metab 1997;17:1191-1201.
22. Adachi K, Takahashi S, Melzer P, Campos KL, Nelson T, Kennedy C, Sokoloff L. Increases in local cerebral blood flow associated with somatosensory activation are not mediated by NO. Am J Physiol 1994;267:H2155-H2162.
23. Takahashi S, Cook M, Jehle J, Kennedy C, Sokoloff L. Preservation of autoregulatory cerebral responses to hypotension after inhibition of nitric oxide synthesis. Brain Res 1995;678:21-28.
24. Tanaka K, Fukuuchi Y, Gomi S, Mihara B, Shirai T, Nogawa S, Nozaki H, Nagata E. Inhibition of nitric oxide synthesis impairs autoregulation of local cerebral blood flow in the rat. Neuroreport 1993;4:267-270.
25. Kobari M, Fukuuchi Y, Tomita M, Tanahashi N, Takeda H. Role of nitric oxide in regulation of cerebral microvascular tone and autoregulation of cerebral blood flow in cats. Brain Res 1994;667:255-262.
26. Okamoto H, Hudetz AG, Roman RJ, Bosnjak ZJ, Kampine JP. Neuronal NOS-derived NO plays permissive role in cerebral blood flow response to hypercapnia. Am J Physiol 1997;272:H559-H566.
27. Hudetz AG, Shen H, Kampine JP. Nitric oxide from neuronal NOS plays critical role in cerebral capillary flow response to hypoxia. Am J Physiol 1998;274:H982-H989.
28. Kelly PA, Buckley CH, Ritchie IM, O'Brien C. Possible role for nitric oxide releasing nerves in the regulation of ocular blood flow in the rat. Br J Ophthalmol 1998;82:1199-1202.
29. Kondo M, Wang L, Bill A. The role of nitric oxide in hyperaemic response to flicker in the retina and optic nerve in cats. Acta Ophthalmol Scand 1997;75:232-235.
30. Schmetterer L, Findl O, Strenn K, Graselli U, Kastner J, Eichler HG, Woltz M. Role of NO in the O_2 and CO_2 responsiveness of cerebral and ocular circulation in humans. Am J Physiol 1997;272:R2005-R2012.
31. Riva CE, Buerk DG. Dynamic coupling of blood flow to function and metabolism in the optic nerve. Neuro-Ophthalmol 1998;20:45-54.

Chapter **34**

NIRS MONITORING OF PILOTS SUBJECTED TO +Gz ACCELERATION AND G-INDUCED LOSS OF CONSCIOUSNESS (G-LOC)

Paul B. Benni[1,2], John K-J. Li[1], Bo Chen[1,2], Joseph Cammarota[3], & David W. Amory[4]

[1]Dept. Biomedical Engineering, Rutgers University, Piscataway, NJ 08854-8014, USA; [2]Dept. Anesthesia, UMDNJ-Robert Wood Johnson, New Brunswick, NJ 08903, USA; [3]EDO M-Technologies, Huntingdon Valley, PA 19006, USA; [4]Institute of Applied Physiology and Medicine, Seattle, WA 98122, USA.

Abstract: With the increasing maneuverability of modern aircraft, there is an increased frequency of pilots losing consciousness due to high +Gz acceleration. This phenomena is defined as "G-induced loss of consciousness" (G-LOC). We used an NIRS system of our design to monitor cerebral oxygenation changes of pilots subjected to high +Gz acceleration and G-LOC. During the +Gz pulse, ΔHbO_2, and ΔTotalHb decreased, with lesser changes of ΔHb. The maximum decrease of ΔHbO_2 and ΔTotalHb usually occurred at the onset of G-LOC. After G-LOC, ΔHbO_2 and ΔTotalHb increased rapidly for the first few seconds, beginning the reactive hyperemic recovery phase. ΔHbO_2 and ΔTotalHb peaked, and then began to decrease towards baseline. The subjects were unconscious for 3--10 seconds after the onset of G-LOC. Upon returning to consciousness, the subjects were disoriented for another 4--11 seconds. NIRS provides an additional means of studying physiological mechanisms leading to and recovery from G-LOC.

Key words: Cerebral oxygenation, +Gz acceleration, G-LOC (G-induced loss of consciousness), NIRS (Near-Infrared Spectroscopy)

1. INTRODUCTION

In 1918, Head reported incidences of pilots losing control of their aircraft from fainting after a high +Gz maneuver [1]. This phenomenon is now termed gravitation-induced loss of consciousness (G-LOC) and has been suspected as one of the causes of fatal aircraft crashes [1, 2]. A G-LOC episode consists of a two-phase incapacitation period. The absolute incapacitation period, or period of unconsciousness, is observed to begin at the point where the pilot's muscle tone becomes flaccid, resulting in the pilot's head slumping to one side, as if the pilot is nodding off to sleep. At first, there are no eventful observations. During the last few seconds of unconsciousness, blood flow to the brain returns, and uncoordinated myoclonic convulsions such as twitching of the arms, legs, and head muscles, are commonly observed [3, 4]. The absolute incapacitation period averages about 12 seconds [4]. Some pilots report dreaming during the brief unconsciousness period [3, 4]. Upon returning to consciousness, the pilot is confused and disorientated for another 12 seconds (average) [4]. This is defined as the relative incapacitation period. Our research expands the scope of previous work [5-10] involving NIRS monitoring of pilots subjected to high +Gz acceleration.

2. METHODS

2.1 Experimental protocols

These experiments were conducted at the U.S. Navy Air Warfare Center – Aircraft Division (NAWCAD) centrifuge facility in Warminster, PA. This study was carried out with the approval of the U.S. Navy G-Acceleration Committee on Human Experimentation and the Centrifuge Medical Response Council. Voluntary, fully informed written consent of the G-acceleration protocol was obtained from each subject. Medical personnel were present at all times during the experiments.

Two different +Gz acceleration profile protocols were used in the experiments: *G-plateau* and *G-pulse*. The *G-plateau* profile is characterized as a single +Gz pulse to induce the subject to G-LOC. The onset rate and offset rate was about 6 +Gz/second, with a +Gz pulse maximum amplitude set at 6 to 10 +Gz, depending on the G tolerance of the subject. The maximum +Gz pulse duration was set at 15 seconds. When the subject was observed going into G-LOC, the centrifuge was immediately decelerated to normal G levels (1.2 +Gz). No advanced technology anti-G suit or anti-G straining maneuvers were used. The subject's attention and anxiety levels were partially diverted by giving them a task of simple addition and subtraction exercises, with required verbal response to the questions. Subjects had to deactivate a G-LOC recovery monitoring system consisting of a beeping signal triggered by the medical monitor upon the onset of G-LOC, which determined total recovery. After each +Gz pulse run, the subjects reported any +Gz acceleration stress symptoms that were experienced.

The *G-pulse* profile consisted of a series of +Gz pulses starting at a predetermined pulse width of 0.25 seconds and commencing 1.0 sec., 2.0 sec., 2.5 sec.... until G-LOC or exhaustion had occurred. All other details are similar to the *G-plateau* profile. Consent by the subject and attending medical personnel was obtained before proceeding to the next +Gz pulse.

NIRS System - The NIRS system used in the +Gz acceleration studies was of our own design, and constructed to survive in 15 +Gz conditions with high +Gz transients. The NIRS optical transducer "probe" consisted of three laser diodes (wavelengths 750 nm, 780 nm, and 810 nm) and a photodetector 5.4 cm away from the laser light source. The laser diodes and the photodetector were incorporated into a flexible material, which was placed directly on the subject without the use of fiber optic light guides. An algorithm based on a multi-wavelength modified Beer-Lambert Law was used to calculate relative hemoglobin concentration changes, ΔHb, ΔHbO_2, and ΔTotalHb. The differential pathlength factor (DPF_{807}) was selected to be 6.09 [11].

3. RESULTS

Nine subjects participated in the NIRS +Gz acceleration study; eight male and one female. NIRS data was recorded from G-LOC episodes resulting from the *G- plateau* protocol (n=11), and *G-pulse* protocol (n=12). The subjects were unconscious for 3--10 seconds after the onset of G-LOC, and upon returning to consciousness, the subjects were disoriented for another 4--11 seconds.

Figure 1 is a representative example from the *G-plateau* protocol. For the measurement of relative changes in the NIRS parameters, a baseline is established at a steady-state point before the +Gz pulse in which ΔHb, ΔHbO_2, and ΔTotalHb are set to zero. As the +Gz pulse acceleration begins, a gradual decrease of ΔHb, ΔHbO_2, and ΔTotalHb is observed, with G-LOC occurring within a second of the maximum decrease of ΔHbO_2 and ΔTotalHb. When G-LOC is observed, the centrifuge is decelerated to a normal G level. After G-LOC, ΔHbO_2 and ΔTotalHb both increased rapidly for the first few seconds. Soon, ΔHbO_2 and ΔTotalHb reached a peak value, and then began to decrease. This observation can be described as the reactive hyperemic response phase. Eventually, the NIRS parameters returned to baseline about 1 minute after G-LOC (not shown in the figure).

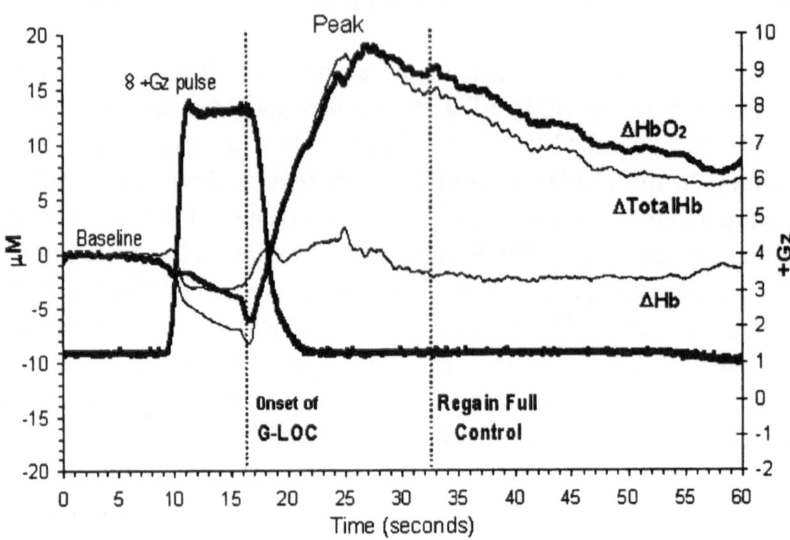

Figure 1. A representative example of the changes in ΔTotalHb, ΔHbO_2, and ΔHb during a 8 +Gz pulse. Event markers indicating the onset of G-LOC, and full recovery are shown.

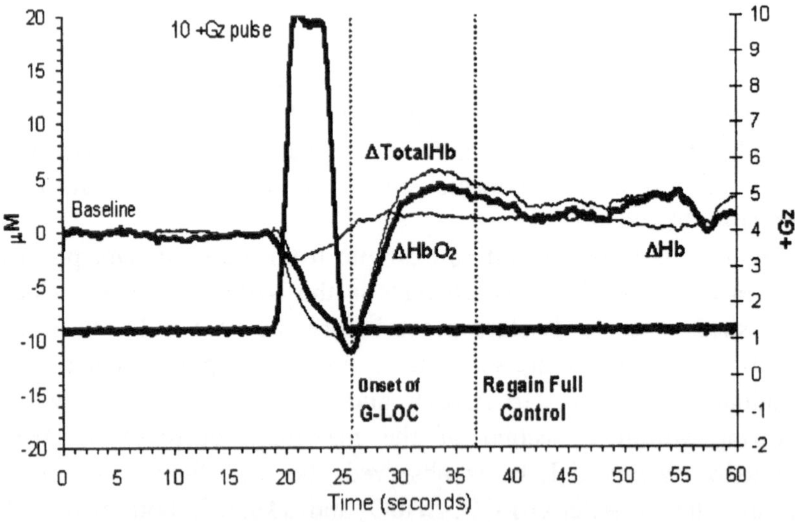

Figure 2. This figure is interesting because G-LOC occurred about 1 second after the completion of the +Gz pulse. ΔHbO_2 and ΔTotalHb continued to decrease until the onset of G-LOC.

Figure 2 is a representative example from the *G-pulse* protocol. This result is very interesting because G-LOC occurred about one second after the end of the +Gz pulse. ΔHbO_2 and ΔTotalHb continued to decrease after the +Gz pulse was completed, with a rise of ΔHb. The hyperemic response, in this case, was not very large. Data from the *G-pulse* protocol indicated that 6 of 12 G-LOCs occurred after the completion of the +Gz pulse, at an average of about 1 second after normal G-levels were restored.

Individual Subject Responses - The NIRS parameters of some subjects (n=5) consistently returned to baseline in 1 to 2 minutes, while other subjects (n=4) consistently took more than 3 minutes to return to baseline. During G-LOC recovery, subjects experienced twitching and tingling sensations in the extremities (n=5), "floating" sensations (n=3), brief dreams (n= 3), mood changes (n=3), hearing loss (n=2), and convulsions (n=1). One subject became upset and frightened from a nightmare after a G-LOC episode.

Analysis of G-LOC Episodes - G-LOC episodes induced by the *G-plateau* protocol were compared to the *G-pulse* protocol G-LOC episodes. The NIRS zero reference baseline was set a few seconds before the onset of the +Gz pulse. The values of the NIRS parameters were tabulated at three different events: 1) *Onset of G-LOC*; 2) *Hyperemic Response Peak* determined by the maximum values of ΔTotalHb and ΔHbO_2 during the hyperemic response recovery phase; and 3) *Total Recovery* of subject (end of relative incapacitation period – when subject deactivated the G-LOC beeper). The onset of G-LOC usually occurred at the maximum decrease of ΔTotalHb and ΔHbO_2. The NIRS parameter values were pooled together and averaged, grouped by the two protocols. The average and standard deviation for ΔTotalHb, ΔHbO_2, ΔHb, and $\{\Delta HbO_2\text{-}\Delta Hb\}$ for the three events are shown in the bar graphs of Figure 3. The averaged values of ΔTotalHb, ΔHbO_2, and $\{\Delta HbO_2\text{-}\Delta Hb\}$ appeared greater for the *G-plateau* protocol during *Event 2: Hyperemic Response Peak* and *Event 3: Total Recovery*, being nearly statistically significant ($p < 0.10$, two-tailed t-test) for ΔTotalHb (Event 2) and ΔHbO_2 (Events 2 and 3). The averaged $\{\Delta HbO_2\text{-}\Delta Hb\}$ values for the *G-plateau* protocol was significantly higher ($p < 0.05$, two-tailed t-test) compared to the *G-pulse* protocol during *Event 3: Total Recovery*. When comparing the averaged total incapacitation period (*Tincap*) of both protocols, the *G-plateau* protocol *Tincap* (15.9 sec.) was significantly higher ($p = 0.00003$) than the *G-pulse* protocol (9.6 sec.). In summary, the results showed that the *G-plateau* protocol generated higher *Tincap* times, as well as higher levels of ΔTotalHb, ΔHbO_2, and $\{\Delta HbO_2\text{-}\Delta Hb\}$ during the recovery phase when compared to the *G-pulse* protocol.

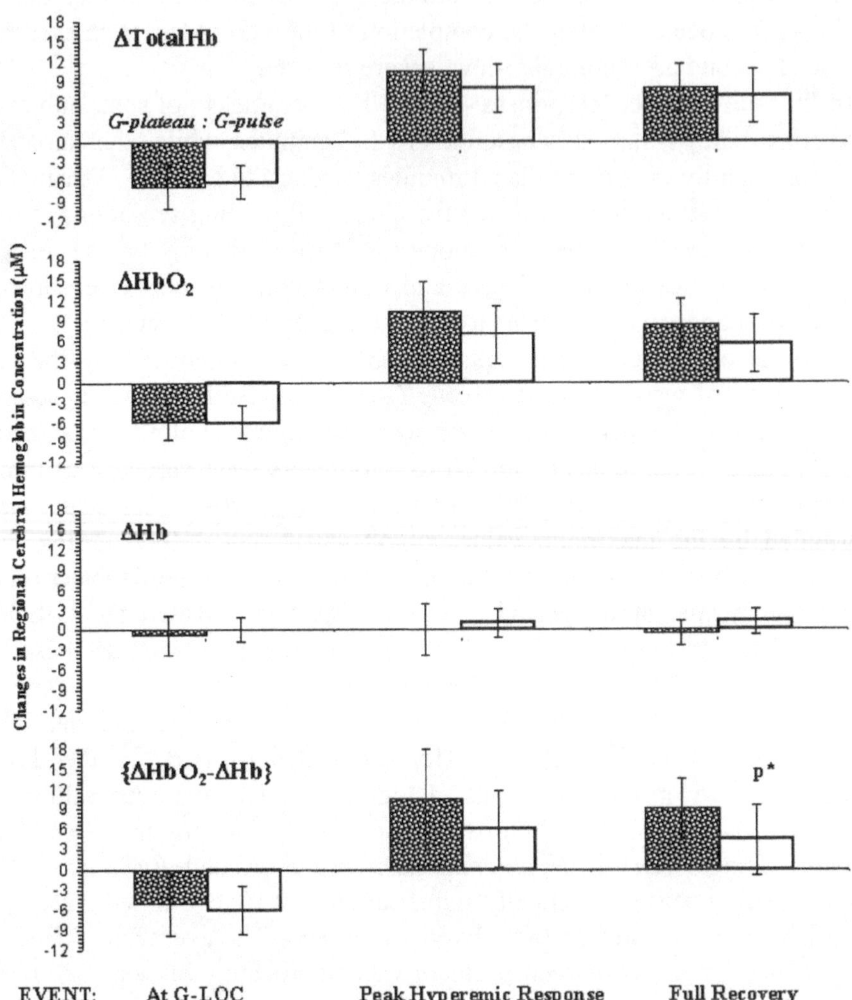

Figure 3. The bar graphs show averaged values with standard deviation of ΔHb, ΔHbO_2, $\Delta TotalHb$ and $\{\Delta HbO_2\text{-}\Delta Hb\}$ during three different G-LOC events divided by the *G-plateau* (n=11) and *G-pulse* (n=12) protocols. Averaged ΔHbO_2, $\Delta TotalHb$ and $\{\Delta HbO_2\text{-}\Delta Hb\}$ appeared greater during the G-LOC recovery phase for the *G-plateau* protocol. Statistical significance ($p < 0.05$, t-test) is indicated by p*.

4. DISCUSSION

Representative changes of ΔHb, ΔHbO_2, and ΔTotalHb due to a +Gz pulse are shown in Figures 1 and 2. For the first 1 to 2 seconds after the onset of the +Gz pulse, both ΔHbO_2 and ΔHb decreased, likely because of the expected drop of arterial and venous pressures in the brain from the high hydrostatic and orthostatic pressures generated during +Gz acceleration. Then, ΔHb appeared to level off and even increase in some cases during the plateau of the +Gz pulse, while ΔHbO_2 continued to decrease. The biphasic response of ΔHb may be due to the pooling of blood in the venous circulation of the lower extremities (orthostatic pressure effect), which may cause the jugular venous pressure to approach zero, causing this high compliant vessel to collapse. As a result, venous return from the brain would be severely reduced. After venous collapse, blood would be trapped in different regions in the brain, and the remaining oxygen available from the oxygenated hemoglobin would be metabolized, producing more deoxyhemoglobin [12].

The delayed G-LOC phenomenon is very interesting because G-LOC occurs after the end of the +Gz pulse. ΔHbO_2 and ΔTotalHb appeared to be continually decreasing at normal G levels. For all cases, the maximum decreased value of ΔHbO_2 occurred within a second of the onset of G-LOC. The causes of G-LOC occurring after the +Gz pulse may be due to an inadequate circulatory reflex response after a certain critical point is reached, when G-LOC is imminent. Optical monitoring studies of the stimulated brain demonstrated that the vascular response appeared 0.5 -- 3 seconds after the stimulus occurred [13-15]. We observed that G-LOC occurred within 2 to 3 seconds after the return to normal G levels for the delayed G-LOC cases. This suggests that the vascular recovery from high hydrostatic and orthostatic pressures generated during +Gz acceleration have a similar latency.

A rapid rise of ΔHbO_2 and ΔTotalHb was observed after the onset of G-LOC. After reaching their peak, ΔHbO_2 and ΔTotalHb began to gradually decrease. ΔHb usually did not change significantly when compared to ΔHbO_2 and ΔTotalHb, which indicated an increase in the volume of oxygenated blood in the cerebral microcirculation due to cerebral vasodilation. The rise in cerebral blood flow and blood volume, which is usually proportional to ΔTotalHb, is commonly known as the reactive hyperemic response. The magnitude and duration of reactive hyperemia usually depends on the magnitude of the tissue oxygen deficit preceding the response. Interestingly, the duration of hyperemia for some subjects, as indicated by elevated ΔHbO_2 and ΔTotalHb, consistently returned to baseline more quickly than others. Also, the hyperemic peak for some subjects was consistently higher. G-LOC dreaming episodes appeared to occur during the rapid increase of ΔHbO_2 and ΔTotalHb, since dreaming is believed to occur during the late absolute incapacitation period [3].

G-pulse vs. G-plateau Induced G-LOCs - The results showed that the *G-plateau* protocol generated a significantly higher total incapacitation period (*Tincap*) during G-LOC, as well as higher levels of ΔTotalHb, ΔHbO_2, and $\{\Delta HbO_2\text{-}\Delta Hb\}$, an index of oxygen saturation [16], during the G-LOC recovery phase when compared to the *G-pulse* protocol. A longer *Tincap* period is more dangerous because the pilot is not in control of the aircraft for a longer duration. In the experimental setting, the differences of *Tincap* may be due to the protocol design. The +Gz pulse of the *G-plateau* protocol is terminated when the onset of G-LOC is observed. The centrifuge takes about 2 to 3 seconds to decelerate to normal G-levels after manual +Gz pulse termination. The +Gz pulse of the *G-pulse* protocol terminated at a finite pulse width, incremented 0.25 to 0.5 seconds longer than the previous +Gz pulse. These findings suggest that high +Gz acceleration exposure, after the onset of G-LOC, extended the total incapacitation time. The higher levels of ΔTotalHb, ΔHbO_2, and $\{\Delta HbO_2\text{-}\Delta Hb\}$ during the G-LOC recovery phase may be due to increased compensation for neuron oxygenation deficit caused by the continued high G-levels after G-LOC. Therefore, the ability to decelerate aircraft to lower or normal G-levels as fast as possible after the onset of G-LOC will likely reduce *Tincap*, allowing the pilot to quickly regain full control of the aircraft.

In conclusion, NIRS provides an additional means of studying physiological mechanisms leading to and recovery from G-LOC.

ACKNOWLEDGMENTS

The authors would like to express their appreciation to the subjects who participated in the study, as well as to the NAWCAD centrifuge operation staff whose assistance was invaluable.

REFERENCES

1. Burton RR. G-Induced Loss of Consciousness: Definition, History, Current Status. Aviat Space Environ Med 1988;59:2-5.
2. Wood EH, Prevention of the Pathophysiologic Effects of Acceleration in Humans: Fundamentals and Historic Perspectives, IEEE Engg in Medicine and Biology, 1991; 10:26-35.
3. Whinnery JE. Observations on the Neurophysiologic Theory of Acceleration (+Gz) Induced Loss of Consciousness. Aviat Space Environ Med 1989; 60:589-93.

4. Whinnery JE, Whinnery AM. Acceleration Induced Loss of Consciousness – A Review of 500 Episodes, Arch Neurol 1990; 47:764-76.
5. Glaister DH. Current and Emerging Technology in G-LOC Detection: Noninvasive Monitoring of Cerebral Microcirculation Using Near Infrared. Aviat Space Environ Med 1988; 59:23-8.
6. Glaister DH, Jobsis-VanderVliet FF. A Near-Infrared Spectrophotometric Method for Studying Brain O_2 Sufficiency in Man During +Gz Acceleration. Aviat Space Environ Med 1988; 59:199-207.
7. Miyamoto Y, Mizumoto C, Shimizu K, Kobayashi A, Nakamura A, Yagura S. Evaluation of near infrared specvtrophotometric method for individual tolerance to +Gz acceleretaion. Reports of the Aeromedical Lab, JASDF 1992; 33 (4): 103-109.
8. Miyamoto Y, Mizumoto C, Shimizu K, Kobayashi A, Nakamura A, Yagura S. Quantification of concentration changes in human cerebral haemoglobin during +Gz acceleration: Preliminary results. Reports of the Aeromedical Lab, JASDF 1993;34: 13-17.
9. Miyamoto A, Nagaoka S, Konishi T, Suzuki H, Watanabe S, Ususi S, Kojima T. 1994. Haemodynamic measurement during parabolic flight. J.J. Aerospace Env. Med. 1994;31: 79-86.
10. Tripp LD, Chelette T, Savul S, Widman R. Female exposure to high G: effects of simulated combat sorties on cerebral and arterial O_2 saturation. Aviat Space Environ Med 1998;69:869-74.
11. Duncan A, Meek JH, Clemence M, Elwell CE, Tyszczuk L, Cope M, Delpy DT. Optical Pathlength Measurement on Adult Head, Calf and forearm and the Head of the Newborn Infant Using Phase Resolved Optical Spectroscopy, Phys Med Biol 1995;40:295-304.
12. Shahed AR, Barber JA, Galindo S, Werchan PM. Rat Brain Glucose and Energy Metabolites: Effect of +Gz (Head-to-Foot Inertial Load) Exposure in a Small Animal Centrifuge. J Cereb Blood Flow Metab 1995;15:1040-6.
13. Nemoto M, Nomura Y, Sato C, Tamura M, Houkin K, Koyanagi I, Abe H. Analysis of optical signals evoked by peripheral nerve stimulation in rat somatosensory cortex: dynamic changes in hemoglobin concentration and oxygenation. J Cereb Blood Flow Metab. 1999;19:246-59.
14. Malonek D, Grinvald A. Interactions between electrical activity and cortical microcirculation revealed by imaging spectroscopy: implications for functional brain mapping. Science. 1996; 272(5261):551-4.
15. Gratton E, Fantini S, Franceschini MA, Gratton G, Fabiani M. Measurements of scattering and absorption changes in muscle and brain. Philos Trans R Soc Lond B Biol Sci. 1997; 352(1354):727-35.
16. Brun NC, Moen A, Borch K, Saugstad O, Greisen G, Near-Infrared Monitoring of Cerebral Tissue Oxygen Saturation and Blood Volume in Newborn Piglets, Am J Physiol 1997;273(2 pt 2):H682-H686.

Chapter 35

CORRELATION OF NIRS DETERMINED CEREBRAL OXYGENATION WITH SEVERITY OF PILOT +Gz ACCELERATION SYMPTOMS

Paul B. Benni[1,2], John K-J. Li[1], Bo Chen[1,2], Joseph Cammarota[3], & David W. Amory[4]

[1]*Dept. Biomedical Engineering, Rutgers University, Piscataway, NJ 08854-8014, USA;* [2]*Dept. Anesthesia, UMDNJ-Robert Wood Johnson, New Brunswick, NJ 08903, USA;* [3]*EDO M-Technologies, Huntingdon Valley, PA 19006, USA;* [4]*Institute of Applied Physiology and Medicine, Seattle, WA 98122, USA.*

Abstract: Pilots commonly experience decreased peripheral vision, confusion & disorientation, and/or unconsciousness when exposed to high +Gz acceleration. We correlated NIRS determined ΔHb, ΔHbO_2, and ΔTotalHb with the resultant +Gz stress symptoms that subjects reported after experiencing a 6 to 10 +Gz amplitude pulse. During the hyperemic response phase following the +Gz pulses, an increase of the averaged peak values of ΔHbO_2 and ΔTotalHb as a function of the severity of the subjects' symptoms was observed. Significant increases were found for the averaged peak values of ΔHbO_2 and ΔTotalHb between high vision loss, confusion and disorientation while remaining conscious (A-LOC), and unconsciousness (G-LOC). The results suggest that the confusion and disorientation associated with A-LOC is physiologically based and that A-LOC is an intermediate +Gz stress symptom between high peripheral vision loss and G-LOC. Like G-LOC, pilots who experience A-LOC symptoms momentarily do not have full control of their aircraft.

Key words: A-LOC, Cerebral oxygenation, +Gz acceleration, +Gz stress, G-LOC (G-induced loss of consciousness), NIRS (Near-Infrared Spectroscopy)

1. INTRODUCTION

+Gz acceleration induced stress is a function of G level and time duration under high G. With either an increase in G level or an increase of time duration under high G, or both, the +Gz stress on the human body will increase. When the circulatory system compensatory reflexes become overwhelmed by hydrostatic and orthostatic pressure changes, the decreased perfusion pressure in the retina and brain will initially cause decreased peripheral vision [1]. As the +Gz stress continues, total vision loss and G-induced loss of unconsciousness (G-LOC) will then result (Figure 1).

It has been observed that episodes of confusion and disorientation during partial vision loss result in transient loss of control of aircraft. Often, the confusion and disorientation were thought to be psychological in origin and not have a physiological basis, similar to G-LOC. Hence, an intermediate phase between consciousness with total awareness and G-LOC has been identified and is defined as Almost Loss of Consciousness (A-LOC) [2]. A-LOC is believed to occur more frequently than G-LOC, based on reports by pilots who experienced brief episodes of confusion, amnesia, apathy, loss of situational awareness, muscle twitching, and weakness in the hands, during aerial combat maneuvers [2]. Our NIRS study will be one of the first physiological based studies that will examine A-LOC symptoms as separate and distinct from consciousness and G-LOC.

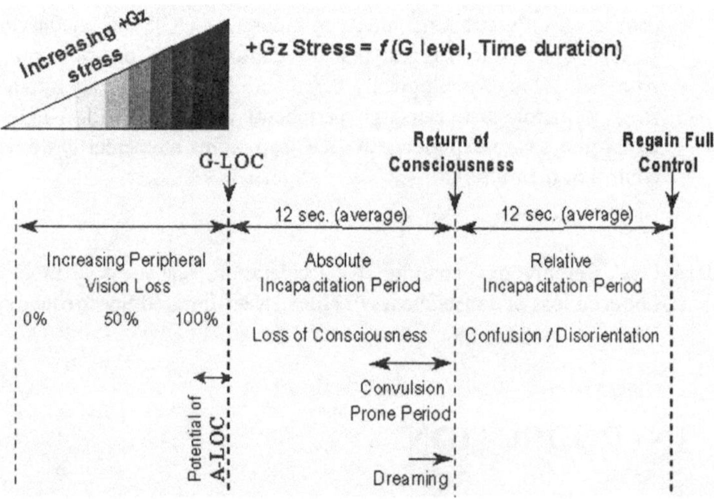

Figure 1. +Gz acceleration induced stress is a function of G level and time duration under high G. With either an increase in G level or an increase of time duration under high G, or both, the +Gz stress on the human body will increase. As +Gz stress increases, there will first be a decrease in peripheral vision, then total vision loss, and finally unconsciousness (G-LOC).

2. METHODS

These experiments were conducted at the U.S. Navy Air Warfare Center – Aircraft Division (NAWCAD) centrifuge facility in Warminster, PA. This study was carried out with the approval of the U.S. Navy G-Acceleration Committee on Human Experimentation and the Centrifuge Medical Response Council. Voluntary, fully informed written consent of the G-acceleration protocol was obtained from each subject. Medical personnel were present at all times during the experiments.

+Gz Acceleration Protocol - The *G-pulse* protocol consisted of a series of +Gz pulses starting at a pulse width of 0.25 seconds with increasing duration (1.0 sec., 2.0 sec., 2.5 sec...) until G-LOC or subject exhaustion had occurred (Figure 2). The onset and offset rate of the +Gz pulse was about 6 +Gz/second. The subject completed verbal mathematical tasks, deactivated the G-LOC recovery beeper upon returning to consciousness, and employed no anti-G suit or anti-G straining maneuvers. After each +Gz pulse run, the subject reported the level of vision loss, determined from a peripheral vision-loss scale inside the centrifuge, as well as any +Gz acceleration stress symptoms that were experienced. For each +Gz pulse, the level of peripheral vision loss that the subject reported was recorded as a percentage (i.e. 100% vision loss = 90 degrees peripheral vision loss or complete blackout). Observed +Gz stress symptoms of confusion and disorientation indicated that A-LOC had occurred. The G-pulse protocol trial ended when G-LOC or subject exhaustion occurred.

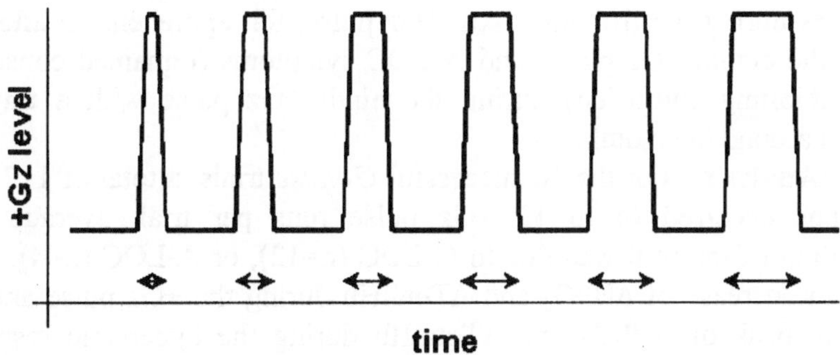

Figure 2. The *G-pulse* Protocol consisted of multiple +Gz pulses, starting with a pulse width of 0.25 sec. for the first pulse, and increased by 0.25 to 1 sec. for the following pulses until G-LOC or subject exhaustion. The +Gz level was set constant at 6, 8, or 10 +Gz for each +Gz pulse, with an onset & offset rate of ~ 6 Gz /second.

NIRS System - The NIRS system used in the +Gz acceleration studies was of our own design, and constructed to survive in 15 +Gz conditions with high +Gz transients. The NIRS optical transducer "probe" consisted of three laser diodes (wavelengths 750 nm, 780 nm, and 810 nm) and a photodetector 5.4 cm away from the laser light source. The laser diodes and the photodetector were incorporated into a flexible material, which was placed directly on the subject without the use of fiber optic light guides. An algorithm based on a multi-wavelength modified Beer-Lambert Law was used to calculate relative hemoglobin concentration changes, ΔHb, ΔHbO_2, and ΔTotalHb. The differential pathlength factor (DPF_{807}) was selected to be 6.09 [3].

3. RESULTS

Nine subjects participated in the NIRS +Gz acceleration study; eight males and one female. Data was recorded from 16 successful trials, consisting of a total of 117 +Gz pulses.

Figure 3 is a representative *G-pulse* protocol result consisting of ten consecutive +Gz pulse runs of increasing pulse width. ΔTotalHb, ΔHbO_2, and $\{\Delta HbO_2\text{-}\Delta Hb\}$ reached their lowest point at the third +Gz pulse (symptom: vision loss 85%) and the last +Gz pulse (symptom: G-LOC). The hyperemic recovery peak of ΔTotalHb and ΔHbO_2 rose steadily for the first few +Gz pulses and then seemed to level off, with a slight rise at G-LOC. ΔHb did not change to a large degree compared to ΔTotalHb and ΔHbO_2. The NIRS parameters returned to the zero reference baseline up to the eighth +Gz pulse, and then slightly shifted negatively below the zero reference baseline for the ninth and tenth +Gz pulse. The subject experienced hearing difficulties during the fifth and sixth +Gz pulse, felt apprehension after the sixth to the eighth +Gz pulse, and A-LOC symptoms (remained conscious while exhibiting confusion) during the ninth +Gz pulse with a tingling sensation around the mouth.

Data Analysis - For the 16 successful *G-pulse* trials, a total of 117 +Gz pulse runs occurred (4 to 13 +Gz pulse runs per trial, average ~7). Termination of the trial was due to G-LOC (n=12), or A-LOC (n=4). The maximum decrease of ΔHbO_2 and ΔTotalHb during the +Gz pulse and the maximum peak of ΔHbO_2 and ΔTotalHb during the hyperemic response recovery phase were determined from each of the +Gz pulses from the individual *G-pulse* trials. The NIRS determined hyperemic response peak was considered an indicator of the degree of cerebral oxygen deficit. ΔHb was considered secondary since this parameter did not change substantially during the +Gz acceleration pulses or the recovery phase. ΔHb values were determined at the time that the respective ΔHbO_2 and ΔTotalHb values were recorded by the expression $\Delta Hb = \Delta TotalHb - \Delta HbO_2$. Then $\{\Delta HbO_2\text{-}\Delta Hb\}$ was also determined. The NIRS values were entered into the data analysis, along with the subject's symptoms for each +Gz pulse. Of the 117 +Gz pulse runs, data from 4 runs exhibited movement artifact and were not included in the data analysis. Data points were placed in one of six categories, which were determined by the subjects' reported +Gz acceleration stress symptoms after each +Gz pulse. The six categories were as follows: 1) *No Vision Loss (VL 0%);* 2) *VL > 0 to 39%;* 3) *VL 40 to 79%;* 4) *VL 80 to 100%;* 5) *A-LOC (confusion / disorientation with 80 to 100% vision loss)*; 6) *G-LOC.* (VL = peripheral vision loss). Categories 4 and 5 attempted to compare 80 to 100% vision loss with and without A-LOC symptoms to determine any differences. The Student's two sample, two tailed t-test was employed between adjacent categories (4 vs. 5, 5 vs. 6, etc) to determine statistical significance ($p < 0.05$).

The results are shown in Figure 4 for ΔTotalHb, ΔHbO_2, ΔHb, and $\{\Delta HbO_2\text{-}\Delta Hb\}$. For each NIRS parameter, the average and standard deviation of the maximum decrease (during +Gz pulse) and maximum increase (peak value during hyperemic response) for each symptom category are shown in the bar graphs along with statistical significance ($p < 0.05$) indicated by p*. Also the comparison of categories 4 and 6, VL 80 to 100% and G-LOC, indicated significant differences of ΔHbO_2 and ΔTotalHb ($p<0.0001$), and $\{\Delta HbO_2\text{-}\Delta Hb\}$ ($p<0.05$).

4. DISCUSSION

The *G-pulse* protocol attempts to correlate the subjects' symptoms to changes in brain oxygenation in response to a series of +Gz pulses with increasing pulse width. The bar graphs of Figure 4 for the NIRS parameters ΔTotalHb, ΔHbO_2, ΔHb, and $\{\Delta HbO_2\text{-}\Delta Hb\}$ show some expected results, as well as some new results. For the hyperemic response phase after the +Gz pulses, a gradual increase of the averaged peak values of ΔHbO_2 and ΔTotalHb as a function of the subjects' symptoms is observed. Interestingly, significant differences ($p < 0.05$, two-tailed t-test) are shown for the peak values of ΔHbO_2, ΔTotalHb, and $\{\Delta HbO_2\text{-}\Delta Hb\}$ between the categories: *Vision loss 80 to 100%* and *A-LOC (with vision loss 80 to 100%)*. This result confirmed that the confusion and disorientation associated with A-LOC is physiologically based, not a psychological situation wherein subjects lose their concentration and focus. Otherwise, the results of the two categories would be expected to be similar. A-LOC probably occurs more frequently than G-LOC [2]. The symptoms of A-LOC are similar to the relative incapacitation period associated with G-LOC, but without the absolute incapacitation period (unconsciousness) associated with G-LOC. Based on analysis of NIRS determined hyperemic response peaks which

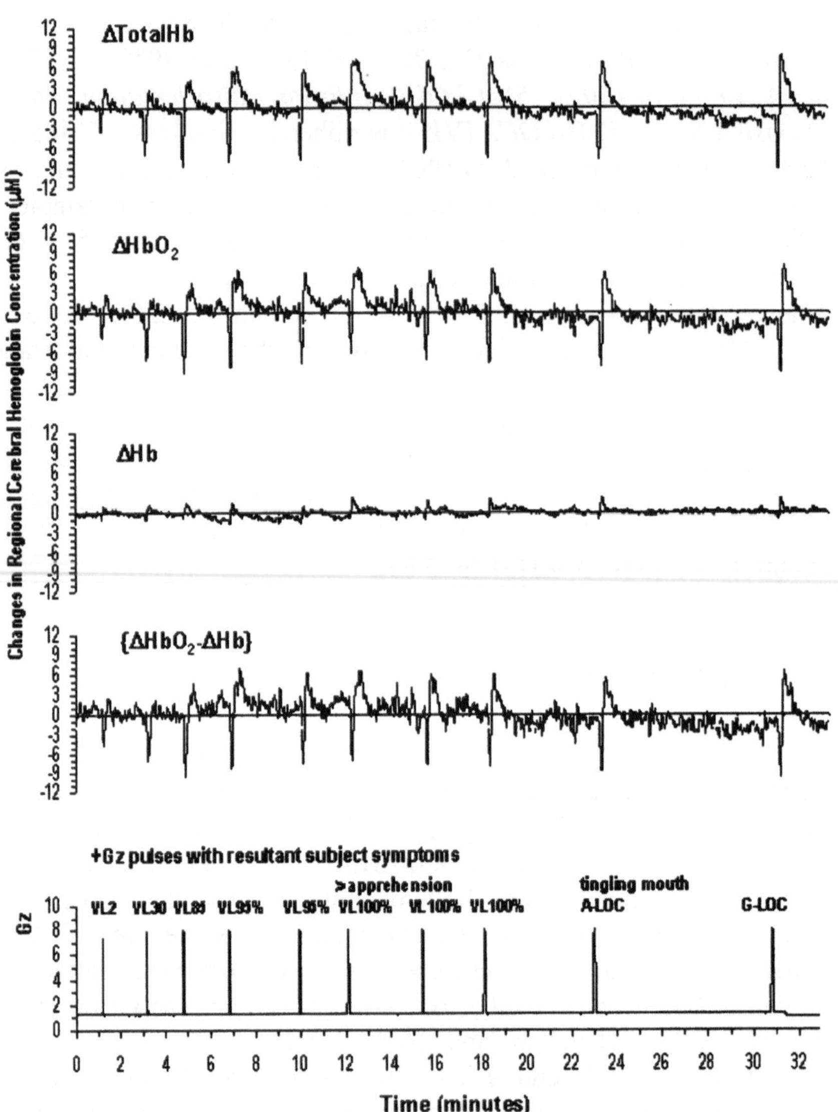

Figure 3. An example of the changes in ΔTotalHb, ΔHbO_2, ΔHb, and {ΔHbO_2-ΔHb} for a subject experiencing multiple +Gz pulses of increasing pulse width during the *G-pulse* protocol. This subject experienced apprehension after the 6th pulse, and a tingling mouth sensation during A-LOC. VL = vision loss.

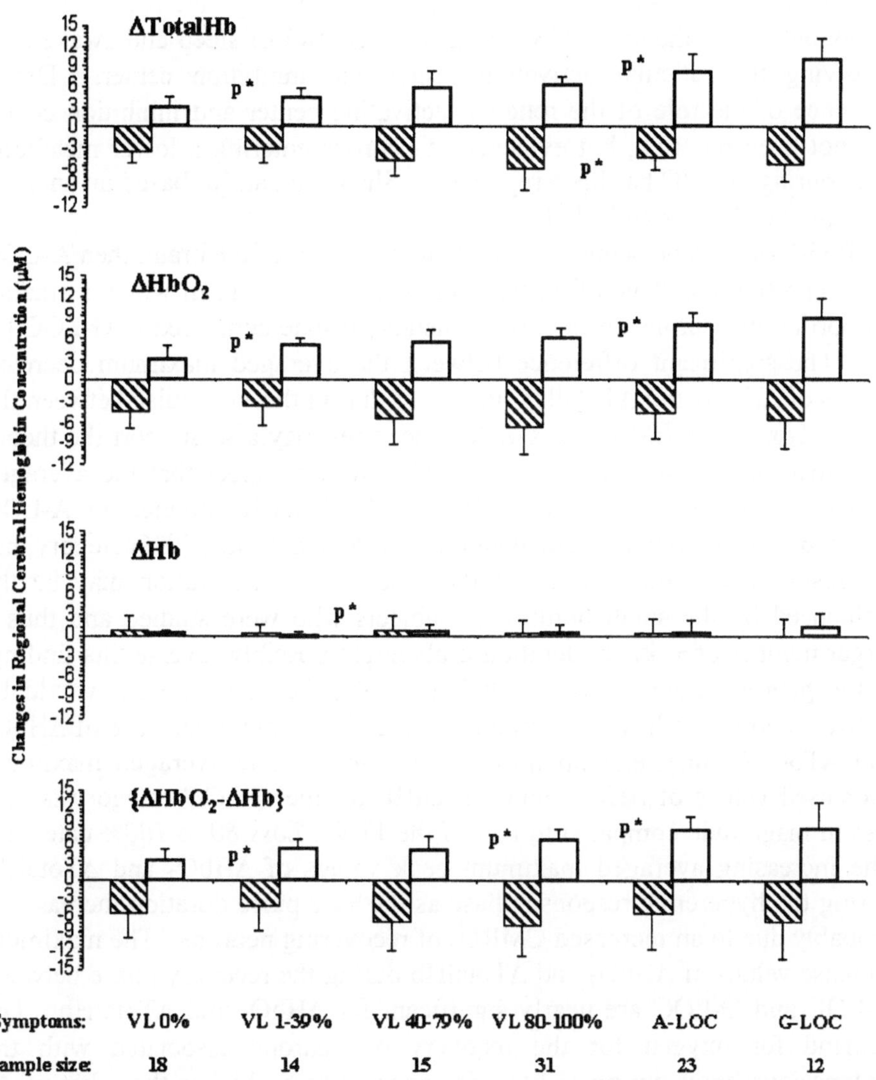

Figure 4. The bar graphs show the average and standard deviation of the maximum decrease (during +Gz pulse) and maximum increase (peak of recovery hyperemic response) of ΔHb, ΔHbO_2, ΔTotalHb, and $\{\Delta HbO_2\text{-}\Delta Hb\}$ categorized by the subjects' symptoms. The six categories are divided by ranges of vision loss (VL), A-LOC, and G-LOC. Significant differences between adjacent categories are denoted by p* ($p < 0.05$, two-tailed t-test).

correlated with +Gz symptoms, it appears that A-LOC is an intermediate +Gz stress response between high vision loss and G-LOC, as shown in Figure 1.

Whinnery [4,5] suggested that G-LOC is a brain protective mechanism analogous to the theory of CNS brain stem control of sleep and awakeness, involving the reticular activation center and inhibition center. Direct evidence of the role of the reticular activating center and inhibition center has not been reported, but evidence of neurons entering a lower metabolic state during G-LOC has been reported by Shahed et. al. [6] based on analysis of brain metabolites and EEG.

If G-LOC is represented by complete inhibition of the brain, then A-LOC may represent the stage of partial inhibition of the brain in which a smaller proportion of neurons enter a lower metabolic state compared to G-LOC [2, 7]. The significant difference between the averaged maximum decrease values of ΔHbO_2 and ΔTotalHb that occur during the +Gz pulse between the *Vision Loss 80 to 100%* and *A-LOC* categories may also support the theory of a metabolic change in the brain. One would expect that the averaged maximum decrease values of ΔHbO_2 and ΔTotalHb attained in A-LOC would be greater in magnitude than the *Vision Loss 80 to 100%* category, but the results show that this is not the case. This observation may be the influenced by the small number of subjects who were studied and thus a larger number of +Gz acceleration trials might possibly reverse this finding. If the neurons enter a lower metabolic state, then less oxygen would be utilized, and thus a lower magnitude averaged maximum decrease of ΔHbO_2 and ΔTotalHb may explain these observations. The averaged maximum decreased values of ΔHbO_2 and ΔTotalHb for the *G-LOC* category is also less in magnitude compared to that of the *Vision Loss 80 to 100%* category. The increasing averaged maximum peak values of ΔHbO_2 and ΔTotalHb during the hyperemic response phase as the +Gz pulse duration increases is probably due to an increased $CMRO_2$ of recovering neurons. The maximum increase values of ΔHbO_2 and ΔTotalHb during the recovery phase between A-LOC and G-LOC are nearly significant for ΔHbO_2 and ΔTotalHb. The demand for oxygen for the recovery of neurons associated with the unconscious brain during G-LOC is expected to be higher than that of A-LOC.

In conclusion, the use of NIRS has provided us with possible mechanisms that lead to the onset of G-LOC and recovery. Increasing duration of +Gz pulses appear to correlate with the magnitude of the hyperemic response and severity of +Gz stress symptoms experienced by pilots.

ACKNOWLEDGMENTS

The authors would like to express their appreciation to the subjects who participated in the study, as well as to the NAWCAD centrifuge operation staff whose assistance was invaluable.

REFERENCES

1. Whinnery JE, Whinnery AM. Acceleration Induced Loss of Consciousness – A Review of 500 Episodes, Arch Neurol 1990;47:764-76.
2. McGowan DG. "ALOC" - Almost Loss of Consciousness and Its Importance to Fighter Aviation. Aviat Space Environ Med 1997;68:632.
3. Duncan A, Meek JH, Clemence M, Elwell CE, Tyszczuk L, Cope M, Delpy DT, Optical Pathlength Measurement on Adult Head, Calf and forearm and the Head of the Newborn Infant Using Phase Resolved Optical Spectroscopy, Phys Med Biol 1995;40:295-304.
4. Whinnery JE. Observations on the Neurophysiologic Theory of Acceleration (+Gz) Induced Loss of Consciousness. Aviat Space Environ Med 1989;60:589-593.
5. Whinnery JE. Theoretical Analysis of Acceleration-Induced Central Nervous System Ischemia. IEEE Engg in Medicine and Biology 1991;10:41-45.
6. Shahed AR, Barber JA, Galindo S, Werchan PM. Rat Brain Glucose and Energy Metabolites: Effect of +Gz (Head-to-Foot Inertial Load) Exposure in a Small Animal Centrifuge. J Cereb Blood Flow Metab 1995;15:1040-1046.
7. Cammarota JP, Onaral B, State Transitions in Physiologic Systems: A Complexity Model for Loss of Consciousness, IEEE Trans Biomed Eng 1998; 45:1017-1023.

Chapter **36**

ALTERED GENE EXPRESSION FOLLOWING CARDIOPULMONARY BYPASS AND CIRCULATORY ARREST

Tatiana Zaitseva, Gregory Schears*, Jin Shen, Jennifer Creed, David F. Wilson and Anna Pastuszko

Departments of Biochemistry & Biophysics, University of Pennsylvania, School of Medicine and Anesthesiology & Critical Care, Children's Hospital of Philadelphia, Philadelphia, PA 19104*

Abstract: This study investigated the effects of normothermic cardiopulmonary bypass (CPB) and circulatory arrest (DHCA) on expression of specific genes in neonatal piglet brain. CPB was performed through the chest at 100 ml/kg/min for 2 hrs at 37°C. In the second group of animals, CPB was begun as described above and then animals were cooled to a nasopharyngeal/brain temperature of 18°C. When the brain temperature reached 18°C, the CPB circuit was turned off. After 60 min of circulatory arrest (DHCA), CPB was resumed at 100 ml/kg/min, and the piglets were rewarmed to a temperature of 36°C. In both groups, the animals remain sedated, paralyzed, mechanically ventilated, and continuously monitored throughout a four hour study period after CPB. Oxygen pressure in the microvasculature of the cortex was measured by oxygen dependent quenching of phosphorescence. The aRNA technique was used to assess mRNA steady-state levels in the brain tissue.

Control oxygen pressure (pre-bypass) was 61 ± 5 Torr and during CPB this decreased to 32 ± 7 Torr on the beginning of bypass and to 36 ± 5 Torr at the end of bypass. During the recovery period, cortical oxygenation steadily decreased, reaching 29 ± 8 Torr at the end of the four hours period. Cortical oxygen decreased during DHCA to near zero and during rewarming and recovery increased to 35 ± 6 Torr.

Measurements of gene expression following CPB revealed significantly increased levels of mRNA for NMDAR1, DARPP-32, CamKII, GluR1, and D1AR. DHCA caused changes similar to those for CPB in levels of mRNA for NMDAR1, DARPP-32, CamKII and GluR1. In contrast, DHCA caused significantly increased levels of mRNA for GluR6 and GABRB1. There was no significant alteration in the level of D1AR following DHCA.

Oxygen Transport to Tissue XXIV, edited by
Dunn and Swartz, Kluwer Academic/Plenum Publishers, 2003

The results showed that DHCA caused much larger alterations in gene expression in the critical metabolic signaling pathways tested than did CPB.

Key words: Cardiopulmonary bypass, brain oxygen, gene expression and newborn piglets

1. INTRODUCTION

In spite of the vast literature on the pathophysiology of bypass, little is known about how this procedure affects brain oxygenation or the mechanistic basis for the neurological deficit often observed following the procedure. Our earlier studies have shown that normothermic bypass and circulatory arrest cause significant decrease in cortical oxygenation, a decrease that could be responsible for brain dysfunction and/or injury [1]. One result of this oxygen deficiency could be altered expression of genes involved in metabolic pathways that sustain cellular homeostasis and function, leading to cellular dysfunction, and thereby long term brain injury, and even cell death. Preliminary data presented in this paper show changes in expression of selected genes that are directly or indirectly associated with dopaminergic system in the brain.

2. METHODS

2.1 Animal model

Newborn piglets, 2-4 days of age were premedicated with ketamine (20 mg/kg IM) and acepromazine (1 mg/kg IM), intubated, and placed on mechanical ventilation (Sechrist Infant Ventilator, model IV - 100 B). Anesthesia was maintained with fentanyl (100 mcg/kg IV bolus, and 50 mcg/kg/hr infusion) and neuromuscular blockade with pancuronium (0.3 mg/kg IV). The ventilator was set with a positive inspiratory pressure of 25 mm Hg and positive and expiratory pressure of 3 mm Hg. Respiratory rate and inspiratory oxygen fraction were titrated and maintained on arterio

$PaCO_2$ of 35 - 45 mm Hg and a PaO_2 of 200 - 300 mm Hg. Sodium bicarbonate (8.5 %) was used to maintain a base excess between -3 and 3 mmol/L. A jugular venous cannula and femoral arterial cannula were placed for collection of samples for measuring oxygen pressure, hemoglobin desaturation, pH and for monitoring blood pressure. The animal's temperature was monitored throughout the study by nasopharyngeal temperature probe (Yellow Spring Instrument, Inc., Yellow Spring, OH), cortical and deep brain temperature probes (Shiley, Inc., 15mm and 30mm myocardial probes). Temperature was maintained at 36°C except for the period of induced hypothermia. The head of animal was placed in a Kopf stereotaxic holder and an incision was made along the midline of the scalp. The scalp was removed to expose the skull and a hole approximately 5 mm in diameter made in the skull over one parietal hemisphere for measuring the oxygen pressure. After one hour of stabilization period CPB was performed. Sham operated animals served as the control. In all experiments, blood pressure, body temperature and respiratory rate were continuously monitored. Blood samples were routinely taken for the measurement of blood pH, $PaCO_2$ and PaO_2 using a Beckman blood gas machine. The cortical oxygen pressure was measured at regular intervals during the experiment. All animal use procedures were in strict accordance with the NIH Guide for the Care and Use of Laboratory Animals and have been approved by the local Animal Care Committee.

2.2 CPB technique and experimental protocol

The protocol and techniques used in this investigation were designed to duplicate those practiced in the clinical setting. The CPB circuit was consist of two Stocker Shiley Roller Pumps (Model 10 - 10-00, Shiley, Inc.), a membrane oxygenator and a biomedical 370 heat exchanger (Biomedicus). The pump was primed with whole donor blood and normal saline solution, to maintain a CPB hematocrit value between 20 - 25 %. Oxygen and carbon dioxide flows were used to maintain blood gases within preset limits using alpha stat blood gas management. For cardiopulmonary bypass, the common carotid artery and interior jugular vein were exposed through a right neck incision. After intravenous heparin (300 units/kg) was administered, an 8 French arterial and venous cannula was inserted to the level of the aortic arch and the right atrium, respectively. CPB was then begun. Full CPB flow was set at 100 ml/kg/min. The animals were maintained on flow CPB for 120 min and then were weaned off CPB for 240 min. For DHCA the brain temperature was decreased to 18°C (as described above) and then the CPB circuit was turned off. After 60 minutes of DHCA, the CPB was resumed at 100 ml/kg/min, and the piglet was be rewarmed as described above.

2.3 Measurements of oxygen by oxygen dependent quenching of phosphorescence

Cortical oxygen pressure was measured by the oxygen dependent quenching of phosphorescence [2,3]. This is a non-invasive optical method in which an oxygen sensitive phosphor (Oxyphor G2) was injected IV at approximately 2 mg/kg. The measurements were made with a frequency domain phosphorometer developed in t his laboratory (PMOD 5000, Oxygen Enterprises, Ltd.). The excitation light (635 nm) was modulated simultaneously at 60 frequencies from 100 to 40,000 Hz carried to the tissue through one branch of a bifurcated light guide and illuminated an approximately 5 mm diameter area. The phosphorescence (790 nm maximum) was returned through the second branch of the bifurcated light guide, filtered to remove the excitation light and the distribution of lifetimes calculated. The distribution of phosphorescence lifetimes was used to calculate the oxygen histogram for the sampled tissue volume. The reported oxygen pressures are the peak value of the histogram.

2.4 Brain preparation for the aRNA amplification analysis

At the end of experimental protocol, the chest of the anesthetized animals were opened and a tube was inserted into the ascending aorta to perform perfusion of the brain. The brains were perfused *in situ* first with 1 L saline and then with 2 L of 4% paraformaldehyde. Perfuse brains were removed and kept in 4% paraformaldehyde for at least 24 h. After fixation, brains were embedded in paraffin and cut into 6-µm sections. Slices was used for aRNA amplification. RNA amplification technique was described in details by Eberwine and coworkers [4,5].

2.5 Statistical evaluation

All values are expressed as means for n experiments ± SD. Statistical significance was determined using one-way analysis of variance with repeated measures by Wilcoxon signed-rank test. $p < 0.05$ was considered statistically significant.

3. RESULTS

3.1 Effect of normothermic bypass on physiological parameters and cortical oxygen pressure in newborn piglets

Table 1. Physiological parameters of newborn piglets during CPB and recovery

		Heart Rate (beats/min)	MABP (Torr)	pHa	$PaCO_2$ (Torr)	PaO2 (Torr)	PcO_2 (Torr)
Control		261±36	102±9	7.44±0.02	34±3	95±18	62±5
Bypass	0 h	159±66**	63±17**	7.55±0.06**	21±3.5**	91±60	33±12**
	1 h	179±35**	71±17**	7.29±0.12**	48±14*	79±18	35±8**
	2 h	191±50*	79±14**	7.38±0.08	37±5	87±35	36±9**
Post-bypass	1 h	200±31*	64±13**	7.23±0.12**	42±17	109±48	31±12**
	2 h	181±28**	55±10**	7.26±0.08**	41±14	97±33	25±11**

The data are the means ± SD for 7 experiments. *p<0.05; **p<0.01

During CPB the pH of the arterial blood decreased at one hour from 7.44 ±0.02 to 7.29 ± 0.12 ($p<0.01$) but after two hours is not significantly different form control. At the end of two the hours recovery period, blood pH decreased to 7.26 ± 0.08 ($p < 0.01$) (Tab 1). The blood pressure decreased from 102 ± 9 Torr (control) to 79 ± 14 Torr ($p < 0.01$) at the end of bypass and to 55 ± 10 Torr ($p < 0.01$) by the end of recovery. Control oxygen pressure (pre-bypass) was 61.8 ± 4.6 Torr and during CPB this decreased to 32.5 ± 11.6 Torr ($p < 0.01$) on the beginning of bypass and to 36.2 ± 9.4 Torr ($p < 0.01$) at the end of bypass. During the recovery period, cortical oxygenation steadily decreased, reaching 25.1 ± 11 Torr ($p < 0.01$) at the end of the two hour period.

Table 2. Physiological parameters of newborn piglets during DHCA and recovery

		Heart Rate	MABP (Torr)	pHa	$PaCO_2$ (Torr)	PaO_2 (Torr)	PcO_2 (Torr)
Control		238±13	102±13	7.38±0.09	55±6	105±12	62±4
Bypass	Pre cooling	268±12	60±7**	7.37±0.04	50±5.6	111±26	40±3**
	On-cooling	164±11*	65±3**	7.53±0.08	29±4.7**	194±49*	32±7**
	On-DHCA	0	NA	NA	14±2.6**	NA	2.5±1**
	Post-cooling	202±10	67±14*	7.39±0.01	34±3.6**	96.8±18	33±15**
Post-bypass	1 hour	194±11	58±11**	7.34±0.05	33±6**	87.8±27	39±9**
	2 hour	203±7	53±5**	7.34±0.06	31±7**	79±38	35±9**

The data are the means ± SD for 3 experiments. *p<0.05; **p<0.01

3.2 Effect of hypothermic circulatory arrest on physiological parameters and cortical oxygen pressure in newborn piglets

As shown in Table 2, pHa, $PaCO_2$ and PaO_2 for newborn piglets during DHCA were not significantly different from those during CPB. There was no detectable blood pressure during DHCA. During rewarming and the two hour recovery period, blood pressure was lower than prebypass (52.7 ± 4.6 Torr, $p < 0.01$). Cortical oxygen decreased during DHCA from 61.9 ± 4 Torr to 2.5±1 Torr ($p < 0.01$) and during rewarming and recovery increased to 34.8 ± 9 Torr ($p < 0.01$).

3.3 Effect of normothermic bypass and hypothermic circulatory arrest on expression of selected genes in striatum of newborn piglets

The changes in selected gene expression in striatum of newborn piglets following normothermic bypass and hypothermic circulatory arrest are shown on Figure 1. Normothermic CPB resulted in increased levels of mRNA for NMDAR1, 62% (p<0.005), DARPP-32, 47% (p<0.005), CamKII, 28% (p<0.005), GluR1, 75% (p<0.005) and D1AR, 23% (p<0.05).

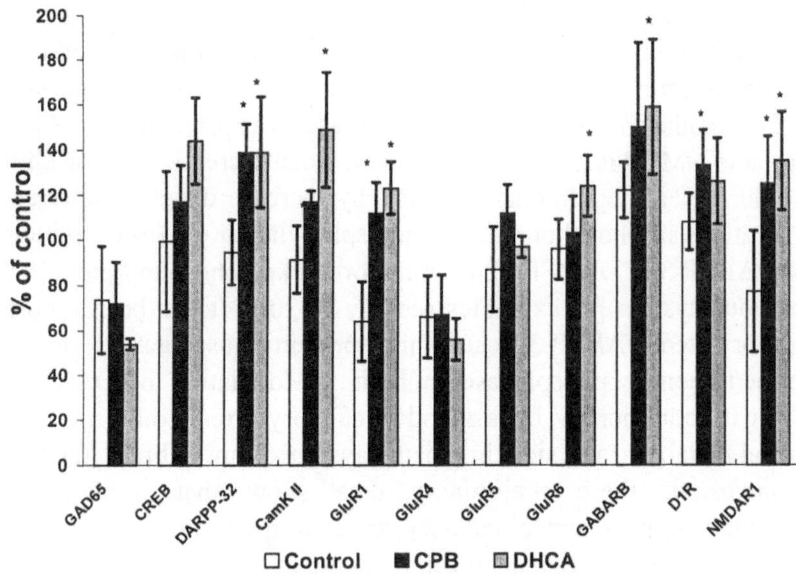

Figure 1. Effect of normothermic cardiopulmonary bypass and circulatory arrest of expression of selected genes in striatum of newborn piglets. The results are means ± SD for 6 control, 3 CPB and 3 DHCA experiments. * Represents statistic significance as described in the text.

DHCA caused increased levels of mRNA for NMDAR1, 75% ($p<0.05$), DARPP-32, 47% ($p<0.01$), CamKII, 63% ($p<0.005$), GluR1, 92% ($p<0.005$), GluR6, 29% ($p<0.05$) and GABRB1, 30%($p<0.05$).

There was also an increase in expression of CREB in both normothermic bypass and DHCA. However, because large variability, the changes were not statistically significant.

4. DISCUSSION

The present study investigated the changes in expression of selected genes in striatum of newborn piglets following normothermic bypass and hypothermic circulatory arrest. The genes which were chosen are directly or indirectly associated with dopamine. Our earlier results had shown that the dopaminergic system in striatum of newborn piglet brain is very sensitive to hypoxia/ischemia and that even small decreases in oxygen pressure causes statistically significant changes in dopamine release/uptake and metabolism [6-9]. Thus, any decrease in brain oxygenation caused by cardiopulmonary bypass can increase extracellular dopamine and affect signal transduction via the D1-like receptors and the intersecting pathway of glutamate-NMDA receptors (see for examples [10-13]. Our results show that normothermic bypass caused a significant increase in D1 like receptors and both circulatory arrest and normothermic bypass caused increase in expression of DARPP-32, GluR1, and NMDAR1. It can be postulated that increase in extracellular dopamine during cardiopulmonary bypass, by increase of gene expression, caused alterations in phosphorylation/dephosphorylation of several proteins, including DARPP-32. DARPP-32 is an important key phosphoprotein in the central nervous system [as examples see 11, 14-20]. It has been reported that phosphorylated DARPP-32 can inhibit protein phosphatase 1 (PP1), a major serine/threonine phosphatase in brain. Modulation of DARPP-32 activity by cardiopulmonary bypass and circulatory arrest could, therefore, affect several metabolic and signaling pathways important to brain function.

In conclusion, these preliminary data show that normothermic cardiopulmonary bypass and circulatory arrest triggers molecular changes that lead to metabolic disturbance in central nervous system of newborn.

ACKNOWLEDGEMENTS

Supported in part by grants HL-58669 and NS-31465.

REFERENCES

1. G Schears , J Shen, J Creed, T Zaitseva, DF Wilson, WJ Greeley, and A Pastuszko. Brain oxygenation during cardiopulmonary bypass and circulatory arrest. Adv. Exptl. Med. Biol. 510; 2003; 510: 325-330.
2. WL Rumsey, JM Vanderkooi, and DF Wilson, Imaging of phosphorescence: a novel method for measuring oxygen distribution in perfused tissue. Science 1988; 241:1649-1651.
3. SA Vinogradov, MA Fernandez-Seara, BW Dugan, and DF Wilson, Frequency domain instrument for measuring phosphorescence lifetime distributions in heterogeneous samples. Rev. Sci. Inst. 72 (8); 2001: 3396-3406.
4. J Eberwine, P Crino, and M Dichter, Single-cell mRNA amplification: implications for basic and clinical neuroscience. Neuroscientist 1995;1:200-211.
5. J Phillips and JH Eberwine, Antisense RNA amplification: a linear amplification method for analyzing the mRNA population from single living cells. A Companion to Methods in Enzymol 1996;10:283-288.
6. A Pastuszko, SN Lajevardi, J Chen, O Tammela, DF.Wilson, and M.Delivoria-Papadopoulos, Effects of graded levels of tissue oxygen pressure on dopamine metabolism in striatum of newborn piglets. J Neurochem 1993;60:161-166.
7. A.Pastuszko, Metabolic response of dopaminergic system during hypoxia in newborn brain. Biochem Med Metab Biol 1994; 51:1-15.
8. Ch-Ch Huang, NS Lajevardi, O.Tammela, A Pastuszko, M Delivoria-Papadopoulos, and DF.Wilson, Relationship of extracellular dopamine in striatum of newborn piglets to cortical oxygen pressure. Neurochem Res 1994; 19: 649-655.

9. M Yonetani, Ch-Ch Huang, SN Lajevardi, A Pastuszko, M Delivoria-Papadopoulos, and DF Wilson, Effect of hemorrhagic hypotension on extracellular level of dopamine , cortical oxygen pressure and blood flow in brain of newborn piglets. Neuroscience Letters 1994;180:247-252.
10. AA Fienberg, N Hiroi, PG Mermelstein, W-J Song, GL Snyder, A Nishi, A Cheramy, JP O'Callaghan, DB Miller, DG Cole, R Corbett, CN Haile, DC Cooper, SP Onn, AA Grace, CC Ouimet, FJ.White, SE Hyman, DJ Surmeier, J-A Girault, EJ Nestler, and P Greengard, DARPP-32: regulator of the efficacy of dopaminergic neurotransmission. Science 1998;281: 838-842.
11. P Greengard, PB Allen, and AC Nairn, Beyond the dopamine receptor: the DARPP-32/protein phosphatase-1 cascade. [Review] Neuron 1999; 23:435-447.
12. P Greengard, AC Nairn, J-A Girault, CC Ouimet, GL Snyder, G Fisone, PB Allen, A Fienberg, and A Nishi, The DARPP-32/protein phosphatase-1 cascade: a model for signal integration. Brain Res Rev 1998; 26:274-284.
13. P Svenningsson, M Lindskog, C Ledent, M Parmentier, P Greengard, BB Fredholm, and G Fisone, Regulation of the phosphorylation of the dopamine- and cAMP-regulated phosphoprotein of 32 kDa in vivo by dopamine D1, dopamine D2, and adenosine A2A receptors. PNAS, 2000;97:1856-1860.
14. GL Snyder, PB Allen, AA Fienberg, CG Valle, RL Huganir, AC Nairn, and P Greengard, Regulation of phosphorylation of the GluR1 AMPA receptor in the neostriatum by dopamine and psychostimulants in vivo. J Neuroscience 2000; 20:4480-4488.
15. J. Flores-Hernandez, S. Hernandez, G.L. Snyder, Z. Yan, A.A. Fienberg, S.J. Moss, P. Greengard, and D.J. Surmeier, D(1) dopamine receptor activation reduces GABA(A) receptor currents in neostriatal neurons through a PKA/DARPP-32/PP1 signaling cascade. J Neurophysi 2000; 83: 2996-3004.
16. J.A. Bibb, G.L. Snyder, A. Nishi, Z Yan, L Meijer, A.A. Fienberg, L.H. Tsai, Y.T. Kwon, J.A. Girault, A.J. Czernik, R.L. Huganir, H.C. Hemmings, Jr., A.C. Nairn, and P. Greengard, Phosphorylation of DARPP-32 by Cdk5 modulates dopamine signalling in neurons. Nature 1999; 402 (6762):588-589, 669-671.
17. A. Nishi, G.L. Snyder, A.A. Fienberg, G. Fisone, A. Aperia, A.C. Nairn, and P. Greengard, Requirement for DARPP-32 in mediating effect of dopamine D2 receptor activation. European Journal of Neuroscience 1999a;11: 2589-2592.
18. A. Nishi, G.L. Snyder, A.C. Nairn, and P. Greengard, Role of calcineurin and protein phosphatase-2A in the regulation of DARPP-32 dephosphorylation in neostriatal neurons. J Neurochem 1999b; 72:2015-2021.
19. M. Lindskog, P. Svenningsson, B.B. Fredholm, P Greengard, and G. Fisone, Activation of dopamine D2 receptors decreases DARPP-32 phosphorylation in striatonigral and striatopallidal projection neurons via different mechanisms. Neuroscience 1999; 88:1005-1008.
20. G.L. Snyder, A.A. Fienberg, R.L. Huganir, and P Greengard, A dopamine/D1 receptor/protein kinase A/dopamine- and cAMP-regulated phosphoprotein (Mr 32 kDa)/protein phosphatase-1 pathway regulates dephosphorylation of the NMDA receptor. Journal of Neuroscience 1998;18:10297-10303.
21. C.C. Ouimet, K.C. Langley-Gullion, and P. Greengard, Quantitative immunocytochemistry of DARPP-32-expressing neurons in the rat caudatoputamen. Brain Research 1998; 808:8-12.

Chapter **37**

DOMINANT EVENTS THAT MODULATE MASS TRANSFER COEFFICIENT OF OXYGEN IN CEREBRAL CORTEX

Fahmeed Hyder[#,%], Ikuhiro Kida[#], Kevin L. Behar[*], Richard P. Kennan[#,+], and Douglas L. Rothman[#,%]
Magnetic Resonance Research Center
Departments of [#]Diagnostic Radiology, [%]Biomedical Engineering, and [*]Psychiatry
Yale University School of Medicine, New Haven, CT, USA
[+]currently at Department of Medicine, Albert Einstein College of Medicine, Bronx, NY, USA

Abstract: Recently, a model of cerebral oxygen delivery was described (*J Appl Physiol* 85:554) which yields a relationship similar to that used to depict substrate transport across the endothelium. Because the endothelium is not a diffusion barrier for oxygen, the permeability surface area product was replaced by an effective mass transfer coefficient term for oxygen, *D*. The cerebral metabolic rate of oxygen utilization (CMR_{O2}) was linked to cerebral blood flow (CBF) and volume (CBV) through properties that modify the vessel-to-tissue oxygen tension giving rise to changes in *D*. Changes in the value of *D* were correlated with changes in CBF, CMR_{O2}, and CBV as measured using NMR methods in a 48 μL volume of the cerebral cortex of anesthetized rats at different levels of activity. We conclude that the changes in total vascular volume (i.e., swelling or shrinking of the capillary bed) contributes <5% to changes in *D*, whereas variations in the number of hematic vs. plasmatic capillaries, or intra-capillary stacking vs. unpacking of erythrocytes, or increase vs. decrease of dissolved oxygen in the tissue (i.e., processes which modify vessel-to-tissue oxygen tension) contribute(s) >95% to changes in *D*.

Key words: glucose, lactate, metabolism, perfusion, permeability, substrate transport

1. INTRODUCTION

Methods for assessing cerebral physiology – e.g., autoradiography, positron emission tomography (PET), magnetic resonance spectroscopy (MRS) and imaging (MRI) – have provided extensive measures of parameters that are linked to the phenomenon of neurovascular-metabolism coupling [1]. The century old hypothesis of Roy and Sherrington [2] conveys the same type of neurovascular-metabolism coupling where the delivery of nutrients is locally adjusted through local regulation of cerebral blood flow (CBF) and volume (CBV) to meet the local metabolic need for oxygen (i.e., CMR_{O2}).

In resting awake and resting anesthetized conditions, steady-state measurements have revealed a direct and stoichiometric-like relationship between local perfusion and metabolism at spatial resolution of about 10^{-2} to 10^{4} μL [3]. However, in the past decade some controversy has been raised about the lack of this type of neurovascular-metabolism coupling within a dynamic range of cerebral activity [4]. For example, when cortical activity is altered from one steady-state level to another, is the degree of coupling between changes in CBF and CMR_{O2} or CBF and CBV maintained?

The literature shows that at steady-state, CBF and CMR_{O2} change in a stoichiometric-like manner from the resting anesthetized condition over a wide dynamic range which includes the resting awake condition [5]. Furthermore, functional PET studies in awake humans have revealed complementary focal changes in CBF and CMR_{O2} during sensory stimulation [5,6]. These studies taken together provide compelling support for the existence of stoichiometric-like coupling between changes in CBF and CMR_{O2} over a wide range of cortical activity, both above and below the resting awake condition. However, presence of similar data regarding CBF and CBV in the literature is scarce [7].

In the last few years, several mathematical models [5,8-10] have attempted to describe macroscopic cerebral oxygen delivery *in vivo*. These models are based upon different assumptions about the microscopic oxygen delivery properties of the capillary bed. The best prediction of the physiological transitions measured above and below the resting awake condition is a model [5] in which the oxygen delivery efficiency of the capillary bed increases with increased flow [11]. A term describing the effective mass transfer coefficient for oxygen through the capillary bed, D, is defined to link changes in CMR_{O2} to changes in CBF and CBV. The effective mass transfer coefficient for oxygen term, D, is theoretically related to the efficiency of oxygen transport from the capillary bed by a ratio Ω which is defined as $(\Delta D/D)/(\Delta CBF/CBF)$. The value of D is hypothesized to be dependent on various characteristics of cerebral vascular volume (which corresponds to CBV), vessel-to-tissue oxygen tension gradient (which is dependent on number of hematic vs. plasmatic capillaries, or intra-capillary stacking vs. unpacking of erythrocytes, or increase vs. decrease of dissolved oxygen in the tissue), and vessel number and geometry (which is associated with vascular morphology). While vascular morphology remains constant during a functional challenge, processes which modify the vessel-to-tissue oxygen tension gradient and CBV may be altered.

Therefore, in this study we determined the relationship between CBF and the value of *D*, and evaluated the primary factors that contribute to changes in the effective mass transfer coefficient for oxygen. Changes in CBV, CBF, and CMR_{O2} were measured using MRI and MRS over a wide range of cortical activity in rats maintained under morphine anesthesia. Based on the previous description of the model [5], we show that the effective mass transfer coefficient for oxygen, *D,* is given by the product of CBV and κ, where κ represents a lumped parameter that depicts the functional characteristics of the capillary bed that influences the vessel-to-tissue oxygen tension. The results suggest that it is the latter (i.e., processes which modify vessel-to-tissue oxygen tension) that contribute mostly to altering the effective oxygen diffusivity in cerebral tissue.

2. METHODS

2.1 Animals and materials

Adult, male, Sprague-Dawley rats (n = 25; 120-250 g; fasted >16 hours) were tracheotomized under halothane (0.7-1.2%) anesthesia and artificially ventilated with a mixture of 70%/30% N_2O and O_2. A femoral artery was cannulated for periodic blood sampling (for measurements of pCO_2, pO_2, pH, and glucose) and continuous blood pressure monitoring. Femoral veins were cannulated for infusions of nicotine hydrogen tartrate, iron oxide contrast agent (AMI-227; Advanced Magnetics Inc., Cambridge, MA), and D-[1-^{13}C]glucose (99 atom %; Cambridge Isotopes, Andover, MA). An intra peritoneal catheter was placed for delivery of *D*-tubocurarine chloride (initial 0.5 mg/kg; supplemental 0.25 mg/kg/30 min) and morphine sulfate (initial 50 mg/kg; supplemental 30 mg/kg/30min). Other details of animal preparation and radio-frequency (RF) probe placement have been described [11,12]. Halothane use was discontinued after the rat was positioned in the magnet and anesthesia was maintained thereafter with morphine. *In vivo* experiments were conducted on rats before and after administration of pharmacological agents to decrease or increase cortical activity relative to the reference condition. Three different levels of brain activity were studied: (A) reference condition of morphine sulfate only, (B) lower activity condition of morphine sulfate plus sodium pentobarbital (initial 30 mg/kg; supplemental 5 mg/kg/30 min), and (C) higher activity condition of morphine sulfate plus nicotine hydrogen tartrate (dose of 4 mg/kg; rate of

16.7 μL/min). Protocols for these experiments have been described [11,13,14].

2.2 *In vivo* MRI and MRS measurements

All *in vivo* MRI and MRS data were obtained on an extremely modified 7 Tesla Bruker Biospec I horizontal-bore spectrometer (Bruker Instruments, Billerica, MA) operating at 300.4 and 75.6 MHz for ^1H and ^{13}C, respectively. An 8 cm diameter ^1H resonator RF transmit coil was used for homogeneous excitation and a 10 mm diameter ^1H RF surface-receiver coil for local reception. This ^1H RF coil arrangement allows better shimming, minimizes sensitivity loss in the receiver coil, and results in high sensitivity. A concentric 20 mm diameter ^{13}C RF surface-coil was used for transmission and decoupling for proton observed carbon edited (POCE) experiments for CMR_{O2} measurements (see below). Multi-slice, gradient-echo coronal images were acquired for the placement of a $7.5\times1.6\times4.0$ mm^3 rectangular volume for the POCE experiments. Imaging parameters were: image matrix = 128×128; in-plane resolution = 156×156 μm^2; slice thickness = 500 μm; repetition time (TR) = 250 ms; echo time (TE) = 20 ms; inversion recovery time (TIR) = 300 ms. The static magnetic field homogeneity of a region of $8\times2\times5$ mm^3 volume in the sensorimotor cortex was manually shimmed prior to data acquisition. The localized MRS pulse sequence has been described [11,12,15]. All CBF and CBV data were acquired with an echo-planar imaging (EPI) method using sequential sampling [16] with coronally orientated multi-slice acquisitions of 1 mm slice separation (image matrix = 32×32; in-plane resolution = 430×430 μm^2; slice thickness = 1000 μm; TR > 5000 ms).

Relative CBV values from the reference condition to lower or higher activity were measured by administration of a high susceptibility contrast agent to enhance blood volume induced MRI signal changes. Spin echo weighted EPI data, acquired with multiple TEs ranging from 10 to 80 ms, were fitted by a single-exponential to give the transverse relaxation rate, R2. Blood volume susceptibility was raised through serial injections of an iron oxide contrast agent (2 mg/kg/0.9 cc bolus), which remains in the intravascular space for several hours [17]. The relative changes in CBV were calculated according to $\Delta CBV/CBV = (\Delta R2^{w} - \Delta R2^{w/o})/(R2^{w} - R2^{w/o})$, where $R2^{w}$ and $R2^{w/o}$ are the rates during the reference condition (A) with and without agents, respectively, and $\Delta R2^{w}$ and $\Delta R2^{w/o}$ are the rate differences as a consequence of transition from the reference condition (A) to altered conditions (B) or (C) with and without agents (see above), respectively. Since the superparamagnetic contrast agent was hypothesized to be distributed uniformly in blood plasma at a constant hematocrit, it was assumed that the derived changes in CBV based on changes in R2 mainly reflected changes in swelling or shrinking of the capillary bed.

CBF data were obtained using a spin echo inversion recovery EPI method [18] with multiple TIRs ranging from 200 to 2200 ms. A single-exponential recovery fit to the multi TIR data was used to create longitudinal relaxation rate, R1, data from two consecutive inversion recovery images corresponding to slice selective (R1s) and non-slice selective (R1n) maps. The value of R1 of blood water (R1b) of $0.5\pm0.1\ s^{-1}$ was obtained from a previous study at 7 Tesla [11]. CBF (in units of mL/g/min) was calculated using the relation $60\times(\lambda/(1+\varepsilon))(R1s - R1n)$, where λ is the brain-blood partition coefficient for water (0.95 mL/g) and ε is a correction coefficient given by ¾(1 R1b/R1n).

Details of the localized POCE have been described previously [11,12,19]. Each *in vivo* POCE time course was converted to a time course of ^{13}C fractional enrichment of C4-glutamate. The details of the metabolic modeling have been described [20]. Best-fits of the metabolic model to the C4-glutamate turnover data were used to obtain the tri-carboxylic acid cycle flux (V_{TCA}). CMR_{O2} (in units of μmol/g/min) was calculated according to $3\times V_{TCA}$ (see refs. [11,19,20]).

2.3 The model

The basis of the cerebral oxygen delivery model has been described [5]. The loss of oxygen from the plasma component of blood during transit, $C_P(\tau)$, is related to the first-order rate constant, k, of oxygen loss during capillary transit of blood

$$dC_T/d\tau = - k'C_P\,(1 - q) \qquad (1)$$

where C_T is the total oxygen in blood, $r = C_P/C_T$ over an elapsed time of $\Delta\tau$, q is related to the dissolved oxygen in the extravascular space in relation to C_P, and k' is related to the spatial gradients of oxygen tension in the blood and the abluminal side of the endothelium. Given that the ratio r is constant over the elapsed time of $\Delta\tau$, it can be shown that

$$C_T(\tau + \Delta\tau) = C_T(\tau)\exp(- kr\Delta\tau) \qquad (2)$$

where k is given by the product of k' and $(1 - q)$ and $\Delta\tau$ is assumed to be a fraction of the capillary bed circulation time, T_c, such that $\Delta\tau = T_c/n$, where $n>1$ and is an integer. Then for the whole circulation time T_c,

$$C_T(T_c) = C_T(0)\exp(- (T_c/(n/2))\Sigma[kr]) \qquad (3)$$

Oxygen extraction fraction (OEF) of the tissue from an infinitesimally thin volume element of blood in an elapsed time of $\Delta\tau$ [21] can be described as

$$\mathrm{OEF}(\tau + \Delta\tau) = 1 - C_{\mathrm{T}}(\tau + \Delta\tau)/C_{\mathrm{T}}(\tau) \quad (4)$$

where $C_{\mathrm{T}}(\tau)$ is the temporal profile of the changing total oxygen content of blood in transit. By integration over the whole circulation time $\mathrm{T_c}$ equation (4) converts to

$$\mathrm{OEF}(\mathrm{T_c}) = 1 - \mathrm{C_T}(\mathrm{T_c})/C_{\mathrm{T}}(0) \quad (5)$$

From equations (3) and (5), OEF through circulation time $\mathrm{T_c}$ is given by,

$$\mathrm{OEF}(\mathrm{T_c}) = 1 - \exp\left(-(\mathrm{T_c}/(n/2))\Sigma[kr]\right) \quad (6)$$

$\mathrm{T_c}$ may be related to the CBV and CBF

$$\mathrm{T_c} = \mathrm{CBV}/\mathrm{CBF} \quad (7)$$

which is based on total vascular volume in a particular neurovascular subunit as defined by the indicator "dilution" method and not the indicator "diffusion" method [22]. Substitution of equation (7) into (6) results in

$$\mathrm{OEF} = 1 - \exp\left(-(\mathrm{CBV}/\mathrm{CBF})/(n/2)\Sigma[kr]\right) \quad (8)$$

where the effective mass transfer coefficient for oxygen term D is defined as

$$D = (\mathrm{CBV}/(n/2))\Sigma[kr] = \kappa\,\mathrm{CBV} \quad (9)$$

κ is a lumped parameter that represents processes of vessels within the tissue all of which affect the vessel-to-tissue oxygen tension, such as conversion from plasmatic to hematic capillaries or *vice versa*, and/or intra-erythrocyte stacking or unpacking. By substitution of equation (9) in (8), OEF can be simply expressed

$$\mathrm{OEF} = 1 - \exp(-D/\mathrm{CBF}) \quad (10)$$

An alternate expression for OEF determined from Fick's relation [23]

$$\mathrm{OEF} = \mathrm{CMR_{O2}}/(\mathrm{CBF}\ C_{\mathrm{a}}) \quad (11)$$

where C_a is the average arterial oxygen concentration in the capillary bed. Combination of equations (10) and (11) shows that oxygen utilization is linked to perfusion

$$\text{CMR}_{O2} = (\text{CBF } C_a)\,(1 - \exp(-D/\text{CBF})) \tag{12}$$

The effective mass transfer coefficient for oxygen is assumed to be coupled to perfusion according to

$$\Omega = (\Delta D/D)/(\Delta\text{CBF}/\text{CBF}) \tag{13}$$

Similar relationships for CMR_{O2} vs. CBF and CBV vs. CBF can be expressed

$$\Psi = (\Delta\text{CMR}_{O2}/\text{CMR}_{O2})/(\Delta\text{CBF}/\text{CBF}) \tag{14}$$

$$\Lambda = (\Delta\text{CBV}/\text{CBV})/(\Delta\text{CBF}/\text{CBF}) \tag{15}$$

If the value of Ω is constant over the range of CBF changes, then equation (13) may be substituted into a relative version of equation (12) to yield

$$\Delta\text{CMR}_{O2}/\text{CMR}_{O2} = \alpha[1 - \exp\{-(D/\text{CBF})(\beta/\alpha)\}]/\text{OEF} - 1 \tag{16}$$

where $\alpha = (1 + \Delta\text{CBF}/\text{CBF})$ and $\beta = (1 + (\Delta\text{CBF}/\text{CBF})\Omega)$. Equation (13) can be expressed as

$$\Omega = (1 + (\Delta\text{CBF}/\text{CBF})^{-1})\,(\omega/\chi) - (\Delta\text{CBF}/\text{CBF})^{-1} \tag{17}$$

where $\omega = \ln[1 - \text{OEF}\,(\gamma/\alpha)]$, $\chi = \ln[1 - \text{OEF}]$, and $\gamma = (1 + (\Delta\text{CBF}/\text{CBF})\Psi)$.

2.4 Simulations

The values of Ψ and Λ were determined experimentally. Equation (17) was used to determine the value of Ω assuming OEF of 0.3 during the resting awake state [5]. The value of $\Delta D/D$ was determined from equation (13) using the calculated value of Ω and the experimental value of ΔCBF/CBF. Equation (9) was expanded as

$$\Delta\kappa/\kappa = (1 + (\Delta\text{CBF}/\text{CBF})\,\Omega)/(1 + (\Delta\text{CBF}/\text{CBF})\,\Lambda) - 1 \tag{18}$$

The extent of CBV and κ contributing to D were expressed as $\zeta = (\Delta CBV/CBV)/(\Delta D/D)$ and $\xi = (\Delta\kappa/\kappa)/(\Delta D/D)$. All results are shown as mean ± standard deviation.

3. RESULTS AND DISCUSSION

3.1 Measurements of CBF, CBV, and CMR_{O2}

We measured changes in CBF, CBV, and CMR_{O2} by *in vivo* MRI and MRS methods in rat sensorimotor cortex at different levels of cortical activity (n = 25). Values of changes in CBF and CBV were obtained (by averaging) from the same region in sensory cortex as the absolute CMR_{O2} measurements (i.e., 48 μL). Fig. 1 shows the relationship between fractional changes in CBF and CMR_{O2} and between CBF and CBV measured in rat sensorimotor cortex over a wide range of cortical activity. The values of Ψ and Λ (equations (14) and (15)) were 0.81±0.08 and 0.03±0.02, respectively. The current results are in good agreement with previous observations [5,11,13,14].

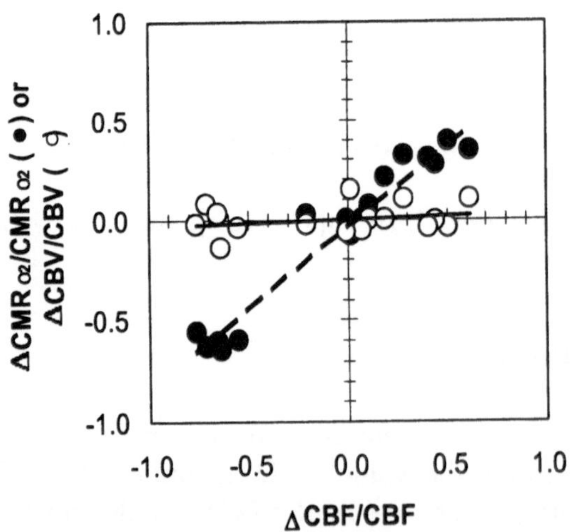

Figure 1. Relative changes CMR_{O2} (filled circles) and CBV (open circles) are plotted against changes in CBF. The best-fits of Ψ and Λ, as defined by equations (14) and (15), are shown by dashed and solid lines with values of 0.81±0.08 and 0.03±0.02, respectively. These results, which are in good agreement with previous studies [5,11,13,14], corresponded to a value of 0.75±0.14 for Ω, as defined by equation (13).

3.2 Changes in *D*

The value of the parameter Ω (equation (13)), which is determined by Ψ and Λ, was equal to 0.75±0.14, which is consistent with an earlier study [5,11]. The effective diffusivity parameter *D* (equation (9)) is equal to the product of κ and CBV, where κ represents a lumped parameter that depicts processes which influence the vessel-to-tissue oxygen tension. Using the relationship given by equation (18), the calculated fractional changes in κ and measured changes in CBV were plotted against changes in *D* (Fig. 2), where it was determined that the lumped parameter κ contributed the most to changes in *D* (98±8%), whereas the contribution of CBV to changes in *D* was much smaller (4±3%).

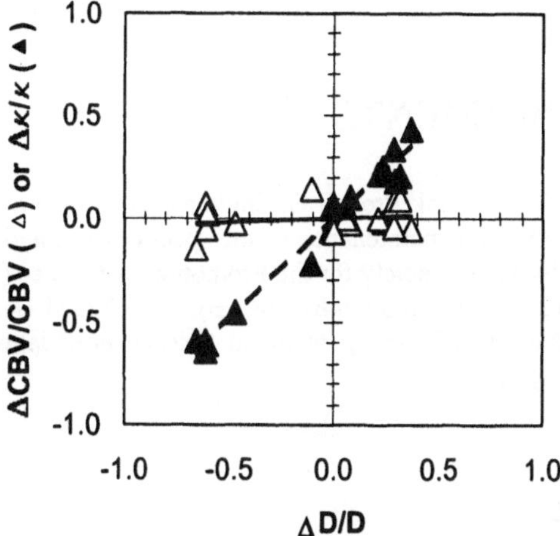

Figure 2. Relative changes in the calculated value of κ (filled triangles) and experimental value of CBV (open triangles) are plotted against calculated changes in *D* (see Fig. 1). The best-fits of ζ and ξ (see Simulations), which describe the extent of CBV and κ contributing to *D* separately, are shown by solid and dashed lines with values of 0.04±0.03 and 0.98±0.08, respectively.

4. DISCUSSION

The changes in CBV (i.e., swelling or shrinking of the capillary bed), and consequently changes in circulation time distributions (e.g., ref. [24]), contribute negligibly (<5%) to changes in *D*, whereas the changes in the capillary bed profile parameter that influences the vessel-to-tissue oxygen tension, κ, must be the dominant contributor (>95%) to changes in oxygen diffusivity, *D*. Therefore, it is suggested that processes such as changes in the number of plasmatic and hematic capillaries, or erythrocyte unpacking against stacking, or alterations in the amount of dissolved oxygen in the tissue, underlie changes in the effective diffusivity of oxygen within cerebral tissue (e.g., ref. [5]).

5. CONCLUSION

While processes such as changes in the number of plasmatic and hematic capillaries [25,26] and erythrocyte unpacking against stacking [27,28] are more easily detected with fluctuations in physiology, changes in the amount of dissolved oxygen in the tissue [29,30] are more difficult to detect. If the amount of dissolved oxygen in cerebral tissue is altered during functional activity, then it is unlikely that oxygen transport from the blood to the brain is diffusion limited [8].

ACKNOWLEDGEMENTS

The authors would like to thank Prof. R.G. Shulman for helpful discussions, T. Nixon, P. Brown, and S. McIntyre for maintenance of the spectrometer, and B. Wang for technical support. Supported by Japan Society for the Promotion of Science (IK); National Institutes of Health: NS-032126 (DLR), NS-034813 (KLB), HD-032573 (KLB), NS-037203 (FH), DC-003710 (FH), MH-067528 (FH); National Science Foundation: DBI-9730892 (FH) and DBI-0095173 (FH).

REFERENCES

1. Shulman RG, Rothman DL, Hyder F (1999) Stimulated changes in localized cerebral energy consumption under anesthesia *Proc Natl Acad Sci USA* 96:3245-3250
2. Roy CS, Sherrington CS. On the regulation of the blood supply of the rat brain. J Physiol (Lond*)* 1890;11:85-108.
3. Sokoloff L. Sites and mechanisms of function-related changes in energy metabolism in the nervous system. Dev Neurosci 1993;15:194-206.
4. Raichle ME. Behind the scenes of functional brain imaging: a historical and physiological perspective. Proc Natl Acad Sci USA 1998;95:576-772.
5. Hyder F, Shulman RG, Rothman DL. A model for the regulation of cerebral oxygen delivery. J Appl Physiol 1998;85:554-564.6. Hyder F, Shulman RG, Rothman DL. Regulation of cerebral oxygen delivery. Adv Exp Med Biol 1999;471:99-109.
7. Kuschinsky W. Capillary perfusion in the brain. Pflugers Arch 1996;432:R42-R46.
8. Buxton RB, Frank LR. A model of the coupling between cerebral blood flow and oxygen metabolism during neural stimulation. J Cereb Blood Flow Metab 1997;17:64-72.
9. Gjedde A. The relation between brain function and cerebral blood flow and metabolism in: *Cerebrovascular Disease* Batjer HH, Ed (Lippincott-Raven, Philadelphia) 1997; pp. 23-40
10. Hudetz AG. Mathematical model of oxygen transport in the cerebral cortex. Brain Res 1999;817:75-83.

11. Hyder F, Kennan RP, Kida I, Mason GF, Behar KL, Rothman DL. Dependence of oxygen delivery on blood flow in rat brain: A 7 Tesla NMR study. J Cereb Blood Flow Metab 2000;20:485-498.
12. Hyder F, Renken R, Rothman DL. *I*n vivo *c*arbon *e*dited *d*etection with *p*roton *e*cho *p*lanar *s*pectroscopic *i*maging (ICED PEPSI): [3,4-$^{13}CH_2$]glutamate/glutamine tomography in rat brain. Magn Reson Med 1999;42:997-1003.
13. Kida I, Hyder F, Kennan RP, Behar KL. Towards absolute quantitation of BOLD functional MRI. Adv Exp Med Biol 1999;471:681-689.
14. Kida I, Kennan RP, Rothman DL, Behar KL, Hyder F. High-resolution CMR_{O2} mapping in rat cortex: a multi-parametric approach to calibration of BOLD image contrast at 7 Tesla. J Cereb Blood Flow Metab 2000;20:847-860.
15. Hyder F, Petroff OA, Mattson RH, Rothman DL. Localized ^{1}H NMR measurements of 2-pyrrolidinone in human brain in vivo. Magn Reson Med 1999;41:889-896.
16. Hyder F, Rothman DL, Blamire AM. Image reconstruction of sequentially sampled echo-planar data. Magn Reson Imaging 1995;13:97-103.
17. Kennan RP, Scaley BE, Gore JC. Physiologic basis for BOLD MR signal changes due to hypoxia/hyperoxia: Separation of blood volume and magnetic susceptibility effects. Magn Reson Med 1997;37:953-956.
18. Schwarzbauer C, Morrissey SP, Haase A. Quantitative magnetic resonance imaging of perfusion using magnetic labeling of water proton spins within the detection slice. Magn Reson Med 1996;35:540-546.
19. Hyder F, Rothman DL, Mason GF, Rangarajan A, Behar KL, Shulman RG. Oxidative glucose metabolism in rat brain during single forepaw stimulation: a spatially localized ^{1}H[^{13}C] nuclear magnetic resonance study. J Cereb Blood Flow Metab 1997;17:1040-1047.
20. Hyder F, Chase JR, Behar KL, Mason GF, Siddeek M, Rothman DL, Shulman RG. Increase tricarboxylic acid cycle flux in rat brain during forepaw stimulation detected with ^{1}H [^{13}C] NMR. Proc Natl Acad Sci USA 1996;93:7612-7617.
21. Crone C. The permeability of capillaries in various organs as determined by the "indicator diffusion" method. Acta Physiol Scand 1963;58:292-305.
22. Zierler KL. Theory of the use of arteriovenous concentration differences for measuring metabolism in steady-state and non steady-state methods. J Clin Invest 1961;40:2111-2125.
23. Siesjo BK (1978) *Brain Energy Metabolism* (New York, Wiley).
24. Rovainen CM, Woolsey TA, Blocher NC, Wang DB, Robinson OF. Blood flow in single surface arterioles and venules on the mouse somatosensory cortex measured with video microscopy, flourescent dextrans, non occluding flourescent beads, and computer-assisted image analysis. J Cereb Blood Flow Metab 1993;13:359-371.
25. Klitzman B, Duling BR. Microvascular hematocrit and red cell flow in resting and contracting striated muscle. Am J Physiol 1979;237:H481–H490.
26. Villringer A, Them A, Lindauer U, Einhaul K, Dirnagl U. Capillary perfusion in rat brain cortex. Circ Res 1994;75:55-62.
27. Mochizuki M. On the velocity of oxygen dissociation of human hemoglobin and red cell. Jpn J Physiol 1966;16:649–657.
28. Kleinfeld D, Mitra PP, Helmchen F, Denk W. Fluctuations and stimulus-induced changes in blood flow observed in individual capillaries in layers 2 through 4 of rat neocortex. Proc Natl Acad Sci USA 1998;95:15741–15746.
29. Gijsbers KJ, Melzack R. Oxygen tension changes evoked in the brain by visual stimulation. Science 1967;156:1392-1393.
30. Ances BM, Buerk DG, Greenberg JH, Detre JA. Temporal dynamics of the partial pressure of brain tissue oxygen during functional forepaw stimulation in rats. Neurosci Lett 2001;306:106-110.

Chapter **38**

RELATIONSHIP BETWEEN REDOX BEHAVIOR OF BRAIN CYTOCHROME OXIDASE AND NEUROLOGICAL PROGNOSIS

Yasuyuki Kakihana, Tamotsu Kuniyoshi, Sumikazu Isowaki, Kazumi Tobo, Etsuro Nagata, Naoko Okayama, Kouichirou Kitahara, Takahiro Moriyama, Takeshi Omae, Masayuki Kawakami, Yuichi Kanmura, *Mamoru Tamura

*Division of Intensive Care Medicine, Kagoshima University Hospital, 8-35-1 Sakuragaoka, Kagoshima 890-8520, Japan; *Biophysics Group, Research Institute for Electronic Science, Hokkaido University, Kita12, Nishi 6, Kita-ku, Sapporo 060, Japan*

Abstract: Currently, no on-line method of assessing cerebral oxygenation is sufficiently accurate to be clinically helpful. In an attempt to find a good predictor of postoperative cerebral outcome, we retrospectively studied the relationship between the redox behavior of cytochrome oxidase (cyt. ox.) during an operation and the neurological prognosis in 83 patients who underwent thoracic aortic surgery. Our data revealed three patterns of change in the redox behavior of cyt. ox. during the operation; the actual pattern exhibited by a given patient showed a highly significant correlation with the neurological prognosis ($p<0.0001$). We conclude that the redox behavior of cyt. ox. during an operation is likely to be a good predictor of postoperative cerebral outcome, which implies that brain tissue oxygen sufficiency can be evaluated by near-infrared measurement of cytochrome oxidase (except for that in local regions far from the monitoring site).

Key words: Cerebral monitoring, Near-infrared spectroscopy, Cytochrome oxidase, Hemoglobin oxygenation, Cardiopulmonary bypass

1. INTRODUCTION

Despite advances in the technology involved in cardiopulmonary bypass (CPB) and separate cerebral perfusion (SCP), as well as in surgical and anesthetic methods, the fact still remains that some patients undergoing cardiac or thoracic aortic surgery suffer an obvious and potentially catastrophic complication during the operative and perioperative periods: namely, neurological dysfunction. The presence of a level of cerebral perfusion that is inadequate to meet the cerebral metabolic demands during CPB has been incriminated as a major factor in this complication. Consequently, during CPB it is not the level of cerebral blood flow per se, but an appropriate balance between cerebral perfusion and cerebral oxygen consumption that is likely to be important. At present, however, no form of on-line analysis for assessing cerebral oxygenation is sufficiently accurate to be clinically helpful.

Near infrared spectroscopy (NIRS) has great potential as a noninvasive way of monitoring changes in hemoglobin oxygenation, blood volume, and the redox state of cytochrome oxidase (cyt. ox.) in the brain. However, the specificity and accuracy of the measurement of the redox state of cyt. ox. are still controversial [1, 2]. In this paper, to evaluate whether the redox state of cyt. ox. is an accurate predictor of postoperative cerebral outcome, we retrospectively studied 83 patients who had undergone thoracic aortic surgery, and we examined the relationship between the redox behavior of cyt. ox. during the operation and the neurological prognosis.

2. METHODS

After institutional approval and informed consent had been obtained, we studied 83 patients (64.3 ± 11.3yr.; 54 male and 29 female) undergoing repair of a thoracic aortic aneurysm (TAA) with cardiopulmonary bypass (CPB) and/or selective cerebral perfusion (SCP) and/or deep hypothermic circulatory arrest (DHCA). Anesthesia was induced intravenously with fentanyl (2-5 μg/kg) and midazolam (0.05-0.1 mg/kg), and intubation was facilitated by the use of vecuronium bromide (0.1 mg/kg). Anesthesia was thereafter maintained with 0.4-1.5% isoflurane in air plus oxygen. Additional doses of fentanyl and vecuronium were given when necessary. A heart-lung machine (HAD-101; Mera, Tokyo, Japan) was used in non-pulsatile flow mode (2.0-3.0 L/min/m^2) together with a membrane oxygenator (SX10R; Terumo, Tokyo, Japan), and $Paco_2$ (uncorrected for temperature) was adjusted to normocapnic levels (α-stat regulation). Nasopharyngeal and rectal temperatures were continuously monitored throughout the operation.

Regional cerebral oxygenation was continuously monitored by NIRS (OM110; Shimadzu Inc., Kyoto, Japan) in all 83 patients. The two probes (light source and detector) of the NIRS system were placed 4 cm apart on one side of the forehead. NIR light was provided through four interference filters (700, 730, 750, and 805 nm) via one light guide, and the transmitted light was detected by the other light guide. Relative changes in the concentration of oxyhemoglobin (oxy-Hb) and deoxyhemoglobin (deoxy-Hb), and in the redox state of cytochrome oxidase (cyt. ox.) were calculated using the algorithm we described previously [2]. As the optical pathlength of the light is obscured because of the values obtained before starting any surgical procedures were taken as the baseline control values.

To test for significant differences among groups, we used the chi-square test for independence ($l \times m$ contingency table). A p value less than 0.05 was considered statistically significant.

3. RESULTS

When, in the entire group of patients, the redox behavior of cyt. ox. during the operation was retrospectively assessed in detail, we realized that there were three different types of cyt. ox. behavior. In terms of time course, we classed these as: (1) no change (type-A), (2) a transient reduction with a subsequent return to the control level (type-B), or (3) a marked and prolonged reduction (type-C) (Fig. 1).

Figure 2 shows the changes in cerebral oxygenation in a case representative of type-A; 37 of the 83 cases (45%) were of this type. The characteristics of the type-A pattern are that the redox state of cyt. ox. remains at its initial level (no change) throughout the operation even though marked changes occur in [oxy-Hb], [deoxy-Hb], and [total-Hb] at the initiation of CPB, and during the periods of hypothermia and rewarming. In the type-B pattern, a transient reduction of cyt. ox. is observed, but recovery to the control level occurs by the end of the operation (Fig. 3). Thirty-nine of the 83 cases (47%) showed this pattern. Figure 4 shows the changes in cerebral oxygenation in a case representative of type-C; only seven of the 83 cases (8%) were of this type. The characteristics of type-C are that the redox behavior of cyt. ox. shows a marked and prolonged reduction during and after the operation.

Thirteen of the 83 patients (15.7%) suffered from postoperative brain injury. This resulted in severe coma (7 cases), hemiparesis (5 cases), or a sight deficit (one case). Six of the 7 patients who exhibited type-C cyt. ox. behavior suffered from severe coma (5 cases) or seizure (one case) after their operation. Five of the 39 type-B patients showed post-operative hemiparesis. Only one of the 37 type-A patients suffered a post-operative deficit (visual loss). The relationship between the occurrence of a brain injury and the type of cyt. ox. behavior was highly significant [$p<0.0001$; chi-square test for independence ($l \times m$ contingency table)] (Table 1).

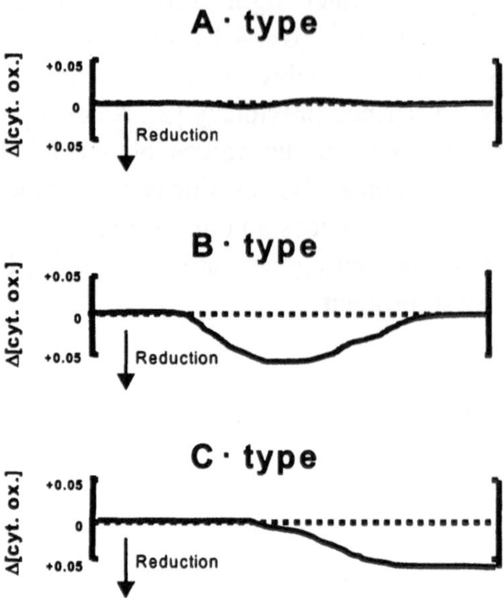

Figure 1. Different patterns of change seen in cytochrome oxidase during operations.

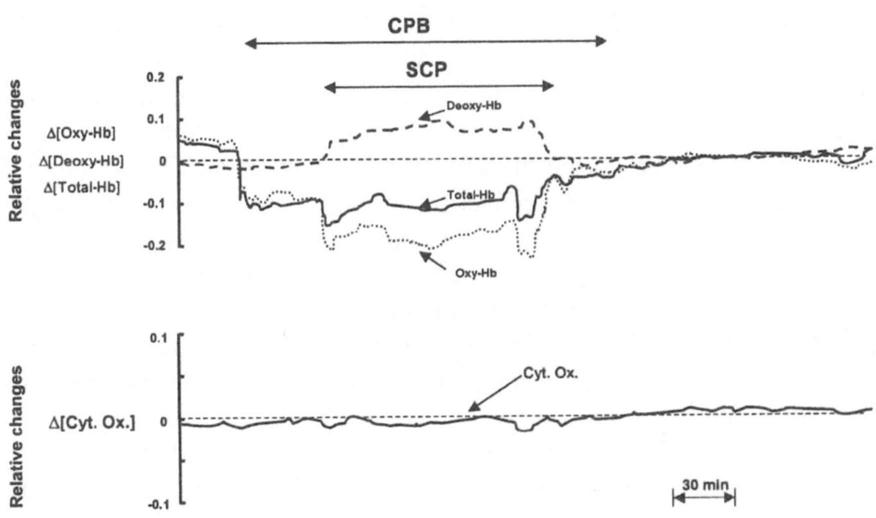

Figure 2. Changes in oxy-Hb, deoxy-Hb, total Hb, and cyt. ox. (measured by NIRS) during cardiopulmonary bypass (CPB) and separate cerebral perfusion (SCP) in an operation to repair the thoracic aortic arch in a 70-year-old male. The redox state of cyt. ox. was kept at the initial level throughout the operation, although there was a marked decrease in hemoglobin oxygenation (type-A). Upward deflections indicate an increase in relative concentration for oxy-Hb, deoxy-Hb, and total Hb, and oxidation of cyt. ox..

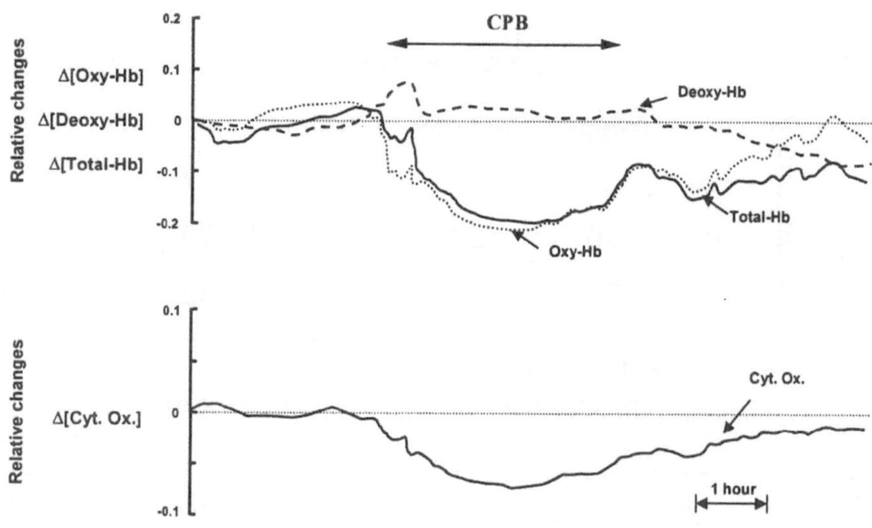

Figure 3. Changes in oxy-Hb, deoxy-Hb, total Hb, and cyt. ox. (measured by NIRS) during cardiopulmonary bypass (CPB) in an operation to repair the ascending aorta in a 74-year-old male. The redox state of cyt. ox. was transiently reduced, but it returned to the initial level at around the end of the operation (type-B). Upward deflections indicate an increase in relative concentration for oxy-Hb, deoxy-Hb, and total Hb, and oxidation of cyt. ox..

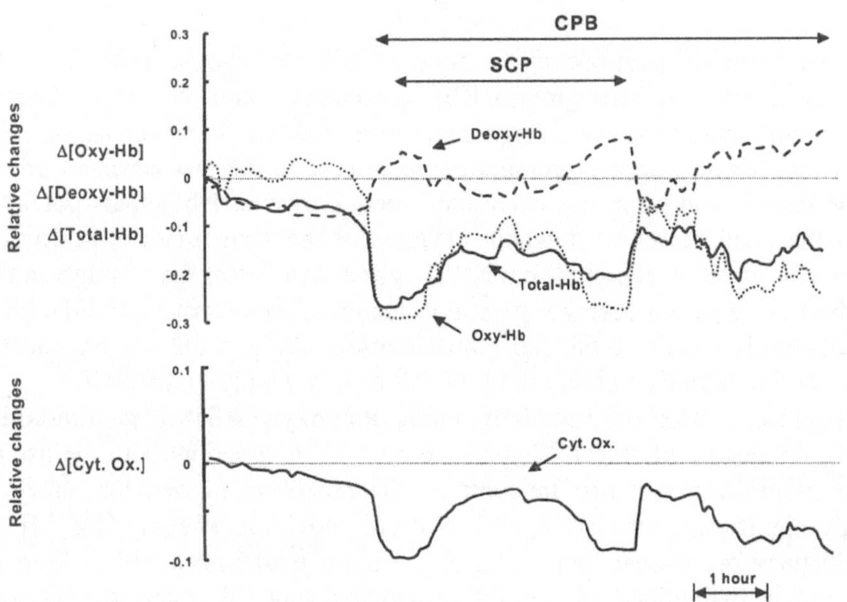

Figure 4. Changes in oxy-Hb, deoxy-Hb, total Hb, and cyt. ox. (measured by NIRS) during cardiopulmonary bypass (CPB) and separate cerebral perfusion (SCP) in an operation to repair the ascending and descending aortae, and the aortic arch in a 70-year-old female. The redox state of cyt. ox. showed a marked and prolonged reduction, and it did not return to the control level during the operation (type-C). Upward deflections indicate an increase in relative concentration for oxy-Hb, deoxy-Hb, and total Hb, and oxidation of cyt. ox.

Type of Cyt. Ox.

Brain injury	A-type	B-type	C-type	
Yes	1	6	6	13
No	36	33	1	70
	37	39	7	83

p<0.0001
(chi-square test for independence)

Table 1. Relationship between brain injury and type of cyt. ox. behavior seen during operations. Brain injury vs. type of cyt. ox. behavior (p<0.0001)(chi-square test for independence, *l*×*m* contingency table).

4. DISCUSSION

Near-infrared spectroscopy, a relative new technique, enables us to monitor changes in hemoglobin (Hb) oxygenation, and the redox state of cytochrome oxidase in the living tissues noninvasively. This technique now finds wide clinical application and its usefulness to the measurement of Hb in the muscle and brain has been confirmed. However Hb signals obtained from the head represented whole changes in the skin, muscle, scalp and brain. Kuroda et al. reported that NIRS system can detect the changes in the cerebral oxygenation during separate occlusion of the external carotid artery, which mainly supply to the skin, muscle and scalp, and the internal carotid artery or the middle cerebral artery, which mainly supply to the brain.

At present, when only cerebral hemoglobin oxygenation is monitored by NIRS, the degree of cerebral hypoxia cannot be assessed without the use of other monitoring systems. In contrast, the reduction of cyt. ox. shows a highly significant correlation with a decreased brain energy state [4]. It would therefore appear that, because the brain works normally as long as cyt. ox. is maintained in the fully oxidized state, the start of cyt. ox. reduction can be used as an alarm, giving notice that the brain condition is metabolically and functionally critical, even though absolute values are lacking [2]. In the present study, one of the 37 type-A patients (visual loss) and five of the 39 type-B patients (hemiparesis) suffered from post-operative brain injury, even though cyt. ox. showed no change (type-A) or a transient reduction followed by a return to its control level at around the end of the operation (type-B). In these cases, local regional cerebral infarction far from the monitoring site was found by computed tomography (CT) scanning and/or magnetic resonance imaging (MRI) after the operation. On the other hand, six of the 7 type-C patients suffered from severe coma or seizure after their operation.

Our data showed that recognizing the type of change in the redox behavior of cyt. ox. (as measured by NIRS) enables us to assess the likely degree of cerebral damage and predict the post-operative cerebral outcome (see Table 1). To judge from our results, we can be fairly certain that a marked and prolonged reduction of cyt. ox. indicates global cerebral hypoxia, resulting in postoperative brain injury. In the clinical field, further investigation will be required to assess (i) whether the continuous monitoring of cyt. ox. will not only help us to predict post-operative cerebral outcome, but also provide information enabling us to assess interventive cerebroprotective regimens, and (ii) whether aggressive cerebral management informed by cyt. ox. monitoring may improve the postoperative neurological prognosis.

From our data, we conclude that the redox behavior of cyt. ox. during an operation is likely to be a good predictor of postoperative cerebral outcome. This implies that the oxygen sufficiency of brain tissue can be evaluated on a global scale by near-infrared measurement of cytochrome oxidase, although changes in local regions far from the monitoring site may elude detection.

REFERENCES

1. Skov L, Greisen G. Apparent cerebral cytochrome aa_3 reduction during cardiopulmonary bypass in hypoxaemic children with congenital heart disease. A critical analysis of *in vivo* near-infrared spectrophotometric data. Physiol Meas 1994;15:447-457.
2. Hoshi Y, Hazeki O, Kakihana Y, Tamura M. Redox behavior of cytochrome oxidase in the rat brain measured by near-infrared spectroscopy. J Appl Physiol 1997;83:1842-1848.
3. Chance B, Leigh JS, Miyake H, Smith DS, Nioka S, Greenfield R, Finander M, Kaufmann K, Levy W, Young M, Cohen P, Yoshioka H, Boretsky R. Comparison of time-resolved and -unresolved measurements of deoxyhemoglobin in brain. Proc Natl Acad Sci USA 1988;85:4971-4975.
4. Matsumoto H, Oda T, Hossain MA, Yoshimura N. Does the redox state of cytochrome aa_3 reflect brain energy level during hypoxia? Simultaneous measurements by near-infrared spectrophotometry and 31P nuclear magnetic resonance spectroscopy. Anesth Analg 1996;83:513-518.

Chapter 39

ARE VEP CORRELATED FAST OPTICAL SIGNALS DETECTABLE IN THE HUMAN ADULT BY NON-INVASIVE NEARINFRARED SPECTROSCOPY (NIRS)?

Frank Syré, Hellmuth Obrig, Jens Steinbrink, Matthias Kohl, Rüdiger Wenzel & Arno Villringer
Neurologische Klinik, Charité, Humboldt-Universität zu Berlin, 10098 Berlin, Germany

Abstract: The potential of near infrared spectroscopy (NIRS) to detect vascular changes in cerebral cortical tissue elicited by functional stimulation has been established (1). The vascular response is considered the result of neuronal activity in the investigated area and forms the basis of imaging techniques such as BOLD-contrast fMRI and PET. In the animal optical methods have been shown to detect optical changes in the illuminated tissue which exactly follow the time course of electrical potential changes, thus optical techniques can potentially assess both the 'fast' neuronal and the 'slow' vascular response. Lately a group of investigators has reported data showing that the 'fast optical signal' is detectable in the adult human in response to a visual stimulus ("optical VEP" termed EROS, i.e. event related optical signal). We failed to reproduce these results, with an almost identical instrumentation and experimental protocol. The negative result is reported in this paper. To check the expected magnitude of such changes we performed a simulation based on data reported in the animal and a human head model of optical properties. The result indicates that changes in optical properties to be expected in a non-invasive approach in the human adult are about three orders of magnitude smaller than those reported previously by the group of Gratton and co-workers (2). Also they are so small that they are below the noise level of the presently available NIRS monitors.

Key words: Fast optical signals, human adults, near-infrared spectroscopy, visually evoked potential

Oxygen Transport to Tissue XXIV, edited by
Dunn and Swartz, Kluwer Academic/Plenum Publishers, 2003

1. INTRODUCTION

Near infrared spectroscopy (NIRS) has been used in many functional activation studies to non-invasively monitor changes in cerebral oxygenation in human adults (1, 3-6). Our group's work has helped to establish a typical pattern of the response in oxygenated and deoxygenated haemoglobin concentration ([oxy-Hb], [deoxy-Hb]) over an activated cortical area. This response consists of an increase in oxygenated accompanied by a decrease in deoxygenated haemoglobin the latter nicely corresponding to the increase in BOLD contrast in fMRI as shown by a approach simultaneously assessing fMRI and NIRS response to a motor paradigm (7). Because the latency of this vascular response is on the order of some 5 seconds (even when a short-lasting stimulus is applied) it will be referred to as "slow response" in this paper. It should be noted that this very response is the basis of modern functional imaging techniques such as fMRI and PET (for review see (8)). The neuronal response to functional stimuli can be recorded in the human by evoked potential (EP) techniques clinically established for example as VEP (visually evoked potential). A technique apt to simultaneously study and the neuronal and the vascular response would be an exquisite instrument to study neuro-vascular coupling non-invasively in the human.

Animal studies on exposed cortex (9) and studies of single neurones (10) showed an optical signal change occurring simultaneously with the electrical potential changes in the activated neurone or cerebral tissue. This response, whose latency is on the order of some hundreds of milliseconds will be termed "fast response" here.

Recently Gratton and co-workers have published NIRS data, claiming that this "fast response" can be detected by non-invasive NIRS in the adult (2, 11). With an intensity-modulated monitor (ISS-Oximeter, USA) the group demonstrated a phase shift corresponding to a delay of the photon mean time of flight occurring about 100 ms after pattern reversal in a set up similar to the clinically well established electrophysiological VEP assessment. The objective of this present study was to test the validity of such an approach.

2. METHODOLOGICAL CONSIDERATIONS

While changes of the vascular parameters ([oxy-Hb] and [deoxy-Hb]) lead to a major effect on the intensity of the reflected light, it has been proposed, that scatter changes in the illuminated tissue, which are a candidate for the origin of the fast response, may require time resolved systems. Time-resolved measurements can be made either in the time-domain (TD) or in the frequency-domain (FD). For time domain measurements the time of flight distribution of the photons travelling through the investigated tissue is assessed. For this end a very short light pulse is injected into the tissue by a light emitting fibre and the temporal distribution of the photons reaching the detector fibre is assessed (time correlated single photon counting, TCSPC). Frequency domain systems using a sinusoidally intensity-modulated light source, rely on the phase shift of the modulation wave between incident and reflected light to estimate the mean time of flight. It was shown that the average run time of the photons is equivalent to the phase shift at a certain modulation frequency (12). In the present study we use the same frequency-domain monitor used for the above mentioned studies by the group claiming to demonstrate the non-invasive 'fast-optical signal' in the adult human (2, 11).

2.1 Description of the monitor

The ISS-Oximeter (ISS-Oximeter, Champaign, IL, USA) is an intensity modulated NIRS-system and was used in a single wavelength, single inter-optode distance mode. Since by default 8 channels are recorded successively the actual time resolution was 4 ms. To reduce noise, however, all 8 channels were binned for the analysis of the data resulting in an effective temporal resolution of 32 ms. All subject were measured at 750 nm, additionally in some subjects measurements at 830 nm were obtained (laser power 63O μW for both wavelengths). To exactly synchronise data acquisition to the reversal of the checkerboard, the stimulus was driven by the ISS-monitor. The checkerboard reversed after the acquisition of 16 data points, resulting in a duration of 512 ms for each reversal, corresponding to 1.95 Hz. The optical probe consisted of a fibre 400 μm in diameter guiding the emitted light to the subject's head and an optic fibre bundle a few mm in diameter collecting the transmitted light. The system returns the intensity changes (ΔDC), the changes in modulation depth (ΔAC) and the phase shift in degrees (ΔPH). The modulation frequency is 112 MHz and the cross-correlation frequency *is* 5 kHz.

2.2 Experimental set-up

2.2.1 Subjects

We tested 12 subjects (one male, age range 20-30) with normal or corrected to normal visual acuity. In all subjects several probe positions were investigated, in three subjects (SN, SS and WC) a "mapping" of the occipital region was performed in successive sessions with at least 12 different probe positions. The positioning was guided by an anatomical MRI in subject SS and WC.

2.2.2 Paradigm

The activation paradigm used, was a circular reversing black and white checkerboard, which was presented on a 17" Monitor The checkerboard reversed at 1.95 Hz (512 ms). The stimulus is well established to maximally stimulate primary visual area V1, since the circular arrangement takes into account the magnification of receptive fields in the visual periphery of the visual field.

2.2.3 Experimental Protocol

To also allow for an analysis of the slow response, which corresponds to the well established vascular response, the data acquisition was arranged in blocks of stimulation (i.e. reversals of the checkerboard), and rest (i.e blank screen). Stimulation periods for the fast response (20.48 s corresponding to 40 reversals) were separated by equally long resting periods. In each session wc acquired 10-20 cycles yielding a total of up to 800 reversals for each probe position.

The subjects were comfortably seated in an EEG-chair. Special care was taken to minimise head movement during the measurement. The experiment was performed in a dark and quiet room, to minimise the artefacts caused by ambient light, the probes and the occiput of the subject were covered with black cloth. In each session only 3-4 different positions were tested with recreational pauses in between to prevent attention differences due to tiring of the subject. During the experiment subjects fixated the centre of the screen marked by a black cross. It is important to note that the three subjects in whom mapping was performed and who are dealt with here in more detail did not report tiring, during the task. Experiments in which the subjects felt loss of attention were repeated after a break. The ECG was co-registered but was not used for additional filtering. The source-detector distance was 3.35 cm.

2.3 Data analysis

To analyse the slow response the data was averaged time-locked to the beginning of the checkerboard reversal. Since only one wavelength was recorded the results do not allow for a differential analysis of the different vascular parameters ([oxy-Hb],[deoxy-Hb]). The changes in DC must therefore be understood as a very general measure of a change in focal cerebral oxygenation. Since the pulse rate (between 60 to 85 min^{-1}) may mimic a visually evoked response due to its spectral vicinity to the stimulation frequency the data were filtered according to the procedure described by Gratton and Corballis (13). To check for the detectability of a fast response data were averaged over 1024 ms (i.e. two reversals) across all stimulation periods in each probe position and each subject individually. Also a FFT was performed across the whole data file.

3. RESULTS

3.1 "Slow response"

In most of the subjects a slow change in the optical parameters was seen in response to the reversal of the checkerboard. This response was most pronounced in the DC-recordings, which correspond to the changes in optical densities as used by the standard continuous wave monitors (14). Examples for the three subjects WC, SS and SN are shown in Figure 1a. The response reaches its maximum 5-10 s after the onset of the reversals. In some recordings it reaches a plateau to decline after the stimulation period. The time course of the changes is in line with previously reported vascular responses over the occipital region elicited by visual stimulation (15-17). Due to the single wavelength approach a differentiation between the different oxygenation parameters was not possible at 750 nm. The response may therefore be taken as a very rough measure of the oxygenation changes in the underlying cerebral region. A decrease in [deoxy-Hb] as expected from previous studies readily explains the increase in DC since at 750 nm the specific absorption coefficient for deoxy-Hb amounts to about three times that of oxy-Hb. On the other hand [oxy-Hb] increases over the stimulated area have been reported to be of at least twice the magnitude of the decrease in [deoxy-Hb], thus it is not surprising that in some locations there was a decrease in DC, most probably resulting from a more pronounced increase in [oxy-Hb].

To compare intensity (DC) and phase (PH) response in the different positions the differences between stimulation and rest were calculated as ‰ of the mean signal and mapped with reference to the inion (see figure 2 for PH (2a) and DC (2b) changes. AC values are not shown but closely follow the DC time course and distribution). For the three subjects shown, there is a distribution of the DC responses in rough analogy to the anatomy with maximal responses bilaterally of the midline.

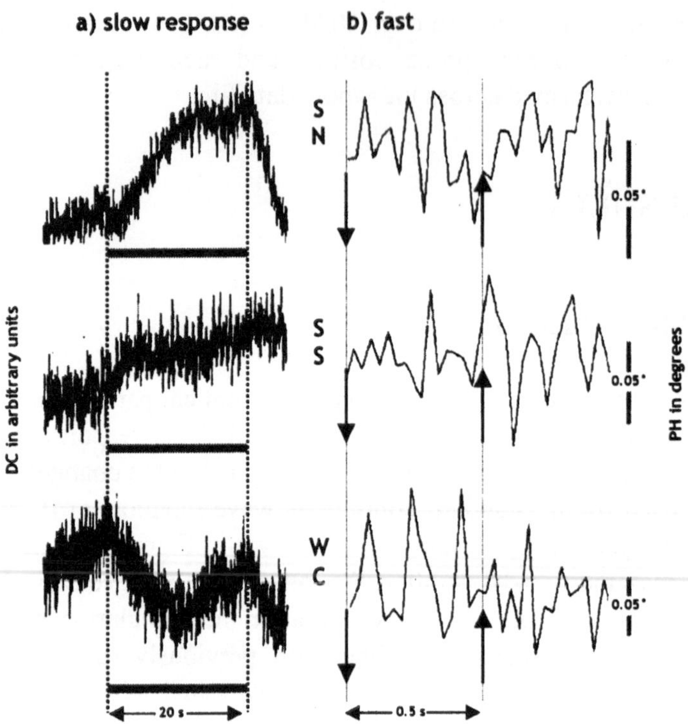

Figure 1. Slow (1a) and fast (1b) response to the stimulus in three subjects (SN, SS, WC). For the slow response the changes in DC show a time course similar to that previously reported for the vascular or oxygenation response. The different response directions are not surprising, since a differentiation between the different vascular parameters is not possible (for explanation see text). 1b shows the averages across all checkerboard reversals in the same subjects (reversal at 0 and 0.5 s). The exact same data are now averaged on a much smaller time scale to check for the potential fast response. However, there is no reproducible feature of the PH changes in response to the checkerboard reversal on this time scale (arrows indicate the reversals of the checkerboard).

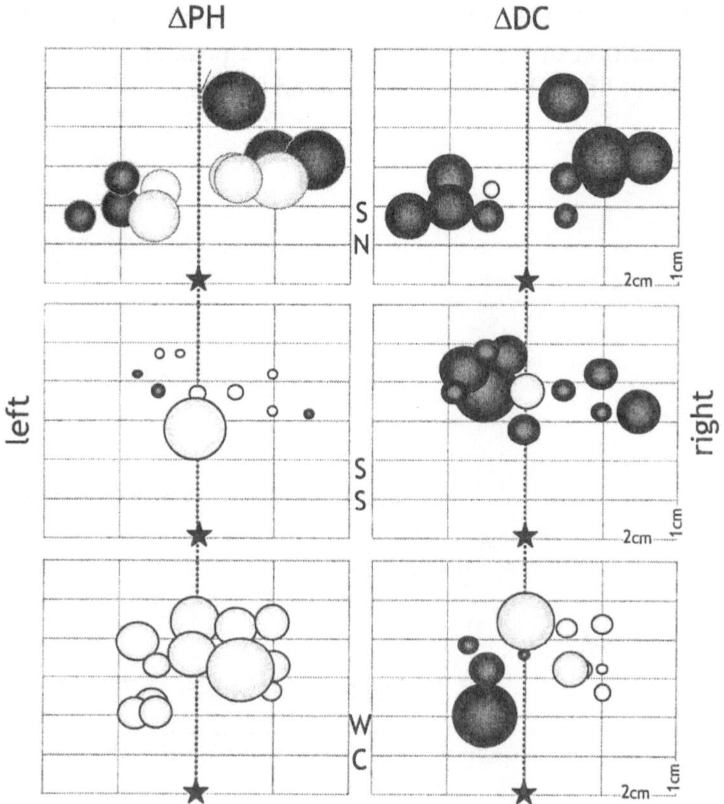

Figure 2. This graph demonstrates the magnitude and distribution of the slow response in the three subjects in whom a successive mapping was performed (SN,SS,WC). The magnitude is represented by the diameter and the colour of the bubbles (dark = positive, light = negative correlation with the stimulus). The location of the bubbles represents the distance of the probes relative to the inion (the star represents the position of the inion, scaling of the grid is 1 cm in y-direction and 2 cm in x-direction). 2a shows the changes in phase (ΔPH), 2b shows the changes in intensity (ΔDC). Note that for the DC changes there are clear maxima on either side of the midline roughly corresponding to the expected location of the bilateral visual fields.

3.2 "Fast response"

There was no clear fast response in any subject or any optode position (see figure 1b). The average over 1024 ms (i.e. two reversals) sometimes showed a single feature most probably related to the heartbeat, which was attenuated by the heart-beat filter. Noise level was so low that the previously reported delay of 10 ps (2) should have clearly shown in the recordings. The DC and AC power-spectra showed a strong peak at the approximate heart rate and a smaller peak at the respiration frequency (0.2

Hz). Around 2 Hz it was essentially featureless (see Figure 3, note the successful attenuation of the heart-rate associated peak around 1 Hz by the filter). Noise level of the PH-signal was around 0.4 ps.

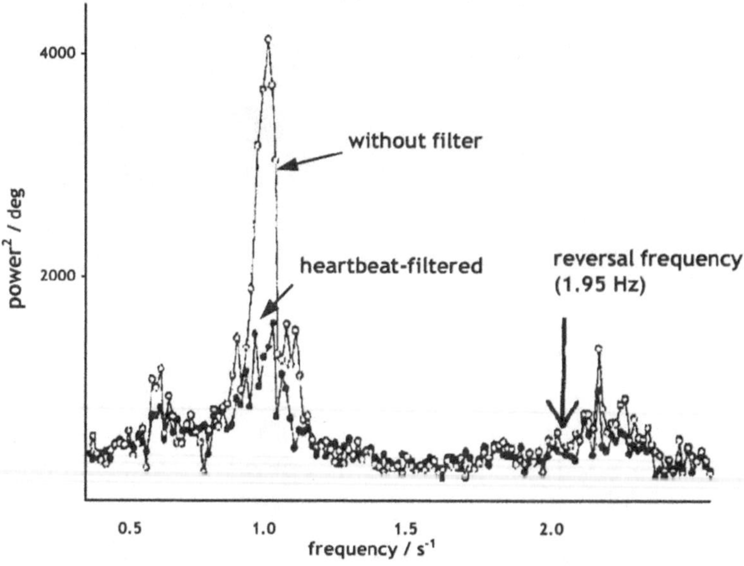

Figure 3. The figure shows the FFT of a representative file of a single subject, in whom a successive mapping was performed (SN). The analysis shows a clear peak around 1Hz i.e. the heart beat frequency. There is a second peak above 2 Hz, which most likely is the first harmonic of the pulse. At the stimulation frequency (arrow) the FFT is essentially featureless. The heart-rate filtering attenuates the peak around 1 Hz (filled circles), but does not change the other features of the spectrum.

4. DISCUSSION

In the present experiment we were not able to reproduce the previously reported results, indicating a detectability of the fast optical response to a visual paradigm ('optical VEP', 'EROS'). Concerning the comparability of the results published by Gratton and co-workers (2, 11) the differences between our and the reported set-up are the following:

- The optical changes were recorded at 750nm (as opposed to 715nm).

- A full field circular checkerboard was used to elicit a response in the visual cortex (as opposed to four quadrant black and white reversing grids).
- The temporal resolution of our measurements was 32 ms (as opposed to 50 ms).
- The optode spacing was 3.35 cm (as opposed to 3.0 cm)
- The probes were vertically fixed to the head (as opposed horizontally).

The differences in the experimental set-up are unlikely to account for our results conflicting with the previously reported data. In fact the stimulus chosen here (full field circular checkerboard) should rather elicit a greater response over the primary visual area due to its adaptation to the retinal representation. Also the area stimulated should extend farther than that stimulated by reversing bars in one quadrant of the visual field only. Concerning the issue of potential inter-individual differences it should be noted, that the subjects in whom we performed a "mapping" of the occipital area, were selected by the excellent intensity of DC transmittance (400-900 counts per data-point). It is unlikely that in subjects with a poorer transmittance a positive result could be found since the detectability is critically dependent on the signal to noise ratio (SNR), see below. Furthermore the distribution of slow response in the DC-signal indicates that the light penetrates into cortical tissue, which was activated by the stimulus. Since the PH-signal is more sensitive to deeper layer changes according to theoretical models (18) the penetration to structures on the rim of the inter-hemispheric cleft can be assumed and thus a potential fast response should not be missed.

4.1 Estimate of the expected magnitude of the "fast response"

To validate our negative result we next addressed the question as to the expected magnitude of the fast optical changes in response to the stimulus. It should be noted here that we in no way question the *existence* of such signals, the pivotal issue is the *detectability* in the human adult by presently available technology.

There is a number of studies on changes in various optical parameters in response to the activation of neuronal tissue (10, 19, 20). In 1991 Stepnoski reported: "We observe an optical signal that is linearly proportional to the change in the membrane potential. Action potentials can be recorded without signal averaging." (10). Rector and co-workers recently published data showing reflectance changes temporally parallel to the electrophysiological response in exposed hippocampus of the cat (21, 22).

By use of a Monte-Carlo simulation we first estimated the magnitude of changes in scatter properties (μs') of the tissue in the latter study. The results were then fed into a second MC simulation applying these changes to the recently published head model by Okada and co-workers (23).

The results indicate, that the expected change is on the order of 0.01 ps for the change in time of flight and a 0.01 % change in the DC-signal, i.e three orders of magnitude smaller than the data reported by Gratton and co-workers (2), and below the sensitivity of the monitor used. From our simulation we infer that methodological alterations substantially increasing SNR are the prerequisite for the detection of fast optical signals non-invasively in the human adult. The unique potential of optical methods to simultaneously monitor vascular and electrophysiological response is certainly worth this effort.

REFERENCES

1. Obrig H and Villringer A. Near-infrared spectroscopy in functional activation studies Can NIRS demonstrate cortical activation? Adv Exp Med Biol 1997; 413:113-127.
2. Gratton G, Corballis PM, Cho E, Fabiani M, Hood DC. Shades of gray matter: noninvasive optical images of human brain responses during visual stimulation. Psychophysiology 1995;32:505-509.
3. Obrig H, Hirth C, Junge-Hulsing JG, Doge C, Wolf T, Dirnagl U, Villringer A. Cerebral oxygenation changes in response to motor stimulation. J Appl Physiol 1996; 81:1174-1183.
4. Boas DA, Gaudette T, Strangman G, Cheng X, Marota JJ, Mandeville JB. The accuracy of near infrared spectroscopy and imaging during focal changes in cerebral hemodynamics. Neuroimage 2001;13:76-90.
5. Meek JH, Elwell CE, Khan MJ, Romaya J, Wyatt JS, Delpy DT, Zeki S. Regional changes in cerebral haemodynamics as a result of a visual stimulus measured by near infrared spectroscopy. Proc R Soc Lond B Biol Sci 1995;261:351-356.
6. Tamura M, Hoshi Y, Okada F. Localized near-infrared spectroscopy and Obrig H, Villringer A. Near-infrared spectroscopy in functional activation studies. functional optical imaging of brain activity. Philos Trans R Soc Lond B Biol Sci 1997;35:737-742.
7. Kleinschmidt A, Obrig H, Requardt M, Merboldt KD, Dirnagl U, Villringer A, Frahm J. Simultaneous recording of cerebral blood oxygenation changes during human brain activation by magnetic resonance imaging and near-infrared spectroscopy. J Cereb Blood Flow Metab 1996;16:817-826.
8. Villringer A, Dirnagl U. Coupling of brain activity and cerebral blood flow: basis of functional neuroimaging. Cerebrovasc Brain Metab Rev 1995;7:240-276.
9. Rector DM, Poe GR, Kristensen MP, Harper RM. Light scattering changes follow evoked potentials from hippocampal Schaeffer collateral stimulation. J Neurophysiol 1997;78:1707-1713.
10. Stepnoski RA, LaPorta A, Raccuia-Behling F, Blonder GE, Slusher RE, Kleinfeld D. Noninvasive detection of changes in membrane potential in cultured neurons by light scattering. Proc Natl Acad Sci U S A 1991;88:9382-9386.

11. Gratton G, Fabiani M. The event-related optical signal: a new tool for studying brain function. Int J Psychophysiol 2001;42:109-121.
12. Sevick EM, Chance B, Leigh J, Nioka S, Maris M. Quantitation of time- and frequency-resolved optical spectra for the determination of tissue oxygenation. Anal Biochem 1991;195:330-351.
13. Gratton G, Corballis PM. Removing the heart from the brain: compensation for the pulse artifact in the photon migration signal. Psychophysiology 1995; 32(3):292-299.
14. Cope M, Delpy DT. System for long-term measurement of cerebral blood and tissue oxygenation on newborn infants by near infra-red transillumination. Med Biol Eng Comput 1988;26:289-294.
15. Colier WN, Quaresima V, Wenzel R, van der Sluijs MC, Oeseburg B, Ferrari M, Villringer A. Simultaneous near-infrared spectroscopy monitoring of left and right occipital areas reveals contra-lateral hemodynamic changes upon hemi-field paradigm. Vision Res 2001;41:97-102.
16. Heekeren HR, Kohl M, Obrig H, Wenzel R, von PW, Matcher SJ, Dirnagl U, Cooper CE, Villringer A. Noninvasive assessment of changes in cytochrome-c oxidase oxidation in human subjects during visual stimulation. J Cereb Blood Flow Metab 1999;19:592-603.
17. Ruben J, Wenzel R, Obrig H, Villringer K, Bernarding J, Hirth C, Heekeren H, Dirnagl U, Villringer A. Haemoglobin oxygenation changes during visual stimulation in the occipital cortex. Adv Exp Med Biol 1997;428:181-187.
18. Firbank M, Okada E, Delpy DT. A theoretical study of the signal contribution of regions of the adult head to near-infrared spectroscopy studies of visual evoked responses. Neuroimage 1998;8:69-78.
19. Cohen LB, Salzberg BM, Grinvald A. Optical methods for monitoring neuron activity. Annu Rev Neurosci 1978;1:171-182.
20. Lipton P. Effects of membrane depolarization on light scattering by cerebral cortical slices. J Physiol (Lond) 1973;231:365-383.
21. Rector DM, Poe GR, Redgrave P, Harper RM. A miniature CCD video camera for high-sensitivity light measurements in freely behaving animals. J Neurosci Methods 1997;78:85-91.
22. Rector DM, Rogers RF, Schwaber JS, Harper RM, George JS. Scattered-light imaging in vivo tracks fast and slow processes of neurophysiological activation. Neuroimage 2001;14:977-994.
23. Okada E, Firbank M, Delpy DT. The effect of overlying tissue on the spatial sensitivity profile of near-infrared spectroscopy. Phys Med Biol 1995;40:2093-2108.

Chapter 40

USING HIGH SPECTRAL AND SPATIAL RESOLUTION BOLD MRI TO CHOOSE THE OPTIMAL OXYGENATING TREATMENT FOR INDIVIDUAL CANCER PATIENTS

Hania A. Al-Hallaq[1], Marta A. Zamora[1], Brian L. Fish[3], Howard J. Halpern[2], John E. Moulder[3], & Gregory S. Karczmar[1]

[1] Department of Radiology, The University of Chicago, Chicago, IL 60637; [2] Department of Radiation Oncology, The University of Chicago, Chicago, IL 60637; [3] Department of Radiation Oncology, Medical College of Wisconsin, Milwaukee, WI 53226

Abstract: We evaluate whether high spectral and spatial resolution (HiSS) BOLD MRI can correctly rank the effects of three tumor-oxygenating treatments on radiosensitivity in BA1112 rhabdomyosarcomas (n = 5). Significant decreases in spectral linewidth predict that treatment with carbogen gas combined with a perfluorocarbon emulsion will increase radiosensitivity more than either treatment alone, which agrees with the known effects of these treatments on hypoxic fraction. High-resolution maps show that tumor response to each treatment is spatially heterogeneous, and that there is a paradoxical response to the treatments in 7 – 12% of tumor pixels. Because HiSS MRI emphasizes changes in necrotic and/or hemorrhagic regions, it is more sensitive to oxygenation changes compared to conventional MRI. These results demonstrate that HiSS MRI is a practical, noninvasive method that could be used to choose the treatment that maximizes the size and extent of increases in tumor oxygenation for individual patients.

Key words: Magnetic resonance imaging; oxygen/blood; radiation sensitizing agents; rats

Abbreviations:

HiSS	High spectral and spatial resolution
PH	peak height
PI	peak integral

1. INTRODUCTION

Since it was discovered that tumor hypoxia may limit response to radiation [1], various treatments have been used to increase tissue oxygen tension (pO_2) in radioresistant tumor regions. Some of these treatments, which rely on increasing oxygen supply to tumors, were shown to effectively reduce hypoxic fractions in experimental tumors [2-4]. However, many *clinical* studies of these adjuvants to radiotherapy have not shown statistically significant increases in tumor response [5]. This may be because the power of the statistical tests was reduced due to both heterogeneity of the tumor populations available for clinical studies and spatial heterogeneity of the response to tumor-oxygenating agents within tumors.

MR imaging (MRI) offers several advantages for studying response to tumor-oxygenating treatments. It is a practical method that allows for noninvasive, repeated measurements in individual tumors. Because MRI uses each tumor as its own control, the statistical power of studies is increased. Finally, MRI provides high spatial resolution (i.e., 200 – 500 microns) which allows assessment of the heterogeneity of response within a tumor.

MRI measurements of tumor oxygenation rely on blood-oxygenation level-dependent (BOLD) contrast, which arises from magnetic susceptibility gradients in and near blood vessels due to paramagnetic deoxyhemoglobin [6]. Correct interpretation of changes in BOLD contrast caused by tumor-oxygenating treatments requires an understanding of *some* of the limitations of the technique. First, BOLD MRI does not provide a quantitative measure of blood oxygenation. In addition, administration of an oxygenating treatment may alter a number of physiologic parameters; these include blood flow, blood volume, and tissue oxygen consumption. These physiologic changes could in turn affect MR image intensity without necessarily leading to increases in tumor oxygenation. Despite these limitations, BOLD MRI could be used to measure *changes in oxygenation*. Also, additional information obtained with MR (when compared to conventional measurements of BOLD) may be used to distinguish changes in oxygenation from other physiologic changes.

Here, we use high spectral and spatial (HiSS) resolution proton MRI to measure changes in BOLD contrast. This technique provides sensitive, more accurate measurements of changes in T_2^* caused by tumor-oxygenating treatments compared to conventional MRI measurements of BOLD. The goal of this work is to determine if HiSS MRI could correctly rank the relative effects of three treatments on radiosensitivity in BA1112 tumors, i.e., whether HiSS data correlate with reductions in hypoxic fractions.

2. METHODS

2.1 Biological models

Studies of BA1112 rhabdomyosarcomas in WAG/Rij rats have shown that infusion of a perfluorocarbon emulsion (PFC) combined with carbogen (95% O_2/5% CO_2) gas breathing significantly increased radiosensitivity relative to either treatment alone [4, 7]. We studied the response of BA1112 tumors, implanted subcutaneously on the lower backs of WAG/Rij rats, to carbogen, PFC, and PFC + carbogen. Tumors (n = 5) were studied at diameters of 0.5 – 1.0 cm and contained areas of significant hemorrhage. Animals were anesthetized by continuous IP administration of a Ketamine hydrochloride (90 mg/kg) and xylazine (5 mg/kg) solution. Blood pressure and body temperature were continuously monitored throughout the experiment. We used perfluoro-15-crown-5-ether, a highly concentrated PFC emulsion (74%w/v, 40% v/v) obtained from Dr. Sotak's lab (Worcester Polytechnic Institute). A catheter implanted into the jugular vein was used for injection of doses of 5 – 7 ml/kg of PFC (i.e., 8 – 11% of blood volume). All experimental protocols involving animals were approved by the University and were consistent with federal, state, and local regulations.

2.2 Magnetic resonance spectroscopic imaging

Data were collected at 4.7 Tesla using a General Electric/Bruker Omega MR scanner. Spectroscopic imaging of the water proton resonance was performed using phase-encoding gradients followed by detection of the water signal in the absence of applied gradients [8]. Spectroscopic images of a single 2-mm-thick slice in the center of the tumor were acquired with in-plane spatial resolution of 0.7 – 1.0 mm (32 x 32 points). A 45° excitation pulse was used with a repetition time of 250 msec. Image acquisition time was approximately 5 minutes. These parameters produced a high-resolution spectrum (3.9 Hz) of the water resonance in each voxel that could be used to accurately measure water signal peak integral (PI) and peak height (PH). Images were continuously acquired as the treatments were administered serially: carbogen, PFC, PFC + carbogen. A control period in which animals breathed room air preceded each treatment. Multiple MR images were acquired during each breathing period in order to improve the stability and reliability of measurements.

3. RESULTS

Changes in T_1 and T_2^* were measured using HiSS images of water signal peak integral and peak height. Under our experimental conditions, PI is sensitive to T_1, which in turn is sensitive to blood flow (via the in-flow effect) and increased oxygenation [9]. PH is proportional to both T_2^* and T_1. Therefore, we determine changes in T_2^* using the parameter:

$$\%\Delta \sim T_2^* = \%\Delta PH - \%\Delta PI = 100 \times \left(\frac{PH_{treatment}}{PH_{control}} - \frac{PI_{treatment}}{PI_{control}} \right).$$

This parameter yields 'changes in approximate T_2^*' with high signal-to-noise ratio. It is positively correlated with increased tumor pO_2 levels ($r = 0.92$, $p < 0.05$) in at least one rodent tumor model [10] and is a measure of changes in linewidth that does not require the assumption of a particular spectral lineshape. When changes in the peak integral are smaller than changes in peak height, PH alone can serve as an approximate, high signal-to-noise indicator of changes in linewidth.

For an aggregate measure of tumor response, $\%\Delta \sim T_2*$ data were averaged over each tumor. Averaged over all five tumors, $\sim T_2^*$ increased 1.1% during carbogen breathing and 3.3% during treatment with PFC + air. The changes in $\sim T_2^*$ increases to 6.5% in response to PFC + carbogen. The $\sim T_2^*$ increases significantly compared to control (by one-tailed, one-sample t-tests) when the tumors are treated with carbogen ($p < 0.02$), PFC alone ($p < 0.05$), and PFC + carbogen ($p < 0.02$), indicating that linewidth decreases significantly. ANOVA analysis shows a difference in $\%\Delta \sim T_2^*$ caused by the three treatments ($p < 0.02$ with blocking on tumors). PFC + carbogen has a significantly larger effect on $\sim T_2^*$ than carbogen or PFC alone, while $\sim T_2^*$ changes caused by carbogen alone are *not* different from those caused by PFC alone ($p < 0.05$ by two-tailed SNK test; [11]).

Because the changes caused by each of the treatments were not homogenous over the entire imaged region, we also studied data on a pixel-by-pixel basis. Data were analyzed to produce maps of response to each treatment; this utilizes the statistical advantage of the images, which allows measurement of changes in each pixel under the various experimental conditions. Response was determined by finding pixels in which mean PH values during treatment were larger than control mean PH values by more than 2.5 times the standard deviation of control PH values (i.e., pixels in which the intensity increased significantly with one-tailed $p < 0.01$). When averaged over all five tumors, 25% of tumor pixels responded to carbogen inhalation, while 40% of pixels responded to treatment with PFC. The percentage of responding pixels increased to 55% when tumors were treated with PFC + carbogen. Thus, the combination of carbogen and PFC enlarges

A B

Figure 1A shows an image of a tumor on the back of a rat. (Back muscle, which was just below the coil, also appears in the image.) Colored pixels, overlaid on grayscale anatomic images, indicate regions of the image where treatment with PFC, carbogen, or PFC + carbogen increased tumor blood oxygenation. ***Yellow*** indicates that carbogen alone increased oxygen levels, ***blue*** indicates that PFC alone increased oxygen levels, ***orange*** indicates that both PFC and carbogen given separately increased oxygen levels, and ***red*** indicates that only the combination of PFC with carbogen increased oxygen levels. This map shows that while certain regions are affected by carbogen only (yellow) or PFC only (blue), combining the two treatments increases oxygen in regions (red) where neither treatment alone produced an effect. **Figure 1B** shows an analogous color map in another tumor. Here, the colors represent pixels that responded *paradoxically* to each of the treatments. The central portion of this tumor responded negatively when treated with either carbogen or PFC.

the area of the tumor in which peak height values are increased when compared to treatment with either carbogen alone or PFC alone ($p < 0.05$ by two-tailed SNK test following significant finding with ANOVA; [11]).

To gain further insight into the mechanism of action of the tumor-oxygenating treatments, we selected the pixels that responded to PFC + carbogen and mapped their response to the three treatments in color. Data from one tumor is shown in Figure 1A. The color map is overlaid on a grayscale anatomic reference image. Yellow indicates that carbogen alone increased PH values, blue indicates that PFC alone increased PH, orange indicates that both carbogen and PFC given separately increased PH, and red indicates that *only the combination of carbogen with PFC* increased PH. In this particular tumor, PFC alone affected a larger area of the tumor than carbogen alone. However, some tumor regions (shown in red) *only* responded to the combination of carbogen with PFC. The patterns depicted by the maps varied greatly from tumor to tumor.

In some tumor regions, significant decreases in PH were observed (one-tailed $p < 0.01$), suggesting a paradoxical increase in deoxyhemoglobin due to decreased tumor oxygenation. Figure 1B shows a color map of a tumor that is *analogous* to that shown in Figure 1A. Here, the colors indicate pixels that responded paradoxically to each of the three treatments. In this tumor, PH decreases occurred in a central region of the tumor and were induced by both carbogen *and* PFC. Decreases in PH occurred in all five tumors. Averaged over all 5 tumors, these decreases occurred in 7% of tumor pixels during carbogen inhalation, 12% during treatment with PFC, and 10% during treatment with PFC + carbogen.

In these tumors, the water spectrum was broad and/or complex (contained multiple resolvable components) due to local magnetic susceptibility gradients caused by extensive hemorrhage. Because HiSS can directly resolve the detailed proton lineshapes in each image voxel, it can monitor changes in *narrow* components of broad, complex signals. Figure 2 shows significant spectral changes caused by carbogen inhalation in one small, hemorrhagic tumor voxel. Different resolvable components of the initially broad spectrum (~ 150 Hz) responded differently to carbogen gas; similar spectral response was also induced by PFC. These resolved components come from water molecules in discrete magnetic environments, which may represent different components of the tumor microenvironment (i.e., water molecules in and around deoxygenated capillaries and venuoles). This type of nonuniform spectral change, which reduces sensitivity to BOLD changes in conventional MRI, occurs in 20 – 45% of tumor voxels studied here and is detected with high signal-to-noise in HiSS datasets.

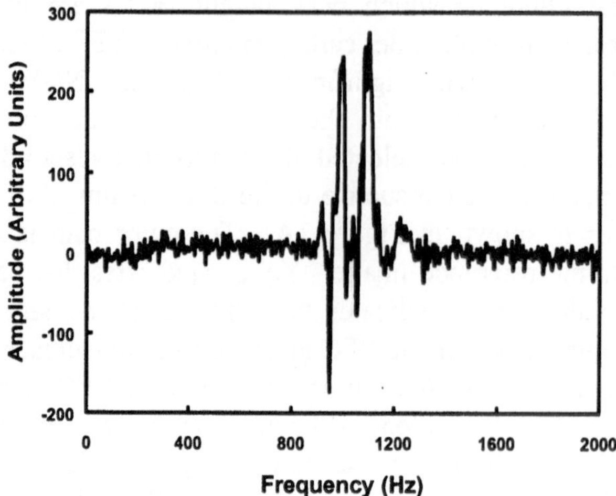

Figure 2 shows significant spectral changes caused by carbogen inhalation in one small, hemorrhagic tumor voxel. Different resolvable components of the initially broad spectrum (~ 150 Hz) responded differently to the treatment. HiSS detects such nonuniform spectral changes with high sensitivity compared to conventional MRI measurements of BOLD contrast.

4. DISCUSSION

Average increases in ~T_2^* caused by PFC + carbogen averaged over the five tumors were significantly larger than the increases caused by carbogen alone or PFC alone. Thus, HiSS MRI predicts that increases in tumor blood oxygenation caused by PFC + carbogen will be greater than increases caused by either carbogen or PFC alone. Since tumor blood oxygen levels are often tightly coupled to tumor tissue oxygenation [12], this predicts a greater radiosensitizing effect of PFC + carbogen. This prediction obtained noninvasively with MR agrees with the known effects of the three tumor-oxygenating treatments on tumor hypoxic fraction in BA1112 tumors.

Several of the advantages provided by MRI are evident in this study. Although radiosensitivity studies normally require large numbers of tumors to obtain statistical significance, we were able to correctly rank the three treatments in order of increasing efficacy with a much smaller number of samples. This is primarily because each tumor could be compared to itself, which increased the statistical power of our study. With HiSS MRI, significant variation in the response of individual tumors to each of the three treatments was detected. Thus, HiSS MRI would allow for easy identification and evaluation of outliers *before* radiotherapy treatment begins. For example, one tumor was extremely responsive to *all three* treatments, while another was less responsive – particularly to PFC or carbogen alone. In a clinical setting, this type of information might be very useful for guiding patient management throughout the course of treatment.

High-resolution maps showed that tumor response to several different oxygenating treatments was highly heterogeneous. Many pixels that did not respond significantly to either PFC alone or carbogen alone responded to the combination. In practice, MR could be used to evaluate combinations of tumor-oxygenating treatments administered serially or simultaneously so that a protocol that causes the largest response in the largest percentage of tumor volume could be designed. Although we cannot detect hypoxic cells directly, it is likely that utilizing MRI to choose the treatment that maximizes the size and extent of increases in oxygenation would reduce the hypoxic fraction. HiSS MRI also detected paradoxical responses to treatment in a small percentage of tumor volumes. Although this could reflect a variety of physiologic changes, including decreased blood volume, this observation is consistent with an 'intratumoral steal effect', i.e., a redistribution of blood flow and blood oxygenation within tumors caused by tumor-oxygenating agents [10]. This suggests that treatments increase radiosensitivity in some tumor regions while decreasing it in other regions. Finally, MR can also provide temporal information so that changes in the effect of tumor-oxygenating treatments over time can be monitored. This temporal information is also important for the design of therapy, particularly for those tumors in which acute hypoxia is important or in which adaptation to oxygenating treatments occurs.

With conventional measurements of changes in oxygenation caused by carbogen inhalation, little or no change is detected in tumor regions with initially low signal intensity [13]. These regions are those where signal has already decayed due to very short T_2^*, and often correspond to hemorrhage. Because of their initially broad linewidths, spectra in these regions are more likely to respond inhomogeneously to oxygenating treatment. Such regions are likely to correspond to areas of poor perfusion or necrosis and may be the most important regions to study because they are likely to contain hypoxic tissue. The response of any viable cells, which could repopulate tumors following radiotherapy, in these regions is very important. Because HiSS resolves detailed lineshapes in each image voxel, it can monitor changes in narrow components of broad, complex spectra. Thus, HiSS MRI may be more sensitive to changes in these regions than conventional GRE imaging. As a result, HiSS MRI may provide more accurate predictions of response to tumor-oxygenating therapies such as carbogen.

Conventional spectroscopic imaging would not be practical in a clinical setting due to the long run times required to image a large volume of tissue with high spectral and spatial resolution. However, new self-shielded gradients make it possible to perform fast spectroscopic imaging with both high spectral and spatial resolution and with reasonable run times [14]. Thus, HiSS MRI is a practical technique that measures response to oxygenating treatment via BOLD contrast with high sensitivity. Routine clinical use of this method to design the optimal oxygenating treatment for each individual patient could have a significant impact on cancer treatment.

ACKNOWLEDGEMENTS

We thank Theodore Karrison, Ph.D. for helpful discussion regarding statistical analysis. We also thank Ms. Kerianne Quanstrom for help in preparing the manuscript.

REFERENCES

1. Thomlinson RH, Gray LH. The histological structure of some human lung cancers and the possible implications for radiotherapy. Br J Cancer 1955;9:539-549.
2. Horsman MR. Nicotinamide and other benzamide analogs as agents for overcoming hypoxic cell radiation resistance in tumours. A review. Acta Oncol 1995;34:571-587.
3. Rojas A. Radiosensitization with normobaric oxygen and carbogen. Radiother Oncol 1991;20(Suppl 1):65-70.
4. Martin DF, Porter EA, Fischer JJ, Rockwell S. Effect of a perfluorochemical emulsion on the radiation response of BA1112 rhabdomyosarcomas. Radiat Res 1987;112:45-53.
5. Saunders M, Dische S. Clinical results of hypoxic cell radiosensitisation from hyperbaric oxygen to accelerated radiotherapy, carbogen and nicotinamide. Br J Cancer 1996;74(Suppl 27):S271-278.
6. Ogawa S, Lee TM. Magnetic resonance imaging of blood vessels at high fields: in vivo and in vitro measurements and image simulation. Magn Reson Med 1990;16:9-18.
7. Rockwell S, Kelley M, Irvin CG, Hughes CS, Porter E, Yabuki H, et al. Modulation of tumor oxygenation and radiosensitivity by a perfluorooctylbromide emulsion. Radiother Oncol 1991;22:92-98.
8. Brown TR, Kincaid BM, Ugurbil K. NMR chemical shift imaging in three dimensions. Proc Natl Acad Sci U S A 1982;79:3523-3526.
9. Oikawa H, Al-Hallaq HA, Lewis MZ, River JN, Kovar DA, Karczmar GS. Spectroscopic imaging of the water resonance with short repetition time to study tumor response to hyperoxia. Magn Reson Med 1997;38:27-32.
10. Al-Hallaq HA, River JN, Zamora M, Oikawa H, Karczmar GS. Correlation of magnetic resonance and oxygen microelectrode measurements of carbogen-induced changes in tumor oxygenation. Int J Radiat Oncol Biol Phys 1998;41:151-159.
11. Zar JH. Biostatistical Analysis. Englewood Cliffs (NJ): Prentice-Hall, Inc.; 1974.
12. Rofstad EK, Fenton BM, Sutherland RM. Intracapillary HbO2 saturations in murine tumours and human tumour xenografts measured by cryospectrophotometry: relationship to tumour volume, tumour pH and fraction of radiobiologically hypoxic cells. Br J Cancer 1988;57:494-502.
13. Robinson SP, Howe FA, Griffiths JR. Noninvasive monitoring of carbogen-induced changes in tumor blood flow and oxygenation by functional magnetic resonance imaging. Int J Radiat Oncol Biol Phys 1995;33:855-859.
14. Kovar DA, Al-Hallaq HA, Zamora MA, River JN, Karczmar GS. Fast spectroscopic imaging of water and fat resonances to improve the quality of MR images. Acad Radiol 1998;5:269-275.

Chapter **41**

ISSUES IN GRE & SE MAGNETIC RESONANCE IMAGING TO PROBE TUMOR OXYGENATION

Franklyn A. Howe, Simon P. Robinson, Loreta M. Rodrigues, Marion Stubbs & John R. Griffiths

Cancer Research UK Biomedical Magnetic Resonance Research Group, Dept. Biochemistry and Immunology, St. George's Hospital Medical School, London, SW17 ORE, England.

Abstract: Tumor oxygenation determines the efficacy of radiotherapy, but there is no non-invasive way to image this parameter. Since gradient recalled echo (GRE) images are sensitive to blood deoxyhæmoglobin concentration ([dHb]) they could have a role in assessing tumor oxygenation. In brain, linear relationships have been demonstrated between brain tissue R_2* relaxation rate and tissue [dHb] or oxygen saturation, but in tumors, vascular and tissue heterogeneity, and the presence of simultaneous oxidative and glycolytic metabolism, complicate the analysis. We have studied the effects of vascular challenge in a rat prolactinoma tumor model by MR imaging and spectroscopy and comment on the implications of these results for calibrating GRE images for blood or tissue pO_2.

Key words: Blood flow, MRI, oxygenation, tumor, vasomodulation

1. INTRODUCTION

Tumor oxygenation determines the efficacy of therapy, and tumor growth and survival depends on angiogenesis, which is regulated by hypoxia [1]. A non-invasive measure of tumour oxygenation would therefore be a valuable tool for *in vivo* studies to predict treatment outcome and to select appropriate therapies. Gradient recalled echo (GRE) and spin echo (SE) MR images are sensitive to blood deoxyhæmoglobin concentration ([dHb]) and could have a role in assessing tumor oxygenation.

GRE and SE methods have been extensively used for functional MRI studies of the brain, and theoretical models developed to describe the contrast mechanisms [2,3]. MR methods are now being used to investigate tumor oxygenation [4,5], changes in response to vascular challenge [6-8] and angiogenesis [9]. Ideally we would like to quantify MR images in terms of blood oxygenation as has been done for SE imaging in the brain [3]. Unlike brain, tumor blood flow and oxygenation is not well regulated and vascular development is chaotic leading to regions of hypoxia [10]. Because tumours exhibit heterogeneity in size and density of blood vessels, a *de novo* theoretical analysis for in tumors is difficult. Furthermore, it is not blood pO_2 but tissue pO_2 that is the radiobiologically important parameter. Oxidative and glycolytic metabolism occur simultaneously in tumors, so the relationship between blood oxygenation and tissue pO_2 is complex [10,11]. We have studied the effects of vascular challenge in a transplanted rat prolactinoma [5-8] and using this and new data we comment here on the implications of these results for calibrating GRE images for blood and tissue pO_2.

2. THEORY

A theoretical description [2] of the contribution of blood oxygenation to the R_2^* ($1/T_2^*$) relaxation rate of brain tissue suggests:

$$R_2^* = A\, bv\, (1-Y)^n \qquad (1)$$

where bv is the blood volume, Y the blood oxygen saturation and A depends on field strength, blood vessel orientation and morphology. The index n is unity for small vessels and 2 for large vessels. Experimentally, correlation of brain tissue R_2^* in terms of blood [dHb] or oxygen saturation have shown linear relationships. Using Near Infra Red Spectroscopy (NIRS) to measure brain [dHb] Punwani [12] found:

$$R_2^* = A\,[dHb] + C \qquad (2)$$

and measuring venous (Y_v) and arterial (Y_a) oxygen saturation Lin [13] found that:

$$\Delta R_2^* = m\,[\alpha\, \Delta Y_a + (1-\alpha)\, \Delta Y_v] + C \qquad (3)$$

where α is the brain tissue arteriolar blood vessel fraction. *A, m* and *C* were the calibration constants determined.

It can be expected that similar relationships will hold for tumors and qualitative correlations have already been demonstrated between changes in tissue pO_2 and those in MR image intensity [4]. However, any quantitative relationship will be complicated by the heterogeneity of tumor tissue structure, which may contain regions of necrosis and oedema, and by the heterogeneous morphology of blood vessels. Therefore it is unlikely that constants for Eqs. (1-3) could be determined to give a standard calibration for all tumors and they may vary even within the tissue of a single tumor. The calibration methods for Eqs. (2) and (3) are also problematic for studying tumors. Firstly, non-localised NIRS measurements will give average tumor $[Hb]$, which cannot necessarily be applied to heterogeneous tissue. Secondly, except for particular experimental situations [11], it is not possible to measure arteriolar and venular blood oxygenation in tumors.

A possible naïve approach to calibrating R_2^* is to modulate blood oxygenation in some way to determine calibration constants for individual tumors directly from MR images. We have used Eq. (3) with the following assumptions. Firstly, the venular blood in tumors is completely desaturated (Y_v=0) and Eq. (3) simplifies to $R_2^*\ (Y) = m\ Y_a + C$. Secondly, that it is possible to somehow completely desaturate the arteriolar tumor blood so $R_2^*(0) = C$. Finally, a change in arterial saturation (ΔY_A) translates into a similar change in the saturation of tumor arteriolar blood saturation and $\Delta R_2^* = m\ \Delta Y_A$.

If these criteria and assumptions are true, then in principle a blood pO_2 map could be created from T_2^* images. In practice, all the assumptions are oversimplifications, but we can learn something about the behaviour of the system from this rudimentary model.

3. METHODS

GH3 prolactinomas were grown in the flanks of Wistar Furth rats. MRS/MRI data were acquired using a 4.7 T SISCO (Spectroscopy Imaging Systems Corporation) instrument with a 3 cm diameter three-turn solenoid. GRE images through the centre of all tumors were acquired with TE/TR/α = 20ms/80ms/45°. Also acquired for some tumors were GRE images at TEs of 12, 20 and 30ms to enable absolute T_2^* images to be calculated and SE images with TE/TR = 20ms/300ms to evaluate flow effects. Changes in R_2^* following vascular challenge were calculated from $\Delta R_2^* = -50 \ln (S_{air}/S_{vasomodulator})$, where S_{air} and $S_{vasomodulator}$ are the average GRE signal intensities taken over the whole tumor in each image. Non-localised ^{31}P spectra were acquired with repetition time TR=3 s. Intracellular pH_i was calculated from the chemical shift difference between P_i and αNTP.

Breathing gases (air, 100% O_2, carbogen (95% O_2/5% CO_2)) were administered via a nose-piece with integral scavenger at 2 l/min. Calcitonin-Gene Related Peptide (CGRP) (39 μg/kg), hydralazine (5mg/kg) and nicotinamide (1000mg/kg) were administered via a tail vein cannula. The iliac artery was cannulated in separate cohorts of Wistar Furth rats and sequential blood samples taken during air and carbogen breathing, and following nicotinamide. A pH/blood gas analyser (Instrumentation Laboratory, Milan) was used to measure blood p_AO_2, p_ACO_2 and pH_A. More complete details of the methods described above can be found in references [5-8].

4. RESULTS

MR and arterial blood gas analysis data are given in Tables 1 and 2. Figure 1 shows the typical heterogeneous T_2* maps that are calculated for a tumor during air and carbogen breathing.

Table 1. ^{1}H MRI and ^{31}P MRS data following a vascular challenge

Vascular challenge	R_2* (s^{-1})	ΔR_2* (s^{-1})	ΔSE (%)	βNTP/Pi	ΔpH_i
CGRP [6]	-	11.2	-	0.5	-0.1
Hydralazine*	-	6.4	-8	0.7	-0.2
Air (21% O_2) [8]	63	0	0	1.0	0
100% O_2 [7]	-	-3.4	-	-	-
Carbogen [6-8]	45	-15	15	1.6	0.1
Nicotinamide*	-	-3.8	0	1.8	0.2

* Unpublished data

Table 2. Arterial blood data following a vascular challenge

Vascular challenge	p_AO_2 (mm Hg)	ΔY_A (%)	ΔpH_A	Glucose (mmol/l)	$p_{blood}O_2$ (mm Hg)
CGRP [6]	-	-	-0.1	-	0
Hydralazine*	-	-	-0.2	-	5
Air (21% O_2) [8]	100	0	0	7	28
100% O_2 [7]	400	8.4	0	-	37
Carbogen [6-8]	400	8.3	-0.15	16	66
Nicotinamide*	-	-	0	11	48

* Unpublished data

The changes in R_2* are as expected for the action of each vascular challenge: breathing high oxygen content gases improves blood oxygenation, reduces [dHb] and thus increases R_2*; CGRP and hydralazine reduce tumor blood flow (TBF) by steal effects, further desaturating the blood and increasing R_2*; the small reduction in R_2* (and implied decrease in [dHb]) with nicotinamide is most likely due to inhibition of arterial vasoconstriction [14]. Changes in the βNTP/P_i ratio are consistent with improved or worsened substrate (oxygen or glucose) supply and show qualitative correlation with the ΔR_2*. Tumor blood oxygen saturation was estimated by assuming that with CGRP tumor blood becomes fully deoxygenated (Y_a=0) and between breathing air and 100% O_2 the tumor ΔY_a is the same as the ΔY_A of arterial blood so that $Y_a = \Delta Y_A\ [\ R_2* - R_2*(O_2)\ ]\ /\ [\ R_2* - R_2*(CGRP)\ ]$. Tumor blood $p_{blood}O_2$ (Table 1) was then calculated using the Hill equation $Y = pO_2^n\ /\ [\ p50^n + pO_2^n\ ]$ with n=2.8, and assuming a p50 of 43mmHg at pH 7.3 for arterial blood and pH 7.2 for arteriolar blood.

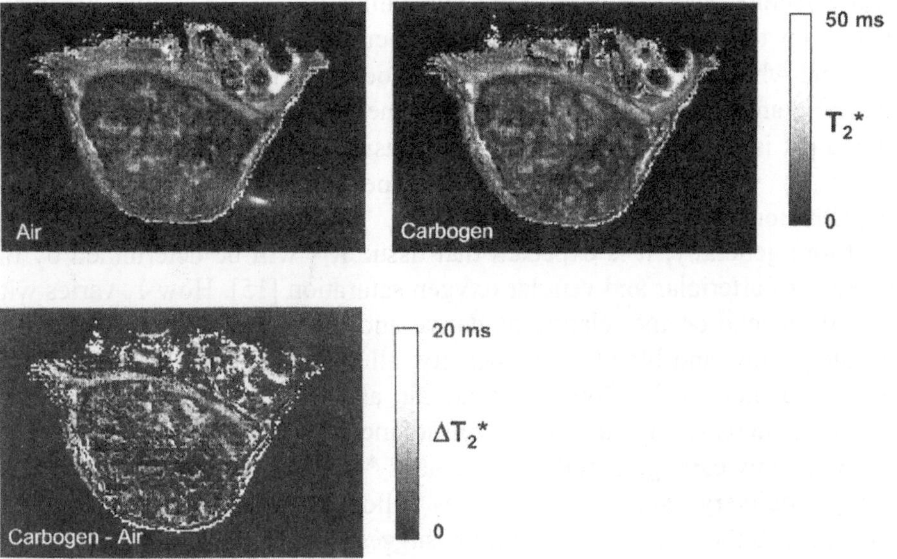

Figure 1. T2* maps from a prolactinoma during air and carbogen breathing and difference map showing regions of increased T2* with carbogen breathing. Note that both the absolute T2* and the difference maps of the tumor tissue are very heterogeneous whereas the body wall muscle has a uniform T2* for both gas breathing regimes. The T2* maps were created using a pixel-by-pixel logarithmic fit to data acquired using a multiple gradient echo sequence with TR=80ms, TE=N*5ms with N=1 to 8. In plane resolution is 0.16 mm with a 1 mm slice thickness

5. DISCUSSION

This simple approach for calibrating the R_2* change gives an average tumor blood pO_2 ($p_{blood}O_2$ in Table 2) similar to arteriolar measurements made in a tumor window model [11]. This suggests that GRE image calibration that includes individual tumor heterogeneity may be possible. How justifiable are the extreme assumptions made in this model?

The constant term in Eq. (3) is determined assuming that tumor blood is completely desaturated. Hydralazine and CGRP, vasomodulators that reduce TBF, produce ΔR_2* values that correlate with the reduction in $\beta NTP/P_i$. However, complete vascular desaturation is unlikely. Furthermore, if the high interstitial pressure found in tumors caused vascular collapse, the reduced blood volume would artefactually reduce the tissue [dHb] and invalidate any calibration. Also, once TBF has been reduced, tumor oxidative metabolism must still be high enough to extract the blood oxygen. To determine the linear coefficient of Eq. (3) the change in tumor arteriolar oxygen saturation must be known, and for the proposed method must be linearly coupled to the change in arterial blood saturation. This depends on many factors, such as host tissue and tumor oxygen extraction fractions, where on the oxyhæmoglobin saturation curve the tumor blood saturation lies and whether there are changes in blood pH. The data in Table 1 show that pH_A and tumor pH_i are altered by some vascular modulations, so tumor blood pH is also likely to change. A perfusion change (indicated by ΔSE in Table 1) will alter tumor [dHb] independently of changes in blood oxygenation.

More generally, it is expected that tissue R_2* will be determined by the average of arteriolar and venular oxygen saturation [15]. How Y_v varies with Y_a will depend on the relative oxidative and glycolytic metabolic rates and on blood flow and blood vessel density, all of which affect how tissue pO_2 relates to blood pO_2. Both nicotinamide and carbogen produce a similar $\beta NTP/P_i$ increase and about a twofold increase in blood glucose levels; however only carbogen produces a large R_2* decrease suggestive of improve oxygen delivery. Since R_2* increases following a TBF reduction due to CGRP, together these observations suggest that the GH3 prolactinoma utilises glucose at least as readily as oxygen. Clearly, R_2* changes will not indicate tissue pO_2 if there is a switch from oxidative to glycolytic metabolism during the vascular challenge.

The calibration of GRE images in terms of blood Y yields the product $A.bv$. However, it is $bv.Y_a$ which will scale with [dHb] and should relate more directly to tissue pO_2. From Eqs. (1) and (2) we can write:

$$\Delta R_2* = A\ \Delta[Hb] = A\ (\Delta bv\ Y + bv\ \Delta Y) \quad (4)$$

and NIRS, for example, can be used to calibrate GRE images [12]. However, the image in Fig. 1 illustrates the problem of applying a global change in average [dHb], obtained by NIRS, to R_2* changes that are clearly heterogeneous. Nevertheless, Eq. (4) demonstrates how ΔR_2* measurements can be used. The term A includes factors relating to blood vessel size, distribution and morphology. If assumed, then modulation of Y yields a ΔR_2* that relates to blood volume. This would imply that non-responding regions have very low blood volumes, consistent with therapeutically resistant hypoxic/necrotic areas, whilst responding regions of the tumour indicate the areas for potential treatment gain [7]. If Y remains constant then the changes in ΔR_2* relate to changes in blood vessel density, due to angiogenic activity for example [9].

In conclusion, although the calibration of GRE images in terms of tissue or blood oxygenation is fraught with difficulties, it clearly has great potential. To validate any quantitative interpretation of GRE MR images of tumors, NIRS and oxygen electrode measurements must be used, but information on tumor metabolism and vascularity is also likely to be needed for a correct interpretation. Nevertheless, qualitative changes observed by MR imaging in response to a physiological manoeuvre, may provide a non-invasive index of the heterogeneity of tumour oxygenation [16].

ACKNOWLEDGEMENTS

This work was supported by the Cancer Research UK Grant No. [SP1971/0402].

REFERENCES

1. Horsman MR. Measurement of tumor oxygenation. Int J Radiat Oncol Biol Phys 1998;42:701-704.
2. Ogawa S, Menon RS, Tank DW, Kim S-G, Merkle H, Ellerman JM, Ugurbil K. Functional brain mapping by blood oxygenation level-dependent contrast magnetic resonance imaging. Biophys J 1993;64:803-812.
3. van Zijl PCM, Eleff SM, Ulatowski JA, Oja JME, Ulug AM, Traystman RJ, Kauppinen RA. Quantitative assessment of blood flow, blood volume and blood oxygenation effects in functional magnetic resonance imaging. Nature Med 1998;4:159-167.
4. Al-Hallaq HA, River JN, Zamora M, Oikawa H, Karczmar GS. Correlation of magnetic resonance and oxygen microelectrode measurements of carbogen-induced changes in tumor oxygenation. Int J Radiat Oncol Biol Phys 1998;41:151-159.

5. Robinson SP, Howe FA, Rodrigues LM, Stubbs M, Griffiths JR. Magnetic resonance imaging techniques for monitoring changes in tumor oxygenation and blood flow. Sem Radiat Oncol 1998;8:197-207.
6. Howe FA, Robinson SP, Griffiths JR. Modification of tumor perfusion and oxygenation monitored by gradient recalled echo MRI and ^{31}P MRS. NMR Biomed 1996;9:208-216.
7. Robinson SP, Collingridge DR, Howe FA, Rodrigues LM, Chaplin DJ, Griffiths JR. Tumour response to hypercapnia and hyperoxia monitored by FLOOD magnetic resonance imaging. NMR Biomed 1999;12:98-106.
8. Howe FA, Robinson SP, Rodrigues LM, Griffiths JR. Flow and oxygenation dependent (FLOOD) contrast in MR imaging to monitor the response of rat tumors to carbogen breathing. Magn Reson Imag 1999;17:1307-1318.
9. Abramovitch R, Frenkiel D, Neeman M. Analysis of subcutaneous angiogenesis by gradient echo magnetic resonance imaging. Magn Reson Med 1998;39:813-824.
10. Vaupel P, Kallinowski F, Okunieff P. Blood flow, oxygen and nutrient supply, and metabolic environment of human tumors: a review. Cancer Res 1989;49:6449-6465.
11. Dewhirst MW, Ong ET, Rosner GL, Rehmus SW, Shan S, Braun RD, Brizel DM, Secomb TW. Arteriolar oxygenation in tumor and subcutaneous arterioles: effects of inspired air content. Br J Cancer 1996;74(suppl XXVII):241-246.
12. Punwani S, Ordidge RJ, Cooper CE, Amess P, Clemence M. MRI measurements of cerebral deoxyhæmoglobin concentration [dHb] – correlation with near infrared spectroscopy (NIRS). NMR Biomed 1998;11:281-289.
13. Lin WL, Paczynski RP, Celik A, Kuppusamy K, Hsu CY, Powers WJ. Experimental hypoxemic hypoxia: changes in R_2* of brain reflect combined effects of changes in arterial and cerebral venous oxygen saturation. Magn Reson Med 1998;39:474-481.
14. Hirst DG, Kennovin GD, Flitney FW. The radiosensitiser nicotinamide inhibits arterial vasoconstriction. Br J Radiol 1994;67:795-799.
15. Dunn JF, Zaim-Wadghiri Y, Pogue BW, Kida I. BOLD MRI vs. NIR spectrophotometry. Will the best technique come forward? In: Hudetz, Bruley, editors. Oxygen Transport to Tissue XX. New York: Plenum, 1998;103-113.
16. Okunieff P. Towards noninvasive human tumor physiologic measurements. Int J Radiat Oncol Biol Phys 1995;33:961-962.

Chapter **42**

FMRI FOR MONITORING DYNAMIC CHANGES IN TISSUE OXYGENATION/BLOOD FLOW: POTENTIAL APPLICATIONS FOR TUMOR RESPONSE TO CARBOGEN TREATMENT

Jianhui Zhong[x,+], W. C. Edmund Kwok[x], & Paul Okunieff[*]
[x]*Department of Radiology,* [+]*Biomedical Engineering, and* **Radiation Oncology, University of Rochester, Rochester, NY, USA*

Abstract: The ability to differentiate between well-oxygenated and poorly-oxygenated tumors may play an important role in selecting an optimal therapeutic regime for tumor treatment of the individual patient. We present preliminary results in the development of a dynamic functional MRI method for mapping tissue oxygenation and blood flow distribution in humans simultaneously. We applied interleaved Blood Oxygenation Level Dependent (BOLD) and Flow-sensitive Alternating Inversion Recovery (FAIR) sequences to detect signals as a subject is inspiring gases of varying oxygen concentration. The method allows quantitation of the spatial distribution and time course of the important physiological functions that are easily registered with high resolution anatomic MR images. It may be used to critically evaluate the efficacy of varying durations of carbogen breathing in tumor patients, and allow a quantitative evaluation of the roles of carbogen and other radiosensitizers as potential adjuncts to radiotherapy and drug therapies.

Key words: oxygenation; blood flow; fMRI; tumor

1. INTRODUCTION

The use of deoxyhemoglobin as an endogenous contrast agent for MRI signal acquisition was first proposed by Ogawa et al [1]. This contrast is based on the difference between the magnetic susceptibility of deoxygenated hemoglobin and that of tissue surrounding the vasculature. Deoxyhemoglobin is diamagnetic, leading to a more efficient magnetic relaxation and faster decay in the MR signal. Whereas oxyhemoglobin is paramagnetic and has similar relaxation properties as that of the surrounding tissue. As a consequence, MRI pulse sequences can be designed to produce the so called blood-oxygenation-level-dependent image contrast, which reflects changes in the supply-and-demand balance of tissue oxygenation, accompanying metabolic changes during physiological processes such as neuronal activation.

Following a phenomenal number of applications for studies of brain somatosensory and cognitive functions using the BOLD functional MR imaging (fMRI) reported over the past 6 years or so, more recent works have focused on the quantitative relationship of the fMRI signal change during activation with underlying physiological parameters. van Zijl et al [2] have derived a functional form of the relaxation rate, R_2, focusing on its dependence on physiological parameters in the brain, which include cerebral blood volume (CBV), cerebral blood flow (CBF), oxygen metabolic rate ($CMRO_2$), blood hematocrit ([Hb]) and the oxygen saturation fraction (Y). Blood volume maps at specific oxygen levels were directly calculated from cat brains during a hypoxic insult. An intravascular contrast agent, Monocrystalline Iron Oxide Nanocompounds (MION), was used by Dennies et al [3] to measure ΔR_2 and ΔR_2*, the differences between relaxation rates pre- and post-contrast agent injection, in a rat glioma model. Since $\Delta R_2^*/\Delta R_2$ increases as vessel size increases, this ratio was used as a measure of the average vessel size within a Region-Of-Interest (ROI) or a single voxel. The technique supports the hypothesis that susceptibility contrast MRI can provide useful quantitative metrics of *in vivo* tumor vascular morphology. Davis, Rosen and coworkers [4] calculated cerebral metabolic rates for the $CMRO_2$ from acquired BOLD/flow data. If $[O]_a$ and $[O]_v$ are oxygen concentrations at arterial and venous sides of the circulation respectively, Fick's equation for conservation of oxygen delivery and uptake suggests that

$$CBF \times ([O]_a - [O]_v) = CMRO_2$$

leading to

$$dHb \sim CMRO_2 / CBF,$$

where dHb is the production of deoxyhemoglobin. A dynamic $CMRO_2$ map can then be calculated from simultaneous measurements of the time course of the BOLD signal (reflecting dHb) and flow-weighted images (CBF).

We used an interleaved BOLD [1] and Flow-sensitive Alternating Inversion Recovery (FAIR) [5] fMRI method to measure the dynamic response of the brain to normoxic, hyperoxic, and hyperbaric conditions. Signals were acquired continuously as a subject was inhaling gases of varying oxygen concentrations. The rational for the design of experiments includes the following:

(1) The delineation of tumors which incorporate host blood vessels that can respond to vasomodulators from those that do not, has significant implications in choice of therapy [6]. The ability to directly monitor the response of tumor blood vessels to variation in respiratory treatment will therefore aid in the timing of radiation treatment and the use of vasomodulators to localize anticancer drugs within the tumor;

(2) Both BOLD and FAIR are completely non-invasive and suitable for dynamic measurements when implemented with fast MR imaging techniques such as echo-planar imaging (EPI) acquisition, and therefore are easily adaptable for patients;

(3) Acquisition parameters can be chosen so that a measure of blood flow rate, and oxygenation can be achieved in an interleaved fashion and used to monitor dynamic changes in blood oxygenation and flow. With the time resolution of a few seconds, and coverage of a large volume including tumor and surrounding normal tissues, comparisons of physiological parameters in different regions under exactly the same conditions can be achieved.

2. METHODS

2.1 Estimate of ΔpO_2 from R^*_2

Even though the BOLD signal in general originates from a combination of physiological sources (e.g. blood oxygenation, blood flow, vascular architecture, etc), parameters can be chosen to primarily detect changes in blood oxygenation and overall magnetic susceptibility effects. We used a pulse sequence with long repetition time, TR, and small flip angle, which allows for an effective elimination of the majority of blood in-flow signals. The signal intensity at low (S^l) and high oxygenation states (S^h) can be written as:

$$S^l = S_0 \exp\left(-R_2^{*l} \times TE\right),$$

$$S^h = S_0 \exp\left(-R_2^{*h} \times TE\right) = S_0 \exp\left(-R_2^{*l} \times TE\right) \times \exp\left(\Delta R_2^* \times TE\right),$$

$$E = \frac{S^h - S^l}{S^l} = \exp\left(\Delta R_2^* \times TE\right) - 1 \approx \Delta R_2^* \times TE,$$

where E is the fractional change in signal intensities, TE is the echo time, S_0 is the spin density, and ΔR_2* is the decrease in relaxation rate R_2* due to the changes in tissue oxygenation. Therefore,

$$\Delta R_2^* = E/TE.$$

If we assume $\Delta R_2^* = k \times \Delta pO_2$, where k is a calibration constant depending on the vascular volume, we obtain

$$\Delta pO_2 = E/(k \times TE).$$

The assumption about the proportional changes in ΔR_2^* and ΔpO_2 is supported by several animal [7,8] and human [9] studies. Lin et al [7] utilizes a rat model of acute normovolemic hemodilution, and found that changes in R_2^* values measured with MR were in close agreement with the expected linear relationship predicted from theory when the measured changes in blood hematocrit were used as independent variables. Prasad et al used BOLD MRI to evaluate intrarenal oxygenation in humans [9], and a good linear relationship was found between the tissue R_2^* and changes in renal oxygenation. Our own measurements inarterial blood pH, pCO_2, and pO_2 in rat undergoing normoxia/hyperoxia [8] also provide supporting evidence that there is minimum contribution of blood flow to the BOLD signal under certain conditions (long repetition time and small flip angle).

2.2 Estimate of blood flow (perfusion) from FAIR

Applications of the FAIR sequence allows direct quantitative measurement of blood flow (perfusion) [5]. From a sequence with a preparation period with either a slice-selected or a non-slice selective inversion pulse with inversion time corresponding to the optimal blood transit time, a "flow-weighted image" can be obtained from the difference between the two images, as follows:

$$\text{flow} - \text{weighted image} = \Delta S = S_{sel} - S_{non-sel}.$$

A perfusion image can then be calculated where image intensity is directly proportional to the cerebral blood perfusion rate in the unit of ml blood per 100g tissue per minute [5]:

$$\text{perfusion } f = \Delta S / (\lambda \cdot TI \cdot S_0 \cdot F(T1, TI, TR))$$

where λ is the blood/tissue partition coefficient (= 0.9 for human brain), T_1 is the spin-lattice relaxation time for the tissue, TR and TI are repetition and inversion time respectively, and

$$F(T1, TI, TR)) = 2\exp(-TI/T1) - \exp(-TR/T1).$$

T_1 and S_0 can be obtained from acquisitions of inversion-recovery images with at least two TIs.

In our interleaved acquisition of the BOLD/FAIR signal under normoxic/hyperoxic conditions, measurements of tissue oxygenation and blood flow responding to changes in the breathing oxygenation concentration can be obtained. Figure 1 displays a representation of the acquisition and data analysis scheme.

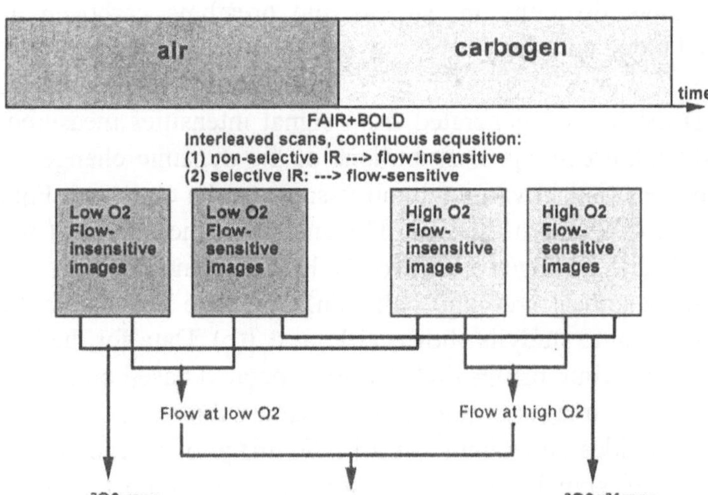

Figure 1. Schematic representation of the experimental design and data analysis for quantitative measurements of changes in cerebral blood flow and oxygenation with carbogen breath and an interleaved BOLD/FAIR acquisition.

3. RESULTS

We used a 1.5T GE Signa scanner for imaging acquisition. In our implementation of BOLD/FAIR pulse sequences, the gradient-recalled echo EPI acquisition was used. Series of images with TR = 4000 ms, TE = 40 ms, TI = 1200 ms were acquired, for which either a slice-selected or a non-slice selective inversion pulse was used. Images acquired with either a slice selective or a non-selective inversion pulse by themselves are BOLD images: the time course of the images reflect changes in tissue oxygenation due to inhaling gases. The difference in image intensity from each image pair was calculated, yielding the flow-weighted image for which the image intensity is proportional to the blood perfusion.

Three normal volunteers (all males) have been studied in multiple sessions of imaging for the brain and leg muscles. The functional MRI protocol was approved by the Human Studies Committee at the School of Medicine and Dentistry of the University of Rochester. Pulse rate, breathing rate and blood oxygen saturation of the subjects were monitored with a pulse oximeter (MR Equipment Corp, Model6500). Oxygen uptake was controlled with gas delivered via a close-fitting face mask during the entire scan. In a typical experiment the subject was asked to breathe normally with air, pure oxygen or carbogen at the same flow rate (~15 l/min). Previous studies have shown that carbogen breathing delivered by this method at similar rates was tolerated without difficulty in patients and breathing carbogen does not influence red blood cell flux [10].

Figure 2 shows results from a brain study of a single subject. Flow/BOLD maps were generated from signal intensities measured during air and carbogen breathing. Distributions of the dynamic changes in blood flow and oxygenation in various brain tissues can be observed. Figure 3(a) displays the time course of the BOLD signals in a measurement where the air was switched to carbogen after two and half min, and then again switched to 100% after another three and half min. The time course for the FAIR signal from the same study is shown in Figure 3(b). Data for the brain gray matter (GM) and white matter (WM) were separated based on segmentation of the T2-weighted images acquired for the same subject.

Figure 4 provides an example of a FAIR image of a brain astrocytoma. Heterogeneity of signal intensity is seen for the tumor: Higher signal intensities are localized to the periphery of the astrocytoma. This is typical where the core of tumors becomes necrotic with reduction in blood flow.

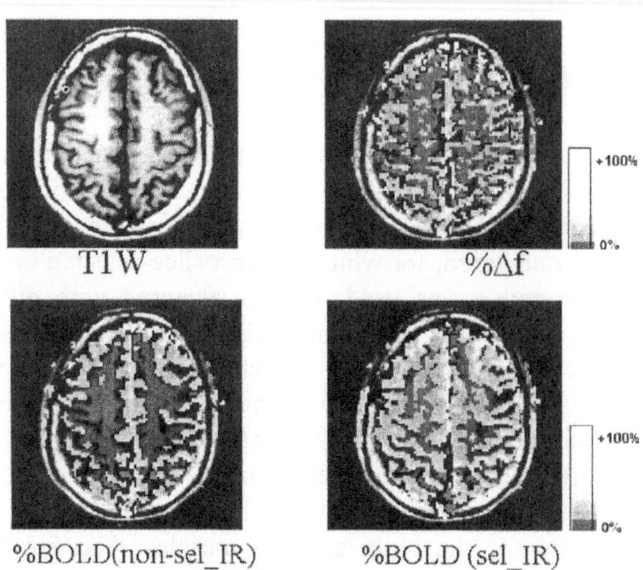

Figure 2. Spatial distribution of changes in the cerebral blood flow and oxygenation with carbogen breath from an interleaved BOLD/FAIR acquisition in a normal volunteer. Percentage changes in the blood flow and BOLD signals are superimposed onto the T1W image of the same location.

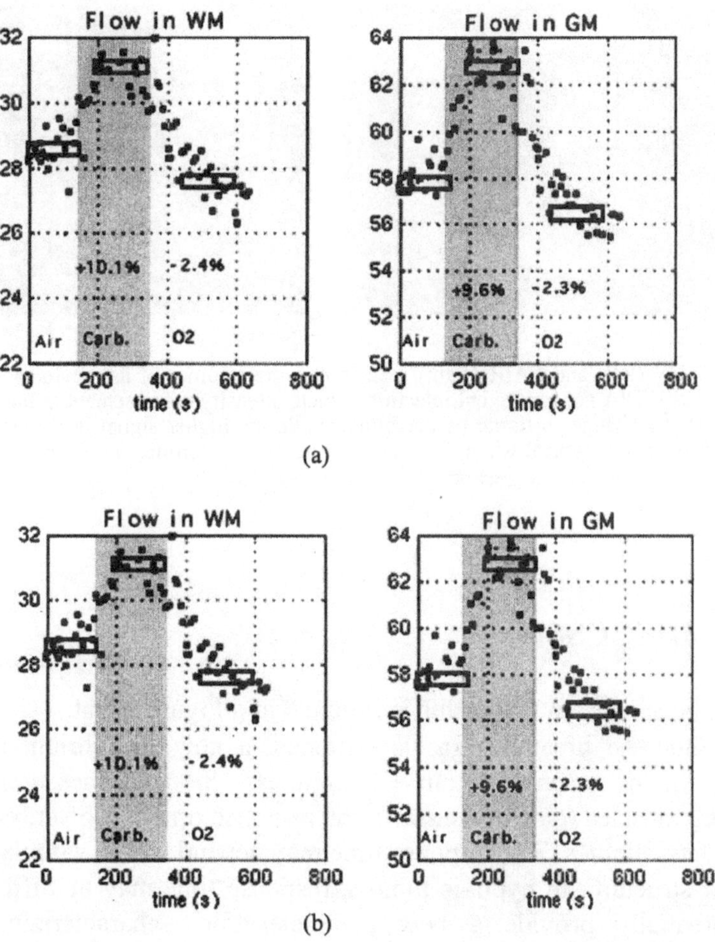

Figure 3. (a) Time course for the BOLD signal in a measurement where the air was switched to carbogen after two and an half min, and then again switched to 100% after another three and half min. (b) The time course for the FAIR signal from the same subject. Data for the brain GM and WM were separated based on segmentation of the T2W images acquired for the same subject. Percentage changes of the corresponding parameters during carbogen or 100% O2 compared with air are calculated from the average values indicated by hollow bars superimposed on the data.

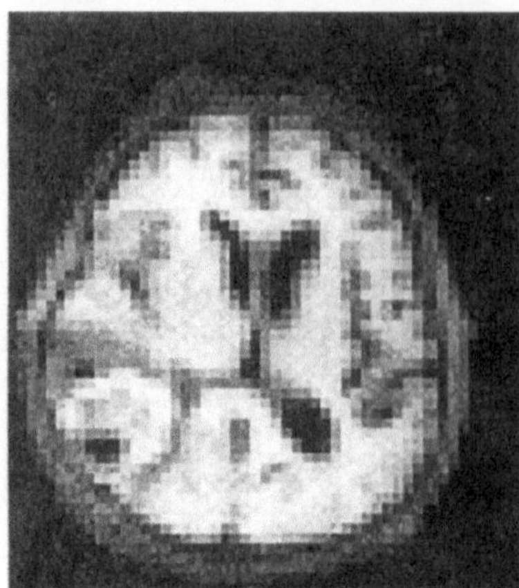 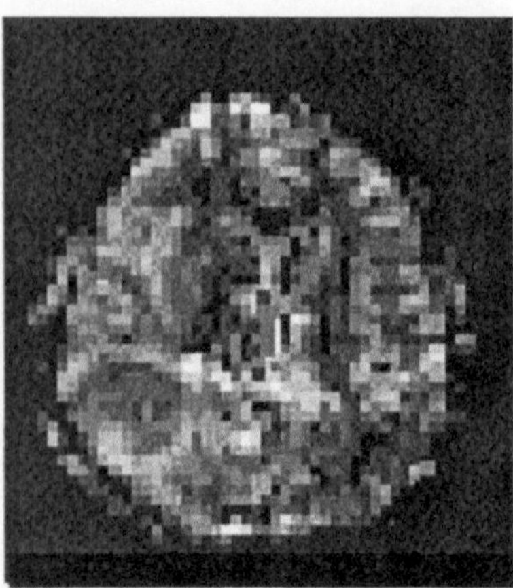

Figure 4. T1W (left) and FAIR (right) images in a brain tumor of astrocytoma. The T1W image post Gd-DTPA contrast agent injection depicts intensity enhancement at the rim of the tumor due to the change in tissue permeability, while the higher signal in the FAIR image along the tumor rim is typical when the center of a tumor is necrotic, and therefore has lower blood flow.

4. DISCUSSION

It can be seen from the results in Figure 3 and Figure 2 that:

(a) When the breathing gas is switched, a new equilibrium in signal intensity can be reached within 1-2 minutes. In other measurements in human leg muscles (data not shown), the response time is slower, but is still within a few minutes. The response time may depend on the vasculature of a particular structure. In hypoxic tumors, response time may be different and may potentially provide a new parameter for characterizing tumor pathology.

(b) The ratio of the flow signals in the brain gray matter (GM) and white matter (WM) is about 2 to 1, which is in the range of other brain perfusion measurements [5,11].

(c) Carbogen is known to be a vasodialator, and it is believed that pure O_2 either has a minimum effect on blood flow or can act as a vasoconstrictor. Our results indeed suggest a large flow-related increase in signal intensity when carbogen is administered, whereas brain tissue either displays a relatively much smaller increase, or even a decrease in flow signal, when pure oxygen is administered. This also correlates with our measurement of pCO_2 in rat brains (data not shown), where pCO_2 was found unchanged for pure O_2 uptake, but increased substantially with the breathing of carbogen. The pH was found to remain constant, but pO_2 increased under both hyperoxic and hypercarbic conditions. Administration of carbogen produces both increases in blood flow and blood oxygenation, whereas hyperoxia predominantly produces an increase in blood oxygenation alone.

(d) The BOLD signal increases when the oxygen concentration is increased, but the increase due to carbogen is greater than that due to O_2. This suggests that when CO_2 is combined with O_2 at the same pressure, an increased flow effectively raises the concentration of the oxyhemoglobin under similar partial pressures of pure O_2. Carbogen therefore may potentially be more effective for the tumor treatment as a radiosensitizer. However, the study of tumors by Thews et al [14] have suggested that at normobaric hyperoxia (1 atm), O_2 and carbogen intake results in approximately the same increase in tumor oxygenation.

(e) The BOLD signal increases in both WM and GM when the oxygen concentration is increased, but the changes in GM are larger than in WM, even though the percentage increase in flow is similar for both. The higher cerebral volume in the GM may facilitate more efficient delivery of oxygen in the GM than in the WM.

Several effects need to be considered when analyzing the fMRI signal change during the varying oxygenation: (a) Increases in arterial blood oxygen concentration result in a decreased deoxyhemoglobin concentration ; (b) If there is an increase in blood flow without proportional changes in oxygen uptake (consumption), there is a decrease in venous deoxyhemoglobin concentration; (c) A secondary effect of vasodilation and increase blood flow is an increase in-flow effect; (d) Density of blood vessels in the tissue determines the deoxyhemoglobin concentration per unit volume of tissue and therefore the magnitude and distribution of the magnetic field inhomogeneity. (a) and (c) cause an increase, and (b) causes a decrease in the BOLD signal. However, (b) and (c) become effective only when carbogen is applied, and can be detected directly via FAIR signal changes.

BOLD contrast has been successfully applied in numerous brain functional imaging studies. In tumors, T2*, resonance signal linewidth, and resonance frequency shift, which all reflect different aspects of the BOLD effect, have been measured to evaluate tumor oxygenation and response to hyperoxia. Since the BOLD phenomena include effects from both oxygenation and flow, whereas the FAIR methods allow independent measure of blood flow, combination of BOLD/FAIR and carbogen/O_2 should allow a better quantitative evaluation of each component. It may allow for example distinguishing areas with hypoxia with regular blood flow (high signals in both BOLD and FAIR) from those hypoxic with low flow (high signals in BOLD but low in FAIR). Furthermore, the BOLD/FAIR combination allows calculations of $CMRO_2$, which completes the description of the supply-and-demand cycle in tumor blood microcirculation. The intrinsic interleaved acquisition mode in this sequence design minimizes the artifacts in the BOLD and flow calculation and permits registration of the results with high-resolution anatomical MR images.

Carbogen has been evaluated as a radiosensitizer in humans but results from randomized controlled clinical trials in the past were found to be disappointing [13]. Lack of quantitation of spatial oxygen distribution in tumors and a non-invasive mechanism to serially measure that distribution efficiently may have limited the design of clinical trials. The fMRI method allows quantitation of spatial and temporal distributions of the important physiological functional information, which can be easily registered with high resolution anatomic information obtained by MRI in the same subject. We will measure the temporal response of tumors to hyperoxic/hyperbaric conditions and determine the relative contributions of blood flow and oxygenation during carbogen delivery. This method can be used to critically evaluate the efficacy of varying durations of carbogen inhalation in tumors, and allow a quantitative evaluation of the roles of carbogen breathing as a possible useful adjunct to radiotherapy and drug therapies. It may also supplement histological classification and clinical staging and grading in order to design the best individualized therapeutic treatment. The combination of a mathematical model, physiological calibration, and pulse sequence design should provide the necessary tools for noninvasive dynamic mapping of blood circulation and oxidative metabolism, essential to the application of fMRI techniques to diseased tissue, including tumors. The technique may also provide tools to seek for answers for the important questions about physiological control of cerebral blood flow.

REFERENCES

1. Ogawa S, Lee TM, Kay AR, Tank DW. "Brain magnetic resonance imaging with contrast dependent blood level oxygenation. Proc Natl Acad Sci (USA), 1990;87:9868-9872.
2. van Zijl PCM, Eleef SM, Ulatowski JA, Oja JME, Ulug AM, Traystman RJ & Kauppinen RA. Quantitative assessment of blood flow, blood volume and blood oxygenation effects in functional magnetic resonance imaging. Nature Medicine, 1998;2:159-157.
3. Dennie J. Mandeville JB. Boxerman JL. Packard SD. Rosen BR. Weisskoff RM. NMR imaging of changes in vascular morphology due to tumor angiogenesis. Magn Reson Med 1998;40:793-799.
4. Davis TL, Kwog KK, Bandettini PA, Weisskoff RM, & Rosen BR. Mapping the dynamics of oxidative metabolism by functional MRI. ISMRM, 1997;151.
5. Kim S-G, Tsekos NV. Perfusion imaging by a flow-sensitive alternating inversion recovery (Fair) techniques: application to functional brain imaging. Magn Reson Med 1997;37:425-435.
6. Robinson SP, Howe FA, Rodrigues LM, Stbbs M, & Griffiths, JR. Magnetic resonance imaging techniques for monitoring changes in tumor oxygenation and blood flow. Seminars in Radiat Onc 1998;8:197-207.

7. Lin W. Paczynski RP. Celik A. Hsu CY. Powers WJ. Effects of acute normovolemic hemodilution on T2*-weighted images of rat brain. Magn Reson Med 1998; 40:857-864.
8. Zhong J, Fulbright RK, Kennan RP, & Gore JC. Quantification of intra- and extravascular contributions to BOLD effects induced by alteration in oxygenation or intravascular contrast agents. Magn Reson Med 1998; 40:526-536.
9. Prasad PV, Chen Q, Edelman RR, Epstein FH. Non-invasive evaluation of intra-renal oxygenation using BOLD MRI. Circulation 1996; 94:3271-3275.
10. Powell, M.E.B, S.A. Hill, M.I. Saunders, P.J. Hoskin, and D.J. Chaplin, Effect of carbogen breathing on tumor microregional blood flow in humans. Radiotherapy and Oncology, 1996;41,225-231.
11. Yang, Y., J.A. Frank, L. Hou, F.Q. Ye, A.C. McLaughlin, and J.H. Duyn, Multislice imaging of qunatitative cerebral perfusion with pulsed arterial spin labeling. Magn Reson Med. 1998; 39:825-832.
12. Thews O, Kelleher DK, and Vaupel P, Tumor oxygenation under normobaric and hyperbaric hyperoxia. Adv Expt Med Biology 1997;428,79.
13. Rubin, Hanley PJ, Keys HM, Marcial V, & Brady NL, Carbogen breathing during radiation therapy. Int J Radiol Oncol Biol Phys 1979;5,1963-1970.
14. Thews O, Kellleher DK, & Vaupel P. Tumor oxygenation under normobaric and hyperbaric hyperoxia. Oxygen Transport to Tissue XIX, ed. Harrison and Delpy, Plenum Press, New York, 1997.

Chapter 43

COMPARISON STUDY OF OXYGEN-INDUCED MRI-SIGNAL CHANGES AND pO_2 CHANGES IN MURINE TUMORS

Lothar Weissfloch[1,5], Michael Peller[2], Juergen Weber[2], Hans-Juergen Feldmann, Reingart Senekowitsch-Schmidtke[3], Karlheinz Tempel[4], Jeffrey A. Coderre[5], Michael Molls[1], and Michael Reiser[3]

[1] *Dept. of Radiation Oncology, Klinikum rechts der Isar, TU Munich, Germany;* [2] *Inst. of Diagnostic Radiology, Klinikum Grosshadem, LMU Munich, Germany;* [3] *Dept. of Nuclear Medicine, Klinikum rechts der Isar, TU Munich, Germany;* [4] *Inst. of Pharmacology, Toxicology and Pharmacy, Faculty of Veterinary Medicine,LMU Munich, Germany;*[5] *Medical Department, Brookhaven National Laboratory, Upton, NY, ,USA*

Abstract: The purpose of this study was to compare the results from oxygen-induced MR-signal intensity changes with polarographic pO_2 measurements in tumors. Balb-c mice with an intramuscular transplanted osteosarcoma were examined. To study the response of tumors to changes in oxygen supply, hyperoxia was induced by breathing pure oxygen for a short period. The examination of each animal started with T2* weighted MRI followed by the pO_2 measurements (Eppendorf Histograph). During oxygen inhalation in all tumors, when the hypoxic tumor fraction drops, both areas of significant MR-signal intensity increase and decrease were observed in each animal.

Key words: Tumor, rodent, hyperoxia, oxygen tension measurement, pO_2, MRI

1. INTRODUCTION

The measurement of tissue oxygenation may be an important factor in tumor treatment planning and prognosis. A non-invasive method has been reported that uses signal intensity changes in T2* weighted magnetic resonance images (MRI) during hyperoxia to differentiate tumor from normal tissue in various tumor models [1, 6]. The exact mechanism of the oxygen-induced MR-signal-changes is not yet completely understood.

Changes in deoxyhemoglobin concentration [2] and in blood flow [3] may be responsible for the signal change. It has been shown that the induced signal intensity changes are reproducible when hyperoxia is repeated [6]. The reproducibility allows a repetition of the experiments for further examination of the effect outside the MRl-unit. A validated method to examine the oxygenation status of tumors is the polarographic oxygen tension (pO_2) measurement [4]. The purpose of this study was to compare the results from oxygen-induced MR-signal intensity changes with pO_2 measurements in tumors.

2. METHODS

10 Balb-c mice (28-32 g) with an intramuscular transplanted OTS 64 osteosarcoma were examined [5]. The average maximum diameter of the tumors was 1.6. $\pm$ 0.3 cm. The mice were anesthetized by an IP-injection of 0.25 ml of a solution of 2.5 ml tiletamine hydrochloride (Tilest500, Parke-Davis), 4 ml xylazine (Rompun, Bayer) and 35 ml 0.9% NaCl. The body temperature of the animals was kept constant at 34° $\pm$ 2° C. To study the response of tumors to changes in oxygen supply, hyperoxia was induced by breathing 100% oxygen during a 10 min interval. The examination of each animal started with the MR experiments followed by the pO_2 measurements. All experiments were performed on a 1.5 Tesla whole-body scanner (Magnetom Vision, Siemens, Germany) using a surface coil. During the experiments T2* weighted FLASH 2D (TE = 22 ms, TR =44 ms, FOV = 80 mm, matrix = 64 x 128, slice thickness = 2 mm) images were acquired every 10 sec.

Maps of signal intensity changes to normoxic conditions were calculated pixel by pixel and tested by applying the unpaired Student's t-test ($p < 0.05$). These maps were used to determine whether signal increase or decrease pre-dominated in the tumors. For this purpose all voxels in the tumor area were sorted by their signal change amplitude in steps of 2% in a range between -80% and +100% signal change. From this data the median relative MR-signal intensity change of the tumor was determined. Additionally, the relative amount of tumor voxels showing either significant increase or decrease were calculated. For the pO_2 measurements an Eppendorf pO_2-Histograph (Eppendorf, Germany) with polarographic needle electrodes of

0.3 mm in diameter were used. The needle probes were moved automatically through the tumor in steps of 0.7 mm [5]. An average number of 130 points were evaluated. Two sets of measurements were done, one before and one after 5 min of 100% oxygen breathing. Median pO_2 changes were calculated and tested by the Wilcoxon-Mann- Whitney test ($p < 0.05$).

3. RESULTS & DISCUSSION

During oxygen inhalation in all tumors, both areas of significant MR-signal intensity increase and decrease were observed in each animal. Signal increase was pronounced in seven of the ten tumors (Table 1). Median signal intensity changes were in those seven animals between +4% and +14%. In three of the animals, the median signal intensity change was between -6% and -8%.

Table 1. Change of median MR signal vs. median pO_2

		Change of median MR signal	
		↑	↓
Change of median pO_2	↑	**3**	1
	↓	1	1
	ns	3	1

The median pO_2 values in the tumor were according to earlier results for those tumors [5]. Four tumors showed significant increase of the median pO_2 values and one tumor a significant decrease. In four cases no significant median change was found. Comparing those animals showing a change of the hypoxic tumor fraction below 5 mm Hg (9 out of 10) and the median MR-signal intensity change (Table 2), there were five animals showing a

Table 2. Change of median MR signal vs. tumor fraction

		Change of median MR signal	
		↑	↓
Change of tumor fraction ($pO_2 < 5$ mmHg)	↑	1	1
	↓	**5**	2
	=	1	0

decrease of the hypoxic tumor fraction during oxygen breathing that also showed a median MR-signal intensity increase. One animal showed an increase in $pO_2 < 5$ mm Hg and a drop in MR-signal during oxygen breathing. The other three animals showed either increase or decrease in both, the hypoxic fraction and the median MR -signal change.

4. CONCLUSION

There seems to be a trend in this tumor model that when the hypoxic tumor fraction drops during oxygen breathing, the MR-signal intensity will increase or vice versa. Physiological processes dependent on oxygen supply in the tumor can be observed with MRI and pO_2 measurements. MRI offers the opportunity to access oxygen dependent processes for each image voxel. This is in contrast to the pO_2 measurements which deliver trends for the whole tumor since only random points can be accessed. Oxygen induced MR-signal intensity changes may yield additional information to the pO_2 measurements on oxygen-dependent physiological processes in tumors. Additional study of this methodology is planned.

ACKNOWLEDGEMENTS

Financial support is gratefully acknowledged from Bayerisches Staatsministerium für Landesentwicklung und Umweltfragen, München, Germany, and Deutsche Forschungsgemeinschaft, Bonn, Germany, as well as from the Medical Department of the Brookhaven National Laboratory, Upton, NY, USA.

REFERENCES

1. Kuperman, V.Y, River J.N, Lewis M.Z et al, Magn Reson Med 1995;33:318.
2. Ogawa S, Lee T.M, Magn Reson Med 1990;14:68.
3. Karczmar G.S, Kuperman V.Y, Lewis M.Z et al, SMR/ESMRMB, 1995. p.1678.
4. Lyng H, Sundfor K, Rofstad E.K, Radiother Oncol 1997; 4:163.
5. Weissfloch L, Auberger T, Wagner F.M et al, Adv Exp Med Biol 1999;338:495.
6. Peller M, Weissfloch L, Stehling M.K et al, MRI, 1998;16:799-809.

Chapter **44**

VENOUS-ARTERIOLAR REFLEX IN HUMAN *GASTROCNEMIUS* STUDIED BY NIRS

Tiziano Binzoni[1], Loan Ngo[1], Massimo Girardis[2], Roger Springett[3], François Terrier[1] & David Delpy[3]

[1]*Depts. of Physiology and Radiology, University of Geneva, Switzerland;* [2]*Cattedra di Anestesiologia e Rianimazione, Università degli Studi, Udine, Italy;* [3]*Dept. of Medical Physics and Bioengineering, University College London, UK.*

Abstract: Heat-up tilting manoeuvre from 0 to 60 degrees induces oxygenated and deoxygenated haemoglobin concentration changes in the human *gastrocnemius*. These changes, measured by NIRS, can only be partially explained by the blood volume displacement due to the gravitational force. In the present study it is demontrated, by a dye dilution technique (indocyanine green), that a reduction in blood flow (venous-arteriolar and/or spinal reflex) is responsible of the limited oxyhaemoglobin concentration increase observed when going from 0 (2.54 ± 0.48 blood flow in arbitrary units, a.u.) to 60 (1.46 ± 0.55 a.u.) degrees. The proposed technique is potentially applicable to the detection of specific pathological aspects of microcirculation, such as arterial occlusion in the leg, diabetes mellitus, and congestive heart failure, where the venous-arteriolar reflex may be affected.

Key words: human, skeletal muscle, blood flow, near infrared spectroscopy.

1. INTRODUCTION

The superficial vascular compliance, in the lower legs, can be assessed by NIRS during a head- up tilting manoeuvre [1]. When such a study is performed, it is observed that when going from -10 to 75 degrees, deoxygenated blood (Hb, ~7 ml l^{-1}) accounts for nearly all the increase in blood volume (~9 ml l^{-1}) observed in the legs (see Figure 1).

To explain this observation, it was hypothesized [1] that the limited increase in oxygenated blood (HbO_2, ~2 ml l^{-1}) content during tilt might be due to a blood flow reduction, probably derived from a venous-arteriolar and/or spinal reflex. For constant metabolic conditions, the blood flow

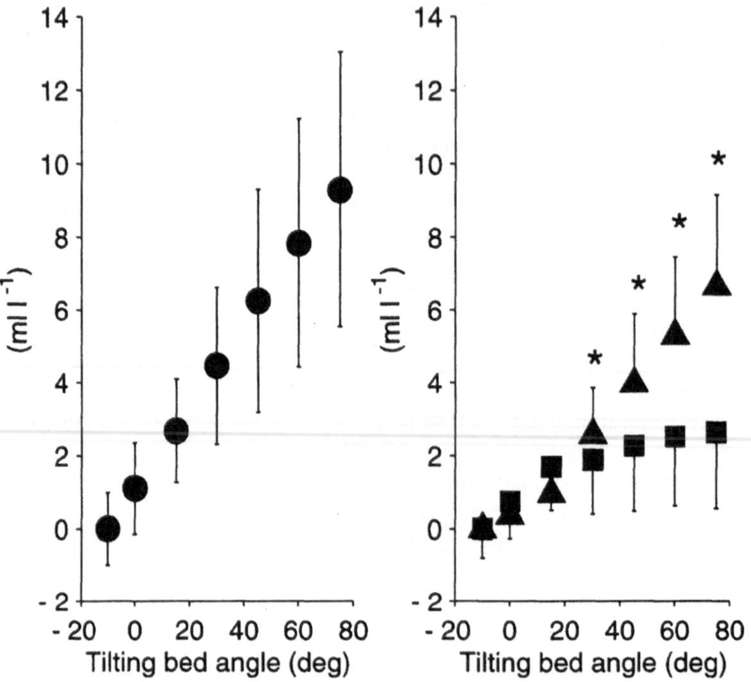

Figure 1. Left) Increase in blood volume as a function of the tilting table position in the *gastrocnemius medialis* of 11 healthy subjects. 0 degrees corresponds to the horizontal position. The vertical bars represent the standard deviation. A mean DPF of 5 was used for all the measurements [5]. Concentration in μM was transformed in ml l^{-1} using: ml l^{-1} = μM x (64000/ 15) x 10^{-4}. **Right)** Increase in oxygenated (squares) and deoxygenated (triangles) blood as a function of the tilting table position. The experimental points above 15 degrees are significantly different (*, $p < 0.05$). (Figure adapted from [1]).

reduction induces an increase in oxygen extraction from haemoglobin and, as a consequence, limits the change in HbO_2.

The aim of the present study was to verify, by NIRS, the hypothesis that the head-up tilting manoeuvre reduces blood flow in human *gastrocnemius*.

2. METHODS

Five healthy volunteers participated at this study (38 ± 8 years, 179 ± 6.2 cm, 72.8 ± 6.3 kg). The protocol was in accordance with the guidelines of the ethical committee of the Medical School of the University of Geneva. The subject was placed on a tilting table (HESS, Dübendorf, Switzerland) in a supine position (see Figure 2).

One index of blood flow (perfusion index, in arbitrary units, a.u.) was assessed successively at 0, 60 and 0 degrees tilt using a dye dilution technique [3] employing indocyanine green (ICG, Cardiogreen, Fluka Chemie, Switzerland), 2 ml (10 mg) of which was injected into the basilic vein in 0.5 sec using an automatic injector (MARK V Plus, Medrad Technology for people, Pittsburgh PA, USA). To study the same region of interest where HbO_2 and Hb measurement are usually performed, the ICG intramuscular dilution curve was measured using a NIRS spectrophotometer. The optical fibers were attached to a white light source (Oriel 77501, Stratford CT, USA) and a CCD based NIR spectrophotometer [4] from which it was possible to measure the near-infrared spectrum over the range 677 to 998 nm. The optodes were positioned, 3 cm apart, on the medial line of the upper third of the right *gastrocnemius medialis*. NIR spectra were measured at 1.5 sec interval over a period of 20-30 min, necessary for the ICG concentration to reach again the baseline levels. Hb, HbO_2 and ICG were calculated using a mean DPF=5 [5]. ICG reference spectrum was obtained using human plasma as a dilution medium. The ICG concentration curve was plotted on a semilogarithmic scale and corrected for recirculation [6] using software specifically written for that purpose (Matlab, The Mathworks Inc., Natick, MA, USA). A new ICG baseline was used for each measurement to eliminated the error induced by the small amount of ICG remaining in the circulation from the previous measurements. The ICG curve was then integrated (A) and the perfusion index was then obtained as: perfusion index=1/A. The blood flow index was assessed after Hb and HbO_2 had reached steady-state values.

Data are reported as mean values ± S.D. Significance of differences was analysed by means of Student's paired *t* test.

Figure 2. Subject lying supine on the tilting table. The tilting table angles can be set electronically from -10° to 75°.

3. RESULTS

In Figure 3. it is shown a typical ICG concentration tracing as a function of time for one subject positioned at 0 degrees. The second maximum appearing in the figure corresponds to the first recirculation of the ICG. The lower decreasing line represents the ICG curve corrected for recirculation [6].

Perfusion index values were 2.54 ± 0.48, 1.46 ± 0.55 and 3.45 ± 1.10 a.u. at 0, 60 and 0 degrees tilt, respectively. As expected, a significant ($p<0.05$) reduction in blood flow was observed at 60 degrees (see Figure 4). The same blood flow behaviour is observed if the data of each subject are normalized at 100 % for the first measurement at 0 degrees.

4. DISCUSSION

The present study demonstrates that head-up tilting modifies blood flow in the leg and the results are compatible with previous measurements performed using ^{133}Xe-technique (for a review: [7]). Therefore, the observed

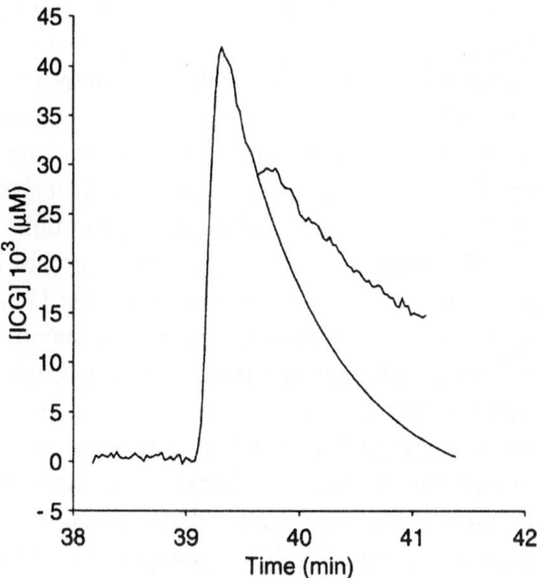

Figure 3. ICG concentration as a function of time for a typical subject. Lower curve corresponds to the ICG curve corrected for recirculation.

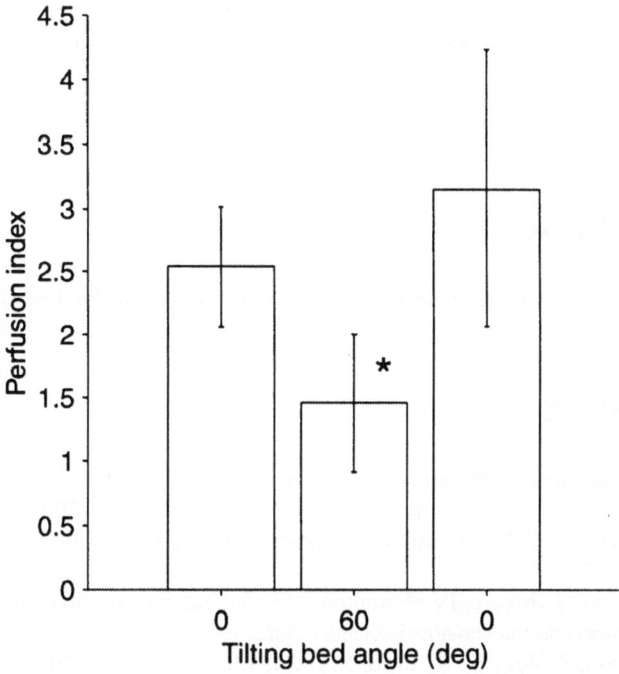

Figure 4. Perfusion index as a function of the tilting table angle (see text).

HbO_2 and Hb changes during head-up tilt are not only determined by the compliance properties of the vascular tree, but also by tissue blood flow modulation. In addition, these results indicate that NIRS may be used to assess, simultaneously, the superficial vascular compliance and the characteristics of the vasomotor response of the lower limbs during a tilting manoeuvre. This information cannot be obtained by classical strain-gauge plethysmography. Thus, the present technique is potentially applicable to the detection of specific pathological aspects of microcirculation, such us arterial occlusion in the leg [8, 9], diabetes mellitus [10], and congestive heart failure [11], where the venous-arteriolar reflex may be affected.

The capability to simultaneously monitor tissue blood volume content and perfusion index might also be extremely useful for the foundamental study of the venous-arteriolar reflex inducing mecanisms in different patho-/physiological conditions. In fact, the local sympathetic venous- arteriolar axon reflex is probably elicited by venous distension (venules) [7], hence the importance of monitoring blood volume changes. The monitoring of blood volume could in this way be used to distinguish between muscle tissue vasoconstriction induced by spinal and/or central sympathetic reflexes on one side and local venous-arteriolar reflex on the other side.

In conclusion, the present results goes far beyond the scope of the study. The data provide a clear demonstration showing that the ability to perform quantitative metabolic studies by NIRS, depends on the capability to measure and ideally to control blood volumes and blood flow changes. If blood volumes and fluxes are not considered, then it is impossible to decide if an HbO_2 or Hb variation is due to an actual metabolic change.

ACKNOWLEDGMENTS

We thank the Swiss National Science Foundation (#31-47075.96) for the financial support.

REFERENCES

1. Binzoni T, Quaresima V, Ferrari M, Hiltbrand E, Cerretelli P. Human calf microvascular compliance measured by near infrared spectroscopy. J Appl Physiol 2000;369-372.
2. Garskell P, Burton AC. Local posture vasomotor reflexes arising from the limb veins. Circ Res 1953;1:27-39.
3. Murray-Kinsman J, Moore JV, Hamilton WF. Studies on the circulation. I. Injection method: physical and mathematical considerations. Am J Physiol 1929;89:322-330.
4. Cope M, Delpy DT, Wray S, Wyatt JS, Reynolds EOR. A CCD spectrometer to quantitate the concentration of chromophores in living tissue utilising the absorption peak of water at 975 nm. Adv Exp Med & Biol 1989;248:33-40.

5. Duncan A, Meek JH, Clemence M, Elwell CE, Tyszczuk L, Cope M, Delpy DT. Optical pathlength measurements on adult head, calf and forearm and head of newborn infant using phase resolved optical spectroscopy. Phys. Med. Biol. 1995;40:295-304.
6. Lassen NA, Perl W. Tracer kinetics methods in medical physiology. Raven Press, New York, 1979.
7. Henriksen O. Sympathetic reflex control of blood flow in human peripheral tissues. Acta Physiol Scand 1991;143 (Suppl. 603):33-39.
8. Henriksen O. Orthostatic changes of blood flow in subcutaneous tissue in patients with arterial insufficiency of the legs. Scand J Clin Lab Invest 1974;34:103-109.
9. Agerskov K, Henriksen O, Tønnesen KH, Lassen NA. Constriction of collateral arteries induced by head-up til in patients with occlusive arterial disease of legs. Cardiovasc Res 1981;15:675-679.
10. Hilsted J. Decreased sympathetic vasomotor tone in diabetic othostatic hypotension. Diabetes 1979;28:970-973.
11. Kassis E, Amtorp O, Skagen K. Sympathetic reflex control of subcutaneous blood flow in patients with congestive heart failure. Clin Sci 1986;70:523-522.

Chapter **45**

MUSCLE OXYGEN CONSUMPTION AT ONSET OF EXERCISE BY NEAR INFRARED SPECTROSCOPY IN HUMANS

Takafumi Hamaoka, Toshihito Katsumura, Norio Murase, Takayuki Sako, Hiroyuki Higuchi, Motohide Murakami, Kazuki Esaki, Ryotaro Kime, Toshiyuki Homma, Akiko Sugeta, Yuko Kurosawa, Teruichi Shimomitsu & Britton Chance*

*Department of Preventive Medicine and Public Health, Tokyo Medical University, JAPAN; *Department of Biochemistry and Biophysics, University of Pennsylvania, PA.*

Abstract: In this study, we tried to continuously measure muscle oxygen consumption (m-VO_2) by near infrared spectroscopy (NIRS) without arterial occlusions. We used an intermittent isometric exercise at high intensity, which elicits a spontaneous occlusion of the blood flow to the muscle due to an increase in intramuscular pressure. Changes in muscle oxygenation and phosphocreatine (PCr) concentration were monitored in 5 subjects during an intermittent isometric exercise (5 sec. contraction/5 sec. relaxation) at 50% of maximum voluntary contraction for 3 minutes. The rate of deoxygenation was measured from the 2nd sec. to the 3rd sec. of each muscle contraction. The rate of deoxygenation at the onset of exercise followed an exponential time course with a time constant of 42.0 ± 12.5 sec. (mean±SD). This value agreed with the time constant of the decrease in PCr (48.2 ± 10.2 sec.). This result suggests that m-VO_2 was successfully monitored with a time resolution of 10 sec. by NIRS during exercise without arterial occlusion.

Key words: Muscle oxygen consumption, near infrared spectroscopy, NMR, phosphocreatine, onset of exercise

1. INTRODUCTION

Near infrared (NIR) spectrometer monitors tissue oxygenation level by measuring optical absorption changes in oxy- and deoxy- fractions of hemoglobin (Hb) and myoglobin (Mb). However, the continuous wave NIR technique (NIR_{CWS}) has its own difficulties in evaluating muscle energy metabolism during exercise. Primarily, the values measured by NIR_{CWS} do

Oxygen Transport to Tissue XXIV, edited by
Dunn and Swartz, Kluwer Academic/Plenum Publishers, 2003

not specifically reflect muscle oxygen consumption, but reflect the balance between muscle oxygen delivery and muscle oxygen consumption. In previous studies [1, 2], an arterial occlusion by pneumatic tourniquet was used to interrupt O_2 supply to muscles in order to dissociate O_2 consumption from O_2 supply in muscles measured by NIR_{cws}. Therefore, the initial rate of muscle deoxygenation during arterial occlusion is an indicator of m-VO_2 [3]. However, measurements of muscle m-VO_2 by NIR_{cws} require repeated brief arterial occlusions. Therefore, disadvantages such as a poor time resolution, occur. In this study, we tried to continuously monitor the rate of deoxygenation by NIR_{cws} without arterial occlusion. In order to achieve this objective, we used an intermittent isometric exercise at a relatively high intensity, which elicits a spontaneous occlusion of the blood flow to the muscle due to an increase in intramuscular pressure. Thus, we hypothesized that the rate of deoxygenation during a sustained muscle contraction, where blood flow is restricted, is an indicator of m-VO_2. The validity of the change in the rate of deoxygenation at the onset of exercise and during exercise was confirmed by measuring the change in phosphorus metabolites, which was determined by ^{31}phosphorus magnetic resonance spectroscopy (^{31}P-MRS).

2. METHODS

2.1 Subjects

Five male subjects (aged 24 to 36 yrs.) were recruited for the experiment from a population of healthy volunteers. None of the subjects were engaged in a physical training program involving forearm muscles at the time of the study. Informed consent was obtained from each subject prior to the experiment. The procedures followed in this study were in accordance with the Helsinki Declaration of 1975, as revised in 1983.

2.2 Experimental design

To examine changes in energy metabolism and Hb/Mb deoxygenation kinetics in muscles, the subjects performed the same procedure twice: once with the ^{31}P-MRS magnet measurement and once with the NIR_{cws}. This dual measurement was made because of the strong magnetic fields in the MRS device which influence the operation of the NIR_{cws}. The measurement site for both ^{31}P-MRS and NIR_{cws} was the right finger flexor muscles, specifically the flexor digitorum superficialis muscles, the most active site in grip contractions. This location was identified during isometric grip contraction, and the specific site of measurement was marked each time for higher precision and reproducibility.

2.3 Experimental protocol

The subjects performed submaximal grip exercise with use of the flexor digitorum superficialis muscle. The subjects performed hand grip contractions at 50% of maximum voluntary contraction (MVC) (5 sec. isometric contractions and 5 sec. relaxation) in the upright position.

After 1 min. of resting arterial occlusion was performed by tourniquet cuff followed by a 2 min recovery, the exercise was performed for 3 min. once with the ^{31}P-MRS and once with the NIR_{cws}.

2.4 NIR_{CWS}

The NIR_{CWS} device (HEO-200, OMRON Inc., Japan) [4] used for this study consisted of a probe and a computerized control system. The basic principle of this NIR_{CWS} device was extensively discussed in a previous paper by Chance et al [5].

The NIR_{CWS} probe contains a light source and optical detector, with the distance of 3.0 cm between light source and detector, providing sensory input for the unit. A pair of two-wavelength LED's, with wavelengths of 760 nm and 840 nm, respectively, were used as the light source. A silicon photodiode was used as the photodetector. The light-emitting intensities were 12 and 4 mW at 760 and 840 nm, respectively. Relative changes in oxygenated hemoglobin (Hb) was measured as follows [4]:

Δoxy-Hb= $\Delta OD_{840} - 0.66\Delta OD_{760}$

Δdeoxy-Hb= $0.80\Delta OD_{760} - 0.59\Delta OD_{840}$

Δtotal-Hb= $0.41\Delta O\Delta_{840} + 0.14\Delta OD_{760}$

where ΔOD_{840} and ΔOD_{760} indicate changes in optical density at 840 and 760 nm, respectively; and Δoxy-Hb, Δdeoxy-Hb, and Δtotal-Hb denote changes in the concentration of oxygenated hemoglobin, deoxygenated hemoglobin, and total hemoglobin, respectively.

The reliability of this method was previously confirmed with *in vitro* experiments [4, 5] and *in vivo* experiments [6]. The depth of the light penetration with the probe in this study was 1-2 cm [7]. As thickness of skin and adipose tissue of the forearm of the experimental subjects was less than 0.2 cm, it can be assumed that the detected signal primarily came from the microvasculature in muscles. The sampling time of the data for NIR_{CWS} was 0.1 sec in this study.

2.5 m-VO2 Measurement

In order to confirm that the 50% MVC used in this study spontaneously occluded the arterial inflow to the muscle, the subjects performed two 10 sec. sustained isometric grip exercises (with and without arterial occlusion) at 50% MVC before the start of the principal exercise. In this preliminary study, the rate of deoxygenation during exercise without arterial occlusion (2.13±1.18 %/sec) was comparable to that with arterial occlusion (2.23±1.26 %/sec). Therefore, in this study, we had confirmed that the rate of muscle deoxygenation during an intermittent exercise at 50% MVC without arterial occlusion is an indicator of m-VO_2 of the muscles. The m-VO_2 was determined from the deoxygenation rate from the 2nd to the 3rd sec. (20 data points) of each intermittent contraction during 3 min. of exercise. The value of m-VO_2 during exercise was expressed relative to the rate of resting muscle oxygen usage.

2.6 ^{31}P-MRS

An NMR spectrometer (Otsuka Electronics Inc.) was used with a 2.0-T superconducting 26-cm-bore magnet. An inductively coupled 3 cm diameter circular surface coil was placed on the forearm, and the coil and arm were held in a fixed position in the magnet by a cradle. The magnetic field homogeneity was optimized by using proton NMR, decreasing the half-line width of the water peak to <0.3 ppm. A radio frequency pulse (60 μsec pulse width at 43.58 Hz) was used for phosphorus signal acquisition and the PCr line broadening was adjusted to 5 Hz to improve the signal-to-noise ratio. The scan repetition time was 2 sec. and the mean FIDs values were obtained every 10 sec.

The FIDs were obtained and Fourier-transformed into spectra and the peak areas were fitted by a non-linear curve-fitting method to calculate the areas under the ß-ATP, PCr, and Pi peaks. To acquire saturation factors for each phosphate compound, a ratio was calculated between each area obtained during the 20 sec. of repetition time (fully relaxed spectra) and those obtained during the 2 sec. pulse intervals. Thereafter, PCr and Pi peak areas were quantified from the ratio of PCr and Pi areas to ß-ATP area with corrections for saturation factor. The muscular pH was calculated from the median chemical shift of Pi peak relative to PCr [8].

3. RESULTS

Figure 1 shows typical changes in oxygenated hemoglobin, deoxygenated hemoglobin, and total hemoglobin in the forearm muscle during intermittent isometric contractions at 50% MVC and recovery after exercise in a subject. Figure 2 shows an enlarged trace of the changes in oxygenated hemoglobin and total hemoglobin in the muscle from the 2nd to the 6th contraction in a subject. Total hemoglobin rapidly decreased in no longer than 1 second and remained stable during the last 4 seconds of each contraction. Figure 3 shows the averaged change in the rate of deoxygenation and PCr in the muscle at rest and during exercise. The rate of deoxygenation during exercise increased 10.7 ± 3.6 fold of resting at its peak value, while the decrease in PCr was 60.0 ± 9.6 % at the end of exercise. The rate of deoxygenation at the onset of exercise followed an exponential time course with a time constant of 42.0 ± 12.5 sec. (mean ± SD). This value agreed with the time constant of the decrease in PCr (48.2 ± 10.2 sec.) as shown at the upper panel in this figure. Figure 4 shows the relationship between the rate of deoxygenation and PCr concentration at the onset of exercise. There was a significant correlation between the change in the rate of deoxygenation and PCr ($r=.96$, $p<0.05$).

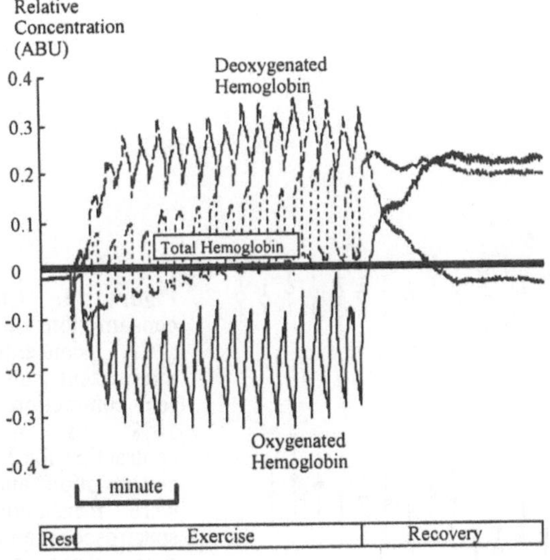

Figure 1. Typical changes in oxygenated, deoxygenated, and total hemoglobin during the intermittent isometric grip exercise (5 sec. contraction and 5 sec. relaxation) at 50% of maximum voluntary contraction for 3 minutes.

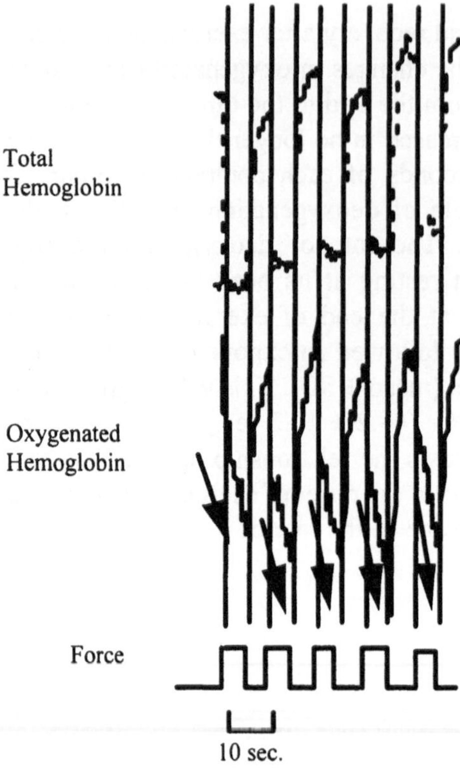

Figure 2. The changes in oxygenated and total hemoglobin at the 2nd, 3rd, 4th, 5th, and 6th isometric contraction (5 sec. contraction and 5 sec. relaxation each) of the 3 minutes of exercise. Total hemoglobin rapidly decreased at the 1st second and remained stable during the last 4 seconds of each contraction. Oxygen consumption was calculated from the rate of decline of oxygenated hemoglobin concentration during the 2nd to 3rd second (20 data points for a 2 second interval) as indicated by the arrows in the figure.

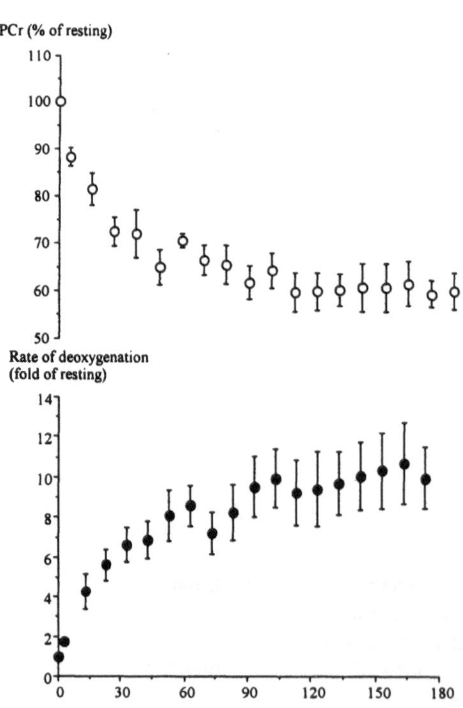

Figure 3. Changes in oxygen consumption and phosphocreatine (PCr) concentration during an intermittent isometric grip exercise (5 sec. contraction and 5 relaxation) at 50% of maximum voluntary contraction for 3 minutes. The oxygen consumption and PCr concentration were measured by near infrared spectroscopy and magnetic resonance spectroscopy, respectively.

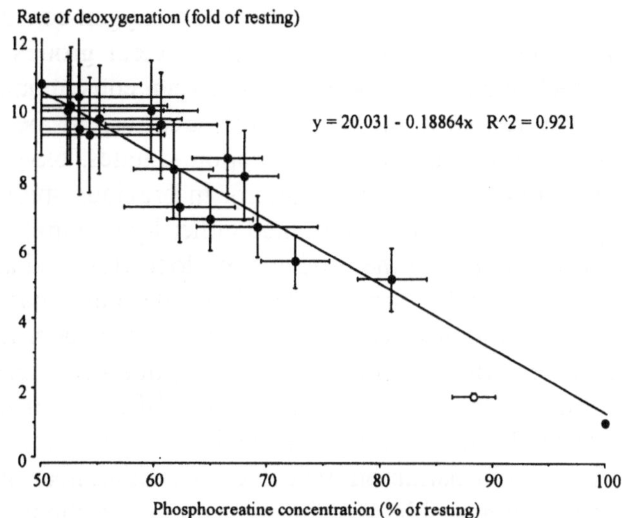

Figure 4. The relationship between oxygen consumption and phosphocreatine (PCr) concentration at rest and during exercise in the human forearm muscle. The first data point at the onset of exercise is the 2.5 sec. for the oxygen consumption and the 5 sec. for PCr. Therefore, the first data point which is indicated by an open circle is not included for the single regression analysis.

4. DISCUSSION

This study is original in that the rate of deoxygenation were continuously monitored with a time resolution of 10 sec. by using NIRS without applying arterial occlusion and that it observes similar on-transit kinetics between the rate of deoxygenation and PCr concentration at the onset of exercise.

The thermodynamic regulation model with assumptions, in which PCr can be directly related to the respiration rate, is reported in the previous paper [9]. We found a significant correlation between PCr and the rate of deoxygenation ($r=0.96$, $P<0.05$) at the onset of exercise and during steady state in this study. This result agreed with a model for the oxidative regulation proposed by Meyer et al. [9]. Furthermore, we also observed the agreement of Tc of the rise in the rate of deoxygenation and Tc of the decrease in PCr. A time resolution for PCr determined by a ^{31}P-MRS measurement in the human muscle is usually 10 to 20 sec. In this study, we achieved a comparable time resolution by using NIRS in place of MRS. This result confirms that m-VO_2 measured by NIRS is valid and provides us with the direct information about muscle oxidative metabolism without arterial occlusion with a high time resolution, which was unable to be determined with the other conventional methodologies.

The on-transit regulation of blood flow and oxygen consumption is one of the major topics in exercise physiology. Several groups have tried to elucidate the mechanism for the kinetics of oxygen consumption at the onset of exercise. The primary interest in this area would be which factor determines the on-transit oxygen consumption kinetics, oxygen delivery or oxygen demand (mitochondrial respiration). In previous studies in humans, oxygen consumption has been determined either by obtaining venous blood samples and then calculating the products of blood flow and arterial-venous oxygen differences or by simply sampling the pulmonary gas. These conventional methodologies have disadvantages such as poor time resolution and time lag between the site of oxygen consumption in muscles and the site of sampling of the pulmonary gas. In this study, NIRS has an advantage over the previous methods with regards to achieving a higher time resolution and collecting a direct information on oxygen consumption near muscles, which might be otherwise impossible with the conventional methods.

5. CONCLUSION

In conclusion, NIR_{CWS} can accurately monitor the rate of deoxygenation during the intermittent isometric exercise with a high time resolution (10 sec.) without applying the brief arterial occlusion to the muscle. In this study, we confirmed that the rate of deoxygenation is an indicator of $m\text{-}VO_2$ and NIR_{cws} provides us with a similar information as the MRS measurement, which is more expensive and less easily accessible.

ACKNOWLEDGMENT

This study was supported in part by a Grant-in-Aid for Encouragement of Young Scientists, the Ministry of Education, Science, Sports and Culture in Japan (T.H., #09780087) and HL 44125 (B.C.).

REFERENCES

1. Cheatle TR, Potter LA, Cope M, Delpy DT, Coleridge Smith PD, Scurr JH. Near-infrared spectroscopy in peripheral vascular disease. Br J Surg 1991;78:405-408.
2. De Blasi RA, Cope M, Ferrari M. Oxygen consumption of human skeletal muscle by near Infrared spectroscopy during tourniquet-induced ischemia in maximal voluntary contraction. Adv Exp Med Biol 1992;317: 771-777.
3. Hamaoka T, Iwane H, Shimomitsu T, Katsumura T, Murase N, Nishio S, Osada T, Kurosawa Y, Chance B. Noninvasive measures of oxidative metabolism on working human muscles by near infrared spectroscopy. J Appl Physiol 1996;81(3): 1410-1417.

4. Shiga T, Yamamoto K, Tanabe K, Nakase Y, Chance B. Study of an algorithm based on model experiments and diffusion theory for a portable tissue oximeter. J Biomed Opt 1997;2(2):154-161.
5. Chance B, Dait TM, Chang C, Hamaoka T, Hagerman F. Recovery from exercise induced desaturation in the quadriceps muscle of elite competitive rowers. Am J Physiol 1992;262:766-775.
6. Shiga T, Tanabe K, Nakase Y, Shida T, Chance B. Development of portable tissue oximeter using near infra-red spectroscopy. Med Biol Eng Comput 1995;33: 622-626.
7. Chance B, Nioka S, Kent J, McCully KK, Fountain M, Greenfeld R, Holtom G. Time resolved spectroscopy of hemoglobin and myoglobin in resting and ischemic muscle. Anal Biochem 1988;174:698-707.
8. Kushmerick MJ and Meyer RA. Chemical changes in rat leg muscle by phosphorus nuclear magnetic resonance. Am J Physiol 1985;248:542-549.
9. Meyer RA. A linear model of muscle respiration explains monoexponential phosphocreatine changes. Am J Physiol 1988;254:548-553.

Chapter **46**

MODELING OF OXYGEN DIFFUSION FROM THE BLOOD VESSELS TO INTRACELLULAR ORGANELLES

Aleksander S. Popel, Daniel Goldman & Arjun Vadapalli
Department of Biomedical Engineering and Center for Computational Medicine and Biology, The Johns Hopkins University School of Medicine, Baltimore, MD 21205

Abstract: We describe recent models of oxygen transport in tissue along the pathway from the hemoglobin molecule to the mitochondria and illustrate their applications. Microvasculature is the major site of exchange between blood and parenchymal cells for gases (O_2, CO_2, CO, NO), nutrients, metabolic products, and drugs. These exchange processes are affected by the architecture of the microvessels and the surrounding cells; distribution of blood flow; transport characteristics of blood, cells, and interstitial space; and rates of various chemical reactions associated with the transport processes. These processes operate at multiple levels of biological organization, from the molecular to the organ levels. Quantitative understanding of molecular transport in cells and tissues, specifically of oxygen transport, is the prerequisite for understanding the mechanisms of many diseases and for designing effective therapies. Mathematical and computational models provide a powerful set of tools for studies of these complex phenomena.

Key words: computational model, microvascular network, transport resistance, capillary transport

1. INTRODUCTION

The first model of oxygen transport was formulated by August Krogh in 1919 [1], as part of his general studies of capillary physiology [2]. In this work, a simple equation was derived for radial distribution of oxygen tension in a tissue unit, later referred to as the Krogh Tissue Cylinder. Following this seminal work, many researchers have extended the model to include detailed oxygen transport in capillaries, pre- and postcapillary vessels, and finally complex microvascular networks. Several general reviews on theoretical aspects of oxygen transport are available [3, 4]. In this paper, we summarize major steps in previous studies and concentrate on the studies of the last decade. Due to space limitations, references are given to either review articles or most recent papers where earlier references can be found.

We focus on the following specific areas: (1) Role of intravascular transport resistance in vessels of different sizes under different flow conditions; (2) Role of pre- and postcapillary vessels and capillaries in oxygen transport; (3) Role of heterogeneities of anatomical and physiological variables (e.g., blood flow, hematocrit, diffusion parameters, mitochondrial density). We conclude by describing oxygen transport in the presence of oxygen carriers in the blood plasma.

2. TRANSPORT RESISTANCES ALONG THE OXYGEN PATHWAY: A REVIEW

It is well recognized that exchange vessels comprise capillaries as well as arterioles and, to a smaller degree, venules [5]. Along the pathway from the hemoglobin molecule to the mitochondrion the oxygen molecule encounters several structures, some of which may be significant barriers to transport. We outline qualitatively the physical characteristics of this pathway and the corresponding transport resistances. For the purpose of this discussion, transport resistance is defined as the ratio of the difference of the oxygen tensions at certain points along the pathway, ΔP, to the flux of oxygen, J. The inverse of the transport resistance is the tissue conductivity or the mass transfer coefficient. In the case of a homogeneous plain layer where free diffusion is the only form of transport, this definition yields the resistance equal to the thickness of the layer divided by the so-called Krogh diffusion coefficient, $K=D\alpha$, where D is the diffusion coefficient and α is the solubility coefficient. We follow the pathway from the Hb molecule to the tissue having in mind the systemic circulation. However, the analysis of the intravascular transport also applies to the pulmonary circulation where the pathway is traversed from the endothelium to the Hb molecule. The pathway is illustrated in Figure 1 for capillary O_2 transport.

First, the O_2 molecule is released from Hb with a finite rate. The molecule then diffuses through the concentrated Hb solution inside the red blood cell; this diffusion is facilitated by the diffusion of oxyhemoglobin. The molecule crosses the red blood cell plasma membrane, which is of negligible resistance, and enters the plasma region. In capillaries, where red blood cells travel in single file, the molecules leaving the cell can cross the plasma region either through a narrow sleeve between the red blood cell and the endothelium or through the gap between the adjacent cells. Because of the relative dimensions of the two domains, the resistance of the gap is much larger. In fact, the hematocrit-dependent resistance of the gap is the major determinant of intracapillary resistance, although the resistance inside the red cell is also significant [6, 7]. In arterioles and venules the O_2 molecule travels through the suspension of the red cell until it reaches the endothelium. Theoretical estimates show that by far the major portion of the resistance resides in the plasma [4]. Recent studies demonstrate the presence of a glycocalyx with its associated protein layer that is substantially thicker than commonly thought – on the order of one micron [8]. Even though the layer should not have a transport resistance for oxygen larger than the equivalent layer of plasma, it nevertheless could have a significant impact on oxygen transport by affecting the shape and distribution of the red blood cells (local hematocrit) in a microvessel [9]. The importance of both factors for oxygen transport in capillaries has been demonstrated, in both systemic [10] and pulmonary capillaries [11].

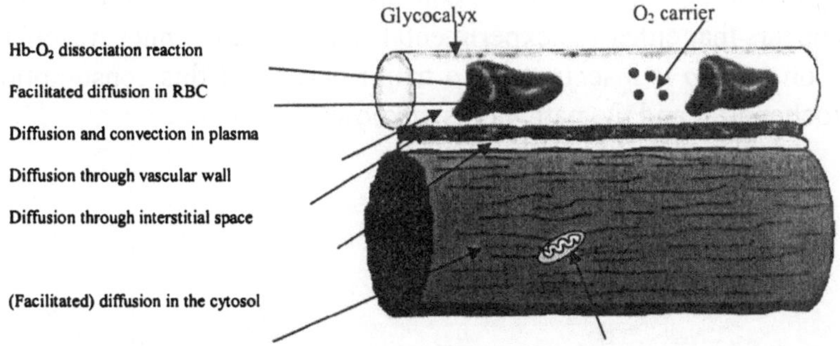

Figure 1. Illustration of the pathway of the oxygen molecule from the Hb molecule to the mitochondrion

Endothelial cells present the next segment of the oxygen transport pathway. In arterioles and venules, the oxygen molecule has to traverse the rest of the vascular wall that comprises smooth muscle cells and connective tissue. The cell plasma membrane is known to have a very small transport resistance to oxygen because the molecule is lipid soluble and the membrane is very thin [12]. The diffusion coefficient in the wall should not be very different from that of the parenchymal cells; however, one study reported a range 1.42-8.73$\times10^{-6}$ cm^2/s for endothelial cell monolayer *in vitro* [13]. We discuss available measurements of intracellular diffusion below, when we discuss parenchymal cells. In addition to cell permeability, cell O_2 consumption is another component of transport resistance – the larger the consumption, the larger the apparent resistance. Our analysis of experimental data on O_2 consumption by endothelial and vascular smooth muscle cells and segments of vascular wall *in vitro* gives values that do not exceed a value that we refer to as the mitochondrial-based maximum O_2 consumption, M_{mt} [14] We estimate $M_{mt}=5\cdot10^{-3}$ ml O_2 ml^{-1} s^{-1} based on the known mitochondrial maximum respiration rate of 0.1 $mlO_2(ml\ mito)^{-1}s^{-1}$ and a volume fraction of mitochondria in vascular cells of approximately 5%. Figure 2 is a summary of these results. Note that all the in vitro O_2 consumption values do not exceed M_{mt}. *In vivo* results for arterial wall also show $M<M_{mt}$[15]. However, a microvascular study suggests that vascular wall consumption exceeds M_{mt} by an order of magnitude [16] as a result of this high consumption the vascular wall, and specifically the endothelial cells, are conjectured to be a major barrier to O_2 transport since most of the oxygen leaving the microvessels would be consumed by the endothelium. To assess this possibility, we estimated, based on the published experimental data, the utilization of O_2 in an extra-mitochondrial pathway associated with nitric oxide (NO) synthesis; our results give values $M_{NO}<8\cdot10^{-4}$ ml O_2 ml^{-1} s^{-1} that are similar to or smaller than O_2 consumption in resting striated muscle. Thus, it appears that either the experimental estimate of the microvascular O_2 consumption *in vivo* is inaccurate or a major portion of this consumption is extra-mitochondrial and the pathway is unknown.

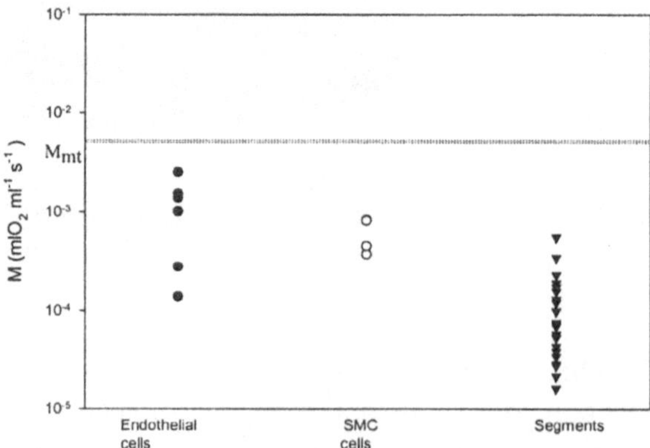

Figure 2. Oxygen consumption of endothelial cells (EC) and smooth muscle cells (SMC) in suspensions and vascular segments *in vitro*.

The dimension of the interstitial space is not well known. The distances estimated from electron-microscopic studies in muscle indicate very small dimensions of the interstitial space, less than a micron. However, estimates of the interstitial space *in vivo* obtained by indicator techniques give values around 20% for the volume fraction of the interstitium, which would result in a larger distance between the vascular wall and plasma membrane of the parenchymal cell. Because of the composition of the interstitium, its permeability to O_2 should be slightly less than that of water and when no O_2 carrier is present.

In parenchymal cells, the permeability to O_2 is usually assumed to be equal to that measured in macroscopic tissue samples, with diffusion coefficients in the range $(1\div2.5)\cdot10^{-5}$ cm^2s^{-1} depending on tissue composition, particularly its lipid content [12]. However, numerous experimental data suggest that diffusion coefficients in isolated single cells are an order of magnitude lower than those in tissues [12, 17]. The volume of interstitial space does not appear to be sufficient to explain the difference. In addition, our theoretical studies suggest that the low values of the muscle-cell Krogh coefficient are inconsistent with muscle performance at maximum oxygen consumption, VO_{2max} (Roy). Thus, additional experimental and theoretical work is necessary to reconcile the differences.

Groebe [18] developed several models of oxygen transport using the Krogh cylinder geometry (i.e., capillary surrounded by a tissue cylinder) and muscle fiber geometry (i.e., cylindrical fiber surrounded by several capillaries). Myoglobin is an oxygen carrier and can significantly facilitate diffusion of oxygen in cells. Groebe [18] made predictions of oxygen distribution in maximally contracting red muscle fibers and concluded that the pO_2 and Mb saturation profiles are essentially flat, both in radial and longitudinal directions, and the pO_2 values are low, a few mm Hg. In his view, the largest drop of the oxygen tension occurred in the region between the red blood cell and the plasma membrane of the muscle fiber, i.e., in the carrier-free region. Importantly, he assumed the value of the Mb diffusion coefficient $D_{Mb}\approx5.5\cdot10^{-7}$ cm^2/s based on experiments of Mb diffusion in protein solutions. However, measurements of Mb diffusion along muscle fibers gave values several times lower [19, 20]. The most pertinent values of D_{Mb} are in the radial direction, where the gradients should be steeper; such measurements in muscle cells are not available. Incorporating these lower values, Roy and Popel [6] predicted a significantly lower effect of Mb facilitation and higher intracellular pO_2 gradients. In contrast, results of Conley et al [20, 21] suggest that Mb plays a role in oxygenation of muscle and heart. Therefore, at present the issue of how much Mb contributes to O_2 diffusion remains open.

The mitochondrial surface area supplied by a single red blood cell is significantly larger than the surface area of the red cell; the area ratio depends on the hematocrit, velocity, and mitochondrial volume fraction [22].

As a result, the transport resistance in the vicinity of mitochondria is negligible and the perimitochondrial pO_2 gradients are small [23]. However, the spatial distribution of mitochondria may play a significant role. It is known that mitochondria are preferentially concentrated near the plasma membrane and in the vicinity of capillaries; they can also form three-dimensional networks. The implications of these arrangements for intracellular transport have not been fully investigated.

As an illustration of transport resistance distribution, we analyze oxygen transport in capillaries with a hematocrit of 30% in a contracting hamster retractor muscle by calculating resistance along each segment and expressing it as a fraction of total resistance between the core of the red blood cell and the mitochondria located furthest from the capillary. We find that 11% of resistance lies in the RBC, 40% in the plasma, 6% in the vascular wall, 1% in the interstitial space, and 42% in the parenchymal cells surrounding the capillary. Clearly, more than half of the total resistance is located outside the muscle fiber. Similar results were obtained in Roy and Popel [6] for skeletal muscle, diaphragm, and heart in several species.

3. MODELING OF OXYGEN TRANSPORT IN MICROVASCULAR NETWORKS

The development of models following the work of Krogh was dominated by the desire to simplify the geometry to a point where a simple solution of the model equations could be obtained. In part, this was due to the lack of proper technology for obtaining accurate three-dimensional anatomic and physiological microvascular data and in part due to the limitations of computing capabilities. Both of these limitations are significantly relaxed at present and we can begin to model transport of oxygen and other substances in biophysically-accurate experiment-based microvascular networks. In this section we give a brief summary of how to formulate such models.

Formulation of a transport model and its implementation consists of the following steps: (1) Define the anatomical structure of the network and surrounding cells; (2) Either specify or model the distribution of blood flow and hematocrit in the network; (3) Formulate the transport equations and solve them numerically; (4) Visualize the spatio-temporal distributions.

The network geometry can be obtained from microvascular casts and cell-vessel geometry can be obtained from appropriately stained fixed tissue samples; this method allows data collection from macroscopic tissue samples and large microvascular networks, but the data are limited to tissue geometry [24]. Alternatively, vascular data can be obtained from *in vivo* microscopic observations [25-27]; with current technology, the size of the tissue sample and especially the depth from the tissue surface are limited, but the methods allow one to obtain physiological data (e.g., red blood cell velocity, vessel hematocrit, pO_2 and Hb-O_2 saturation).

Once the network geometry is specified, the velocity of red blood cells and plasma and local hematocrit can be specified using *in vivo* data. Such data are scarce though and are not readily available for large microvascular networks. Alternatively, a model of blood flow can be formulated that can be applied to blood networks of general geometry, validated against experimental data and then applied to a variety of networks to predict the hemodynamic parameters. A continuum model (i.e., not considering individual red cells) of flow *in vivo* has been formulated by Pries et al [25] and validated against experimental data in rat mesentery. The model includes an expression of vessel apparent viscosity and its hematocrit as functions of vessel diameter and discharge hematocrit. In addition, the model includes a relationship between red blood cell fluxes in daughter branches of arteriolar bifurcations as a function of volumetric blood flow rates into the branches, referred to as the "bifurcation law." The *in vivo* apparent viscosity in small vessels, including capillaries, is significantly greater than the corresponding in vitro viscosity obtained in glass tubes. The difference is attributed primarily to the presence of a glycocalyx at the endothelial surface and is corroborated by theoretical studies [9, 28]. As an alternative to the steady-state continuum model, a time-dependent model, which describes oscillations of rheological origin in the network, has been formulated by considering "groups" of red blood cells [29].

In the preceding section we have indicated that the particulate nature of blood is a major determinant of intravascular transport, primarily because a significant part of the transport resistance resides in the plasma due to its low solubility for oxygen. Thus, this feature has to be taken into account in any model of blood-tissue oxygen transport. One way of achieving this is to track every red blood cell in the network and solve the corresponding unsteady three-dimensional transport equations. However, for networks larger than tens of capillaries this task becomes impossible because of computational complexity. Another possibility is to use a continuum approach in dealing with red blood cells and employ mass transfer coefficients between the red blood cells and the surrounding plasma and the red blood cells and capillary wall that could be calculated separately in a detailed model of capillary transport.

The general equations for intravascular and tissue transport can be found in [3, 4]. The presence in the plasma of a hemoglobin-based oxygen carrier was modeled in [30], and of perfluorocarbon emulsion in [31]. Here we present examples of computer simulations in capillary networks in a striated muscle. In Figure 3 we present examples of pO_2 distribution in and around a blood capillary with parachute-shaped red blood cells at vessel hematocrits of 15%. The distribution is shown at a location close to the end of the capillary. In the standard case shown in the upper panel, red blood cells act as localized sources of oxygen. In the lower panel Hb solution is introduced at a concentration of 7 g/dl, with the Hb kinetic parameters similar to those of the Hb inside the red blood cells. In this case, the capillary acts more like a continuous source of oxygen.

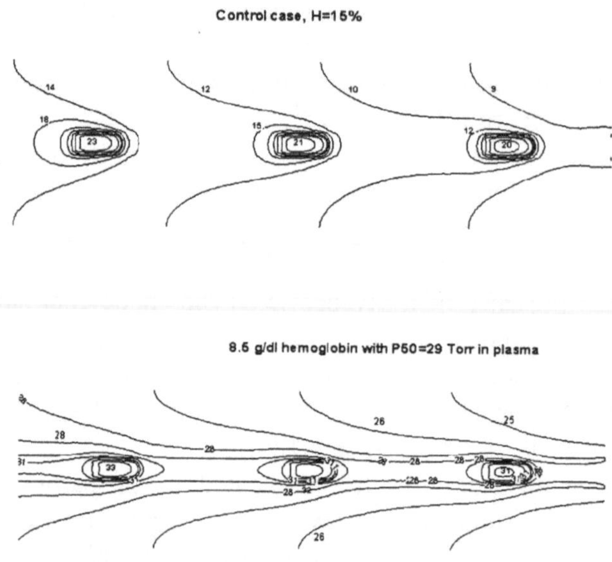

Figure 3. Comparison of PO_2 distribution in a capillary segment having a tube hematocrit of 15%. The upper panel corresponds to the case where there is no hemoglobin and the bottom panel has hemoglobin with concentration of 8.5 g/dl.

Figure 4 presents an example of the oxygen distribution around a capillary network consisting of two modules having countercurrent blood flows. The network structure of a single capillary module, measured *in vivo* in the laboratory of C.G. Ellis, University of Western Ontario, was used for blood flow calculations. For oxygen transport calculations, the original module was combined with a copy that had been rotated 180° about the *x*-axis (vertical axis in Figure 4). Concurrent modules and large individual networks have also been studied using our computational model.

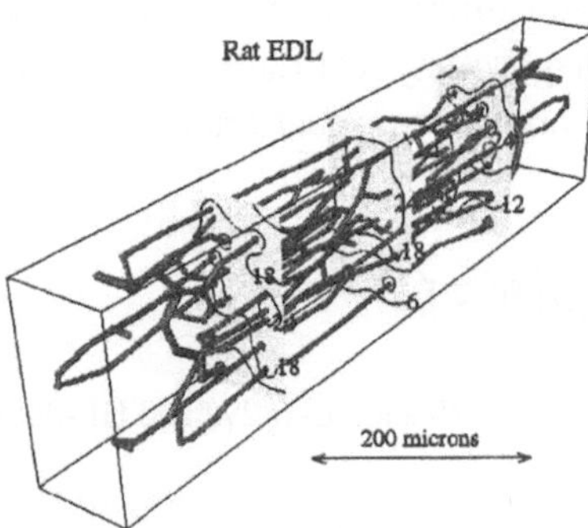

Figure 4. pO_2 distribution around a capillary network constructed from experimental data. Flow in the module on the right front side is into the page. Shown on cross-sections taken at three locations along the tissue block are the contour lines for calculated pO_2 equal to 6, 12, 18, and 24 mm Hg.

4. TRANSPORT OF OTHER GASES

Other gases, such as carbon dioxide, carbon monoxide, and nitric oxide, also play an important role in regulation of the microcirculation and in cellular energetics; therefore, it is important to quantify their transport. General transport equations for these gases can be formulated analogous to the transport equations for oxygen. Compartmental models describing coupled transport of O_2 and CO_2 have been presented recently [32, 33]; extension of these models to complex microvascular networks needs to be accomplished. A model of nitric oxide diffusion between endothelial and smooth muscle cells has also been formulated [34, 35].

5. SUMMARY

In this brief overview, we have presented current knowledge of oxygen transport along the pathway from the hemoglobin molecule to mitochondria and methods of computational modeling of this transport. We have identified several unresolved issues that require further experimental and theoretical efforts. Among these are the magnitude of precapillary transport and oxygen utilization in the vascular wall, the role of myoglobin in facilitating oxygen diffusion, and the role of intravascular and extravascular heterogeneities. The development of computational models in recent years has allowed for simulation of the transport of oxygen and other substances in biophysically-accurate experiment-based microvascular networks. As the models become more accurate and are rigorously validated against experimental data, they will become valuable in clinical applications.

ACKNOWLEDGEMENT

This work was supported by NIH Grant HL-18292 and HL-09595.

REFERENCES

1 Krogh A. The number and distribution of capillaries in muscles with calculations of the oxygen pressure head necessary for supplying the tissue. J Physiol (London) 1919;52:409.
2 Krogh A. The Anatomy and Physiology of Capillaries, New Haven: Yale University Press, 1922;Pages.
3 Popel AS. Theory of oxygen transport to tissue. Crit Rev Biomed Eng 1989;17:257-321.
4 Hellums JD, Nair PK, Huang NS, Ohshima N. Simulation of intraluminal gas transport processes in the microcirculation. Ann Biomed Eng 1996;24:1-24.
5 Ellsworth ML, Ellis CG, Popel AS, Pittman RN. Role of Microvessels in oxygen supply to tissue. News Physiol. Sci. 1994;9:119-123.
6 Roy TK, Popel AS. Theoretical predictions of end-capillary PO2 in muscles of athletic and nonathletic animals at VO2max. Am J Physiol 1996; 271:H721-737.
7 Eggleton CD, Vadapalli AR, Roy TK, Popel AS. Calculations of intracapillary oxygen tension distributions in muscle. Math Biosci 2000; 167:123-143.
8 Henry CB, Duling BR. Permeation of the luminal capillary glycocalyx is determined by hyaluronan. Am J Physiol 1999;277:H508-514.
9 Secomb TW, Hsu R, Pries AR. A model for red blood cell motion in glycocalyx-lined capillaries. Am J Physiol 1998;274:H1016-1022.
10 Wang CH, Popel AS. Effect of red blood cell shape on oxygen transport in capillaries. Math Biosci 1993;116:89-110.
11 Hsia CC, Johnson RL, Jr., Shah D. Red cell distribution and the recruitment of pulmonary diffusing capacity [see comments]. J Appl Physiol 1999;86:1460-1467.
12 Dutta A, Popel AS. A theoretical analysis of intracellular oxygen diffusion. J Theor Biol 1995;176:433-445.
13 Liu CY, Eskin SG, Hellums JD. The oxygen permeability of cultured endothelial cell monolayers. Adv Exp Med Biol 1994;345:723-730.
14 Vadapalli A, Pittman RN, Popel AS. Estimating oxygen transport resistance of the microvascular wall. Am J Physiol 2000;279:H657-H671.
15 Buerk DG, Goldstick TK. Arterial wall oxygen consumption rate varies spatially. Am J Physiol 1982;243:H948-958.
16 Tsai AG, Friesenecker B, Mazzoni MC, Kerger H, Buerk DG, Johnson PC, Intaglietta M. Microvascular and tissue oxygen gradients in the rat mesentery. Proc Natl Acad Sci U S A 1998;95:6590-6595.
17 Baranov VI, Belichenko VM, Shoshenko KA. [The oxygen diffusion coefficient in isolated skeletal muscle fibers]. Fiziol Zh SSSR Im I M Sechenova 1991;77:29-34.
18 Groebe K. An easy-to-use model for O_2 supply to red muscle. Validity of assumptions, sensitivity to errors in data [published erratum appears in Biophys J 1995 69:288]. Biophys J 1995;68:1246-1269.

19 Papadopoulos S, Jurgens KD, Gros G. Diffusion of myoglobin in skeletal muscle cells--dependence on fibre type, contraction and temperature. Pflugers Arch 1995;430:519-525.

20 Conley KE, Jones C. Myoglobin content and oxygen diffusion: model analysis of horse and steer muscle. Am J Physiol 1996;271:C2027-2036.

21 Conley KE, Ordway GA, Richardson RS. Deciphering the mysteries of myoglobin in striated muscle. Acta Physiol Scand 2000;168:623-34.

22 Honig CR, Gayeski TE, Federspiel W, Clark A, Jr., Clark P. Muscle O_2 gradients from hemoglobin to cytochrome: new concepts, new complexities. Adv Exp Med Biol 1984;169:23-38.

23 Clark A, Jr., Clark PA, Connett RJ, Gayeski TE, Honig CR. How large is the drop in pO_2 between cytosol and mitochondrion? Am J Physiol 1987;252:C583-587.

24 Mathieu-Costello O. Comparative aspects of muscle capillary supply. Annu Rev Physiol 1993;55:503-525.

25 Pries AR, Secomb TW, Gessner T, Sperandio MB, Gross JF, Gaehtgens P. Resistance to blood flow in microvessels in vivo. Circ Res 1994;75:904-915.

26 Hudetz AG. Blood flow in the cerebral capillary network: a review emphasizing observations with intravital microscopy. Microcirculation 1997; 4:233-252.

27 Ellis CG, Ellsworth ML, Pittman RN, Burgess WL. Application of image analysis for evaluation of red blood cell dynamics in capillaries. Microvasc Res 1992;44:214-225.

28 Damiano ER. The effect of the endothelial-cell glycocalyx on the motion of red blood cells through capillaries. Microvasc Res 1998;55:77-91.

29 Kiani MF, Pries AR, Hsu LL, Sarelius IH, Cokelet GR. Fluctuations in microvascular blood flow parameters caused by hemodynamic mechanisms. Am J Physiol 1994;266:H1822-1828.

30 Page TC, Light WR, Hellums JD. Prediction of microcirculatory oxygen transport by erythrocyte/hemoglobin solution mixtures. Microvasc Res 1998;56:113-126.

31 Eggleton CD, Roy TK, Popel AS. Predictions of capillary oxygen transport in the presence of fluorocarbon additives. Am J Physiol 1998; 275:H2250-2257.

32 Sharan M, Popel AS, Hudak ML, Koehler RC, Traystman RJ, Jones MD, Jr. An analysis of hypoxia in sheep brain using a mathematical model. Ann Biomed Eng 1998;26:48-59.

33 Ye GF, Jaron D, Buerk DG, Chou MC, Shi W. O2-Hb reaction kinetics and the Fahraeus effect during stagnant, hypoxic, and anemic supply deficit. Ann Biomed Eng 1998;26:60-75.

34 Vaughn MW, Kuo L, Liao JC. Estimation of nitric oxide production and reaction rates in tissue by use of a mathematical model. Am J Physiol 1998;274:H2163-2176.

35 Vaughn MW, Kuo L, Liao JC. Effective diffusion distance of nitric oxide in the microcirculation. Am J Physiol 1998;274:H1705-1714.

Chapter 47

MUSCLE REOXYGENATION RATE AFTER ISOMETRIC EXERCISE AT VARIOUS INTENSITIES IN RELATION TO MUSCLE OXIDATIVE CAPACITY

Ryotaro Kime, Toshihito Katsumura, Takafumi Hamaoka, Takuya Osada, Takayuki Sako, Motohide Murakami, Sang Yong Bae[*†], Koji Toshinai[*], Shukoh Haga[*] & Teruichi Shimomitsu

Dept of Prev. Med. and Pub. Health, Tokyo Medical University, 6-1-1 Shinjuku, Tokyo 160-8402 Japan; [*]Inst. of Health and Sports Sci., University of Tsukuba, Ibaraki, Japan;
[†] Deceased

Abstract: The purpose of this study was to determine whether the reoxygenation rate (Reoxy-rate) immediately after static exercise at various submaximal intensities would be related to muscle oxidative capacity. Seven healthy male subjects performed isometric handgrip exercise for 10 sec at 30%, 60% and 90% of maximal voluntary contraction (MVC). The Reoxy-rate and muscle oxygen consumption during exercise (muscleVO_{2EX}) were monitored by near infrared continuous wave spectroscopy (NIRcws). The muscle oxidative capacity was evaluated by the time constant for phosphocreatine resynthesis (PCrTc) using ^{31}-phosphorus magnetic resonance spectroscopy (^{31}P-MRS). The Peak blood flow of brachial artery after exercise (BABFpeak) was measured using Doppler ultrasound. There was no correlation between PCrTc and Reoxy-rate at 30% and 60% MVC. In contrast, Reoxy-rate at 90% MVC was positively correlated to PCrTc ($r=0.825$, $p<0.05$). The muscle VO_{2EX} increased 5.9, 8.8 and 12.6-fold of the resting on average at 30%, 60% and 90% MVC, respectively, and the muscle VO_{2EX} at 90% MVC was significantly higher than that at 30% and 60% MVC. On the other hand, BABFpeak increased only just 1.9, 2.4 and 2.7-fold of the resting on average at 30%, 60% and 90% MVC, respectively (Fig. 4). These results suggest that the higher oxidative capacity muscle shows slower muscle reoxygenation after 10 sec isometric exercise at 90% MVC because the Reoxy-rate after this type of exercise may be influenced more by muscle VO_2 than by O_2 supply. In contrast, 60%MVC and lower exercise intensities may not be severe enough to influence the muscle VO_2 dependent Reoxy-rate.

Key words: Reoxygenation rate, NIRcws, ^{31}P-MRS, Muscle oxidative capacity

1. INTRODUCTION

Near infrared continuous wave spectroscopy (NIRcws) is a noninvasive method of evaluating changes in oxygenation levels in tissue. NIRcws has been used to measure muscle oxygenation, blood volume, muscle oxygen consumption and blood flow in the case of evaluating muscle metabolism [1-7]. The NIRcws technique, however, has difficulties in measuring energy metabolism during exercise. This is because absolute concentration changes of hemoglobin and/or myoglobin (Hb/Mb) cannot be determined due to unquantifiable biophysical quantities (i.e., optical path length) [8, 9]. Because of this limitation, several studies have measured the relationship between reoxygenation rate (Reoxy-rate) and muscle oxidative capacity [1,6, 7]. However, the results are not consistent. Some studies reported that there was no correlation between the muscle oxidative capacity and the Reoxy-rate after isometric exercise [4, 7]. In contrast, McCully et al [6] reported that the kinetics of the post dynamic exercise recovery for both muscle oxygenation and phosphocreatine (PCr) levels produced the same time constant. These divergent results, which appear to be due to slight differences in the exercise paradigm, suggest that muscle reoxygenation may be influenced more by O_2 supply after dynamic exercise than after isometric exercise [10].

It is well known that the magnitude of post-exercise hyperemic response is lower after short-term and/or isometric exercise than after long-term and/or dynamic exercise [10, 11]. Therefore, when the hyperemic response is limited by short-term isometric exercise, Reoxy-rate after the exercise may be more influenced by muscle VO_2 than O_2 supply. We hypothesized that in the case of high intensity exercise, when O_2 demand is increased, high oxidative capacity muscle (HOCmuscle) could utilize greater O_2 than low oxidative capacity muscle (LOCmuscle) even during the recovery phase. Hence, it is expected that HOCmuscle may exhibit slower reoxygenation after short duration, high intensity isometric exercise (Fig. 1). In contrast, however, with low intensity exercise, when O_2 demand is not fully increased, muscle VO_2 would not enough to increase as much as the muscle VO_2 effects would significantly influence to Reoxy-rate. The primary goal of this study was to test the hypothesis that the Reoxy-rate immediately after short-term isometric exercise at various intensities would be related to muscle oxidative capacity. We further evaluated muscle VO_2 and brachial arterial blood flow (BABF) separately because the Reoxy-rate is determined by the O_2 balance.

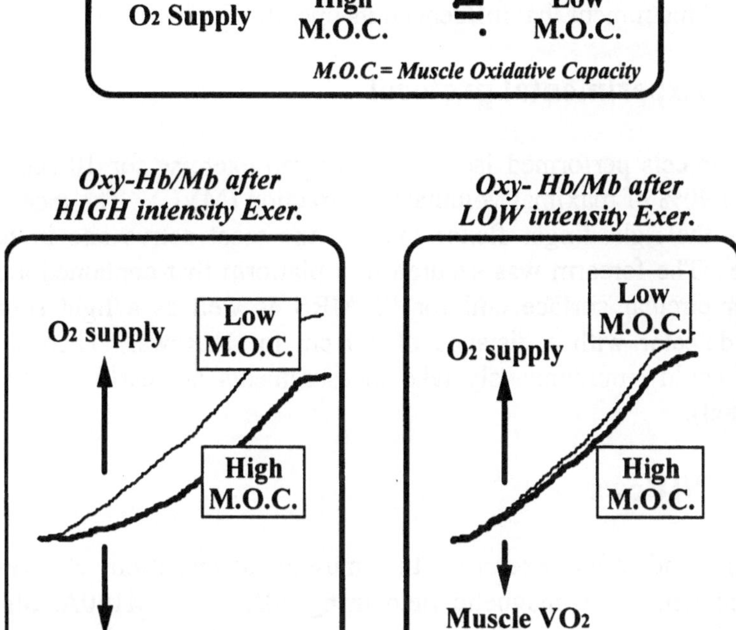

Figure 1. Illustration of our hypothesis.

2. METHODS

2.1 Subjects

Seven healthy male volunteers were studied (five moderately physically active, one well-trained triathlete, and one former Japanese windsurfing champion, mean age ± S.D.: 28.3 ± 2.5 yr.). All subjects were briefed about the experimental protocol, and informed consent was obtained before the test.

2.2 Experimental design

The muscle oxygenation level was measured by NIRcws while phosphocreatine (PCr) concentration was evaluated simultaneously by ^{31}P-MRS. The subject performed the same experiment twice: once with the

NIRcws and ^{31}P-MRS and once with Doppler ultrasound only. This dual measurement was made because of the strong magnetic fields in the Doppler ultrasound instrument that influenced its operation.

2.3 Experimental protocol

All subjects performed isometric handgrip exercise for 10 sec at 30%, 60% and 90% of maximal voluntary contraction (MVC). The measurement site was the right finger flexor muscle, the most active site in handgrip exercise. The forearm was secured to a platform that contained a 3.0 cm-diameter circular surface coil for ^{31}P-MRS as well as a light source and optical detector with a distance of 3.0 cm for NIRcws. Therefore, both devices could simultaneously take measurements at nearly identical sites (Sako et al).

2.4 NIRcws

During and after exercise, the muscle oxygenation changes were measured within the magnetic field using NIRcws (OM100A, Shimadzu, Kyoto, Japan). The basic operation of this NIRcws device has been extensively explained in a previously published paper [1, 2]. In this study, three-wavelength optical fibers, with wavelength of 780, 805 and 830 nm, were used as the light source. Changes in oxygenated Hb/Mb (Oxy-Hb/Mb) and total Hb (Total-Hb) were calculated with the least-squares method using data from the changes in light absorbance at different wavelength [2].

The reoxygenation rate (Reoxy-rate) was determined by the rate of oxy-Hb/Mb recovery immediately after exercise up until 4 sec at various intensities.

Hamaoka et al [12] recently reported that muscle VO_2 during exercise (muscle VO_{2EX}) can be monitored by NIRcws during isometric exercise without arterial occlusion because intramuscular pressure causes a sufficient transient occlusion of the blood flow to the muscle. To confirm that the arterial inflow was completely interrupted during isometric exercise, the isometric exercise was performed at each intensity exercise with and without spontaneous arterial occlusion. These results showed that arterial inflow was interrupted at all intensities because the deoxygenation rates with or without artificial arterial occlusion were similar. Further, the restriction of the brachial artery inflow during isometric exercise was also confirmed at all intensities by Doppler ultrasound measurement. Therefore, the deoxygenation rate during the last 4 sec of exercise was defined as the muscle VO_{2EX} at each intensity exercise.

The fat layer thickness varies greatly in humans and the optical densities of Oxy-Hb/Mb and Total-Hb cannot be compared between individuals. Therefore, to evaluate differences in muscle oxygenation between each subject, arterial occlusion was used [3, 5]. The muscle oxygenation level during and after exercise was expressed using the overall change from the resting pre-occluded to minimum oxygenation level during arterial occlusion.

2.5 ^{31}P-MRS

^{31}P-MRS signals were obtained by a NMR spectrometer (Otsuka Electronics) with a 2.0-T superconducting 26-cm-bore magnet. Pulse repetition time was 2 sec. Five pulses were averaged to obtain a free induction decay, therefore, each spectrum was obtained over a period of 10 sec. All ^{31}P-MRS spectra were fitted by a non-linear, curve-fitting method to calculate the areas under the ß-ATP, PCr, and Pi peaks. The area of the PCr peaks during recovery from 90%MVC isometric handgrip exercise was integrated and fit to a single-exponential curve, and the time constant of PCr recovery was determined by the following equation:

$$[PCr] = PCr_0 + \quad PCr\,(1-e^{-kt})$$

where PCr_0 is the initial [PCr], PCr is the difference between final and initial [PCr], t is time, k is the rate constant, and $1/k$ is the time constant. This time constant was defined as the indicator of the muscle oxidative capacity. The intracellular pH was calculated by the median chemical shift between the Pi and PCr peaks [13].

2.6 Doppler ultrasound instrument

BABF was determined using a Doppler ultrasound instrument (SONOS 1500, HP77035A, Tokyo, Japan) [14]. A two-dimensional-imaging echography analysis was conducted with a duplex scanner fitted with a 7.5-MHz linear probe to estimate the brachial artery diameter. Velocity analysis by pulsed Doppler flowmetry was carried out during the experimental protocol. The probe was fixed to the skin over the brachial artery just proximal to the forearm. Peak blood velocity was calculated by integration of the outer envelope of the maximal velocity values in the flow profile and was determined as the mean value of the three successive cardiac cycles for each protocol. BABF was calculated by multiplying the cross sectional area of the brachial artery [area = $\pi \times (\text{diameter}/2)^2$] with the angle-corrected, time and space-averaged, and amplitude (signal intensity)-weighted mean blood velocity (Vmean); e.g., BABF = Vmean × area.

Heart rate (HR) was determined from the R-R intervals of the electrocardiogram.

2.7 Statistics

One-way analysis of variance (ANOVA) was used to compare the parameters for exercise at different intensities with Scheffe's post hoc comparisons. The interdependence between Reoxy-rate and time constant for PCr recovery was expressed by the Spearman rank correlation coefficient. A 5% level was considered significant.

3. RESULTS

3.1 PCr resynthesis rate

Tc for PCr resynthesis and minimum value of intracellular pH were 28.9 ± 9.1 sec. and 6.8 ± 0.04 (mean ± SD), respectively. All seven subjects were regarded as having little restriction of mitochondrial oxidative phosphorylation because all of their minimum pH values were above 6.8 [15]. Some studies have demonstrated that a linear relationship exists between PCr resynthesis rate and mitochondrial enzyme activities in human skeletal muscle [7, 16]. Therefore, PCrTc in the experiment should be considered as an indicator of muscle oxidative capacity.

3.2 Muscle oxygen dynamics

A typical example of the changes in oxy-Hb/Mb during and after exercise at 90% MVC is shown in Fig. 2. After static handgrip exercise, total and oxy-Hb/Mb rose exponentially. The Reoxy-rates of each of the workloads were 5.9 ± 2.4, 5.5 ± 1.6, and 7.1 ± 4.1 %/sec on average at 30%, 60%, and 90% MVC, respectively. Fig. 3 shows the relationship between the Reoxy-rates and the PCrTc at each exercise intensity. There was no correlation between the PCrTc and Reoxy-rate at 30% and 60% MVC. In contrast, significant positive correlation was found between the Reoxy-rate at 90% MVC and the PCrTc ($r=0.825$, $p<0.05$).

At the beginning of exercise, total-Hb dropped rapidly and then remained stable while oxy-Hb/Mb continued to decline linearly during exercise (Fig. 2). Muscle VO_{2EX} was calculated by the rate of decline of oxy-Hb/Mb during the 4 seconds where the total-Hb remained stable. The muscle VO_{2EX}

increased 5.9, 8.8 and 12.6-fold of the resting on average at 30%, 60% and 90% MVC, respectively (Fig. 4). Moreover, the muscle VO_{2EX} at 90% MVC was significantly higher than that at 30% and 60% MVC.

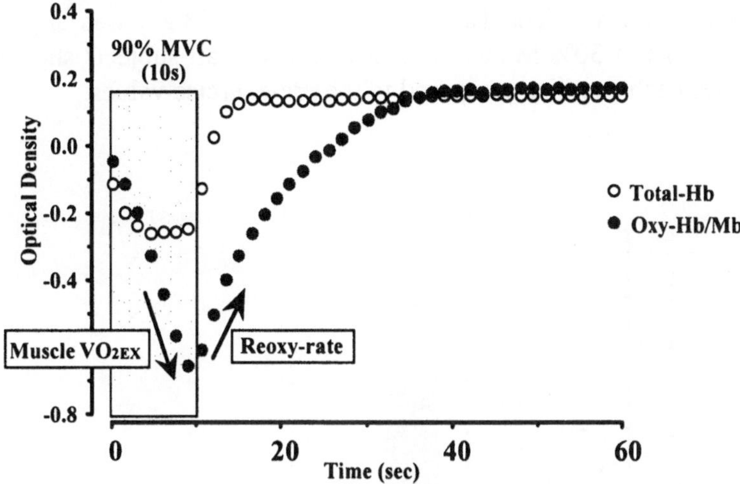

Figure 2. Typical example of changes in Oxy-Hb/Mb and Total-Hb during and after exercise at 90% MVC.

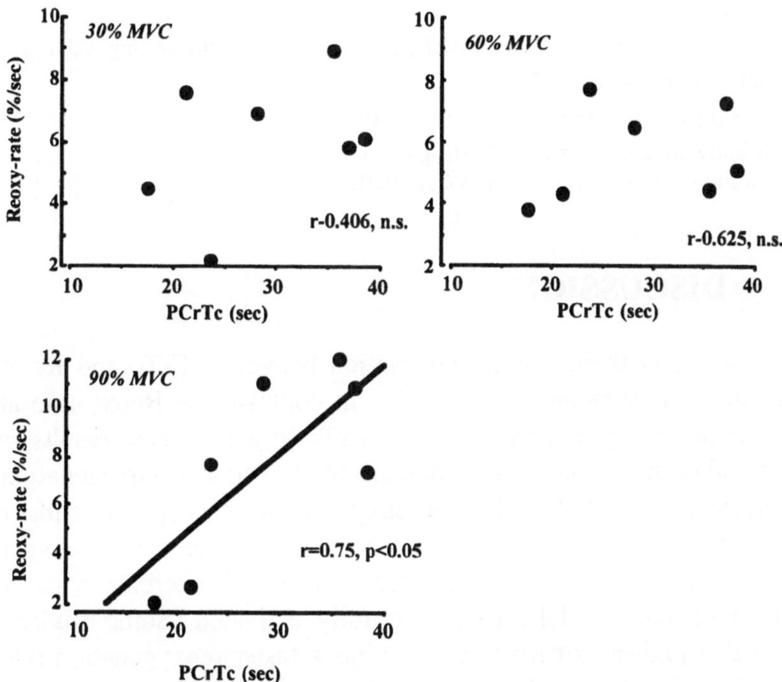

Figure 3. Relationship between the Reoxy-rate at each intensity exercise and the time constant for PCr recovery.

3.3 Brachial arterial blood flow

Peak values of brachial arterial blood flow (BABFpeak) increased 1.9, 2.4 and 2.7-fold of the resting on average at 30%, 60% and 90% MVC, respectively (Fig. 4). The BABFpeak after 90% MVC was significantly higher than that at 30% MVC. The heart rates of all subjects showed little change after each exercise compared to the pre-exercise values.

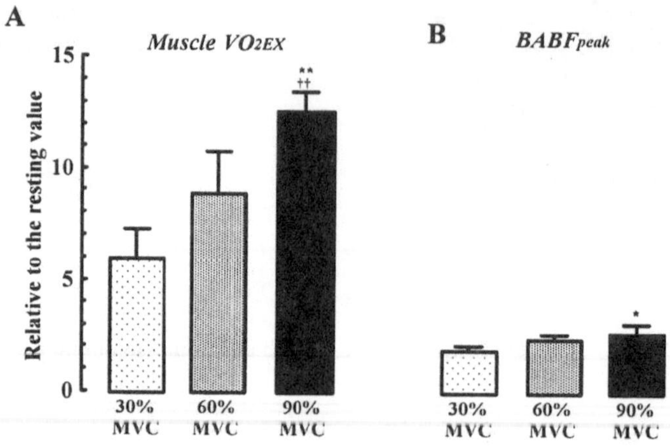

Figure 4. Muscle VO_{2EX} (A) and $BABF_{peak}$ (B) relative to the resting value at various intensities (Mean ± SE).

** Significantly different from 30% MVC; $p<0.01$.

* Significantly different from 30% MVC; $p<0.05$.

†† Significantly different from 60% MVC; $p<0.01$.

4. DISCUSSION

We found that there was no correlation between PCrTc and Reoxy-rate after exercise at 30% and 60% MVC. In contrast, the Reoxy-rate at 90% MVC was positively correlated to the PCrTc (Fig 3). These results suggest that the higher the muscle oxidative capacity, the slower reoxygenation from the exercise at 90% MVC. These findings support our hypothesis that higher muscle oxidative capacity muscle may exhibit a slower muscle oxygenation recovery after short duration, high intensity isometric exercise, when muscle VO_2 is high and O_2 delivery is not fully activated. Some studies have reported that higher oxidative capacity has a faster reoxygenation rate after exercise [1, 6]. To our knowledge, this is the first report that a higher muscle oxidative capacity exhibited a slower reoxygenation rate after exercise.

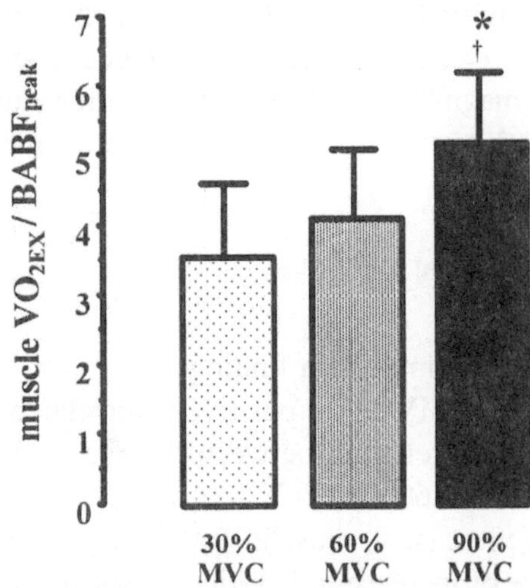

Figure 5. Muscle VO_2 / $FABF_{peak}$ ratio at various intensities.
* Significantly different from 30% MVC; $p<0.05$.
† Significantly different from 60% MVC; $p<0.05$.

Reoxy-rate is defined as the balance between the muscle VO_2 and O_2 supply. As mentioned earlier, according to our hypothesis, the O_2 balance (Reoxy-rate) after short-term isometric contraction is more influenced by muscle VO_2 than O_2 supply. This study found much greater increased muscle VO_{2EX} than $BABF_{peak}$, especially at 90%MVC (Fig. 4). Also, the ratio of muscle VO_{2EX} / $BABF_{peak}$ was 3.6 ± 1.0 (30%MVC), 4.1± 1.0 (60%MVC) and 5.2 ± 1.0 (90%MVC), respectively (Fig. 5) and the ratio of muscle VO_{2EX} / $BABF_{peak}$ was significantly higher at 90%MVC than 30% and 60%MVC ($p<0.05$). These results suggest that O_2 supply increase after exercise may be suppressed by short-term isometric exercise. Additionally, with high intensity exercise, increased O_2 demand promotes greater O_2 extraction from capillaries in HOCmuscle than LOCmuscle. These may cause delayed reoxygenation after 90%MVC in HOCmuscle

BABF was blocked by increased intramuscular pressure, and Total-Hb was stable during isometric exercise. Therefore, O_2 inflow to the exercising muscle is completely interrupted during isometric exercise. These results suggest that the availability of O_2, but not of convective O_2 transport, may limit muscle VO_2 during isometric exercise. Bangsbo et al [17] found that O_2 extraction increased after a few seconds of exercise. This finding suggests that the intracellular PO_2 may decrease within a few seconds, which may induce a greater O_2 extraction at the onset of blood reperfusion. Some studies reported that HOCmuscle produces a more rapid adjustment in muscle VO_2 [18]. Therefore, the O_2 gradient from capillary to myocyte at the onset of reperfusion may be higher in HOCmuscle than in LOCmuscle after high intensity isometric exercise. These data suggest that the mitochondrial driving force and the magnitude of Mb deoxygenation during exercise may be important factors influencing the Reoxy-rate after isometric exercise.

5. CONCLUSION

This study produced four important findings.

1. Reoxy-rate at 90% MVC was positively correlated to the PCrTc.
2. There was no correlation between PCrTc and Reoxy-rate after exercise at 30% and 60% MVC.
3. The degree of increase in muscle VO_{2EX} was much greater increased than the degree of increase in $BABF_{peak}$, especially at 90%MVC.
4. The ratio of muscle VO_{2EX} / $BABF_{peak}$ was significantly higher at 90%MVC than 30% and 60%MVC.

These results support our hypothesis that higher oxidative capacity muscle exhibits a slower muscle oxygenation recovery after short duration, high intensity isometric exercise when muscle VO_2 is high and O_2 delivery is not fully activated. In contrast with low intensity exercise, when O_2 demand is also not fully increased, the Reoxy-rate may not be correlated to muscle oxidative capacity because muscle VO_2 may not be sufficient enough to where the muscle VO_2 effects would significantly influence the Reoxy-rate.

ACKNOWLEDGEMENTS

The authors would thank Joohee Im for all her help and Eric Sell in wiring English manuscript.

REFERENCES

1. Chance B, Dait TM, Chang C, Hamaoka T, and Hagerman F. Recovery from exercise induced desaturation in the quadriceps muscle of elite competitive rowers. Am J Physiol 1992;262:766-775.
2. Homma S, Eda H, Ogasawara S, and Kagaya A. Near-infrared estimation of O_2 supply and consumption in forearm muscles working at varying intensity. J Appl Physiol 1996; 80:1279-1284.
3. Hamaoka T, Iwane H, Shimomitsu T, Katsumura T, Murase N, Nishio S, Osada T, Kurosawa Y, and Chance B. Noninvasive measures of oxidative metabolism on working human muscles by near-infrared spectroscopy. J Appl Physiol 1996;81:1410-1417.

4. Hamaoka T, Mizuno M, Katsumura T, Osada T, Shimomitsu T, and Quistorff B. Correlation between indicators determined by near infrared spectroscopy and muscle fiber types in humans. Jpn J Appl Physiol 1998; 28 (5): 243-248.
5. Sako T, Hamaoka T, Higuchi H, Kurosawa Y, Katsumura T. Validity of NIR spectroscopy for quantitatively measuring muscle oxidative metabolic rate in exercise. J Appl Physiol 2001;90:338-344.
6. McCully KK, Iotti S, Kendrick K, Wang Z, Posner JD, Leigh JS, and Chance B. Simultaneous in vivo measurement of HbO_2 saturation and PCr kinetics after exercise in normal humans. J Appl Physiol 1994;77:5-10.
7. Mizuno M, Hamaoka T, Osada T, Shimomitsu T, Katsumura T, Quistorff B. Correlation between mitochondrial enzyme activities and the rate of hemoglobin deoxygenation at onset of exercise in human gastrocnemius muscles. Med Sci Sports Exerc 1999;31: S275.
8. Homma S, Fukunaga T, and Kagaya A. Influence of adipose tissue thickness on near infrared spectroscopic signals in the measurement of human muscle. J Biomed Opt 1996; 1:418-424.
9. Niwayama M, Yamamoto K, Kohata D, Hirai K, Kudo N, Hamaoka T, Kime R, Katsumura T. A 200-cahnnel imaging system of muscle oxygenation using CW near-infrared spectroscopy. IEICE Trans Inf & Syst 2002; E85-D: 115-123.
10. Huonker M, Halle M, Keul J. Structural and functional adaptations of the cardiovascular system by training. Int J Sports Med 17 1996; 3:S164-S172.
11. Saltin B, Radegran G, Koskolou MD, Roach RC. Skeletal muscle blood flow in humans and its regulation during exercise. Acta Physiol Scand 1998;162:421-436.
12. Hamaoka T, Katsumura T, Sako T, Higuchi H, Murakami M, Esaki K, Kime R, Homma T, Kurosawa Y, Shimomitsu T and Chance B. A continuous measurement of oxidative rate by near infrared spectroscopy in human muscles. Med Sci Sports Exerc 1999;31: S246.
13. Kushmerick MJ, and Meyer RA. Chemical changes in rat leg muscle by phosphorus nuclear magnetic resonance. Am J Physiol 1985;248:542-549.
14. Osada T, Katsumura T, Hamaoka T, Inoue S, Esaki K, Sakamoto A, Murase N, Kajiyama J, Shimomitsu T, and Iwane H. Reduced blood flow in abdominal viscera measured by Doppler ultrasound during one-legged knee extension. J Appl Physiol 1999; 86:709-719.
15. Takahashi H, Inaki M, Fujimoto K, Katsuta S, Anno I, Niitsu M, and Itai Y. Control of the rate of phosphocreatine resynthesis after exercise in trained and untrained human quadriceps muscles. Eur J Appl Physiol 1995;71:396-404.
16. McCully KK, Fielding RA, Evans WJ, Leigh JS, Posner JD. Relationship between in vivo and in vitro measurements of metabolism in young and old human calf muscles. J Appl Physiol 1993;77:2740-2747.
17. Bangsbo J, Krustrup P, Gonzalez-Alonso J, Boushel R, Saltin B. Muscle oxygen kinetics at onset of intense dynamic exercise in human. Am J Physiol 2000;279:R899-906.
18. Phillips SM, Green HJ, MacDonald MJ, Hughson RL. Progressive effect of endurance training on VO_2 kinetics at the onset of submaximal exercise. J Appl Physiol 1995;79: 1914-1920.

Chapter **48**

OXYGEN DIFFUSION COEFFICIENT AND OXYGEN PERMEABILITY OF METMYOGLOBIN SOLUTIONS DETERMINED IN A DIFFUSION CHAMBER USING A NON-STEADY STATE METHOD

Jolanda P.W.M. Lamers-Lemmers, Louis JC Hoofd & Berend Oeseburg
237 Department of Physiology, University Medical School PO Box 9101, 6500 HB Nijmegen, The Netherlands.

Abstract: Non-steady state measurements of oxygen diffusion through various model layers can be performed using a diffusion chamber that was described earlier [1, 2]. A closer analysis of these measurements showed that they not only yield the oxygen diffusion coefficient (DO_2) of the diffusion layer, but also the oxygen permeability (***P***O_2). In this study DO_2 and ***P***O_2 have been determined in solutions of metmyoglobin (metMb) with concentrations varying between 5 and 40 g/dL at 25 °C. Both DO_2 and ***P***O_2 decreased with increasing metMb concentration. This decrease was comparable to the values reported for DO_2 in protein solutions by Kreuzer and Hoofd [3]. Using this diffusion chamber for non-steady state measurements, oxygen diffusion coefficients, oxygen permeability and oxygen solubility (aO_2 = ***P***O_2 / DO_2) of various model layers could be determined..

Key words: Diffusion, Metmyoglobin, Model, Oxygen

Abbreviations: αO_2: oxygen solubility c: protein concentration
cHb: hemoglobin concentration cMb: metmyoglobin concentration
DO_2: oxygen diffusion coefficient metMb: metmyoglobin
pO_2: oxygen partial pressure ***P***O_2: oxygen permeability
$t_{0,silicon}$: delay time for diffusion through gas chambers and silicon membrane
$t_{0,silicon+metMb}$: delay time for diffusion through gas chambers, silicon membrane, and layer of metMb solution
$t_d = (\sqrt{t_{0,silicon}} - \sqrt{t_{0,silicon+metMb}})^2$: layer delay time for diffusion through layer of metMb solution

1. INTRODUCTION

Oxygen diffusion through various model layers can be studied by performing non-steady state [1] and steady state measurements [2] using a diffusion chamber. The first yield the oxygen diffusion coefficient (DO_2) and the latter yield the oxygen permeability ($\boldsymbol{P}O_2$). However, an improved analysis of the non-steady state measurements shows that these also yield $\boldsymbol{P}O_2$ [4 (full paper in preparation)].

In this study DO_2 and $\boldsymbol{P}O_2$ of metmyoglobin (metMb) solutions were determined using both the non-steady state and the steady state method. These values will be used in an additional study of myoglobin solutions to correct for the "passive", e.g. non-facilitated, diffusion through these solutions.

2. METHODS

The diffusion chamber was similar to the one used in a former study [2], consisting of a top gas chamber, a bottom gas chamber, and a special holder for the model layer (figure 1). Both gas chambers could be flushed separately with gases of different composition and the temperature could be maintained constant within 0.03 °C throughout the measurement.

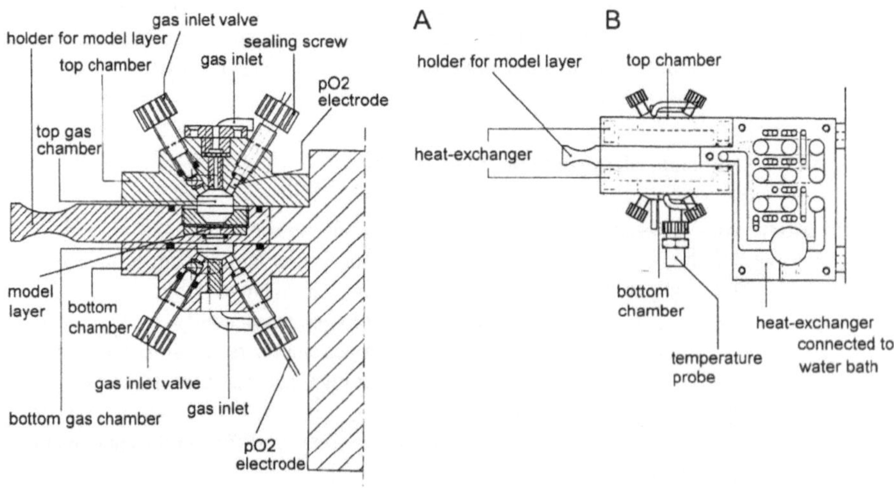

Figure 1. The diffusion chamber set-up
A: diffusion chamber with special holder for the model layer
B: built-in circulating system for temperature control of the diffusion chamber

The model layer in this study was composed of a MEM-100 silicon membrane (Membrane Products Corporation, Salt Lake City USA) with a metMb solution in 20 mM Tris-HCl buffer (pH 8.0) on top.

Using the non-steady state method, the oxygen diffusion coefficient (DO_2) and the oxygen permeability ($\boldsymbol{P}O_2$) of metMb solutions were determined. The non-steady state measurement was performed by first flushing both gas chambers with 100% nitrogen gas until steady state was reached. At time t=0 sec the bottom chamber was closed and the top chamber was flushed with an oxygen containing gas mixture. Figure 2, upper panel, shows the recordings of a typical non-steady state measurement for a silicon membrane, where the oxygen partial pressures in the top and bottom chambers are plotted against time (note the different scales!). After a certain delay the oxygen partial pressure (pO_2) in the bottom chamber started to increase, because O_2 diffuses through the model layer into the bottom chamber. The increase became essentially after some time. The linear increase was fitted using a linear fit and its intercept with the horizontal time axis ($pO_2 = 0$) was determined. This delay time t_0 represented the delay as a consequence of the diffusion through the gas chambers and the model layer. t_0 was determined for both the silicon membrane ($t_{0,silicon}$) as well as for the silicon membrane with the metMb solution on top ($t_{0,silicon+metMb}$). The layer delay time t_d was then calculated as $t_d = (\sqrt{t_{0,silicon}} - \sqrt{t_{0,silicon+metMb}})^2$ and DO_2 of the metMb solution was calculated from this layer delay time as $DO_2 = {}^1/_6\, L^2 / t_d$, where L was the thickness of the layer of metMb solution [4].

$\boldsymbol{P}O_2$ was calculated from the increase of pO_2 in the bottom chamber (dp_{bottom}/dt), the slope of the linear part of the curve. The oxygen flux J, which is the amount of oxygen diffusing through a unit area, was derived from the value of dp_{bottom}/dt, using the ideal gas law [5]:

$$\frac{dn}{dt} = J \times A = \frac{d}{dt}\left(\frac{p \times V}{R \times T}\right)$$

(n = amount of oxygen in moles, t = time, A = surface area of the layer, p = partial pressure, V = chamber volume, R = gas constant, T = temperature)

The oxygen permeability $\boldsymbol{P}O_2$ of the layer was then calculated from the oxygen flux and the oxygen pressure difference across the model layer (Δp):

$$J = \boldsymbol{P}O_2 \frac{\Delta p}{L}$$

Because $\boldsymbol{P}O_2 = DO_2 \cdot \alpha O_2$, the oxygen solubility (αO_2) of the diffusion layer could also be calculated.

The oxygen permeability ($\boldsymbol{P}O_2$) of metMb solutions was also determined using a steady state method as described previously [5] in order to compare these values with the values determined using the non-steady state method.

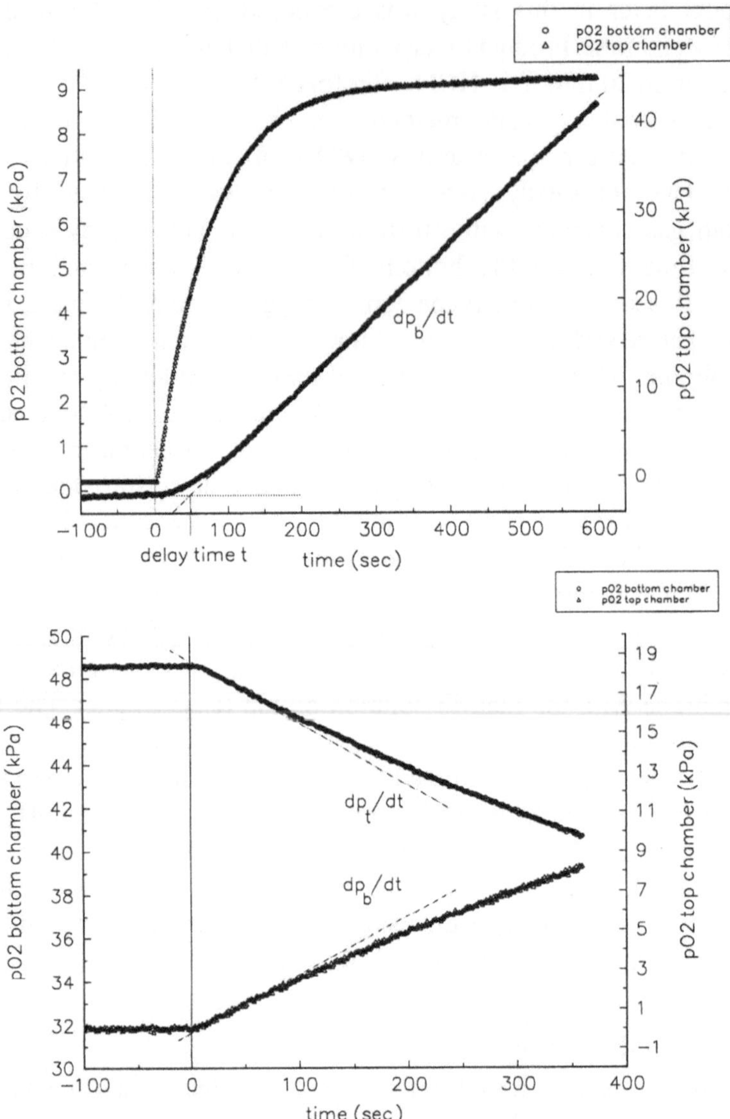

Figure 2. Recordings of a typical non-steady state measurement (upper panel) and a typical steady state measurement (lower panel). Note the different scales!

Figure 2, lower panel, shows the recordings of a typical steady state measurement, where the oxygen partial pressures in the top and bottom chambers are plotted against time (again note the different scales!). At the beginning of the experiment the gas chambers were flushed with gases of different composition until steady state was achieved. At t=0 sec the valves of the gas inlets were closed. Because O_2 diffuses through the model layer,

pO_2 in the top chamber started to decrease and pO_2 in the bottom chamber started to increase. The values for dp_{top}/dt and dp_{bottom}/dt were determined as the initial decrease or increase in pO_2, using a least squares fit method. These initial changes should reflect the preceding steady state. Again the oxygen flux J was derived from these values, using the ideal gas law. Calculating the oxygen permeability $\boldsymbol{P}O_2$ of the layer is more complicated in this case [5], but essentially similar to the calculation shown above.

3. RESULTS

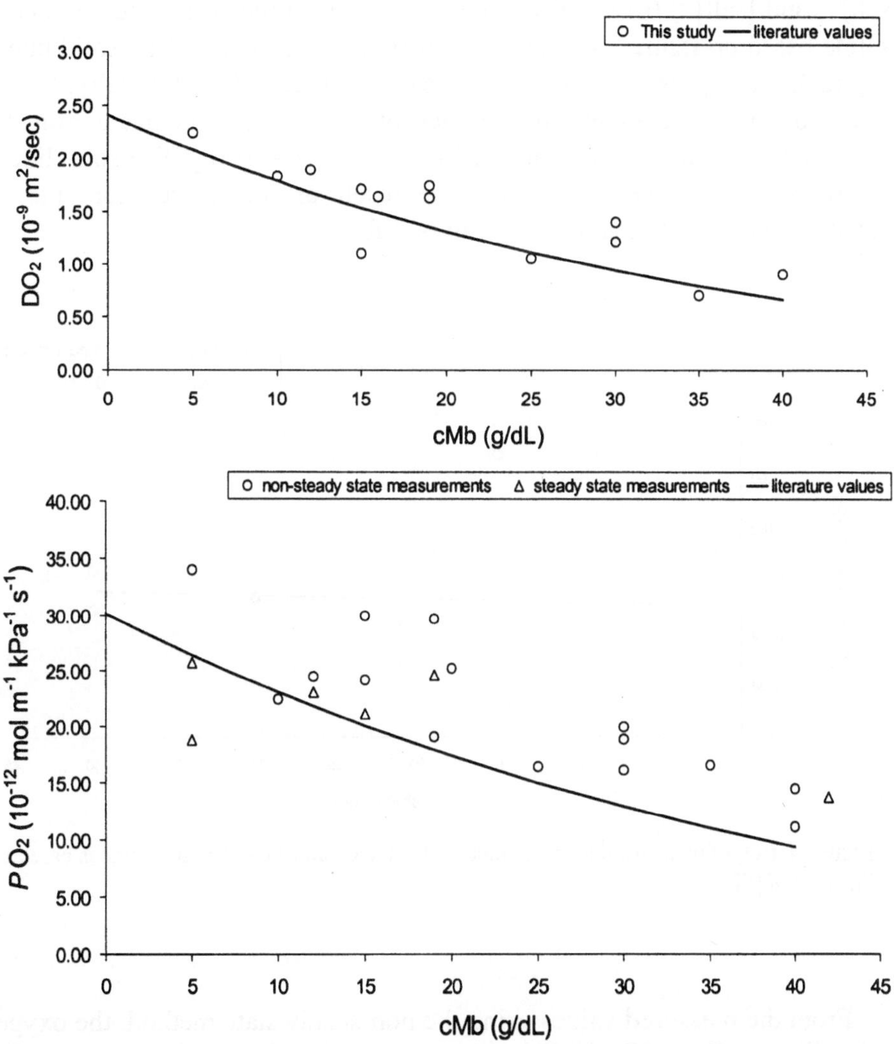

Figure 3. Oxygen diffusion coefficients DO_2 (upper panel) and oxygen permeabilities $\boldsymbol{P}O_2$ (lower panel) of metMb solutions. The solid lines are based on literature values [3, 6, 7] (see text).

DO_2 and $\boldsymbol{P}O_2$ have been determined in solutions of metMb with concentrations varying between 5 and 40 g/100 mL at 25 °C. Figure 3 shows the results of these measurements. DO_2 and $\boldsymbol{P}O_2$ decreased with increasing metMb concentration. The measurements were compared with literature compilations of DO_2 against protein concentration from Kreuzer & Hoofd [3], where $DO_2 = 2.07 \cdot 10^{-9} \cdot (1\text{-}c/100) \cdot 10^{-c/119}$ (c = protein concentration in g/dL). These literature values were corrected for the absence of saline in the present study, as follows. First, $\boldsymbol{P}O_2$ (= $DO_2 \cdot \alpha O_2$) was calculated from the DO_2 values and $\alpha O_2 = \alpha_0 \cdot (1+0.00312 \cdot cHb) - \Delta\alpha[salt]$ [6], with $\alpha_0 = 1.25 \cdot 10^{-5}$ mol L^{-1} kPa^{-1}, cHb = hemoglobin concentration, $\Delta\alpha = 3.2 \cdot 10^{-6}$ mol kPa^{-1} (mol salt)$^{-1}$, and [salt] = 0.15 mol/L. Then, $\boldsymbol{P}O_2$ was corrected for the absence of saline based on figure 1 of the paper of Breepoel *et al.* [7]. The solid line in figure 3, lower panel, represents the result of this calculation. Finally, DO_2 (= $\boldsymbol{P}O_2$ / αO_2), corrected for the absence of saline, was calculated from the corrected $\boldsymbol{P}O_2$ and the calculated αO_2 (with [salt] = 0) [6]. The solid line in figure 3, upper panel, represents the result of this calculation. For all calculations we assumed that c = cHb = cMb.

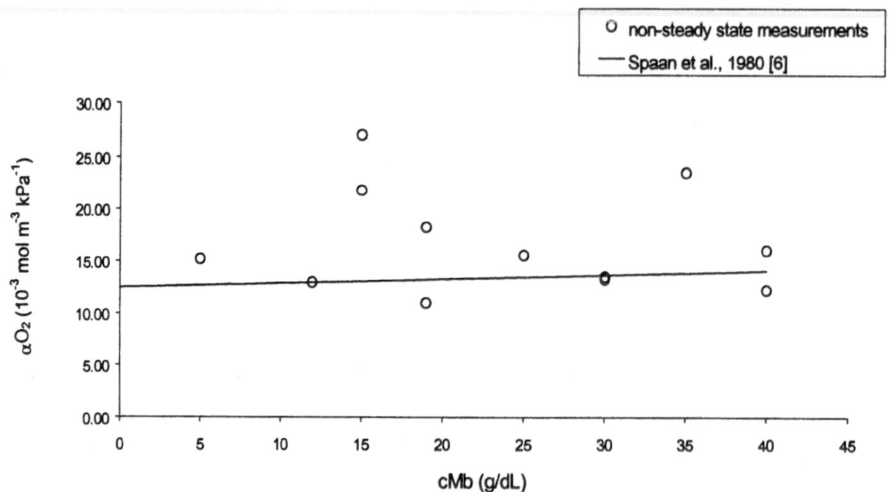

Figure 4. αO_2 of metMb solutions calculated from DO_2 and $\boldsymbol{P}O_2$. The solid line is based on Spaan *et al.*[6].

From the measured values using the non-steady state method, the oxygen solubilities ($\alpha O_2 = \boldsymbol{P}O_2$ / DO_2) were calculated and shown in figure 4. The straight line in this graph represents the calculated αO_2 according to the above formula for protein solutions with [salt] = 0 [6].

4. DISCUSSION

The increase in pO_2in the bottom chamber, p_b in figure 2, is not exactly linear, but for lower times will be approached exponentially [4 (full paper in preparation]:

$$p_b = \text{Const} \times \boldsymbol{PO}_2\left[t - t_0 + \frac{12}{\pi^2} t_0 \sum_{k=1}^{\infty} \frac{2(-)^{k-1}}{\pi^2 k^2} \exp\left(-\frac{\pi^2 k^2 t}{6t_0} \right) \right]$$

where $t_0 = L^2/(6DO_2)$ and Const is a constant evaluated from the experimental conditions. For larger times, the increase will slow down because the driving force across the layer decreases due to an increasing bottom chamber back pressure. So, for the linear evaluation a portion of the line was chosen between 2 t_0 and 6 t_0. When the above theoretical formula is fitted to a straight line, this yields $0.99t-0.97t_0$, an underestimation of the slope of 1% and of t_0 of 3%. However, also the top chamber gas pressure does not change immediately. This will not influence the first appearance of O_2 in the bottom chamber - O_2 will diffuse into the layer instantly after its appeareance - but the amount of O_2 will increase, resulting in increasing slope. The latter effect will lead to an overestimation in slope and t_0. The net effect is not clear, but seems to be minor since none of the evaluated experiments showed any statistically significant deviations from a linear increase.

Possibly, a statistical fit to a theoretically better defined function than the straight line could lead to more confident results.

It should be noted, however, that the data for DO_2 presented here are from the difference in t_0 between two measurements: with and without liquid layer. Consequently, it is expected that effects of methodological error are minimized.

In particular for the membrane without liquid layer, as shown in figure 2, there is no steady state of pO_2 in the top chamber during the evaluated time part - as is not for the bottom chamber. Therefore, the analysis of $\boldsymbol{PO}_2$ was done assuming time-averaged chamber pO_2s over the evaluated period.

The results in figure 3 show that DO_2 and $\boldsymbol{PO}_2$ decrease with increasing metMb concentration. DO_2 (m^2 sec^{-1}) determines the speed of O_2 diffusing through the model layer whereas $\boldsymbol{PO}_2$ (mol m^{-1} kPa^{-1} sec^{-1}) determines the amount of O_2 that diffuses through the model layer. Therefore, DO_2 and $\boldsymbol{PO}_2$ are expected to decrease with increasing metMb concentration, because the viscosity of the solution increases with increasing metMb concentration.

The decrease in DO_2 in this study is comparable to the values reported for DO_2 in protein solutions by Kreuzer and Hoofd [3], assuming that c = cMb in the evaluations of DO_2 and αO_2 [3, 6]. This comparison applies to the whole range of protein concentrations.

$\boldsymbol{P}O_2$ also decreases with metMb concentration, but this decrease seems to be somewhat slower than the decrease discussed for DO_2. The decrease for $\boldsymbol{P}O_2$ is roughly comparable to the literature data, which are calculated from DO_2 and αO_2 of hemoglobin [3, 6] and corrected for the absence of saline [6, 7].

The values of $\boldsymbol{P}O_2$ from the non-steady state measurements are comparable with the values of $\boldsymbol{P}O_2$ from the steady state measurements (see figure 3, lower panel, circles and triangles, respectively), indicating that there is no essential difference between the results of these two methods.

The relation between DO_2 and $\boldsymbol{P}O_2$ is the oxygen solubility αO_2 (= $\boldsymbol{P}O_2$ / DO_2), which is defined as the amount of O_2 that is solved per m^3 solution per kPa partial pressure (mol m^{-3} kPa^{-1}). The calculated values for αO_2 showed no significant increase or decrease with metMb concentration, contrary to hemoglobin where αO_2 does increase slightly with hemoglobin concentration [6]. However, there is no significant difference between the calculated values for metMb in this study and the values for hemoglobin, so a slight increase in αO_2 of metMb could not be excluded based on our data.

In conclusion, this diffusion chamber can be used for non-steady state measurements. From these non-steady state measurements the oxygen diffusion coefficients, oxygen permeabilities, and oxygen solubilities of various model layers can be determined. In this way, information about oxygen diffusion through various model layers can be obtained using a relatively simple set-up.

ACKNOWLEDGEMENTS

This research was supported by Grant number 902-19-109 by NWO (Netherlands Organization for Scientific Research)

REFERENCES

1. van Dijk AMC, Hoofd LJC, Oeseburg B. Diffusion coefficient of oxygen in various model layers as determined by analysis of time-dependent diffusion. Adv Exp Med Biol 1996; 388:327-331.
2. Lamers-Lemmers JPWM, Hoofd LJC, Otte-Höller I, de Waal RMW, Oeseburg B. Measurements of oxygen diffusion through cultured endothelial cell monolayers. Oxygen Transport to Tissue XXI, Adv Exper Med Biol 471. New York: Plenum Press, 1999; 691-695.
3. Kreuzer F, Hoofd LJC. Facilitated diffusion of oxygen in the presence of hemoglobin. Respiration Physiology 1970; 8:280-302.

4. Hoofd L, Lamers-Lemmers J. Die Bestimmung von DO_2 an Diffusionsschichten auf einen Träger. In: Abstract Book 33rd Atmungsphysiologische Arbeitstagung, Nijmegen: January 1999; 10
5. Hoofd L, de Koning J, Kreuzer F, Lamboo A. Determination of permeabilities for two gases from recording the partial pressure of one gas. Pflügers Arch 1986; 407:320-326
6. Spaan JAE, Kreuzer F, van Wely FK. Diffusion coefficients of oxygen and hemoglobin as obtained simultaneously from photometric determination of the oxygenation of layers of hemoglobin solutions. Pflügers Arch 1980; 384:241-251
7. Breepoel PM, de Koning J, Hoofd L. Diffusion of oxygen in methemoglobin solutions. Dependence on salt concentration. Biochem Biophys Res Comm 1982; 3:848-850

Chapter 49

PINACIDIL-INDUCED OPENING, LIKE GLIBENCLAMIDE-INDUCED CLOSURE OF CARDIAC K_{ATP} CHANNELS, PROTECTS CARDIAC FUNCTION AGAINST ISCHEMIA IN ISOLATED, WORKING, ERYTHROCYTE PERFUSED RAT HEARTS

Roger J. Legtenberg[1], Ralph J.F. Houston[1], Paul Smits[2] & Berend Oeseburg[1]
[1]Department of Physiology 237, University of Nijmegen, The Netherlands; [2]Department of Pharmacology 233, University of Nijmegen, The Netherlands

Abstract: Glibenclamide-induced closure of ATP-dependent potassium (K_{ATP}) channels decreases coronary blood flow during normoxic and post-ischemic conditions. We have found that post-ischemic cardiac function is improved after glibenclamide treatment. Our theory was that this is a result of higher intracellular calcium concentrations due to reduction in ischemia-mediated hyperpolarization of the myocardial cell membrane. We hypothesized therefore that opening K_{ATP} channels would reduce post-ischemic function in our isolated, erythrocyte perfused, working rat heart model. During treatment with 1 or 12 μmol.L^{-1} pinacidil (protein unbound concentration) both before and after 12 minutes global ischemia coronary blood flow increased 2-3 fold compared with vehicle, while cardiac functional recovery post-ischemically was improved with both concentrations. Because closing and opening cardiac K_{ATP} channels both improve post-ischemic function, our calcium theory above can be discounted. The protective effect of glibenclamide may possibly be ascribed to metabolic effects such as preservation of ATP levels during ischemia.

Key words: Coronary blood flow, diabetes mellitus, glibenclamide, isolated rat heart, K_{ATP} channels, pinacidil

Abbreviations: CBF: coronary blood flow; K_{ATP} channel: ATP-dependent potassium channel

1. INTRODUCTION

Closing or opening of ATP-dependent potassium (K_{ATP}) channels in the cardiovascular system has been frequently studied. In diabetes mellitus type 2, K_{ATP} channel blockers, sulfonylurea derivatives, have been in use for over two decades. Their target organ are the pancreatic islets but it seems that these drugs also have extrapancreatic effects, for example in the cardiovascular system. It is now clear that closure and opening of the K_{ATP} channel mediates coronary blood flow (CBF) during normoxic and post-ischemic situations. In anesthetized dogs and isolated rat hearts basal CBF was reduced after treatment with glibenclamide, a K_{ATP} channel blocker [1,2,3], even at clinically relevant concentrations [4, In press], while opening the channel leads to a marked increase in CBF [5]. Closure of K_{ATP} channels also prevents the hypoxic vasodilation response [6] and diminishes the post-ischemic vasodilation response [7].

During ischemia, endogenous opening of cardiac K_{ATP} channels occurs, resulting in a potassium efflux, thereby hyperpolarizing the cell membrane. This results in a shortening of the action potential duration, with subsequently less contractility during the ischemic period, and in this way results in protection of the myocardium [8]. Therefore, it has been postulated that closing the K_{ATP} channels should result in a decrease in post-ischemic function, while opening the channels exogenously should result in improvement of post-ischemic function. This could be of great importance for the treatment of diabetes mellitus type 2 because these patients have a 3-fold higher risk of cardiovascular accidents, so maybe the K_{ATP} blocking drugs can increase the risk or worsen the outcome of an cardiovascular accident. We and others have shown that using the K_{ATP} channel blocker glibenclamide, post-ischemic function was improved [4, In press,9]. Others, however did show detrimental effects on post-ischemic [10]. Opening of cardiac K_{ATP} channels should result in improvement of function, by augmentation of the endogenous protective mechanism. This has been confirmed in various studies [11,12,13].

A major shortcoming in most of the studies concerning closure and opening of K_{ATP} channels is the use of unrealistically high concentrations which reduces clinical relevance of their studies, especially in the context of the treatment of diabetes mellitus. Also the majority of the studies are limited by the use of a Langendorff preparation over a working heart. Most studies did not have protein in their perfusate, while in the *in vivo* situation blockers and openers of K_{ATP} channels are highly bound to protein, e.g. glibenclamide >99%. Furthermore in these animal studies the normal kinetics of endogenous adenosine (from ATP breakdown during ischemia) was not taken into account, because of using a buffer solution as perfusate.

Adenosine is quickly taken up by erythrocytes. Another disadvantage of using buffer solutions is the vasodilation occurring due to lower oxygen capacity, and therefore CBF effects are difficult to determine. There also exists a great deal of disagreement about the adequacy of oxygenation in buffer perfused hearts, and one may conclude that it is on the brink of hypoxia [14].

Closing cardiac K_{ATP} channels results in an increase in intracellular calcium levels due to the reduction in hyperpolarization of the myocardial membrane. It may be that this higher level can result in improvements of post-ischemic function, especially when the ischemic interval is short, so possibly no calcium overload can occur. Because we have shown that closing cardiac K_{ATP} channels results in improved post-ischemic function in our isolated, erythrocyte perfused, working rat heart model, we hypothesized that pinacidil, a K_{ATP} channel opener, would result in a decreased post-ischemic recovery in this model.

2. METHODS

2.1 Reagent and experimental groups

Pinacidil was kindly provided by Leo Pharmaceutical Products, Copenhagen, Denmark. Three groups were studied: vehicle (n=5), 1 $\mu mol.L^{-1}$ (n=3) and 12 $\mu mol.L^{-1}$ (n=2) pinacidil. The solutions were infused as closest to the heart as possible by a syringe with a computer-controlled pump (1:10). The vehicle composition in the syringe was as follows: 1 $mmol.L^{-1}$ NaOH, 0.4% DMSO, 0.4% NaCl and 1.5% bovine serum albumin (fraction V, ICN Biomedicals B.V., Zoetermeer, The Netherlands). The pinacidil concentrations were non-protein bound concentrations.

2.2 Animal model

The animal procedures described were approved by the Faculty Ethical Committee on Animal Research. A male Wistar rat was anesthetized with diethyl ether, the heart excised via thoractomy and placed in ice-cold buffer. The aorta was cannulated and Langendorff perfusion with buffer was started within 8-10 min after heart removal. Subsequently the left atrium was cannulated, pacing-wires were attached and after checking for leaks, the system was switched to working configuration with erythrocyte suspension. Fluid columns were used to maintain the preload pressure at 15 mm Hg and the afterload pressure at 100 mm Hg. The buffer (95% O_2 and 5% CO_2) and erythrocyte suspension (18% O_2, 8% CO_2 and rest N_2) were equilibrated using membrane oxygenators (Minimax Plus, CB3381, Medtronic, Inc, Minneapolis, U.S.A.). The composition of the modified Krebs-Henseleit buffer solution was as follows: NaCl 118, $CaCl_2$ 3.0, KCl 4.7, $NaHCO_3$ 25, $MgSO_4$ 1.2, KH_2PO_4 1.2, NaEDTA 0.5, Glucose 11.1 mmol.L^{-1}. For the erythrocyte suspension albumin was added to the same buffer solution to 1.5% followed by adding washed bovine erythrocytes to a hematocrit of 0.25. The free calcium concentration in this composition was 1.6 mmol.L^{-1}. The temperature of the heart and perfusate was maintained throughout the experiment at normothermia of a rat (38 °C). A more detailed description of the perfusion set-up has already been published [15].

2.3 Experimental protocol & statistical analysis

First performance was assessed during at least 8 min. by changing the preload pressure from 15 to 20 and 10 mm Hg and holding the afterload pressure constant at 100 mm Hg. Secondly, 10 min infusion of drug was performed, followed by 12 min normothermic global ischemia. Thereafter reperfusion began during 20 min with infusion of drug. Finally, performance was assessed in the same way as the first time.

The results are presented as means with standard error of the mean (SEM). Differences in CBF and cardiac functional recovery were statistically tested using one-way analysis of variance (ANOVA) followed by a Bonferroni comparison test. Statistical analysis was performed using Instat (version 3.00, Graphpad Software Inc.). It should be noted however, that with a group size of three, in order to reach a power of 0.9 the expected difference must be around six times the standard deviation. A p-value lower than 0.05 was considered to be significant.

3. RESULTS

Figure 1 shows the CBF response to vehicle, 1 and 12 μmol.L^{-1} pinacidil. The CBF significantly increased after first treatment with 1 and 12 μmol.L^{-1} pinacidil to 264 % (17 %, $p<0.001$) and 240 % (14 %, $p<0.001$) respectively, compared to vehicle, and after ischemia at the end of reperfusion CBF also increased to 252 % (15 %, $p<0.001$) and 209 % (8 %, $p<0.001$) with 1 and 12 μmol.L^{-1} pinacidil, respectively compared to vehicle. After washout of drugs CBF was significantly higher with 12 μmol.L^{-1} pinacidil compared to vehicle.

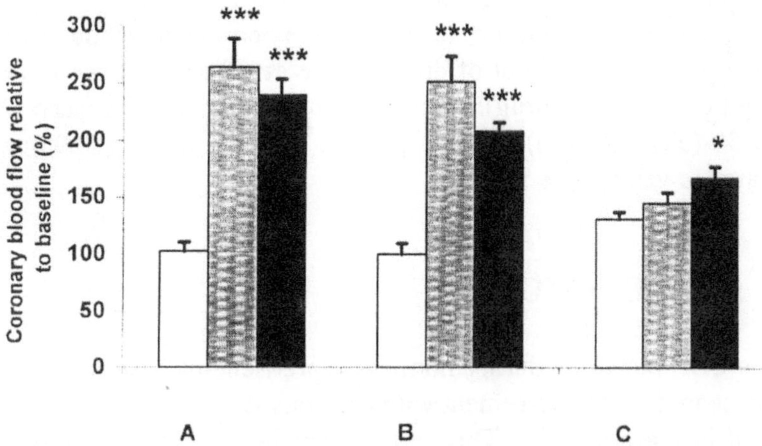

Figure 1. Coronary blood flow levels relative to baseline levels (measured during first performance assessment) after first treatment with pinacidil (A), after reperfusion with treatment with pinacidil (B) and during second performance assessment (C). Vehicle is represented by white bars, 1 $\mu mol.L^{-1}$ pinacidil by gray bars and 12 $\mu mol.L^{-1}$ pinacidil by black bars ($p<0.05$=*, $p<0.01$=**, $p<0.001$=***, compared to control).

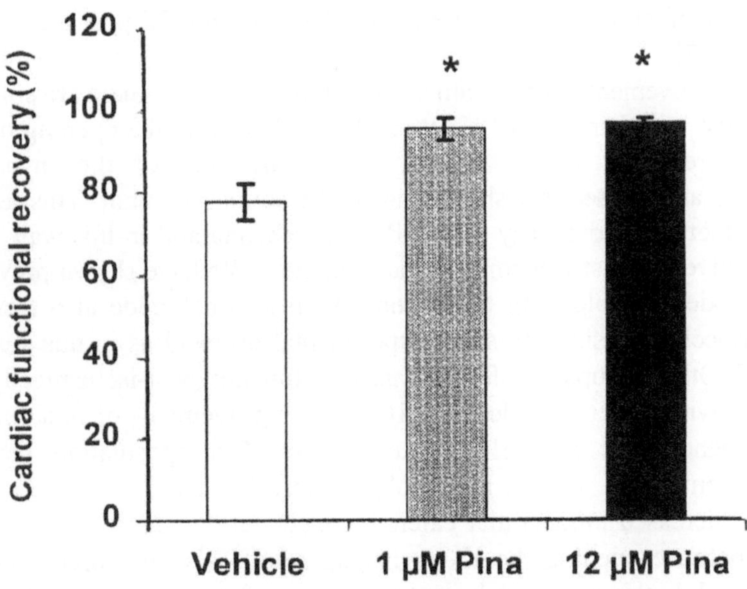

Figure 2. Cardiac functional recovery (see text) after treatment with vehicle, 1 and 12 $\mu mol.L^{-1}$ pinacidil (pina, $p<0.05$=*, compared to control).

In Figure 2, the cardiac functional recovery after treatment with vehicle and pinacidil is shown. The cardiac functional recovery was evaluated by dividing the cardiac output (aorta flow + coronary blood flow) post-ischemically (after washout of drug) by the cardiac output pre-ischemically (before treatment with drug). A significant improvement was seen with both 1 (96 % (3%, $p<0.05$)) and 12 $\mu mol.L^{-1}$ (97 % (1%, $p<0.05$) pinacidil compared to vehicle (78 % (4%)).

4. DISCUSSION

The CBF increases during normoxia (pre-ischemic) caused by pinacidil treatment are in close agreement with previous studies [5]. Pinacidil caused a 2-3 fold increase in CBF. This can be explained by relaxation of vascular smooth muscle cells due to hyperpolarization of the membrane. We have shown previously that glibenclamide treatment results in a marked CBF decrease [4, In press]. Other groups also showed similar effects [1,2,3], and it was concluded that K_{ATP} channels in the cardiovascular system are important in maintaining CBF during physiological situations [1]. These decreases in CBF were even seen at therapeutic anti-diabetic concentrations [4, In press]. After a period of global ischemia, the CBF remains high following the reactive hyperemia during pinacidil treatment, Again blocking the K_{ATP} channel has the opposite effect: decrease in CBF post-ischemically [4, in press,7].

The improvements of function seen with pinacidil treatment are similar to those found by others [11,12,13. It can be explained by the opening of K_{ATP} channels, resulting in K^+ efflux, hyperpolarization of the myocardial membrane and subsequent shortening of the action potential. This leads to less contractions and energy saving during ischemia and in this way, it will be protective for post-ischemic cardiac function. We have shown previously in our model that blocking the channel with glibenclamide also results in cardioprotectivity using the same experimental protocol as in this study [4, In press]. Other groups also found improved function post-ischemically after treatment with glibenclamide [9,13,16]. The improvements of function seen with glibenclamide are difficult to explain. Our speculation that this improvement is due to less hyperpolarization, thus more contractions as a result of increased intracellular calcium concentrations does not seem to be correct because opening the K_{ATP} channel results also in improvement of function, while this results in hyperpolarization of the membrane. Therefore, other explanations must be found. It has been proposed that preservation of intracellular high-energy phosphate compounds, e.g. ATP, during ischemia by glibenclamide treatment is the explanation for the improved post-ischemic function [9]. Another explanation may be the attenuation of intracellular lactate accumulation during ischemia, because of glibenclamide treatment, thereby less intracellular acidosis occurs, so protecting the ischemic heart [13].

Because the CBF response to pinacidil is actually near maximal vasodilation at control conditions already, it is very important that erythrocytes are in the artificial plasma perfusing the isolated heart, like in the present study. This is because buffer perfused hearts have a CBF which is also near maximal vasodilation, due to the lower oxygen capacity of the solution due to the lack of hemoglobin, meaning that increases seen with pinacidil in studies using buffer perfused hearts may have been masked or have been underestimated. Due to the pharmacological interaction of both K_{ATP} channel blockers and openers with protein, the use of albumin in the perfusate can contribute to improvement of clinical relevance of the results and therefore this was also included in our perfusate.

The present study shows that using pinacidil, the CBF is markedly increased and post-ischemic function is also improved. Therefore, we conclude that the cardioprotective effect of glibenclamide post-ischemically is not due to electrophysiological effects (less hyperpolarization) but speculate that it may be due to metabolic effects, like preservation of intracellular ATP levels during ischemia.

ACKNOWLEDGEMENTS

The authors would like to thank Jos Evers and Maarten de Jong for their invaluable biotechnical assistance during the experiments. Prof. dr. S.H. Skotnicki and Prof. dr. F.W.A. Verheugt are acknowledged for their valuable advice. This study was supported by Dutch Diabetes Foundation (grant 97-105).

REFERENCES

1. Samaha FF, Heineman FW, Ince C, Fleming J, Balaban RS. ATP-sensitive potassium channel is essential to maintain basal coronary vascular tone in vivo. Am J Physiol 1992;262:C1220-C1227.
2. Imamura Y, Tomoike H, Narishige T, Takahashi T, Kasuya H, Takeshita A. Glibenclamide decreases basal coronary blood flow in anesthetized dogs. Am J Physiol 1992;263:H399-H404.
3. Randall MD. The involvement of ATP-sensitive potassium channels and adenosine in the regulation of coronary flow in the isolated perfused rat heart. Br J Pharmacol 1995;166:3068-3074.
4. Legtenberg RJ, Houston RJF, Smits P, Oeseburg B. Hemodynamic changes caused by glibenclamide in isolated, working, rat heart. Adv Exper Med Biol (In press)

5. D'Alonzo AJ, Darbenzio RB, Hess TA, Sewter JC, Sleph PG, Grover GJ. Effect of potassium on the action of the KATP modulators cromakalim, pinacidil, or glibenclamide on arrhythmias in isolated perfused rat heart subjected to regional ischaemia. Cardiovasc Res 1994;28: 881-887.
6. Daut J, Rudolph WM, von Beckerath N, Meherke G, Gunther K, Meinen LG. Hypoxic vasodilation of coronary arteries is mediated by ATP-sensitive potassium channels. Science 1990;247:1341-1344.
7. Aversano T, Ouyang P, Silverman H. Blockade of the ATP-sensitive potassium channel modulates reactive hyperemia in the canine coronary circulation. Circ Res 1991;69:618-622.
8. Nichols CG, Ripoll C, Lederer WJ. ATP-sensitive potassium channel modulation of the guinea pig ventricular action potential and contraction. Circ Res 1991;68:280-287.
9. Tosaki A, Hellegouarch A. Adenosine triphosphate-sensitive potassium channel blocking agent ameliorates, but the opening agent aggravates, ischemia/reperfusion-induced injury. JACC 1994;23:487-496.
10. Auchampach JA, Maruyama M, Cavero I, Gross GJ. Pharmacological evidence for a role of ATP-dependent potassium channels in myocardial stunning. Circulation 1992;86:311-319.
11. Grover GJ. Protective effects of ATP-sensitive postassium channel openers in experimental myocardial ischemia. J Cardiovasc Pharmacol 1994;24(Suppl.4):S18-S27.
12. Hearse DJ. Activation of ATP-sensitive potassium channels: a novel pharmacological approach to myocardial protection? Cardiovasc Res 1995;30:1-17.
13. Docherty JC, Gunter HE, Kuzio B, Shoemaker L, Yang L, Deslauriers R. Effects of cromakalim and glibenclamide on myocardial high energy phosphates and intracellular pH during ischemia-reperfusion: 31P NMR studies. J Mol Cell Cardiol 1997;29:1665-1673.
14. Olders J, Turek Z, Evers J, Hoofd L, Oeseburg B, Kreuzer F. Comparison of tyrode and blood perfused working isolated hearts. Adv Exper Med Biol 1990;277:403-413.
15. Olders J, Boumans T, Evers J, Turek Z. An experimental set-up for the blood perfused working isolated rat heart. Adv Exp Med Biol 1990;277:151-160.
16. Ali K, Morimoto M, Fukaya Y, Furukawa Y, Iida F. Beneficial effects of the ATP sensitive K^+ channel opener cromakalim on recovery of cardiac function from cardioplegic arrest in isolated perfused rat hearts. Asia Pac J Pharmacol 1993;8:173-179.

Chapter **50**

EFFECT OF IRRADIATION ON ENZYMES OF THE CAPILLARY BED IN RAT VENTRICLES

Ming Gao, Hiroki Shirato, Kazuo Miyasaka & Tomiyasu Koyama*

*Department of Radiology, Hokkaido University, School of Medicine, 060 Sapporo, Japan; *064-0821 Sapporo, N-1 W-25 C,I. Ms516, Japan*

Abstract: The effect of localized irradiation on the enzyme activity in rat cardiac capillaries was examined in experiments in which the arteriolar and venular portions of the capillary bed were distinguished by the double-staining method. This method shows that the endothelial cells of the former contain alkaline phosphatase (AP) and those of the latter, dipeptidylpeptidase IV (DPP). At both 1 week and 3 weeks after irradiation with 20 Gy, staining for AP was reduced but staining for DPP was unchanged. The loss of enzyme from the arteriolar portions may be a consequence of the greater radiosensitivity of tissues exposed to high oxygen tension, or it may indicate that AP is less stable than DPP when exposed to irradiation.

Key words: heart, x-ray irradiation, arteriolar and venular capillary portion, capillary density, endothelial enzymes

1. INTRODUCTION

In the early period after X-irradiation the enzyme activity of alkaline phosphatase (AP) decreases in capillaries without decrease in the total number of open capillaries [1,2]. AP, however, distributes in the arteriolar capillary portion and not in the venular capillary portion in rat hearts [3-5]. Since capillaries are non-uniform tubes located between arterioles and venules, different enzymes are distributed along with capillaries. In normal adult Wistar rats dipeptidylpeptidase IV (DPP)-positive capillaries form 70% of the total present in cross-section of cardiac tissue. A double staining method for studying arteriolar and venular capillaries has been used in previous investigations of changes in capillarity in rat cardiac tissues [4-6].

Oxygen Transport to Tissue XXIV, edited by
Dunn and Swartz, Kluwer Academic/Plenum Publishers, 2003

Extensive validation of the techniques was made by double staining of cardiac tissues, to which microspheres had been infused through coronary artery, indicates that AP-positive capillaries adjoin the arterioles and DPP-positive capillaries, on the venules [7,8]. Thus, AP- and DPP-positive capillary portions can be designated arteriolar and venular capillary portions, respectively. Changes in the proportions of arteriolar and venular capillary portions have been observed in rat hearts exposed to pathophysiological stresses, supporting the usefulness of the double-staining method [9-13]. A question therefore arose as to whether both endothelial enzymes AP and DPP were equally affected by X-irradiation.

In the present work the double-staining method has been applied to irradiated cardiac tissues of rat, and the changes in stainability of the capillaries have been studied.

2. MATERIALS AND METHODS

Thirty 7-week-old male Wistar rats were used. They were provided with standard rat chow and water ad libitum, and all procedures were performed according to the institutional guidelines for care and use of laboratory animals, School of Medicine, Hokkaido University. Rats were anaesthetized by intraperitoneal injection of sodium pentobarbiturate (40 mg/kg). Fifteen rats were each placed in a 6 mm-thick lead housing with an 8 mm diameter hole in the top through which irradiation of 20 Gy was directed onto the heart from a TOSHIBA KXC18 instrument emitting 180 kV X-rays filtered by 0.5 mm Cu and Al at a dose rate of 2 Gy/min. The rats were rapidly decapitated with a guillotine 1 hour, 1 week or 3 weeks after irradiation. Fifteen anaesthetized control rats were placed in the housing without any irradiation. They were randomly divided into three groups and killed concomitantly with the irradiated rats at the three time points. The heart was removed, dipped in O.C.T. Compound^{R} (Miles Inc. USA), frozen in liquid nitrogen and stored at –80°C.

Using a cryotome, tissue cross-sections were cut from the widest part of the ventricle at 16μm, four to six sections, each 80 μm apart, being obtained from each left ventricle. The sections were double stained as described previously [4,5], then covered with Crystal Mount (Biomeda, CA, USA) [9]. Arteriolar capillary portions whose endothelial cells express alkaline phosphatase stained blue, venular portions containing dipeptidylpeptidase IV stained red, and the intermediate portions containing both enzymes stained violet. When capillaries are followed longitudinally on double-stained cardiac sections, a transition from blue- to violet- to red-stained portions can be seen.

Microscopic observations reveal cross-section of capillaries in the outer and inner muscle layers and longitudinal ones in the middle layer. In each of four sections per rat four visual fields were arbitrarily selected in the inner muscle layer close to the endocardium, i.e. subendocardium and morphological measurements were made in these chosen fields which covered 0.0615 mm^2. The numbers of blue, violet and red elements were separately counted and the results from the four sections were averaged. Between 100 and 150 capillaries were present in each visual field, hence more than 1600 capillaries were examined in each rat.

All data are given as means±SD. Comparisons between groups were carried out using Mann-Whitney U-test. P-values < 0.05 were considered significant.

3. RESULTS

No significant differences between the control and treated groups were found in body weight, whole heart weight, or the weight of the left ventricle, at 1 hour, 1 week or 3 weeks after irradiation.

A significantly reduced number of blue-stained capillaries were present while many red-stained capillaries remained at 1 week and 3 weeks after irradiation, compared with control group. Stainings of the capillaries of the longitudinal sections were often interrupted probably because of the loss in the AP-sensitivity in the irradiated heart.

The densities and proportions of each type of capillary are listed in Table 1; there were no significant changes in the capillary network 1 hour after irradiation. AP-positive capillaries decreased significantly in number and in proportion at 1 week and 3 weeks after irradiation. The decreases are reflected in the decrease in the total stained capillary density, compared with controls. On the other hand, DPP-positive capillary portions showed no significant changes in density or proportion at either time point. C:M ratio was slightly but significantly decreased compared with controls at 1 week and at 3 weeks.

4. DISCUSSION

The main finding in the present study was the selective decrease in the AP-staining without significant changes inDPP-staining. It is not possible from our study to distinguish whether the decrease in AP-staining is due to death of endothelial cells or to inactivation of AP. However, it suggests that

Table 1. Body weight (BW), left ventricular weight (LVW) of X-irradiated and control rats and numbers of capillary of three types in the left ventricular tissue at three time points after X-irradiation.

	BW (g)	LVW (mg)	Number of three Capillary Types ($/mm^2$)			
			Arteriolar	Intermed.	Venular	Total
Control	181±3	429±29	438±69	173±28	1451±127	2062±107
1 h	180±5	409±12	316±129	165±30	1422±123	1903±41
P	n.s.	n.s.	n.s.	n.s.	n.s.	n.s.
Control	255±9	519±41	388±55	251±112	1356±219	1995±91
1 wk	253±9	524±65	140±33	231±102	1365±86	1736±93
P	n.s.	n.s.	<0.01	n.s.	n.s.	<0.01
Control	260±6	531±95	289±45	210±50	1578±129	2077±57
3 wks	258±8	593±114	138±18	200±24	1349±146	1687±150
P	n.s.	n.s.	<0.025	n.s.	n.s.	<0.025

P n.s. n.s. <0.025 n.s. n.s. <0.025

Table 2. Left ventricular weight to body weight ratio (LVW/BW) of X-irradiated and control rats, capillary proportion (%) and capillary to myocyte ratio (C:M) at three time points after X-irradiation.

	LVW/BW	Proportion of Capillary three Types			C:M
		Arteriolar	Interm.	Venular	
Control	2.38±0.18	21.3±4.3	8.3±0.91	70.4±3.5	1.02±0.01
1 h	2.27±0.07	18.4±3.0	7.3±3.6	74.2±7.7	1.03±0.05
P	n.s.	n.s.	n.s.	n.s.	n.s.
Control	1.94±0.06	19.7±1.0	11.8±5.8	68.4±8.8	1.01±0.03
1 wk	2.04±0.19	10.0±2.3	11.4±4.3	78.5±6.4	0.96±0.03
P	n.s.	<0.01	n.s.	<0.01	<0.05
Control	1.90±0.24	14.7±2.5	10.6±2.5	74.7±4.6	1.04±0.05
3 wks	2.00±0.07	8.2±1.6	11.9±3.4	79.9±2.8	0.96±0.04
P	n.s.	<0.025	n.s.	n.s.	<0.05

the enzymes of the endothelial cells of capillaries on the arteriolar side are more radiosensitive than those on the venular side of capillaries.

The preferential decrease in AP-staining may be an *in vivo* example at the capillary level of the way in which oxygen tension can affect sensitivity to irradiation; this topic was elegantly reviewed by Streffer [14]. Subsequently, Benderitter et al. [15] investigated from the viewpoint of the production of oxygen-related radicals. The oxygen tension at the beginning of the arteriolar capillary portion is the same as that in the arteries, i.e. nearly 100 mmHg. The values reported for oxygen tension in the coronary sinus are remarkably low because of the high oxygen consumption of beating cardiac tissues, ranging from 19 to 23 mm Hg in resting dogs [16-18]. Working on isolated hepatocytes Alati et al. [19] found that at this level of low oxygen tension radio sensitivity was 60% of that at high oxygen tension. It is possible that radiosensitivity might be similarly low in the venular capillary portions in rat hearts as in dog hepatocytes. The arterial tension corresponds to 84% of the radiosensitivity on the same curve (Fig. 6 of [19]). Capillary oxygen tension decreases sharply near arterioles but more slowly in the venular capillary portions [20]. It seems possible that the lower oxygen tension on the venular side may partially explain why DPP activity on the venular capillary portions are spared.

Unlike the AP activity, DPP activity was not affected by the present experiments. DPP protein or mRNA may be less sensitive to X-irradiation than AP. Reports on differences in the stability of AP and DPP activity are conflicting. In a study of cultured endothelial cells from rat coronary microvessels, Koyama et al. [7] found that AP stainability was reduced by repetition of passages, while the DPP stainability was unaffected. On the other hand, it has been reported on the basis of the double-staining method that in free muscle grafts in the rat DPP activity was lost from the capillaries 1 or 2 days after grafting, but AP activity was preserved for 6 days, suggesting AP to be more stable than DPP [21].

As yet, we have no decisive evidence for the possible cause of the selective reduction in the number of AP-positive capillary portions after irradiation. Further work is needed to establish whether AP is intrinsically more sensitive to radiation than DPP or whether the higher oxygen tension on the arteriolar side renders it more vulnerable to radiation. In future study X-irradiation should be made in rats breathing oxygen or hypoxic gas mixture. In addition, the present decrease in AP-staining could be due to loss of enzyme activity or to a loss of arteriolar endothelial cells. To resolve this it will be necessary to compare the radiosensitivity of AP and DPP activities in cultured cardiomyocytes.

ACKNOWLEDGMENTS

The authors wish to express their thanks to Prof. Dr. Mikinori Kuwabara and Dr. Daiji Endo, School of Veterinary Medicine for their kind permission and instruction to use the X-ray emission instrument.

REFERENCES

1. Lauk S. Endothelial alkaline phosphatase activity loss as an early stage in the development of radiation-induced heart disease in rats. Radiat Res 1987;110: 118-128.
2. Schultz-Hector S, Balz K, Bohm M, Ikehara Y, Rieke L. Cellular localization of endothelial alkaline phosphatase reaction product and enzyme protein in the myocardium. J Histochem Cytochem 1993; 41:1813-21.
3. Lojda Z. Studies on dipeptidyl(amino)peptidase IV (glycyl-proline naphthylamidase). II blood vessels. Histochemistry 1979; 59: 153-166.
4. Batra S, Rakusan K. Geometry of capillary network in volume overloaded rat heart. Microvasc Res 1991;42:39-50.
5. Batra S, Rakusan K. Capillary network geometry during postnatal growth in rat hearts. Am J Physiol 1992; 262:H636-H640.
6. Batra S, Gao M, Koyama T, Horimoto M, Rakusan K. Microvascular geometry of the rat heart arteriolar and venular capillary regions. Jpn Heart J 1992;33:817-828.
7. Koyama T, Gao M, Uede T, Batra S, Itoh K, Ushiki T, Abe K. Different enzyme activities in coronary capillary endothelial cells. Oxygen transport to Tissue XVIII, ed. By Nemoto & LaManna, Plenum Press, New York, 1997;359-364.
8. Koyama T, Xie, ZL, Gao M, Juzuki J, Batra S. Adaptive changes in the capillary network in the left ventricle of rat heart. Jap J Physiol 1998;48:229-241.
9. Gao M, Batra S, Koyama T. Capillary geometry in the subendocardium of the left ventricle from young rats subjected to exercise training. In: Progress in Microcirculation Research, eds Niimi H, Oda T, Sawada T, Xiu R-J. Pergamon Press, London, 1994; 187-191.
10. Xie Z, Gao M, Koyama T. The capillary of left ventricular tissue of rats subjected to coronary artery occlusion. Cardiovasc Res 1997;33: 671-676.
11. Suzuki J, Gao M, Xie Z, Koyama T. Effects of the β_2-adrenergic agonist Clenbuterol on capillary geometry in cardiac and skeletal muscles in young and middle-aged rats. Acta Physiol Scand 1997;161:317-326.
12. Xie ZL, Gao M, Batra S, Koyama T. Remodeling of capillary network in left ventricular subendocardial tissues induced by intravenous vasopressin administration. Microcirculation 1997;4:261-266.
13. Xie Z, Gao M, Koyama T. Effects of transient coronary occlusion on the capillary network in the left ventricle of the rat. Jpn J Physiol 1997;47:537-543.
14. Streffer C, Van Beuningen D. Cellular radiobiology. In: Radiopathology of organs and tissues, ed by Scherer E, Streffer C, Trott KR. Berlin, Heidelberg, New York, London, Paris, Tokyo, Hong Kong, Barcelona, Budapest, 1991; 1-22.
15. Benderitter M, Maingon P, Abadie C, Assem M, Maupoil V, Briot F, Horiot JC, Rochette L. Effect of in vivo heart irradiation on the development of antioxidant defenses and cardiac functions in the rat. Radiat Res 1995;144:64-72.
16. Feigl EO. Control of myocardial oxygen tension by sympathetic coronary vasoconstriction in the dog. Circ Res 1975;37:88-95.

17. McKenzie JE, Steffen RP, Haddy FJ. Relationships between adenosine and coronary resistance in conscious exercising dogs. Am J Physiol 1982;242:H24-29.
18. Heyndricks GR, Muylaert P, Pannier JL. Alpha-adrenergic control of oxygen delivery to myocardium during exercise in conscious dogs. Am J Physiol 1982;242:H805-809.
19. Alati T, Van Cleeff M, Jirtle R. Radiosensitivity of parenchymal hepatocytes as a function of oxygen concentration. Radiat Res 1989;118:488-501.
20. Opitz E, Schneider M. Uber die Sauerstoffversorgung des Gehirns und den Mechanismus von Mangelwirkungen. Ergeb Physiol Biolog Chem Exper Pharmakol. 1950;46:126-260.
21. Grim M, Mrazkowa O, Carlson BM. Enzymatic differentiation of arterial and venous segments of the capillary bed during the development of free muscle grafts in the rat. Am J Anat 11986;77:149-159.

Chapter **51**

THE REDOX STATE OF CYTOCHROME OXIDASE IN BRAIN *IN VIVO*: AN HISTORICAL PERSPECTIVE

Joseph C. LaManna
Department of Neurology, Case Western Reserve University School of Medicine, 10900 Euclid Avenue (BRB 525), Cleveland, OH 44106-4938, USA

Abstract: Recent evidence suggests that cytochrome oxidase is partially reduced under resting conditions in the brain. Previous data, recorded over the past 30 years from intact brain using optical methods in the visible wavelength range, are consistent with this observation. These older data, while not conclusive in themselves, support the overall conclusions. The historical perspective on the experiments and controversies illustrates a number of useful principles. The first is that new methods tend to produce new observations that may be difficult to reproduce due to the uniqueness of the instrumentation. The second is that any new and different observations cannot be assimilated without an acceptable theoretical framework and, without assimilation can have little impact. Finally, the mechanisms which might explain why cytochrome oxidase may be more reduced than previously thought are still not fully developed and, therefore, the physiological significance of such reduction is not known.

Key words: Brain Cytochrome redox state, Cytochrome oxidase, Cytochrome a,a_3, Brain oxygen metabolism, Nitric oxide

1. INTRODUCTION

The redox state of cytochrome oxidase in the mammalian brain has been continuously debated over the past 25 years without resolution. A definitive answer to the question would be especially useful because of its implications for understanding the fundamental normal brain metabolic physiology that underlies the rapidly developing new (and some not so new) non-invasive techniques that are being applied to human subjects, with the hope of providing clinically useful tools for the study of wide ranging brain pathophysiological conditions. Techniques such as positron emission tomography (PET), functional magnetic resonance imaging (fMRI), and near-infrared spectroscopy (NIR) offer potentially powerful methods for investigating brain function, but their usefulness is currently limited by a lack of appreciation of the basic mechanisms that link blood flow, metabolism and neuronal function. Much current work is underway and there is an ever-growing and impressive accumulation of knowledge on this issue that promises·to rectify the situation fairly soon. On the other hand, there exists an older body of literature on the subject that may have been forgotten, or may be only of historical interest, having been superseded by newer methods. Nevertheless, it might of value to re-examine this previous literature to gain some insight into how the current concepts were developed and from where the contentious positions might have arisen.

2. HISTORICAL BACKGROUND

2.1 MacMunn and Hoppe-Seyler

The first full report of heme-like absorption bands in tissue was by MacMunn in 1884 [1, 2]. An extensive account and analysis of MacMunn's work was provided by Keilin [3]. It is of direct relevance to our discussion that MacMunn made his observations from intact tissue using the direct-vision prism spectroscope ocular that Sorby had devised in 1867. Sorby had also mentioned observing similar absorption bands from snail tissues using this instrument [4]. MacMunn had observed the multiple absorption bands in many tissues of vertebrates and invertebrates, and for the general case he assigned the bands to something he called "histohaematin" and in the specific case of muscle, to "myohaematin".

These observations had little impact on later work, mainly due to the overwhelming critical conclusions of Hoppe-Seyler, whose main arguments rested on his assignment of the absorption bands to "tissue hemoglobin" (i.e., myoglobin) and residual surface blood hemoglobin. Hoppe-Seyler had been the first to observe the absorption spectrum of blood [5] and had given the names to "oxyhemoglobin" and "hemoglobin" [6], (although it was Stokes who first had realized the oxygen dependence of the two blood colors [7]). Hoppe-Seyler's laboratory was trying to develop methodology for determining blood volume in tissue and the presence of an intrinsic unknown absorber would interfere with the method [8]. He clearly demonstrated the myoglobin spectrum in muscle and thus claimed there was no myohaematin. He did not examine the tissue with a low dispersion microspectroscope however, and did not attempt to duplicate MacMunn's observations. He also conveniently brushed aside the argument that the histohaematin bands were clearly observable in invertebrate tissues that contained no hemoglobins.

MacMunn on the other hand could not admit to the presence of myoglobin, insisting the signal was from histohaematin, in addition, he had difficulty explaining how the complex multiple banded absorption spectrum could arise from the single molecule of histohaematin. In the end he erroneously concluded that histohaematin was an evolutionary intermediate of hemoglobin. It was not possible to overcome the problems of differing instrumentation and lack of a reasonable theoretical explanation, and so the observations of MacMunn languished for 35 years.

2.2 Keilin and Warburg

Keilin, while involved in research experiments on the respiratory system of a horse gut parasite, saw the same absorption bands that MacMunn had previously reported. He recognized that the absorption spectrum was a combination of at least 3 related molecules which he called cytochromes (cell colors). He argued that the bands were coming from an oxygen using enzyme, apparently identical to the *atmungsferment* that was being touted by Warburg as the enzyme responsible for tissue oxygen consumption. There was much acrimonious debate between the two scientists, as recounted in Keilin's book [3] and in Krebs' biography of Warburg [9]. The debate spurred much high quality research in the 1930's and 1940's that eventually resulted in the recognition of the cytochromes as the mitochondrial oxygen consuming enzymes. Some interesting additional landmarks should be noted as deriving from these extended experimental studies. Warburg made use of the photoactivation action spectrum that he had previously developed [10] and Keilin observed redox changes in the cytochromes of living wax moth wing muscles in vivo.

In this case, the experimental observations were established through the use of accepted instrumentation (Keilin used the Zeiss version of the microspectroscope ocular, and the grating-based Hartridge reversion microspectroscope [11]). It was also crucial that the development of an understanding of the mechanisms of oxidative phosphorylation provided a solid theoretical framework for the optical observations. In fact, these optical methods were developed to help explicate the role of the mitochondrial respiratory chain in energy metabolism.

2.3 The years of 1950 – 1990

A new wave of scientific investigation for this field was triggered in the early 1950's by the development of the precise spectrophotometric instruments created by Britton Chance at the Johnson Foundation in Philadelphia [12], especially through their application to mitochondrial studies [13]. The story of the development of these optical methods has been briefly recounted [14]. This instrumentation was so successful for in vitro work that attempts were soon made to apply it to in vivo studies as well. Initial investigations in skeletal muscle by Frans Jöbsis (now Jöbsis van der Vliet) were quite successful [15, 16], but soon discrepancies began to appear in data recorded from intact organs with high oxidative energy capacity, with and without an intact blood supply. The major departure from theoretical expectation was found in the reports of a more reduced cytochrome oxidase redox state than was predicted from the in vitro experimental evidence. Many of these studies also demonstrated activity-induced cytochrome oxidase oxidations in these tissues. Jöbsis brought back the idea of the action spectrum as a way of using a very sensitive 2 wavelength instrument to obtain spectral information. He also used the term "equibestic" wavelength to indicate a reference wavelength where the absorption differences between Hb and HbO_2 were the same as at the sampling wavelength used to detect cytochrome redox state [17]. This term was based on "isosbestic point" (an IUPAC term, see P. Müller, *Pure Appl. Chem.*, **66**, 1077-1184 (1994) http://www.chem.qmw.ac.uk/iupac/gtpoc/), coined by Thiel in 1924, according to the Oxford English Dictionary, to indicate a wavelength of light at which two absorbers of an acid-base pair are equal, and derived from the Greek *isoj* = "equal", and sbest-oj = "extinguished". [Note: the Jöbsis term should have been "equisbestic" to be consistent with the Greek derivation.] The usefulness of the concept has faded with the growth of multiple wavelength analyses and full spectrum instrumentation.

Early observations interpreted as reduced cytochrome oxidase in tissues other than brain included heart [18-21], gut [22, 23] and carotid body [24]. These observations, plus similar observations in brain, led to the hypothesis that intact tissues with high metabolic rates and specialized for ion transport had partially reduced cytochrome oxidase [17].

In the brain, the evidence for partially reduced cytochrome oxidase was based on observations that an optical signal with wavelength characteristics similar to cytochromes a,a_3 could be detected to change in both oxidized and reduced directions during increased and decreased oxygen delivery, and to oxidize during neuronal activation. Identification of cytochrome a,a_3 was usually done by 2 wavelength action spectra between 580nm and 630nm. These action spectra all clearly show a shoulder in the region of 600nm - 610nm, usually on a significantly inflected baseline [17, 25] that was most likely due to HbO_2 [26].

3. LIGHT SCATTERING AND HEMOGLOBIN

The validity of the observations of reduced cytochrome a,a_3 made by instruments using wavelengths in the visible spectrum was never satisfactorily demonstrated. The technical issues, where artifact might cause misinterpretation, are fundamentally due to light scattering and the strong absorption by hemoglobin.

3.1 Light scattering

The major effects of light scattering are to produce shifts in the peak wavelength, and increases in absorption peak amplitudes and widths [27-30]. The effects are predictable and various successful compensation strategies have been used. Light scattering changes have also been used as signals of pathophysiological processes in brain tissue [31], and this has spawned an entirely new field of study using intrinsic light scattering properties of tissue.

3.2 Strong absorption by hemoglobin

As a result of the strong absorption of hemoglobin, the light path taken by photons through the tissue can be greatly affected. The major outcome of this effect is that the volume of tissue sampled by a light beam in an optical instrument varies with wavelength of light and with large shifts in blood volume. Thus, it is nearly impossible to guarantee good reference spectra for differential methodology. The appearance or disappearance of an absorption peak may, therefore, simply reflect the sampling of more or less tissue volume, without any required change in redox state.

The non-linear effect on pathlength with wavelength means that the shapes of the absorption peaks will also be altered in ways for which it is not easy to compensate. This is a problem that has not yet been solved satisfactorily and continues to interfere with interpretation of visible and NIR derived data.

Attempts to overcome the large hemoglobin influence included the in situ freeze trapping method, whereby the tissue is frozen rapidly and the milled brain surface examined spectrally by a small diameter scanning beam [32]. Although these authors clearly discounted the possibility that cytochrome oxidase could be significantly reduced, they were unable to exclude the possibility that cytochrome oxidase was as much as 20% reduced.

The changes in blood volume and oxygenation state produce absorption changes that are large (>10 times) compared to the expected cytochrome signal. Efforts to remove blood by saline or blood substitute solutions have been inconclusive in answering the redox question. Blood substitutes show a nearly fully oxygenated cytochrome, but tissue oxygen is also very high and vascular flow is much higher in the blood free preparations [33-35].

The spectral effects of light scattering and hemoglobin on brain optical measurements were well studied by Lübbers in Dortmund [36]. He had developed very well-designed fast spectral scanning photometers that could be applied to intact tissues such as the brain [37-40]. In the absence of hemoglobin, in the saline perfused guinea pig brain, he showed that the cytochrome spectrum was observable [41]. In that study, cytochrome a,a_3 was found to be 40% reduced at 37°C. This was logically attributed to the expected tissue hypoxia from oxygen delivery deficiency in the absence of the carrier function of hemoglobin. These investigators then dropped the temperature as low as 18°C to observe 95% oxidation of cytochrome a,a_3. If the data are examined in detail, it might be concluded that this study is compatible with a partially reduced cytochrome for the following reasons. The metabolic reduction produced by the anesthetic (nembutal) is at least 1/3. The saline was equilibrated with 95% O_2 and 5%CO_2, thus blood flow was at least double. Thus, the oxygen delivery, even at 37°C was most likely within 50% of the blood perfused control. Therefore, the relevant temperature for comparison, based on a Q_{10} of at least 2, would be 27°C. The authors report cytochrome as 80% oxidized at 24°C. These conclusions then are in substantial agreement that at reasonable delivery and consumption rates, cytochrome might be partially reduced.

In the isolated, whole toad brain preparation, in the absence of hemoglobin, Moffett and Jöbsis showed that cytochrome a,a_3 became more oxidized with electrical activation [42]. This was confirmed with an action spectrum. The oxidative transient was recorded over a range of temperatures, and exhibited a Q_{10} in the physiological range [43].

3.3 Conclusions?

To attempt a reasonable resolution from the visible data we can say that the 2 wavelength method cannot be successfully used to monitor the cytochrome oxidase redox state in vivo in the visible portion of the spectrum. But, 3 wavelength and spectral analysis methods applied to the visible spectrum have shown that cytochrome a,a_3 becomes oxidized with neuronal activation [25, 26], and that cytochrome oxidase may be as much as 15-20% reduced in the steady state, resting brain [25, 32, 41].

4. THEORETICAL FRAMEWORK FOR PARTIALLY REDUCED CYTOCHROME OXIDASE

While the above conclusions based on data from visible spectrophotometry cannot be considered definitive in themselves, they are compatible with some recent observations using near-infrared spectrophotometry (NIR), (with the caveat that there may be important differences in reactions between the iron and copper centers). Cooper et al did not detect any evidence of partial reduction of cytochrome oxidase within a maximum error of 15%, under the most generous conditions [44]. But small (as much as 10%) oxidations were observed with increased carbon dioxide ventilation [45] and in response to visual stimulation [46]. If these observations hold up and are confirmed definitively, there will still be a need to place them into a physiological context within a solid theoretical framework.

4.1 Heterogeneity of brain mitochondria

In vitro studies put the Km for brain mitochondria at about 0.1:M oxygen (about 0.07 torr) [47-49], and measurements have shown that cytochrome oxidase would be more than 90% oxidized in these preparations [48]. In vitro methodological inconsistencies that might explain an artificially low Km have been ruled out [50]. However, it could be that brain mitochondria are somehow different --good optical studies of brain mitochondria are lacking-- or that the environment of brain mitochondria is not reproduced in in vitro studies. Further studies would help to better define the relationship among cytochrome oxidase redox state, oxygen tension and oxygen consumption [51-53]. This relationship is by no means clear, especially under conditions of oxygen limitation and changes in phosphate potential [54], and Jöbsis, at one time, proposed [53]that there might be an additional site of respiratory control between cytochrome oxidase and oxygen [55].

The redox state of cytochrome oxidase is affected by other uncontrolled variables [56] and active-state mitochondria may have an effective Km in the physiologically relevant range [57]. Another possibility is that brain tissue may contain a unique cytochrome-like heme protein that perhaps acts as an optical sensor analogous to that found in the carotid body [24, 58] Thus, in vivo conditions may not be equivalent to in vitro conditions and the kinetic assays may not be relevant to the physiological state [57]. Other variations of this argument include suggestions of the existence of mitochondrial sub-populations, such as neuronal and glial, for example [59].

4.2 Heterogeneity of oxygen availability

It may be that the cytochrome oxidase of brain tissue exhibits the common Km, but that the intracellular peri-mitochondrial pO_2 is well below 1 torr. The overall partially reduced cytochrome oxidase redox state measured using volume averaging optical methods might then represent an average between tissue volumes with fully oxidized cytochrome and tissue volumes of reduced cytochrome. These low oxygen regions would presumably occupy 15-20% of the tissue. Very low oxygen tensions are difficult to ascertain in vivo, but the histogram distribution of tissue oxygen tension allows the possibility of the existence of brain tissue regions of very low oxygen [60, 61].

4.3 Nitric oxide interactions

The potential role of NO was assessed, and perhaps too quickly, rejected [46]. Certainly, a simple mechanism can be rejected. It has been pointed out that the increase of NO that is produced with increased neuronal activity would produce a reduction of cytochrome oxidase and this would be incompatible with the observed oxidation of cytochrome oxidase during functional stimulation. On the other hand, NO plays a much more complicated role in the brain. First of all, it must be considered that there is much evidence that NO is an important contributor to control of local blood flow. NO mediates the vasodilation after neuronal activation [62, 63]. NO might be a coupling factor that allows neurons to control their own blood supply [64]. A significant source of the vasodilating NO is most likely through specific neuronal pathways. NOS positive neurons are closely associated with blood vessels [65]. These neurons may be local interneurons, or may arise from other brain regions, such as the pontine nucleus locus ceruleus. The optically detected hemoglobin changes that accompany neuronal activation were attenuated when the locus ceruleus was lesioned [66]. Neuronal activation would then be accompanied by increased local blood flow that would result in increased oxyhemoglobin and oxygen. The combination of these would act to oxidize cytochrome with a time course reminiscent of the vascular response, well after and out of proportion to any neuronal NO production. If the steady state represents a balance between NO production and tissue oxygen tension, then a more reduced cytochrome is favored. Oxygen levels are low based on histogram analysis and NO levels are in the range to compete for cytochrome. Activation, as said before, would disturb this balance by an overwhelming provision of oxygen, thus activation produces oxidation. The incoming HbO_2 would react at diffusion-limited rates with NO [67] and oxygen delivery would be enhanced through the NO-mediated decreased hemoglobin oxygen affinity [68]. Thus, the cytochrome oxidase oxidation during activity would represent quenching of NO and increased O_2 provision, while the steady state baseline redox level would be more reduced because of the interaction of NO with mitochondrial metabolism and through direct competition with oxygen for cytochrome oxidase [69-71]. It has been suggested that these interactions among NO, oxygen and cytochrome oxidase have the effect of permitting cytochrome oxidase to act as an oxygen sensor [72].

REFERENCES

1. MacMunn CA. On myohaematin, an intrinsic muscle pigment of vertebrates and invertebrates, on histohaematin, and on the spectrum of the supra-renal bodies. J Physiol (Lond) 1884;5:xxiv-xxvi.
2. MacMunn CA. Researches on myohaematin and the histohaematins. Phil Trans R Soc Lond B 1886;177:267-298.
3. Keilin D. The History of Cell Respiration and Cytochrome. Cambridge: Cambridge University Press; 1966.
4. Sorby HC. On a definite method of qualitative analysis of animal and vegetable colouring-matters by means of the spectrum microscope. Proc Roy Soc Lond 1867;15:433-455.
5. Hoppe F. Über das Verhalten des Blutfarbstoffes im Spectrum des Sonnenlichtes. Virchows Archiv A Pathol Anat 1862;23:446-449.
6. Hoppe-Seyler F. Über die chemischen und optischen Eigenschaften des Blutfarbstoffes. Virchows Archiv A Pathol Anat 1864;29:233-235.
7. Stokes GG. On the reduction and oxidation of the colouring matter of the blood. Proc Roy Soc Lond B 1864;13:355-364.
8. Levy L. Über Farbstoffe in den Muskeln. Hoppe-Seyler Z Physiol Chem 1889;13:309-325.
9. Krebs HA. Otto Warburg. Oxford: Clarendon Press; 1981.
10. Warburg O, Negelein E. Über die photochemische Dissoziation bei intermittierender Beichtung und das absolute Absorptionsspektrum des Atmungsferments. Biochem Z 1928;202:202-228.
11. Hartridge H. A spectroscopic method of estimating carbon monoxide. J Physiol (Lond) 1912;44:1-21.
12. Chance B. Rapid and sensitive spectrophotometry. III. A double beam apparatus. Rev Sci Instrum 1951;22:634-638.
13. Chance B, Williams GR. The respiratory chain and oxidative phosphorylation. Adv Enzym 1956;17:65-134.
14. Chance B. Optical method. Annu Rev Biophys Biophys Chem 1991;20:1-28.
15. Jöbsis FF. Spectrophotometric studies on intact muscle. II. Recovery from contractile activity. J Gen Physiol 1963;46:929-969.
16. Jöbsis FF. Spectrophotometric studies on intact muscle. I. Components of the respiratory chain. J Gen Physiol 1963;46:905-928.
17. Jöbsis FF, Rosenthal M, LaManna JC, Lothman E, Cordingley G, Somjen G. Metabolic activity in epileptic seizures, In: Ingvar D, Lassen N, editors. Brain Work, Alfred Benzon Symposium VIII. Copenhagen: Munksgaard; 1975;185-196.

18. Ramirez J. Oxidation-reduction changes of cytochromes following stimulation of amphibian cardiac muscle. J Physiol (Lond) 1959;147:14-32.
19. Ramirez J. Oxidation-reduction changes of cytochromes in lobster heart. Biochim Biophys Acta 1964;88:648-650.
20. Hassinen IE, Hiltunen K. Respiratory control in isolated perfused rat heart. Role of the equilibrium relations between the mitochondrial electron carriers and the adenylate system. Biochim Biophys Acta 1975;408:319-330.
21. Snow TR, Kleinman LH, LaManna JC, Wechsler AS, Jöbsis FF. Response of cytochrome a,a3 in the in situ canine heart. Basic Res Cardiol 1981;76:289-304.
22. Hersey SJ, Jöbsis FF. Redox changes in the respiratory chain related to acid secretion by the intact gastric mucosa. Biochem Biophys Res Commun 1969;36:243-250.
23. Mandel LJ, Moffett DF, Jobsis FF. Redox state of respiratory chain enzymes and potassium transport in silkworm mid-gut. Biochim Biophys Acta 1975;408:123-134.
24. Mills E, Jöbsis FF. Mitochondrial respiratory chain of carotid body and chemoreceptor response to changes in oxygen tension. J Neurophysiol 1972;35:405-428.
25. Jöbsis FF, Keizer JH, LaManna JC, Rosenthal M. Reflectance spectrophotmetry of cytochrome aa3 in vivo. J Appl Physiol 1977;43:858-872.
26. LaManna JC, Sick TJ, Pikarsky SM, Rosenthal M. Detection of an oxidizable fraction of cytochrome oxidase in intact rat brain. Am J Physiol 1987;253:C477-C483.
27. Latimer P. Apparent shifts of absorption bands of cell suspensions and selective light scattering. Science 1958;127:29-30.
28. Butler WL, Norris KH. The spectrophotmetry of dense light-scattering material. Arch Biochem Biophys 1960;87:31-40.
29. Butler WL. Absorption of light by turbid materials. J Opt Soc Am 1962;52:292-299.
30. Longini RL, Zdrojkowski R. A note on the theory of backscattering of light by living tissue. IEEE Trans Biomed Engn 1968;BME-15:4-10.
31. Grinvald A, Lieke E, Frostig RD, Gilbert CD, Wiesel TN. Functional architecture of cortex revealed by optical imaging of intrinsic signals. Nature 1986;324:361-364.
32. Bashford CL, Barlow CH, Chance B, Haselgrove J, Sorge J. Optical measurements of oxygen delivery and consumption in gerbil cortex. Am J Physiol 1982;242:C265-C271.
33. Piantadosi CA, Jöbsis-VanderVliet FF. Spectrophotometry of cerebral cytochrome a,a3 in bloodless rats. Brain Res 1984;305:89-94.
34. Sylvia AL, Piantadosi CA. O2 dependence of in vivo brain cytochrome redox responses and energy metabolism in bloodless rats. J Cereb Blood Flow Metab 1988;8:163-172.
35. Lee PA, Sylvia AL, Piantadosi CA. Effect of fluorocarbon-for-blood exchange on regional cerebral blood flow in rats. Am J Physiol 1988;254:H719-H726.
36. Wodick R, Lübbers DW. Quantitative evaluation of reflection spectra of living tissue. Hoppe-Seyler Z Physiol Chem 1974;355:583-594.
37. Thews G, Lübbers D. Ein schnellregistrierendes Absorptionsspektralphotometer. Zeit ang Physik 1955;7:325-331.
38. Lübbers D, Niesel W. Der Kurzzeit-Spektralanalysator. Ein schnellarbeitendes Spektralphotometer zur laufenden Messung von Absorptions- bzw. Extinktionsspektren. Pflüg Arch 1959;268:286-295.
39. Niesel W, Lübbers DW, Schneewolf D, Richter J, Botticher W. Double beam spectrometer with 10 ms recording time. Rev Sci Instrum 1964;35:578-581.
40. Lübbers DW, Wodick R. The examination of multicomponent systems in biological materials by means of a rapid scanning photometer. Appl Opt 1969;8:1055-1062.
41. Heinrich U, Hoffman J, Lübbers DW. Quantitative evaluation of optical reflection spectra of blood-free perfused guinea pig brain using a non-linear multicomponent analysis. Pflüg Arch 1987;409:152-157.

42. Moffett DF, Jöbsis FF. Response of toad brain respiratory chain enzymes to ouabain, elevated potassium, and electrical stimulus. Brain Res 1976;117:239-255.
43. LaManna JC, Rosenthal M, Novack R, Moffett DF, Jöbsis FF. Temperature coefficients for the oxidative metabolic responses to electrical stimulation in cerebral cortex. J Neurochem 1980;34:203-209.
44. Cooper CE, Delpy DT, Nemoto EM. The relationship of oxygen delivery to absolute haemoglobin oxygenation and mitochondrial cytochrome oxidase redox state in the adult brain: a near-infrared spectroscopy study. Biochem J 1998;332:627-632.
45. Quaresima, V., Springett, R., Cope, M., Wyatt, J. T., Delpy, D. T., Ferrari, M., and Cooper, C. E. Oxidation and reduction of cytochrome oxidase in the neonatal brain observed by in vivo near-infrared spectroscopy. Biochimica Biophysica Acta 1366(3), 291-300. 1998.
46. Heekeren HR, Kohl M, Obrig H, Wenzel R, von Pannwitz W, Matcher SJ et al. Noninvasive assessment of changes in cytochrome-c oxidase oxidation in human subjects during visual stimulation. J Cereb Blood Flow Metab 1999;19:592-603.
47. Clark JB, Nicklas WJ, Degn H. The apparent Km for oxygen of rat brain mitochondrial respiration. J Neurochem 1976;26:409-411.
48. Starlinger H, Lubbers DW. Polarographic measurements of the oxygen pressure performed simultaneously with optical measurements of the redox state of the respiratory chain in suspensions of mitochondria under steady-state conditions at low oxygen tensions. Pflugers Arch 1973;341:15-22.
49. Oshino N, Sugano T, Oshino R, Chance B. Mitochondrial function under hypoxic conditions: the steady states of cytochrome a+a3 and their relation to the mitochondrial energy states. Biochim Biophys Acta 1974;368:298-310.
50. Chance B. Early reduction of cytochrome c in hypoxia. FEBS Lett 1988;226:343-346.
51. Wilson DF, Erecinska M, Drown C, Silver IA. The oxygen dependence of cellular energy metabolism. Arch Biochem Biophys 1979;195:485-493.
52. Wilson DF, Rumsey WL, Green TJ, Vanderkooi JM. The oxygen dependence of mitochondrial oxidative phosphorylation measured by a new optical method for measuring oxygen concentration. J Biol Chem 1988;263:2712-2718.
53. Jobsis FF. Oxidative metabolism at low PO 2. Fed Proc 1972;31:1404-1413.
54. Bienfait HF, Jacobs JM, Slater EC. Mitochondrial oxygen affinity as a function of redox and phosphate potentials. Biochim Biophys Acta 1975;376:446-457.
55. Muraoka S, Slater EC. The redox states of respiratory-chain components in rat-liver mitochondria. II. The "crossover" on the transition from state 3 to state 4. Biochim Biophys Acta 1969;180:227-236.
56. Cooper CE, Matcher SJ, Wyatt JS, Cope M, Brown GC, Nemoto EM et al. Near-infrared spectroscopy of the brain: relevance to cytochrome oxidase bioenergetics. Biochem Soc Trans 1994;22:974-980.
57. Cooper CE. The steady-state kinetics of cytochrome c oxidation by cytochrome oxidase. Biochim Biophys Acta 1990;1017:187-203.
58. Lahiri S, Ehleben W, Acker H. Chemoreceptor discharges and cytochrome redox changes of the rat carotid body: role of heme ligands. Proc Natl Acad Sci U S A 1999;96:9427-9432.
59. Pysh JJ, Khan T. Variations in mitochondrial structure and content of neurons and neuroglia in rat brain: An electron microscopic study. Brain Res 1972;36:1-18.
60. Lübbers DW. Oxygen delivery and microcirculation in the brain, In: Manabe, Zweifach, Messmer, editors. Microcirculation in Circulatory Disorders. Tokyo: Springer-Verlag; 1988;33-50.

61. Sick TJ, Lutz PL, LaManna JC, Rosenthal M. Comparative brain oxygenation and mitochondrial redox activity in turtles and rats. J Appl Physiol 1982;53:1354-1359.
62. Faraci FM, Breese KR. Nitric oxide mediates vasodilatation in response to activation of N-methyl-D-aspartate receptors in brain. Circ Res 1993;72:476-480.
63. Dirnagl U, Lindauer U, Villringer A. Role of nitric oxide in the coupling of cerebral blood flow to neuronal activation in rats. Neurosci Lett 1993;149:43-46.
64. Millar J. The nitric oxide / ascorbate cycle: How neurons may control their own oxygen supply. Med Hypoth 1995;45:21-26.
65. Schottler F, Collins JL, Fergus A, Okonkwo D, Kassell NF, Lee KS. Structural interactions between NOS-positive neurons and blood vessels in the hippocampus. NeuroReport 1996;7:966-968.
66. Harik SI, LaManna JC, Light AI, Rosenthal M. Cerebral norepinephrine: Influence on cortical oxidative metabolism in situ. Science 1979;206:69-71.
67. Liu XP, Miller MJ, Joshi MS, SadowskaKrowicka H, Clark DA, Lancaster JR. Diffusion-limited reaction of free nitric oxide with erythrocytes. J Biol Chem 1998;273:18709-18713.
68. Kosaka H, Seiyama A. Physiological role of nitric oxide as an enhancer of oxygen transfer from erythrocytes to tissues. Biochem Biophys Res Comm 1996;218:749-752.
69. Brown GC, Cooper CE. Nanomolar concentrations of nitric oxide reversibly inhibit synaptosomal respiration by competing with oxygen at cytochrome oxidase. FEBS Lett 1994;356:295-298.
70. Brown GC. Nitric oxide regulates mitochondrial respiration and cell functions by inhibiting cytochrome oxidase. FEBS Lett 1995;369:136-139.
71. Torres J, Darley-Usmar V, Wilson MT. Inhibition of cytochrome c oxidase in turnover by nitric oxide: Mechanism and implications for control of respiration. Biochem J 1995;312:169-173.
72. Clementi E, Brown GC, Foxwell N, Moncada S. On the mechanism by which vascular endothelial cells regulate their oxygen consumption. Proc Natl Acad Sci U S A 1999;96:1559-1562.

Chapter **52**

A QUANTITATIVE STUDY OF OXYGEN AS A METABOLIC REGULATOR

Krishnan Radhakrishnan[1], Joseph C. LaManna[2] & Marco E. Cabrera[3]
[1]ICOMP, NASA Glenn Research Center, Cleveland, OH 44135 USA; [2,3]Case Western Reserve University, School of Medicine, Cleveland, Ohio 44106 USA; [3]Rainbow Babies and Children's Hospital, Cleveland, Ohio 44106 USA

Abstract: An acute reduction in oxygen delivery to a tissue is associated with metabolic changes aimed at maintaining ATP homeostasis. However, given the complexity of the human bioenergetic system, it is difficult to determine quantitatively how cellular metabolic processes interact to maintain ATP homeostasis during stress (e.g., hypoxia, ischemia, and exercise). In particular, we are interested in determining mechanisms relating cellular oxygen concentration to observed metabolic responses at the cellular, tissue, organ, and whole body levels and in quantifying how changes in tissue oxygen availability affect the pathways of ATP synthesis and the metabolites that control these pathways.

In this study, we extend a previously developed mathematical model of human bioenergetics, to provide a physicochemical framework that permits quantitative understanding of oxygen as a metabolic regulator. Specifically, the enhancement—sensitivity analysis—permits studying the effects of variations in tissue oxygenation and parameters controlling cellular respiration on glycolysis, lactate production, and pyruvate oxidation. The analysis can distinguish between parameters that must be determined accurately and those that require less precision, based on their effects on model predictions. This capability may prove to be important in optimizing experimental design, thus reducing use of animals.

Key words: metabolism, metabolic control, oxygen, regulation, sensitivity analysis

1. INTRODUCTION

Under conditions of adequate oxygen (O_2) supply, ATP synthesis is derived mainly from oxidative phosphorylation [1]. During acute reduction in oxygen delivery to a tissue, ATP homeostasis can be maintained temporarily through changes in tissue stores (ATP, O_2, and phosphocreatine, PC), as well as through accelerated rates of glycolysis and adjustments in the relative rates of carbohydrate and fatty acid oxidation, to provide reducing equivalents for oxidative phosphorylation [2, 3]. However, the relative contribution of each of these energetic pathways and their regulation depend on the metabolic state of the cell and the degree of reduction in oxygen delivery [4]. Given the complexity of the human bioenergetic system and its components, it is difficult to determine quantitatively how cellular metabolic processes interact to maintain ATP homeostasis during stress, such as hypoxia or ischemia [1, 5]. In this paper, we extend a previously developed mathematical model of the human bioenergetic system [6] by including sensitivity analysis [7], to facilitate study of the effects of key tissue biochemical species and cellular kinetic parameters on metabolic transport and conversion processes at the cellular, organ system, and whole body levels.

2. METHODS

The present work uses an existing multi-compartment, mathematical model of the human bioenergetic processes that provides a physicochemical framework for analyzing the role of oxygen in the regulation of lactate dynamics during hypoxia or ischemia [6]. The human body is described as a bioenergetic system consisting of four metabolically distinct tissue/organ subsystems—alveolar space, splanchnic bed, skeletal muscle, and other tissues—that exchange materials with the blood (i.e., 8 blood-tissue interactions in all). A typical tissue/organ subsystem consists of 25 interacting metabolic pathways among 12 biochemical species. The model is based on dynamic mass balances for glycogen, glucose, pyruvate, lactate, oxygen, and carbon dioxide, as well as for three other molecular pairs, which are coupled to most reactions: NAD^+ and NADH; ADP and ATP; and PC and creatine (CR). Other substrates participating in the reactions are alanine, fatty acids, and glycerol.

Mathematical descriptions of the transport and reaction processes involved in energy metabolism constitute sets of coupled, nonlinear, first-order ordinary differential equations (ODE's). To study the dynamic response of each subsystem to stimuli, we use the packaged code LSODE [8] to solve the ODE's describing the temporal evolution of metabolite levels in all tissue subsystems and the arterial blood compartment.

The solution of the ODE's describing energy metabolism depends on a number of input parameters, such as initial conditions and rate coefficients, as well as physiological and thermodynamic properties. The input parameters are often not known exactly, since they depend on our current knowledge of underlying biochemical and physiological mechanisms, which are described mathematically in the model. Application of sensitivity analysis to the mathematical model of the system can help design experiments that increase our understanding of these mechanisms and hence result in their improvement. This process fine-tunes the model by discarding unimportant reactions and adjusting rate coefficient parameters to match experimental data. The sensitivity analysis procedure requires the solution of an additional set of ODE's that describes the changes in tissue and arterial compartment concentrations (and reaction rates) due to changes in kinetic parameters and/or initial metabolite concentrations. The relationships between model predictions (i.e., outputs) and problem parameters (i.e., inputs—initial concentrations, rate coefficients, etc.) help determine the effects of uncertainties or changes in input parameters on the predictions, which ultimately are compared with experimental observations in order to validate the model.

Sensitivity analysis can identify parameters that must be determined accurately because of their large effect on the model predictions, as well as parameters that need not be known with great precision because they have little or no effect on the solution. Hence, we applied this approach to investigate the effects of changes in initial concentrations of muscle and arterial substrates (e.g., glucose) and other biochemical species (e.g., ADP, NADH, PC, and O_2), as well as the effects of changes in key kinetic parameters (e.g., k_{max} for oxidative phosphorylation) on selected model outputs (e.g., blood lactate concentration) during normal and reduced blood flow to skeletal muscle.

We computed sensitivity coefficients representing the changes in model outputs due to variations in initial concentrations and parameter values. In particular, we present at a simulation time of 30 min the normalized sensitivity coefficients of selected model outputs with respect to important model inputs (Table 1) at various levels of muscle ischemia. The calculation procedures were adapted from the NASA Lewis kinetics and sensitivity analysis code, LSENS [9,10].

3. RESULTS

To illustrate the predictive value of the ischemia model, we compared results of our computer simulations to experimental observations from occlusion studies in humans [11]. The computed changes in muscle glycogen, glucose, pyruvate, lactate, and phosphocreatine concentrations closely corresponded to those observed experimentally (Figure 1).

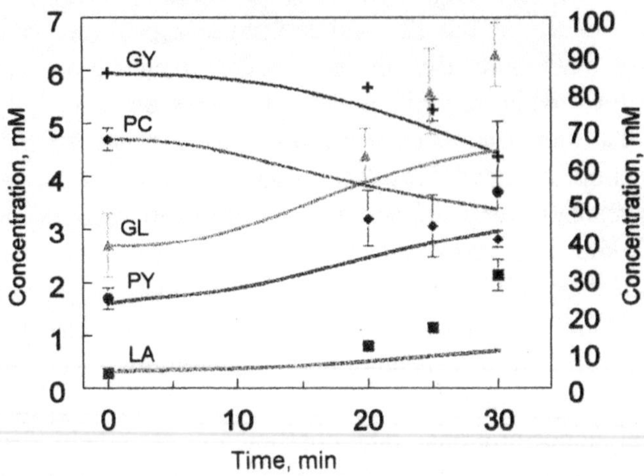

Figure 1. Comparison between model predictions and experimental data from occlusion studies in humans [11] of muscle glycogen (GY), phosphocreatine (PC), glucose (GL), pyruvate (PY), and lactate (LA) concentrations. Solid lines represent model simulations corresponding to a muscle blood flow of 0.3 L/min (i.e., severe muscle ischemia). Curves are shifted to match baseline values obtained experimentally. Pyruvate concentrations are displayed an order of magnitude larger than the actual values. Experimental data are represented by the following solid symbols: GY (+), PC (♦), GL (▲), PY (•) and LA (■). The left-hand scale applies to glucose and pyruvate and the right-hand scale to glycogen, phosphocreatine, and lactate.

Among the species examined (Table 1), those that had the largest effect on selected model outputs, as evidenced by the normalized sensitivity coefficients, were ADP and PC. The rate coefficients with the largest effects on model outputs were k_{max} for oxidative phosphorylation and k_{max} for ATP hydrolysis. Figure 2 shows at a time of 30 min the normalized sensitivity coefficients of glycogen, glucose, pyruvate, and lactate with respect to the initial concentration of ADP. The normalized sensitivity coefficients of muscle NADH/NAD$^+$ and lactate/pyruvate (LA/PY) ratios, as well as of the rates of glycolysis and lactate production, with respect to k_{max} of oxidative phosphorylation are presented in Figure 3. The rates of lactate accumulation in muscle and arterial blood were the model outputs that displayed the greatest sensitivity to most of the input parameters. For example, for a skeletal muscle blood flow of 0.9 L/min the parameters with the largest effect on the rates of lactate accumulation and the normalized sensitivity coefficients were: (a) k_{max} and K_m for ATP hydrolysis (10^4 and 10^3); (b) k_{max} and K_m for oxidative phosphorylation (10^4 and 10^2); (c) initial muscle ADP and PC concentrations (10^3 and 10^3); (d) initial muscle glucose concentration (10^2); (e) initial muscle O_2 concentration (10^1); and (f) k_{max} for NAD^+ reduction (10^1).

Table 1. Model Inputs and Outputs Considered in the Sensitivity Analysis

Model Inputs		Model Outputs
Initial Concentrations	Kinetic Parameters	
Muscle Glycogen	k_{max} for Glycolysis	Tissue Glycogen
Muscle and Arterial Glucose	k_{max} for Lactate Production	Tissue and Blood Glucose; Tissue Pyruvate and Lactate
Muscle Pyruvate	k_{max} for Lactate Oxidation	Tissue and Blood O_2 and CO_2
Muscle and Arterial Oxygen	k_{max} for Oxidative Phosphorylation	Muscle ADP, ATP, NAD^+, NADH, PC, and CR
Muscle Lactate	K_m for Oxidative Phosphorylation	Muscle NADH/NAD^+
Muscle ADP	k_{max} for ATP Hydrolysis	Muscle and Blood LA/PY
Muscle NADH	K_m for ATP Hydrolysis	Rates of Glycolysis, Lactate Production, and Oxidative Phosphorylation
Muscle PC	k_{max} for NAD^+ Reduction	Net Rates of Muscle and Blood Lactate Accumulation

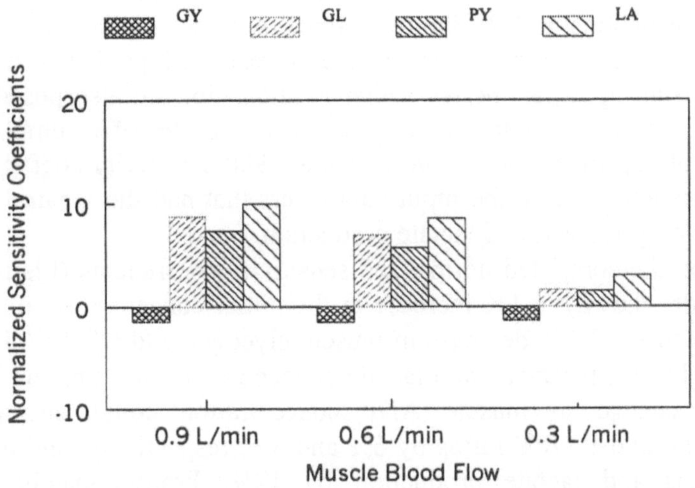

Figure 2. Normalized sensitivity coefficients of muscle glycogen (GY), glucose (GL), pyruvate (PY) and lactate (LA) concentrations with respect to the initial concentration of ADP in muscle. Normal muscle blood flow (0.9 L/min), moderate ischemia (0.6 L/min), severe ischemia (0.3 L/min); time = 30 min.

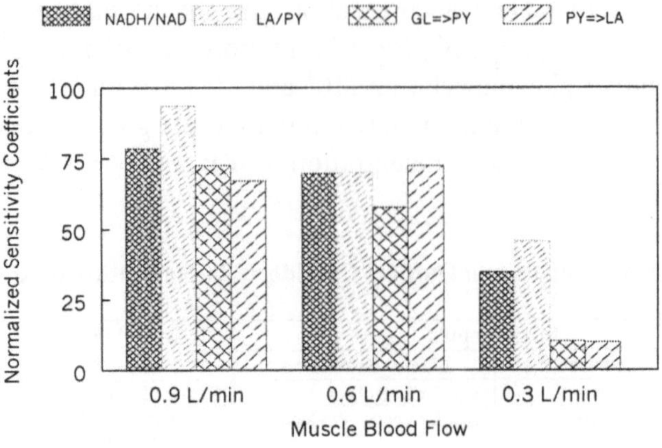

Figure 3. Normalized sensitivity coefficients of muscle NADH/NAD$^+$ and lactate/pyruvate (LA/PY) ratios, as well as of the rates of glycolysis (GL=>PY) and lactate production (PY=>LA), with respect to the parameter representing muscle mitochondrial density in the mathematical representation of muscle oxygen consumption as a Michaelis-Menten function of oxygen concentration {i.e., VO_2 muscle = $k_{max}[O_2]/(K_m + [O_2])$}. Normal muscle blood flow (0.9 L/min), moderate ischemia (0.6 L/min), severe ischemia (0.3 L/min); time = 30 min.

4. DISCUSSION

Sensitivity analysis was applied to a previously developed mathematical model of human bioenergetics [6], to permit quantitative understanding of oxygen as a metabolic regulator [12,13]. This model consists of 42 ordinary differential equations and 72 input parameters, which upon application of sensitivity analysis produces up to 3,024 relationships between model predictions and input parameters. Other relationships are also possible; for example, sensitivities of concentration ratios and of rates of accumulation of species (Table 1). In this study, we computed 900 sensitivity coefficients in all, but focus attention on the input parameters that had the greatest impact on the rates of glycolysis and lactate production.

Based on the computed normalized sensitivity coefficients (Figures 2-3, not all results shown), a 1% increase in the initial concentration of muscle ADP can induce a 1.5% decrease in muscle glycogen and a 7-10% increase in muscle glucose, pyruvate, and lactate concentrations at a time of 30 min. Also, this change in muscle ADP concentration could increase the NADH/NAD$^+$ and LA/PY ratios by 6% and 3%, respectively, and the rates of glycolysis and lactate production by 12%. Proportionately similar normalized sensitivity coefficients were obtained with respect to initial muscle [PC], although the actual magnitudes were approximately twice the values obtained with respect to initial muscle [ADP].

Initial arterial and muscle oxygen concentration did not appear to have a large effect on any of the metabolic reactions and thus on the concentrations of the biochemical species in the range of muscle blood flows studied. Model outputs also displayed low sensitivity to changes in muscle glycogen, glucose, pyruvate, lactate, and NADH. However, the normalized sensitivity coefficients with respect to the model parameter representing mitochondrial density—k_{max} for oxidative phosphorylation—were very large for most model outputs. In particular, muscle NADH/NAD$^+$, LA/PY, and the rates of glycolysis and lactate production could all increase by ~70% if oxidative phosphorylation were decreased by 1%. The rates of lactate accumulation in muscle and blood showed an even greater sensitivity to changes in oxidative phosphorylation.

These results, obtained by means of sensitivity analysis, are in good agreement with those previously obtained by actually changing the values of selected model parameters [14]. In both cases, we observed that the tissue oxygen concentration has to be extremely low to affect directly most of the reactions involved in energy metabolism. On the other hand, both sensitivity analysis and parameter-variation studies have illustrated the high sensitivity of the rates of glycolysis and lactate formation to reductions in oxygen consumption. It has been demonstrated that mitochondrial respiration is dependent on oxygen, as well as on the NADH/NAD$^+$ and ADP/ATP ratios [13]. In the present model, muscle oxygen consumption was assumed to be an explicit function of oxygen concentration and indirectly coupled to the rates of ATP formation and NADH oxidation [6]. Although tissue oxygen concentration has to be extremely low (near its critical value) to reduce the rate of oxygen consumption by one-half, changes in oxygen delivery that result in a reduction in the tissue oxygen consumption rate by as little as 1% can significantly affect lactate metabolism through the effects on ADP and NADH.

5. CONCLUSION

Sensitivity analysis can be applied in conjunction with any mathematical description of a physiological/ biological system. The main advantage of sensitivity analysis is that it can identify parameters that must be determined accurately because of their large effect on the model predictions, as well as parameters that need not to be known with great precision because they have little or no effect on the solution. This capability may prove to be important in optimizing the design of experiments, thereby reducing use of animals.

Sensitivity analysis can also be used to study the metabolic effects of reduced oxygen delivery to cardiac muscle due to local myocardial ischemia, as well as the effects of acute hypoxia on brain metabolism. Other important applications include identification of quantitatively relevant pathways and biochemical species within an overall mechanism, when examining the effects of a genetic anomaly or pathological state on energetic system components and whole system behavior.

ACKNOWLEDGEMENTS

This work was supported in part by NASA Glenn Research Center through Grant NCC3-622 and a Venture Fund Award, and by NHLBI Grant HL58653-01A1.

REFERENCES

1. Gutierrez G. Cellular energy metabolism during hypoxia. Crit Care Med 1991; 19:619-626.
2. Katz A, Sahlin K. Role of Oxygen in Regulation of Glycolysis and Lactate Production in Human Skeletal Muscle. Exerc Sports Sci Rev 1990; 18:1-28.
3. Gutierrez G. The Relationship of Tissue Oxygenation to Cellular Bioenergetics, in Gonzalez NC and Fedde MG (eds): Oxygen Transfer from Atmosphere to Tissues. New York, NY, Plenum Press, 1988; 183-205.
4. Connett RJ, Honig CR, Gayeski TEJ, Brooks GA. Defining hypoxia: a systems view of VO2, glycolysis, energetics, and intracellular PO2. J Appl Physiol 1990; 68:833-842.
5. Katz A. G-1,6-P2, glycolysis, and energy metabolism during circulatory occlusion in human skeletal muscle, Am J Physiol 1988; 255:C140-C144.
6. Cabrera ME, Saidel GM, Kalhan SC. Role of O2 in Regulation of Lactate Dynamics during Hypoxia: Mathematical Model and Analysis. Ann Biomed Eng 1998; 26:1-27.
7. Radhakrishnan K. Combustion Kinetics and Sensitivity Analysis Computations. In: Oran ES, Boris JP, editors. Numerical Approaches to Combustion Modeling. Washington DC: AIAA, 1991; 83-128.
8. Radhakrishnan K, Hindmarsh AC. Description and Use of LSODE, the Livermore Solver for Ordinary Differential Equations. Washington DC, NASA Reference Publication 1327, 1993.
9. Radhakrishnan K. LSENS. A General Chemical Kinetics and Sensitivity Analysis Code for Homogeneous Gas-Phase Reactions. I. Theory and Numerical Solution Procedures. Washington DC, NASA Reference Publication 1328, 1994.
10. Radhakrishnan K. LSENS, The NASA Lewis Kinetics and Sensitivity Analysis Code, Reston, VA: American Institute of Aeronautics and Astronautics 99-2394, 1999.
11. Chasiotis D, Hultman E. The effect of circulatory occlusion on the glycogen phosphorylase-synthetase system in human skeletal muscle. J Physiol 1983; 345:167-173.
12. Wilson DF, Erecinska M, Drown C et al. Effect of oxygen tension on cellular energetics. Am J Physiol 1977; 233:C135-C140.
13. Wilson DF, Owen CS, Holian A. Control of mitochondrial respiration: a quantitative evaluation of the roles of cytochrome c and oxygen. Arch Biochem Biophys 1977; 182:749-762.
14. Cabrera ME, Saidel GM, Kalhan SC. A model analysis of lactate accumulation during muscle ischemia. J Crit Care 1999; 14:151-163.

Chapter **53**

THE OXYGEN DEPENDENCY OF CEREBRAL OXIDATIVE METABOLISM IN THE NEWBORN PIGLET STUDIED WITH ^{31}P NMRS AND NIRS

Roger J Springett[1], Marzena Wylezinska[3], Ernest B. Cady[3], Veronica Hollis[2], Mark Cope[2] & David T. Delpy[2]

[1]Department of Radiology, Dartmouth College, Hanover, New Hampshire, USA. Department of Medical Physics & Bioengineering, University College London[2] and University College Hospitals NHS Trust[3], London WC1E 6JA, UK.

Abstract: Mean cerebral saturation and changes in the oxidation state of the Cu_A centre of cytochrome oxidase were measured by near infra-red spectroscopy simultaneously with phosphorous metabolites and intracellular pH measured using ^{31}P NMR spectroscopy during transient anoxia (inspired oxygen fraction = 0.0 for 105 seconds) in the newborn piglet brain. By collecting high quality ^{31}P spectra every 10 seconds, it was possible to resolve the delay between the onset of anoxia and the fall in PCr and to show that the Cu_A centre of cytochrome oxidase reduced simultaneously with the fall in PCr. From these observations it is concluded that, at normoxia, oxygen tension at the mitochondrial level is substantially above a critical value at which oxidative metabolism becomes oxygen dependent.

Key words: NIRS, ^{31}P NMRS, cytochrome oxidase, hypoxia

1. INTRODUCTION

Enzymatic studies of flash-frozen rat brain at different levels of hypoxeamia have shown that, at normoxia, cerebral phosphocreatine (PCr) concentration is independent of arterial oxygen tension (p_aO_2) but that with increasingly severe hypoxeamia, there is a fall in PCr before there is a fall in cerebral metabolic rate for oxygen ($CMRO_2$) [1]. In the brain where phosphorylation potential is buffered by the creatine kinase reaction, a fall in the PCr concentration is indicative of a fall in the phosphorylation potential. Although the results from in-vitro mitochondrial models show that the redox state of the components of the electron transport chain are independent of oxygen tension at high oxygen tension [2], it is difficult to predict the response of the electron transport chain at normoxia in-vivo because it is not possible to measure mitochondrial oxygen tension directly. However, when phosphorous metabolites have been measured simultaneously with redox state of cytochrome c and NADH, it has been shown that cytochrome c reduces simultaneously with the fall in phosphorylation potential [3,4].

Near infrared spectroscopy (NIRS) is a non-invasive technique which has the potential to measure changes in the redox state of the Cu_A centre of cytochrome oxidase as well as changes in haemoglobin concentration and saturation. However, the Cu_A signal is controversial with different algorithms producing contradictory results: some algorithms show continuous changes in redox state with hypoxia [5], and some showing that redox state is independent of cerebral oxygenation at normoxia [6,7]. Modelling has shown that full spectra NIRS measurement is more precise in separating the Cu_A component from the haemoglobin components [8] and has the added advantage that absolute quantification of the optical pathlength and absolute deoxyhaemoglobin can be obtained by using 2nd differential spectroscopy [9]. Absolute oxyhaemoglobin can be obtained by assuming the oxyhaemoglobin concentration falls to zero during brief anoxia so allowing mean cerebral saturation ($SmcO_2$) to be determined.

The aim of this study was to simultaneously measure NIRS parameters and phosphorous metabolites with ^{31}P NMR spectroscopy in order to validate the Cu_A signal and determine the $SmcO_2$ at which oxidative metabolism becomes oxygen dependent. Previous studies [10,11,12] attempting to correlate redox states with phosphorous metabolites using ^{31}P NMR spectroscopy have usually been performed by collecting phosphorous spectra at varying inspired oxygen fraction or p_aO_2. However, this technique requires p_aO_2 to be maintained at low values for many minutes in order to obtain high quality ^{31}P spectra which can lead to circulatory failure resulting in partial cerebral ischaemia, a more profound cellular acidosis, and severe energy failure (compare Tsuji et al. [10] and Matsumoto et al. [11] with the data presented here). Such changes can lead to cellular oedema and a change in the tissue scattering coefficient which may not be accounted for by the NIRS algorithm used to separate attenuation changes into chromophore concentration changes and hence may lead to spurious changes in the redox signal. In order to circumvent this problem, brief anoxias were performed in a neonatal piglet model and the phosphorous spectra collected with a high temporal resolution during the onset of anoxia. The neonatal piglet model is particularly suited to this procedure because the neonatal heart is resistant to hypoxia. In order to obtain high quality spectra, the anoxia was repeated six times in each piglet and the phosphorous spectra from equivalent time points of each anoxia were summed.

2. METHODS

The animal studies were performed in accordance with current UK Animals (Scientific Procedures) Act 1986. Six piglets, born at term but less than 24 hours old and weighing 1.72±0.14 (SD) Kg, were sedated with Midazolam and anaesthetised with 2% isoflurane. A tracheotomy was performed and the piglets were artificially ventilated with an oxygen and nitrogen gas mixture with the inspired oxygen fraction was set to 0.4. Heart rate and mean arterial blood pressure were monitored from a catheter sited in the umbilical arterial artery using a strain-gauge pressure transducer. Rectal temperature was maintained at 38.5°C using a heated water mattress.

The piglets were placed in a custom-made pod in the bore of a 7-Tesla Bruker Avance spectrometer (Karlsruhe, Germany). The head was fixed in a stereotaxic frame which also held a 25mm diameter double-tuned (^{31}P and ^{1}H), inductively-coupled surface coil against the intact scalp over the parietal lobes and two end-on optodes 35mm apart symmetrically about the midline approximately 1cm posterior to the eyes. In this configuration, the NIR optodes and ^{31}P surface coil were expected to probe approximately the same volume of cerebral tissue.

In order to obtain high quality ^{31}P spectra with a high temporal resolution, it was necessary to use a 2s recovery time (T_R) so that 4 free induction decays (FIDs) could be collected in 10s and then to repeat the anoxia six times in each piglet summing the free induction decays at equivalent time points from each anoxia. The length of the anoxia was chosen to be 105s as pilot studies had showed that this period produced complete reduction of the Cu_A centre of cytochrome oxidase compared with longer anoxias but was sufficiently short that a full haemodynamic and metabolic recovery was made on return to normoxia.

The pulse sequence consisted of a DANTE pulse train [13] of 500 10μs pulses with a 200μs repetition time to reduce signals from bone and phospholipids, followed by a square acquisition pulse (180° flip angle at the centre of the coil). The 24 FIDs were summed, exponentially weighted with an 8Hz filter, Fourier transformed and phase corrected. Relative concentrations of phosphorous metabolites were obtained by fitting Lorenzian peaks to the real component of the phosphorous spectra using a non-linear iterative curve fitting method and a quartic background. The high temporal resolution precluded the acquisition of fully relaxed ^{31}P spectra so that metabolite concentration is expressed as percent of baseline values. Intracellular pH (pH_i) was calculated from the frequency shift between the PCr and inorganic phosphate (P_i) peaks [14].

Light from a stabilised tungsten halogen light source was filtered with a 610nm long-pass filter and transmitted to the piglet head with a glass optic fibre bundle (3.3mm diameter). Transmitted light was collected with a second fibre bundle and focused onto the slits of a 0.27m spectrograph (270M, Instruments SA, Lonjumeau, France) equipped with a 300g/mm grating blazed at 1000nm. NIR spectra between 650 and 980nm were collected contiguously every second on a CCD detector (Wright Instruments, Enfield, London) with the shutter held open. The pixel bandwidth was 0.32nm and the slits set to give a signal of ≈100,000 electrons per digital conversion at 800nm. Spectral resolution was between 2.5 and 3.7nm. Decomposition of the attenuation spectra into changes in HbO_2, Hb and Cu_A oxidation state corrected for optical pathlength were performed as described previously [15] and absolute deoxyhaemoglobin was measured using the 2nd differential technique [9]. Baseline absolute HbO_2 was back-calculated by assuming that HbO_2 fell to zero at the depth of the anoxia and mean cerebral saturation ($SmcO_2$) was calculated from absolute HbO_2 and absolute Hb.

Analysis of variance (ANOVA) was used to compare repeated measurements and all data are presented as mean ± SD (n=6 animals). The significance of changes was determined using a paired Student's *t*-test; the criterion for significance was $P<0.05$

3. RESULTS

Heart rate, mean arterial blood pressure, rectal temperature, arterial pO_2 and arterial pCO_2 were in the normal range and did not change significantly between baseline periods before each anoxia. Likewise, $SmcO_2$, Cu_A redox state, PCr concentration and pH_i did not change significantly in the baseline periods before each anoxia. Plasma glucose and lactate were in the normal range but increased significantly between the anoxias from 5.0±0.9 and 1.5±0.5mM to 5.6±1.3 and 2.5±0.7mM respectively.

Figure 1 shows time courses of $SmcO_2$ (top panel), redox state of the Cu_A centre of cytochrome oxidase (upper middle panel), PCr concentration (lower middle panel) and intracellular pH (bottom panel) time-normalized so that time zero indicates the first data point at which $SmcO_2$ has fallen below baseline. In brief, the fall in arterial p_aO_2 leads to a fall in $SmcO_2$, a reduction in the Cu_A centre of cytochrome oxidase, a fall in the concentration of PCr and an acidification of the intracellular space. However, there is a delay between the onset of anoxia and the reduction of Cu_A and fall in PCr and a then further delay to the first fall in pH_i.

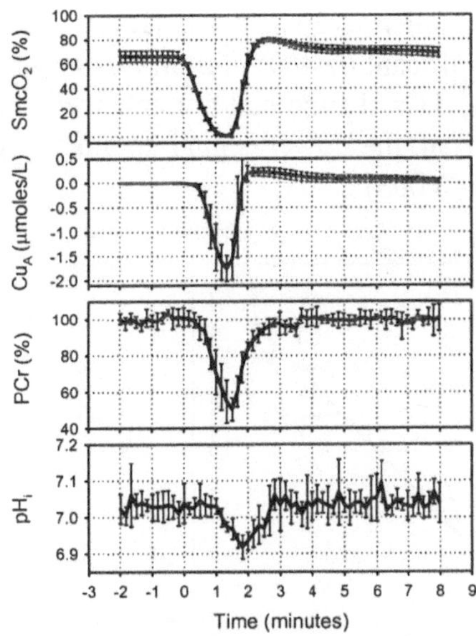

Figure 1. Time courses of mean cerebral saturation (top panel), changes in the oxidation state of the Cu_A centre of cytochrome oxidase (upper middle panel), Phosphocreatine concentration (lower middle panel) and intracellular pH (bottom panel). The results are expressed as mean±SD (n=6 piglets) and time-normalized so that time zero indicates the first data point at which HbO_2 has fallen below baseline.

The nadir of the hypoxia occurs at 90s and, on reoxygenation, $SmcO_2$, Cu_A and PCr rapidly return towards baseline but pH_i continues to fall for a further 30s before beginning to return towards baseline. The difference in the period of zero inspired oxygen and period of hypoxia observed in the brain is due to gas mixing in the ventilation tubing and lungs. The hypoxaemia triggers an increase in CBF which, during the return to normoxia, leads to a rise in $SmcO_2$ above baseline levels and $SmcO_2$ reaches a maximum of (79.7±1.9%) at 2 minutes 40s and then returns to baseline levels over the subsequent 5 minutes. The Cu_A centre hyperoxidises to 0.23±0.08 µmoles/L at 2 minutes 10s and then returns to baseline over the same time course as $SmcO_2$. On reoxygenation, there is a rapid increase in PCr towards baseline levels over the first 30 seconds post reoxygenation, but the increase slows and it is not until approximately 4 minutes post anoxia that PCr reaches baseline whereas pH_i returns to baseline before 3 minutes post anoxia.

During the anoxia, the concentration of P_i increased (data not shown) mirroring the decrease in the concentration of PCr but with a lower signal to noise. There was no observable change in the concentration of ATP (data not shown).

The baseline value of $SmcO_2$ was 66.2±4.6%. The first observable change in Cu_A occurred at 30s when Cu_A had fallen to –0.14±0.11 µmoles/L and when $SmcO_2$ had fallen to 30.2±6.4%. However, due to the inter-piglet variation, the first significant ($p<0.05$) fall in Cu_A occurred at 40s ($p=0.024$). At this time point the PCr change had not achieved significance ($p=0.069$) probably as a result of the lower signal to noise but PCr did achieve significance at 50s ($p=0.003$).

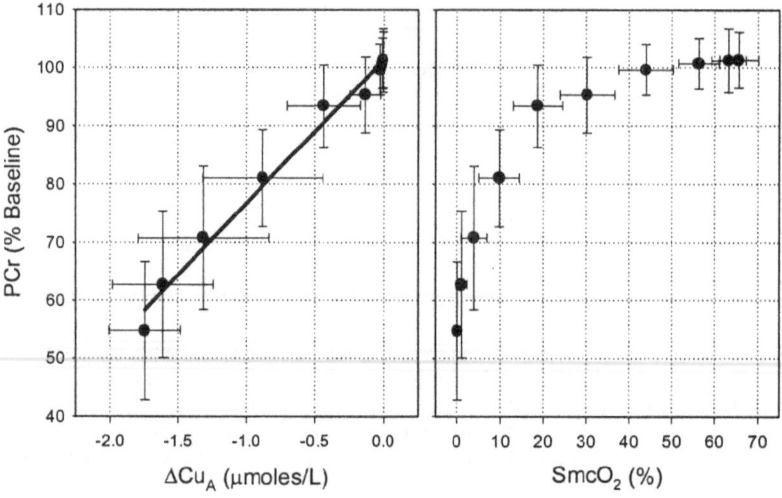

Figure 2. Phosphocreatine concantration plotted against mean cerebral saturation (right panel) and against change in Cu_A oxidation state (left panel) during the onset of anoxia. The results are expressed as mean±SD (n=6 piglets).

The right panel of figure 2 shows the PCr concentration plotted against $SmcO_2$ during the onset of anoxia. It can be clearly seen that there is no change in the PCr concentration until $SmcO_2$ falls to 30%. The left panel of figure 2 plots the PCr concentration against the change in Cu_A oxidation state. This plot shows that PCr and Cu_A change simultaneously and that, at least over this range of PCr in which there was no noticeable change in ATP, there is a strong linear relationship between Cu_A redox state and PCr concentration.

4. DISCUSSION

The stability of the systemic parameters, the NIRS parameters and the NMRS parameters between anoxias shows that the anoxias were highly reproducable and that the technique of summing equivalent time points from each anoxia was valid. The rise in plasma lactate concentration was consistent with repeated anoxias and probably systemic in origin and not cerebral. Although plasma lactic acid can be in part used as a substrate by the neonatal mammalian brain [16], the increase in lactate was small and did not affect the Cu_A redox state, PCr concentration or pH_i.

The Cu_A centre of cytochrome oxidase is the primary electron acceptor from cytochrome c and lies on the cytosolic side of the enzyme close to the docking site with cytochrome c [17,18]. In-vitro mitochondrial work has shown that the Cu_A centre feels the same membrane potential as cytochrome c [19] and is in redox equilibrium with cytochrome c during coupled turnover [20]. These results confirm that the phosphorylation potential and Cu_A redox state are independent of cerebral oxygenation at normoxia and that the Cu_A centre of cytochrome oxidase reduces simultaneously with the fall in PCr concentration. This is further evidence [15] that the Cu_A signal is accurately measuring oxidation changes of the Cu_A centre of cytochrome oxidase in this model and under these conditions and that interference by haemoglobin is negligible. However, care should be taken when using the Cu_A signal as a surrogate marker for PCr because the oxidation state of the Cu_A centre of cytochrome oxidase is dependent on parameters other than oxygen tension and phosphorylation potential. For instance, at two minutes post anoxia there is an hyperoxidation of the Cu_A signal even though $SmcO_2$ and PCr are still below baseline values.

It is the oxygen gradient between the vasculature and mitochondria which provides the driving force for the diffusion of oxygen to the mitochondria. In the steady state, the rate of diffusion of oxygen to the mitochondria is equal to $CMRO_2$. If there is no change in cerebral perfusion or $CMRO_2$ then the oxygen gradient between mitochondria and vasculature must be fixed so that, as vasculature pO_2 decreases so must mitochondrial pO_2 (p_mO_2). Using this model, the observation that Cu_A redox state is independent of cerebral oxygenation would imply that, at normoxia, p_mO_2 is above a critical value at which the Cu_A centre becomes oxygen dependent.

This critical value of p_mO_2 is reached when $SmcO_2$ falls from a baseline value of $\approx$66% to $\approx$30% (see Results). Assuming the haemoglobin saturation curve of Kokholm [21], a P_{50} of piglet haemoglobin of 3.7KPa and an average vasculature pH of 7.4, saturations of 66% and 30% corresponds to a mean pO_2 of $\approx$4.8 and $\approx$2.7KPa respectively. This analysis would therefore suggest that the p_mO_2 is approximately 2.1KPa above the critical p_mO_2 at normoxia. These figures should be regarded as approximations because the assumption that mean vasculature pO_2 can be calculated from $SmcO_2$ is not exact since both the pH and CO_2 tension change from arterial to venous compartment. However, this analysis would strongly suggest that p_mO_2 is substantially above zero.

When the ATP generated by glycolysis has been hydrolysed, glycolysis is a proton generating system. These protons can either be consumed by the TCA cycle and oxidative phosphorylation or they can be exported from the cell as lactic acid. At normoxia when homeostatis is achieved, proton consumption and proton production occur at equal rates. Figure 1 shows that there is a rapid acidification of the intracellular space which could indicate the point at which oxygen tension limits oxygen consumption. This interpretation would be consistent with the observation that oxygen consumption is inhibited at lower values of p_aO_2 than affects the PCr concentration [1]. Inspection of figure 1 shows that pH_i begin to fall when $SmcO_2$ is 3.9±3.0% corresponding to a pO_2 of ≈0.9KPa. However this interpretation of the decrease in pH_i is complicated because physiochemical buffering would tend to obscure small changes in pH, the transfer of phosphate from PCr to ADP is a proton consuming reaction which would tend to mask small increases in production of protons, and increases in blood flow triggered by the hypoxaemia could wash out carbon dioxide increasing pH_i. Furthermore, the fall in the phosphorylation potential would result in small increases in ADP which would activate phosphofructokinase one of the major regulatory enzymes of glycolysis. This would have the effect of increasing glycolysis to values above the basal rate resulting in the slow accumulation of protons even at constant $CMRO_2$. In fact, increases in lactate are often detectable by ^1H NMR spectroscopy even when there are no detectable changes in PCr [22].

ACKNOWLEDGEMENTS

We are grateful to Hamamatsu Photonics KK for financial assistance.

REFERENCES

1. Siesjo BK. Brain Energy metabolism, John Wiley & Sons, Chichester, 1978.
2. Sugano T, Oshino N, Chance B. Mitochondrial functions under hypoxic conditions. The steady states of cytochrome c reduction and of energy metabolism. Biochim. Biophys. Acta 1974;347:340-358.
3. Wilson DF, Erecsinska M, Drown C, Silver IA. Effects of oxygen tension on cellular energetics. Amer. J. Physiol. 1977;233: C135-140.
4. Wilson DF, Erecsinska M, Drown C, Silver IA. The oxygen dependence of cellular energy metabolism. Arch. Biochem. Biophys. 1979;195:485-493.
5. Hampson NB, Camporesi EM, Stolp BW, Moon RE, Shook JE, Griebel JA, Piantadosi CA. Cerebral oxygen availability by NIR spectroscopy during transient hypoxia in humans. J. Appl. Physiol. 1990;69:907-913.

6. Hoshi Y, Hazeki O, Kakihana Y, Tamura M Redox behavior of cytochrome oxidase in the rat brain measured by near-infrared spectroscopy. J Appl Physiol 1997;83:1842-1848.
7. Cooper CE, Matcher SJ, Wyatt JS, Cope M, Brown GC, Nemoto EM, Delpy DT. Near-infrared spectroscopy of the brain: relevance to cytochrome oxidase bioenergetics. Biochem Soc Trans 1994;22:974-980.
8. Matcher SJ, Elwell CE, Cooper CE, Cope M Delpy DT. Performance comparison of several published tissue near-infrared spectroscopy algorithms. Anal. Biochem. 1994;227: 54-68.
9. Matcher SJ, Cope M, Delpy DT. Use of the water-absorption spectrum to quantify tissue chromophore concentration changes in near-infrared spectroscopy. Phys. Med. Biol. 1994;39:177-196.
10. Tsuji M, Naruse H, Volpe J, Holtzman D. Reduction of cytochrome aa3 measured by near-infrared spectroscopy predicts cerebral energy loss in hypoxic piglets. Pediatr. Res. 1995;37 253-259.
11. Matsumoto H, Oda T, Hossain MA, Yoshimura N. Does the redox state of cytochrome aa3 reflect brain energy level during hypoxia? Simultaneous measurements by near infrared spectrophotometry and 31P nuclear magnetic resonance spectroscopy. Anesth Analg 1996;83:513-518.
12. Gyulai L, Chance B, Ligeti L, McDonald G, Cone J. Correlated in vivo ^{31}P-NMR and NADH fluorometric studies on gerbil brain in graded hypoxia and hyperoxia. Am. J. Physiol. 1988;254: C699-C708.
13, Morris P., Freeman R. Selective excitation in Fourier transform nuclear magnetic resonance. J. Magn. Reson. Imaging 1978;29:433-462.
14. Moon RB, Richards JH. Determination of intracellular pH by ^{31}P magnetic resonance. J. Biol. Chem. 1973;48: 7276-7278.
15. Cooper CE, Cope M, Springett R, Amess PN, Penrice J, Tyszczuk L, Punwani S, Ordidge R, Wyatt J, Delpy DT Use of mitochondrial inhibitors to demonstrate that cytochrome oxidase near-infrared spectroscopy can measure mitochondrial dysfunction non-invasively in the brain J. Cereb. Blood Flow Metab. 1999;19:27-38.
16. Medina JM, Tabernero A, Tovar JA, Martin-Barrientos. Metabolic fuel utilization and pyruvate oxidation during the postnatal period. J Inherit Metab Dis 1996;19:432-42.
17. Iwata S, Ostermeier C, Ludwig B and Michel H. Structure at 2.8Å resolution of cytochrome c oxidase from Paracoccus Denitrificans. Nature 1995;376: 660-669.
18. Tsukihara T, Aoyama H, Yamashita E, Tomizaki T, Yamaguchi H, Shinzawa-Itoh K, Nakashima R, Yaono R and Yoshikawa S. Structure of metal sites of oxidized bovine heart cytochrome c oxidase at 2.8Å. Science 1995; 269: 1069-1074.
19. Rich PR, West IC and Mitchell P. The location of Cu_A in mammalian cytochrome c oxidase. FEBS Lett. 1988; 233: 25-30.
20. Morgan JE, Wikström M. Steady-state redox behaviour of cytochrome c, cytochrome a, and Cu_A of cytochrome oxidase in intact rat liver mitochondria. Biochemistry 1991; 30: 984-958.
21. Kokholm G. Simultaneous measurements of blood pH, pCO_2, pO_2 and concentrations of hemoglobin and its derivates--a multicenter study. Scand J Clin Lab Invest Suppl 1990; 203:75-86.
22. Gyulai L, Schnall M, McLaughlin AC, Leigh JS, Chance B (1987) Simultaneous 31P and 1H-nuclear magnetic resonance studies of hypoxia and ischaemia in the cat brain J. Cereb. Blood Flow Metab. 7:543-551.

Chapter 54

EFFECT OF MYOGLOBIN INACTIVATION ON INTRACELLULAR GRADIENTS OF NADH FLUORESCENCE AT CRITICAL MITOCHONDRIAL OXYGEN SUPPLY[1]

Eiji Takahashi, Hiroshi Endoh, Mizue Ishikawa, Machiko Kishi & Katsuhiko Doi

Department of Physiology, Yamagata University School of Medicine, Yamagata 990-9585, Japan

Key words: myoglobin facilitated oxygen diffusion, NADH fluorescence, single cardiomyocyte, metmyoglobin

1. PURPOSE

In single isolated cardiomyocytes of rats, we examined effects of functional inhibition of cytosolic myoglobin by nitrite on mitochondrial oxidative metabolism at critical mitochondrial oxygen supply.

2. METHODS AND RESULTS

Mitochondrial oxidative metabolism was assessed in single isolated cardiomyocytes of rats by quantitative measurement of mitochondrial NADH fluorescence with a subcellular spatial resolution [1]. Incubation of cardiomyocytes with 5 mM $NaNO_2$ for > 5 min abolished changes in light absorption at 434 nm on aerobic-to-anaerobic transitions,

[1] Full text version of this article has been published in *Appl. Cardiopulmonary Pathophysiol.* 9: 368-373, 2000.

implying that cytosolic myoglobin was completely converted to the ferric myoglobin (functional inhibition). Five mM $NaNO_2$ decreased the oxygen consumption of quiescent coupled cardiomyocytes at nonlimiting pO_2 by approximately 30%, while it was unchanged in cardiomyocytes treated with an uncoupler of oxidative phosphorylation (1 μM carbonyl cyanide *m*-chlorophenylhydrazon, CCCP). In coupled quiescent cardiomyocytes with a low oxygen consumption rate, lowering the extracellular pO_2 from 22 mm Hg to 1.8 mm Hg elevated the NADH fluorescence from 0.50±0.12 (fraction to the anoxic NADH fluorescence, photo bleaching not compensated) to 0.85±0.12. This relationship was not affected by 5 mM $NaNO_2$ (0.48±0.08 and 0.87±0.09 for extracellular pO_2 of 22 mmHg and 1.8 mm Hg, respectively).

Lowering the extracellular pO_2 to 15 – 30 mm Hg while stimulating mitochondrial respiration by CCCP produced significant radial gradients of the NADH fluorescence (Fig.1A). Because we have previously demonstrated the presence of quite steep radial gradients of myoglobin oxygen saturation in cardiomyocytes in the same experimental condition [2], these gradients of the NADH fluorescence may imply the critical dependency of mitochondrial oxygen metabolism on intracellular diffusional oxygen supply to mitochondria. We then repeated the same experiment after treating the cardiomyocytes with 5 mM $NaNO_2$. Functional inhibition of myoglobin at the critical mitochondrial oxygen supply further elevated the NADH fluorescence compared to the control cell with the intact myoglobin (Fig. 1B).

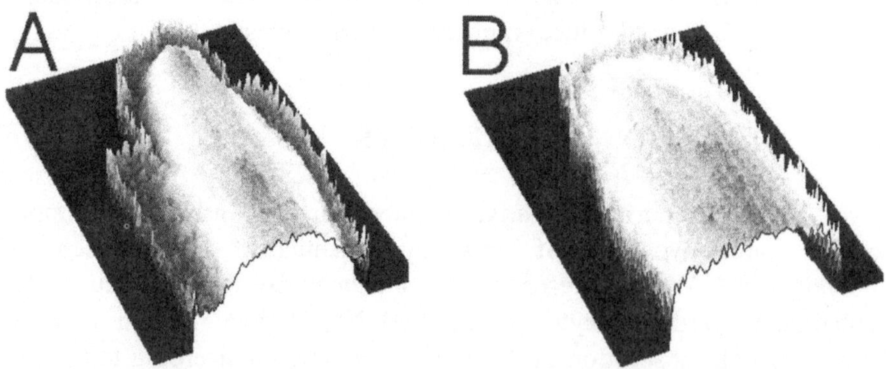

Figure 1. Three-dimensional presentation of the mitochondrial NADH fluorescence in single isolated cardiomyocytes of rats. Intensity of the NADH fluorescence is represented in shades of gray. The images were cross-sectioned near the center of the cell so that the radial NADH fluorescence profile can be seen. Extracellular pO_2 was 22 mm Hg. Mitochondrial respiration was stimulated by 1 μM CCCP. A: control. B: 5 mM $NaNO_2$.

3. DISCUSSION

In CCCP uncoupled cardiomyocytes, we previously demonstrated, by the microspectrophotometry of myoglobin, significant radial gradients of intracellular oxygen supply [2]. Unfortunately, this technique cannot be used in $NaNO_2$ treated cardiomyocytes, because it uses in situ ferrous myoglobin as an oxygen probe. Autofluorescence of mitochondrial NADH is a sensitive indicator of mitochondrial oxygen supply at critical pO_2. Thus, mitochondrial NADH was used to assess the mitochondrial oxygen supply in cardiomyocytes without functioning myoglobin.

Functional inhibition of cytosolic myoglobin by 5 mM $NaNO_2$ did not affect the NADH fluorescence in quiescent coupled cardiomyocytes with a very low oxygen flux, but the treatment certainly elevated the NADH fluorescence in uncoupled cardiomyocytes with a ~7 times higher oxygen flux. Thus, present results that functional inhibition of myoglobin resulted in the relative suppression of mitochondrial oxidative metabolism suggest a physiological role of myoglobin at critical oxygen supply to mitochondria, most likely the myoglobin facilitated intracellular oxygen diffusion [3].

REFERENCES

1. Takahashi, E., Endoh, H. and Doi, K. Intracellular gradients of O_2 supply to mitochondria in actively respiring single cardiomyocyte of rats. Am J Physiol 1999; 276:H718-H724.
2. Takahashi, E., Sato, K., Endoh, H., Xu, Z.L. and Doi, K. Direct observation of radial intracellular pO_2 gradients in a single cardiomyocyte of the rat. Am J Physiol 1998;275: H225-H233.
3. Wittenberg, J.B. Myoglobin-facilitated oxygen diffusion: role of myoglobin in oxygen entry into muscle. Physiol Rev 1970; 50:559-636.

Chapter **55**

THE EFFECTS OF ANESTHESIA ON CEREBRAL TISSUE OXYGEN TENSION: USE OF EPR OXIMETRY TO MAKE REPEATED MEASUREMENTS

Harold M. Swartz[1], Satoshi Taie[2], Minoru Miyake[2], Oleg Y. Grinberg[1], Huagang Hou[1], Hisham El-Kadi[3] & Jeff F. Dunn[1]

[1]*Department of Diagnostic Radiology, Dartmouth Medical School, 7785 Vail Building Hanover, NH 03755 USA;* [2]*Department of Anesthesiology and Emergency Medicine, Kagawa Medical School, 1750-1, Miki-cho, Kita-gun, Kagawa 761, 07, Japan;* [3]*Boston Medical Center, Dept. of Anesthesia, Dowling 6 North, One Boston Medical Center Place, Boston, MA 02118-2392 USA.*

Abstract: While very useful data can be obtained from measurements of pO_2 within various compartments of the vascular system, such measurements do not necessarily provide accurate information on the pO_2 in the brain. Anesthetics can significantly affect the tissue pO_2 in the brain by several mechanisms involving both delivery and utilization. Electron Paramagnetic Resonance (EPR or ESR) oximetry has the potential for non-invasively carrying out repeated direct measurements of pO_2 in tissues during the course of anesthesia. In this paper we describe the use of EPR oximetry for studying the influence of anesthesia on tissue pO_2, and present illustrative results from experiments with five different anesthetics in rats. The results indicate that the tissue O_2 can be measured directly using EPR oximetry, and data can be obtained non-invasively during the course of anesthesia.

Key words: anesthesia, pO_2, EPR oximetry, rats, brain

1. INTRODUCTION

General anesthesia is needed in many animal experiments in order to carry the studies out in an effective and appropriate manner. Anesthesia also is required for many clinical interventions. In both the experimental and operating room environments, the tissue oxygenation (the partial pressure of oxygen in tissues, P_tO_2) is a critical variable. Anesthesia would be expected to affect tissue oxygenation because it can alter many physiological parameters, especially cardiovascular, pulmonary, and metabolic parameters.

While many of these physiological parameters can be measured, the relationships between them and tissue oxygenation are not straightforward. It is known, for example, that hyperventilation can increase arterial oxygenation and hemoglobin saturation at the same time as it causes a reduction in cerebral oxygenation [1]. It is also known that the steady state value for P_tO_2 can be quite different, depending on the anesthesia being used [2-6]. The effects of anesthetics on these parameters would be expected to vary with the type of anesthetic, the route of administration, and the method of administration. Therefore, it seems very desirable to directly measure the effects of anesthesia on tissue oxygenation. In addition to providing important information for experimental studies in animals, the data obtained also should have some direct relevance to the effects of anesthesia on the P_tO_2 of the brain and other tissues during clinical anesthesia.

The measurement of brain P_tO_2 is now becoming possible in the clinic [2,4,5,7,8], although the techniques are rather complex and difficult. These capabilities appear to be useful. For example, measurements of P_tO_2 highlighted localized hypoxic regions in patients under conditions where more standard methods such as jugular bulb oximetry did not indicate the presence of hypoxia [8]. Results such as these make it increasingly apparent that it is important to determine the influence of anesthesia on P_tO_2 and therefore techniques that can do this facilely would be desirable.

Electron paramagnetic resonance (EPR or completely equivalently, ESR) oximetry is a noninvasive (after implantation of the paramagnetic material) technique that can measure P_tO_2 level in tissue repeatedly (with time resolution of less than 10s) and non-invasively [9]. Recently the development of new stable paramagnetic particles such as fusinite, India ink, and lithium phthalocyanine and the development of low frequency EPR instruments (1 GHz or lower) have enabled rapid progress to be made in the field of *in vivo* EPR oximetry [10-13].

The technique is based on the effect that the paramagnetism of molecular oxygen has on other paramagnetic species. It can affect the relaxation times and the distribution of the unpaired electrons, thereby affecting the EPR spectra. It is quite feasible to measure the effects of pO_2 on the EPR spectra of each type of paramagnetic materials and use this information to measure P_tO_2 by EPR. A variety of paramagnetic materials that are especially sensitive to oxygen have been developed. These include both soluble and insoluble materials. The various types of oxygen-sensitive paramagnetic materials have quite different quantitative responses to changes in the pO_2, which makes it feasible to select a paramagnetic material that has an appropriate sensitivity in the range of P_tO_2 that is likely to occur under a particular circumstance. Because many of the measurements of pO_2 are based on the broadening effect of oxygen on the EPR spectrum, the method is especially sensitive for the low values of pO_2 that are likely to be of particular biomedical interest.

The physical and chemical properties of the oxygen-sensitive paramagnetic materials vary considerably. The soluble free radicals that are used most frequently for oximetry are the nitroxides and derivatives of triphenyl methyl radicals. The soluble nitroxides are widely distributed within the animal; the charge and lipophilicity of the soluble free radicals are the principal factors that determine the details of their distribution. Due to metabolism of to non-paramagnetic derivatives and excretion, these usually disappear within minutes to hours. The soluble materials may be especially useful for imaging but due to the relatively modest changes in their spectra from the presence of oxygen, their sensitivity to small differences is limited.

The particulate oxygen sensitive paramagnetic materials cannot readily be used for imaging, but are very useful for spectroscopy. Because they are insoluble and usually quite inert, they can remain at the site at which they are placed and report on P_tO_2 for periods of up to a year or more in some tissues. Their responsiveness to oxygen is independent of factors such as pH, oxidants, reductants, and other paramagnetic materials except for paramagnetic gases such as O_2 and NO. The particulates based on carbonaceous materials such as India ink, coals, and charcoals can have very large oxygen-induced changes in their EPR spectra and therefore are especially useful for measuring low concentrations of oxygen. Lithium phthalocyanine (LiPc) is especially useful for measuring higher concentrations of oxygen such as occur in brain, because the effects of oxygen on their EPR spectra are linear and of a magnitude that can readily be measured accurately through the range of 0-60 torr or more [14]. LiPc crystals equilibrate with local P_tO_2 within 30 sec, and the responses of the line width to changes in P_tO_2 are stable for at least 30 days in the brain [15]. The high density of unpaired spins, combined with a narrow intrinsic line width of LiPc (~15 m gauss in the absence of oxygen), allow measurements of pO_2 in the brain by one or more crystals with a diameter of approximately 0.2 mm. Because of these characteristics, combined with its stability on the CNS, LiPc is often the material of choice for EPR oximetry in the brain.

2. MATERIALS AND METHODS

The experimental animals were male Sprague-Dawley rats weighing 350-450 g with 6 rats per group with 5 different anesthetic agents: ketamine / xylazine (100 mg/10 mg/kg IM, pentobarbital (80 mg/kg IP), Urethane/alpha chloralose (1250 mg/50 mg/kg, isoflurane (2.2%), and halothane (1.5%).

Approximately 7 days prior to the experiments with the anesthetics, LiPc crystals that were synthesized in our laboratory were placed directly via a spinal needle into the brain at a depth of 3 mm from the surface of the skull through 1 mm drilled holes located 4.5 mm from the midline and 1 mm in front of the bregma, using stereotactic techniques. After an adequate level of anesthesia was achieved, a tracheostomy tube was inserted and positive pressure ventilation started with continuous monitoring of inspiratory and expiratory pO_2 and pCO_2 and concentrations of inhalation anesthetics. A polyethylene arterial catheter was placed in the left femoral artery for continuous monitoring of BP and periodic blood gas measurements. After each sample was withdrawn, 0.5 ml of saline containing 0.006% heparin was infused into the catheters to prevent clotting; to avoid dehydration an additional 1.0 ml/hr of saline was infused. Body temperature was maintained at 37° on a heated pad (temperature monitored by a rectal probe). FIO_2 was maintained at 33% during surgical manipulation and during the pO_2 measurement in the brain (approximately 30 minutes). The ventilatory volume was continuously adjusted to maintain the measured pCO_2 at 35-40 torr. At the end of the experiment, the rats were sacrificed and it was confirmed that LiPc was implanted into a suitable site of the cerebral cortex. Gross and microscopic examination of the tissue around the implanted LiPc confirmed the absence of significant local perturbations from the LiPc.

The spectra of LiPc were obtained using an EPR spectrometer constructed in our laboratory with a low frequency (1.2 GHz, "L-band") microwave bridge. [16]. The rat was placed in the magnet and the head was positioned so that the brain was directly under the extended loop resonator that was adjusted to obtain the maximum signal from the LiPc in the cerebral cortex. Typical settings for the spectrometer were: incident microwave power, 10 mW; magnetic field center, 425 gauss; scan range, 1 gauss; modulation frequency, 27 kHz. Modulation amplitude was set at less than one-third of the EPR line width. Scan time was 2 min. 3-5 scans were usually averaged to achieve a better signal to noise ratio for the spectra.

3. RESULTS & DISCUSSION

The different anesthetics resulted in very different patterns of response of the P_tO_2 in the brain (Table 1). The values of P_tO_2 in the brain were lower for the injectable anesthetics. These variations could not be explained simply on the basis of the arterial pO_2, pCO_2, ventilation, blood pressure, or heart rate. For example, the mean blood pressure was the lowest in the halothane group, while the associated value for P_tO_2 was one of the highest.

Table 1. Effects of anesthetic agents on physiological variables

Drugs	n	P_tO_2 (mm Hg)	MBP (mm Hg)	PaO_2 (mm Hg)	$PaCO_2$ (mm Hg)	pH
KT/XYL	5	6.7±1.9	86.4±3.8	115.9±9.6	35.8±2.5	7.43±0.03
PB	5	13.9±3.3	106.1±2.4	116.2±15.4	38.9±2.6	7.44±0.01
UT/CH	6	16.0±4.5	93.1±1.2	127.1±6.2	38.8±1.7	7.40±0.03
HT	5	22.6±1.1*	66.2±1.2	121.8±18.1	39.5±1.9	7.38±0.01
IF	5	44.6±5.1*°×+	69.5±1.5	131.8±18.2	36.3±2.2	7.39±0.01

KT/XYL, Ketamine/xylazine; PB, Pentobarbital; UT/CH, Urethane/chloralose; HT, Halothane; IF, isofluorane.
Mean different $p<0.05$ from KT/XYL *, from PB °, from UT/CH ×, from HT + (1 way ANOVA). Values are means ± SE, FiO_2=33%.

The P_tO_2 values under halothane anesthesia are similar to those previously reported, when the animals were breathing spontaneously [7,15]. The current results show higher values of P_tO_2 when using isoflurane or halothane vs. injectable anesthetics, a result that also is consistent with a previous report from this laboratory [15]. The current study, however, shows a much greater difference in brain P_tO_2 between halothane and isoflurane. It may be that the level of anesthesia was different in the current study as increased desflourane (similar to isoflurane) causes metabolic depression and graded increases in P_tO_2 [4].

Histological studies have shown that there is little or no inflammation, reactivity or hemorrhage 3 days after the insertion of the LiPc. In addition, viable cells are in immediate contact with the LiPc and healthy appearing vasculature occurs in close proximity [17,18].

Different techniques for measuring tissue oxygenation have different strengths and weaknesses [19]. We note here some of the potential advantages of the technique used in this study. For many purposes, the possibility of allowing for tissue recovery after implanting the oxygen sensitive material provides a significant advantage over the more acute nature of microelectrode studies. Very important for the study of anesthetics is the fact that measurements can be made on non-anesthetized (restrained) animals, allowing one to obtain control values from an unanesthetized group. The measurements are from the same site in the brain and therefore can provide useful data tracking changes over a period of time. The measurements can be made repeatedly as needed on the same day and/or on subsequent days.

A key consideration in any measurement of P_tO_2 or $[O_2]$ is what volume is being sampled [20]. The extracellular site of the EPR material in this study results in the measurements being from the interstitium of the brain, rather than the circulatory system. With this type of EPR oximetry, the LiPc crystal is responding to the P_tO_2 in the immediate vicinity of the crystal. The volume of the brain that this represents beyond that immediately adjacent to the crystal depends on whether there are gradients in the local P_tO_2. Such considerations, of course, are common to virtually all types of measurements of oxygen in tissues.

Because the crystal or implant is in the range of 200 μm, the volume being studied can be very heterogeneous in P_tO_2. Lubbers clearly showed that there can be large gradations of P_tO_2 over microns [21]. Due to the volume occupied by the EPR-sensitive material, we propose that the values obtained with LiPc are similar to values of mean P_tO_2 as measured by a microelectrode.

In summary, EPR oximetry appears to be well suited for repeated direct measurements of P_tO_2. This capability is well suited to study the influence of anesthetics on P_tO_2. The values of P_tO_2 in the brain are not inferable from usual physiological measurements and therefore it seems useful to measure these directly. The data in this report indicate that inhaled anesthetics are associated with higher values of P_tO_2 than are injected anesthetics.

ACKNOWLEDGEMENTS

This study was supported by NIH Grants PO1 GM51630 and R01 CA67431, and used the facilities of the EPR Center for Viable Biological Studies, supported by NIH Grant P41 RR11602.

REFERENCES

1. Miller AT Jr., Curtin KE, Shen AL; Suiter CK. Brain oxygenation in the rat during hyperventilation with air and with low O_2 mixtures. Am J Physiol 1970;219:798-801.
2. Dings J., Meixensberger J., Amschler J, Roosen K. Continuous monitoring of brain tissue pO_2: a new tool to minimize the risk of ischemia caused by hyperventilation therapy. Zentralblatt fur Neurochirurgie 1996;57:177-183.
3. Liu KJ., Bacic G, Hoopes PJ, Jiang J, Du HK, Lou LC, Dunn JF, Swartz HM. Assessment of cerebral pO_2 by EPR oximetry in rodents: Effects of anesthesia, ischemia, and breathing gas. Brain Res 1995;685:91-98.
4. Hoffman, WE., Charbel, FT, Edelman, G. Desflurane increases brain tissue oxygenation and pH. Acta Anaesth Scand 1997a;41:1162-1166.

5. Hoffman WE, Charbel FT, Edelman G, Ausman JI. Thiopental and desflurane treatment for brain protection. Neurosurg 1998;43:1050-1053.
6. Murr R, Schurer Berger S, Enzenbach R, Peter K, Baethmann A. Effects of isoflurane, fentanyl, or thiopental anesthesia on regional cerebral blood flow and brain surface PO_2 in the presence of a focal lesion in rabbits. Anesth Analg 1993;77:898-907.
7. Charbel FT, Hoffman WE, Misra M, Hannigan K, Ausman JI. Cerebral interstitial tissue oxygen tension, pH, HCO_3, CO_2. Surg Neurol 1997;48:414-417.
8. Hoffman WE., Charbel FT, Abood C, Ausman, JI. Regional ischemia during cerebral bypass surgery. Surg Neurol 1997b;47:455-459.
9. Swartz HM, Walczak T. Developing *In Vivo* EPR Oximetry for Clinical Use. Adv Exp Med Biol 1998;454:243-252.
10. Swartz HM. *In vivo* oximetry. Berliner, LJ, editor. In: Biological Magnetic Resonance: Volume 20 *In Vivo* EPR (ESR) theory and applications. New York: Plenum Publishing. 2000.in press.
11. Swartz HM, Clarkson, RB. The measurement of oxygen *In Vivo* using EPR techniques. Phys Med Biol 1998;43:1957-1975.
12. Swartz HM, Halpern H. EPR studies of living animals and related model systems. (*In Vivo* EPR). In: Berliner, LJ, editor. Spin Labeling: The Next Millennium, New York: Plenum. 1998;367-404.
13. Swartz HM, Liu KJ, Goda F, Walczak T. India ink: A potential clinically applicable EPR oximetry probe. Magn Reson Med 1994;31:229-232.
14. Liu KJ, Gast Moussavi M, Norby SW, Vahidi N, Walczak Wu, M; Swartz HM. Lithium phthalocyanine: A probe for electron paramagnetic resonance oximetry in viable biological system Proc Nat Acad Sci USA 1993;90:5438-5442.
15. Liu KJ, Hoopes PJ, Rolett EL, Beerle BJ, Azzawi A, Goda F, Dunn JF; Swartz HM. Effect of anesthesia on cerebral tissue oxygen and cardiopulmonary parameters in rats. Adv Exp Med Biol 1997;411:33-39.
16. Nilges MJ, Walczak T, Swartz HM. 1 GHz *in vivo* ESR spectrometer operating with a surface probe. Phys. Med. 1989;5:195-201.
17. Hoopes PJ, Liu KJ., Bacic G, Rolett EL, Dunn JF, Swartz HM. Histological assessment of rodent CNS tissues to EPR oximetry probe material. Adv Exp Med Biol 1995;411:13-21.
18. Rolett EL, Azzawi A, Liu KJ, Yongbi, MN, Swartz HM, Dunn, JF. Critical oxygen tension in rat brain: a combined 31P-NMR and EPR oximetry study. Am J Physiol 2000. In press.
19. Swartz HM, Dunn JF. Measurements of oxygen in tissues: overview and perspectives on methods. (this volume).
20. Swartz HM, Dunn JF, Grinberg O, O'Hara J, Walczak, T. What does EPR oximetry with solid particles measure--and how does this relate to other measures of pO_2?. Adv Exp Med Biol 1997;428:663-670.
21. Lubbers D.W. The meaning of the tissue oxygen distribution curve and its measurement by means of Pt electrodes. Prog in Resp Res 1969;3:112-123.

Chapter **56**

CAPILLARIZATION AND VASCULAR ENDOTHELIAL GROWTH FACTOR EXPRESSION IN HYPERTROPHYING ANTERIOR LATISSIMUS DORSI MUSCLE OF THE JAPANESE QUAIL

Hans Degens[1], Rebecca K. Anderson[2] & Stephen E. Alway[3]
[1]*Dept. Physiology, University of Nijmegen, 6500 HB Nijmegen, The Netherlands;* [2]*Dept. Biomedical Engineering, University of Florida, 231 Aerospace Building, Gainesville, FL 32611 USA;* [3]*Laboratory of Muscle, Sarcopenia, and Muscle Diseases, Div. Exercise Physiology, West Virginia University School of Medicine, Morgantown, WV 26506-9227 USA*

Abstract: Hypertrophy may increase the diffusion distances from capillaries to the interior of the muscle fibers. We hypothesized that capillary proliferation occurs during hypertrophy, which is accompanied by an up-regulation of vascular endothelial growth factor (VEGF). Hypertrophy of the left anterior latissimus dorsi muscle of Japanese quail (2–3 months old) was induced by 1-4 week stretch-overload. Capillarization was analyzed in cross-sections stained for ATPase. VEGF expression was determined with RT-PCR. Initially, hypertrophy was not accompanied by increases in fiber cross-sectional area (FCSA), but after 1 week the average FCSA did increase. The capillary to fiber ratio was decreased after 1 week, but returned to control values in subsequent weeks. This indicates that capillary proliferation occurred, because this model is characterized by extensive fiber proliferation. The absence of any significant change in VEGF mRNA levels indicates that increased levels of VEGF mRNA are not crucial for capillary proliferation during muscle hypertrophy.

Key words: capillary domains, RT-PCR, VEGF

1. INTRODUCTION

Both proliferation and hypertrophy of muscle fibers contribute to stretch-induced hypertrophy in the anterior latissimus dorsi (ALD) muscle of the quail [1,2]. In the absence of capillary proliferation, the hypertrophy would result in increased diffusion distances from the capillaries to the interior of the muscle fibers, thus compromising the oxygenation of the muscle. This may ultimately result in anoxic areas and affect muscle function. However, both in muscles from rats [3,4] and chickens [5,6] capillary proliferation takes place during hypertrophy to compensate for the increase in diffusion distances.

Vascular endothelial growth factor (VEGF) is a specific mitogen for endothelial cells. It plays a crucial role in angiogenesis both during embryonic development and in several conditions with angiogenesis in the adult organism [7]. Also in quail embryos VEGF expression in several tissues, including skeletal muscle, coincided temporally and spatially with angiogenesis [8]. In adult skeletal muscle, VEGF expression correlates with capillary proliferation in chronically stimulated muscles [9,10]. Consequently, it may well be that stretch will also induce an increased VEGF expression in hypertrophying skeletal muscle preceding or accompanying capillary proliferation. In addition, since an adequate capillary supply is important for muscle function, it may be that capillary proliferation precedes or concurs with the increase in fiber size and fiber proliferation during hypertrophy.

The aim of the present study was to determinc thc time course of capillary proliferation and VEGF expression during the development of stretch-induced hypertrophy. We hypothesized that an increase in VEGF expression precedes or occurs simultaneously with capillary proliferation, which in turn precedes or coincides with muscle fiber hypertrophy. To evaluate this we used quail in which hypertrophy of the ALD muscle was induced by stretch-overload as described previously [1,2].

2. METHODS

2.1 Animals and materials

Japanese quail of 2-3 months and weighing on average 179 g were used. The animals were housed in groups at a 12 h light : 12 h dark cycle. Food and water were provided *ad libitum*. The procedure to obtain stretch-induced hypertrophy of the ALD muscle has been described previously [1]. Briefly, a weight corresponding to 10 – 14% of the body weight was placed around the left wing. The right wing served as an internal control. The birds were randomly selected to carry the weight for 1, 2, 3 or 4 weeks. Then the birds were sacrificed by an overdose of pentobarbital. ALD muscles were excised and the mid-belly was positioned between sections of the pectoralis muscle for histochemistry. The remaining parts of the ALD were used for RNA isolation. All tissue was quickly frozen in liquid nitrogen, and stored at -80°C until use. The Animal Review and Use Ethical Committee of the University of South Florida approved all procedures.

2.2 RT-PCR

RNA was isolated from the ALD muscles using TriReagent [11,12]. RT-PCR was then performed using 1-2 μg of the total RNA as described previously [12]. The GAPDH primers were described earlier [12]. The upper and lower primers for VEGF were 5'-CGGGTTGCTGCGGCGATG-A-3' and 3'-GACCCTTCCCCTTTCCTCGCTTT-5', respectively. The annealing temperature was 54.5°C and the PCR went for 33 cycles with both primer sets in the same tube. GAPDH served as an internal control. The ratio of VEGF:GAPDH was identified by densitometry from ethidium bromide stained 1.5% agarose gels (ImageQuant; Molecular Dynamics, Sunnyvale, CA). The identity of the PCR product was verified by determination of the size of the DNA fragments after digestion with the restriction enzyme AluI.

2.3 Histochemistry

Cross-sections (12 μm) of the ALD were cut at -20°C. The sections were stored at -80°C until staining for ATPase to depict capillaries [see 5].

2.4 Analysis

Muscle capillarization was analyzed from photomicrographs of the ALD by the method of capillary domains [3,13]. Using a digitizing tablet, fiber outlines were read into the computer as contour coordinates and capillary locations as coordinates of the capillary centers. The capillary density (CD) was defined as the number of capillaries per square millimeter of tissue (muscle fibers plus intercellular space). Capillary domains (DOM) were defined as the area surrounding a capillary delineated by equi-distant boundaries from adjacent ones and their surface area and radius (R) calculated. DOM areas have a log-normal distribution. Therefore, the logarithmic standard deviation of the DOM areas estimates the heterogeneity of capillary spacing. Fiber cross-sectional areas (FCSA) were derived from complete fiber contours on these photographs. The DOM to fiber ratio was calculated and presented as the capillary to fiber ratio (C/F) and the number of DOMs overlapping a fiber is given as DAF (DOMs around a fiber), which is similar to the commonly used parameter 'capillaries around a fiber'. The local capillary fiber ratio (LCFR) was defined as the sum of the fractions of each DOM area overlapping the fiber. DAF or LCFR divided by the FCSA provides the capillary density for the particular fiber, DAF/FCSA and the capillary fiber density (CFD) respectively.

2.5 Statistics

Differences with controls were identified with a paired t-test. Group means of 1, 2, 3 and 4 weeks overload were compared by ANOVA. Transformation of the data was performed where appropriate. Differences were considered significant at $p < 0.05$.

Table 1 Animal and muscle characteristics, and parameters of capillary supply in the anterior latissimus dorsi muscle of the quail after 0, 1, 2, 3 or 4 weeks stretch-overload.

Weeks	0	1	2	3	4
ALD/BW ($mg\ g^{-1}$)	0.30 ± 0.02 (28)	0.47 ± 0.04$^{*,\&}$ (8)	0.59 ± 0.05* (6)	0.81 ± 0.05* (8)	0.84 ± 0.06$^{*,@}$ (6)
MRNA ($mg\ g^{-1}$)	0.77 ± 0.06 (21)	0.74 ± 0.19 (2)	0.86 ± 0.11 (5)	0.75 ± 0.21 (5)	1.38 ± 0.22 (5)
CT %	19 ± 1 (28)	25 ± 4 (8)	22 ± 6 (6)	22 ± 3 (8)	30 ± 3 (6)
R (μm)	22.0 ± 0.4 (28)	28.6 ± 2.3 (8)	29.9 ± 1.6* (6)	28.8 ± 1.3* (8)	29.2 ± 1.6* (6)
LogSD	0.157 ± 0.003 (28)	0.173 ± 0.011 (8)	0.161 ± 0.014 (6)	0.166 ± 0.011 (8)	0.191 ± 0.019 (6)
LCFR	1.23 ± 0.05 (28)	0.75 ± 0.09 (8)*	0.95 ± 0.13 (6)	1.19 ± 0.11 (8)	0.84 ± 0.16 (6)
CFD ($N\ mm^{-2}$)	703 ± 25 (28)	479 ± 92 (8)	383 ± 39* (6)	435 ± 31* (8)	421 ± 34* (6)
DAF	4.74 ± 0.10 (28)	3.68 ± 0.24* (8)	4.06 ± 0.29 (6)	4.65 ± 0.25 (8)	3.90 ± 0.43 (6)
DAF/FCSA ($N\ mm^{-2}$)	2986 ± 97 (28)	2638 ± 350 (8)	2286 ± 188$^{*,\&}$ (6)	2749 ± 519 (8)	3951 ± 484$^{\$}$ (6)

BW: body weight; ALD: anterior latissimus dorsi muscle; CT: connective tissue; R: radius of the Krogh cylinder surrounding a capillary, or estimation of maximal diffusion distance from a capillary; LogSD: logarithmic standard deviation of the capillary domain areas, as an index of heterogeneity of capillary spacing; LCFR: local capillary to fiber ratio; CFD: LCFR per fiber cross-sectional area; DAF: domains supplying a fiber, similar to capillaries around a fiber; DAF/FCSA: DAF per fiber cross-sectional area; mean ± SEM (n); *: different from 0 week; @: different from 1 week; $: different from 2 week; &: different from 4 weeks

3. RESULTS

The ALD hypertrophied throughout the entire 4-week period as reflected in a continuing increase of the ALD/BW ratio (Table 1). This was at least partly a result of increases in the FCSA (Fig. 1a). The standard deviation of the FCSA also increases over time (Fig. 1b), following roughly the same pattern as the increase in fiber number observed previously [1,2]. This may be an indirect indicator of fiber hyperplasia. The amount of non-contractile tissue did not show a significant change (Table 1).

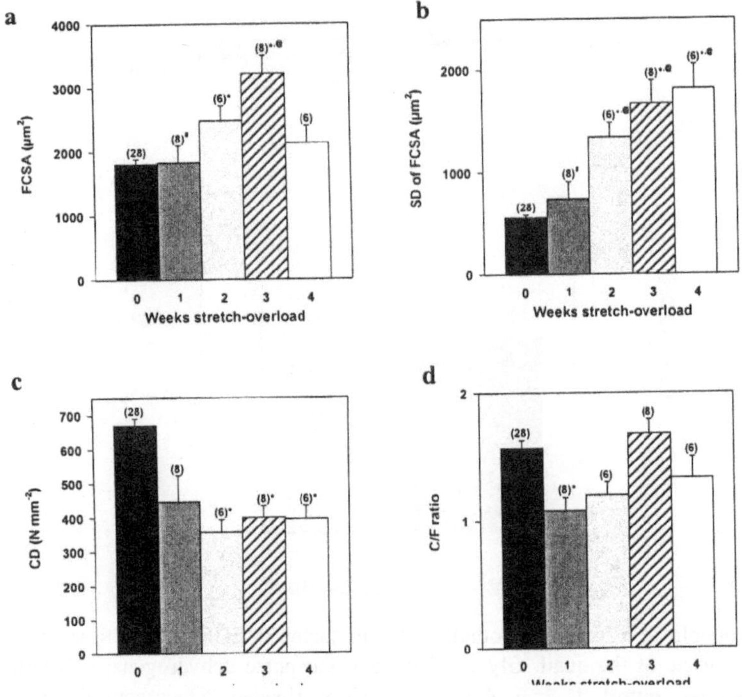

Figure 1. a) Fiber cross-sectional area (FCSA), **b)** standard deviation of the FCSA, **c)** capillary density (CD) and **d)** capillary to fiber ratio (C/F) in 0, 1, 2, 3 or 4 weeks stretch-overloaded anterior latissimus dorsi muscle of the quail. Number of observation between parentheses; *: different form 0 weeks; @: different from 1 week; #: different from 3 weeks.

The decrease in CD (Fig. 1c) during the first week was caused by a combination of increases in FCSA (Fig. 1a), fiber proliferation, as reflected by an increase in the SD of the FCSA (Fig. 1b), and a decrease in C/F (Fig. 1d). The absence of a significant decrease in C/F after 2 weeks combined with the continuing fiber hyperplasia [1], as indicated by a progressive increase in the SD of the FCSA, indicates capillary proliferation. The capillary proliferation was adequate to prevent a significant increase in the heterogeneity of capillary spacing, but lagged behind the muscle hypertrophy, since CD did not return to control levels. As a result the diffusion distances were increased in stretch-overloaded muscles (Table 1). These patterns are confirmed by other parameters of capillary supply (see Table 1).

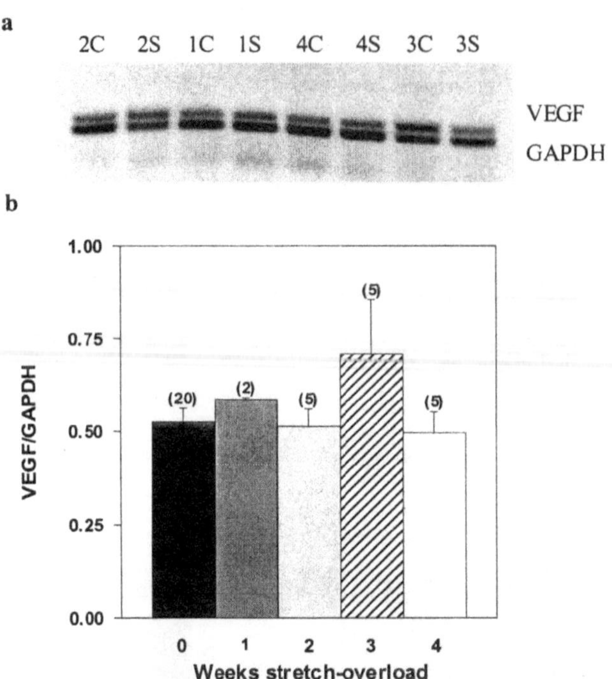

Figure 2. a) Example of vascular endothelial growth factor (VEGF) RT-PCR in anterior latissimus dorsi muscle of the quail. Glyceraldehyde-3-phosphate dehydrogenase (GAPDH) served as an internal control. 1, 2, 3, 4 week stretch-overloaded (S) or contralateral (C) anterior latissimus dorsi muscle muscle of the quail, **b)** VEGF/GAPDH ratios in 0, 1, 2, 3, 4 week stretch-overloaded muscles.

The relative mRNA content did not change during stretch-overload (Table 1). The VEGF/GAPDH ratio was also unchanged (Fig. 2). However, in 3 preliminary experiments of 1-week stretch-overloaded ALD muscles with another primer set for VEGF, the VEGF/GAPDH ratio was increased in all of them. This was also found in 1 of the 2 with the normal primer set. Inclusion of these preliminary data did, however, not result in significance.

4. DISCUSSION

The primary new finding of the present study is the occurrence of capillary proliferation during the development of stretch induced hypertrophy, which is not accompanied or preceded by an increased relative presence of VEGF mRNA.

Stretch-overload resulted in a significant hypertrophy of the ALD muscle. This hypertrophy is at least partly the result of fiber hypertrophy, but also fiber hyperplasia is known to contribute significantly [1,2]. Fiber hyperplasia already occurs during the first week of the development of hypertrophy [2] and results in a 52% increase in fiber number after 4 weeks [1]. This fiber proliferation results in a significant increase in the SD of the FCSA [1] and the progressive increase in the SD of the FCSA in the present study thus indirectly reflects fiber proliferation.

Compensatory hypertrophy in the rat plantaris muscle [3,4] and stretch-induced hypertrophy of chicken ALD muscles [5,6] are accompanied by capillary proliferation, which lags behind the increases in FCSA. In those studies the occurrence of capillary proliferation was indicated by an increase in the C/F ratio. In the present study the C/F ratio was decreased after 1 week, but similar to the control situation later on. The C/F ratio as such thus does not seem to indicate capillary proliferation. However, the maintained C/F ratio combined with the increase in fiber number indicates that capillary proliferation does occur in this model of hypertrophy. Also in this model the proliferation lags behind the magnitude of hypertrophy, but only during the first 2 weeks, whereafter the capillary proliferation appears to be sufficient to prevent further decreases in CD.

Previous studies have shown increases in C/F that are preceded by increases in VEGF mRNA [10] and VEGF protein [9] after chronic electrical stimulation. However, we did not find an up-regulation of VEGF preceding or coinciding with the capillary proliferation during the development of hypertrophy of the ALD. However, during chronic stimulation the metabolic demand and muscle blood flow are increased and the muscle may at least temporarily become hypoxic. Each of these factors is an important stimulus for VEGF expression and capillary proliferation [14,15]. During hypertrophy, however, muscular activity [5] and blood flow [16] are not increased. Apparently, the stimuli for capillary proliferation during chronic electrical stimulation and the development of hypertrophy are different. Accordingly, it has been suggested that capillary proliferation during stretch-induced hypertrophy may be caused directly by stretch-induced deformations of the endothelial cells, causing an increased intracellular [Ca^{2+}], which is an important signal for the stimulation of proliferation [15,16].

Although we did not find an increased expression of VEGF mRNA as expressed per mg of muscle tissue, the absolute amount of VEGF mRNA is increased. In addition, our data does not exclude the possibility that the VEGF protein level is increased despite an unchanged VEGF mRNA concentration. In line with this suggestion, it has been reported that both VEGF transcription and translation can be regulated [17]. Also, the sensitivity of the endothelial cells may have increased (e.g. an increased number of VEGF receptors). Clearly, these possibilities deserve further investigation.

5. CONCLUSION

In conclusion, our study confirms that capillary proliferation occurs during the development of hypertrophy. Our data do not support, however, the hypothesis that an increased expression of VEGF is crucial for this capillary proliferation.

ACKNOWLEDGEMENTS

We are grateful for the technical assistance of Dr. Dawn A. Lowe, Jo Ann Moore, Paul Llobet and James Ignatius. This work was supported by the National Institute on Aging Grant AG-17143.

REFERENCES

1. Alway SE, Winchester PK, Davis ME, Gonyea WJ. Regionalized adaptations and muscle fiber proliferation in stretch-induced enlargement. J Appl Physiol 1989;66:771-781.
2. Carson JA, Alway SE, Yamaguchi M. Time course of hypertrophic adaptations of the anterior latissiumus dorsi muscle to stretch overload in aged Japanese Quail. J Geront 1995;50A:B391-B398.
3. Degens H, Turek Z, Hoofd LJC, van 't Hof MA, Binkhorst RA. The relationship between capillarisation and fibre types during compensatory hypertrophy of the plantaris muscle in the rat. J Anat 1992;180:455-463.
4. Plyley MJ, Olmstead BJ, Noble EG. Time course of changes in capillarization in hypertrophied rat plantaris muscle. J Appl Physiol 1998;84:902-907.
5. Holly RG, Barnett JG, Ashmore CR, Taylor RG, Mole PA. Stretch-induced growth in chicken wing muscles: a new model of stretch hypertrophy. Am J Physiol 1980;238:C62-C71.
6. Snyder GK. Capillary growth in chick skeletal muscle with normal maturation and hypertrophy. Resp Physiol 1995;102:293-301.
7. Ferrara N, Davis-Smyth T. The biology of vascular endothelial growth factor. Endocr Rev 1997;18:4-25.
8. Aitkenhead M, Christ B, Eichmann A, Feucht M, Wilson DJ, Wilting J. Paracrine and autocrine regulation of vascular endothelial growth factor during tissue differentiation in the quail. Dev Dyn 1998;212:1-13.

9. Annex BH, Torgan CE, Lin P, Taylor DA, Thompson MA, Peters KG, Kraus WE. Induction and maintenance of increased VEGF protein by chronic motor nerve stimulation in skeletal muscle. Am J Physiol 1998;274:H860-H867.
10. Škorjanc D, Jaschinski F, Heine G, Pette D. Sequential increases in capillary proliferation and mitochondrial enzymes in low-frequency-stimulated rabbit muscle. Am J Physiol 1998;274:C810-C818.
11. Chomczynski P, Sacchi N. Single step method of RNA isolation by acid guanidinium thiocyanate-phenol-chloroform extraction. Anal Biochem 1987;162:156-159.
12. Lowe DA, Lund T, Alway SE. Hypertrophy-stimulated myogenic regulatory factor mRNA increases are attenuated in fast muscle of aged quail. Am J Physiol 1998;275:C155-C162.
13. Hoofd L, Turek Z, Kubat K, Ringnalda BEM, Kazda S. Variability of intercapillary distance estimated on histological sections of rat heart. Adv Exp Biol 1985;191:239-247.
14. Breen EC, Johnson EC, Wagner H, Tseng H-M, Sung LA, Wagner PD. Angiogenic growth factor mRNA responses in muscle to a single bout of exercise. J Appl Physiol 1996;81:355-361.
15. Hudlicka O, Brown M, Egginton S. Angiogenesis in skeletal and cardiac muscle. Physiol Rev 1992;72:369-417.
16. Egginton S, Hudlicka O, Brown MD, Walter H, Weiss JB, Bate A. Capillary growth in relation to blood flow and performance in overloaded rat skeletal muscle. J Appl Physiol 1998;85:2025-2032.
17. Akiri G, Nahari D, Finkelstein Y, Le S-Y, Elroy-Stein O, Levi B-Z. Regulation of vascular endothelial growth factor (VEGF) expression is mediated by internal initiation of translation and alternative initiation of transcription. Oncogene 1998;17:227-236.

Chapter 57

EXPRESSION OF PROLIFERATING CELL NUCLEAR ANTIGEN IN RAT HEARTS SUBJECTED TO TRANSIENT ISCHEMIA FOLLOWED BY REPERFUSION

Tomiyasu Koyama*, Zhonglin Xie, Jun'ichi Suzuki & Kazuhiro Abe
*Third Dept. of Anatomy, School of Medicine, Hokkaido University, 060 Sapporo, Japan; *064-0821 Sapporo, N-1 W-25 C,I. Ms516.*

Abstract: Early mechanisms involved in improving capillarity and oxygen transport to cardiac tissue exposed to transient coronary ischemia followed by reperfusion were studied in rats. Under ether anaesthesia, the left coronary artery was mechanically occluded for 3 min after which it was released, and the rats allowed to recover. After 2, 24 or 48 h the rats were sacrificed and the hearts frozen in liquid nitrogen. Frozen cross-sections were stained immunohistochemically for proliferating cell nuclear antigen (PCNA) and for the growth factors, VEGF and bFGF. No reaction for PCNA was seen in sections of sham-operated hearts but an inhomogeneous reaction occurred in annular structures in the occluded hearts at 48 h reperfusion. The stain appeared to be located in proliferating nuclei, and in the cytosol of endothelial cells. It is suggested that PCNA is stimulated by the increase in growth factors that is known to occur within 2 h after the end of the coronary occlusion. It is concluded that the increase in capillarity, indicated by the nuclear proliferation of endothelial cells, will improve the transport of oxygen to the cardiac tissues.

Key words: heart, ischemia/reperfusion, proliferating cell nuclear antigen, capillarity, vascular growth factors

1. INTRODUCTION

In previous studies on rat hearts the effects on ventricular capillarity produced by transient ischemia have been investigated. Ischemia was induced either by intravenous injection of vasopressin [1] or by mechanical occlusion of the left coronary artery [2] and was followed by reperfusion. Both procedures produced an increase in capillary density in the left ventricle tissue within one month; this change would increase the oxygen supply to the ventricular tissues. These long-term effects on capillarity were preceded by expression of basic fibroblast growth factor (bFGF) and vascular endothelial growth factor (VEGF) in the early phase of reperfusion, immunohistological staining for both proteins being increased within 2 h of the vasopressin injection or of the onset of reperfusion following the mechanical occlusion. While it is possible that these growth factors are involved in endothelial cell proliferation, direct evidence of proliferation and of capillary neogenesis has not been available.

In the present immunohistochemical investigation we examined the expression of proliferating cell nuclear antigen (PCNA), which is involved in DNA synthesis and functions as a cofactor for DNA polymerase-δ [3]. PCNA has been used as a marker for regenerating cells in brain tissue [4], proliferation of tumor cells [5], and uterine endothelial cells [6] and of changes in capillarity in thyroid-treated rat hearts [7]. PCNA staining can thus be expected to provide evidence for endothelial cell proliferation in rat hearts subjected to mechanical ischemia/reperfusion.

2. METHODS AND MATERIALS

Twelve male 9-week-old Wistar rats given standard rat chow and water ad libitum were used as ischemia/reperfused group (n=9) or the Sham-operated (n=3) group. All procedures followed the institutional guidelines for care and use of laboratory animals. Rats were anesthetized with ethyl ether, intubated with a polyethylene tube and mechanically ventilated with a rodent ventilator using room air, being added intermittently as a low concentration of ethyl ether as necessary. The surgical procedures and method of coronary occlusion have been described previously [2]. In brief, a left thoracotomy was performed via the fourth intercostal space, and the pericardium opened. The left atrial appendage was raised and a small curved needle, threaded with fine silk, was passed through the ventricular myocardium to encompass the left coronary artery. The thread was then tied in an overhand knot (an occluder). Two other threads were tied to the main knot (releasers) [8]. The ligature could be tightened or loosened by pulling on the relevant threads. The left coronary artery was occluded for 3 min, and the ligature then released. In the Sham-operated group a thread was looped around the coronary artery but not tightened. After the treatment the left lung was inflated, the pneumothorax evacuated and the chest closed. Three rats of the occluded group, concomitantly one Sham-operated rat,

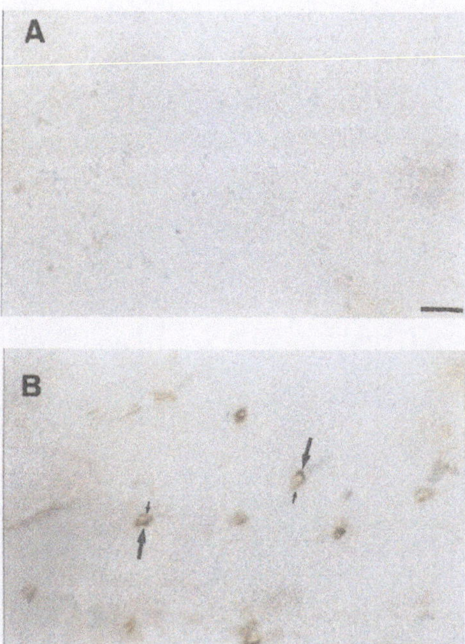

Figure 1. Micrographs of transverse sections of rat hearts. Sections stained brown by an immunohistochemical reaction for PCNA, and counterstained with methyl green. (A) Section of the subendocardium of a Sham-operated rat heart. (B) Similar section from heart subjected to a 3-min coronary occlusion followed by a 48 h reperfusion. Thick arrows: rings with area of strong stain; thin arrows: rings with fainter but uniform staining. Horizontal bar - 20 μm.

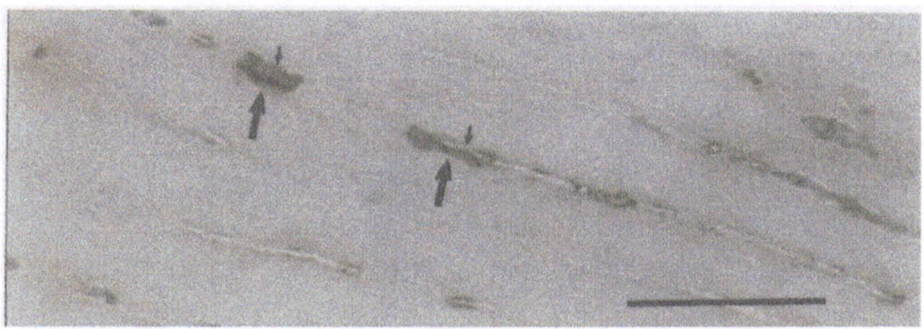

Figure 2. Micrograph similar to those in Figure 1 but the fields were selected from the middle layer of the left ventricular wall, where the plane of section of the tissue is longitudinal rather than transverse. Note the heavily stained elongated elements (thick arrows) in continuity with lightly stained areas (thin arrows). Horizontal bar: 20 μm.

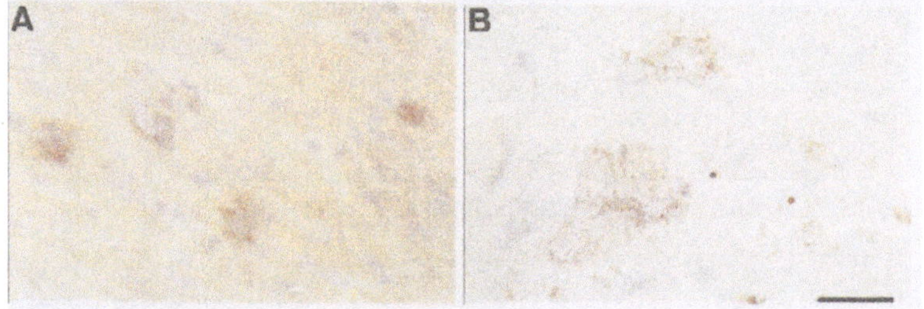

Figure 3. Immunohistochemical staining for VEGF (A) and bFGF (B), visualized brownish red with anti-mouse IgG and aminoethylcarbasol (AEC) with a counterstaining using hematoxylin solution. Sections were obtained from the same hearts (48 h reperfused) as in Figure 1. Horizontal bar: 20 μm.

were sacrificed by decapitation with a guillotine at 2, 24 or 48 h for each time point after the operation. The heart was removed, dipped in optimum cutting temperature compound (O.C.T. compound, Miles Inc., USA) and frozen in liquid nitrogen at –80C°.

Using a cryotome, tissue sections 6 µm thick, were cut transversely from the upper part of the ventricle. They were processed immunohistochemically for PCNA using a mouse monoclonal anti-PCNA antibody (DAKO-PCNA: PC10, USA). The procedure was as described by Heron and Rakusan [7] except that the dilution was 1:50; the reaction was visualized with diaminobenzidine which gave a brown coloration. Methyl green was then used to counterstain nuclei light green. Staining for VEGF and bFGF was as described previously [1,2] using anti-mouse IgG and aminoethylcarbasol.

3. RESULTS

In the Sham-operated hearts no PCNA staining was apparent at either 24 or 48 h (Figure 1A). Hearts from the ischemic group showed some weak reaction at 24 h but at 48 h brown-stained annular structures, presumed to be capillaries, were present among the counterstained nuclei in the subendocardial layer (Figure 1B). Some of the rings (indicated by thin arrows) stained uniformly lightly but others (thick arrows) showed a crescent-shaped thickening that appeared to be a darkly stained nucleus; staining in the rest of the ring was fainter. It is noteworthy that the staining was exclusive to the annular structures, suggesting that no cell types other than capillary endothelial cells were proliferating as a consequence of ischemia.

Figure 2 shows PCNA staining in the middle layer of the ventricular wall. Since cardiomyocytes run circumferentially in this layer, the capillaries are likely to be cut longitudinally rather than transversely. In these sections, elongated areas of dark staining (thick arrows) are continuous with more diffuse areas of faint staining (thin arrows).

Staining for VEGF and bFGF at 48 h is shown in Figure 3. VEGF is associated with capillaries, and bFGF is diffusely distributed over cardiomyocytes. At 24 h also a clear staining was obtained but not shown.

4. DISCUSSION

Since PCNA staining is associated with nuclei, we expected any reaction in proliferating capillaries to be confined to crescent-shaped elements corresponding to cross-sections of endothelial cell nuclei. Contrary to this expectation, we frequently found that the stain in the strongly reacting crescent was continuous with a semi-circle of fainter staining in the transverse staining. These annular elements are similar to the PCNA-expressing capillaries in rat endometrium, whose identity was confirmed by lectin staining [6]. Furthermore, where tissue was cut longitudinally we observed well-stained elongated elements continuous with areas of fainter stain. This observation seems consistent with the micrograph of brain shown by Liu and Chen [4], where faint staining of cytosol is seen along with the capillaries together with strongly stained nuclei in capillary endothelial cells. Two forms of PCNA with different solubilities have been identified in 3T3 cells [3]. One form is highly soluble and the other tightly bound to sites of DNA replication; the soluble appears to be distributed in cytosol. We suggest that those circular images seen in some sections that were uniformly stained with no dark crescent probably represent capillary endothelial cells, the proliferating nuclei of which were out of the plane of section.

Heron and Rakusan [7] found a marked increase in the number of nuclei expressing PCNA in endothelial cells, cardiomyocytes, fibroblasts and other cells including smooth muscle in rat hearts exposed to one month of hypothyroidism followed by 6 days of hyperthyroidism. In our study, if cell types other than capillary endothelium were expressing PCNA, the stain structure would appear solid rather than ring-like. Since virtually no solidly stained elements were present in the transverse sections, it seems probable that it was the endothelial cells only that were proliferating. The brief period of occlusion was probably too short to induce proliferation in other cells.

In our previous studies [2] we demonstrated expression of VEGF and bFGF proteins as early as 2 h and as late as 30 days after reperfusion; our present finding of expression at 24 h and 48 h is consistent with our original observations.

In conclusion, the transient increase in oxygen demand caused by ischemia/reperfusion appears to be met by an increase in capillarity. It seems likely that the growth factors, which are known to be stimulated soon after the ischemia, are involved in the proliferation of endothelial cells indicated by the rise in PCNA. The finding that cell proliferation appears to be confined to the capillary endothelium indicates the important role of the capillaries in the recovery of the ischemic tissue.

REFERENCES

1. Xie ZL, Gao M, Batra S, Koyama T. Remodeling of the left ventricular network in left ventricular subendocardial tissue induced by intravenous vasopressin administration. Microcirculation 1997;4: 261-266.

2. Xie ZL, Gao M, Koyama T. Effects of transient coronary occlusion on the capillary network in the left ventricle of rat. Jpn J Physiol 1997;47:537-543.
3. Bravo R, Macdonald-Bravo H. Existence of two populations of cyclin/proliferating cell nuclear antigen during the cell cycle:association with DNA replication sites. J Cell Biol 1987;105:1549-1554.
4. Liu HM, Chen HH. Correlation between fibroblast growth factor expression and cell proliferation in experimental brain infarct: Studied with proliferating cell nuclear antigen immunohistochemistry. J Neuropathology and Experimental Neurology 1994;53: 118-126.
5. Louis DN, Edgerton S, Thor AD, Hedley-Whyte ET. Proliferating cell nuclear antigen and immunohistochemistry in brain tumors.: A comparative study. Acta Neuropathol 1991;81:675-679.
6. Goodger AM, Rogers PAW. Uterine endothelial cell proliferation before and after embryo implantation in rats. J Reprod Fert 1993;99:451-457.
7. Heron MI, Rakusan K. Proliferating cell nuclear antigen (PCNA) detection of cellular proliferation in hypothyroid and hyperthyroid rat hearts. J Mol Cell Cardiol 1995;27:1393-1403.
8. Himori N & Matsuura A. A simple technique for occlusion and reperfusion of coronary artery in conscious rats. Am J Physiol 1989;256:H1719-1725.

Chapter **58**

FIBROBLAST GROWTH FACTORS (FGFS) INCREASE BREAST TUMOR GROWTH RATE, METASTASES, BLOOD FLOW, AND OXYGENATION WITHOUT SIGNIFICANT CHANGE IN VASCULAR DENSITY

Paul Okunieff[1], Bruce M. Fenton[1], Lurong Zhang[2], Francis G. Kern[3], Timothy Wu[4], J. Robert Greg[4] & Ivan Ding[1]

[1]Department of Radiation Oncology, University of Rochester, Rochester, NY; [2]Department of Cellular and Molecular Biology, Georgetown University, Washington, DC; [3]Southern Research Institute, Birmingham, A; [4]Radiation Oncology Branch, National Cancer Institute, Bethesda, MD

Abstract: Breast tumors expressing no detectable FGFs (MCF-7) were compared with tumors transfected with FGF4 or FGF1 (FGF4/MCF-7 or FGF1/MCF-7), and with MDA-MB-435, which produce endogenous FGF2. Tumor blood flow was measured by ^{133}Xe diffusion, oxygen distribution was measured by Eppendorf pO_2 histography, and vascular density was measured by CD31 staining. Tumors that overexpress angiogenic factors grew at a rate far exceeding that of MCF-7. The FGF producing tumors also had much higher metastatic rates to lung. Tumor blood flow was significantly higher in the two FGF-transfected xenografts compared with the parent MCF7. Median tumor pO_2 was also higher, and tumor oxygenation was preserved even for large tumors. The vascular density as determined by CD31 staining, however, was not markedly increased in tumors overexpressing angiogenic factors. We found that angiogenic factors preserve and augment neovascular function, thus facilitating tumor growth and progression.

Key words: angiogenesis, hypoxia, fibroblast growth factor, blood flow, breast cancer

1. INTRODUCTION

The aggressive malignant phenotype of breast tumors has been attributed to the tumors' ability to generate neovasculature (1, 2). Angiogenic growth factors are the primary signaling peptides that lead to angiogenesis of both normal and neoplastic tissues (2). Among angiogenic factors, fibroblast growth factors (FGFs) are extremely effective at stimulating angiogenesis, facilitating metastases, and promoting tumor growth in both animal and clinical tumor studies (3-7). Neither the mechanism of these angiogenic-related growth factor mediated properties, nor the effect of angiogenic factors on tumor physiology are completely understood. Important questions remain regarding the relative growth effects of these angiogenic factors on the tumor cells themselves compared with their regulation of functioning vasculature. We therefore measured and compared the rate of tumor growth, lung metastasis, and physiological parameters including total blood vessels, blood flow, and oxygenation, in a series of human breast tumor xenografts growing in nude mice. Tumors expressing little or no detectable FGFs (MCF-7) were compared with tumors transfected with FGF4 or FGF1 (FGF4/MCF-7 or FGF1/MCF-7), and with MDA-MB-435, which naturally over-expresses FGF2. Tumor blood flow was measured by ^{133}Xe diffusion, oxygen distribution was measured by Eppendorf pO_2 histography, and vascular density was measured by CD31 staining. We also measured tumor growth rate, tumor take rate, and lung metastasis in these models.

2. METHODS

2.1 Tumor formation in nude mice

MCF-7, MDA-MB-435, FGF1/MCF-7 and FGF4/MCF-7 cells were grown in cell culture dishes in DMEM with 5% fetal bovine serum, and gently scraped into the same medium without serum under sterile conditions. Viable cells were counted using Trypan blue exclusion and a hemocytometer. Cells were harvested by first washing the plates with PBS containing 10 mM EDTA, after which cells were allowed to detach from the dish. Cells were then centrifuged and resuspended into an appropriate volume of medium to give 5×10^6 viable cells/0.2 ml for injection. Five-week old nude mice (Athymic NCR nu/nu, National Cancer Institute, Frederick, MD) were injected with 5×10^6 cells/site into the mammary fat pad. Tumors of different sizes were chosen at appropriate times after inoculation to determine tumor size-dependent changes in physiological parameters. Tumor volume, tumor take rate, and lung metastasis were measured and recorded after sacrificing mice. Guidelines for the humane treatment of animals were followed as approved by the University Committee on Animal Resources.

2.2 Tumor blood flow (^{133}Xe clearance)

^{133}Xe ($T_{1/2}$ 5.3 day; γ 81 KV) dissolved in 0.9% NaCl solution was injected directly into the tumors in order to study tumor blood flow. The total volume injected was 10 μl, which contains 60 to 150 mCi and resulted in maximum counting rates of 100,000 to 300,000/min. The mouse was placed in a plastic jug approximately the same size as the diameter of the lead collimator (the collimator shields extraneous radiation from the body of the animal). The collimator has a 1.5 cm diameter bore and is 10 cm in length. The bore exposes the 1.5 inch NaI crystal detector. The tumor was set in front of the collimator bore at an appropriate distance and geometry. The output of the detector was processed with a multichannel analyzer, with counts summed at 3 to 10 sec intervals. The radioactivity from each animal was recorded for at least eight min (range 8 to 60 min, median 18 min), which represents at least one, but usually several, half-times. These lines were plotted semi-logarithmically and fitted by linear regression for computation of the half-times ($T_{1/2}$) of radioisotope clearance (clearance is a single exponential function). The half-times were converted to rates of blood flow as determined by the following formula: blood flow (ml/100 g/min) = 100 L (0.693) ($T_{1/2}$), where the partition coefficient, L, was determined for samples of the tumors under investigation. The reproducibility of the method is excellent based on repeated studies of single tumors and repeated analysis of data sets. All data analysis was done by a single observer.

2.3 Tumor pO_2 measurements

Tumor pO_2 was determined using a polarographic oxygen electrode system (Eppendorf, Germany, Model 6650). The measurement of tumor pO_2 levels using the Eppendorf pO_2 histograph is a well established procedure for quantifying both human and animal tumor oxygen distributions (8, 9). Estimates of tumor pO_2 were obtained in anesthetized animals using a fine needle Eppendorf electrode probe (needle diameter 270 μm). To measure pO_2, the needle is inserted up to a depth of about 1 mm into the tumor and then moved automatically through the tissue in 0.7 mm increments, followed each time by a 0.3 mm backward step prior to measurement. From three to seven repeated insertions are performed in each tumor, with 7-15 measurements per track, depending on tumor volume. The relative frequency of the pO_2 measurements is automatically calculated and displayed as a histogram, and tumors are characterized on the basis of median pO_2. The forward motion is chosen such that the net motion will not overlap on a preceding needle location. The reverse motion reduces pressure on the lanced tumor.

2.4 Immunohistochemical staining for CD31

CD-31 immunohistochemical staining was carried out on 5 μm thick paraffin embedded tissue sections. Rabbit polyclonal antibody against CD31 was used. Slides were deparaffinized and rehydrated followed by blocking of endogenous peroxidase activity with 0.3% hydrogen peroxide in methanol for 30 min. Incubation with a 1: 1000 dilution of the primary antibody was performed at 4°C overnight, followed by incubation with 1:200 dilution of a biotinylated horse anti-rabbit IgG for 60 min at room temperature and 1:100 dilution of peroxidase-coupled avidin according to the manufacturer's protocol (ABC kit, Vector, Burlingame, CA). The specimens were then stained with 50 μg/ml 3,3'-diaminobenzidine (Sigma Chemical Co., St Louis, MO) in 0.05 m Tris buffered saline containing 0.01% hydrogen peroxide for 10 min. Slides were counter-stained with 1% methyl green solution for 2 min.

3. RESULTS

3.1 Tumor growth rate and metastases were increased by FGF overexpression

Four human breast tumor lines were tested in this study as shown in Table 1. MCF-7 is an estrogen-dependent, non-metastatic line, and produces little or no FGF mRNA. MCF-7 tumors usually grew slowly in nude mice and formed tumors of only 36-80 mm^3 two months after inoculation. However, the two MCF-7-derived transfectants, FGF1/MCF-7 and FGF4/MCF-7, as well as MDA-MB-435, grew quickly and had a relative large tumor size after two months (280-1440 mm^3) as shown in Table 1. After inoculation of $5x10^6$ cells, only 35% of MCF-7 tumors were detectable, and none of the seven tumors that did grow produced metastases to lung. The other three tumor xenografts had higher tumor take rates (65-85%) and frequently produced metastases (31-47% of tumors produced metastases).

TABLE 1. Tumor growth, lung metastasis and FGFs mRNA expression in four human breast carcinoma xenografts.

Cells	Endogenous FGFs mRNA	Tumor-take rate (%)	Tumor Size (mm^3)	Lung metastasis frequency
MCF-7	None	7/20 (35)	36-80	0/7
MBA-MB-435	FGF2	17/20 (85)	280-968	8/17 (47)
FGF1/MCF-7	FGF1	14/20 (70)	360-1440	6/14 (43)
FGF4/MCF-7	FGF4	13/20 (65)	240-1320	4/13 (31)

3.2 Vascular density did not correlate with FGF overproduction

As shown in Figure 1, FGF1/MCF-7 formed a purple-blue bloody tumor mass (Figure 1a) compared with other tumors, such as FGF4/MCF-7 (Figure 1d). Histologically, FGF1/MCF-7 tumors had very active angiogenesis, and enhanced vascularity was noted in both densely cellular tumor tissue and the surrounding stroma regions (Figure 1b and 1c). FGF4/MCF-7 tumors, however, were densely cellular with active tumor cell proliferation and less vascularity (Figure 1e and 1f). MCF7 tumor cells grew with a glandular structure surrounded by a fibrous capsule (Figure 2). MDA-MB-435, like the transfected tumor models, is an undifferentiated adenocarcinoma with almost no glandular structure (Figure 2). Vascular density was assessed immunohistochemically using antiCD-31 antibody. The number and density of vessels was similar based on CD31 staining in all four tumor models (Figure 3).

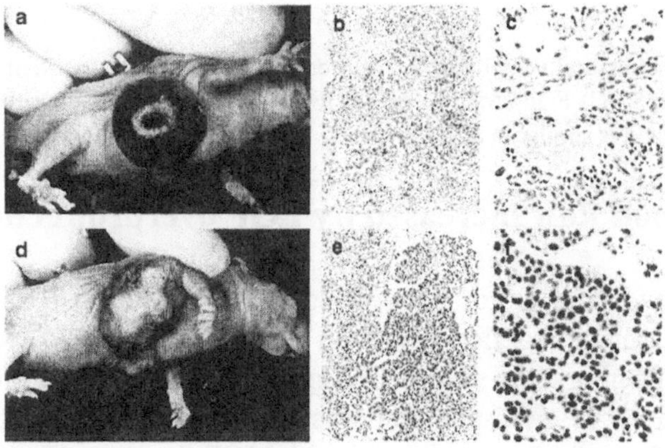

Figure 1. MCF-tumors transfected with FGF1 (a,b,c) or FGF4 (c,d,e. FGF1 transfected tumors had a particularly impressive vascular engorgement. All tumor models had a similar level of vascular density.

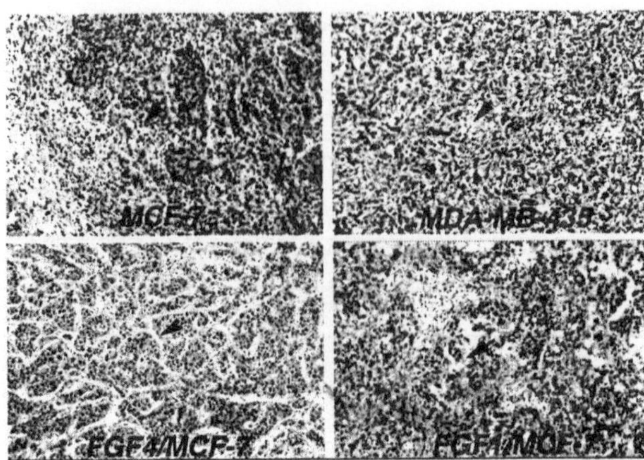

Figure 2. Tumor histology of MCF-7, MDA-MB-435 (FGF2 overexpressing tumor), FGF4/MCF-7 (FGF4 transfected tumor), and FGF1/MCF-7 (FGF1 transfected tumor). MCF-7 tumors had a more glandular histology than angiogenesis factor overexpressing tumor models.

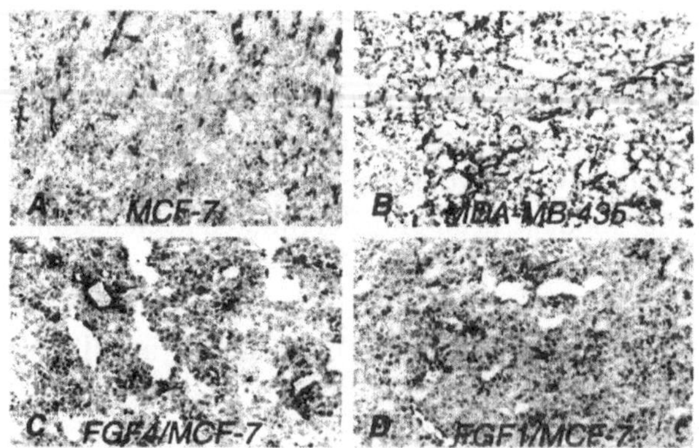

Figure 3. CD31 structural vessel staining of breast cancers grown in nude mice. (A) MCF-7, (B) MDA-MD-435, (C) FGF4/MCF-7, (D) FGF1/MCF-7. Number of positive staining cells per high power field is similar in all.

3.3 Blood flow increased by FGF overproduction

It is well known that tumor neovasculature is chaotic and frequently non-functional. Vascular function was then determined using ^{133}Xe washout in another set of tumors. As shown in Figure 4, both FGF transfected tumors had increased ^{133}Xe clearance that was particularly obvious at large tumor sizes (100-500 mm^3) when compared with MCF-7 tumors (Figure 4). Tumor perfusion in all three FGF overexpressing tumor models was often much higher than the low levels seen in MCF-7.

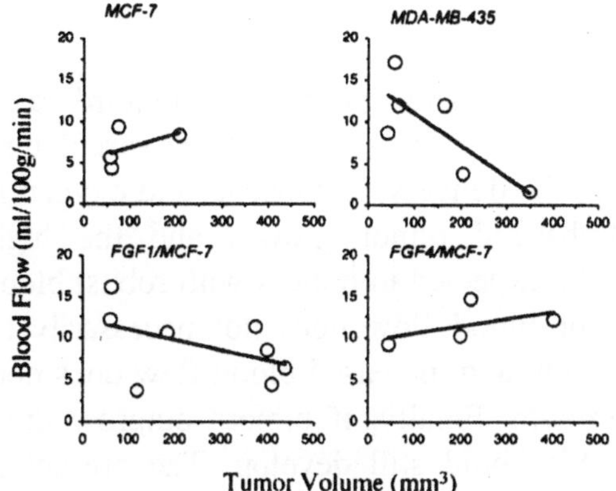

Figure 4. 133Xe tumor blood flow measurements as a function of tumor size. The volume scale (x-axis) varies between tumor models due to size variation at the 2 month end-point. MCF-7 blood flow decreased with increasing tumor volume. FGF-overexpressing tumors maintained near normal flow, even for large tumors.

3.3.1 Mean pO_2 was increased by FGF overproduction

In a separate set of experiments, we also measured tumor median pO_2 levels using polarographic oxygen electrodes (Figure 5). Here again there appeared to be maintenance of vascular function by tumors overexpressing FGFs, evidenced by relatively higher median pO_2 values even for tumors of large size.

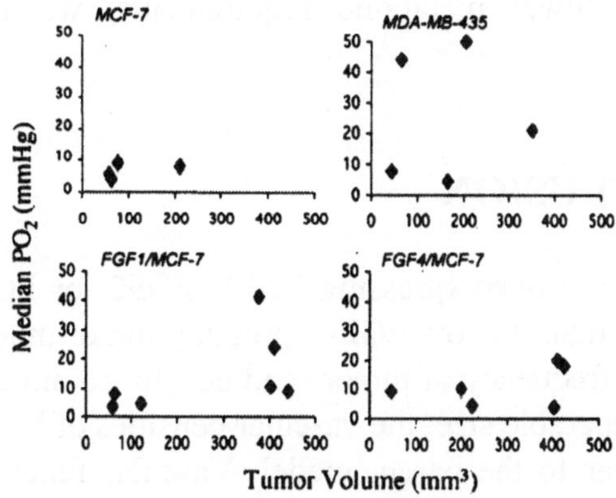

Figure 5. Median pO_2 of tumor versus tumor size. Median pO2 was highly variable between tumors of similar size and histology. MCF-7 tumors, however, consistently had low median pO_2 values compared with FGF1/MCF-7 and FGF4/MCF-7 tumors.

4. DISCUSSION

Breast tumor progression and metastases require a number of physiological processes including access to adequate vasculature. Likewise, the tumor growth requires the delivery of nutrients and oxygen for activated cellular metabolism. Enhanced tumor growth and the higher rate of metastases therefore can be expected in tumors with robust blood perfusion. However, increased tumor blood flow does not necessarily result in the enhancement of oxygenation, and increased blood flow does not necessarily result in faster tumor growth. Finally, if tumors outgrow the functioning vasculature, tumor hypoxia could still develop. The present studies help delineate the major effects of FGFs on tumor microenvironment and vascular physiology.

Our study also demonstrated that overexpression of angiogenic growth factors results in a more malignant phenotype in breast tumors. This result supports the work of several other laboratory and clinical investigations (4, 5, 10, 11). The loss of normal glandular architecture is consistent with FGFs stimulating tumor proliferation. Similarly, overexpression of FGFs by tumor cells also likely stimulates tumor endothelial cell proliferation.

Our data suggest that overexpression of FGF1, FGF2 or FGF4 produces breast cancer tumors with less glandular architecture, faster growth rates, higher frequency of lung metastases, but no significant alteration of structural vascular density. The relatively stable vascular density occurs despite a preservation of tumor oxygenation and flow compared with MCF-7 tumors of large size. The physiology underlying this effect could result from (1) lower metabolic activity of FGF-overexpressing tumors causing lower consumption, or (2) better functional vasculature in FGF-overexpressing tumors leading to better nutrient delivery. Since faster growing tumors are unlikely to have lower metabolic requirements, we favor the second explanation.

5. CONCLUSION

The breast tumors overexpressing FGF1, FGF2, or FGF4 all had more rapid growth rates than the low FGF-expressing tumor model. These tumors produced a higher frequency of tumors and developed more lung metastases. For tumors of comparable size, the vascular densities of FGF-overexpressing tumors were similar to the parent model. Vascular function and perfusion were, however, preserved even for large tumor sizes among tumors expressing high levels of angiogenic factors. Therefore tumors that overexpress fibroblast growth factors are more aggressive tumors in part through the production of a better functioning vascular network, and not a total structural vascular network.

ACKNOWLEDGEMENTS

REFERENCES

1. Folkman J. Angiogenesis inhibitors generated by tumors, Mol Med 1995; 1:120-122.
2. Folkman J. A new family of mediators of tumor angiogenesis, Cancer Invest 2001;19:754-755.
3. Kurebayashi J, McLeskey SW, Johnson MD, Lippman ME, Dickson RB, Kern FG. Quantitative demonstration of spontaneous metastasis by MCF-7 human breast cancer cells cotransfected with fibroblast growth factor 4 and LacZ, Cancer Res 1993;53:2178-2187.
4. McLeskey S W, Kurebayashi J, Honig SF, Zwiebel J, Lippman ME, Dickson RB, Kern FG Fibroblast growth factor 4 transfection of MCF-7 cells produces cell lines that are tumorigenic and metastatic in ovariectomized or tamoxifen-treated athymic nude mice, Cancer Res 1993;53:2168-2177.
5. Zhang L, Kharbanda S, Chen D, Bullocks J, Miller DL, Ding IY, Hanfelt J, McLeskey S W, Kern FG. MCF-7 breast carcinoma cells overexpressing FGF-1 form vascularized, metastatic tumors in ovariectomized or tamoxifen-treated nude mice, Oncogene 1997;15:2093-108..
6. Zhang L, Kharbanda S, McLeskey SW, Kern FG. Overexpression of fibroblast growth factor 1 in MCF-7 breast cancer cells facilitates tumor cell dissemination but does not support the development of macrometastases in the lungs or lymph nodes, Cancer Res 1999; 59:5023-5029.
7. Nguyen, M. Angiogenic factors as tumor markers, Invest New Drugs 1997;15:29-37.
8. Vaupel P, Schlenger KH, Hoeckel M, Okunieff P. Oxygenation of mammary tumors: from isotransplanted rodent tumors to primary malignancies in patients, Adv Exp Med Biol 1992;316:361-371.
9. Vaupel P and Hockel M. Blood supply, oxygenation status and metabolic micromilieu of breast cancers: characterization and therapeutic relevance, Int J Oncol 2000;17:869-879.
10. Compagni A, Wilgenbus P, Impagnatiello MA, Cotten M, Christofori G. Fibroblast growth factors are required for efficient tumor angiogenesis, Cancer Res 2000;60:7163-7169.
11. Hajitou A, Deroanne C, Noel A, Collette J, Nusgens B, Foidart J M, Calberg-Bacq C M. Progression in MCF-7 breast cancer cell tumorigenicity: compared effect of FGF-3 and FGF-4, Breast Cancer Res Treat 2000;60:15-28.

Chapter 59

FGF1 AND VEGF MEDIATED ANGIOGENESIS IN KHT TUMOR-BEARING MICE

Ivan Ding, Weimin Liu, Jianzhong Sun, Scott F. Paoni, Eric Hernady, Bruce M. Fenton, & Paul Okunieff

Department of Radiation Oncology, University of Rochester, Box 647, Rochester, NY, 14642

Abstract: Isotransplants of murine fibrosarcoma (KHT) cells were inoculated i.m. into the hind limbs of 6-8 week-old female C3H/HeJ mice. Intratumoral injection of FGF1 or VEGF proteins decreased hypoxic marker uptake in murine fibrosarcoma KHT. Reduction of tumor hypoxia did not correlate with mRNA expression of transcription factors in tumors. Likewise, there was no significant alteration in either apoptotic frequency or the mRNA levels of 10 apoptotic-related molecules in FGF1- or VEGF-treated tumors. mRNA expression for MCP-1, IL-1ß, IL-18, and IL-1Ra, however, were decreased in the tumors following FGF1 or VEGF treatment. Among the normal tissues tested (brain, kidney, liver, spleen, and lung), basal mRNA levels for cytokines and chemokines varied. Intratumoral injection of FGF1 or VEGF (6 daily intra-tumor injections of 6μg/mouse) did not alter most cytokine or chemokine mRNA expression in spleen and lung. In summary, alteration of tumor oxygenation by local administration of angiogenic growth factors may be mediated by cytokine/chemokine production in the tumor.

Key words: VEGF, FGF, fibrosarcoma, cytokine and chemokine, apoptosis, transcription factors, gene expression

1. INTRODUCTION

Tumor angiogenesis requires at least four steps (1, 2): (i) local secretion of proteinase or cytokines/chemokines, (ii) degradation of the extracellular matrix of tumor stroma surrounding tissues, (iii) chemotaxis of endothelial cells and other cell types toward an angiogenic stimulus, and (iv) proliferation of endothelial cells and formation of tumor-specific blood vessels. A variety of tumor-related growth factors/cytokines have been

Oxygen Transport to Tissue XXIV, edited by
Dunn and Swartz, Kluwer Academic/Plenum Publishers, 2003

identified as potential positive regulators of angiogenesis, including fibroblast growth factors (FGFs) and vascular endothelial growth factor (VEGF) (3). Some of these factors are able to directly stimulate endothelial cell growth, while others lack direct stimulatory effects on endothelial cells, and are thought to require the paracrine release of growth factors from other cell types, such as tumor fibroblasts, infiltrating inflammatory cells, and tumor cells (1, 2, 4-6). In this study, we investigated the effects of exogenous FGF1 and VEGF on KHT tumor growth and hypoxia, and explored the underlying molecular mechanisms. Using FGF1- or VEGF-treated and control mice, mRNA expression for 7 transcriptional factors, 17 cytokines, 9 chemokines, and 10 apoptotic-related molecules were determined in tumors and five normal organs (brain, liver, lung, kidney, and spleen) by RNase protection assay. Our goal was to further understand the role of these two angiogenic growth factors in the regulation of tumor oxygenation by alteration of gene expression for cytokine/chemokine, apoptosis, and transcriptional factors.

2. METHODS

2.1 Tumor models and angiogenic growth factors treatment

Isotransplants of murine fibrosarcoma (KHT) cells were inoculated i.m. into the hind limbs of 6-8 week-old female C3H/HeJ mice (Jackson Laboratories, Bar Harbor, ME). Tumors were selected for treatment when they reached volumes of 100 and 300 mm^3 [as measured by calipers and the formula: volume = (Diameter^3/6)]. 6 μg of either FGF1 or VEGF (Pepro Tech Inc., Rocky Hill, NJ) was injected intratumorally once a day for 6 days. Control mice were given the same volume of saline as controls. Mice were sacrificed 24 hours after the final injection. Tumor, liver, spleen, lung, kidney, and brain from tumor bearing mice were removed, and tissue RNA was isolated. Guidelines for the humane treatment of animals were followed as approved by the University Committee on Animal Resources.

2.2 EF5/Cy3 hypoxic marker

Localized areas of tumor hypoxia were assessed in frozen tissue sections by immunohistochemical identification of sites of 2-nitroimidazole metabolism. A pentafluorinated derivative (EF5) of etanidazole was injected i.v. one hour before tumor freezing. Protein conjugates of EF5 have been previously used to immunize mice from which monoclonal antibodies were developed. These antibodies are extremely specific for the EF5 drug adducts that form when the drug is incorporated by hypoxic cells (7). Regions of high EF5 metabolism in tumors (hypoxic regions) were visualized immunochemically using a fluorochrome (Cy3) conjugated to the ELK3-51 antibody (8).

2.3 mRNA expression for transcriptional factors, apoptotic molecules, and cytokines/chemokine measured by RNase protection assay

Tumor and normal tissues from each treatment group were removed and immediately frozen. The RNA was then isolated by pulverizing the frozen tissue and dissolving it in TRIzol reagent (MRC, Cincinnati, OH). RNase protection was then performed using established multi-probe template sets (PharMingen, San Diego, CA). The MCK5 chemokines set includes Lymphotactin, Rantes, Eotaxin, MIP-1α, MIP-1β, MIP-2, IP-10, MCP-1, and TCA-3. The MCK3 inflammatory cytokine set includes tumor necrosis factors α and β, transforming growth factors β1, β2, β3, interferon β and γ, interleukine 6, and lymphotoxin. The MCK2 interleukine set includes IL-1α, IL-1β. IL-1Ra, IL-6, IL-8, IL-12, IL-18, and MIF. The transcriptional factors set includes c-jun, JunB, JunD, c-fos, FosB, Fra-1, and Fra-2. The apoptosis-related molecular set includes Caspase 8, FAS, FASL, FADD, FAP, FAF, Trail, TRADD, and TNFRp55. Two internal controls, L32 and GAPDH, were used. The mRNA expression levels for each sample were quantified by phosphorimaging (HP Company, Meriden, CT).

2.4 Statistical methods

mRNA expression from saline and FGF1 or VEGF treated mice was evaluated using the unpaired Student's t-test or Mann-Whitney Rank Sum test. Differences were considered significant for $P < 0.05$.

3. RESULTS

3.1 Effects of FGF1 and VEGF intratumoral injection on tumor sections

Six daily intratumoral injections of either FGF1 or VEGF decreased the EF5 staining in tumor sections compared with saline-treated controls (Figure 1a-c). This phenomenon appears tumor size-dependent (Ding: Unpublished data). Although angiogenic growth factor-mediated improvement of tumor oxygenation could be the result of alteration in local tumor blood flow, other mechanisms may also be involved in this process, such as up- or down-regulation of angiogenic-related cytokines and/or chemokines. Angiogenic growth factors have been shown to lessen cell death from radiation by decreasing radiation-induced apoptosis. In our study, intratumoral injection of FGF1 and VEGF slightly increased tumor cell apoptosis, but not significantly (Figure 1d-f). We also examined mRNA expression of ten apoptosis-related molecules and seven transcriptional factors in these tumors. As shown in Figure 2, administration of FGF1 or VEGF in KHT tumors did not significantly alter mRNA expression of either the transcriptional factors (Figure 2a) or the apoptosis-related molecules (Figure 2b). Similarly, FGF1 and VEGF did not affect either the transcriptional factors or gene expression of the apoptosis-related molecules in five major normal organs (Figure 2c).

3.2 Effects of FGF1 and VEGF intratumoral injection on normal tissues

To explore the effects of intratumoral injection of VEGF or FGF1 on cytokine and chemokine mRNA expression in normal tissues, mRNA was determined for seventeen cytokine and nine chemokine samples of brain, liver, kidney, lung, and spleen tissue. As shown in Figure 2d, although different organs had varied baseline levels of each of the molecules, intratumoral injection of VEGF affected interleukine mRNA expression in kidney, liver, and brain. VEGF upregulated both liver and kidney IL-18 mRNA 1.6 fold, IL-1Ra mRNA 1.4 fold, and IL-1β mRNA 1.3- and 1.4-fold, respectively. Injection of FGF1 and VEGF did not alter expression of any of the TGFβ isoforms. The other cytokines, including TNFα, TNFβ, IFNβ, and IFNγ were also unaffected. Finally, injection of angiogenic growth factors also altered liver and tumor chemokine mRNA levels. As summarized in Table 1, both FGF1 and VEGF increased liver Rantes mRNA expression (1.7 fold and 2.8 fold, respectively). VEGF also upregulated mRNA expression of two C-C family member chemokines in liver: MCP1 (3.7 fold) and TCA-3 (2.2 fold). Both VEGF and FGF1 also decreased tumor MCP-1 mRNA expression.

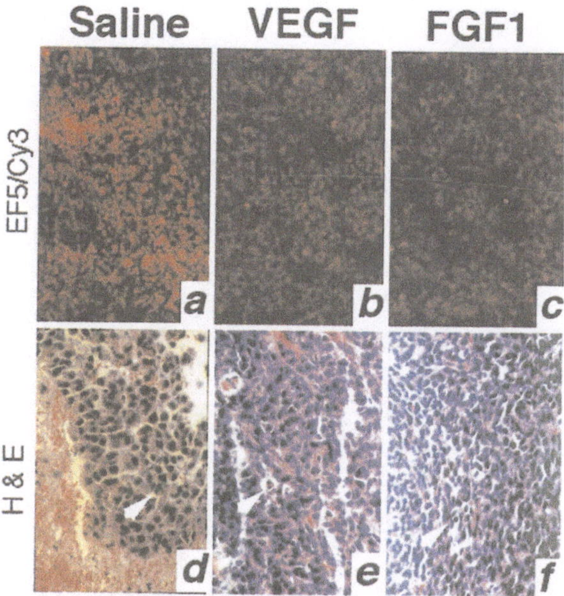

Figure 1. KHT tumor sections from EF5/Cy3-stained slides (a-c), and H&E stained slides (d-f). Tumor-bearing C3H mice were treated with six doses of saline (a,d), VEGF (b,e), or FGF1 (c,f). Both VEGF and FGF1 treated tumors had a decreased EF5/Cy3 staining, indicating a decrease in cellular hypoxia. Tumor apoptotic frequency was not affected by FGF1 or VEGF treatment.

Table 1. Chemokine mRNA Expression in Brain, Kidney, Liver, Spleen, Lung and KHT tumorinf C3H mice Treated with FGF1 and VEGF

	Brain	Kidney	Liver	Spleen	Lung	Tumor
MCP-1	-	++	++↑	+++	-	+++↓
Rantes	-	+	++↑	++++	+++	+ ↑
TCA-3	-	-	+++↑	-	-	-
MIP-2	-	-	-	-	-	++
Lymphotactin	-	-	-	+	-	-

(+) to (++++)represent the basal levels of mRNA expression of chemokine gene;
(-), no detectable mRNA expression; ↑, induction of gene expression after angiogenic growth factor Treatment.

4. DISCUSSION

The effects of angiogenic growth factors on tumor angiogenesis and growth have been studied extensively (4). Active tumor cell proliferation is dependent on the tumor blood supply (2). It is believed that tumors larger than 1 mm^3 in volume require angiogenesis for further growth. In theory, local or systemic administration of angiogenic growth factors could improve blood vessel formation, and thus may alter tumor oxygenation. Our data support this notion. However, the underlying molecular mechanisms for angiogenic growth factor-mediated tumor pathophysiology are not clear (2). Most angiogenic growth factors not only stimulate endothelial cell proliferation, but also regulate tumor cell growth and differentiation. The process leading to tumor proliferation is either direct, acting on both endothelial and tumor cells, or indirect, regulating production of other growth factors, such as cytokines and chemokines. An example of the second mechanism comes from the work of Seghezzi et al (9). They recently reported that both exogenous FGF2 administration and expression of endogenous FGF2 result in increased VEGF gene expression. They also concluded that endothelial cell-derived VEGF is a major autocrine mediator of FGF2-induced angiogenesis.

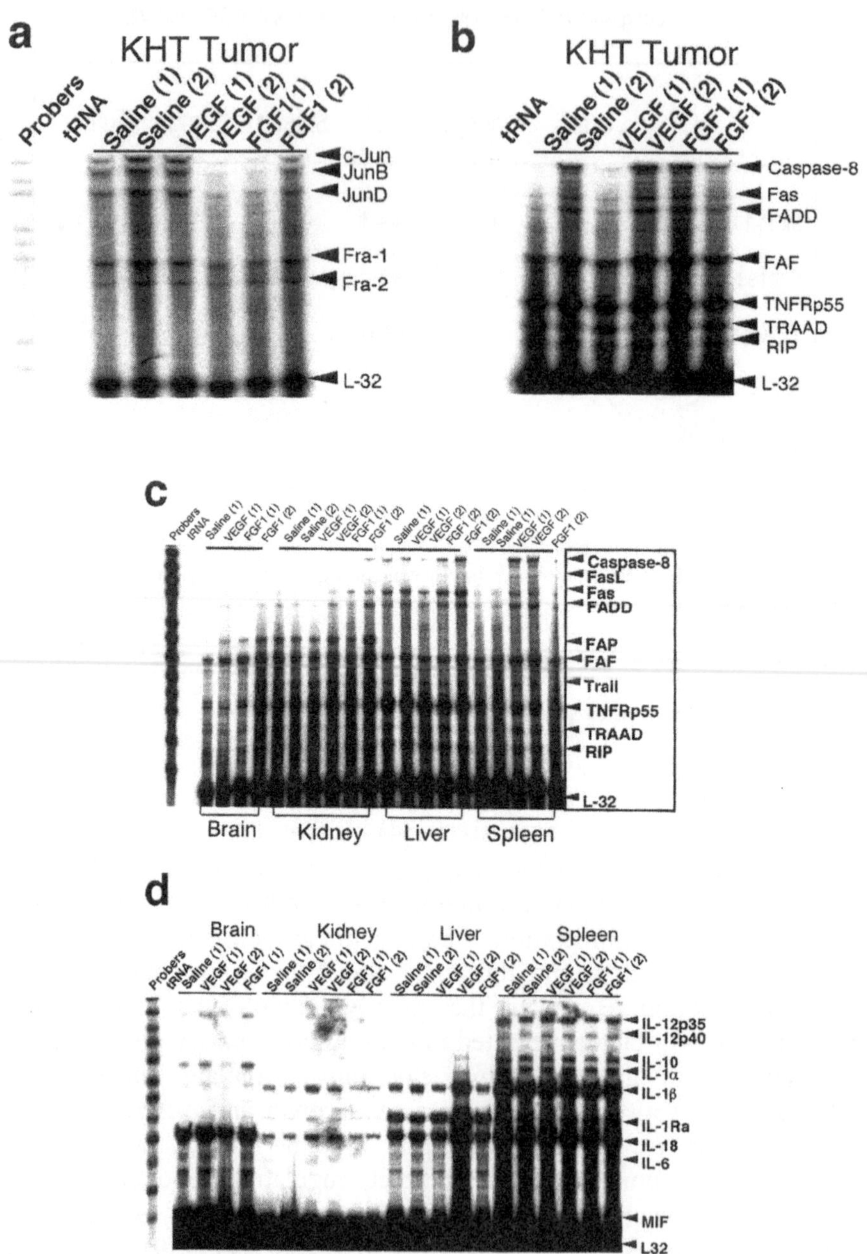

Figure 2. Total RNA from brain, liver, spleen, lung, kidney, and tumor were extracted from KHT tumor-bearing mice. Mice were treated with six doses of FGF1, VEGF, or saline and sacrificed 24 hours later. RNase protection assay was utilized to determine mRNA expression of transcriptional factors in tumor (a), apoptosis-related molecules in tumor (b) and normal tissue (c), and cytokine in normal tissue (d).

5. CONCLUSION

In this study, we explored the molecular mechanisms of angiogenic growth factor-mediated angiogenesis in KHT tumor-bearing mice treated with FGF1 and VEGF. Our findings indicate the following: (1) FGF1 and VEGF improved KHT tumor oxygenation without altering tumor transcriptional factors, tumor cell death-related molecule mRNA expression, or tumor cell apoptotic frequency. However, decreased tumor MCP1 and interleukin mRNA expression may indirectly affect local tumor angiogenesis; (2) Local tumor administration of FGF1 and VEGF altered cytokine or chemokine mRNA expression in some normal organs, with the greatest effects seen in kidney and liver. Increased C-C (MCP-1, Rantes and TCA-3) and C-X-C (IP-10) mRNA expression generated by normal organs in animals with high intratumor cytokines may also participate in local tumor angiogenesis and immunoreactivity.

ACKNOWLEDGEMENTS

This work was supported in part by National Cancer Institute P01-CA11051-25A2 and National Cancer Institute R01CA52586.

REFERENCES

1. Couffinhal T, Silver M, Zheng LP, Kearney M, Witzenbichler B, and Isner JM. Mouse model of angiogenesis. Am J Pathol 1998;152:1667-1679..
2. Folkman J. Angiogenesis in cancer, vascular, rheumatoid and other disease. Nat Med 1995;1:27-31.
3. Andre T, Chastre E, Kotelevets L, Vaillant JC, Louvet C, Balosso J, Le Gall E, Prevot S, Gespach C. Tumoral angiogenesis: physiopathology, prognostic value and therapeutic perspectives. Rev Med Interne 1998;19:904-913.
4. Battegay EJ. Angiogenesis: mechanistic insights, neovascular diseases, and therapeutic prospects. J Mol Med 1995;73:333-346.
5. Folkman J. Fighting cancer by attacking its blood supply. Sci Am 1996;275:150-4
6. Folkman J. New perspectives in clinical oncology from angiogenesis research. Eur J Cancer 1996;32A:2534-2539.
7. Lord EM, Harwell L, and Koch CJ. Detection of hypoxic cells by monoclonal antibody recognizing 2- nitroimidazole adducts. Cancer Res 1993;53:5721-5726.
8. Fenton BM, Paoni SF, Lee J, Koch CJ, and Lord EM. Quantification of tumour vasculature and hypoxia by immunohistochemical staining and HbO2 saturation measurements. Br J Cancer 1999;79:464-471.
9. Seghezzi G, Patel S, Ren CJ, Gualandris A, Pintucci G, Robbins ES, Shapiro RL, Galloway AC, Rifkin DB, and Mignatti P. Fibroblast growth factor-2 (FGF-2) induces vascular endothelial growth factor (VEGF) expression in the endothelial cells of forming capillaries: an autocrine mechanism contributing to angiogenesis. J Cell Biol 1998;141:1659-1673.

Chapter **60**

HIF-1α AND VEGF EXPRESSION AFTER TRANSIENT GLOBAL CEREBRAL ISCHEMIA

Paolo Pichiule, Faton Agani, Juan C. Chavez, Kui Xu & Joseph C. LaManna
Departments of Anatomy and Neurology, Case Western Reserve University, School of Medicine, Cleveland, OH 44106-4938. USA

Abstract: Hypoxia inducible factor-1α (HIF-1α) and vascular endothelial growth factor (VEGF) expression were studied in rat cerebral cortex after reversible global cerebral ischemia produced by cardiac arrest and resuscitation. Immunoblot analysis showed a significant induction of HIF-1α protein after 1 hour of recovery from cardiac arrest which remained elevated for at least 12 hours. Upregulation of VEGF mRNA and protein were also observed but this was delayed in comparison to the HIF-1α response. $VEGF_{188}$ and $VEGF_{164}$ mRNA levels were increased at 12- 48 h of recovery from cardiac arrest but returned to basal expression after 7 days. Changes in $VEGF_{120}$ mRNA expression did not reach statistical significance. Correspondingly, VEGF protein levels increased by about double at 24 and 48 hours of recovery but returned to basal levels after 7 days. These results suggest that cardiac arrest and resuscitation triggers HIF-1α induction, which might be at least in part responsible for the stimulation of VEGF expression.

Key words: cardiac arrest and resuscitation, hypoxia inducible factor, vascular endothelial growth factor, vascular permeability factor, oxygen-regulated genes

1. INTRODUCTION

Almost all the rats subjected to 12 minutes of cardiac arrest can be initially resuscitated. However, about half of them will not survive for more than a few days. One of the most likely reasons for this secondary mortality is metabolic failure due to vasogenic cerebral edema that develops over the first few days after reperfusion. The causes of vasogenic cerebral edema remain unknown, but VEGF, otherwise known as vascular permeability factor could play a major role [1,2]. Analysis of the gene encoding VEGF identified a hypoxia-response element containing HIF-1 binding sites required for its transcriptional activation in response to hypoxia [3]. HIF-1 is a heterodimeric transcription factor consisting of HIF-1α and HIF-1β subunits. Upon decrease in oxygen tension, this complex regulates the expression of several genes involved in maintenance of oxygen homeostasis. Whereas HIF-1α serves as heterodimerization partner for various others transcription factors, HIF-1α is unique to HIF-1 and its expression is tightly regulated by cellular O_2 concentration. Moreover, it has been shown that HIF-1α protein but not mRNA levels are upregulated in response to hypoxia [4,5]. Several studies have demonstrated in vivo regulation of HIF-1α expression by hypoxic-related conditions. For instance, increased myocardial expressions of VEGF and HIF-1α protein were observed in fetal sheep subjected to chronic anemia [6]. Expression of HIF-1α protein was also induced in ischemic retina and this showed temporal and spatial correlation with increased VEGF expression [7].

In this study, we investigated the effect of transient global cerebral ischemia on HIF-1 protein expression and whether increased HIF-1α level was correlated with VEGF upregulation. Induction of HIF-1α and subsequent transactivation of target genes such as VEGF could play a key role in the brain response to ischemic and reperfusion stress.

2. MATERIALS AND METHODS

2.1 Induction of global cerebral ischemia

Reversible global cerebral ischemia was achieved by a modification of the cardiac arrest model described by Crumrine and LaManna [8]. This protocol was reviewed and approved by our Institutional Animal Care and Use Committee.

Cardiac arrest was induced in conscious male Wistar rats by a rapid sequential intra-atrial injection of d-tubocurare (0.3 mg) and ice-cold KCl (0.5M, 0.12ml/100g of body weight). Resuscitation efforts began after 7 min of arrest. For this purpose rats were mechanically ventilated accompanied by cardiopulmonary resuscitation and intravenous saline injections. Once spontaneous heartbeat returned, a small dose of epinephrine (1μg) was administered as a bolus intravenously to achieve a mean arterial blood pressure of at least 80 mm Hg. The duration of ischemia was targeted for ~12 min and it was defined as the period between the decrease of blood pressure to zero and its return to 80% of pre-arrest value. Ventilation was adjusted to achieve normoxia and normocapnia until the rats gained spontaneous respiration. Control animals were subjected to the same surgical procedure, except that they did not undergo cardiac arrest and resuscitation. Rats were sacrificed after different periods of recovery from cardiac arrest. Brains were quickly removed and cortex dissected and their cortices frozen at –80°C until used for RT-PCR and Western blot analysis.

2.2 RNA extraction and VEGF reverse transcriptase polymerase chain reaction (RT-PCR)

Total RNA was extracted from brain cortex using RNAgents Isolation System (Promega Corp.) according to the manufacturer's instructions. RNA (1 μg) was reverse transcribed using AMV reverse transcriptase and oligo(dT) primer at 48°C for 45 minutes. The cDNA was amplified by PCR for 30 cycles (94°C for 30 seconds, 60°C for 1 min, 68°C for 2 min) using *Tfl* DNA polymerase,VEGF sense (5'-CCATGAACTTTCTGCTCTCTTG-3') and antisense (5'-GGTGAGAGGTCTAGT TCCCGAA-3') primers. These primers mapped nucleotides -2 to +20 of the 5' end coding region and nucleotides 606 to 627 of the untranslated 3' end region of the rat cDNA sequence [9]. As an internal control, co-amplification of β-actin was performed using the following primers: sense (5'-AACCCTAAGGCCAACCGTGAAA-3') and antisense (5'-TCATGAGGTAGTCTGTCAGGTC-3'). PCR products were electrophoresed in 2% agarose gel, visualized by ethidium bromide staining and the signals were quantified using a FluorImager system (Molecular Dynamics). Individual data were normalized to b-actin and expressed as percentage increase relative to control values. To verify the identity of the PCR products, they were gel purified, subcloned and sequenced

2.3 HIF-1α and VEGF Western blot analysis

Brain cortex samples were homogenized in ice-cold buffer (0.02 M HEPES, pH 7.5, 1.5 mM $MgCl_2$, 0.2mM EDTA, 0.1M NaCl) supplemented with 0.2 mM dithiothreitol and protease inhibitors (0.4 mM phenylmethylsulfonyl fluoride and 2 μg/ml each of leupeptin, pepstatin and aprotinin). Homogenate was centrifuged at 16000g for 30 min and the resulting supernatant used for Western blot analysis. Protein concentration was determined by Bradford protein assay (Bio-Rad Inc.) using bovine serum albumin as a standard. Samples were subjected to electrophoresis in SDS-polyacrylamide gel and transferred to nitrocellulose membrane using standard procedures. The membrane was blocked with 10% non-fat milk powder and incubated for 2 hours with HIF-1α (1:500, Novus Biologicals) or

VEGF (1:400, Santa Cruz Biotech.) antibody. The respective horseradish peroxidase-conjugated antibody was used with enhanced chemiluminescence reagent (ECL, Amersham) to visualize primary antibody binding. For comparison, crude nuclear extracts of Hepa cells exposed to 1% or 20% O_2 were used.

2.4 Statistical analysis

Data were expressed as the percentage increase of the post-resuscitation samples with respect to control samples. The data were reported as means ± SD. One sample t-tests were used to determine if ischemic: control ratio of VEGF protein density and mRNA level were significantly greater than 1. Comparison among VEGF ratios at various time-points were assessed by ANOVA followed by Tukey correction. In all the cases, $p<0.05$ was considered significant.

3. RESULTS

3.1 Physiological variables

The major effects of 12 minutes of global ischemia induced by cardiac arrest and resuscitation in the rat have been reported previously [8, 10]. These include initial arterial acidosis, hemoconcentration and hypertension that returned to pre-arrest levels between 30 and 180 min of reperfusion. Central changes in energy metabolites were rapidly reversed. The rats regained spontaneous respiration within 6 hours after resuscitation and usually regained consciousness before 24 hours. Generally by 36 hours the resuscitated rats were able to move and feed themselves.

3.2 HIF-1α protein expression

HIF-1α protein levels were studied at various time intervals of recovery from cardiac arrest in rat brain cortex. As previously described, HIF-1α protein was detected as a series of isoforms with apparent molecular mass of ~110-130 kDa [11]. Levels of HIF-1α were very low in sham-operated cortex but it rapidly increased after cardiac arrest and resuscitation. An approximately ten-fold increase was detected at one hour of recovery, remaining significantly elevated at 48 hours of recovery (Figure1).

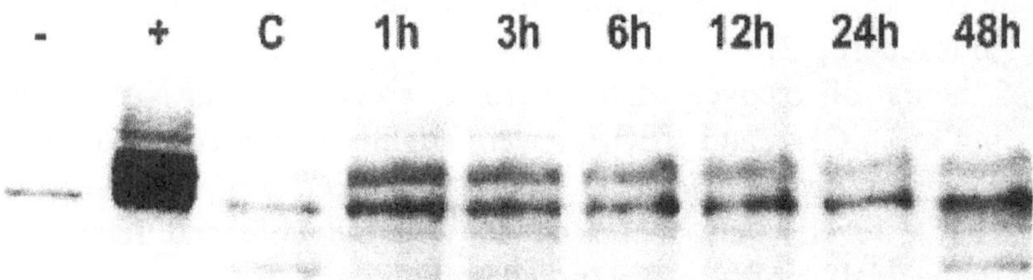

Figure 1. Western Blot showing HIF-1α protein levels in brain cortex of sham operated (c) and post-ischemic rats (1-48h recovery from cardiac arrest). Nuclear extracts of Hepa cells exposed to 20 % (-) and 1 %(+) of O_2 were used as controls.

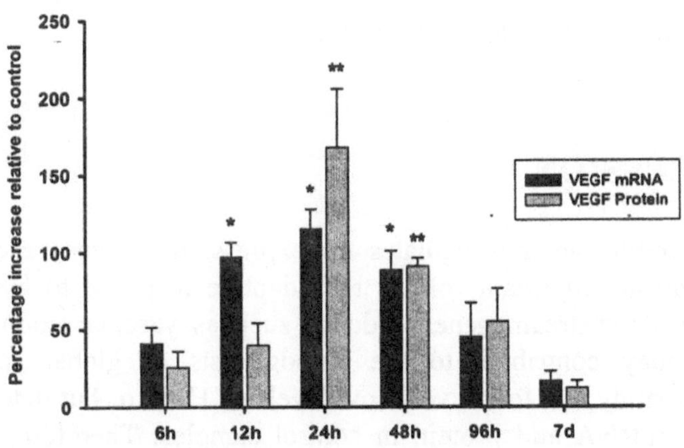

Figure 2. Time course of VEGF $_{164}$ mRNA and VEGF protein expression after cardiac arrest and resuscitation in brain cortex. RT-PCR individual data were normalized to β-actin and expressed as percentage increase relative to control values. Similar pattern of changes was found for $VEGF_{188}$ while $VEGF_{120}$ upregulation did not reach statistical significance. Protein data were expressed as percentage increase of optical density relative to control values. Data represent mean± S.D. from 3 experiments for mRNA analysis and 3-8 experiments for protein analysis (*, ** $p<0.05$ versus control).

3.3 VEGF expression

Three major VEGF isoforms corresponding to $VEGF_{188}$, $VEGF_{164}$ and $VEGF_{120}$ were found in brain cortex by RT-PCR analysis. $VEGF_{164}$ was the predominant isoform and $VEGF_{206}$ was not observed, which is in agreement with a previous report [12].

After 6 hours of recovery from cardiac arrest, the mRNA levels corresponding to $VEGF_{188}$ and $VEGF_{164}$ were slightly increased but did not reach statistical significance. However, after 12h of recovery, $VEGF_{188}$ and $VEGF_{164}$ mRNA levels were increased by about double and remained elevated at 24 h and 48 h of recovery ($p<0.05$, $n=3$, Figure 2).

The polyclonal VEGF antibody used for Western blot detected a single major band at ~23 kDa which is consistent in size with $VEGF_{164}$ monomer. Immunoreactivity was observed in control samples as well as in post-ischemic brain samples. VEGF protein levels were significantly elevated only after 24 h and 48 hours of recovery ($p<0.05$, $n=5$). After 7 days of recovery, both mRNA and protein levels returned to basal expression (Figure 2). Expression of $VEGF_{120}$ mRNA followed a similar pattern but it did not reach statistical significance (data not shown).

4. DISCUSSION

Hypoxia inducible factor-1 regulates the expression of genes encoding proteins that play an important role in the adaptive response to hypoxia. Several of these downstream gene products, such as vascular endothelial growth factor, may contribute to the pathogenesis of global cerebral ischemia. In this study, we found very low levels of HIF-1α, but detectable levels of VEGF mRNA and protein in control samples. Therefore, under normoxic conditions, VEGF expression might be regulated by HIF-1 independent mechanisms. In addition, we showed that HIF-1α was significantly induced and preceded VEGF upregulation after a short period of global cerebral ischemia.

It is important to mention that during the first hours of recovery, the animals were mechanically ventilated first with 100 % O_2 (2 hours) and then with room air until they recovered spontaneous respiration. Despite the fact that at 1h of reflow, rats showed high values of PaO_2, severe cerebral hypoperfusion at that time [8] resulted in tissue hypoxia that could be responsible for the HIF-1α induction. Nevertheless, if arterial oxygen tension and cerebral metabolism were at control levels after 6 h of recovery, it would be unlikely that there was a continued brain hypoxia. Thus, the initial ischemia-induced metabolic stress must have activated other mechanisms that preserve HIF-1α levels in the brain.

Cerebral ischemia, with both oxygen and glucose insufficiency, is expected to result in increased VEGF expression. In addition, reperfusion of ischemic tissue involves generation of reactive oxygen intermediates that are also considered to induce VEGF expression *in vitro* and *in vivo* [13]. Upregulation of VEGF levels after cardiac arrest and resuscitation may be due to HIF-1 transcriptional activation and/or an increase in mRNA stability. However, other HIF-1 independent pathways cannot be excluded.

The fact that HIF-1α is elevated, followed by VEGF, and at the same time edema is occurring, suggests that metabolic stress occurring during global ischemia or during reperfusion activates mechanisms that might result in the secondary mortality. Thus, interfering with the mechanisms that alter VEGF might prove useful in promoting the long-term survival after cardiac arrest and resuscitation.

ACKNOWLEDGEMENTS

This study was supported by NIH grant NS 37111.

REFERENCES

1) Dobrogowska DH, Lossinsky AS, Tarnawski M, Vorbrodt AW. Increased blood-brain barrier permeability and endothelial abnormalities induced by vascular endothelial growth factor. J Neurocytol 1998;27:163-173.
2) Strugar J, Rothbart D, Harrington W, Criscuolo GR. Vascular permeability factor in brain metastases: correlation with vasogenic brain edema and tumor angiogenesis. J Neurosurgery 1994;81:560-566.
3) Forsythe JA, Jiang B-H, Iyer NV, Agani F, Leung SW, Koos RD and Semenza GL. Activation of vascular endothelial growth factor gene transcription by hypoxia-inducible factor 1. Mol Cell Biol 1996;16:4604-4613
4) Semenza GL. Hypoxia-inducible factor 1 and the molecular physiology of oxygen homeostasis. J Lab Clin Med 1998;131:207-214.
5) Wenger RH, Gassmann M. Oxygen(es) and the Hypoxia-Inducible Factor-1. Biol Chem 1997;378:609-616.
6) Martin C. Cardiac hypertrophy in chronically anemic fetal sheep: increased vascularization is associated with increased myocardial expression of vascular endothelial growth factor and hypoxia-inducible factor 1. Am J Obstet Gynecol 1998; 178: 527-534.
7) Ozaki H, Yu AY, Della N, Ozaki K, Luna JD, Yamada H, Hackett SF, Okamoto N, Zack DJ, Semenza GL, Campochiaro PA. Hypoxia inducible factor-1alpha is increased in ischemic retina: temporal and spatial correlation with VEGF expression. Invest Ophthalmol Vis Sci 1999;40:182-189.

Chapter **61**

RELATIONSHIP BETWEEN THE GENE EXPRESSION OF C-FOS AND DEGREE OF HYPOXIA IN RAT BRAIN, AS REVEALED BY NEAR-INFRARED SPECTROSCOPY

Yasutomo Nomura, Masataka Kinjo* & Mamoru Tamura*
*Deptpartment of Bio-System Engineering, Faculty of Engineering, Yamagata University, Yonezawa 992-8510 Japan, *Laboratory of supramolecular Biophysics, Research Institute for Electronic Science, Hokkaido University, Sapporo 060-0812 Japan*

Abstract: Hypoxic induction of c-*fos* was studied in rat brains as a function of the cerebral oxygenation state using near-infrared spectroscopy by which the hemoglobin oxygenation state and redox state of mitochondrial cytochrome oxidase can be monitored noninvasively. Following reoxygenation after hypoxia, the expression of c-*fos* and MAP2 mRNAs was determined by reverse transcription-coupled PCR. The expression of MAP2 remained unchanged throughout all conditions from 21 to 8% FiO_2. Under the mildly hypoxic conditions, c-*fos* mRNA was not induced. Hemoglobin was partially deoxygenated but cytochrome oxidase remained fully oxidized. Severe hypoxia, where cytochrome oxidase was reduced, caused a significant induction of c-*fos* mRNA. At this stage, the oxygen concentration in cerebral tissue fell to lower than 10^{-7} M. These data suggest that the decline in oxidative phosphorylation might be a trigger for the induction of c-*fos* mRNA.

Key words: c-*fos* mRNA, cytochrome oxidase, hemoglobin, hypoxia, immediate early genes, near infrared spectroscopy

1. INTRODUCTION

Mammalian cells are critically dependent on oxygen for survival. The molecular mechanisms by which cells sense and respond to a reduction in oxygen are not known, although the findings on the signal transduction pathway have been reported by several authors [1, 2]. A typical example is that hypoxia stimulates transcription of a number of genes crucial to survival in the hypoxic state, as well as apoptosis. The enhancement of transcription of some genes by hypoxia is mainly mediated by both activator protein-1 (AP-1) and hypoxia-inducible factor-1 (HIF-1) [1]. The c-Fos protein is a component of the hypoxia-induced AP-1 complex and c-*fos* gene transcription is induced via a mitogen-activated protein kinase-dependent pathway [2]. However, it is not clear whether these transcription factors are activated by hypoxia-induced changes in energy metabolism or by oxygen shortage directly through an oxygen sensor. To answer this, we attempted to correlate the gene-expression and intracellular oxygenation state in cerebral tissue *in situ*. We selected c-*fos* in the present study because it is known as one of the immediate early genes inducible by hypoxia. Previously, Hazeki and Tamura [3], and then Hoshi *et al.* [4] reported quadruple wavelength analysis of the hemoglobin oxygenation state and the redox state of copper in cytochrome oxidase in rat brain by using near-infrared spectrophotometry, which could successfully differentiate the redox state of cytochrome oxidase from the hemodynamic changes. Measurement of the redox state allows noninvasive estimation of the intracellular oxygen concentration and oxidative phosphorylation level *in vivo*. In the present study, we determined the dependence on the hypoxia-induced c-*fos* transcription of the redox state of cytochrome oxidase using near infrared spectrophotometry.

2. METHODS

2.1 Animals

Adult male Wistar rats (180-200g) were anaesthetized with urethane (1.25g/kg ip). Tracheotomy, and femoral venous and artery catheterizatoin were performed. After being paralyzed with pancuronium bromide (0.1 mg/kg i.v.), rats were ventilated with a positive-pressure rodent respirator. The tidal volume and respiratory rate were adjusted to maintain normal p_aCO_2 (35-40 mm Hg). Skin and muscle overlying the calvaria were removed. We simultaneously measured the mean arterial pressure with near infrared spectrophotometry. Following a 15-min control period, hypoxia was induced by lowering the FiO_2 from 21% to 6-18% for 1h. Unless otherwise stated, at the end of the hypoxic period, the FiO_2 was raised to 21% and maintained at that level throughout the 1-h period of reoxygenation. At the end of reoxygenation, the animals were ventilated with 100% nitrogen to determine the full scale for the optical signals. The brain was rapidly removed from the cranium. The cerebral cortex was dissected in cold saline and frozen by liquid nitrogen. Sham-operated animals, ventilated with 21% O_2 for 2h, served as a control.

2.2 Near infrared spectroscopy

A four wavelength (700, 730, 750, and 805 nm) near infrared spectrophotometer (NIR spectrophotometer, Unisoku, Japan) was used in this study. The changes in the oxygenation state of hemoglobin and in the oxidation state of cytochrome oxidase in the brain were calculated from the changes in absorbance at the above four wavelengths according to the equation derived by Hazeki and Tamura [3] and expressed as a percentage of the full-scale value. The calculated values were averaged ones throughout the whole brain tissue.

2.3 Reverse transcription-coupled PCR

Total cellular RNA of the frozen cortex was prepared with TRIZOL reagent (GibcoBRL Life Tech., USA) . Reverse transcription (RT) of the total RNA and then PCR were performed by RNA PCR Kit (AMV) Ver.2.1 (Takara Biomed., Japan) according to the manufacturer's manual. RT-PCR was carried out in a programmable heating block, Model PC-700 (Astec, Japan). The temperature profile used for RT was 30 min incubation with oligo-dT primer at 60°C and 5 min denaturation at 99°C. For PCR, the following profile was 1 min denaturation at 96°C, 1 min annealing at 68°C, and 2 min elongation at 72°C for both c-*fos* and MAP2. In the case of PCR, the tube was subjected to 21 cycles with primers. We verified that products were indeed obtained in the exponential phase of the amplification. The locations of the PCR primers were between the following positions (5'-3'): c-*fos* (F:338-361; R:861-838), MAP2 (F:4716-4740; R:5226-5202) [5]. PCR solution was subjected to 1.4% agarose gel electrophoresis before SYBR gold staining (Molecular Probe, USA).

3. RESULTS

3.1 Effects of hypoxia on expression of c-fos mRNA

After 60 min of reoxygenation following the ventilation with 10% FiO_2 for 60 min, we observed a substantial increase in c-*fos* mRNA in the cortex.

The increase was detected within 30 min and persisted for as long as 90 min. In contrast to c-*fos*, MAP2 mRNA as an internal standard, remained unchanged throughout 90 min. Thus we could estimate the hypoxic induction of c-*fos* mRNA using the protocols in which rats were ventilated with 21% oxygen for 1 h after hypoxia. As shown in Figure 1A, the c-*fos* expression depended on the degree of hypoxia. From 21% to 14% FiO_2, the c-*fos* mRNA showed only a faint band on the gel. The maximal effect was obtained with 10-8% FiO_2. Animals did not survive with FiO_2 lower than 8%.

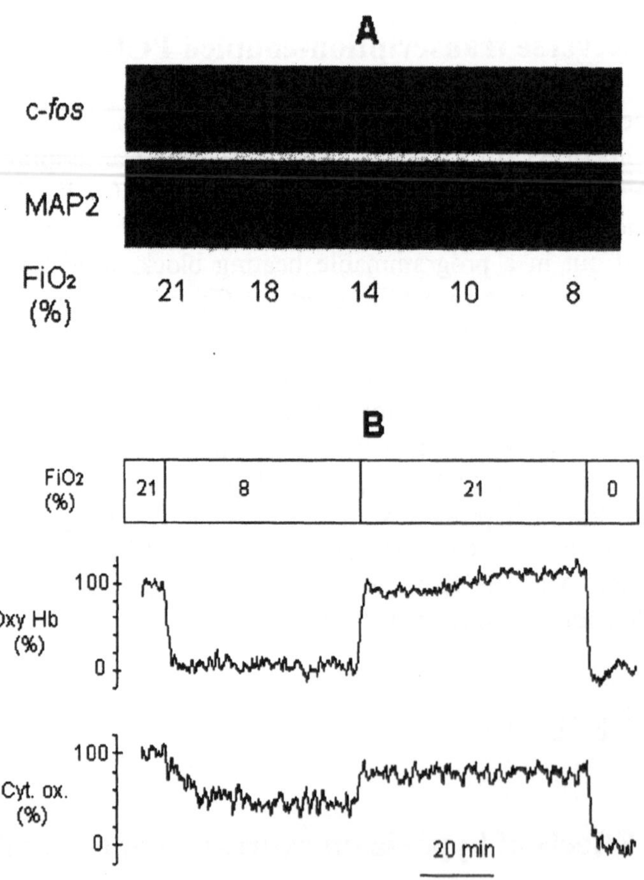

Figure 1. A. mRNA analysis of hypoxia induced c-fos expression. c-fos expression in the different FiO2. Each rat was killed after the end of 1 h reoxygenation following 1 h hypoxia. The expression of MAP2 was also indicated under each condition. **B.** Typical traces of hemoglobin oxygenation and the redox state of cytochrome oxidase in severe hypoxia.

3.2 Effects of hypoxia on cerebral oxygenation state

From the redox response of cytochrome oxidase, we classified animals into two groups, those with mild and severe hypoxia. In the former, hemoglobin was partially deoxygenated (30-80%), but cytochrome oxidase remained fully oxidized. In severe hypoxia, cytochrome oxidase was substantially reduced under the employed hypoxic conditions. As shown in Figure 1B, hemoglobin was almost completely deoxygenated just after lowering FiO_2 to 8%, and cytochrome oxidase started to be reduced. The redox state of cytochrome oxidase became stable after 30 min of hypoxia and became 40%. After the reoxygenation, the rats were ventilated with 100% nitrogen to determine the full scale for the optical signals, where cytochrome oxidase was fully reduced and hemoglobin was completely deoxygenated.

Table 1. Cerebral oxygenation state under various hypoxic conditions and the induction of c-*fos* mRNA

Group Animal No.	FiO2 (%)	Oxy Hb (%)	Cyt. ox. (%)	c-fos
Mild hypoxia				
1	18	77.1	97.8	-
2	18	75.3	95.9	-
3	14	62.1	118	-
4	14	28.8	104	-
Mean (SD)		60.8 (22.3)	104 (10.1)	
Severe hypoxia				
1	14	62.9	67.6	+
2	10	45.4	65.6	+
3	10	3.9	51.2	+
4	8	5.8	46.7	+
5	8	8.4	35.2	+
Mean (SD)		25.3 (27.1)	53.3(13.3)	

Grades of expression, -:no induction, +:substantial induction

3.3 Expression of c-fos and cerebral oxygenation state

Hemoglobin was deoxygenated in both mild and severe hypoxia (Table 1). The c-*fos* mRNA expression tended to be induced in the progressive deoxygenation of hemoglobin. When 62.9% of hemoglobin was oxygenated, c-*fos* mRNA was substantially increased but the mRNA was not induced in a rat with 28.8% of oxygenated hemoglobin. In contrast to hemoglobin deoxygenation, the reduction of cytochrome oxidase was correlated strictly with the induction of c-*fos* mRNA. The expression of c-*fos* mRNA was not induced for any rats in which cytochrome oxidase was not reduced under hypoxic conditions. Inversely, when cytochrome oxidase was substantially reduced, c-*fos* mRNA increased without exception. The mean oxidation state of cytochrome oxidase was 53.3%. In the present protocols, cytochrome oxidase was substantially reduced under the conditions of 10-8% FiO_2.

4. DISCUSSION

Hypoxic induction of c-*fos* mRNA was correlated with the reduction of cytochrome oxidase more strictly than with the deoxygenation of hemoglobin. In near infrared spectrophotometry, we can noninvasively measure the redox state of mitochondrial cytochrome oxidase within cells as an internal marker for the degree of hypoxia. Therefore, we should note the redox state of cytochrome oxidase in order to determine the intracellular activities such as the synthesis of mRNA. In 10% and 14% FiO_2, the redox state markedly changed. It was difficult to control the degree of hypoxia *in vivo* by changing FiO_2 since each animal showed a different response of systemic circulation. However, as shown in Table 1, near infrared spectrophotometry has the advantage of discriminating mild from severe hypoxia concomitant with c-*fos* mRNA induction under the same hypoxic conditions of 14% FiO_2. The expression of c-*fos* mRNA was not induced in this mild hypoxia of 14% FiO_2. Although the cerebral oxygenation state in the present method was the mean value of the whole brain, the regional differences are small according to simultaneous measurement with the regional blood oxygenation level dependent MRI signal [6]. Thus, we can correlate the hypoxic induction of c-*fos* mRNA in the cerebral cortex with the mean oxygenation in near infrared spectrophotometry.

The redox state of copper A, one of the redox centers of cytochrome oxidase, is measured in near-infrared spectrophotometry [7]. The redox in isolated mitochondria is dependent only on the oxygen concentration [8]. Using the calibration *in vitro*, the present data demonstrated that c-*fos* mRNA was not induced with an intracellular oxygen concentration of above 10^{-6}M without reduction. When half of the cytochrome oxidase was reduced at 10^{-7}M, the expression of c-*fos* mRNA was elevated. Considering the findings of Matsunaga *et al.* who showed that there was a parallel relationship between the reduction of copper A and the decrease in ATP content *in vivo* [7], the decline in oxidative phosphorylation can be correlated with the cortical c-*fos* mRNA induction. This suggests that a decrease in ATP might be the trigger for the induction of c-*fos* mRNA. This is supported by a previous work in which fructose 1, 6 phosphate, known to maintain the ATP level in brain slices during hypoxia, suppresses the c-*fos* mRNA induction [9].

Using an O_2-dependent phosphorescence quenching method, Murphy *et al.* reported that hypoxic insult in which the cortical pO_2 was 3-10 mmHg induced expression of hsp72 mRNA in regions of both white and gray matter, with strong expression occurring in the cerebral cortex [10]. This pO_2 was comparable to the value expected from the present hemoglobin oxygenation state. In addition to the mean pO_2 in the blood vessels, our near-infrared spectrophotometry can directly monitor the intracellular oxygen concentration. The present method allows estimation of the intracellular oxygen concentration activating the molecular mechanism by which cells sense a reduction in oxygen.

Fos protein is involved in the adaptive response to neural activity such as convulsions, through alteration of receptor numbers due to membrane depolarization and enhanced firing rates [11]. However, using near-infrared spectrophotometry, Hoshi and Tamura reported that pentylenetetrazole caused the half reduction of cerebral cytochrome oxidase [12]. To determine whether c-*fos* mRNA after seizure is induced by cellular hypoxia or by other mechanisms as above, we are now systematically comparing the relationship between the dose response of c-*fos* induction and the redox state of cytochrome oxidase. Furthermore, using a fluorocarbon-perfused brain model in which the cerebral tissue without hemoglobin can be adjusted to various redox states, we plan to determine the relationship in detail [7].

5. CONCLUSION

Using near-infrared spectrophotometry, the redox state of cytochrome oxidase was measured under hypoxic conditions in which c-*fos* mRNA was substantially induced. The present data indicated that the expression of c-*fos* mRNA remained unchanged above the intracellular oxygen concentration of 10^{-6} M, but was markedly induced at 10^{-7} M.

ACKNOWLEDGEMENTS

This study was supported in part by Grants-in-Aid for Scientific Research (C) (2) (14580761) from the Ministry of Education, Science, Sports and Culture of Japan.

REFERENCES

1. Damert A, Ikeda E, Risau W. Activator-protein-1 binding potentiates the hypoxia-inducible factor-1-mediated hypoxia-induced transcriptional activation of vascular-endothelial growth factor expression in C6 glioma cells. Biochem J 1997; 327: 419-423.
2. Muller JM, Krauss B, Kaltschmidt C, BaeuerlePA, Rupec RA. Hypoxia induces c-fos transcription via a mitogen-activated protein kinase-dependent pathway. J Biol Chem 1997;272:23435-23439.
3. Hazeki O, Tamura M. Near infrared quadruple wavelength spectrophotometry of rat head. Adv Exp Med Biol 1989;248:71-76.
4. Hoshi Y, Hazeki O, Kakihana Y, Tamura M. Redox behavior of cytochrome oxidase in the rat brain measured by near-infrared spectroscopy. J Appl Physiol 1997;83: 842-1848.
5. Ferhat L, Khrestchatisky M, Roisin MP, Barbin G. Basic fibroblast growth factor-induced increase in zif/268 and c-fos mRNA levels is Ca2+ dependent in primary cultures of hippocampal neurons. J Neurochem 1993;61:1105-1112.
6. Kida I, Yamamoto T, Tamura M. Interpretation of BOLD MRI signals in rat brain using simultaneously measured near-infrared spectrophotometric information. NMR Biomed 1996;9:333-338
7. Matsunaga A, Nomura Y,Kuroda S, Tamura M, Nishihira J, Yoshimura N. Energy-dependent redox state of heme a + a3 and copper of cytochrome oxidase in perfused rat brain in situ. Am J Physiol 1998;275:C1022-C1030.
8. Hoshi Y, Hazeki O, Tamura M. Oxygen dependence of redox state of copper in cytochrome oxidase in vitro. J Appl Physiol 1993;74:1622-1627.
9. Hasegawa K, Litt L, Espanol MT, Gregory GA, Sharp FR, Chan PH. Effects of neuroprotective dose of fructose-1,6-bisphosphate on hypoxia-induced expression of c-fos and hsp70 mRNA in neonatal rat cerebrocortical slices. Brain Res 1997;750:1-10.
10. Murphy SJ, Song D, Welsh FA, Wilson DF, Pastuszko A. The effect of hypoxia and catecholamines on regional expression of heat-shock protein-72 mRNA in neonatal piglet brain. Brain Res 1996;727:145-152.
11. Morgan JI, Cohen DR, Hempstead JL, Curran T. Mapping patterns of c-fos expression in the central nervous system after seizure. Science 1987;237:192-197.
12. Hoshi Y, Tamura M. Dynamic changes in cerebral oxygenation in chemically induced seizures in rats: study by near-infrared spectrophotometry. Brain Res 1993 603:215-221.

Chapter **62**

BLOOD VOLUME CHANGES ARE CONTROLLED CENTRALLY NOT LOCALLY - A NEAR-INFRARED SPECTROSCOPY STUDY OF ONE LEGGED AEROBIC EXERCISE

Chris E. Cooper & Caroline Angus
University of Essex, Essex, United Kingdom

Abstract: It is well known that blood flow increases in an exercising limb to match increases in oxygen consumption. However, it is less clear what effects occur in the opposite limb. We performed a one legged incremental cycling protocol and used near-infrared spectroscopy to measure changes in muscle blood volume and oxygenation in the exercising and non-exercising leg. As expected during exercise the exercising leg was deoxygenated relative to the non-exercising leg. However, there were similar increases in blood volume to both legs during the exercise, and during the post-exercise recovery period similar volume *and* oxygenation increases were seen in both legs. We conclude that blood volume increases may be signalled locally, but the effect is expressed globally. Previous studies have demonstrated a training effect in the non-exercising leg following one legged aerobic exercise. The large haemodynamic changes in the non-exercising leg observed here may be partially responsible for the cross-training effect.

Key words: blood flow, exercise, near infrared spectroscopy, muscle, haemoglobin, oxygen consumption

Abbreviations: NIRS; Near-Infrared Spectroscopy

1. INTRODUCTION

Following injury to a limb, athletes often experience a reduction in both central and peripheral factors of aerobic fitness, together with a loss of strength and atrophy in the injured limb [1]. Athletes can maintain the central parameters of aerobic fitness by adopting a one-legged training program during their rehabilitation [1]. Several studies have shown that a local training response can also be measured in the non-exercising leg of individuals that have followed either a one-legged aerobic exercise [2] or isometric and isokinetic [3] training program. This indicates that a relationship exists between the trained and untrained limbs and a cross-transfer or cross-education effect occurs between the trained and untrained sides. Some [3] but not all [4, 5] studies have also shown an improvement in the muscular strength, power and endurance muscles in the untrained limb.

It has also been shown that training one limb causes an increase in the local peak blood flow of both the trained and untrained sides [6]; however, there have been few studies of the haemodynamic changes in the exercising or resting limbs during the one-legged exercise training process itself. Such studies should shed light on both the mechanism of the cross-training effect and the extent to which local and global factors affect blood flow and blood volume changes in the leg. In this study we used Near-Infrared Spectroscopy (NIRS) to measure muscle blood volume and oxygenation changes in the exercising and non-exercising legs during a one-legged aerobic exercise protocol. Surprisingly we found that, although the exercising limb deoxygenated more, there was a very similar increase in blood volume to both limbs.

2. METHOD

2.1 Subjects

Seven men and seven women volunteered to participate in this study. The subjects were all students at the University of Essex, who engaged in exercise regularly but were not following any specific cycling training program. The test procedure and possible risks were explained to each subject before they provided written informed consent. The study was approved by the University of Essex ethics committee. Subjects participated in two testing sessions within a one-week period: they followed the same protocol but exercised their dominant leg during the first session and their non-dominant leg during the second.

2.2 Test protocol

The subjects completed a one-legged incremental exercise test to fatigue. Subjects were asked to maintain a pedal rate of 60 rpm on a cycle ergometer (Monark 824E, Varberg, Sweden). Several modifications were made to allow one-legged exercise to be performed on the ergometer: the foot was secured on the pedal with a toe clip, a counterweight was used on the vacant pedal to facilitate an even pedalling motion (3.2 kg for males and 2.0 kg for females) and the resting leg was placed on a chair beside the ergometer. The initial resistance was set at 90W (males) and 60W (females) and increased every two minutes by 24W and 18W for males and females respectively. This procedure was repeated until the subject reached volitional fatigue and could no longer maintain the specified pedal rate or until no increase in VO_{2peak} was elicited. On termination of the test the subject remained in a seated position on the ergometer and further data were collected during a passive recovery phase lasting approximately two minutes.

Gas exchange measurements were recorded using an Oxycon Beta, breath by breath respiratory gas analyser (Mijnhardt, Bunnik, The Netherlands) for the measurement of inspiratory and expiratory volumes and flows. Heart rate data was collected using a standard heart rate monitor (Polar Vantage NV). Both gas exchange measurements and heart rate were monitored continuously throughout the test period.

2.3 Near-infrared spectroscopy measurements

NIRS measurements were recorded with a probe (Micro Run-Man 96, Philadelphia, PA) placed on both the exercising and resting legs to record data simultaneously during the test. The probes were placed on the vastus lateralis muscles, 150mm from the knee as previously described by Belardinelli et al. [7]. The probe consists of two tungsten lamps placed 6cm apart and two silicone diodes (equidistant from the lamps) that absorb light in the NIR region. The source:detector separation was therefore 3 cm. The two detectors have 20 nm band pass filters one centred at 760 nm and the other at 850 nm. The absorbance changes at 760 and 850 nm were converted to micromolar changes in the concentration of oxyhaemo(myo)globin and deoxyhaemo(myo)globin using simultaneous equations. Previously published absorbance spectra for oxyhaemoglobin and deoxyhaemoglobin [8] were corrected for the wavelength dependence of the optical pathlength [9] and the resultant changes divided by the average optical pathlength for the male and female leg [10]. Although this procedure results in more quantitative data than the muscle "blood volume" and "oxygenation" changes in the standard manufacturers algorithm the essential features of this study were independent of which algorithm was used.

As it is not possible to differentiate between the absorption characteristics of Hb and Mb in the NIR region (700-1000nm) the oxygenation changes observed are likely to be a combination of haemoglobin and myoglobin. However, as the myoglobin concentration will not change in the course of this study, volume changes are essentially entirely due to haemoglobin changes. Thus the data were analysed as micromolar changes in muscle "oxygenation" changes (oxygenated haemo(myo)globin minus deoxygenated haemo(myo)globin) and muscle "blood volume" (oxygenated haemo(myo)globin + deoxygenated haemo(myo)globin).

2.4 Statistical analysis

The physical characteristics of the men and women were compared using independent t-tests. Paired sample t-tests were used to compare muscle oxygenation and blood volume changes in the exercising and resting legs at selected time points during the test. Results were considered to be significant at the 0.05 level of confidence.

3. RESULTS

The maximal physiological responses of the subjects during the exercise test are shown in table 1. All physiological variables except peak heart rate and respiratory quotient were significantly higher in males than females. Heart rate reached significantly higher levels in females and there was no significant difference in peak respiratory quotient values between males and females.

Table 1. Peak physiological responses to one-legged incremental exercise

MEAN (± SD)	MALES	FEMALES
Age (years)	22.2 ± 3.1	19.9 ± 0.6
Height (m)	1.82 ± 0.07	1.64 ± 0.09 *
Weight (kg)	82.3 ± 8.5	64.1 ± 9.5 *
Length of test(s)	498 ± 118	400 ± 96 *
HR_{peak} (bpm)	168 ± 15	181 ± 8 *
$\dot{V}O_2$ (ml/kg/min)	38.3 ± 5.3	33.9 ± 5.1 *
$\dot{V}CO_2$ (ml/kg/min)	42.3 ± 6.7	39.2 ± 6.9 *
Ve (l/min)	101.7 ± 23.4	76.8 ± 15.4 *
Respiratory Quotient	1.14 ± 0.08	1.18 ± 0.07

*Significant gender difference for that variable ($P < 0.05$). Seven males and seven females each performed two tests as described in the methods section, one exercising their dominant leg and the other exercising their non-dominant leg.

Typical responses in muscle oxygenation and blood volume in both the exercising and resting legs are shown in figures 1 and 2 respectively. In this instance the dominant leg was exercised and the test terminated at 495s. In the exercising leg the muscle oxygenation showed an initial decrease from the baseline level within the first minute of the test commencing. The muscle remained deoxygenated until the test ended, when there was a dramatic rise in oxygenation, as oxygen consumption levels returned to normal but blood flow to the muscle remained high. In contrast, in the resting leg the muscle was hyperoxygenated compared to baseline throughout the test, and this effect was enhanced during the recovery period. In the exercising leg, following an initial fall at the onset of exercise, blood volume rose during the remainder of the test. In contrast, the blood volume levels in the resting leg increased continuously immediately from the start of the test.

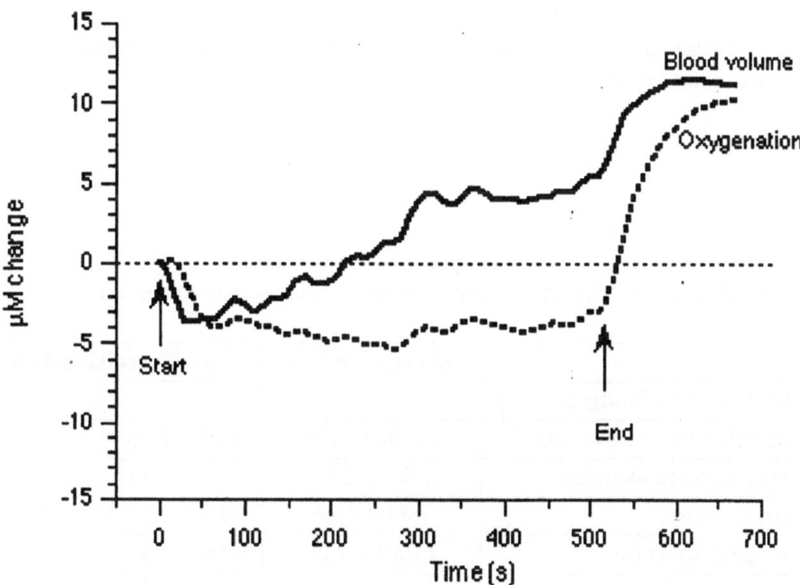

Figure 1. Muscle blood volume and oxygenation changes during one-legged incremental cycling exercise in a typical male: Exercising leg.

Table 2 shows the data from all the subjects. As expected the exercising leg remained deoxygenated relative to the non-exercising leg throughout the test protocol. However, the blood volume changes were less distinguishable.

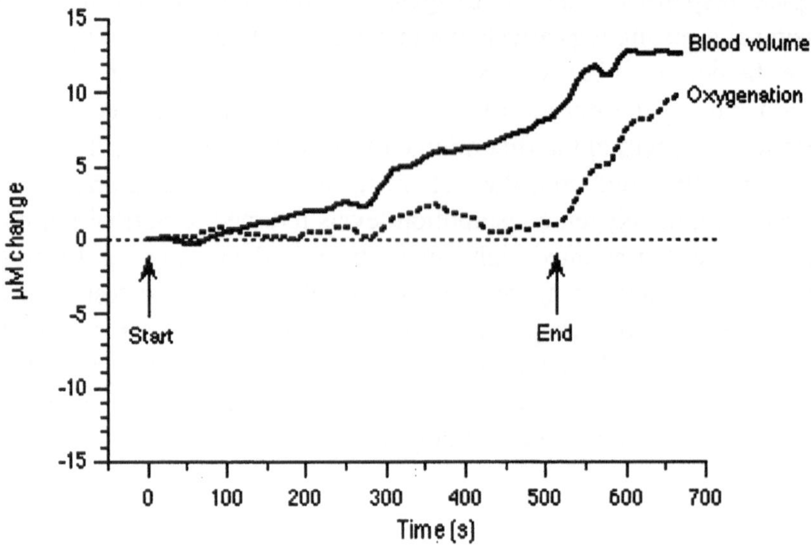

Figure 2. Muscle blood volume and oxygenation changes during one-legged incremental cycling exercise in a typical male: Non-Exercising leg.

Table 2. Haemodynamic responses to one-legged incremental exercise

	Exercising Leg	Non-Exercising Leg
Blood Volume Changes (µM)		
1 min. after start of exercise	-1.96 ± 2.38	0.16 ± 1.28 *
1/2 way through exercise	3.01 ± 2.72	2.62 ± 2.11
End of exercise	7.43 ± 5.66	5.85 ± 3.81 *
2 min. after exercise	11.49 ± 7.00	9.56 ± 4.85 *
Oxygenation Changes (µM)		
1 min. after start of exercise	-.4.04 ± 3.32	0.32 ± 1.34 *
1/2 way through exercise	-2.01 ± 2.87	1.46 ± 2.10 *
End of exercise	0.75 ± 3.67	1.91 ± 3.35 *
2 min. after exercise	9.63 ± 7.76	6.66 ± 4.88 *

*Significant difference for that variable (n=28, $P < 0.05$). Seven males and seven females each performed two tests as described in the methods section, one exercising their dominant leg exercising and the other exercising their non-dominant leg.

Half way through the test the blood volume increases were identical in the two limbs, although there was a significant (30%) difference in the blood volume increase in the two legs by the end of the exercise. During the recovery phase large muscle blood volume and oxygenation increases were seen in both legs; surprisingly the increases in the exercising leg were only on average 15-20% higher than in the non-exercising leg. These results suggest that there is a significant increase in blood flow to both legs during one-legged aerobic exercise.

4. DISCUSSION

Immediately at the onset of exercise there is a reduction in muscle blood volume and oxygenation in the exercising leg, probably due to vasoconstriction [11]. However, blood volume rises subsequently and oxygenation does not decrease further; increases in oxygen delivery can therefore compensate for the increased oxygen consumption as the power increases during the rest of the exercise test. This is consistent with cardiac output not being limiting in one-legged aerobic exercise. At the end of the exercise the large increase in muscle oxygenation is consistent with increased muscle blood flow relative to the baseline.

Thus the results in the exercising leg during this test were largely as predicted from a number of previous NIRS studies [7, 12, 13]. What *is* surprising, however, is how similar the haemodynamic changes are in the non-exercising leg. As expected the non-exercising leg is always more oxygenated than the exercising leg during the test. However, the blood volume rise during exercise is identical in both legs. Furthermore, during the recovery period muscle blood volume *and* muscle oxygenation are almost equally raised in the two limbs (although on average oxygenation and volume are higher in the exercising limb during recovery, for many individuals the recovery data in the non-exercising limb is indistinguishable from that in the exercising limb). We are left with the surprising conclusion that muscle blood flow, at least during the recovery, must be almost identical in both limbs. We are currently testing this hypothesis with direct measurements of blood flow during this period [14]. If true, it would suggest that, although local factors in the exercising limb may signal an increase in blood flow, the dominant effect is expressed centrally and not locally, resulting in large increases in blood flow to both limbs.

The results in this study may have implications for the theory of cross-transfer and it's underlying mechanisms. The existence of a cross-transfer effect as a result of one-legged training has been shown previously; Saltin et al. [2] showed a training response in the untrained leg following a one-legged aerobic training program and Kannus et al. [3] showed a training response in the untrained leg following a one-legged isometric and isokinetic training program.

Several theories have been suggested as the underlying mechanisms behind cross-transfer. When a one-legged *aerobic* training program has been followed it is suggested that cross-transfer occurs as a result of improved oxidative cell metabolism and more economical use of the muscles. Different mechanisms for cross-transfer are suggested after a one-legged *strength* training program has been followed; in this case it has been suggested that there may be an increase in motor units per time and the use of the active motor units may be more economical or effective [15, 16]. Yet the *mechanism* as to how these effects are transferred from the exercising to the non-exercising limb is unclear. Training effects can be "driven" by cardiac output changes or by changes in the peripheral vasculature. This current study demonstrates that there are large haemodynamic changes in the non-exercising leg that are almost equivalent to those in the exercising leg. Thus in principle the training effect could be triggered by these local changes in blood volume, rather than by the global rise in cardiac output. However, it is more likely that changes in blood flow and/or oxygen consumption in the non-exercising limb would be responsible for the training effect. NIRS measurements of blood flow and oxygen consumption require venous and arterial occlusions respectively. It is easier to perform these in the arm, rather than the leg; therefore we are currently performing experiments using one-armed exercise, measuring training effects and comparing them to changes in blood flow and oxygen consumption in the non-exercising limb.

This study has shown that some (perhaps many?) central and peripheral responses to exercise are "blind" to the limb that is exercising; in effect the message from the brain is to increase the muscle blood volume in both limbs. When cast in this light it is perhaps not surprising that long-term training responses can be measured in the non-exercising limb. Although non-exercising, the limb is certainly not "resting" from a haemodynamic viewpoint. Thus by exercising the uninjured limb during their rehabilitation, an athlete may be able to prevent a deterioration in muscle function in an injured limb.

ACKNOWLEDGEMENTS

CEC is grateful for a Wellcome Trust University Award.

REFERENCES

1. Bell G, Neary P, Wenger. HA. The influence of one-legged training on cardiorespiratory fitness. J. Orthopedic and Sports Physiotherapy 1988;10:8-11.
2. Saltin B, Nazar K, Costill DL, Stein E, Jansson E, Essen B, Gollnick PD. The nature of the training response; Peripheral and central adaptations to one-legged exercise. Acta Physiol Scand 1976;96:289-305.
3. Kannus PD, Alosa D, Cook L, Johnson RJ, Renstrom P, Pope M, Beynnon B, Yassuda K, Nichols C, Kaplan M. Effect of one-legged exercise on the strength, power and endurance of the contralateral leg. Eur J Appl Physiol 1992:117-126.
4. Young A, Stokes M, Round JM, Edwards RHT. The effect of high-resistance training on the strength and cross-sectional area of the human quadriceps. Eur J Clin Invest 1983;13:411-417.
5. Tesch PA, Karlson J. Effects of exhaustive, isometric training on lactate accumulation in different muscle fiber types. Int J Sports Med 1984;5:89-91.
6. Yasuda Y, Miyamura M. Cross transfer effects of muscular training on blood flow in the ipsilateral and contralateral forearms. Eur J Appl Physiol 1983;51:321-329.
7. Belardinelli R, Barstow TJ, Porszasz J, Wasserman K. Skeletal-Muscle Oxygenation During Constant Work Rate Exercise. Med Sci Sports Exerc 1995;27:512-519.
8. Matcher SJ, Elwell CE, Cooper CE, Cope M, Delpy DT. Performance Comparison of Several Published Tissue Near-Infrared Spectroscopy Algorithms. Anal Biochem 1995;227:54-68.
9. Essenpreis M, Elwell CE, van der Zee P, Arridge SR, Delpy DT. Spectral dependance of temporal point spread functions in human tissues. Applied Optics 1993;32:418-425.
10. Duncan A, Whitlock TL, Cope M, Delpy DT. A multiwavelength, wideband, intensity modulated optical spectrometer for near infrared spectroscopy and imaging. SPIE 1993;1888:248-257.
11. Taylor JA, Joyner MJ, Chase PB, Seals. DR. Differential control of forearm and calf vascular resistance during one-leg exercise. J Appl Physiol 1989;67:1791-1800.
12. Belardinelli R, Barstow TJ, Porszasz J, Wasserman K. Changes in skeletal muscle oxygenation during incremental exercise measured with near infrared spectroscopy. Eur J Appl Physiol 1995;70:487-492.
13. Bhambhani YN, Buckley SM, Susaki T. Detection of ventilatory threshold using near infrared spectroscopy in men and women. Med Sci Sport Exer 1997;29:402-409.
14. De Blasi RA, Ferrari M, Natali A, Conti G, Mega A, Gasparetto A. Noninvasive measurement of forearm blood flow and oxygen consumption by near-infrared spectroscopy. J Appl Physiol 1994;76:1388-1393.
15. Hakkinen K. Neuromuscular and hormonal adaptations during strength and power training. A review. J Sports Med Phys Fitness 1989;29:9-26.
16. Coyle EF, Feiring DC, Rotkis TC, Cote RWd, Roby FB, Lee W, Wilmore JH. Specificity of power improvements through slow and fast isokinetic training. J Appl Physiol 1981;51:1437-1442.

Chapter 63

A MODELING INVESTIGATION TO THE POSSIBLE ROLE OF MYOGLOBIN IN HUMAN MUSCLE IN NEAR INFRARED SPECTROSCOPY (NIRS) MEASUREMENTS

Louis Hoofd, Willy Colier & Berend Oeseburg
237 Department of Physiology, University of Nijmegen, Geert Groote Plein 21, 6525 EZ Nijmegen, The Netherlands

Abstract: Near Infrared Spectroscopy (NIRS) analyzes infrared light having traveled through tissue, for its oxygenation status. The main chromophore analyzed is hemoglobin (Hb), but in muscle tissue also myoglobin (Mb) is present. Since NIRS cannot discern between these two species experimentally, we did model calculation studies using general data for human muscle. Where such data were not directly available, we derived these from analogous data or straightforward assumptions. Consequently, conclusions have to be drawn cautiously. Solid conclusions are, that myoglobin is an important factor with red muscle, and that it is always partly desaturated, significantly depending on workload. Here, both deoxygenated Hb and Mb as detected by NIRS varied between 0.04 and 0.13 mol/m^3, while the variation in Mb saturation (53-86%) even exceeded that of Hb (63-84%).

Key words: muscle, myoglobin, near infrared spectroscopy, tissue oxygenation

1. INTRODUCTION

Near Infrared Spectroscopy (NIRS) makes use of oxygen-dependent absorption in the near infrared light band. The main species detected is hemoglobin (Hb), and the changes in absorption at different wavelengths are interpreted in terms of changes in oxyhemoglobin (O_2Hb) and deoxyhemoglobin (HHb). In red muscle tissue, however, also the similar protein myoglobin (Mb) is present. The absorption characteristics of this species are at least very similar to those of Hb, so that there is insufficient experimental basis to draw conclusions about its possible part in the NIRS signal.

We developed mathematical models of muscle tissue oxygenation that allow calculation of all the species involved [1, 2]. These models were applied to human muscle tissue, and overall amounts of O_2Hb, HHb, O_2Mb and HMb calculated, for a first impression about the relative importance of Mb in the NIRS signal.

2. METHODS

The mathematical model applied is the Multicapillary Model [2], which allows calculation of pO_2 at any location in the muscle tissue. Since Mb is incorporated in this model, from these pO_2s local myoglobin saturations sMb can be calculated - under the assumption of local equilibrium between pO_2 and sMb. Likewise, from the pO_2s in the capillary blood, local Hb saturations sHb can be determined. Tissue averaged values of these species were assumed to be representative for the influence on NIRS measurements.

For this first investigation, model input data were chosen from textbook data (37°C) or generally accepted values, if possible. Tissue data were: oxygen permeability (or Krogh's diffusion coefficient - note the misleading naming) $\wp O_2 = 1.18\ 10^{-11}$ mol m^{-1} kPa^{-1} sec^{-1} [3], oxygen solubility αO_2= 0.024 L/L/atm [4] leading to a diffusion coefficient $DO_2 = \wp O_2/\alpha O_2 = 1.12\ 10^{-9}\ m^2/sec$, myoglobin p_{50}= 0.7 kPa [5] and a facilitation pressure [6] p_F= 2 kPa consistent with a myoglobin content cMb of 0.28 mol/m^3 assuming DMb/DO_2= 0.075 [5]. The latter is a fairly high value but within the wide range of cMb found in the literature. Blood data were: oxygen capacity 200 mL/L, p_{50}= 3.56 kPa and Hill's n= 2.72, arterial pO_2= 13 kPa, capillary radius 3.5 μm and capillary length 1000 μm (generally accepted values).

For the model, a representative cross section of the tissue is needed with its individual capillary locations. Since no such data were available for human muscle, we generated a pattern based on a realistic capillary density and assuming a lognormal distribution with a domain area log standard deviation of 0.13 as found in animal muscle [7]. This cross section is shown in figure 1, 86 capillaries in a 500 μm by 350 μm field. The block of tissue calculated was 500 μm thick, half the capillary length, allowing capillary staggering as described in the figure legends.

For the relation with NIRS, also assumptions had to be made about the amount of blood detected by the measuring method, since it is generally

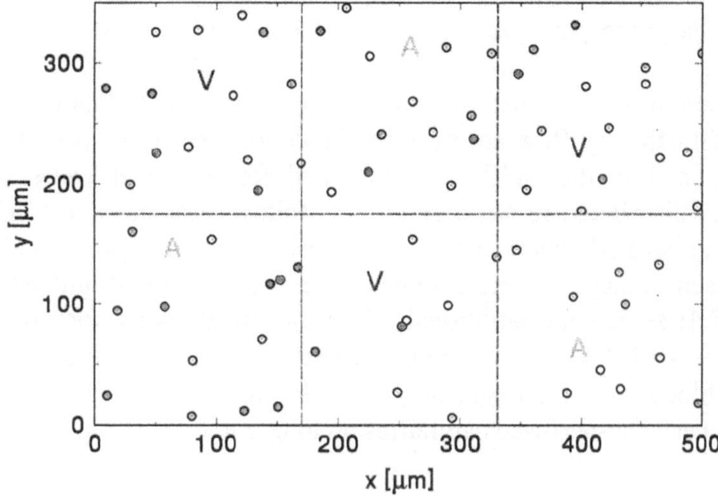

Figure 1. Cross section through muscle used for the calculations. Circles represent capillaries, in zones A starting from an arteriolar pO_2, in zones V from 'half-way pO_2' and draining into a venule. The dots in the capillaries represent flow distribution for 20% maximum work.

assumed that larger vessels absorb all the incoming infrared light and consequently do not contribute to the NIRS signal at the detector. About 5% of muscle volume is blood, whereas from our data we calculated 1.9% capillary blood. Therefore, we assumed that arteriolar and venular blood portions both were half the amount of capillary blood, leaving about 25% of the blood undetected.

Table 1. Conditions for the four calculated cases

Case	$\dot{V}O_2$	Flow	EP
	L/min	L/min	kPa
MAX	0.822	5.63	0.21
40% max	0.329	2.87	0.10
20% max	0.164	1.95	0.05
REST	0.018	0.34	0.035

Four situations were considered, maximum aerobic work (MAX), 40% and 20% of this maximum, and resting state (REST). Data are as in Table 1, based on experimental data for knee extensors, where MAX= 55W [8], and assuming a resting flow of 6% of the maximum (generally accepted value). Values for the Extraction Pressures EP (the oxygen pressure drop due to intracapillary transport) were calculated according to [9]. Maximum aerobic work seems not a realistic situation for NIRS but it served as a 'baseline' here, where it was assumed that all capillaries are open in that case. Data for individual capillary flow are missing. Thus, identical flow in each capillary was assumed in the MAX situation, and for the other cases flow was randomly distributed in steps of 10% as follows: for the ith capillary, flow percentage was ρ_i^k where ρ was a random number between 0 and 1 and k was chosen so that the overall flow had the correct value. Rounding off ρ_i^k to steps of 10% had the additional advantage of allowing for non-perfused capillaries (0% flow), as happens in reality. Such a capillary still might contain blood, but also might not. To make this effect visible, we assumed 25% of these non-perfused capillaries no longer to contain blood.

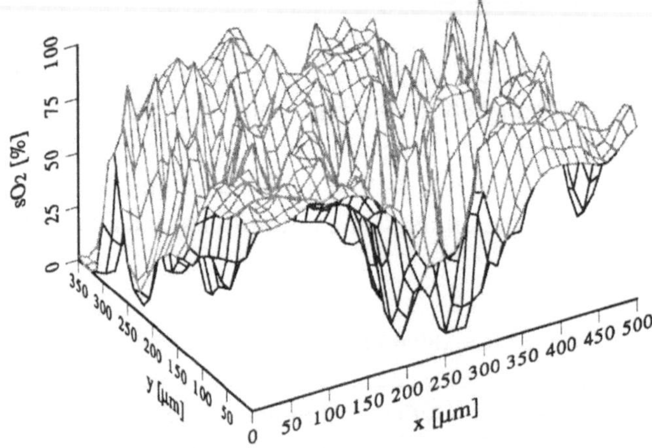

Figure 2. Myoglobin saturation sMb (vertical axis) as a function of location, in a cross section as depicted in figure 1.

3. RESULTS

For the 'maximum work' situation, we calculated considerable variation in myoglobin saturation locally, as shown in figure 2 for one of the cross sections in the tissue block. Averaged sMb was 53% and average pO_2 1.7 kPa in this case - note the discrepancy with the myoglobin saturation curve

(p_{50}= 0.7 kPa) caused by the skew distributions of both pO_2 and sMb. Venous blood saturation s_vHb was calculated from average end-capillary sHb and average blood saturation was the mean of venous and arterial saturation since it was assumed that arteriolar and venular blood portions were equal amounts. For the submaximal cases, end-capillary sHb had to be weighted over capillary flow. The results are gathered in Table 2 and the NIRS relevant results are shown in figure 3. It is clear, that both amounts of and changes in each of the four relevant species are significant.

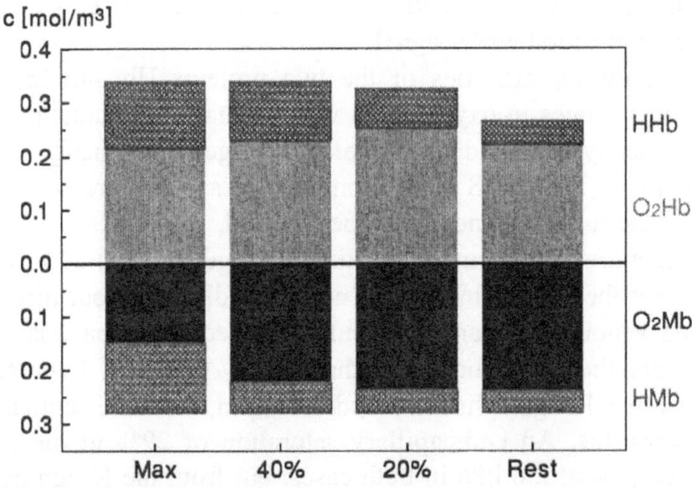

Figure 3. Bar diagram representation of the calculated results relevant for NIRS measurements, showing the relevant overall concentrations of HHb, O_2Hb, O_2Mb and HMb as calculated for the four situations

Table 2. Calculated results for the four work situations

Work sit.	sMb %	cO_2Mb mol/m^3	cHMb mol/m^3	s_vHb %	cO_2Hb mol/m^3	cHHb mol/m^3	open caps	Oxy mol/m^3	Deoxy mol/m^3
Max.	53	0.149	0.131	29	0.215	0.125	86	0.364	0.256
40%	80	0.224	0.056	38	0.230	0.110	86	0.454	0.166
20%	85	0.238	0.042	57	0.252	0.075	73	0.490	0.117
Rest	86	0.239	0.041	70	0.222	0.044	11	0.461	0.084

4. DISCUSSION

Although the above calculation was based on a number of data assumptions, the results are that clear that valid overall conclusions can be drawn. For example, even if only half of the assumed Mb were present, O_2Mb and HMb still are non-negligible. Mb saturation in the MAX state will be low, even for higher capillary densities. Details are more data choice dependent, e.g., the small difference in sMb between the 20% and REST cases results from the different data approaches and easily could be much more significant (the 20% case in fact is an extrapolation of the data of [8]).

Hb concentration in blood is much larger than Mb concentration in muscle tissue, but blood vessel volume is only a low percentage of tissue so that the total amounts of Hb and Mb can be comparable. This means a high Mb amount in the total NIRS signal.

Although the O_2 affinities of the two proteins Hb and Mb are quite different, the changes in oxygenation were in many situations parallel. This means that mostly a marked amount of Mb oxygenation change is present in the NIRS signal. The NIRS absorptions for Hb and Mb are not discernable, so that only the summed amount can be detected.

It is imperative that a model of tissue oxygenation is applied for such estimates. For the blood, things will not or hardly come out different since arterial and venous sHb can be calculated based on global mass transport equations. For the tissue, however, that is very different. Representing the tissue block by a Krogh cylinder of radius 26 μm, the same capillary density, clearly shows this. An end-capillary saturation of 29% in the MAX case means blood pO_2 of 2.6 kPa in both cases, but from the Krogh cylinder we calculated a pressure drop of only 1 kPa to the 'dead corner' boundary leading to a minimum tissue pO_2 of 1.6 kPa so a minimum sMb of 70%. The average sMb would be around 80%. This is way off the results from the multicapillary model, 53% average sMb.

It depends on the interpretation of the NIRS signal whether the Mb changes the conclusions derived from the measurement or not. For instance, when the NIRS signal is interpreted in terms of a change in oxygen, it often does not. An example is, obtaining muscle O_2 consumption in an occlusion experiment [10, 11, 12]. Then, it does not matter whether the O_2 comes from the Hb or from the Mb. An example of the opposite is, when the NIRS signal is used to determine blood volume. Then, O_2 released from or bound to Mb is misinterpreted as related to the Hb in blood.

Indirect effects, however, such as changing the light path due to changing absorption by Mb, cannot be excluded. For evaluating that, a better model including light scattering is needed.

REFERENCES

1. Hoofd L. Calculation of oxygen pressures in tissue with anisotropic capillary orientation. I: Two-dimensional analytical solution for arbitrary capillary characteristics. Math Biosci 1995;129:1-23.
2. Hoofd L. Calculation of oxygen pressures in tissue with anisotropic capillary orientation. II: Coupling of two-dimensional planes. Math Biosci 1995;129:25-39.
3. Krogh A. The number and distribution of capillaries in muscles with calculations of the oxygen pressure head necessary for supplying the tissue. J Physiol 1919;52:409 415.
4. Altman PL, Gibson JF & Wang CC. Handbook of Respiration. Dittmer DS, Grebe RM, editors. Philadelphia & London: Saunders, 1958;6-9.
5. Federspiel WJ. A model study of intracellular oxygen gradients in a myoglobin containing skeletal muscle fiber. Biophys J 1986;49:857-868.
6. Hoofd L. Updating the Krogh model – assumptions and extensions. In: Egginton S, Ross HF, editors. Oxygen Transport in Biological Systems, Soc Exper Biol Seminar Series 51. Cambridge: Cambridge University Press, 1992;197-229.
7. Turek Z, Hoofd L & Rakusan K. Myocardial capillaries and tissue oxygenation. Can J Cardiol 1986;2:98-103.
8. Andersen P, Saltin B. Maximal perfusion of skeletal muscle in man. J Physiol 1985;366:233-249.
9. Hoofd L, Bos C. Extraction pressures calculated for rat heart and dog skeletal muscle and application in models of tissue oxygenation. In: Harrison DK, Delpy DT, editors. Oxygen Transport to Tissue XIX, Adv Exper Med Biol 428. New York: Plenum Press, 1998;679-685.
10. Beekvelt MCP van, Colier WNJM, Engelen BHM van, Wevers RA & Oeseburg B. Validation of measurement protocols to assess oxygen consumption and blood flow in the human forearm by near infrared spectroscopy. SPIE 1998;3194:133-144.
11. Binzoni T, Colier W, Hiltbrand E, Hoofd L & Cerretelli P. Muscle O2 consumption by NIRS: a theoretical model. J Appl Physiol 1999;87:683-688.
12. Colier WNJM, Meeuwsen IBAE, Degens H & Oeseburg B. Determination of oxygen consumption in muscle during exercise using infrared spectroscopy. Acta Anaesth Scand 1995;39:150-155.

Chapter **64**

TISSUE HYPOXIA DURING BACTERIAL SEPSIS IS ATTENUATED BY PR-39, AN ANTI-BACTERIAL PEPTIDE

Philip E. James[1], Melanie Madhani[2], Chris Ross[3], Linda Klei[4], Aaron Barchowsky[4], and Harold M. Swartz[2]

[1] Department of Cardiology, Wales Heart Research Institute, University of Wales College oj Medicine, Wales, U.K.; [2]EPR Center, Radiology Department, Dartmouth Medical School, USA; [3] Department of Anatomy and Physiology, College of Vetinary Medicine, Kansas State University, USA.; [4] Department of Pharmacology and Toxicology, Dartmouth Medical School, USA.

Abstract: Endotoxin (a lipopolysaccharide (LPS) component of the Gram negative bacterial cell wall) induces sepsis in laboratory animals and is the cause of septic shock in patients. Tissues often develop necrotic regions, particularly in kidney and liver, thought to be directly the result of endotoxin-induced release of nitric oxide (NO). These studies investigated the potential of PR-39, an anti-bacterial peptide, as an alternative treatment for sepsis. Our rationale for these experiments was based on the knowledge that PR-39 inhibits the superoxide-producing NADH/NADPH-oxidase system, and also inhibits NOS. In a mouse model of sepsis, we carried out EPR measurements of liver pO_2 and NO simultaneously *in vivo*. Physiological parameters were also measured in these animals (blood pressure, heart rate). NO levels in blood were measured by EPR analysis of red blood cell nitrosyl-hemoglobin. We found PR-39 alleviated endotoxin-induced liver hypoxia 6 hrs after treatment. Tissue NO was higher in the PR-39+LPS group compared to LPS alone. Circulating levels of NO were the same in these groups. Taken together, these results suggest PR-39 is effective in improving survival following a septic episode. The exact mechanism is unclear, but increased NO as a result of decreased superoxide production seems to play an important role in alleviating tissue hypoxia.

Oxygen Transport to Tissue XXIV, edited by
Dunn and Swartz, Kluwer Academic/Plenum Publishers, 2003

1. INTRODUCTION

Endotoxin (LPS), a lipopolysaccharide component of the Gram negative bacterial cell wall induces sepsis in laboratory animals and is the cause of septic shock in patients [1]. The septic episode is usually characterised by a severe decrease in mean arterial blood pressure and altered tissue perfusion, leading to multiple system organ failure [2]. Tissues often develop necrotic regions, particularly in kidney and liver. These effects are thought to be directly the result of endotoxin-induced release of nitric oxide (NO) at early (within minutes) and especially late (>4hr) stages during a septic episode [3]. These times correspond to induction of constitutive (cNOS) and inducible (iNOS) forms of the NO producing enzyme, respectively [4]. Inhibition of NO synthesis, however, only incompletely alleviates septic events; the decrease in MABP can be prevented and the lifespan prolonged, but animals eventually die and suffer enhanced liver necrosis. Treatment of septic animals with superoxide dismutase can be effective in reducing the liver necrosis, but decreased MABP and the resulting organ failure are usually lethal [5].

Our studies investigated the potential of PR-39, an anti-bacterial peptide, as an alternative treatment for sepsis. Our rationale for these experiments was based on the knowledge that PR-39 inhibits the superoxide-producing NADH/NADPH-oxidase system, and also inhibits NOS. Proline rich (PR-)peptide 39 is an antibacterial peptide that was first isolated from pig intestines [6], and more recently from other sources such as human neutrophils [7]. It has been shown to exhibit antibacterial activities, including killing bacteria by halting protein and DNA synthesis [8], as a NADPH oxidase inhibitor [9] and also as a neutrophil chemoattractant [10]. Furthermore, it has been suggested that PR39 may participate in wound repair and inflammation [11].

We have demonstrated that PR-39 decreased liver myeloperoxidase activity and also expression of iNOS in a mouse model of sepsis (unpublished data). Taken together with our finding that PR-39 prevents postischemic microvascular dysfunction by abolishing leukocyte NADPH-oxidase activation and production of reactive oxygen species *in vivo* in rats [12]; we assessed what affect PR-39 might have on tissue levels of NO and pO_2. In a mouse model of sepsis, we carried out simultaneous EPR measurements of liver pO_2 and NO *in vivo*. Physiological parameters were also measured in these animals (blood pressure, heart rate). NO levels in blood were measured by EPR analysis of red blood cell nitrosyl-hemoglobin at liquid nitrogen temperature. PR-39 (or vehicle only) was given at the onset of the septic episode.

2. MATERIALS AND METHODS

2.1 EPR oximetry

Gloxy (an oxygen-sensitive, paramagnetic material was injected directly into the liver. Typically, two 200 *u*m pieces were injected through a 22 G needle, implanting the gloxy at the desired tissue location so as to report on the local pO_2. The spectral characteristics and oxygen-sensitivity of gloxy have been published previously [13]. In brief, the EPR spectral linewidth obtained from gloxy is broadened by oxygen. The rate of change in linewidth is particularly sensitive at lower oxygen tensions (i.e. <30 mm Hg) making it particularly suitable for making measurements of pO_2 from tissue [14]. The peak-to peak linewidth was measured using a spectral simulation package (EWVoight software) and the pO_2 calculated by calibrating against known oxygen tensions.

2.2 *In vivo* spin-trapping of nitric oxide

Mice were given (subcutaneously, lower leg muscle) ferrous sulfate (50 mg/Kg) and disodium citrate dihydrate (250 mg/Kg) in saline, and then DETC (250 mg/Kg) was administered intra-peritoneally. Mice were then placed supine between the EPR magnet poles so that the organ of interest was within the active volume of the resonator. EPR spectra were recorded repeatedly (every 2 min.) from 30 min. to 90 min. post-DETC/Fe injection. The maximum signal amplitude detected was used as a measure of $NO^{\cdot}$ level in each mouse, reflecting the rate of $NO^{\cdot}$ production over the investigation period. In order to measure the level of $NO^{\cdot}$ at a particular time point during sepsis, animals were injected with DETC/Fe at different times relative to the stimulus, allowing approximately 60 min for achieving peak EPR signal intensity.

2.3 Animals

The experiments were performed on anaesthetised spontaneously breathing male Balb/C mice (approx. 20 g body weight). Animals were anaesthetised using isofluorane (1.5%) inhalation via a cone mask. Core body temperature (measured rectally) was maintained at 37^0C by using a heated operating table. Hemodynamic parameters were monitored via canulation of the left carotid artery with PE 10 tubing. This was linked via a small pressure transducer to an IBM driven Biopac system for recording of heart rate (HR) and mean arterial blood pressure (MABP).

2.4 EPR measurement of tissue nitric oxide and pO_2

A loop gap resonator ("whole body" resonator) 3.0 cm in diameter, with a 2 cm active region with two bridged gaps (designed and constructed in our laboratory) was used in conjunction with an EPR spectrometer equipped with a microwave bridge operating at 1.1 GHz.

Because the signals from gloxy and NO-Fe-$(DETC)_2$ did not overlap, we were able to monitor these signals simultaneously (14). Spectrometer conditions, however, were generally optimised for either $NO^{\cdot}$ or pO_2 because over modulation of the EPR spectrum of gloxy could, potentially, give rise to over-estimation of the pO_2 reported. Under these conditions, alternate pO_2 or NO measurements had to be made. A modulation amplitude of 0.8 Gauss was typically used for measurements of NO-Fe-$(DETC)_2$ to enable comparisons of signal amplitude to be made, whereas the modulation amplitude was adjusted for recording of accurate linewidth from Gloxy (approximately one third the observed linewidth was used as rule of thumb).

2.5 Septic model and PR-39 treatment

Endotoxin (1mg/20g body weight) was injected intraperitoneally in 100 ul pyrogen-free saline. Control animals received the same volume of saline only. Treatment groups were also set up in which animals were given either PR-39 (0.41mg/Kg body weight; 100ul in saline given i.p) at the same time as LPS, or PR-39 alone.

3. RESULTS

NO data measured from mouse liver after treatment with LPS, LPS+PR-39, PR-39 in saline, or controls (saline alone) are shown in Figure 1. LPS caused an increase in liver NO, peaking at 5 hrs post treatment, and remaining elevated at 6 and 7 hrs. Levels of NO remained similar to baseline values in both saline, and PR-39 treated controls. Co-treatment of animals with LPS and PR-39 resulted in a further increase in liver NO at 6 hrs which was greater than that observed in LPS treated animals (LPS=0.664+/-0.164; LPS+PR39=1.176+/-0.215; p=0.005; n=8 for each).

Measurements of liver pO2 were also undertaken in these same animals. We were specifically interested in the 6 hr time point following LPS treatment, because this corresponded to maximal liver hypoxia in LPS treated animals [14]. The pO2 results 6 hrs following treatment are summarized in Figure 2, and show how liver hypoxia was alleviated when

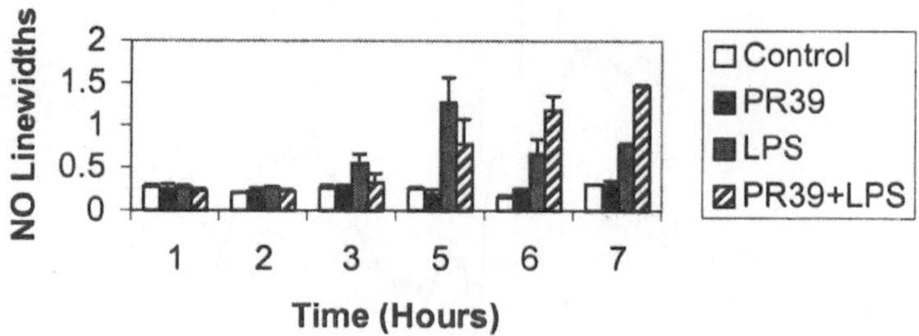

Figure 1. Liver NO-Fe-(DETC)2 in the various treatment groups as a function of time post treatment. n=8 for each group.

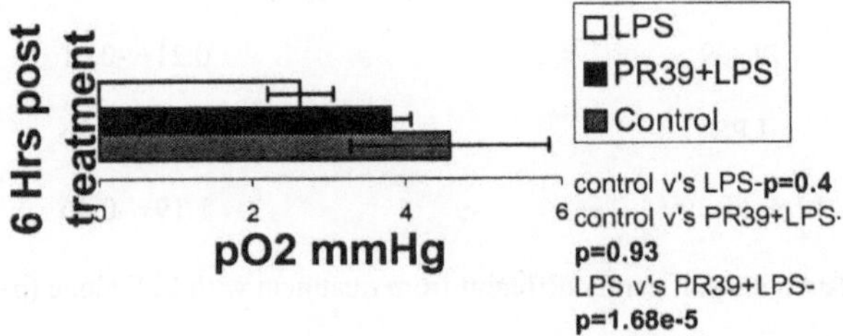

Figure 2. Liver pO_2 after 6 hrs treatment. n=6 for each group.

3A

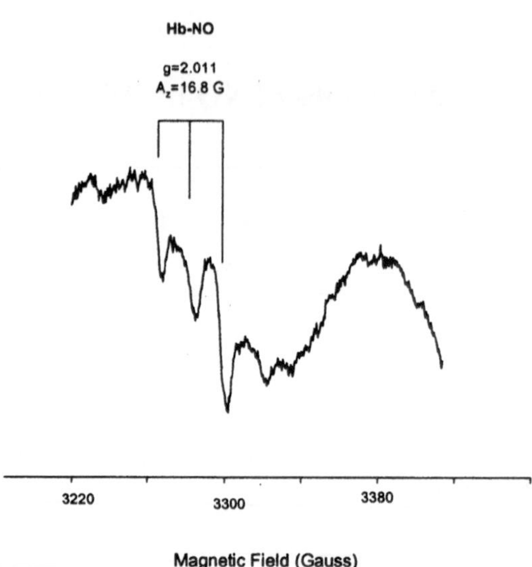

3B Table summarizing Hb-NO data from mice in various treatment groups

Treatment Group	Hb-NO (a.u.+/-s.d.; n=4 each)
CONTROL	0.27+/-0.09
PR-39	0.21+/-0.01
LPS	1.34+/-0.36
LPS+PR39 *	1.19+/-0.46

*** - data not significantly different from treatment with LPS alone (p=0.294).**

Figure 3. Measurement of circulating nitrosylated Hemoblobin (Hb-NO). A typical spectrum recorded from a red blood cell sample obtained from a septic mouse 6 hrs following LPS. EPR measurements were carried out on a Varian E4 spectrometer operated at 9.5 GHz with 100-KHz modulation frequency at liquid nitrogen temperature, with samples held in a quartz finger dewar.

PR39 was given simultaneously with LPS (control at 6 hrs = 4.56+/-1.28 mm Hg; LPS = 2.64+/-0.60 mm Hg; LPS+PR39 = 3.80+/-0.13 mm Hg).

Mean arterial blood pressure (MABP) remained stable throughout the investigation period in control and PR39 treated animals. LPS treatment resulted in a characteristic fall in MABP, reaching minimum values at 6 hrs. This was particularly evident after 4 hrs LPS treatment. Similar results were observed from animals given LPS and PR39 simultaneously. Despite no significant difference in LPS-induced hypotension, the survival rate was increased in the LPS+PR39 group (from 7-8 hrs in LPS treated animals to 17 hrs with LPS+PR39; p= 0.036).

Measurement of nitrosyl hemoglobin was undertaken in red blood cell samples from mice in the various treatment groups. The results are summarized in Figure 3 (and table). We found high levels of circulating nitrosyl-hemoglobin in both LPS and LPS+PR39 groups, but no significant difference between these groups (p=0.294; n=5 each).

4. DISCUSSION AND SUMMARY

We found:

1. LPS induced liver hypoxia at times corresponding with increased levels of tissue NO.
2. LPS-induced liver hypoxia was alleviated by PR39 treatment.
3. Levels of liver NO at 6 hrs in animals given PR-39+LPS were higher than in animals given LPS alone (despite marked inhibition of iNOS expression in the former).
4. Survival improved with PR-39 treatment and histological assessment of liver tissue showed little evidence of tissue necrosis.

These results show that PR-39 is able to reduce LPS-induced tissue hypoxia. The exact mechanism is not yet completely understood, but inhibition of superoxide and NO production (as evidenced by reduced myeloperoxidase activity and iNOS expression), taken together with previous *in vivo* data demonstrating inhibition of NADPH-oxidase activity by PR-39 *in vivo* [12] are important. The overall decrease in damaging oxidising species, simultaneously increasing the level of bio-available NO at the cellular/tissue level, is likely to play an important role in improving tissue oxygenation and survival in this sepsis model.

ACKNOWLEDGEMENTS

Supported by the Superfund Basic Research Program grant, NIEHS at NIH, ES 07373 and "Measurement of pO_2 in Tissues *In Vivo* and *In Vitro*," NIGMS at NIH, GM 51630.

REFERENCES

1. Cobb, J.P. and Danner, R.L. Nitric oxide and septic shock. JAMA 1996;275:1192-1196.
2. Wright, C.E., Rees, D.D. and Moncada, S. Protective and pathological roles of nitric oxide in endotoxic shock. Cariovascular Research 1992; 26:48-57.
3. Nathan, C.F., Stuehr, D.J. Does endothelium-derived nitric oxide have a role in cytokine-induced hypotension? J Natl Cancer Inst *1990;*82:726-728.
4. Nathan, C., Xie, Q-W. Nitric oxide synthases: roles, tolls, controls. Cell 1994;78:915-918.
5. Di Silvio, M., Nussler, A.K., Geller, D.A., and Billiar, T.R. A role for nitric oxide in liver inflammation and infection. In:Nitric Oxide: Principles and Actions (Lancaster, J. Jr., ed), Academic Press, 1996; 219-236.
6. Agerbeth, B., Lee, J., Bergman, T., Carlquist, M., Boman, H. G., Mutt, V., and Jornvall, H.: Amino acid sequence of PR39 – Isolation from pig intestine of a new member of the family of proline-argine-rich antibacterial peptides. European Journal Biochemistry. 1991;202: 849-854.
7. Shi, J., Ross, C.R., Chengappa, M.M., and Blecha, F. Identification of a proline-rich antibacterial peptide from neutrophils that is analogous to PR-39, and antibacterial peptide from the small intestine. J Leukocyte Biol 1994; 56:807-811.
8. Boman, H. G., Agerbeth, B., and Boman, A.: Mechanism of action on Escherichia coli of cecropin P1 and PR39, two antibacterial peptides from pig intestine. Infection and Immunity. Biochemical and Biophysical research communication.1993;61:29782984.
9. Shi, J., Ross, C. R, Leto, T. L, and Blecha, F.: PR29, a proline-rich antibacterial peptide that inhibits phagocyte NADPH oxidase activity by binding to Src homolgy 3 domains of p47phox . Preoceedings National Academy of Science, USA.1996; 93:6014-6018.
10. Huang, H., Ross, C. R, and Blecha, F.: Chemoattractant properties of PR39, a neutrophil antibacterial peptide. Journal of Leukocytes Biology 1997;61:624-628.
11. Gallo, R. L, Ono, M., Povsic, T., Page, C., Eriksson, E.: Syndecans, cell surface heparan sulfate proteoglycans, are induced by a proline-rich antibacterial peptide from wounds. Proceedings National Academy of Science, USA 1994;91:11035-11039.
12. Korthuis, R.J., Gute, D.C., Blecha, F., Ross, C.R. Pr-39, a proline/arginine-rich antimicrobial peptide, prevents postischemic microvascular dysfunction. Am J Physiol 1999;277:H1007-H1013.
13. James, P.E., Grinberg, O.Y., Goda, F., O'Hara, J.A., and Swartz, H.M. Gloxy: an oxygen-sensitive coal for accurate measurement of low oxygen tensionis in biological systems. Mag Res Med 1997; 38:48-58.
14. James, P.E., Miyake, M., and Swartz, H.M. Simultaneous measurement of NO and pO_2 from tissue by *in vivo* EPR. Nitric oxide; Chemistry and Biology 1999; 3:292-301.

Chapter **65**

EFFECTS OF THE CONTRAST MEDIUM IOPROMIDE ON RENAL HEMODYNAMICS AND OXYGEN TENSION IN THE DIABETIC RAT KIDNEY

Fredrik Palm[1,2], Per-Ola Carlsson[1], Angelica Fasching[1], Olof Hellberg[3], Anders Nygren[2], Peter Hansell[1] & Per Liss[2]

[1]Department of Medical Cell Biology, [2]Department of Diagnostic Radiology, [3]Department of Internal Medicine, Biomedical Center, Box 571, SE-751 23, Uppsala University, Uppsala, Sweden

Abstract: We investigated the effects of the contrast medium (CM) iopromide on regional renal blood flow and oxygen tension (pO_2) in the streptozotocin (STZ)-induced diabetic Wistar Furth rats. *Results.* In normoglycemic rats, CM injection induced a transient decrease followed by an increase in renal cortical blood flow (CBF), whereas CBF increased directly in the diabetic animals. Renal outer medullary blood flow (OMBF) increased in controls, while it decreased in the diabetic animals following CM injection. In control rats a marked initial decrease in OM pO_2 following injection of CM was observed. In animals diabetic for 4 weeks only a slight decrease was seen, whereas in 9-week diabetic animals a persistent increase was recorded. *Conclusions.* An altered oxygen tension and hemodynamic response to CM was found in diabetic rats. If these disturbances may contribute to the development of renal dysfunction by CM in the diabetic rat kidney remains to be elucidated.

Key words: Clark-type microelectrode; diabetic kidney; renal blood flow; renal oxygen tension

Abbreviations of medical terms:

CBF	cortical blood flow
CM	contrast medium
mTAL	medullary thick ascending limb
OMBF	outer medullary blood flow
pO_2	oxygen tension
STZ	streptozotocin

1. INTRODUCTION

The development of renal dysfunction on the basis of insulin-dependent diabetes is the most common cause of end-stage renal disease [1]. The mechanisms causing this decrease in renal function in diabetic patients remain unknown. However, a number of both morphological and functional disturbances have been reported. Histological changes described in diabetic kidneys involve glomerular base membrane thickening [2], renal enlargement [3], diffuse glomerulosclerosis [4] and an increased mesangial matrix area [5]. Proteinurea [6], increased renal plasma flow [7] and increased metabolism [8] are some of the occurring functional disturbances. In this study, we investigated if the response to contrast medium (CM) with respect to renal blood flow and oxygen tension (pO_2) in the diabetic rat kidney is different from that of the non-diabetic rat kidney.

2. MATERIALS AND METHODS

2.1 Animals

The experiments were performed on inbred male Wistar Furth rats weighing 230–300 g, purchased from M&B A/S (Ry, Denmark). Throughout the study, the animals had free access to tap water and standard rat chow (R3, Ewos, Södertälje, Sweden). All experiments were approved by the local animal ethics committee at Uppsala University, Sweden.

2.2 Induction of diabetes

Diabetes was induced by intravenous injection of streptozotocin (STZ, 55 mg/kg BW) (Pharmacia-Upjohn, Kalamazoo, MI, USA) 4 or 9 weeks prior to the experiments. Blood glucose concentrations were determined with test reagent strips (MediSense, Bedford, MA, USA) from blood samples obtained from the cut tip of the tail. Animals were considered diabetic if blood glucose concentration rose 12 mM 48 hours after injection. The blood glucose concentrations and the weight of the animals were monitored throughout the study.

2.3 Surgical procedures

The animals were anesthetized with an intraperitoneal injection of thiobutabarbital (80 mg/kg BW) (Inactin®, Research Biochemicals

International, Natick, MA, USA), placed on an operating table maintained at 37°C and tracheostomized. Polyethylene catheters were placed in the right femoral artery and both femoral veins. The arterial catheter was used for monitoring blood pressure (Statham P23dB, Statham Laboratories, CA, USA). Animals with a mean arterial blood pressure < 80 were excluded from the study. The catheter in the left femoral vein was used for infusion of CM; the right vein was used for continuous infusion of saline to make up for body fluid loss. The urinary bladder was catheterized for draining and the urine was discarded.

The left kidney was exposed by a left subcostal flank incision, immobilized in a plastic cup and embedded in pieces of cotton wool soaked in saline. The surface of the kidney was covered with paraffin oil (Apoteksbolaget, Gothenburg, Sweden). Throughout the experiment, the temperature of the kidney was monitored with a thermocouple probe (CT D85, Ellab, Copenhagen, Denmark). The left ureter was catheterized to prevent stasis.

2.4 Measurements of renal pO_2 and blood flow

Renal pO_2 was measured with a polarographic technique using modified Clark-type microelectrodes (Unisense, Aarhus, Denmark) [9] and applied to the kidney as described by Liss et al [10]. A linear correlation was obtained between pO_2 and electric current. The latter was measured by picoamperometry (University of Aarhus, Aarhus, Denmark). The electrodes were calibrated in water saturated with $Na_2S_2O_2$ or air at 37°C. The microelectrode was inserted into the renal medulla by means of a micromanipulator under a stereomicroscope.

The renal outer medullary blood flow (OMBF) and cortical blood flow (CBF) were measured by laser-Doppler flowmetry (PF 4001-2, Perimed, Stockholm, Sweden). A needle probe (411, tip 0.45 mm o.d., Perimed, Stockholm, Sweden) was inserted into the renal medulla using a micromanipulator, and a second probe was placed on the surface of the kidney to simultaneously record the CBF. Renal outer medullary pO_2, OMBF and CBF, blood pressure, and body and kidney temperature were continuously recorded with a MacLab Instrument (AD Instruments, Hastings, UK) connected to a Macintosh Power-PC 6100.

2.5 Experimental protocol

After the surgical procedures, the animals were allowed a 60-minute recovery period. Following 60 minutes of control measurements, CM was injected and the measuring parameters were recorded during the subsequent 60 minutes.

The CM iopromide (1600 mg I/kg BW) (Ultravist® 300 mg I/ml, Schering, Berlin, Germany) or volume-matched saline infusion was given over a period of 2 minutes. All solutions were administered at room temperature.

2.6 Statistical analysis

All values are given as means ± 1 standard error of the mean (SEM). Between groups with different treatments, analysis of variance (ANOVA) followed by Fisher's protected least significant differences (PLSD) test was used (Statview, Abacus Concepts, Berkeley, CA, USA). For all comparisons, $P < 0.05$ was considered statistically significant.

3. RESULTS

3.1 Renal CBF

In the control rats, CM caused a transient decrease followed by an 18 ± 6% increase in CBF (n=6; Table 1). In the 4-week diabetic group (n=7), a transient increase of 10 ± 4% in CBF was seen, whereas the animals diabetic for 9 weeks (n=3) responded with a 20 ± 7% increase which remained throughout the experiment (Table 1).

3.2 Renal OMBF

The control rats responded with a 40 ± 14% increase in OMBF after CM injection. The 4-week diabetic group, by contrast, responded with an 16 ± 9% decrease in renal OMBF (Table 1). In the 9-week diabetic rats, OMBF initially decreased by 15 ± 4%, which was followed by a 12 ± 6% increase (Table 1).

3.3 Renal outer medullary pO_2

In non-diabetic control rats, injection of CM initially reduced pO_2 by 62 ± 10% (Table 1). In the 4-week diabetic group, outer medullary pO_2 initially decreased by 28 ± 10% followed by a 14 ± 7% increase, whereas in the animals diabetic for 9 weeks, pO_2 paradoxically increased by 30 ± 22% immediately after the injection (Table 1).

Table 1

Renal blood flow and oxygen tension before and after intravenous injection of the contrast medium iopromide in non-diabetic and streptozotocin-induced diabetic rats diseased for 4 and 9 weeks.

	Cortical blood flow (% of control)					Outer medullary blood flow (% of control)					Outer medullary oxygen tension (mmHg)				
Time after CM injection (min)	Control	1	2	3	60	Control	1	2	3	60	Control	1	2	3	60
Non-diabetic animals	100	99±15	84±13	92±16	118±6	100	104±7	107±5	109±4	140±14	29±6	11±3	12±2	14±3	31±3
4-week diabetic animals	100	110±4	109±5	108±6	96±4*	100	102±6	95±9	92±6*	84±9*	29±3	28±3*	22±3*	21±3	33±2
9-week diabetic animals	100	120±7	118±9	115±8	107±10	100	103±4	91±4	85±4*	112±6*	23±4	30±5*	25±5*	22±5	35±4

All values are expressed as means ± SEM for 3–7 experiments.

* $P<0.05$ compared with corresponding non-diabetic animals.

4. DISCUSSION

In the present study, we demonstrate that the effects of injection of the CM iopromide on outer medullary pO_2 and regional renal blood flow in the diabetic rat kidney are both qualitatively and quantitatively different from the response in the non-diabetic rat kidney. Our previous studies [10–11] support the present findings of an initial decrease in pO_2 after CM injection in the outer medulla in normal rats. The most plausible explanation for this decrease in pO_2 is that the CM induces an increase in sodium load through osmotic forces in the tubules, which causes an increased reabsorption in the medullary thick ascending limb (mTAL) of the loop of Henle [12–13]. This increased reabsorption is due to an increased activity of the Na-K-ATPase in the mTAL, causing increased oxygen consumption in this region [14]. This reabsorption accounts for about 80% of the "workload" of the kidney. The reasons for the altered response in the diabetic kidney are unclear. Possible mechanisms underlying the different response include vascular and metabolic disturbances in the diabetic kidney.

It has previously been reported that an increase in the total renal blood flow immediately follows CM administration in diabetic animals [15]. When measuring renal blood flow in diabetic Wistar Furth rats in the present study, we found different responses to CM, compared with non-diabetic animals, in different parts of the kidney (i.e. the cortex and medulla). In the diabetic kidney, the CBF increased whereas the OMBF decreased. The opposite was seen in the non-diabetic kidney. An increase in OMBF after injection of CM is in contrary to our prior studies in Lewis-DA and Sprague-Dawley rats [10-11, 16-17]. The reason for this opposite effect in Wistar Furth rats is unclear.

Previous reports have suggested several possible explanations for the increase in OMBF in non-diabetic animals. One is that the hypertonic CM may induce a release of nitric oxide [18]. Other potent vasodilators such as adenosine [19] and prostaglandin E_2 [20] have been reported to increase in the urine after CM has been administered. It may be speculated that the altered vascular response seen in the diabetic kidney might be an effect of vascular dysfunction in medullary blood vessels. Previous investigations have demonstrated such vascular dysfunction in the diabetic kidney [21]; others have proposed that an altered nitric oxide system may be involved [22–23].

The metabolic disturbances present in diabetic animals *per se* may also affect the renal tissue pO_2 and vascular blood perfusion. Hyperglycemia has been reported to disturb the NAD/NADH balance and thus, cause a state of pseudohypoxia [23]. The pseudohypoxic state has in several organs been described to be associated with vascular hyperperfusion [24].

To conclude, a decrease in renal outer medullary pO_2 following CM injection was evident in control rats; by contrast, diabetic animals demonstrated an increase in pO_2. Moreover, an altered hemodynamic response in the renal cortex and medulla of diabetic animals was seen following CM administration. To what extent these disturbances may contribute to an increased risk of developing renal dysfunction upon CM administration in the diabetic rat kidney remains to be elucidated

ACKNOWLEDGEMENTS

The skilled technical assistance of Astrid Nordin is gratefully acknowledged. Financial support for this study was provided by the Marcus and Amalia Wallenberg Foundation, the Konsul Thure Foundation, the Knut and Alice Wallenberg Foundation and the Swedish Medical Research Council (proj. No. 10840 to PH).

REFERENCES

1. US Renal Data System. URDS Annual Data Report. National Institutes of Health, National Institute of Diabetes and Digestive Kidney Disease, 1992.
2. Wehner H, Hohn O, Faix-Schade U, Huber H, Walzer P. Glomerular changes in mice with spontaneous hereditary diabetes. Lab Invest 1972;27:331–340.
3. Mogensen CE, Anderson MJ. Increased kidney size and glomerular filtration rate in untreated juvenile diabetes. Normalization by insulin treatment. Diabetologia 1975;11: 221–224.
4. Kimmelstiel P. Structural Basis of Renal Disease. New York, NY, Harper & Row, 1968, p 468.
5. Østerby R. Early phases in the development of diabetic glomerulopathy. Acta Med Scand 1974;574(suppl): 3–82.

6. Carrie BJ, Mayers BD. Proteinurea and functional characteristics of glomerular barrier in diabetic nephropathy. Kidney Int 1980; 17:669–676.
7. Ditzel J, Junker K. Abnormal glomerular filtration rate, renal plasma flow and renal protein excretion in recent and short-term diabetics. Br Med J 1972; 2:13–19.
8. Körner A, Eklöf A-C, Celsi G, Aperia A. Increased renal metabolism in diabetes. Diabetes 1994; 43:629–633.
9. Revsbech NP. An oxygen microsensor with a guard cathode. Limnol Oceanogr 1989; 34(2): 474–478.
10. Liss P, Nygren A, Revsbech NP, Ulfendahl H. Intrarenal oxygen tension measured by a modified Clark electrode at normal and low blood pressure and after injection of x-ray contrast media. Pflügers Arch 1997; 434: 705–711.
11. Liss P, Nygren A, Ulfendahl HR, Eriksson U. Injection of low and iso-osmolar contrast medium decreases oxygen tension in the renal medulla. Kidney Int 1998; 53: 698–702.
12. Giebisch G, Klose RM, Windhager EE. Micropuncture study of hypertonic sodium chloride loading in the rat. Am J Physiol 1964; 206: 687–693.
13. Caldicott WJ, Hollenberg NK, Abrams HL. Characteristics of response of renal vascular bed to contrast media. Evidence for vasoconstriction induced by renin-angiotensin system. Invest Radiol 1970; 5:539–547.
14. Ullrich KJ, Pehling G. Active sodium transport and oxygen consumption in the outer medulla of the kidney. Plügers Arch 1958; 267:207–217.
15. Bakris GL, Lass NA, Glock D. Renal hemodynamics in radiocontrast medium-induced renal dysfunction: a role for dopamine-1 receptors. Kidney Int 1999; 56:206–210.
16. Nygren A, Ulfendahl, HR. Hansell P. Erikson U. Effects of intravenous contrast media on cortical and medullary blood flow in the rat kidney. Invest Radiol 1988; 23: 753-761.
17. Liss P. Nygren A. Hansell P. Effects of intravenous contrast media on cortical and medullary blood flow in the rat kidney. Acta Radiol 1999; 40:521-527.
18. Agmon Y, Peleg H, Greenfeld Z, Rosen S, Brezis M. Nitric oxide and prostanoids protect the renal outer medullary from radiocontrast toxicity in the rat. J Clin Invest 1994;94: 1069–1075.
19. Arend LJ, Bakris GL, Burnett JC, Megerian C, Spielman WS. Role of intrarenal adenosine in renal hemodynamic response to contrast media. J Lab Clin Med 1987;110: 406–411.
20. Cantley LG, Spokes K, Clark B, McMahon E, Carter J, Epstein FH. Role of endothelin and prostaglandins in radiocontrast-induced renal artery contraction. Kidney Int 1993;44: 1217–1223.
21. Idee JM, Beaufils H, Bonnemain B. Iodinated contrast media-induced nephropathy: pathophysiology, clinical aspects and prevention. Fundam Clin Pharmacol 1994; 8:193–206.
22. Galle J, Wanner C. Impact of nitric oxide on renal hemodynamics and glomerular function: modulation by atherogenic lipoproteins? Kidney Blood Press Res 1996; 19:2–15.
23. Williamson JR, Chang K, Frangos M, Hasan KS, Ido Y, Kawamura T, Nyengaard JR, Van den Enden M, Kilo C, Tilton R. Hyperglycemic pseudohypoxia and diabetic complications. Diabetes 1993; 42: 801-813.
24. Pugliese G, Tilton RG, Williamson JR. Glucose-induced metabolic imbalance in the pathogenesis of diabetic vascular disease. Diabetes Metab Rev 1991; 7:35–59.

Chapter **66**

POSTOCCLUSIVE REACTIVE HYPEREMIA IN HEALTHY VOLUNTEERS AND PATIENTS WITH PERIPHERAL VASCULAR DISEASE MEASURED BY THREE NONINVASIVE METHODS

Tomaž Jarm[1], Rudi Kragelj[1], Adam Liebert[2], Piotr Lukasiewitz[2], Tatjana Erjavec[3], Marjeta Prešeren--Štrukelj[3], Roman Maniewski[2], Pavle Poredoš[4] & Damijan Miklavčič[1]

[1] University of Ljubljana, Faculty of Electrical Engineering, Trzaska 25, SI-1000 Ljubljana, SLOVENIA; [2] Institute of Biocybernetics and Biomedical Engineering, Trojdena 4, Warsaw 02-109, POLAND; [3] Institute of the Republic of Slovenia for Rehabilitation, Linhartova 51, SI-1001 Ljubljana, SLOVENIA; [4] Trnovo Hospital of Internal Medicine, Riharjeva 24, SI-1000 Ljubljana, SLOVENIA

Abstract: Postocclusive reactive hyperemia (PORH) was evaluated in three healthy volunteers and in three patients with different etiologies and suffering from peripheral arterial occlusive disease (PAOD). Three noninvasive methods were used: transcutaneous oximetry ($TcPO_2$), near-infrared spectroscopy (NIRS), and laser Doppler flowmetry (LDF). Changes in perfusion and oxygenation of tissue were measured on foot before, during, and after arterial occlusion on thigh. Numerical parameters were derived from measured signals for quantification of the PORH response. Results of all three methods provided distinction between healthy volunteers and patients. The experimental optical techniques of NIRS and LDF demonstrated more clearly than the well-established $TcPO_2$ method the difference between healthy volunteers and patients. The dynamics of the PORH response proved to be a better indicator of peripheral vascular disorder than the amplitude of responses.

Key words: Human subjects; Laser Doppler flowmetry; Near-infrared spectroscopy; Noninvasive measurement methods; Oxygenation of tissue; Peripheral arterial occlusive disease.

Abbreviations: HbO_2: concentration of oxygenated hemoglobin; LDF: laser Doppler flowmetry; NIRS: near-infrared spectroscopy; PAOD: peripheral arterial occlusive disease; PORH: postocclusive reactive hyperemia; $TcPO_2$: transcutaneous oximetry

1. INTRODUCTION

Different invasive and noninvasive methods based on different physical principles have been developed for assessment of tissue oxygenation and blood perfusion. In recent years much effort has been put in development of optical techniques, which enable noninvasive monitoring of tissue oxygenation and perfusion [1,2].

Near-infrared spectroscopy (NIRS) is a relatively new optical technique for noninvasive monitoring of tissue oxygenation and hemodynamics. The technique is based on relative transparency of human tissue for the near-infrared light (wavelengths roughly between 700 and 1000 nm) and on the oxygenation- or oxidation-dependent absorption of light by certain compounds (chromophores) involved in transport and consumption of oxygen in tissue, such as hemoglobin, myoglobin and cytochrome-c-oxidase. In its most fundamental implementation NIRS can be used to quantify the changes in concentration of these chromophores. This is accomplished by applying the modified Beer-Lambert law to the measured changes in absorption of near-infrared light at different wavelengths [1,3,4].

Laser Doppler flowmetry technique (LDF) is a noninvasive optical method for monitoring of skin blood flow [2,5,6]. It is based on the principle of changes of the frequency of light (the Doppler shift) when scattered by moving structures in cutaneous and subcutaneous layers of tissue. Since the predominant moving particles at this level are erythrocytes in capillaries and small vessels the detected frequency shift is a carrier of information about the skin blood flow. The recorded signal is related to perfusion, which is defined as the product of the concentration of blood cells and their speed averaged over the measured volume. The method yields relative perfusion data in arbitrary units. As is the case with the basic implementation of the NIRS method, the LDF method requires a provocation to induce changes in perfusion (or blood oxygenation in case of NIRS), which can be measured and then used for assessment of tissue oxygenation and perfusion status.

In contrast to experimental methods of NIRS and LDF the transcutaneous oximetry ($TcPO_2$) has already been established in clinical environment as a tool for assessment of tissue oxygenation [1]. It measures the diffusion of dissolved oxygen from the extracellular space through heated skin to a sensor, which is attached to the surface of the skin. In principle it is a polarographic method. The electric current generated by the diffusion of oxygen to the negatively polarized sensor is in theory proportional to the partial pressure of oxygen in the local area under the sensor. Absolute values of pO_2 can thus be measured. But many conditions have to be met for the measurements to be truly representative of tissue oxygenation.

In the presented preliminary study we measured changes in tissue oxygenation and perfusion in foot of three healthy volunteers and three patients with peripheral arterial occlusive disease (PAOD). The arterial occlusion on thigh and the subsequent postocclusive reactive hyperemia (PORH) were used as a provocation. PORH is a reproducible transient increase in blood flow after the release of arterial occlusion [6]. It is also one of the well-known tests in clinical practice for evaluation of the functional aspects of arterial blood flow in extremities.

NIRS and LDF methods were used in the study simultaneously with $TcPO_2$ method, which is a standard method for evaluation of tissue oxygenation. The aims of the study were: a) to examine the applicability of near-infrared spectroscopy and laser Doppler flowmetry for quantification of PORH, (b) to determine parameters which could be used to evaluate the level of PAOD or other vascular disorders in lower extremities and (c) to compare the three methods from the point of applicability for detecting differences between healthy and diseased subjects.

2. MATERIALS AND METHODS

2.1 Subjects

Three healthy volunteers and three patients suffering from PAOD in lower extremity were included in this preliminary study. Both groups were age-matched. The basic patient data is presented in Table 1. The study was approved by the ethical committee of the host institution (Institute of the Republic of Slovenia for Rehabilitation).

2.2 Near-infrared spectroscopy

The $NIRO_2X$-2 instrument (Keele University, U.K.) was used. The device measures light attenuation changes at the nominal wavelengths 775, 800, 845, and 904 nm which enables measurement of concentration changes of oxygenated and deoxygenated hemoglobin by application of the modified Beer-Lambert law. The emitting and the receiving optode were attached to the top and to the bottom of foot respectively (Figure 1) using a custom-made silicone strap. The geometrical interoptode distance was within the 3.5--4.0 cm range as measured by a caliper gauge individually on all subjects after the measurement.

Table 1. Basic subject data. Description of healthy volunteers (H1--H3) and patients with PAOD (P1--P3)

subject	age (years)	blood pressure* (mmHg)			PAOD†	comments‡			
		arm	thigh	ankle		Smok.	Diab.	Alco.	Hyp.
H1	59	120/80	140	130	-	no	no	no	no
H2	75	120/70	130	110	-	no	no	no	no
H3	65	135/70	140	125	-	no	no	no	no
P1§	74	130/80	150	95	IIa	no	yes	no	yes
P2§	62	135/75	200	50	IIb	yes	no	yes	yes
P3§	68	135/70	185	95	IIa	yes	no	yes	no

* The pressures given are systolic/diastolic for arm and systolic for thigh and ankle.
† Clinical stage of peripheral arterial occlusive disease according to the Fontaine classification (IIa = claudication distance > 200 m; IIb = claudication distance < 200 m).
‡ Smok = smoker; Diab = diabetic; Alco = alcoholic; Hyp = hypertension.
§ All patients had one leg amputated (below knee in P1 and above knee in P2 and P3) as a result of PAOD and the measurements were performed on the foot of the remaining leg.

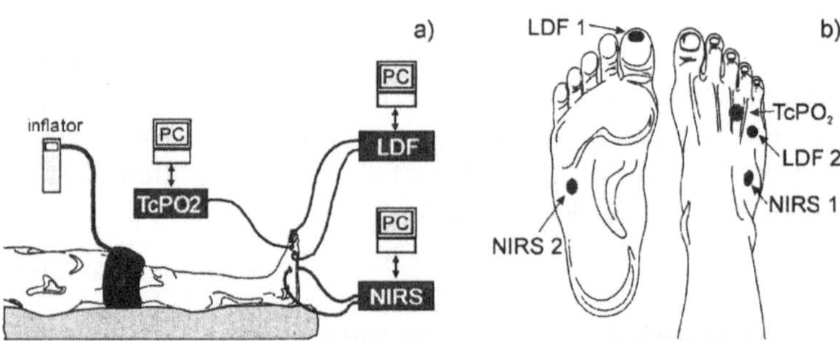

Figure 1. Experimental setup (a) and placement of probes on foot for NIRS, LDF, and $TcPO_2$ instruments (b). LDF 1 and LDF 2 belong to two channels of the laser Doppler instrument and NIRS 1 and NIRS 2 correspond to placement of the emitting and the receiving optode, respectively, for the one-channel NIRS instrument.

2.3 Laser Doppler flowmetry

The two-channel instrument MBF3/D (Moor Instruments, U.K.) was used. The standard probe (distance between fibers 0.5 mm, fiber diameter 0.2 mm) was attached to the lower surface of the big toe and the other probe was placed on the top of the foot as shown in Figure 1. Wavelength of the light used in this instrument is 810 nm.

2.4 Transcutaneous oximetry

The TCM3 instrument (Radiometer, Denmark) was used. The probe was attached to the upper surface of the foot between the third and the fourth digit (Figure 1). Skin under the sensor was heated to 43 degrees Celsius. Stabilization of the pO_2 readings took between 15 and 30 minutes.

2.5 Measurement protocol

Subjects were in supine position and were lying comfortably on a cushioned bed during the experiment. The room temperature was 23 degrees Celsius. A large contoured thigh cuff (Hokanson, U.S.A.) was placed on the leg above the knee. Probes of the three instruments were attached to the foot at locations shown in Figure 1. After this initial manipulation the subjects were left to rest for 20 to 30 minutes before the experiment. During this period stable baseline values for all three methods were established. Following this rest period the cuff was inflated to a pressure of 30 mmHg above the individual systolic pressure using the standard TD312 cuff inflator (Hokanson Inc., USA). Inflation of the cuff to the final pressure took approximately 45 seconds. After the maximum pressure had been reached the arterial occlusion thus obtained was maintained for five minutes and was then followed by a rapid release of the cuff. The measurement continued for about 10 minutes until the preocclusion conditions were restored as demonstrated by a return of all measured signals to vicinity of their initial preocclusion values.

3. RESULTS

Different parameters were calculated from NIRS, LDF, and $TcPO_2$ signals in order to evaluate changes in perfusion and oxygenation during PORH. Their definition is presented in Figure 2 and given as follows.

NIRS signal (change in oxygenated hemoglobin concentration HbO_2) and LDF signal (perfusion):

t_R (seconds): time of recovery; the time interval between release of the cuff and the moment when HbO_2 (or LDF) signal reaches the initial preocclusion level for the first time.

t_M (seconds): time to maximum; the time interval between release of the cuff and the moment of the maximum hyperemic response for HbO_2 and LDF signals.

MR (%): maximum hyperemic response; the amplitude difference for HbO_2 and LDF signals between the lowest level reached at the end of arterial occlusion and the maximum level reached during hyperemia after release of the cuff. This difference is expressed as a percentage of the total decrease of HbO_2 and LDF signals during arterial occlusion.

$TcPO_2$ signal (partial pressure of oxygen):

pO_2 (mmHg): the absolute value of pO_2 at rest immediately prior to inflation of the cuff.

t_H (seconds): half time; the time interval between release of the cuff and the moment when 50% of the rest pO_2 value is restored

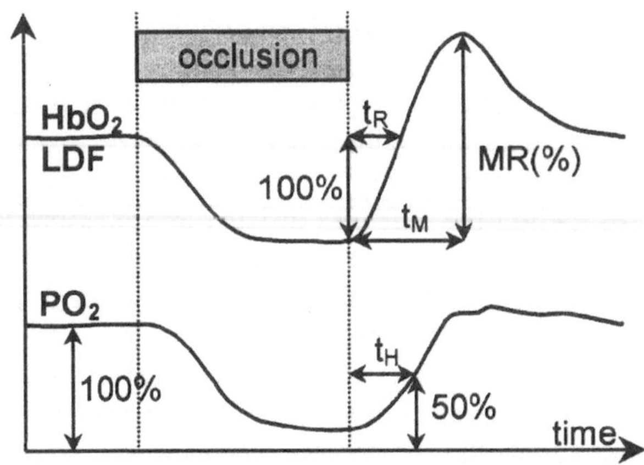

Figure 2. Parameters derived from the HbO_2, the LDF perfusion, and the pO_2 signals. The same type of parameters were calculated for the HbO_2 and the LDF signals.

Values of evaluated parameters for all subjects are given in Table 2 and presented in Figures 3 and 4. The results clearly show that the hyperemic response was significantly slower in patients with PAOD than in healthy subjects as demonstrated by larger t_R and t_M values for both HbO_2 and LDF with the LDF method. Hyperemia after release of the cuff could not be observed by the $TcPO_2$ method since there was no significant increase of the pO_2 value above the initial pre-occlusion level in either healthy subjects or patients. The rest pO_2 value however was higher and the recovery after occlusion demonstrated by parameter t_H was faster in healthy subjects than in patients (Figure 4a,b).

Table 2. The results -- values of evaluated parameters as defined in Figure 2 for healthy volunteers (H1--H3) and patients with PAOD (P1--P3)

subject	NIRS (HbO_2)			LDF (perfusion)			$TcPO_2$ (pO_2)	
	t_R (s)	t_M (s)	MR (%)	t_R (s)	t_M (s)	MR (%)	rest pO_2 (mmHg)	t_H (s)
H1	6	23	420	1	11	530	52	90
H2	11	45	127	1	25	210	48	105
H3	11	56	200	2	33	280	37	95
P1	105	150	83	111	111	97	37	150
P2	92	216	177	144	174	123	32	170
P3	36	78	88	48	110	150	34	105

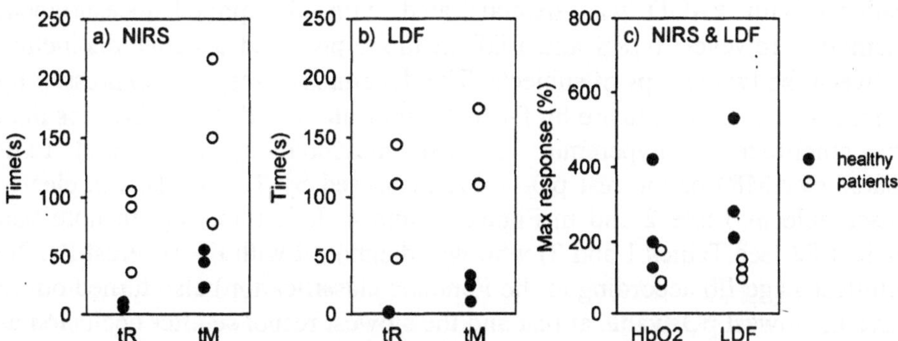

Figure 3. Values of evaluated parameters for NIRS and LDF measurements for healthy volunteers (solid circles) and patients (open circles): a) time of recovery t_R and time to maximum response t_M for the HbO_2 signal; b) time of recovery t_R and time to maximum response t_M for the LDF perfusion signal; c) the amplitude of the maximum response (MR) for the HbO_2 and LDF perfusion signals. See Figure 2 for definitions of the parameters.

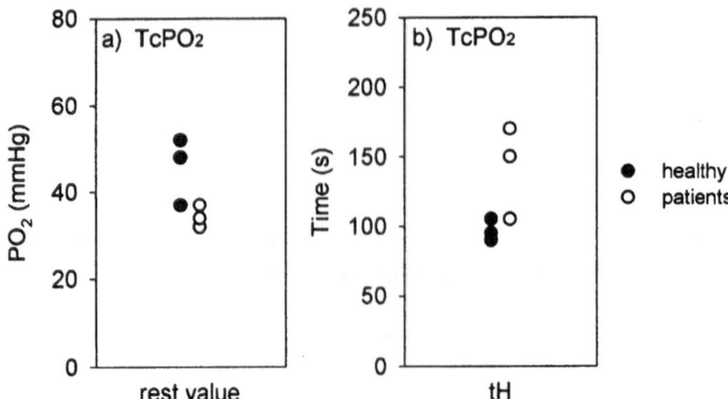

Figure 4. Values of evaluated parameters for $TcPO_2$ measurements for healthy volunteers (solid circles) and patients (open circles): a) absolute pO_2 values at rest before the occlusion; b) half time t_H. See Figure 2 for definitions of the parameters.

4. DISCUSSION AND CONCLUSIONS

The difference in hyperemic response between healthy volunteers and patients with PAOD was demonstrated with all applied measurement methods. However, NIRS and LDF methods provided a better distinction between the two groups of subjects. The dynamics of response expressed by parameters t_R, t_M, and t_H are by far better indicators of these differences than the amplitude of hyperemic response measured by NIRS and LDF (parameter MR) or the rest pO_2 value measured by $TcPO_2$. This is clearly observable in Table 2 and in Figures 3 and 4. It is important to note that patient P2 (see Tables 1 and 2) who was diagnosed with the severest PAOD (clinical stage IIb according to the Fontaine classification) also turned out to have the lowest pO_2 value at rest and the slowest response after occlusion as demonstrated by values of parameters t_R (LDF), t_M (NIRS and LDF) and t_H ($TcPO_2$). The amplitude of response MR did not provide this distinction of patient P2.

The three different methods used in the study measure different variables, which reflect oxygenation of tissue under observation. These variables are related to each other even though the anatomical sites at which measurements are made as well as the sampling volumes of tissue are different for each method. This preliminary study included only a small number of subjects but in general it appears that both experimental optical methods could be suitable for evaluation of the level of peripheral arterial occlusive disease and that they could supplement the more established $TcPO_2$ method. This is also supported by another study in which we have shown that the repeatability of measurements is better with NIRS than with $TcPO_2$ method (unpublished observations). In any case the provocation protocol and signal analysis will have to be improved and more patients with various diagnosed stages of PAOD will have to be included in the study in order to support our preliminary findings.

ACKNOWLEDGEMENT

This study was carried out as a part of the Slovenian-Polish Scientific and Technological Cooperation Joint Project.

REFERENCES

1. Benaron DA, Benitz WE, Ariagno RL, Stevenson DK. Noninvasive methods for estimating in vivo oxygenation. Clin Pediatr 1992;31:258-273.
2. Oberg PA. Laser Doppler flowmetry. Critical Rev Biomed Eng 1990;18:125-163.
3. Jöbsis FF. Noninvasive infrared monitoring of cerebral and myocardial oxygen sufficiency and circulatory parameters. Science 1977;198:1264-1267.
4. Brazy JE. Near-infrared spectroscopy. Clin Perinatol 1991;18:519-534.
5. Fagrell B. Peripheral vascular disease in Laser Doppler Blood Flowmetry. In: Shepherd AP, Oberg PA, editors. Laser Doppler Flowmetry. Kluwer Academic, 1990.
6. de Mul FFM, Koelink MH, Kok ML, Harmsma PJ, Greve J, Graaf R, Aarnouds JG. Laser Doppler velocimetry and Monte Carlo simulations on models for blood perfusion in tissue. Appl Opt 1995;34:6595-6611.

Chapter **67**

ROLE OF MYOGLOBIN IN REGULATING RESPIRATION

Thomas Jue and Youngran Chung
Department of Biological Chemistry, School of Medicine, Unviersity of California Davis, Davis, CA 95616-8635 USA

Abstract: The ^{1}H NMR Val E11 signal provides a unique opportunity to observe carbon monoxide (CO) inhibition of Mb in the *in vivo* myocardium and to assess the functional role of Mb in regulating respiration. Upon carbon monoxide infusion, the MbO_2 Val E11 signal at -2.76 ppm gradually disappears, and a new signal at -2.26 ppm, corresponding to MbCO, emerges. These signals yield the intracellular partial pressure of both O_2 and CO and the extent of Mb inactivation, since CO binds more tightly to Mb than O_2. Although contractile function decreases slightly to a steady state level, it shows no dose dependence on pCO. Up to 80% MbCO saturation, the contractile function remains at the steady state level. Neither the PCr concentration nor the oxygen consumption rate is significantly perturbed. Above 80% MbCO saturation, the oxygen consumption rate starts to decline. The experimental observations raise provocative questions about the functional role of Mb in the cell.

Key words: myoglobin, nmr, respiration, bioenergetics, heart

1. INTRODUCTION

The abundance of myoglobin in myocyte has always raised unsettling questions about its functional role [1, 2]. In the orthodox view, Mb acts as an oxygen storage protein, ready to compensate any cellular oxygen deficit or as a facilitator of oxygen diffusion. That view originates partly from the high tissue Mb concentration in seals and whales. During a dive the oxygen stored in Mb can help maintain oxidative metabolism. However, in rat heart, the Mb concentration is almost two orders of magnitude lower and can sustain the normal heart function for only a few seconds.

Because the O_2 storage role appears limited in many mammalian tissues, researchers have posited an alternative function for Mb, which is predicated and the low solubility of oxygen in the physiological milieu. Since O_2 binding affinity of Mb is high, its O_2 carrying capacity is also higher than free O_2 and can contribute significantly to facilitate intracellular O_2 transport, provided that Mb diffusion is sufficiently rapid. In fact as the pO_2 decreases, Mb plays an increasingly significant role, as *in vitro* experiments have convincingly demonstrated [3].

In vivo doubts persist about the actual function of Mb. Mb diffusion in the cell appears insufficient to contribute significantly to the intracellular O_2 flux [4-6]. Furthermore, Mb inactivation experiments have yielded equivocal results [7-9]. Some experiments indicate that Mb inactivation produces a change in the high energy phosphate signals, as expected, if Mb contributes significantly in facilitating O_2 diffusion [10, 11]. Others, however, show no alteration in the ^{31}P spectra [9]. These disparate observations arise partly from the model system differences. They also arise from the uncertain extent of Mb inactivation. Most techniques to date only assume the extent of Mb inactivation in the myocardium, such as nitrite oxidation or CO binding.

^{1}H NMR can now measure the extent nitrite inactivation of Mb by following a reporter metMb signal at –3.9 ppm [9, 12]. In these experiments, the infused nitrite concentration must exceed >10 mM before any noticeable Mb oxidation appears [9]. The stoichiometry is much greater than previously reported Mb inhibition *in vivo* and is far greater than the stoichiometry predicted from *in vitro* experiments [7-11]. CO can also inactivate Mb by binding more tightly to the heme than O_2. We report herein that the distinct MbCO signal of the γ CH_3 Val E11 at -2.26 ppm is now detectable in the myocardium, separated from the corresponding MbO_2 signal at -2.76 ppm [13, 14]. Increasing the partial pressure of CO (pCO) in the perfusate enhances the MbCO signal intensity and at the same time depresses the MbO_2 signal intensity.

The ability to monitor the extent of CO inactivation of Mb sets the stage for experiments to investigate the function of Mb *in vivo*. Up to 80% saturation of MbCO, the myocardial oxygen consumption (MVO_2) still remains constant, while rate pressure product (RPP) shows no dose dependent effect. The ^{31}P NMR spectra reveal that PCr, ATP, and Pi levels are undisturbed. Along with the recent reports of a viable myoglobinless mouse model, the presented experimental results provoke a re-examination of Mb function in the cell [15-17].

2. METHODS

2.1 Animal preparation and heart perfusion

Male Sprague-Dawley rats (350-400 g) were anesthetized by an intraperitoneal injection of sodium pentobarbital (60 mg/kg) and heparinized (1000 U/kg body weight). The heart was quickly isolated and placed in an ice cold buffer solution until aortic cannulation. It was then perfused in a retrograde mode, as prescribed by a modified Langendorff model, and was maintained at 35 °C with a Lauda MT-3 water bath and temperature jacketed reservoirs and tubings. A peristaltic pump (Rainin Rabbit) maintained a constant, non-recirculating perfusion flow of 18 ml/min. A saline-filled latex balloon inserted in the left ventricle monitored the heart rate (HR) and left ventricular pressure (LVP) via a strain gauge transducer (Statham P23XL) connected to an oscillographic recorder (Gould RS 3200). The balloon volume was adjusted to give an end-diastolic pressure of 6-8 mm Hg. Rate pressure products were calculated from heart rate (HR) times the left ventricular developed pressure (LVDP).

The perfusion medium was a modified Krebs-Henseleit buffer containing (in mM) 118 NaCl, 4.7 KCl, 1.2 KH_2PO_4, 1.8 $CaCl_2$, 20 $NaHCO_3$, 1.2 $MgSO_4$, and 15 glucose. The perfusate was first oxygenated with 95% O_2-5% CO_2 and then passed through a home built lucite mixing chamber, comprising of gas permeable Dow Corning silastic tubing (i.d. 0.058 in, o.d. 0.077 in) wrapped around a heat exchanger. Different gas mixtures equilibrated with the in flowing perfusate just before it entered the heart. The perfusate passed through both 5 μm and 0.45 μm Millipore filters. The other experimental conditions were similar to ones previously reported (5,14).

2.2 Perfusate oxygen measurement

The heart was placed in an NMR tube and isolated with a Teflon plug with holes to permit perfusate overflow. Approximately 50% of the perfusate was withdrawn via a polyethylene (PE) catheter inserted close to the pulmonary artery. A Yellow Springs Instrument (YSI) 5300 meter monitored the perfusate oxygen concentration with two YSI 5331 oxygen electrodes in a temperature jacketed chamber (one for inflow and the other for outflow perfusate). The remaining 50% of the perfusate exited the chamber above the Teflon plug as an overflow.

Parallel bench experiments determined empirically the O_2 loss in the tubing and adjusted the measured pO_2 value to reflect the venous value proximal to the heart. In the first set of measurements, the tubing lines from the heart chamber to the oxygen electrode were kept as short as possible. Independent measurements determined a small O_2 loss/ length of tubing in such an arrangement. In the second set of measurements, the tubing length matched the NMR experimental conditions. The results from the two sets of measurements formed a calibration curve that adjusted the observed outflow O_2 level to approximate the venous O_2 (5,6,14). The CO loss was assumed to be similar.

2.3 Carbon monoxide infusion protocol

Carbon monoxide was introduced into the perfusion buffer in a stepwise manner. Two flow meters control the flow of 95% O_2/5%CO_2 and 95%CO/5%CO_2 gases, which entered a temperature jacketed gas mixing chamber and equilibrated with the perfusate passing through 50 feet of gas permeable Dow Corning silastic tubing (i.d. 0.058 in, o.d. 0.077 in). The CO flow rate varied stepwise at 0.0, 0.1, 0.3, 0.5, and 1.0 liter/min, while the 95% O_2/5% CO_2 flow rate remained constant at 5.0 liter/min. After correcting for loss in the tubings, the resulting pCO values at the catheter tip were 0.0, 12.6, 36.3, 58.4, and 107.0 torr respectively. Oxygen measurement of perfusate exiting the gas mixing chamber confirmed that equilibration was essentially complete within 2 min after adjusting the CO flow rate at each step.

After the last CO infusion step, the heart was reperfused with oxygenated buffer. The RPP and MVO_2 returned to values observed in control hearts, where no CO was introduced. In the control hearts the RPP declined approximately 10% over the course of experiment. In the CO treated myocardium experiments, the control and the oxygen reperfusion data determined the baseline drift, which was consistent with the drift during the control period. All the CO induced changes in RPP were then normalized against this extrapolated baseline.

2.4 Curve fitting and statistical analysis

Linear regression analysis, using a least squares method (Sigma Plot, Jandel Scientific) determined the correlation coefficient, slope, and intercept. Errors were noted as standard error. Student's t test indicated statistical significance when $p < 0.05$.

2.5 NMR

An AMX 400 MHz Bruker spectrometer recorded $^1H/^{31}P$ signals with a 20 mm ^1H-(X) probe, where X represented nuclei from ^{15}N to ^{31}P. A modified 1331 binomial pulse sequence suppressed the H_2O line and selectively excited the MbO_2 and MbCO Val E11 resonances at -2.76 and -2.26 ppm [13, 18]. The ^1H 90° pulse was 65 μs, calibrated against the perfusate H_2O signal. Observing the oxy- and carbon monoxy- Mb signals required a 40 ms acquisition time and a 45° pulse. The spectral width was set at 8065 Hz; the data block size was 512. Six thousand (6,000) transients were averaged for a typical ^1H spectrum, requiring 5 minutes of signal accumulation. The FIDs were then zerofilled to 2K and multiplied by a digital window function. A non-linear spline fit (Bruker UXNMR algorithm then smoothed the baseline. All spectral lines were referenced to the H_2O resonance at 4.67 ppm at 35°C. The chemical shift was in turn calibrated against TSP as 0 ppm. The integrated area of the Val E11 signal at 18 ml/min flow rate was normalized as 100 % MbO_2 saturation. For the ^{31}P spectra, a typical spectrum utilized a 45° pulse angle, a 0.5 s repetition time, and 512 scans per block (4.3 min). The ^{31}P 90° pulse was 72 μs, calibrated against a 0.1M phosphate solution. Spectral width was set at 6494 Hz; the data size was 4K. FIDs were apodized with an exponential function to improve the ^{31}P signal to noise ratio. The ^{31}P signals were referenced to phosphocreatine (PCr) as 0 ppm and apodized with a 15 Hz exponential function. PCr, ATP, and Pi levels were determined from integrated areas of the PCr and ß-ATP, and Pi signals, respectively. The areas were then normalized to the control values. The Pi chemical shift reflects the pH_i, which was estimated from the equation $pH = pK + \log(\frac{\delta_A - \delta_o}{\delta_o - \delta_B})$, where pK = 6.9, $\delta_A = \delta_{ppm}$ of $[H_2PO_4]^{-1}$ at 3.290 ppm, $\delta_B = \delta_{ppm}$ of $[HPO_4]^{-2}$ at 5.805, and $\delta_o = \delta_{ppm}$ of Pi referenced to PCr δ_{ppm} as 0 ppm.

3. RESULTS

The myocardial response to increasing pCO is reflected in the ^1H and ^{31}P spectra, fig. 1A-B. During the control period, the ^1H NMR spectra from well oxygenated myocardium, perfused with 95 % O_2/ 5 % CO_2 saturated buffer flowing at 18 mL/min, exhibit a distinct γ-CH_3 Val E11 signal of MbO_2 at -2.76 ppm, fig. panel 1A. Under these perfusion conditions, the signal reflects the fully saturated state [13, 18, 19]. As the pCO of the perfusate increases, the MbO_2 signal decreases, while the corresponding MbCO peak at -2.26 ppm increases, fig 1b-c. At pCO 12.6 torr, the MbO_2 signal decreases to 53.5% of control level, while the MbCO signal increases correspondingly, fig. 1b. Increasing the pCO to 58.4 torr decreases further the MbO_2 signal to 20.3% of control level, while MbCO peak rises to 84.9%, fig 1c. Upon reperfusion with 95% O_2/5% CO_2 saturated buffer, the MbO_2 signal recovers as the MbCO signal falls, fig 1d. In contrast the ^{31}P NMR spectra show no response to the infused CO, fig 1a'-d'. Despite the increasing pCO, the high energy phosphate signals, ATP, PCr, and Pi , show no alteration, and pH remains constant at 7.4.

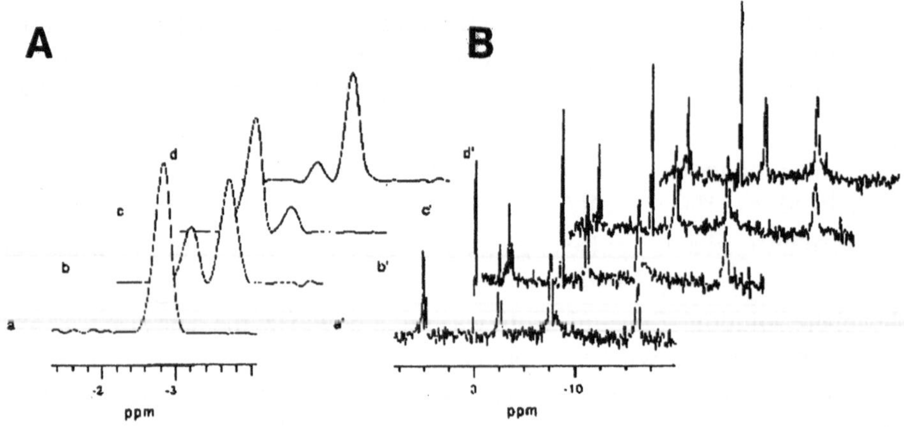

Figure 1. ^{1}H/^{31}P NMR spectra from perfused rat myocardium perfused with CO buffer.

The *in vivo* MbCO off rate kinetics is shown in fig. 2. At pCO of 58.4 torr 84.9 % of intracellular myoglobin is sequestered as MbCO. Upon reperfusion with oxygenated buffer, the MbCO declines to 50 % of its original intensity within 10 minutes. After 40 minutes, the MbCO signal is no longer detectable, whereas the MbO_2 signal has returned to its control level. The time for equilibration and dead space volume clearance is under 2 min, a time frame sufficient for the PCr signal to recover fully (unpublished observations).

The MVO_2 and RPP response to varying pCO is shown in fig 3. Oxygen consumption remains constant up to 76.8% MbCO saturation. At 87.6% MbCO saturation, MVO_2 shows a significant decline (34.0 ± 1.3 μmol/min/g dry wt). In contrast RPP has already dropped significantly at 53.5% MbCO saturation (27,436 ± 2483 mm Hg/min) and remains at this depressed level up to 87.6% MbCO saturation.

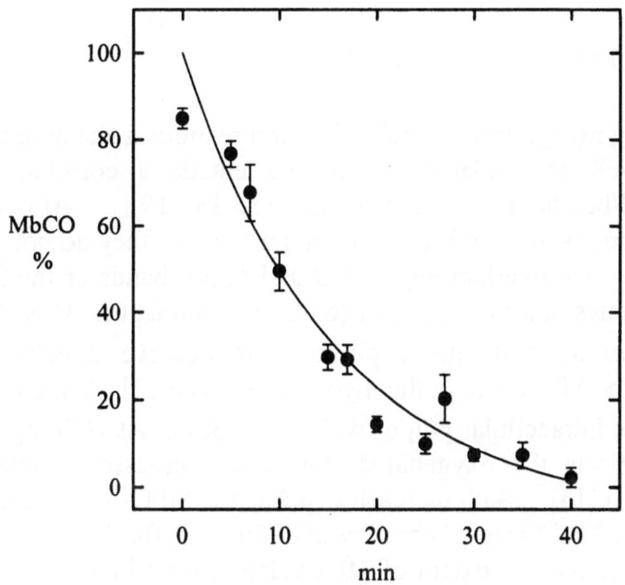

Figure 2. Graph of CO kinetics of MbCO during reperfusion with O_2.

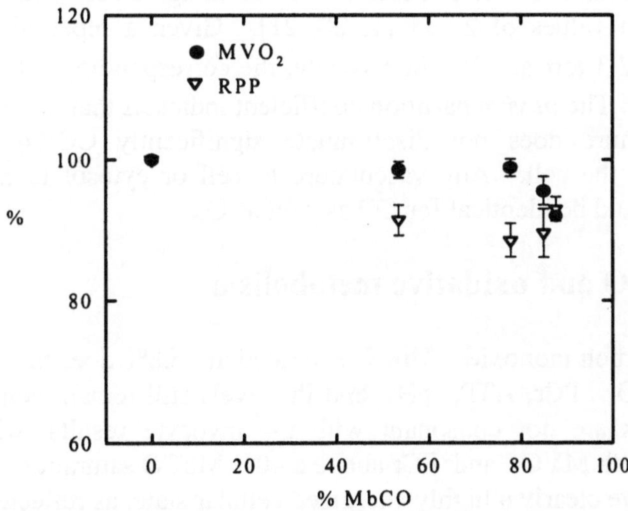

Figure 3. Graph of MVO2 and RPP as the function MbCO saturation in perfused rat myocardium.

4. DISCUSSION

4.1 CO inactivation of Mb

Analyzing myoglobin's cellular function requires a characterization of its ligand as well as oxidation states and entails a correlation with the physiological/biochemical response [9, 13, 18, 19]. Although optical techniques can distinguish MbO_2 saturation *in vitro*, they do not successfully discriminate the overlapping MbCO and MbO_2 bands or the Fe (II) from the Fe (III) states in a beating heart (6,22). In contrast the ^1H NMR CH_3 Val E11 signal offers a unique opportunity to observe directly the MbO_2, MbCO, and metMb states in the myocardium. The CH_3 Val E11 signal can mark both the intracellular pO_2 as well as the pCO. At -2.76 ppm the signal of MbO_2 reflects the oxygenated state and decreases its intensity upon deoxygenation [13]. With increasing pCO, the MbO_2 signal declines as the corresponding MbCO signal emerges at -2.26 ppm, fig. 1A. A reporter signal at -3.9 ppm reflects the extent of Mb oxidation from Fe (II) to Fe (III) state [9, 12].

At 37°C the intracellular partition coefficient, $P = \frac{[MbCO]pO_2}{[MbO_2]pCO}$, between CO and O_2 in myocardium is 36, in agreement with the myocyte and solution values of 20-35 [1, 20, 21]. Given a reported intracellular $[pO_2]_{50}$ of 2.3 torr at 37°C in myocyte, the corresponding $[pCO]_{50}$ is then 0.06 torr [1]. The *in vivo* partition coefficient indicates that in perfused heart, the vasculature does not discriminate significantly CO/O_2 delivery or transport to the cell. Any vasculature to cell or cytosol to mitochondria gradient would be identical for CO as well as O_2.

4.2 CO and oxidative metabolism

With carbon monoxide, MbCO saturated at 76.8% does not significantly impair MVO_2. PCr, ATP, pH, and Pi levels still remain constant. These observations are not consonant with the myocyte results, which note a decline in both MVO_2 and PCr above a 40% MbCO saturation threshold [9-11, 21]. Quite clearly a highly energized cellular state, as reflected by the ^{31}P spectra, does not prevent the decline in contractile function. After the initial drop in the developed pressure, RPP remains at a constant level and shows no dose dependent response to MbCO saturation. At 53.5% MbCO saturation, RPP has already declined significantly from 29,846 ±1093 to 27,436 ± 2483 mm Hg/min but does not decline further as MbCO saturation increases to 76.8%.

Although the RPP has fallen, neither respiration nor oxidative phosphorylation MVO_2 and PCr appears perturbed. Under the reported experimental conditions, Mb inactivation does not disrupt intracellular O_2 transport. Even under a range of physiological conditions, Mb inactivation with CO does not appear to impair any cellular function.

5. CONCLUSION

Recent myocardium experiments with nitrite inhibition of Mb have suggested a direct role for Mb in mediating respiration. However the experiments have not quantitated the relationship between the extent of Mb inhibition with the cellular response and cannot exclude nitrite, not Mb inactivation as the cause for the observed experimental results. The present study has utilized the ^{1}H NMR Val E11 signal of Mb to map the extent of CO inhibition in order to test whether Mb has a role in regulating cellular respiration. It has established the NMR methodology to CO inactivation of Mb in the cell.

With CO, contractile function drops immediately to a steady state level at 90% of the control level. Oxygen consumption remains constant until MbCO reaches 84.9%, while the high energy phosphate levels are unperturbed throughout the entire experiment protocol. The results do not support a significant role of Mb in facilitating oxygen diffusion in normoxic myocyte.

ACKNOWLEDGEMENTS

We gratefully acknowledge the funding from NIH GM 57355 (TJ) and AHA CA 92-08 Affiliate (YC)

REFERENCES

1. Antonini E, Brunori M. Hemoglobin and Myoglobin in Their Reactions with Ligands. Elsevier/North Holland: Amsterdam, 1971.
2. Wittenberg JB.Myoglobin-facilitated oxygen diffusion: Role of myoglobin in oxygen entry into muscle. Physiol Rev 1970;50:559-636.
3. Wittenberg BA, Wittenberg JB.Transport of oxygen in muscle. Ann Rev Physiol 1989; 51:857-878.
4. Jurgens KD, Peters T, Gros G.Diffusivity of myoglobin in intact skeletal muscle cells. Proc Natl Acad Sci 1994;91:3829-3833.
5. Papadopoulos S, Jurgens KD, Gros G.Diffusion of myoglobin in skeletal muscle cells - dependence on fibre type, contraction and temperature. Pflugers Arch-Eur J Physiol 1995; 430:519-525.

6. Wang D, Kreutzer U, Chung Y, Jue T.Myoglobin and hemoglobin rotational diffusion in the cell. Biophys J 1997;73:2764-2770.
7. Taylor DJ, Mathews PM, Radda GK.Myoglobin-dependent oxidative metabolism in the hypoxic rat heat. Resp Physiol 1986;63:275-283.
8. Cole RP, Wittenberg BA, Caldwell PRB.Myoglobin function in the isolated fluorocarbon-perfused dog heart. Am J Physiol 1978;234:H567-H572.
9. Chung Y, Xu D, Jue T.Nitrite oxidation of myoglobin in perfused myocardium: implication for energy coupling in respiration. Am J Physiol 1996;271:H687-H695.
10. Doeller J, Wittenberg B.Myoglobin function and energy metabolism of isolated cardiac myocytes: Effect of Sodium Nitrite. Am J Physiol 1991;261:H53-H62.
11. Gupta R, Wittenberg B. ^{31}P NMR studies of isolated adult heart cells: effect of myoglobin inactivation. Am J Physiol 1991;261:H1155-H1163.
12. La Mar GN, Budd DL, Smith KM, Langry KC.Nuclear magnetic resonance of high-spin ferric hemoproteins. Assignment of proton resonances in met-aquo myoglobins using deuterium-labeled hemes. J Am Chem Soc 1980;102:1822-1827.
13. Kreutzer U, Wang DS, Jue T.Observing the ^{1}H NMR signal of myoglobin val e11 in myocardium: an index of cellular oxygenation. Proc Natl Acad Sci USA 1992;89:4731-4733.
14. Patel DJ, Kampa L, Shulman RG, Yamane T, Wyluda BJ.Proton nuclear magnetic resonance studies of myoglobin in H20. Proc Natl Acad Sci USA 1970;67:1109-1115.
15. Garry DJ, Ordway GA, Lorenz JN, Radford NB, Chin ER, Grange RW, Bassel-Duby R, Williams RS.Mice without myoglobin. Nature 1998;395:905-908.
16. Glabe A, Chung Y, Xu D, Jue T.Carbon monoxide inhibition of regulatory pathways in myocardium. Am J Physiol 1998;274:H2143-H2151.
17. Godecke A, Flogel U, Zanger K, Zhaoping D, Hirchenchain J, Decking UKM, Schrader J.Disruption of myoglobin in mice induces multiple compensatory mechanisms. Proc Natl Acad Sci USA 1999;96:10495-10500.
18. Chung Y, Jue T.Cellular Response to Reperfused Oxygen in Postischemic Myocardium. Am J Physiol 1996;271:H1166-H1173.
19. Kreutzer U, Jue T.Critical intracellular oxygen in the myocardium as determined with the ^{1}H NMR signal of myoglobin. Am J Physiol 1995;268:H1675-H1681.
20. Coburn RF, Forman HJ.Carbon monoxide toxicity. In: Farhi LE, Tenney SM, editors. Handbook of physiology: the respiratory system IV. American Physiological Society: Bethesda, 1987; pp. 439-456.
21. Wittenberg BA, Wittenberg JB.Myoglobin-mediated oxygen delivery to mitochondria of isolated cardiac myocytes. Proc Natl Acad Sci USA 1987;84:7503-7507.

Chapter **68**

OXIDATIVE DEFENSES IN THE SEA BASS, *DICENTRARCHUS LABRAX*

Giulia Guerriero, Alessandra Di Finizio and Gaetano Ciarcia
Department of Zoology, Federico II University, via Mezzocannone 8, 80134 Napoli – Italy

Abstract: A study on the oxidative defenses during larval growth and under stress conditions was carried out in the bred sea bass, *Dicentrarchus labrax*. A high-pressure liquid chromatography (HPLC) method was used for the quantitative determination of vitamins C and E; glutathione peroxidase activity was measured by an enzymatic assay. Vitamin E was measured in the seminal fluid, eggs, embryos and larvae. Vitamins C and E, and glutathione peroxidase activity were measured in adults of *Dicentrarchus labrax* under normal conditions and subjected to hypoxia. Vitamin E content was high in seminal fluid, eggs, and embryos and at the early stage of larval development. It decreased slowly, but steadily, throughout the larval growth. In adults exposed to hypoxia, vitamins C and E levels were significantly lower with respect to the control group. Glutathione peroxidase levels showed a decrease in the hypoxia-subjected group, although the values were not significant.

Key Words: *Dicentrarchus labrax,* Vitamin E, Vitamin C, glutathione peroxidase, hypoxia

1. INTRODUCTION

Oxidative damage reflects an imbalance between the production of oxidants and removal or scavenging of those oxidants. The antioxidants neutralize via enzymatic and non-enzymatic mechanisms the toxic effects of the free radicals, acting at different levels both within the cell and in the extra cellular fluids [1-4]. Oxidative damage is considered to play an important role in cells and tissues in fast proliferation such as in growing organism [5]. Antioxidative defenses are essential to survive, for the gamete functionality and for the embryonic development [6]. In the germinal cells free radicals cause modifications of the plasmatic membrane that are believed to impair the fertilization [7]. Furthermore, free radicals seem to be implicated in the block of the segmentation of the zygote and the embryonic growth [8]. The vitamin E is essential for the reproduction and post-natal growing in different mammals [9]. The presence of vitamin E in the human seminal liquid seems to be sufficient to prevent the peroxidative damage of spermatozoa [10]. Recently, papers have focused on the role played by the antioxidant enzymes during the early stages of embryonic development and in the adults of fish [8, 11-16]. However, no data are available for the sea bass, *Dicentrarchus labrax*, a widely utilized species in aquaculture. In this study we provide information on vitamin E, vitamin C and glutathione peroxidase levels in the gametes, embryos, larvae at different stages of development and in the adults of *Dicentrarchus labrax* in natural conditions and under stress during hypoxia.

2. MATERIALS AND METHODS

From five sexually mature females (1.20 ± 0.07 Kg) and males (0.80 ± 0.03 Kg) of *Dicentrarchus labrax* during the reproductive phase (January-February) milt and eggs were collected by hand stripping. Gametes, after elimination of the celomatic liquid, were divided into two aliquots. One aliquot was immediately frozen in the liquid nitrogen. The other aliquot was left in the tank for the fertilization. Living embryos at an early stage of development, not fertilized eggs, and embryos at an early stage of development with limited surviving recognizable by their opaque colour, were collected and immediately frozen. They were easily recognizable according to the following characteristics: 1) living embryos (Panel 1.A), at an early stage of development floated at the surface of the tank; 2) not fertilized eggs laid at the bottom of the tank; 3) and embryos not vital at an early stage of development with limited surviving (low survival embryo), floated halfway the tank. An aliquot of vital embryos was left in the tank to allow larval development. Larvae at hatching (Panel 1. B) and hatched (Panel 1.C) were collected and immediately every 24-hour from the first to the 14th day of development. Automatic feeders according to the standard feeding tables for *Dicentrarchus labrax* fed the larvae [17].

Sexual mature males and females of *Dicentrarchus labrax* were anaesthetized with 2-phenoxyethanol (Sigma) weighed and labeled at branchial operculum with colored links. Sea bass with similar body weight were gathered into five groups (5 sea bass each) for the experimental treatments (hypoxia). The sea bass were placed in a different tanks with similar characteristics (capacity: 8 Kg/m^3, oxygen: 8 mg/l; salinity: 7 ± 0.5

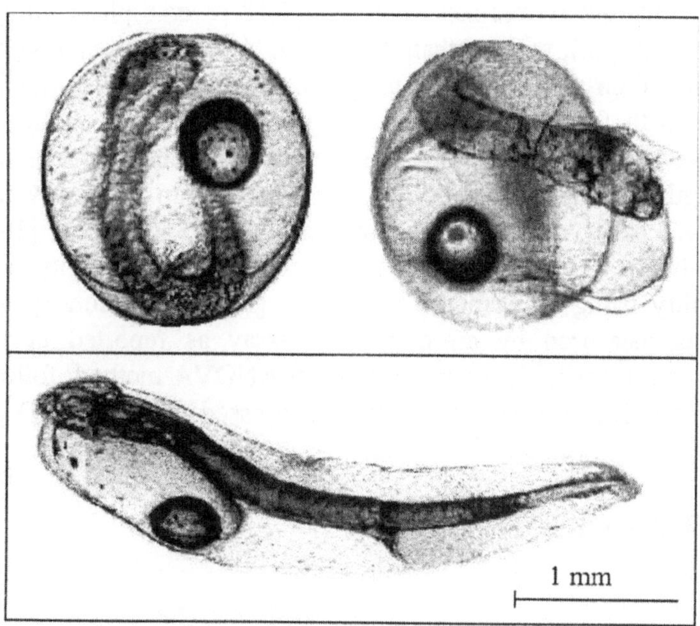

Panel 1. *Dicentrarchus labrax* exemplar of (**A**) egg with embryo; (**B**) embryo at hatching; (**C**) yolk sac larva, 1 day old.

g/l; temperature: 15-18°C; photoperiod: natural) and supplied with continuos light and adequate amounts of running fresh water. For twenty days all animals were fed with a free vitamin C and E meal. After this period, the animals were exposed to hypoxia-recovery conditions. Hypoxia (oxygen: 1.2 mg/l) was induced by reduction of water amount and getting 15 animals without labeling in the tanks After 40 min of hypoxia, the normal value of O_2 (8 mg/l) was restored by introduction of therapeutic oxygen and addition of fresh water. After 60 min of recovery, all animals were anaesthetized, weighed and bled. Control groups (normoxia) were kept for each experimental group.

Blood was drawn by a heparinized capillary inserted into the heart. Plasma samples were obtained after centrifugation at 800-g for 15 min, at 4°C. Samples were stored at –20°C for glutathione peroxidase assay and in liquid nitrogen for vitamin E and vitamin C determinations.

Vitamin E and C were measured by HPLC, a Waters LC 625 with UV/VIS 486 spectrophotometer. Briefly, samples were saponified and extracted with hexane, then dried under a stream of nitrogen and reconstituted in methanol. As a control procedure, in a different set of samples β-sitosterol was added at a known concentrations to evaluate the possible oxidation of vitamin E. α-tocopherol and a known solution of vitamin C were respectively used as internal standards for the vitamin E and vitamin C. Vitamin E was separated on a C18 reverse phase column (10 mm, 8 mm x 100 cm), with methanol: water 95% as a mobile phase; vitamin C was separated on a C18-NH_2 column, (5 mm, 4 mm X 250 cm), C18-NH_2 precolumn, with 0.005 KH_2PO_4, acetonitrile 27/75 as a mobile phase. Eluted peaks were detected by UV/VIS Waters 486 spectrophotometer, set at a wavelength of 292 nm and 270 nm for vitamin E and vitamin C respectively. Peaks were integrated with a Waters 860 system [18]. For method validation we evaluated the recovery, the linearity and the reproducibility using a standard procedure [19]. Glutathione peroxidase activity was measured by an enzymatic assay as reported in [8, 20]. Numerical data were analyzed by a one-way ANOVA method, followed by Duncan's multiple range test. Values were expressed as means ± SD.

3. RESULTS

3.1 Vitamin E and seminal liquid, eggs, dead eggs, not fertilized eggs and embryos of *Dicentrarchus labrax*

Vitamin E levels in seminal liquid, eggs and embryos are reported in Table 1. Vitamin E levels are similar in seminal liquid, eggs, not fertilized eggs and embryos at an early stage of development. On the contrary, values are significantly lower ($P<0.05$) in dead eggs and low survival embryos.

Table 1. Vitamin E concentration (µg/gr fresh tissue) in seminal liquid, eggs, death eggs, not fertilized eggs, embryos at an early stage of development, and embryos with few chances of surviving of *Dicentrarchus labrax*. Each value is the mean ±S.D. of three different assays. Level of significance versus the mean values observed in other samples reported: * $P < 0.05$

	Vitamin E (µg/gr.)
Seminal liquid	36.2 ± 8.2
Eggs	32.1 ± 6.8
Dead eggs	17.5 ± 4.3 *
Not fertilized eggs	32.1 ± 6.8
Embryos at an early stage of development	29.1 ± 4.1
Embryos with few chances of surviving	15.1 ± 5.3 *

3.2 Vitamin E and larvae of *Dicentrarchus labrax*

Vitamin E levels during the first fourteen days of the larval development are reported in Figure 1. Vitamin E levels dropped sharply during the first four days of development and decreased slowly but regularly during the following days of development.

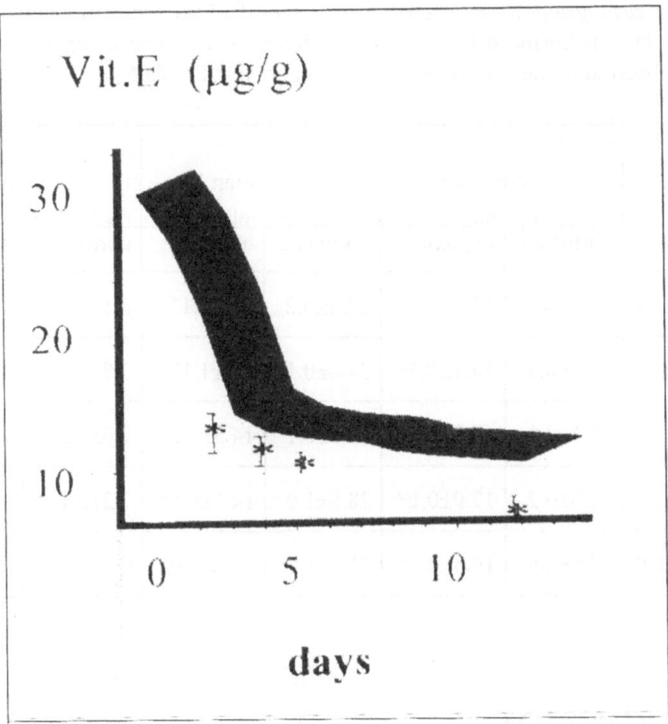

Figure 1. Vitamin E levels during the first fourteen days of the larval development of *Dicentrarchus labrax*. * represent vitamin E content in dead larvae. Values reported are the mean of three different experiments. Symmetric bars represent standard deviations.

3.3 Vitamin E, Vitamin C and Glutathione Peroxidase and plasma of adult *Dicentrarchus labrax*

The plasma levels of vitamin E, vitamin C and glutathione peroxidase activity in the sea bass under normal conditions and exposed to hypoxia-recovery (60 min) are shown in Table 2 . A significant decrease ($P<0.05$) in both vitamin E and vitamin C levels was detected in those animals under hypoxia with respect to those kept in normal conditions. Glutathione peroxidase activity was lower in hypoxia-exposed animals, although such a peroxidase activity was lower in hypoxia-exposed animals, although such a decrease was not statistically significant. Differences in weight, vitamin E, vitamin C and glutathione peroxidase activity among groups under normal conditions were not statistically significant.

Table 2. Vitamin C, vitamin E and glutathione peroxidase plasma levels in adults of *Dicentrarchus labrax*, in normal conditions (normoxia) and under hypoxia. Mean (±S.D.) body weights, for each group, were obtained from 5 animals. Each value of vitamin C, E and glutathione peroxidase is the mean ±S.D. of three different assays Level of significance versus the mean value observed in the normoxia's samples reported: * P < 0.05

Groups	Weight (gr)	Vitamin C (ug/ml)		Vitamin E (ug/ml)		Glutathione peroxidase (ug/min mg prot.)x10^{-3}	
		normoxia	hypoxia	normoxia	hypoxia	normoxia	hypoxia
I	249.2±1.4	20.0±0.3	12.0±0.3*	22.8±0.2	9.1±0.1*	8.1±1.4	7.3±2.1
II	261.3±2.7	22.1±0.6	14.7±0.2*	24.8±0.1	7.2±1.1*	7.9±2.2	6.9±1.1
III	281.7±1.3	25.4±0.3	17.0±0.2*	28.1±0.1	14.5±1.2*	7.6±1.3	6.8±0.5
IV	285.9±1.1	26.0±0.2	17.9±0.1*	28.8±1.0	14.1±0.8*	7.2±2.1	6.5±0.6
V	298.8±2.0	27.1±0.5	19.5±0.1*	31.1±1.0	12.2±1.6*	6.8±0.9	6.0±1.0

4. DISCUSSION

In this paper we report vitamin E concentrations in seminal liquid, eggs, embryos and during the first 14th days of the larval development of the sea bass *Dicentrarchus labrax*. Moreover, vitamin C and vitamin E levels and glutathione peroxidase activity were measured in the adult *Dicentrarchus labrax*, under normal and stress conditions.

Our data suggest that vitamin E is involved in the larval development of *Dicentrarchus labrax*. Indeed, we found high levels of vitamin E in vital eggs, both fertilized and not fertilized, in contrast with the low levels in death eggs. During the first four days of larval development vitamin E levels dropped abruptly, most likely an indication that the endogenous reserve of vitamin E, probably confined to the yolk sac, was rapidly utilised by the growing larvae. As reported by [14,15], the rate of the utilization of endogenous defenses, and in particular the non-enzymatic defenses, is quicker than the rate of their replacement. The decrease of vitamin E in the larvae of *Dicentrarchus labrax* may be related with an acceleration of the metabolic processes and an increase of the genesis of reactive oxygen species [21,22]. Vitamin C and vitamin E levels and glutathione peroxidase activity decreased in adults of *Dicentrarchus labrax* subjected to hypoxia-recovery. Their reduction could be interpreted in light of their utilization of such molecules to counteract the increase of free radicals as a consequence of the stressing conditions. Although these results do not prove that glutathione peroxidase activity is coupled to vitamin E recycling by its ability to regenerate the vitamin C, they suggest that a relationship between vitamin C, vitamin E and glutathione peroxidase is likely to exist [23-26]. In conclusion the results presented here provide evidence that Vitamin E could play an important role during larval development in *Dicentrarchus labrax.* Further, both vitamins C and E and glutathione peroxidase might be a parameter for the evaluation of oxidative processes in the sea bass.

ACKNOWLEDGEMENTS

We wish to thank Dr. Giovanni Cozzolino and the staff of the Stazione Sperimentale di Acquacoltura Termica (Torrevaldaliga-Civitavecchia, Roma, Italy) for providing logistic help and field assistance during the collecting trips.

REFERENCES

1. Halliwell B, Gutteridge JMC. The antioxidant of human extracellular fluids. Arch Biochem Biophys 1990;280:1-8.
2. Sies H. Strategies of antioxidant defense. Eur J Biochem 1993; 215:213-219.
3. Halliwell B. Free radicals and antioxidants: A personal view. Nutrition Reviews 1994;52:253-265.
4. Halliwell B., Murcia MA, Chirico S, Aruoma O I. Free radicals and antioxidants in food and in vivo: what they do and how they work. Critical Review in Food Science and Nutrition 1995;35:7-20.
5. Riley JCN, Behrman HR. Minireview oxygen radicals and reactive oxygen species in reproduction. Proc Soc Exp Biol Med 1991;198:781-791.
6. Aten RF, Duarte KM, Behrman HR. Regulation of ovarian antioxidant vitamins, reduced glutathione, and lipid peroxidation by luteinizing hormone and prostaglandin f-2α. Biol Reprod 1992;46: 401-407.
7. Aitken JR, Clarkson JS, Fishel S. Generation of oxygen species, lipid peroxidation, and human sperm function. Biol. Reprod.1989;40: 183-197.
8. Nars-Esfahani MH, Johnson MH. Quantitative analysis of cellular glutathione in early preimplantation mouse embryos developing *in vivo* and *in vitro*. Human Reproduction 1992;7:1281-1290.

9. Goodman-Gilmann A, Theodore WR, Nies AS, Almer T. In: Zanichelli, editor. Le basi farmacologiche della terapia: Vitamina E. Bologna,1992;1460-1464.
10. Baker GHW, Brindle J, Irvine SD, Aitken JR. Protective effect of antioxidant on the impairment of sperm motility by activated polimorphonuclear leukocytes. Fertility and Sterility 1996;65:411-419.
11. Wilhelm Filho D, Boveris A. Antioxidant defences in marine fish II. Elasmobranchs. Comp Biochem Physiol 1993;106C:415-418.
12. Wilhelm Filho D, Giulivo C., Boveris A. Antioxidant defences in marine fish I. Teleosts. Comp Biochem Physiol 1993;65C:409-413.
13. Rudneva T. Change in the activity of antioxidant enzymes in the processes of early ontogenesis of various species of black sea fishes. UK Biokhi Zh 1995; 67:92-95.
14. Umaoka N. Effects of oxygen toxicity on early development of mouse embryo. Molec Reprod Develop 1992;31:28-33.
15 Surai PF, Noble RC, Speake BK. Tissue-specific differences in antioxidant distribution and susceptibility to lipid peroxidation during development of the chick embryo. Biochim Biophys Acta 1996; 1304:1-10.
16. Aceto A, Amicarelli F, Sacchetta P, Dragani B, Bucciarelli T, Masciocco L, Miranda M, Di Ilio C. Developmental aspects of detoxifying enzymes in fish (Salmo iridalus). Free Rad Biol Med 1994;21:285-294.
17. Saroglia M, Ingle E, Del Prete M, Prata L, Venzi L. Tecniche di acquacoltura. Le tecnologie di allevamento. Saroglia M, Ingle E, editors. Edagricole 1992;71-162.
18. Huo, JZ, Nelis, HJ, Lavens, P.; Sorgeloos, P, De Leenheer, AP. Determination of vitamin E in aquatic organisms by high performance liquid chromatography with fluorescence detection. Analytical Biochem 1996;242:123-128.
19. Passey RB, Bee DE, Caffo A, Erikson JM. Evaluation of the linearity of quantitative analytical methods. NCCLS Document EP6-P, 6:8-11.
20 Akshes A, LR. Naja. Catalase, glutathione peroxidase and superoxide dismutase in different fish species. Comp Biochem Physiol 1981;69B:893-896.
21. Umaoka N. Effects of oxygen toxicity on early development of mouse embryo. Molec Reprod Develop 1992;31:28-33.
22. Mani-Ponset, Guyot, E, Draw JP, Connes R. Utilisation of yolk reserves during post-embryonic development in three teleosts an species: the sea bream Sparus aurata, the sea bass Dicentrarchus labrax and the pike-perch Stizostedion lucioperca. Marine Biology 1996;126:539-547.
23. Meydani M, Evans W, Handelman G, Fielding RA, Meydani SN, Fiatarone MA, Blumberg JB, Cannon JG. Antioxidant response to exercise-induced oxidative stress and protection by vitamin E. Ann NY Acad Sci 1992;30:363-364.
24. Sharma MK, Buettner GR. Interaction of vitamin C and vitamin E during free radical stress in plasma: an ASR study. Free Rad Biol Med 1993;14:649-653.
25. Hamre K, Waagbo R, Berge RK, Oyvind L. Vitamins C and E interact in Juvenile atlantic salmon. Free Rad Biol Med 1997;22:137-149.

Chapter 69

RHEOLOGIC DISSIMILARITIES IN FEMALE AND MALE BLOOD: POTENTIAL LINK TO DEVELOPMENT OF CARDIOVASCULAR DISEASES

Marina V. Kameneva[a,b], Mary J. Watach[a], and Harvey S. Borovetz[c]

[a]McGowan Institute for Regenerative Medicine, [b]Department of Surgery and [c]Department of Bioengineering, University of Pittsburgh, Pittsburgh, Pennsylvania, USA

Abstract: Oxygen Delivery Index (ODI) was introduced as the ratio of red blood cell concentration (hematocrit) to blood viscosity. The ODI can be considered an indirect characterization of oxygen transport to organs and tissues. ODI was obtained for 98 healthy donors (47 pre-menopausal women and 51 age-matched men). In this population ODI levels were found to be significantly lower ($p<0.001$) in male blood (7.7±0.3 vs.8.4±0.5 in female blood). Average ODI obtained for 15 cardiac patients (all males) was found to be significantly lower than that for healthy men. In red blood cell suspensions with the same hematocrit, ODI was found to decrease when plasma viscosity was increased via an increase in protein concentration. Additionally, it was found that ODI measured for samples of blood over a wide hematocrit range, obtained by dilution with autologous plasma, possessed the highest values at the hematocrit levels 30 to 40%. The decreased oxygen transport might contribute to the significantly higher morbidity and mortality from cardiovascular diseases for men compared to pre-menopausal women. ODI may be a useful parameter for evaluation of risk of development of cardiovascular disorders.

Key words: Blood viscosity, cardiovascular, gender, hematocrit, oxygen delivery, risk factors

1. INTRODUCTION

The mortality of men from coronary heart disease (CHD) and, especially, from myocardial infarction is significantly higher than the mortality of women during their reproductive years [1]. There are several speculations regarding the reason for this phenomenon. The most commonly accepted hypothesis is that the higher level of some circulating hormones that produce a vasodilatory effect on coronary arteries and the aorta [2-6] protect women of reproductive age. However, the increase in risk of heart disease was found to be essentially the same in women with surgically induced menopause regardless of ovary removal [7]. Another assumption is that the greater incidence of heart disease in men and post-menopausal women is due to higher levels of stored iron in these two groups compared with pre-menopausal women [7, 8]. However, many epidemiological studies report a negative or no association between CHD and various indicators of body iron [9].

We hypothesize that the significant difference between pre-menopausal women and age-matched men in morbidity and mortality from cardiac diseases is associated with menstruation. Due to regular physiologic bleeding there is a significantly reduced red blood cell (RBC) concentration (hematocrit) and blood viscosity in female compared to male blood [10]. In addition, due to monthly blood loss female blood is expected to contain more young and fewer old RBCs. In fact, it has been shown that the age distribution of RBCs is significantly different in males versus females [11]. The authors demonstrated that female blood has about 80% more young RBCs and 85% fewer old RBCs than male blood. Our recent study revealed a statistically significant difference in most hemorheological parameters between male and pre-menopausal female blood. The most pronounced differences between male and female blood were found for hematocrit (45.3±2.3% vs. 40.0±1.8%, $p<0.001$), asymptotic (5.8±0.5 centipoise [cP] vs. 4.8±0.4 cP, $p<0.001$) and low shear (55.4±11.1 cP vs. 35.2±4.8, $p<0.001$) blood viscosity (both measured at the original hematocrit) and erythrocyte sedimentation rate measured at the standardized hematocrit, Ht=40% (11.7±3.4 mm/hr vs. 8.2±2.6 mm/hr, $p<0.005$) (12). Thus, the comparison of male and female blood at the same standardized hematocrit revealed that male blood has significantly higher low shear viscosity and erythrocyte sedimentation rate reflecting higher RBC aggregation. In addition male blood had lower RBC deformability ($p<0.001$) than did the blood of pre-menopausal women [12]. Viscosity of plasma did not differ significantly (1.74±0.08 cP in male blood vs. 1.73±0.09 cP in female blood, $p>0.05$). The differences in mechanical properties of male RBCs compared to female RBCs are most likely associated with the aforementioned difference in age distribution of RBCs in blood of men versus pre-menopausal women [13].

The superior rheological properties of female blood compared to male blood may provide an increased tissue oxygen delivery and this, in turn, may be a reason for the reduced risk of cardiovascular diseases in pre-menopausal women as compared to age-matched men. We introduced Oxygen Delivery Index (ODI) derived from hemorheological parameters as the ratio of hematocrit to blood viscosity [14]. The ODI can be considered an indirect characterization of oxygen transport to organs and tissues. This is by virtue of the fact that the ODI accounts for bulk transport of oxygen via the red blood cells (hematocrit). The inverse relationship between bulk flow and fluid viscosity is also incorporated in ODI. In the present study we compared ODI in blood samples drawn from 47 pre-menopausal women, 51 age-matched men and 15 cardiac patients. Furthermore, we calculated ODI for the blood samples at the same hematocrit, but different plasma protein concentration and, thus, different plasma viscosity. In addition, ODI was evaluated for blood viscosity measured in the same samples of blood over a wide range of hematocrit obtained by dilution of blood with autologous plasma. Based on the latter data optimal hematocrit levels were estimated for ODI.

2. METHODS

Hematocrit level and blood and plasma viscosity were determined in blood samples obtained from 98 healthy volunteers (students and lab personnel) including 47 pre-menopausal women (age 26.1±5.0 years) and 51 age-matched men (age 27.3±5.6 years). Smokers and pregnant women were excluded from this study. Blood samples were obtained from 15 cardiac patients (age 42.3±12.6, all males). Blood samples were obtained in accordance with study protocols approved in advance by the University of Pittsburgh Institutional Review Board (IRB).

All blood samples were anticoagulated with ethylenediamine-tetraacetic acid (EDTA). Our preliminary tests have shown that the hematocrit levels in the EDTA blood samples are normally about 0.5% lower than hematocrit measured in blood anticoagulated with dry heparin. Hematocrit was determined using a microhematocrit centrifuge (IEC MB Centrifuge, International Equipment Company, Needham Heights, MA). Asymptotic blood viscosity was measured in the same EDTA blood samples by a capillary viscometer (Cannon Instrument Co., State College, PA) at room temperature. Plasma viscosity was measured by Couette rheometer (Contraves Low Shear 30) also at room temperature. We have studied ODI as a function of plasma viscosity by varying the concentration of plasma proteins. Here blood was centrifuged (CR 4-12 Centrifuge, Jouan, Inc., Winchester VA) at 3000 rpm for 15 minutes. Plasma was collected by aspiration and saved, the buffy coat was carefully removed and discarded. RBCs were washed three times with a standard phosphate buffered saline (PBS, 0.01 M phosphate buffer, 0.0027 M potassium chloride and 0.0137 M sodium chloride, pH 7.4, osmolality - 290 mOsm/kg at room temperature, Sigma Chemical Co., St. Louis, MO) and resuspended either in plasma; plasma mixed with PBS; or plasma mixed with albumin and fibrinogen solutions in various proportions. All prepared suspensions had the same hematocrit and osmolality.

Osmotic pressure was measured by a microosmometer (mOSMETTE, Model 5004 Automatic Osmometer, Precision Systems Inc., Natick, MA). Total plasma protein concentration was measured with a desktop clinical refractometer (KERNCO Clinical Refractometer 11F10, El Paso, TX). Suspension viscosity was measured by a capillary viscometer at room temperature.

Oxygen Delivery Index is calculated according to the formula:

$$ODI = \frac{Hematocrit}{Asymptotic\ Blood\ Viscosity}$$

All data are presented as the mean ± standard deviation. Statistical analyses (unpaired Student t-tests) were performed to test for differences in ODI values between male blood and female blood and between male blood and cardiac patient's blood. A value of $p < 0.05$ was assumed to indicate statistical significance.

3. RESULTS

In the studied healthy population the ODI levels are found to be significantly lower in male blood than in blood samples obtained from pre-menopausal females (7.7±0.3 in males vs. 8.4±0.5 in females, $p<0.001$). Figure 1 represents the entire collection of the ODI data for 47 females and 51 males plotted versus hematocrit. In addition, Figure 1 shows values of the asymptotic blood viscosity used for the calculation of ODI. One can see that the majority of female blood samples possess much lower viscosity and much higher ODI than do male blood samples.

Despite the relatively low hematocrit level (35.5±4.5%), the ODI level (7.3±0.5) obtained for 15 cardiac patients (data not shown) is found to be significantly ($p<0.001$) lower than that derived for healthy men. These patients have average asymptotic blood viscosity of 4.9±0.5 cP and average plasma viscosity of 1.96±0.21 cP.

Figure 2 represents blood viscosity and ODI measured in the same blood for different values of hematocrit, which were prepared by dilution with autologous plasma. One can see that the ODI had the highest values at hematocrit levels between 30 and 40%.

Figure 3 shows blood viscosity and ODI as functions of plasma viscosity. The ODI parameter is calculated from the viscosity measured in the samples with the same hematocrit and different plasma protein concentration and, thus, different plasma viscosity. ODI is found to decrease when plasma viscosity increased secondary to an increase in protein concentration.

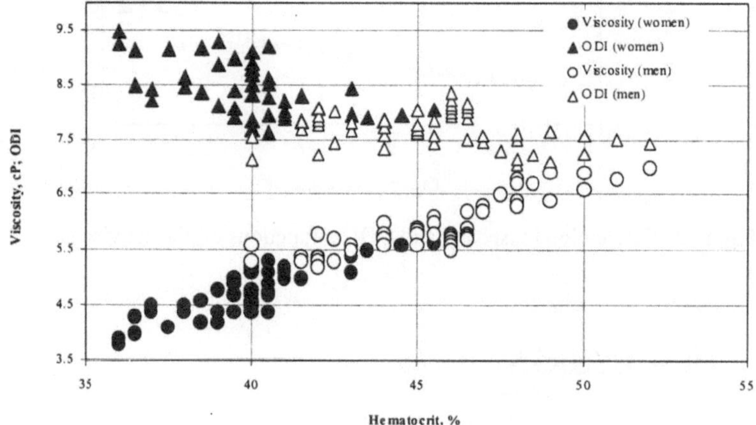

Figure 1. Asymptotic blood viscosity and oxygen delivery index (ODI) vs. hematocrit for 47 pre-menopausal women and 51 age-matched men

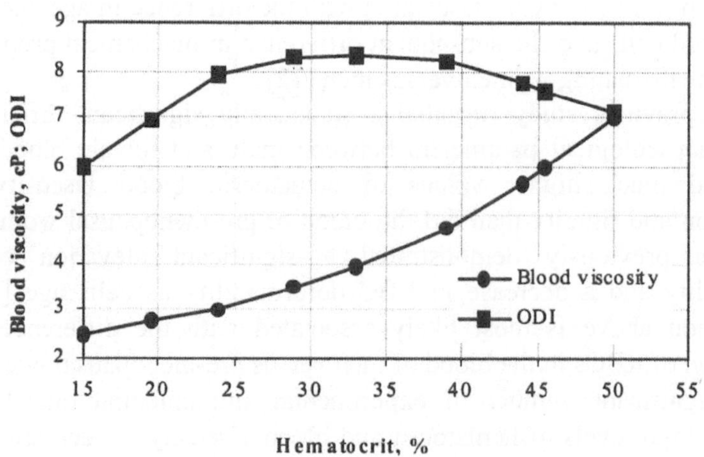

Figure 2. Blood viscosity and ODI measured as a function of hematocrit.

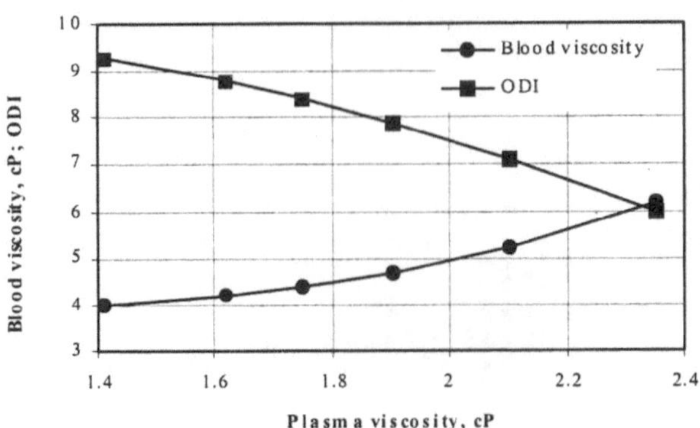

Figure 3. Whole blood viscosity and ODI as functions of plasma viscosity.

4. DISCUSSION

Coronary heart disease is the leading cause of death in the United States. However, the mortality of CHD in men, especially, from myocardial infarction is significantly higher than the mortality in women during their reproductive years [1]. We hypothesize that the reduced morbidity and mortality from cardiovascular diseases in women of reproductive age in comparison to men may be associated with the difference in age distribution of red blood cells and the subsequent difference in mechanical properties of blood of menstruating women versus men [12].

Our previous study revealed a statistically significant difference in most hemorheological parameters between male and female blood. Male blood had much higher values of hematocrit, blood viscosity, RBC aggregation and rigidity than did the blood of pre-menopausal women [12]. Since we previously demonstrated a significant elevation of RBC aggregability and a decrease in RBC deformability as cells age [12], the phenomenon above is most likely associated with the difference in age distribution of RBCs in the blood of men versus pre-menopausal women.

A significant number of experimental and epidemiological studies show that high levels of hematocrit and blood viscosity, a decrease in RBC deformability, and an increase in the ability of RBCs to aggregate are all important hemorheological risk factors for cardiovascular disease [15, 16]. We introduced the Oxygen Delivery Index (ODI) based on rheological properties of blood [14], which is defined as the ratio of hematocrit to blood viscosity. Since the characteristic shear rates in arterial flow range between 150 to 2000 s^{-1} (17) the asymptotic blood viscosity is used for calculation of the Oxygen Delivery Index. Blood viscosity has a strong effect on the resistance to blood flow. Viscosity of blood in turn strongly depends on hematocrit, plasma viscosity, RBC deformability and aggregation. In the population of healthy donors studied here, ODI levels are found to be significantly ($p<0.001$) lower in male blood than in the blood of pre-menopausal females. Since male and female blood have the same plasma viscosity [12] this difference could be explained by the higher viscosity of the male blood due to higher hematocrit and lower RBC deformability. To prove the former we determined ODI at various concentrations of RBCs in suspension. Results of this study reveal that in the blood of a healthy donor the optimal hematocrit, with the corresponding highest ODI levels, lies between 30 and 40%. Remarkably, the ODI level in cardiac patients is found to be significantly lower than that for healthy men despite the much lower hematocrit in the former (35.5±4.5% vs. 45.3±2.3%). These patients have an average asymptotic blood viscosity of 4.9±0.5 cP and average plasma viscosity of 1.96±0.21 cP. Viscosity of healthy donor (male and female) plasma is approximately 1.7 cP (12). In additional experiments we found that in the blood samples with the same hematocrit ODI level decreases when plasma viscosity is increased via an elevation in protein concentration. Thus the low ODI levels in the cardiac patients' blood can be explained by elevated plasma viscosity. An adequate oxygen supply is essential for the normal functioning of tissue. A decreased oxygen delivery may contribute to the development of cardiovascular disease. In the present study we found that the ODI levels are significantly lower in male blood than in blood of pre-menopausal females. The major factors accountable for this phenomenon are the higher hematocrit and increased blood viscosity, and the fewer number of young and larger number of old RBCs in male blood compared to the blood of pre-menopausal females.

Regular blood donations may reduce the hematocrit and blood viscosity, improve rheological properties of blood and increase the volume rate of oxygen delivery in men. Several recent epidemiological studies support our conclusions regarding the advantage of blood donation. Finnish researchers report that among 2,682 middle-aged men followed for about 5 years, the risk of heart attack is 86 percent lower among blood donors [18]. An epidemiological study performed by Kansas University Medical Center researchers on nearly 4,000 people found that the non-smoking men in the study who had donated blood in the previous three years, are at almost twice lower risk to suffer a heart attack or stroke compared to those who had never donated blood [19].

In conclusion, Oxygen Deliver Index is found to be lower in populations with a higher risk of Coronary Heart Disease (CHD) and, especially, in cardiac patients. This suggests that ODI may be a useful parameter for assessing risk of CHD.

REFERENCES

1. Vital Statistics of the United States 1990, Vol.II, Mortality, U.S. Department of health and human services. Hyattsville, Maryland, 1994.
2. Stampfer MJ, Colditz GA, Willet WC. Menopause and heart disease. Ann. NY Acad Sci 1990; 592:193-203; discussion 257-262.
3. Gura T. Estrogen: key player in heart disease among women. Science 1995; 269,771-773.
4. Weiner CP, Lizasoain I, Baylis SA, Knowles RG, Charles IG, Moncada S. Induction of calcium-dependent nitric-oxide synthases by sex-hormones. Proc Natl Acad Sci 1994; 91:5212-5216.
5. Hishikawa K, Nakaki T, Marumo T, Suzuki H, Kato R, Saruta T. Up-regulation of nitric-oxide synthase by estradiol in human aortic endothelial-cells. FEBS Lett 1995; 360:291-293.
6. Yue P, Chatterjee K, Beale C, Poolewilson PA, Collins P. Testosterone relaxes rabbit coronary-arteries and aorta. Circulation 1995; 91:1154-60.
7. Sullivan JL. Iron and the sex difference in heart disease risk. Lancet 1981;1:1293-1294.
8. Sullivan JL. The sex difference in ischemic heart disease. Perspectives Biol Med 1983; 26:657-671.
9. Meyers DG. The iron hypothesis - does iron cause atherosclerosis? Clin. Cardiology 1996;19:925-929.
10. Grigorian SS, Kameneva MV. Role of the hematocrit index in the development of hemodynamic disorders. Soviet Phys Dokl 1985; 30:750-752.
11. Micheli V, Taddeo A, Vanni AL, Pecciarini L, Massone M, Ricci MG. Distribuzione in gradiente di densita' degli eritrociti umani: differenze lagate al sesso. Boll Soc Italiana Biol Speriment 1984; LX(3):665-671.
12. Kameneva MV, Garrett KO, Watach MJ, Borovetz HS. Red blood cell aging and risk of cardiovascular diseases. Clinical Hemorheology and Microcirculation 1998; 18:67-74.
13. Kameneva MV, Antaki JF, Borovetz HS, Griffith BP, Butler KC, Yeleswarapu KK, Watach MJ, Kormos RL. Mechanisms of red blood cell trauma in assisted circulation. Rheologic similarities of red blood cell transformations due to natural aging and mechanical stress. ASAIO Journal 1995; 41:M457-460.
14. Kameneva MV, Watach MJ, Litwak P, Antaki JF, Butler KC, Thomas D, Taylor LP, Borovetz HS, Kormos RL, Griffith BP. Chronic animal health assessment during axial ventricular assistance: importance of hemorheological parameters. ASAIO Journal 1999; 45:183-188.
15. Lowe GDO, editor. Clinical Blood Rheology. CRC Press, Boca Raton, Florida, 1988.
16. Koenig W, Ernst E. The possible role of hemorheology in atherothrombogenesis. Atherosclerosis 1992; 94:93-107.
17. Whitmore RL. Rheology of circulation. Oxford: Pergamon Press; 1968.
18. Tuomainen TP, Salonen R, Nyyssonen K Salonen JT. Cohort study of relation between donating blood and risk of myocardial infarction in 2682 men in eastern Finland. British Medical Journal 1997; 314:793-794.
19. Meyers DC, Strickland D, Maloley PA, Seburg JJ, Wilson JE, McManus BF. Possible association of a reduction in cardiovascular events with blood donation. Heart 1997; 78:188-193.

Chapter **70**

PRELIMINARY STUDIES OF THE APPLICATION OF NEAR INFRARED SPECTROSCOPY IN THE DIAGNOSIS OF DEEP VEIN THROMBOSIS

Lino K. Korah, Frederick D. Scott[#], G. Melville Williams*, & Kyung A. Kang

*Chemical Engineering Department, University of Louisville, Louisville, Kentucky;[#]Chemical and Biochemical Engineering Department, University of Maryland, Baltimore County, Baltimore, Maryland; *Vascular Surgery Department, Johns Hopkins Hospital, Baltimore, Maryland*

Abstract: A feasibility study for diagnosing deep vein thrombosis utilizing near-infrared continuous wave spectroscopy was performed, as a real-time, non-invasive, and inexpensive method. The probe contains two light sources and two detectors with optical filters that monitor reflected light at wavelengths 760 and 850 nm to measure the changes in the amount of deoxyhemoglobin and oxyhemoglobin, respectively. These changes and the blood volume changes are recorded while the subject performs a series of light leg exercises. The test protocol is designed to determine the muscle tissue blood volume capacity, rate of blood filling (venous valve functionality test), and efficiency to promote one-directional venous flow from the leg to heart. The subject pool consists of the patients with leg deep vein thrombosis (DVT) diagnosed by the Johns Hopkins Hospital Vascular Surgery Department and of normal subjects as the control. Abnormal venous systems showed distinct characteristics: high blood volume in the leg; high rate of blood filling while the subject stands upright; and the inability to decrease the blood volume during the muscle contraction. The NIR device proved to be an inexpensive, effective, and portable device that can detect DVT in the leg in real-time.

Key words: NIR spectroscopy, Deep Vein Thrombosis, Blood Volume, oxy- and deoxy-hemoglobin

Abbreviations: DVT - Deep vein thrombosis, EVF - Ejection volume fraction, MRM - Micro Run Man, NIR - Near infrared, RVF - Residual volume fraction

1. INTRODUCTION

Deep vein thrombosis (DVT) is a condition that abnormal blood clots form in deep veins. It usually occurs in the leg possibly due to the stagnation of blood (depletion of anti-coagulant) caused by gravity. Some of the causes of DVT include anticoagulant (e.g. Protein C, anti-thrombin III) deficiency, major surgeries, accidents, and pregnancy. When formed clots dislodge from the original DVT site and travel through the circulatory system, they can block the blood vessels in vital organs causing heart attack, lung embolism, stroke, etc. Approximately 2 million Americans are diagnosed with DVT each year. Each year 600,000 patients develop pulmonary emboli and 60,000 die of related complications [1]. Therefore, it is important to detect DVT in an early stage. Once DVT has developed, constant monitoring of the patient is also important to prevent its reoccurrence.

Current methods for detecting DVT in the leg include invasive (venography) and non-invasive methods (plethysmography and ultrasound). *Venography* is the gold standard for DVT diagnosis. The process consists of injecting a radioactive dye in the leg vein through the foot. Then an X- ray is taken to see how far the dye has spread in the leg circulatory system. This process is painful, expensive, and needs a highly trained person to perform it. In some cases, repeated use of venography has even led to the development of DVT [2]. *Plethysmography* is performed by placing two sets of electrodes around the patient's calf and an oversized blood pressure cuff around the thigh. Changes in venous filling are produced by inflating the thigh cuff to obstruct venous return on the calf while the blood volume reaches the baseline. If a thrombus is present venous emptying is delayed. This method is effective and cost-efficient for detecting proximal DVT, however, not very accurate. *Ultrasound imaging* of the venous system is obtained with high-resolution equipment to produce two-dimensional images by real-time computation of reflected signals from an array of ultrasound. Ultrasound is highly effective in the thigh but the instrument is expensive. Plethysmography and ultrasound are less effective in the calf DVT detection and can sometimes lead to the misdiagnosis [3].

Micro Run Man–96 (MRM-96) utilizes safe and non-ionizing light in the near infrared (NIR) region. In this study, this instrument was tested for an inexpensive, effective, safe, and non-invasive way of detecting DVT and also evaluating the functionality of leg venous valves.

2. INSTRUMENT

MRM-96 (Figure 1; NIR Sales; Philadelphia, PA), a continuous wave spectrometer consists of a probe and a monitor.

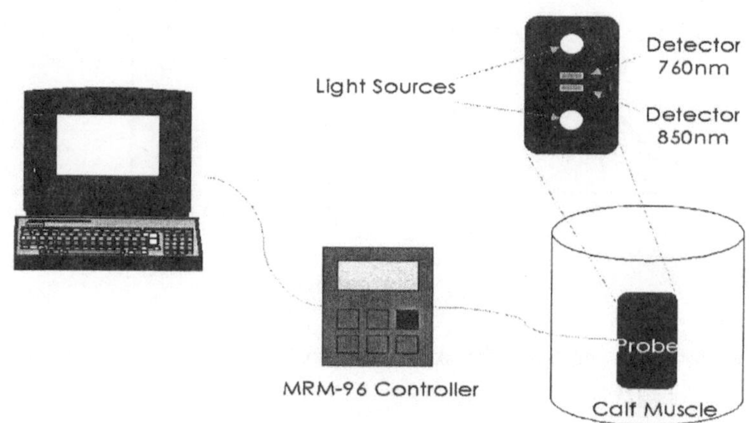

Figure 1. A schematic diagram of Micro Run Man-96 used for detecting DVT in the leg.

This instrument uses the two wavelengths in the near infrared (NIR) region, 760 and 850 nm, which corresponds to mainly deoxy- and oxy-hemoglobin, respectively (Figure 2). As the amount of the oxy- and deoxy-hemoglobin in the monitored area changes, the amount of light being absorbed changes as well. The sum of oxy- and deoxy- hemoglobin correspond to the total blood volume. The light intensity measured by the detector is converted to electrical signals and then to digital signal in the main unit. MRM-96 gathers this data and through the mathematical manipulations, values of the deoxy- and oxy- hemoglobin, and blood volume are obtained. These values are then displayed on the LCD and stored in the unit. A personal computer can also be connected for real-time display of profiles and the data can be stored for later analysis or review [4].

3. METHOD

The changes in the oxy- and deoxy- hemoglobin concentration are monitored at the medial head of the gastrocnemius muscle in the leg, during

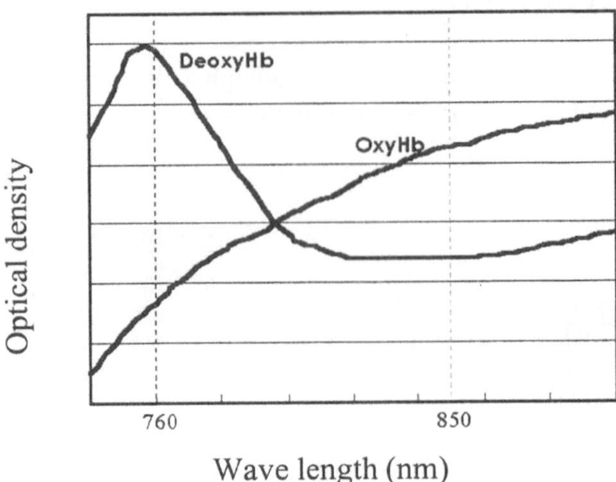

Figure 2. The absorption spectra of the oxy- and deoxy- hemoglobin in the near infrared region.

a light exercise protocol. The test protocol combined the leg elevation and calf raise. The calf raise was selected for the exercise protocol for the following reasons: to promote one-directional blood flow; to isolate the muscle that is monitored; to create a light stress that can generate a recovery effect; and to select a light exercise routine for the patient without any harmful effect.

The overall protocol is divided into six stages: (1) baseline at supine position; (2) leg elevation; (3) standing; (4) exercise; (5) standing recovery; (6) supine position with leg elevation. The first stage was to calibrate the resting levels of oxy- and deoxy- hemoglobin values to zero. A baseline is needed because each individual has unique blood oxygenation and blood volume values at rest. During leg elevation step (stage 2), the leg is elevated at an angle of 17° from the floor in order to create gravity assisted venous flow. 17° angle was used because this provided enough gravity to empty the blood from the calf muscle and an angle was used instead of the distance from the ground since the length of the leg of each subject was different. The standing position (stage 3) creates the condition in which the venous flow is reversely directed by the gravity. By observing the rate of the blood filling, the functionality of the valves in the veins can also be tested. The secondary baseline is obtained once the blood volume in the leg stabilizes. The next step, exercise consists of two sub-stages (stage 4). In the first sub-stage, the subject performs one calf raise causing the muscle to squeeze the blood out of the tissue. Thus, there is a decrease in the blood volume while the body weight is on the toes. Once the reading becomes stable, the individual performs 10 calf raises at a rate of one calf raise every two seconds, to empty the blood from the muscle at the maximum. The standing recovery step (stage 5) generates the recovery characteristic that is a result of the systemic response in the circulatory system after exercise. There is an increase of blood volume to re-oxygenate the muscle. In the final leg elevation recovery step (stage 6), the individual lies back in the original position and elevates the leg at the same angle as the previous leg elevation. This is to test the ability of the leg to return to the original condition.

4. RESULTS AND DISCUSSION

The age range of the patients and the normal subjects was between 20 and 60. There were eleven normal subjects and six DVT patients. The patient pool consisted of two female (one African American and one Caucasian) and four male patients (three Caucasians and one African American). The normal subject pool consisted of six males and five females. In the male category one was African American, three were Caucasian and two were Orientals. Female normal subjects consisted of one African American, two Caucasians, and two Orientals. The readings of both right and left legs of all subjects were used for analysis. From analyzing the results, it was found that the normal subject values showed a rather consistent trend independent of the gender, age, race, or individuals athletic ability. Since this is a preliminary study with a few subjects, only the major differences in the profile between the normal and the patient were evaluated. In general, patients have problems with only one leg. Therefore, we also compared profiles from both legs of a subject. The normal leg of the patient showed the same trend as that of the normal subject. The leg with the DVT (diagnosed by the Vascular Surgery Department, Johns Hopkins Hospital) showed values different from normal legs, regardless of the age, gender, or race.

The profile for a typical normal subject is shown in Figure 3.

After the baseline is established when the subject elevates the leg during the leg elevation position (Figure 3; stage 2), there is a decrease in the blood volume level. This is because gravity pulls the blood away from the leg muscle and directs it to the heart. For the third stage, when the subject stands up, there is an increase in the blood volume in the calf muscle because the gravity is now reversed and direct the blood to the calf muscle. In the

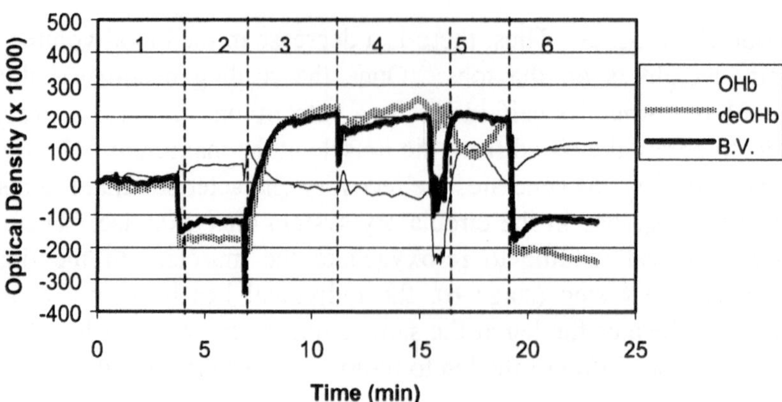

Figure 3. A typical profile for normal subjects.

fourth stage, while the subject is performing a single calf raise, there is a distinguished decrease in the blood volume by muscle contraction. Here, the contraction promotes blood flow leaving the muscle to the vein. There is more decrease in the blood volume while the subject performs the ten calf raises (stage 5). The muscle helps to remove the venous blood nearly to the maximum from the muscle. The blood volume remaining in the muscle at this point can indicate the status of the blockage in the vein. The oxygenation of tissue also decreases due to the exercise. During the resting stage (stage 6), there is a slight increase in the blood volume because the body is responding to the exercise by supplying more oxygen.

A typical patient profile is shown as below in Figure 4.

Compared to the normal subject profile, in the second stage while the subject elevates the leg, there is no significant decrease in the blood volume. Even though the gravity assists the blood in the vein to flow in the direction from the leg to the hip, it cannot flow out of the system due to the blockage in the vein. When the patient stands up (stage 3) blood flow reverses much more rapidly than normal subjects because, due to the venous

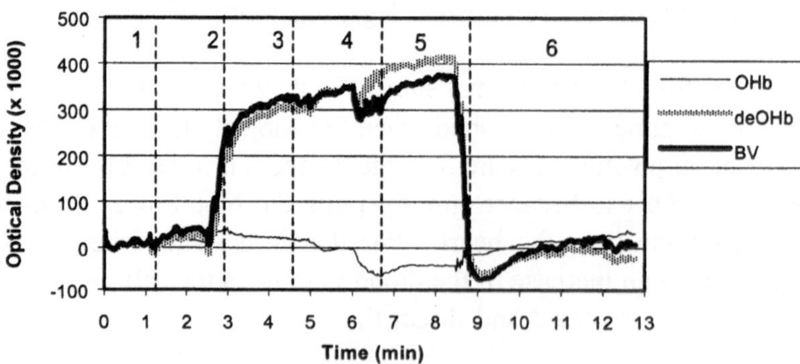

Figure 4. A typical profile for a patient.

valve damage, there is little resistance to the reverse flow. In the fourth stage, although the muscle contraction helps the vein to promote the blood flow out of the muscle, it cannot flow out well due to the blockage in the vein.

5. PARAMETERS SELECTED TO CHARACTERIZE DVT IN LEGS

For a systemic analysis of the profiles, four major parameters that provide the most distinction between normal and patient groups are selected: the muscle capacitance, the rate of filling, the ejection fraction (EVF), and the residual volume fraction (RVF). The *muscle capacitance* is the maximum blood volume change in the calf muscle. It is the difference between the second baseline (standing) and leg elevation. The *rate of the filling* measures the slope of the blood volume increase occuring during the filling of the blood into the calf muscle (stage 3). The *ejection volume* is the blood volume decrease by a single calf raise. The *ejection volume fraction* is the fraction of the ejection volume to the muscle capacitance. The *residual volume* is the amount of the blood left in the muscle after the ten repetitions of the calf raises. *Residual volume fraction* is the fraction of residual volume to the muscle capacitance.

The *rate of filling* values for the normal and patient subjects is shown in Figure 5. For a normal subject, this value is below 500 since it takes some time for the calf muscle to refill with the blood, because the venous valve prevent the blood reversing. The slope of filling is much higher for the patients possibly due to the damaged valves in the vein, thereby rapid blood filling in the calf muscle. Patient profile typically showed a value above 800. Normal subjects 16 and 17, who do athletic activities daily, showed the values slightly higher than the other normal values (Figure 5).

The ejection volume fraction (EVF) is shown in Figure 6. The normal subjects value was usually above 20, since the muscle contraction assists the vein to transport blood from calf to heart. The range of EVF values for the patient stayed between 0~20. These values are lower because of the blockage in the leg vein, which creates a larger muscle capacitance (denominator) and a smaller ejection volume (numerator).

Residual volume fraction (RVF) is shown in Figure 7. The normal values ranged between 0 and 60, while the patient values tend to be above 75. Due to the blockage in the vein and the valve dysfunction, the muscle contraction does not help to decrease the blood volume from the leg muscle, unlike in the leg of a normal subject. Note here again that, the normal subject 16 and 17 (athletes) have slightly higher values.

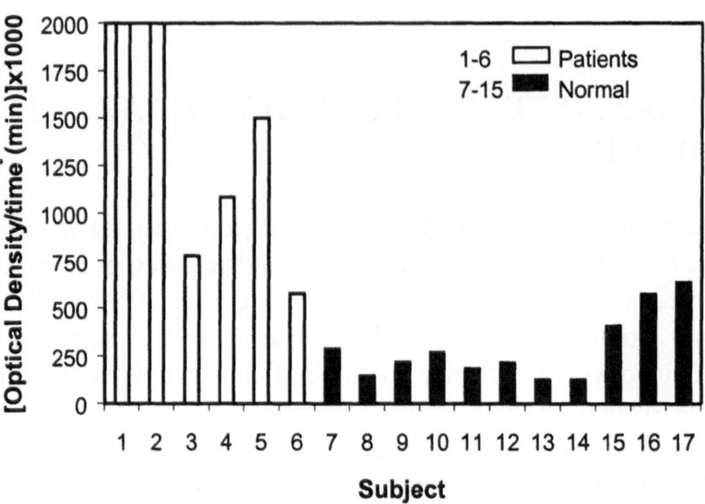

Figure 5. Rate of filling for the patients and normal subjects. Note that the patient 1 and 2 had a very large value (about 4000) and is shown as 2000.

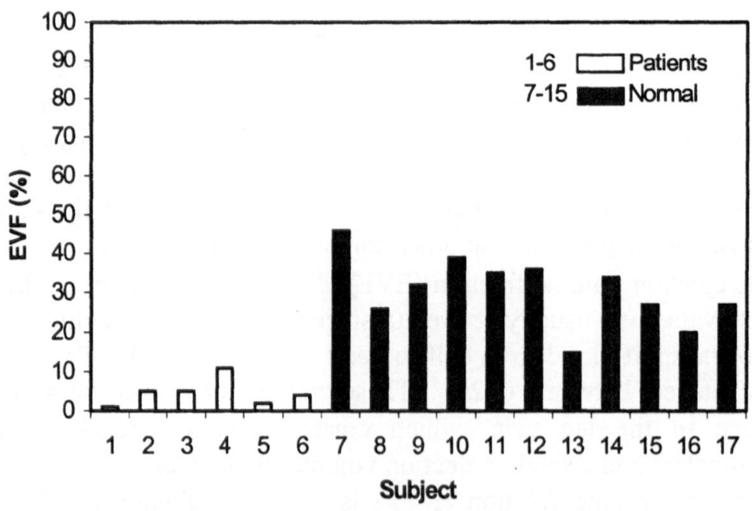

Figure 6. Ejection Volume Fraction of the tested subjects.

While analyzing these parameters, it was also noted that most right-handed patients had DVT's in their left legs. It was also seen that the left legs of most right-handed normal people yielded slightly higher values for the rate of filling and RVF and lower EVF values compared their right legs. This may support the fact that a reason for DVT is due to the stagnation of blood caused by the less mobility.

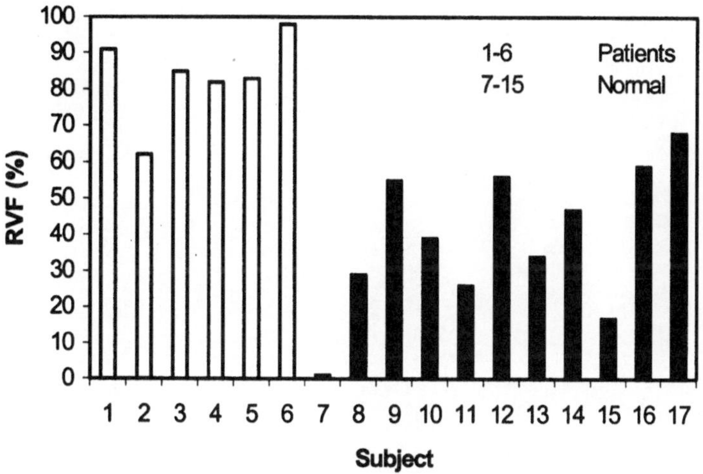

Figure 7. Residual volume fraction of the subjects.

6. CONCLUSIONS

From the blood flow profiles of eleven normal subjects and six patients, it can be concluded that the Micro Run Man-96 can be used as an effective way to detect DVT in the leg and can be used as an initial diagnostic tool for DVT. This instrument also provides the extent of venous valve functionality, while other DVT diagnostic techniques are not capable of providing this type of prognostic information. This instrument and the test protocol are user friendly, non-invasive, and inexpensive.

ACKNOWLEDGEMENTS

The author would like to thank NIR Sales and Dr. Britton Chance for loaning the Micro Run Man-96. The author would also like to thank the nurses at Johns Hopkins Hospital for their help and cooperation in obtaining the patient data. Likewise, author would like to thank all the subjects who had participated in the studies.

REFERENCES

1. Deep Venous Thrombosis and Thrombophebitis. http://www.emedicine.com/EMERG/topic122.htm
2. Wells, P.S. and Ginsberg, J.S. DVT and Pulmonary Embolism: Choosing the night diagnostic tests for patients at risk. Geriatrics, 1995; 50:3-4.
3. Deep Vein Thrombosis and Detection. http://www.amhrt.org/
4. Guide to Using MRM-96 and Nircom. Pennsylvania, 1997; 1-2.

Chapter **71**

CEREBROVASCULAR RESPONSE TO ACUTE METABOLIC ACIDOSIS IN HUMANS

Marjo van de Ven[1], Willy N.J.M. Colier[3], Bregina T.P. Kersten[3], Berend Oeseburg[3] & Hans Folgering[2]

[1]Department of Pulmonology, Rijnstate Hospital Arnhem, PO 9555, 6800 TA Arnhem, The Netherlands;[2]Department of Pulmonology Dekkerswald, University of Nijmegen, PO Box 9001, 6560 GB Groesbeek, The Netherlands;[3]Department of Physiology,Faculty of Medical Sciences, University of Nijmegen, PO Box 9101, 6500 HB Nijmegen, The Netherlands.

Abstract: Objectives: Evaluation of the cerebrovascular response ($\Delta CBV/\Delta Paco_2$) during baseline metabolic conditions and acute metabolic acidosis. Methods: 15 healthy subjects, 5 m, 10 f, 56±10 yrs were investigated. For acidification, NH_4Cl was given orally. CBV was measered using Near Infrared Spectroscopy (OXYMON) during normo-, hyper- and hypocapnia. Results: Acute metabolic acidosis was realised: mean ΔBE -2.7 $mEq.L^{-1}$ ($p<0.001$) with mean $\Delta Paco_2$ –0.2 kPa ($p<0.01$). During normo-, hyper- and hypocapnia, CBV values of 3.51, 4.82 and 2.55 $mL.100g^{-1}$ were calculated during baseline metabolic conditions and 3.70, 4.86 and 2.63 $mL.100g^{-1}$ during acute metabolic acidosis. The $CBV/Paco_2$ response showed a hockeystick configuration with the point of infliction around normocapnia. $\Delta CBV/\Delta Paco_2$ reactivity from normo- to hypercapnia and from normo- to hypocapnia was calculated; no significant differences in $\Delta CBV/\Delta Paco_2$ were found in both metabolic conditions. Conclusion: Cerebrovascular reactivity to CO_2 does not alter during acute metabolic acidosis.

Key words: Ammonium-Chloride-pharmacology, carbon dioxide, cerebral blood volume, NIRS

1. INTRODUCTION

Metabolic acid-base disorders resulting from ingesting nonvolatile acid are buffered in both the extracellular (ECF) and intracellular fluid (ICF)[1]. The CO_2/HCO_3^- buffersystem is the principal ECF buffer. When a nonvolatile acid is added to the body, buffering of H^+ occurs by binding HCO_3^- resulting in a lowered plasma $[HCO_3^-]$. Respiratory compensation ensues, the induced metabolic acidosis is corrected by the kidneys by reabsorbing more HCO_3^- from the urine and increased ammonium excretion (enhanced production of new HCO_3^-). In the cerebral compartment, the shifts in acid-base balances depend mainly on the permeability of the blood brain barrier.

Ammoniumchloride (NH_4Cl) has been shown to be effective at inducing acute metabolic acidosis[2,3]. The resultant decrease in pH causes a systemic (metabolic) acidosis. The blood-brain barrier will prevent [H+] ions to diffuse into the brain tissue and therefore the pH of the brain-ECF will not alter initially. Subsequently, the cerebral vessels hardly dilate in response to an increase in the hydrogen ion concentration of the arterial blood. However, they are very sensitive to CO_2 changes. Cerebrovascular reactivity can be measured as the change of cerebral blood volume (CBV) to a change of $PaCO_2$[4,5]. We wanted to investigate respiratory changes with and without a metabolic acidosis induced by the administration of oral NH_4Cl and compare cerebrovascular reactivity, expressed as $\Delta CBV/\Delta PaCO_2$ during both conditions. The influence of an altered concentration of [H+] in arterial blood on the cerebrovascular autoregulation can be monitored and is valuable for understanding central chemoreceptor activity[6-8].

2. METHODS

2.1 Subjects

Fifteen healthy volunteers –5 male and 10 female, mean age 56 years (range 46 to 68 years) -participated in the study. None of them were on medication. At least two hours prior to the experiments, they had to abstain from caffeinated drinks and cigarettes. A short physical examination was done with a supplementary brief dynamic lungfunction test. All volunteers gave informed consent. The study was approved by the ethical committee of the Department of Pulmonology Dekkerswald, University of Nijmegen.

2.2 Administration of NH_4CL

The administered dose of NH_4Cl was based on the assumption that ECF varies between 23 $\pm$ 3%[9] and 30%[10] of total bodyweight. This value (in litres) multiplied by the target change in BE of 2 mEq provides the total amount of NH_4Cl necessary to reach the desired degree of acidification. Multiplication of this value by the molecular weight of NH_4Cl (53.5), gives the dosage in mg. The dose was given at t=0 and a repeated dose after 60 min to approximate the t½ of 1 - 1¼ hours[11]. Taking these doses, an altered base-excess (ΔBE) of at least 2 $mEq.L^{-1}$ can be achieved for at least 3 hours[12].

2.3 NIRS

NIRS is an optical technique, based on the principle that light in the NIR region is absorbed by chromophores of oxyhemoglobin [O_2Hb] and deoxyhemoglobin [HHb][13]. Changes in light absorption are converted into concentration changes of these chromophores. Owing to light scattering in the tissue, only changes in concentration in the chromophores from an arbitrary zero can be determined. A modified Lambert-Beer law gives the relationship between the changes in concentration and absorption. The difference between [O_2Hb] and [HHb] is defined as the oxygenation-index [OI].

The OXYMON (Departments of Physiology and Instrumentation, University of Nijmegen, the Netherlands) uses a high sampling rate of 10 Hz[14]. Changes in light absorption are measured at three different wavelengths: 905, 845 and 770 nm. Optodes were strapped to the forehead, approximately 2 cm from the midline. A distance of 5.5 cm between transmitting and receiving fibers was used. This optode distance ensures deep enough penetration of the near infrared light into the brain to exclude most of the exta-cranial circulation from the detected signal[15]. A pathlength factor of 6.0 was used. Data were stored on disk for off-line analysis and calculation of CBV.

2.4 CBV measurements

To quantify CBV, a slight decrease in arterial saturation (ΔSaO_2) of 5% is necessary. This was performed by lowering the FiO_2. ΔSaO_2 was measured with a pulse oximeter (N200 Nellcor Puritan Bennett, St. Louis, USA), in beat-to-beat mode. When stable readings of [O_2Hb], [HHb] and SaO_2 at a lowered saturation level were seen, a FiO_2 of 21% was given to restore normoxia. When the change in arterial SaO_2 was related to the change of [OI], absolute values of CBV can be calculated [16,17], under the assumption that blood flow, blood volume and oxygen consumption remain constant. Steady state conditions are needed to obtain reproducible values of [OI] and SaO_2 at two levels. A steady state condition was defined as a condition with stable values of [O_2Hb], [HHb] and [tHb], SaO_2, mean arterial pressure (MAP), endtidal CO_2 ($P_{ET}CO_2$), heart rate (HR) and respiratory rate (RR) for at least 1 minute. Values of [OI] and SaO_2 at the applied desaturation level and during normoxia were obtained from a 20 s average.

2.5 Experimental setup

The setup was described previously[4]. The subjects were in supine position during the whole experiment. SaO_2 and HR were monitored with a pulse-oximeter, with the sensor attached to the right-frontal forehead. The optodes of the NIRS were strapped to the left side of the forehead. The subjects breathed through a facemask with valves for in- and expiratory gasmixture. Changes in inspiratory gasmixture of oxygen, nitrogen and carbon dioxide were induced by means of a computer controlled massflow system (Bronckhorst Hitec, Veenendaal, The Netherlands). The FiO_2 was monitored continuously using an oxygen (OM-11 Beckman, Fullerton, CA., USA). Fast changes in inspiratory gasmixture could be performed; the aimed changes were reached within one breath.The expiratory port of the mask was connected to a capnograph (N200 Nellcor Puritan Bennett, St. Louis, USA) to monitor $P_{ET}CO_2$ (kPa), respiratory rate (min^{-1}). The analogue data were linked directly to the NIRS computer for real time display and storage simultaneously with the NIRS data.

Hypercapnia was induced by giving adequate amounts of CO_2 in the inspired air; hypocapnia could be performed by making the subject increase the respiratory rate and/or tidal volume. Arterialized capillary bloodgas samples were taken during (each) respiratory condition, from a warmed fingertip. Samples were analyzed within 2 minutes (Synthesis 25, Instrumentation Laboratory SpA, Italy). The [Hb] was determined on the same analyzer. A metabolic control state is defined as a condition without applied metabolic changes. During control state, the three respiratory conditions were measured. A calculated dose of oral administrated NH_4Cl was given at the end of the control measurements. One hour after ingestion of the first dose of NH_4Cl and immediately after the second dose, respiratory changes were repeated to determine the same parameters during acute metabolic acidosis.

2.6 Statistical analysis

Each subject was measured twice during the respiratory conditions. Duplicate values of CBV and p_aCO_2 were averaged afterwards.

During the whole experiment, values of $P_{ET}CO_2$, HR, RR and MAP were recorded continuously; they were expressed as means with standard deviation during each CO_2 challenge. All former variables, including CBV and capillary p_aCO_2 values measured during hyper- and hypocapnia, were compared to the values determined during normocapnia using a paired *t*-test. A paired *t*-test. was also used to compare all variables (CBV, p_aCO_2, MAP, HR, RR) measured during acidification to variables measured during baseline (control) conditions. Linear regression analysis was performed on the slopes of CBV/p_aCO_2 of the individual subjects. The control state was compared to the results from acidification. The level of statistical significance was set at $p<0.01$. All values were reported as mean ± SD.

3. RESULTS

This study shows the cerebrovascular response to CO_2 exposed to two different metabolic conditions. Table 1 shows the results of the measurements of CBV, capillary p_aCO_2, HR and RR during metabolic baseline conditions (controls) and after administration of NH_4Cl.

Table 1. Mean values (with SD) during normo-, hyper- and hypocapnia in baseline metabolic conditions (control) and after NH_4Cl administration.

n=15	Age (yrs)	CBV* (ml.100g-1) Mean(SD)	$PaCO_2$† (kPa) Mean(SD)	HR‡ (min^{-1}) Mean(SD)	$PetCO_2$ (kPa) Mean(SD)	RR§ (min^{-1}) Mean(SD)
control						
normocapnia	56(7)	3.51(0.71)	5.14(0.36)	72(10)	4.33(0.42)	14(3)
hypercapnia		4.82(1.12)	5.63(0.29)	71(10)	5.57(0.46)	15(4)
hypocapnia		2.55(0.72)	3.80(0.24)	70(8)	3.21(0.14)	19(4)
NH_4Cl						
normocapnia		3.65(0.56)	4.94(0.39)	73(10)	4.47(0.34)	15(3)
hypercapnia		4.86(0.70)	5.58(0.32)	74(11)	5.80(0.25)	16(3)
hypocapnia		2.67(0.85)	3.78(0.21)	73(9)	3.24(0.12)	18(5)

Cerebral Blood Volume
† capillary p_aCO_2
‡ Heart Rate
§ Respiratory Rate

The mean dose of NH_4Cl was 2250 (SD 400) mg. During the entire experiment of metabolic acidosis, capillary arterialized blood samples showed a mean ΔBE of -2.7 (1.3) $mEq.L^{-1}$ and ΔpH of –0.04 (0.02) (both $p<0.001$). Capillary p_aCO_2 was significantly decreased (mean $\Delta PaCO_2$ 0.20 (0.17) kPa) after NH_4Cl ingestion ($p<0.001$).

During metabolic control state, an increase of p_aCO_2 of +0.52 (0.27) kPa was reached during hypercapnia. Hypocapnia resulted in a decreased p_aCO_2 of –1.35 (0.39) kPa. Similar respiratory changes are effected in the course of metabolic acidosis eventuating in an increase of Δp_aCO_2 of 0.64 (0.25) and a decrease of 1.16 (0.43) kPa towards hyper- and hypocapnia, respectively. CBV values were increased during hypercapnia and decreased during hypocapnia ($p<0.001$), compared to normocapnic values (table 1).

Figure 1 shows the data of all subjects in three different respiratory conditions; mean values of p_aCO_2 during hypo-, normo- and hypercapnia with concomitant CBV values are plotted from each individual during both metabolic conditions. Linear regression analysis in the control condition gave the following equation: CBV = 1.01 * p_aCO_2 - 1.39 ($r=0.69$, $p<0.0001$). During the condition of metabolic acidosis, the CBV shows a not significant different slope and intercept: CBV = 1.05 * p_aCO_2 – 1.29 ($r=0.67$, $P<0.0001$) (CBV: $mL.100g^{-1}.kPa^{-1}$; p_aCO_2: kPa). Values of RR are the same in both metabolic conditions. In contrast, HR is significantly increased during acidosis ($P<0.01$) compared to the control condition.

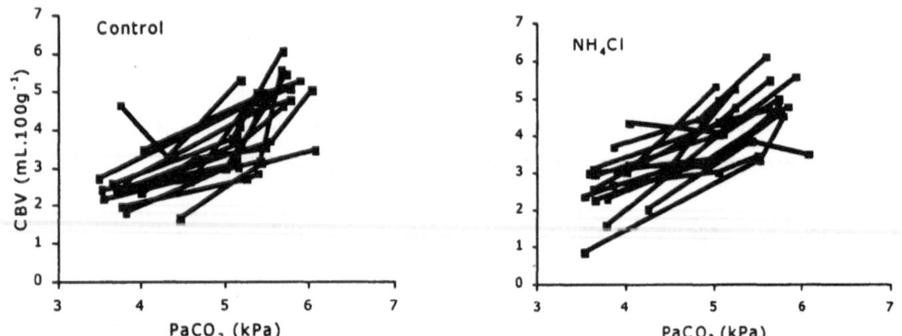

Figure 1. CBV/$PaCO_2$ reactivity: data of all subjects (n=15) during acute metabolic acidosis and control condition.

We determined the individual $\Delta CBV/\Delta p_aCO_2$ reactivity for the hypercapnic and for the hypocapnic response, taking the normocapnic situation as a starting point (Figure 2). Due to one blood sample error to determine capillary arterialized p_aCO_2, the outcome of only 14 subjects is shown in this histogram. Mean slopes in the hyper- and hypocapnic ranges were calculated, 2.60 (1.32-3.88) $mL.00g^{-1}.kPa^{-1}$ and 0.83 (0.64-1.02) $mL.100g^{-1}.kPa^{-1}$ (95% confidence) respectively, during control condition. After NH_4Cl administration, mean hyper- and hypocapnic slopes show not significantly different, values of 2.07(1.39-2.76) and 0.72 (0.413-1.034) $mL.100g^{-1}.kPa^{-1}$, respectively.

4. DISCUSSION

We investigated cerebrovascular reactivity to acute changes in p_aCO_2 during metabolic control condition and acute metabolic acidosis using NIRS.

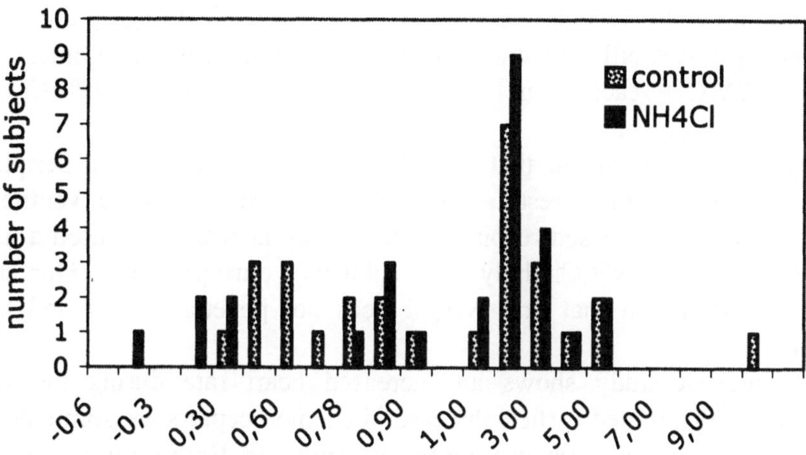

Figure 2. Individual $\Delta CBV/\Delta Pa_{CO_2}$ reactivity for the hypercapnic and hypocapnic response, during control condition and after oral administration of NH_4Cl.

The use of CBV, instead of CBF, may be discussed. Using $[O_2Hb]$ as a tracer to measure CBF, NIRS cannot provide the standards to obtain reproducible values of CBF [18,19]. CBV however, can be measured reproducible [17,20]. Furthermore, the use of CBV in stead of cerebral blood flow (CBF) eliminates the problems related to the mean transit time [21]. CBV values of the present study during normocapnia (3.51 ± 0.71 $mL.100g^{-1}$) are in good agreement with other studies: 2.85 ± 0.97 $mL.100g^{-1}$ [17].

In the present study, a linear relationship was seen to both hyper- and hypocapnia, but with significant different slopes. This confirms the findings of Edvinsson et al.[22] cerebrovascular responsiveness to hypocapnia is much lower than to hypercapnia. Interpretation of the CBV response to p_aCO_2 may therefore be divided into a reactivity from normo- to hypercapnia and normo- to hypocapnia.

Absolute values of CBV and cerebrovascular ($\Delta CBV/\Delta p_aCO_2$) responsiveness did not differ significantly during both metabolic conditions. This means that possible ventilatory responses to non-volatile acid, will not be modulated by changes in cerebral blood volume or blood flow. So it seems that the location of the chemoreceptor modulating vasoresponsiveness is at the interstitial side of the blood brain barier and in this way not sensitive to H^+ changes in the blood. These sensors are however sensitive to changes in Pa_{CO_2} since CO_2 easily diffuses through the blood brain barrier. In the present study, the small but significantly lowered p_aCO_2 did not lead to a lowered CBV. This study does not allow conclusions whether the vascular response at the brain side is affected either by CO_2 itself, or by H^+ from the bicarbonate equilibrium.

Data on cerebrovascular reactivity, expressed as CBV/p_aCO_2, are scarce. Brun et al.[5] used NIRS to study mechanically ventilated preterm infants and found a CBV/CO_2 reactivity in the same range (0.89 $mL.100g^{-1}$, 95% CI 0.63-1.26) as in the present study.

Tojima et al.[3] stated that NH_4Cl primarily stimulates the peripheral chemoreceptor as their results show an augmented hypoxic ventilatory response and an increased respiratory frequency at rest. They used a much higher dose (8 gr NH_4Cl daily in 6 subjects), during 3 days. Our study supports this notion that oral NH_4Cl does not penetrate the blood brain barrier.

The present study shows an increased heart rate during metabolic acidosis. As mentioned earlier, the carotid chemoreceptors are stimulated by NH_4Cl, resulting in a compensating hyperventilation. Both influences tend to depress the primary cardiac response to chemoreceptor stimulation and thereby accelerate the heart[23].

We conclude that an acute metabolic acidosis, induced by oral administrated NH_4Cl, did not change absolute values of cerebral blood volume, nor cerebrovascular reactivity to CO_2. It may hypothesised that cerebrovascular reactivity is not predominantly controlled by fixed $[H^+]$-ions in the blood.

ACKNOWLEDGMENTS

We would like to thank Ing. MC van der Sluijs for his technical support and J Evers for his assistance during the experiments. This study is granted by the Dutch Astma Foundation (96.09).

REFERENCES

1. Stanton BA, Koeppen BM. Role of kidneys in the regulation of acid-base balance. In: Berne RM, Levy MN, eds. Physiology Mosby, Inc., St. Louis. 1998; pp. 763-776.
2. Lerche D, Katsaros B, Lerche G, Loeschke HH. Vergleich der Wirkung verschiedener Acidosen (NH_4Cl, $CaCl_2$, Acetazolamid) auf die Lungenbelüftung beim Menschen. Pflügers Arch 1960; 270:450-460.
3. Tojima H, Kunitomo F, Okita S, Yuguchi Y, Tatsumi K, Kimura H, Kuriyama T, Watanabe S, Honda Y. Difference in the effects of acetazolamide and ammonium chloride acidosis on ventilatory responses to CO_2 and hypoxia in humans. Jpn J Physiol 1986;36: 511-521.
4. van de Ven MJT, Colier WNJM, Kersten BTP, Oeseburg B, Folgering H. Cerebral blood volume responses to acute $Paco_2$ changes in humans, assessed with near infrared spectroscopy. Adv Exp Med Biol 1999; 471:199-207.

5. Brun NC, Greisen G. Cerebrovascular responses to carbon dioxide as detected by near-infrared spectrophotometry: comparison of three different measures. Pediatr Res 1994;36: 20-24.
6. Javaheri S, Herrera L, Kazemi H. Ventilatory drive in acute metabolic acidosis. J Appl Physiol 1979; 46:913-918.
7. Teppema L. Effects of metabolic arterial pH changes on medullary ECF-pH, CSF-pH and ventilation in peripherally chemodenervated cats with intact blood-brain barrier. Respir Physiol 1984;58:123-136.
8. Wolff CB. Long term respiratory control - Metabolic acidosis. The physiological control of respiration Pergamon Press Ltd., U.K., 1993; pp. 529-539.
9. Baarends EM. Peak exercise response in relation to tissue depletion in patients with chronic obstructive pulmonary disease. Eur Respir *J* 1997; 10: 2807-2813.
10. Rose BD. The total body water and the plasma sodium concentration. Clinical physiology of acid-base and electrolyte disorders. McGraw-Hill International Editions, Singapore.1989; pp. 211-215.
11. Martindale. Ammoniumchloride (NH_4Cl), drug information. In: Reynolds JEF, eds. The extra pharmacopoeia. London Royal Pharmaceutical Society, 1996; pp. 1063.
12. van de Ven MJT, Colier WNJM, Oeseburg B, Folgering H. Induction of acute metabolic acid/base changes in humans. Clin Physiol 1999; 19:290-293.
13. Jöbsis FF. Noninvasive, infrared monitoring of cerebral and myocardial oxygen sufficiency and circulatory parameters. Science 1977; 198:1264-1267.
14. Van der Sluijs MC, Colier WNJM, Houston RJF, Oeseburg B. A new and highly sensitive continuous wave near infrared spectrophotometer with multiple detectors. SPIE 1997;3194: 63-82.
15. Harris DN, Cowans FM, Wertheim DA, Hamid S. NIRS in adults--effects of increasing optode separation. Adv Exp Med Biol 1994; 345:837-840.
16. Wyatt JS, Cope M, Delpy DT, Richardson CE, Edwards AD, Wray S, Reynolds EO. Quantitation of cerebral blood volume in human infants by near-infrared spectroscopy. J Appl Physiol 1990; 68:1086-1091.
17. Elwell CE, Cope M, Edwards AD, Wyatt JS, Delpy DT, Reynolds EO. Quantification of adult cerebral hemodynamics by near-infrared spectroscopy. J Appl Physio*l* 1994; 77:2753-2760.
18. Newton CR, Wilson DA, Gunnoe E, Wagner B, Cope M, Traystman RJ. Measurement of cerebral blood flow in dogs with near infrared spectroscopy in the reflectance mode is invalid [see comments]. J Cereb Blood Flow Metab 1997;17:695-703.
19. van de Ven MJ, Colier WN, Walraven D, Oeseburg B, Folgering H. Cerebral blood flow in humans measured with near infrared spectroscopy is not reproducible. *Adv Exp Med Biol* 1999;471:749-758.
20. van de Ven MJT, Colier WNJM, van der Sluijs MC, Walraven D, Oeseburg B, Folgering H. Can cerebral blood volume be measured reproducibly with an improved near infrared spectroscopy system? *J Cereb Blood Flow Metab* 2001; 21:110-113.
21. Shockley RP, LaManna JC. Determination of rat cerebral cortical blood volume changes by capillary mean transit time analysis during hypoxia, hypercapnia and hyperventilation. *Brain Res* 1988; 454:170-178.
22. Edvinsson L, MacKenzie ET, McCulloch J. Fundamental responses of the cerebral circulation. Cerebral blood flow and metabolism. Raven Press, New York. 1993; pp. 553-580.
23. Berne RM, Levy MN. Special circulations. In: Berne RM, Levy MN, eds. Physiology Mosby, Inc., St. Louis. 1998; pp. 478-501.

Chapter **72**

FREE FLAP MONITORING IN PLASTIC AND RECONSTRUCTIVE SURGERY

John A. Pickett[1], Maureen S. Thorniley[2], Nigel Carver[3] & Deric P. Jones[4]

[1]Clinical Physics Group and [3]Department of Plastic & Reconstructive Surgery, Bart's & The London NHS Trust, London EC1A 7BE; [2]Department of Instrumentation and Analytical Science, UMIST, PO Box 88, Manchester, M60 1QD; [4]Department of Medical Electronics, St. Bartholomew's & The Royal London Medical School, Queen Mary & Westfield College, Charterhouse Sq., London EC1M 6BQ.

Abstract: Free flaps are regularly used in plastic and reconstructive surgery but have a significant failure rate due to vessel thrombosis in the re-established arterial or venous circulation. A monitor of flap perfusion and oxygenation would allow the early detection of progressing flap ischaemia, hastening the required intervention and maximising the chances of salvaging the flap. A dual wavelength spectrophotometer has been designed and constructed which can monitor haemodynamic events in flaps during surgery and postoperatively. Eleven patients undergoing free flap surgery were studied. Measurements were made during surgery before division of the vessels and during and after microvascular anastomosis. Significant changes in all parameters were observed on reperfusion of the flaps after anastomosis or tourniquet ischaemia. Abnormal reperfusion in one flap and subsequent ischaemic events in two others were identified.

Key words: Surgery, ischaemia, spectroscopy, perfusion, flap, vascular-occlusions.

Abbreviations: Deoxyhaemoglobin Hb; Light emitting diode LED; Oxygenation Index HbD = $[HbO_2]$-[Hb]; Oxyhaemoglobin HbO_2; Total haemoglobin HbT; TRAM flap - transverse rectus abdominis musculocutaneous.

1. INTRODUCTION

Free flaps are regularly used in plastic and reconstructive surgery. However, some 10% of flaps require re-exploration in the first 48 hours. Salvage rates for these flaps can be as high as 83% if re-exploration is timely [1]. A reliable monitor of flap perfusion would increase the likelihood of a successful re-exploration by reducing the delay between the onset of ischaemia and the return to the operating theatre.

A new monitoring device has been constructed which uses an inexpensive light emitting diode (LED) and photodiode probe to measure changes in concentration of oxy- and deoxy-haemoglobin (HbO_2 and Hb) in free flaps. These variables are used to compute an index of ischaemic change, HbD (HbO_2-Hb) and total haemoglobin, HbT (HbO_2+Hb). The device performed well in a comparison with a commercial near-infrared spectrometer using a forearm occlusion model [2]. Characteristic repeatable patterns for different types of vascular occlusions were obtained [2] which were in agreement with earlier studies [3,4].

Our objective was to extend the work and assess its use in surgery. We used the device on patients undergoing plastic and reconstructive surgery following either completion of microvascular anastomosis or release of tourniquet.

2. METHODS

The device, previously described in detail [2], employs an inexpensive dual wavelength LED, operating at 660 and 880 nm and a photodiode detector. These components were obtained from a neonatal pulse oximeter probe (Johnson and Johnson Medical)[5]. The components were mounted in a reflectance configuration, approximately 1cm apart in a stainless steel shell. The output from the photodiode is amplified and demultiplexed into red and infra-red channels and the resulting signals low pass filtered [6]. These signals form the input to a 12-bit analogue-to-digital converter connected to a laptop computer. A "LabView" program is used to calculate and display the concentration changes of HbO_2, Hb, HbT and HbD[2].

Eleven patients undergoing reconstructive surgery were consented for study, eight following mastectomy, two following excision of carcinomas (one of the ear and one of the neck) and one to reconstruct the chest wall after surgery for necrotising fasciitis. Measurements were made during reperfusion of free flaps (TRAM or radial forearm) in all eleven patients.

Three flaps were reperfused twice. The reflectance probe was placed on the flap surface on a skin island before the flap was raised. The maximum change in each variable (peak response) was measured during reperfusion of the flap (immediately after the release of the arterial and venous clamps).

Reperfusions were also measured after release of a tourniquet on the upper arm when a radial forearm flap was being raised. In this case the arm had been exsanguinated prior to raising the flap and was reperfused for approximately forty minutes before division of the vessels.

3. STATISTICAL ANALYSIS

The maximum change in Hb, HbO_2, HbD and HbT (peak response) was measured during reperfusion of the flap (immediately after the release of the arterial and venous clamps). The mean and standard error of the mean were calculated, Mean $\pm$ SEM.

4. RESULTS

A positive response in all variables was observed on reperfusion of the successful flaps. This is consistent with an increase in blood volume in the flap. Reperfusion was characterised by an increase in HbO_2 (0.075$\pm$ 0.013), Hb (0.030 $\pm$0.006), HbD (0.054 $\pm$ 0.010) and in HbT (0.098 $\pm$ 0.016) (mmolar, Mean $\pm$ SEM) n=13. The majority of the increased blood volume was oxygenated. The observed increase in HbD, haemoglobin oxygenation index is indicative that there is far more oxyhaemoglobin than deoxyhaemoglobin. In the one failed flap measured to date no response was recorded in any variable on release of the vascular clamps.

Figures 1 to 4 show typical recordings made on a free TRAM flap (Figures 1 and 2) and a free radial forearm flap (Figures 3 and 4).

Figure 1 shows reperfusion of a flap after completion of the microvascular anastomoses and release of the vascular clamps. This is followed by a period when the flap was covered and undisturbed while the surgical team worked on the donor site on the abdomen. Some diathermy interference is visible after event 3 but then this is followed by a slow increase in Hb and HbT, with a slowly decreasing HbO_2 concentration, these factors resulted in a fall in the haemoglobin oxygenation index, HbD. The traces were interpreted as progressing venous congestion and the surgeon was alerted.

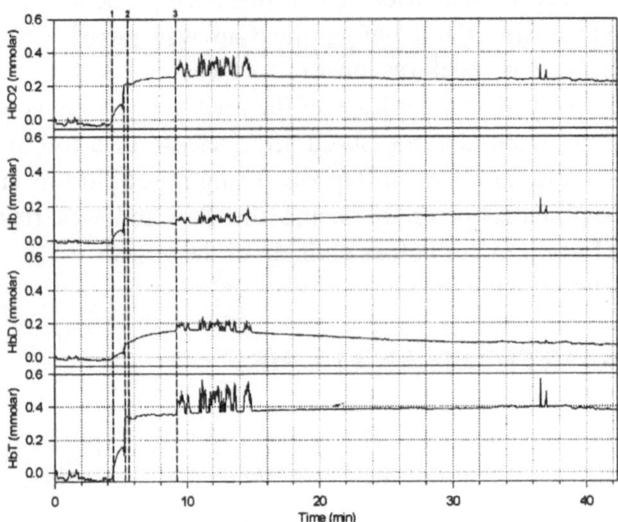

Figure 1. Reperfusion of a Free TRAM Flap. This shows reperfusion of a flap after completion of the microvascular anastomoses and release of the vascular clamps (event 1). The operating microscope was moved away (event 2) and the flap was then covered while the surgical team worked on the donor site on the abdomen and some diathermy interference is visible after event 3.

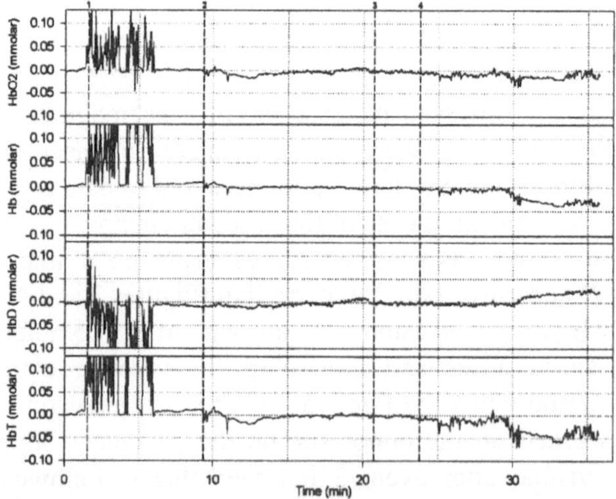

Figure 2. Reinvestigation of a Free TRAM Flap. In this trace is shown the investigation of the congestion shown in Figure 1. The first event marker indicates the presence of diathermy interference. The second event marker shows when the flap was uncovered and inspected. A significant amount of clotted blood was removed from the flap (event 3) and its surroundings, and a kink in the vessels supplying the flap was corrected by repositioning the flap (event 4).

In Figure 2 the investigation of this congestion is shown. The second event marker shows when the flap was uncovered and inspected. A significant amount of clotted blood was removed from the flap (event marker 3) and its surroundings, and a kink in the vessels supplying the flap was corrected by repositioning the flap (event marker 4). Shortly after event 4 a clear decrease in the total haemoglobin and deoxygenated haemoglobin concentrations was observed with a concomitant increase in HbD. HbO_2 remains relatively steady.

Figures 3 and 4 show two reperfusions on the same free radial forearm flap. When raising the flap on the left forearm the arm was exsanguinated and a tourniquet applied to the upper arm to maintain the bloodless operating field. Event 1 in figure 3 shows the release of the tourniquet after the flap had been raised and the vessels prepared, but before the vessels were divided. Reperfusion was accompanied by a large increase in blood volume which was mainly composed of oxygenated haemoglobin. After the reperfusion the flap was covered and allowed to recover from the period of ischaemia before the vessels were divided and the flap moved to its new position. Event 2 marks the start of the work to separate the flap from the donor site. The flap was uncovered and the vessels were clipped starting with the arterial clip at event 3. Motion artifact noise on the HbO_2, Hb and HbT traces makes any trend difficult to see but the HbD trace, where the noise from the HbO_2 and Hb recordings cancel out, shows a clear decrease. The first $1^1/_4$ hours of figure 4 shows the flap monitored while the microvascular anastomoses were made. Significant noise and baseline shifts

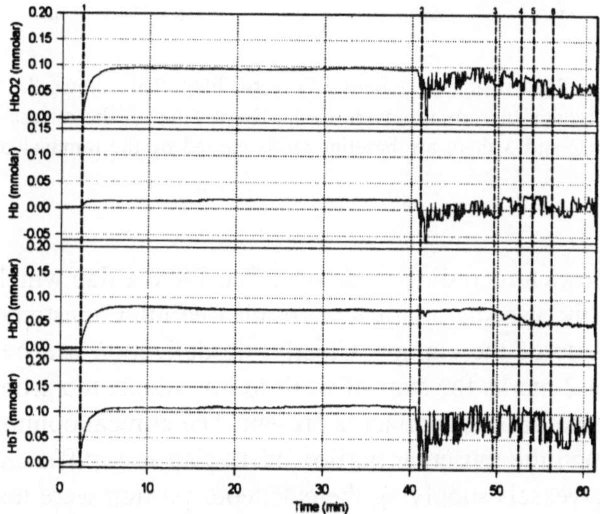

Figure 3. First Reperfusion of a Radial Forearm Flap. Event 1 shows the release of the tourniquet after the flap had been raised and the vessels prepared, but before the vessels were divided. Event 2 marks the start of work to divide the vascular pedicle. Events 3-6 mark the application of the first arterial clip and subsequent arterial and venous clips.

can be seen due to movement and repositioning of the flap and the moving shadows of the surgeons hands under the intense lights of the operating microscope. Event 1 marks the release of the vascular clamps. After the initial increase in HbO_2 on release of the vascular clamps there is a baseline shift caused by the removal of the operating microscope lamp (event 2). Although this has a significant effect on the HbO_2 and Hb traces, this is subtracted out to leave a relatively smooth HbD trace. The microscope was subsequently replaced to investigate a leak in the arterial anastomosis and was eventually removed again at event 3.

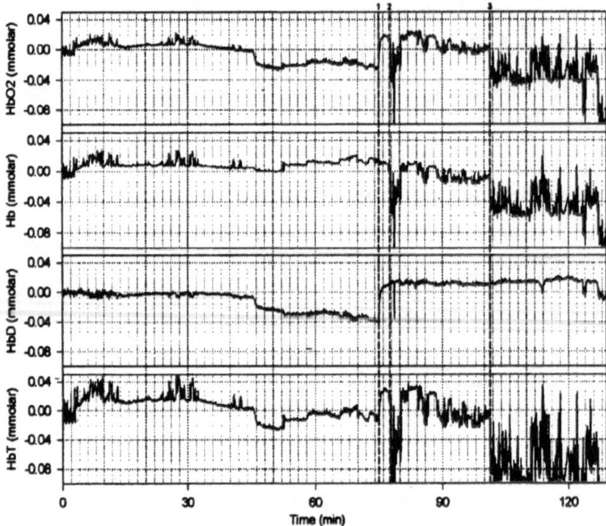

Figure 4. Second Reperfusion of a Radial Forearm Flap. This shows the flap monitored while the microvascular anastomoses were made. Event 1 marks the release of the vascular clamps. At events 2 and 3 there are baseline shifts caused by the removal of the operating microscope lamp.

Figure 5 shows a recording made on a free TRAM flap which was started shortly before the vascular clamps were released after anastomosis. Event 1 marks the release of the clamps. Clearly, no deflection can be seen in any variable. Event 2 marks the start of work to investigate the problem with the flap, so significant motion artifact is present. The clinical opinion of this flap was that although the muscular portion of the flap was sufficiently perfused, the perforating vessels supplying the cutaneous portion were inadequate and hence the distal portion where we were monitoring did not reperfuse. The flap was completed and the patient was sent to recovery, keeping the flap warm and the blood pressure up, but sadly the flap eventually failed and had to be removed.

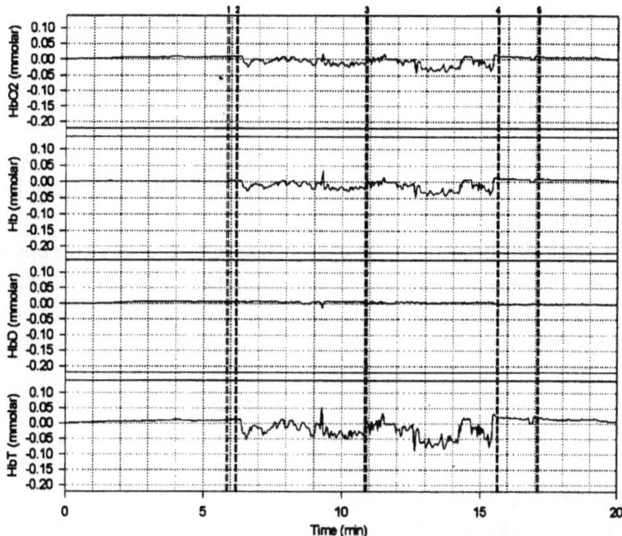

Figure 5. Free TRAM Flap-Abnormal Reperfusion. Event 1 marks the release of the vascular clamps. Event 2 marks the start of work to investigate the problem with the flap, so significant motion artifact is present. Event 3 indicates the use of diathermy at the donor site and events 4 and 5 mark repositioning of the flap by the surgeon.

5. CONCLUSIONS

It has already been shown that NIRS yields repeatable characteristic patterns for different types of vascular occlusions [3,4] and that NIRS is useful in monitoring flaps [3]. We have previously shown that this device is also effective in discriminating between venous and total occlusions in the same way as current NIR instruments [2]. In addition it has the advantages of being inexpensive, portable and has a sterilisable probe which can be applied directly to a flap during surgery. The results presented here show that the device can readily monitor disturbances in perfusion following free flap surgery, and have already proved valuable during surgery. Trials of extended monitoring in the post-operative period are being undertaken.

ACKNOWLEDGEMENTS

The authors would like to thank the EPSRC (Dr Thorniley) for Financial Support.

REFERENCES

1. Furnas H, Rosen JM. Monitoring in microvascular surgery Ann Plast Surg 1991;26:265-272.
2. Pickett JA, Thorniley MS, Balogun E, Jones DP. A dual wavelength spectrophotometer for use in Plastic surgery: Comparison with a Hamamatsu NIRO-500 Instrument. Oxygen Transport to Tissues 1999;XXI:723-730.
3. Thorniley MS, Sinclair JS, Green CJ. The use of Near Infra-red spectroscopy for assessing flap viability during reconstructive surgery. Brit J Plast Surg 1998;51:218-226.
4. Irwin MS, Thorniley MS, Dore C, Green CJ. Near infra-red spectroscopy:a non-invasive monitor of perfusion and oxygenation within the microcirculation. Brit J Plast Surg 1995;48:14-226.
5. Pickett JA, Amoroso P, Nield DV, Jones DP. Optical Technique for continuous monitoring of flap viability in plastic surgery In: Bridging Disciplines for Biomedicine Proc of the 18th Ann Int Conf of the IEEE Eng in Med Biol Society, Amsterdam, Netherlands. (IEEE Piscataway, NJ,USA)1999 Abstract p46, CDROM #520.
6. Pickett JA, Amoroso P, Nield DV, Jones DP. Pulse oximetry and PPG measurements in plastic surgery In: Magnificent Milestones and Emerging Opportunities in Med Eng Proc of the 19th Ann Int Conf of the IEEE Eng in Med and Biol Society, Chicago, Illinois, USA. (IEEE Piscataway, NJ,USA) 1999 CDROM paper pp2330-2332.

Chapter 73

INHIBITION OF MITOCHONDRIAL RESPIRATION DURING EARLY STAGE SEPSIS

Nathan A. Davies*, Chris E Cooper*, Ray Stidwill° & Mervyn Singer°
**Dept. Biological Sciences, University of Essex, Colchester, Essex CO4 3SQ, UK*
°Bloomsbury Institute of Intensive Care Medicine, University College London Medical School, London WC1E 6JJ, UK

Abstract: It is known that nitric oxide (NO) is produced in response to a septic insult such as bacterial invasion and that overproduction of NO can have serious debilitating consequences. The mechanism by which NO causes damage at the cellular level is less clear. We have therefore studied the response to a septic insult in an anaesthetised spontaneously breathing Sprague-Dawley rat model. Six rats were given either an intravenous infusion of bacterial cell wall lipopolysaccharide (LPS, 5mg/kg) or saline control over 1 hour. For electron paramagnetic resonance (EPR) studies, blood samples were collected every hour for a further two hours and liver tissue samples were collected post-mortem. Measurement was also made of PaO_2, blood pressure, base deficit, aortic and renal blood flow and hepatic microvascular pO_2 (using porphyrin phosphoresence). Tissue samples were also collected for mitochondrial complex activity analysis. After the administration of LPS blood pressure, blood flow and microvascular PO_2 were diminished and the base deficit increased. In addition a clear difference was observed by EPR between control and insulted blood and tissue samples. A large heam-nitrosyl signal is observed as well as an increase in the signal at g=1.94, corresponding to the iron-sulphur centres of complex I becoming more reduced. However no significant difference was observed for any of the mitochondrial complex activities. The effect of the NO produced was to depress the circulatory variables and increase base deficit, combined with a reduced oxygen consumption this implies an impairment of normal aerobic respiration. This was supported by increased iron-sulphur signals observed by EPR indicating a blockage in the mitochondrial redox chain with the subsequent accumulation of electrons. As no effect was observed in the mitochondrial complex activities this indicates that this inhibition is reversible in early stage sepsis. We conclude that nitric oxide produced in response to a septic insult can inhibit mitochondria causing an impairment of oxygen utilisation by aerobic respiration.

Oxygen Transport to Tissue XXIV, edited by
Dunn and Swartz, Kluwer Academic/Plenum Publishers, 2003

Key Words: Nitric oxide, sepsis, mitochondria, electron paramagnetic resonance spectroscopy, oxygen utilisation.

1. INTRODUCTION

Sepsis is defined as the systemic response to an infectious insult. Whereas infections can often be localized, sepsis initiates a systemic inflammatory response affecting distant organs. This is then a precipitating factor in multiple organ dysfunction syndrome (MODS), the extreme case of which is multiple organ failure (MOF) which is the commonest cause of death in intensive care units (ICU) [1].

Macrophage/monocytes, endothelium, neutrophils and platelets are all activated as part of the inflammatory response. Activated neutrophils and macrophages produce nitric oxide (NO) in response to such stimuli. NO binds to guanylate cyclase, and the subsequent release of cyclic GMP induces blood vessel dilation. NO is produced by one of three isoforms of the enzyme nitric oxide synthase (NOS), including the neuronal NOS (nNOS or NOS1), the inducible or inflammatory NOS (iNOS or NOS2), and the endothelial (eNOS or NOS3) [2]. nNOS and eNOS are calcium/calmodulin-dependent enzymes expressed constitutively, while iNOS activity is independent of intracellular calcium concentration and is expressed in response to stimuli. It has been suggested that NO displays a cytoprotective role during inflammatory injury [3]; however, its reactivity combined with a relatively long life span enables it to reach a variety of cellular targets, giving it cytotoxic potential [4]. NO is employed by macrophages as a defence mechanism to combat invading pathogens such as bacteria [5]. It has been shown in the rat that iNOS expression is induced in response to a septic insult, with a consequent rise in NO production [6].

A feature of early sepsis is a reduced oxygen usage characterized by a rise in mixed venous oxygen saturation. This has previously been attributed to the shunting of oxygenated blood away from nutrient capillaries. Although there is considerable evidence of neutrophil- and platelet-induced capillary blockage [7], significant anatomic shunting has yet to be demonstrated in sepsis. An alternative explanation is that an impairment of oxygen utilization accounts for the rise in venous pO_2. It has been previously suggested that mitochondrial respiration may be affected either by NO directly, or by a reactive oxygen species (ROS) generated from NO, such as peroxynitrite. Peroxynitrite is formed on reaction of NO with superoxide [8] (scheme 1) and has been shown to irreversibly inhibit electron transport complexes in isolated mitochondria [9]. NO has also been shown to reversibly bind with high affinity to the haem a_3-Cu_B binuclear centre of cytochrome oxidase, the terminal electron acceptor of the mitochondrial electron transport chain [10]. Inhibition of the electron transport chain causes the concentration of superoxide to increase [11], thus providing the possibility of increased peroxynitrite levels.

$NO + O_2^- \longrightarrow ONOO^-$ Scheme 1.

Electron paramagnetic resonance (EPR) spectroscopy is a technique particularly suited to deriving information about the environment of transition metals in biological systems. In addition to the transition metals observed in blood (e.g. Fe in hemoglobin) each of the mitochondrial electron transport complexes contain metal centers with distinct signals observable at cryogenic temperatures (< 20 K). This technique can be used to examine the cellular effects on mitochondria in early stage sepsis. Endotoxin is a lipopolysaccharide derived from the cell wall of Gram-negative bacteria. It is a potent activator of the various cellular and humoral pathways involved in septic shock.

We have studied early phase endotoxaemia in the rat to examine the time course of NO production by EPR spectroscopy and to evaluate whether mitochondrial electron transfer complexes are affected in early stage sepsis.

2. METHODS

2.1 Animal model studies

Two groups of six male Sprague-Dawley rats with body weight 280g (± 20 g) were used in this study and were allowed access to food and water *ad libitum* prior to experiments. Approval for this study was obtained from the Home Office UK [according to the Animals (Scientific Procedures) Act, 1986].

2.2 Surgical procedure

The method used is an adaptation of that described previously by Rosser et al. [12]. Two groups of six rats were anaesthetised with 5% isoflurane and placed supine on a heated operating table to maintain rectal temperature at 37°C. Throughout the experiments the animals were allowed to breathe spontaneously and anaesthesia was maintained by 1% isoflurane (increased to 2-2.5% during highly stimulating periods of surgery). Arterial blood pressure was monitored continuously through the connection of a catheter placed in the carotid artery connected to a pressure transducer. At 30 minute intervals the arterial blood gas status (pO_2, pCO_2, pH and base deficit) was determined using a pH/blood gas analyser (ABL4, Radiometer, Copenhagen,

Denmark). A second catheter was placed into the right jugular vein for infusion of fluid and drugs. A 1-ml bolus of 0.9% saline containing an oxygen-quenched phosphorescent probe (0.9% Pd-meso-tetra-(4-carboxyphenyl) porphine), (Medical Systems Corp., Greenvale, New York, USA.) was given after placement of these lines, followed by a continuous saline infusion of $2.5\ \text{ml} \cdot \text{h}^{-1} \cdot 100\text{g body wt}^{-1}$. A midline laparotomy was performed to gain access to the abdominal vasculature and bladder. Doppler flow probes were placed on the left renal artery (1mm, J reflector-1RB) and on the infrarenal abdominal aorta (2mm, J reflector with sliding gate-2SB) and were connected to a flow monitor (T206; monitor and probes from Transonics, Ithaca, NY). A cannular was inserted surgically into the bladder, via the laparotomy, through an incision in the avascular area at its dome to ensure complete drainage of urine from the bladder. A mercury-in-glass thermometer was inserted into the rectum. Finally a fibreoptic cable from a phosphorimeter (OxySpot, Medical Systems, Greenvale, NY) was placed 5mm above the left ventral lobe of the liver to measure microvascular tissue oxygen. Each group was then given either an intravenous infusion of bacterial cell wall lipopolysaccharide (LPS, 5mg/kg) or saline control over 1 hour (17.5ml/kg/hr).

For electron paramagnetic resonance (EPR) studies, arterial blood samples were collected every hour, venous blood samples were collected at time 0 & 3 hours and liver tissue samples were collected post-mortem.

2.3 EPR spectroscopy

Blood samples collected were placed into quartz EPR tubes and frozen immediately in liquid nitrogen. Tissue samples collected post-mortem were roughly diced (1mm^3) and packed into open-ended quartz EPR tubes before freezing in liquid nitrogen. EPR spectra were recorded in a 4103 T_M cylindrical cavity on a Bruker EMX spectrometer.

2.4 Mitochondrial complex analysis

Respiratory chain enzymes were assayed using a Hewlett-Packard 8453 diode array spectrophotometer. Control and test samples were assayed at the same time to match conditions. Methods for complex analysis have been described previously- complex I, complex II-III, complex III and complex IV[13].

2.5 Statistical evaluation

Data are expressed as mean ± standard error. Statistical significance between control and test groups was established by ANOVA with post-hoc Fisher PLSD tests. Significance was accepted at the 0.05 level.

3. RESULTS

3.1 Haemodynamics

After the administration of endotoxin, blood pressure (Fig. 1.top) rose slightly; it then declined rapidly after 90 minutes and was significantly different from the control after 120 minutes ($p < 0.01$). Aortic and renal blood flows declined steadily from the start of the experiment for both the control and treated groups (data not shown). However, after 60 minutes an increase in the rate of decline for the endotoxin treated group was seen resulting in a significant difference between the groups ($p < 0.01$).

Whereas the blood pH of the two groups did not alter during the time course of the experiment (due to compensatory hyperventilation), a substantial change was seen in the base deficit. The control group did not vary during the time course of the experiment. However, in the endotoxin group, the base deficit worsened from the beginning of the experiment (Fig. 1. middle). The two groups were significantly different at 30 minutes ($p < 0.01$) after which the groups continued to diverge.

The microvascular oxygen concentration exhibited a decline in both groups, although the rate was greater for the endotoxin treated animals. After 30 minutes the groups were significantly different ($p < 0.01$) and continued to diverge thereafter.

3.2 EPR

EPR spectra showed a clear difference between the control and endotoxin treated liver samples (Fig.2). In addition to a haem nitrosyl signal detected in the g2 region of the spectra, a variation in the signal intensities from mitochondrial complex iron-sulphur centers was also observed (Table 1).

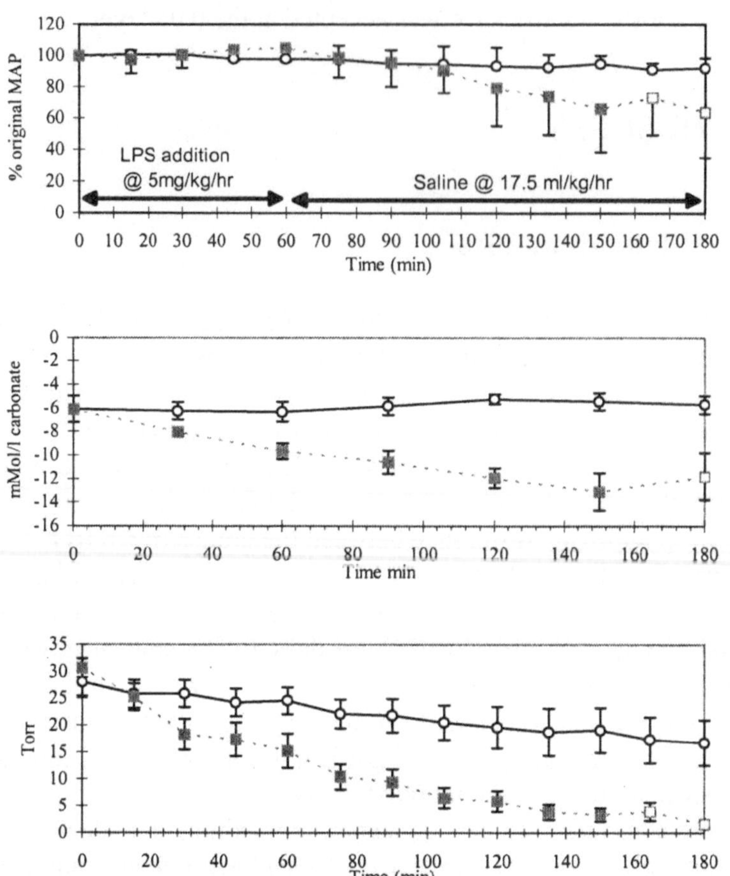

Figure 1. Physiological measurements during endotoxaemia. Arterial blood pressure (top), base deficit (BXS; middle), and microvascular oxygen tension (bottom) measured by porphyrin phosphoresence, time 0 is taken from when endotoxin (LPS) or placebo was started. Control, n=6 —O— , endotoxin treated, n=6 - -■- - (Two animals died at 150 min, n=4 thereafter, - -□- -) values shown mean ± S.E.M.

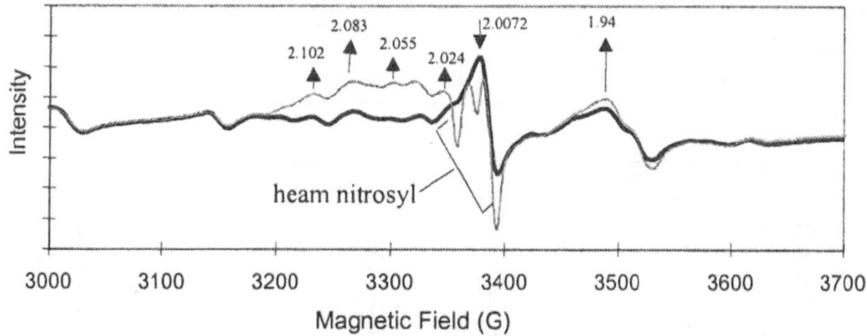

Figure 2. EPR analysis of liver tissue. X-band EPR spectra of control (▬) and endotoxin treated (······) tissue. Arrows show the direction of intensity change between Fe-S clusters with the corresponding g value. Spectra shown are an average of 6. Conditions were:- Temperature, 10K; microwave frequency, 9.470 GHz; time constant, 40.96ms; microwave power, 50.41 μW.

Table 1

Contributing metal centres	g factor			Cluster type	Redox state observable	Signal change
Complex I	g_z	g_x	g_y			
N1	2.24	1.94	~1.91	[2Fe-2S]	Reduced	↑
N2	2.055	1.93	1.93	[4Fe-4S]	Reduced	↑
N3	2.083	~1.94	1.860	[4Fe-4S]	Reduced	↑
N4	2.102	~1.94	1.880	[4Fe-4S]	Reduced	↑
Complex II						
S1	2.024	1.94	~1.91	[2Fe-2S]	Reduced	↑
S3		2.0072		[3Fe-4S]	Oxidised	↓
Complex III						
Rieske Centre.	2.024	1.94	1.782	[2Fe-2S]	Reduced	↑

The Fe-S centers observable in the reduced state gained in signal intensity and the center observed in the oxidized state reduced in intensity (see Table 1). This implies that the mitochondrial electron transport chain becomes more reduced.

The 3 hour EPR spectrum of venous blood showed a large triplet signal at g2 characteristic of a haem nitrosyl in the endotoxin treated sample, whereas the 3 hour control sample did not differ from the sample taken at time 0. A similar result was observed in arterial blood though the nitrosyl signal was not so pronounced.

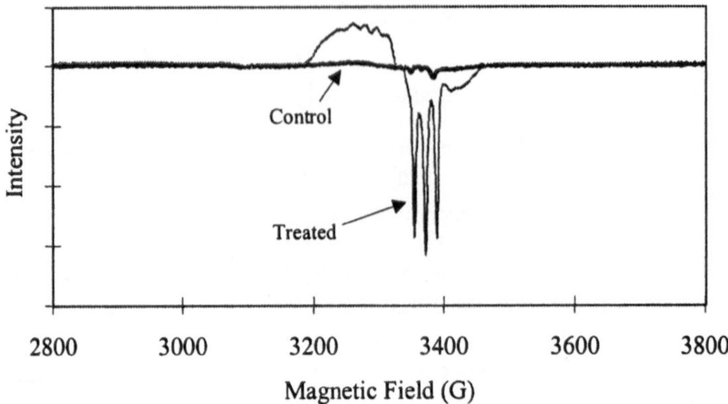

Figure 3. EPR analysis of venous blood. X-band EPR spectra showing the difference between control (▬) and endotoxin treated (—) samples of venous blood taken at T=3 hours. Conditions as for Figure 2.

In the endotoxin treated rats, the g6 ferric haem signal increased progressively with time in both the arterial and venous blood samples. This indicates that a higher proportion of the haemoglobin is present in the met Fe^{III} form [14]. An 800% increase occurred in this signal over the relatively short time course of this experiment. A small increase also occurred in the control samples; however, this change was slight with only a 30% increase of the original signal intensity observed (Fig. 4).

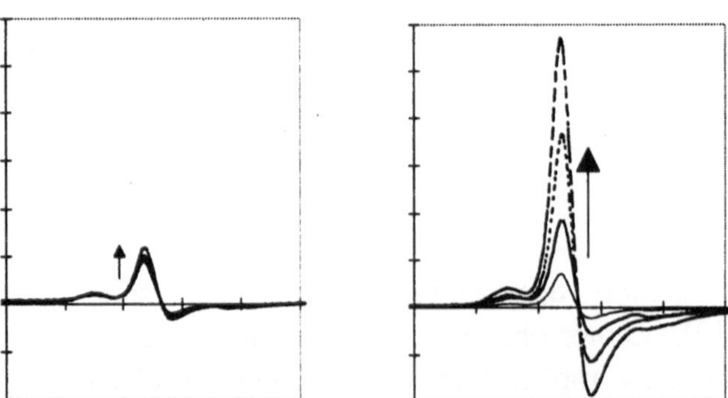

Figure 4. EPR analysis of arterial blood. X-band EPR spectra showing the difference between control and endotoxin treated samples on the arterial blood ferric haem g6 signal with time. Samples were collected at time 0, 1, 2 and 3 hours. Conditions were:- Temperature, 10K; microwave frequency, 9.470 GHz; time constant, 81.92ms; microwave power, 3.181 mW.

3.3 Mitochondrial complex analysis

When normalized to citrate synthase activity or total protein content, the mitochondrial complex activity analysis showed no significant difference between the control and endotoxin treated samples (citrate synthase is used as a standard mitochondrial marker enzyme).

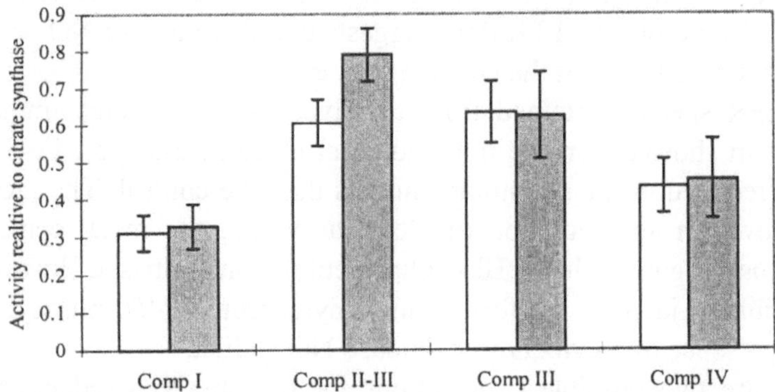

Figure 5. Mitochondrial complex activity analysis. Complex activities normalised to citrate synthase activity. Groups are control □ and endotoxin treated , values expressed mean ± S.E.M., n=6 in each case.

4. DISCUSSION

Endotoxaemia caused a depression of circulatory variables compared to the control group. Despite continual fluid resuscitation there was a fall in blood pressure and blood flows. While blood oxygen tensions remained high the fall in hepatic microvascular oxygen implies either a failure of blood to reach tissues or increased O_2 extraction. It is more likely to be the former as previous endotoxic models have shown elevated tissue O_2 [12].

The increasing metabolic acidosis as demonstrated by rising base deficit with compensatory hypocapnia suggests an imbalance between glycolysis and oxidative phosphorylation, as would occur, for example, during anaerobic respiration.

The formation of a large Hb-nitrosyl and the progressive rise of the g6 ferric haem signal in the blood samples taken from the endotoxic group shows that there are high levels of NO being formed. Met-Hb can be formed from NO in two ways: (1) from the breakdown of a haem-nitrosyl complex, from Fe^{II}-NO to Fe^{III} (met Hb) and nitrate (scheme 2); and (2) NO binding to the ferrous-oxy complex to produce nitrate and met Hb (scheme 3).

Fe^{II}-NO + O_2 $\longrightarrow$ Fe^{III} + NO_3^- Scheme 2

Fe^{II}-O_2 + NO $\longrightarrow$ Fe^{III} + NO_3^- Scheme 3

It is unclear where the NO is formed in response to the insult; previous studies have suggested that iNOS expression is responsible for increased NO. However, the reported time taken for iNOS expression in quantities capable of producing the observed response would be several hours[6]. The fact that this signal increase is seen as early as the first hour, with the emergence of a nitrosyl triplet in the g2 region, and the increase in the g6 met-Hb signal shows there is a rapid response to the presence of the endotoxin. Therefore the EPR data suggests that there is also NO production from constitutive forms of the enzyme (e.g. eNOS).

The EPR spectra obtained from the liver tissue collected immediately post-mortem showed that the mitochondrial electron transport Fe-S centers are more reduced in the endotoxic animals than the control. The increase in signal shows an accumulation of electrons within the metal centers. The haem-nitrosyl signal includes Hb-NO present within the tissue However; it is also possible to identify the formation of cytochrome P450-nitrosyl and the possible presence of a cytochrome oxidase-NO adduct.

There are two possible explanations for the observed reduction of the electron transport chain. Inhibition of cytochrome oxidase by NO would block the terminal electron acceptor of the mitochondrial redox chain causing the observed effect. The same effect would be observed if the tissue became hypoxic due to a fall in intracellular PO_2. It is not trivial to distinguish between these two possibilities, though further analysis and deconvolution of the haem-nitrosyl signals may shed light on whether NO is bound to cytochrome oxidase.

Significantly, no change is seen in isolated mitochondrial complex activities indicating that if an inhibition is occurring, then it is reversible. It is therefore unlikely that peroxynitrite is a significant factor over the time scale of this experiment as this causes irreversible inhibition.

However, in an ongoing study in patients with septic shock, mitochondrial complex activity was diminished in prolonged sepsis (unpublished results). It is probable that the time course of the animal studies presented here is insufficient to observe such damage occurring. We are currently designing a study to test longer time scales, and whether the length of exposure to the insult is a factor in irreversible mitochondrial damage.

5. CONCLUSION

In endotoxaemia there is a depression of circulatory variables, an increased base deficit and increased nitric oxide production. The response time to the insult is too short to be accounted for by iNOS induction, suggesting that NO is produced by constitutive isoforms of the enzyme in the early phase response. EPR data shows evidence for intracellular haem nitrosyl signals and reduced mitochondrial electron transfer. We suggest that a combination of increased NO production and/or reduced intracellular pO_2 leads to an inhibition of mitochondrial oxygen consumption. No effect is observed in the isolated mitochondrial complex activities, indicating that this inhibition is reversible in early stage endotoxaemia.

ACKNOWLEDGEMENTS

We would like to thank the Wellcome Trust, the University of Essex (Research Promotion Fund) and the Charlene Noble Charitable Trust for financial support.

REFERENCES

1. Singer M: Management of multiple organ failure: guidelines but no hard-and-fast rules. J. Antimicrob Chemoth1998;41:103-112.
2. Moncada S, Palmer RMJ, Higgs EA: Nitric Oxide: Physiology, Pathophysiology, and Pharmacology. Pharmacol Rev 1991;43:109-142.
3. Kim YM, de Vera ME, Watkins SC, Billiar TR: Nitric oxide protects cultured rat hepatocytes from TNF α-induced apoptosis by inducing heat shock protein 70 expression. J Biol Chem 1997;272:1402-1411.
4. Kronke KD, Fehsel K, Kolb-Bachofen V: Nitric oxide: Cytotoxicity versus cytoprotection- How, Why, When, and Where? Nitric Oxide 1997;1:107-120.
5. Moilanen E, MoilanenT, Knowles R, Charles I, Kadoya Y, AlSaffar N, Revell PA, Moncada S.: Nitric oxide synthase is expressed in human macrophages during foreign body inflammation. Amer J Path 1997;150:881-887.
6. Knowles RG, Merrett M, Salter M, Moncada S: Differential Induction of Brain, Lung and Liver Nitric-Oxide Synthase By Endotoxin in the Rat. Biochem J 1990; 270:833-836.
7. Cipolle MD: Secondary organ dysfunction. From clinical perspectives to molecular mediators. Crit Care Med 1993;9:261-298.
8. Koppenol WH, Moreno JJ, Pryor WA, Ischiropoulos H, Beckman JS: Peroxynitrite, a cloaked oxidant formed by nitric oxide and superoxide. Chem Res Tox 1992;5 834-842.
9. Cassina A, Radi R: Differential inhibitory action of nitric-oxide and peroxynitrite on mitochiondrial elextron transport. Arch Biochem Biophys 1996;328:309-316.
10. Torres J, Darley-Usmar V, Wilson MT: Inhibition of cytochrome *c* oxidase in turnover by nitric oxide: mechanism and implications for control of respiration. Biochem J 1995; 312:169-173.
11. Poderoso JJ, Carreras MC, Lisdero C, Riobo N, Schopfer F, Boveris A: Nitric-oxide inhibits electrontransfer and increases superoxide radical production in rat-heart mitochondria and submitochondrial particles. Arch Biochem Biophys 1996;328:85-92.
12. Rosser DM, Stidwill RP, Jacobson D, Singer M: Oxygen tension in the bladder epithelium rises in both high and low cardiac output endotoxemic spsis. J App Physiol 1995;79:1878-82.

13. Ragan CI, Wilson MT, Darley-Usmar VM, Lowe PN: Sub-fractionation of the mitochondria. In: Darley-Usmar VM, Rickwood D, Wilson MT, eds. Mitochondria, A Practical Approach. Oxford: IRL Press, 1987; 89-105.
14. Peisach J, Blumberg WE, Ogawa S, Rachmilewitz EA, Oltzik R: The Effects of Protein Conformation on the Heme Symmetry in High Spin Ferric Heme Proteins as Studied by Electron Paramagnetic Resonance. J Biol Chem 1971;246:3342-3355.

AUTHOR INDEX

SUBJECT INDEX